Five Millennium Catalog of Solar Eclipses: –1999 to +3000

2nd Edition

Fred Espenak and Jean Meeus

July 2021

Five Millennium Catalog of Solar Eclipses: –1999 to +3000

Second Edition

Five Millennium Catalog of Solar Eclipses: –1999 to +3000 (2nd Edition)
is published through Astropixels Publishing.

For more information, contact the publisher:

Astropixels Publishing
P.O. Box 16197
Portal, AZ 85632

Astropixels Publishing Website: *astropixels.com/pubs*

This book may be ordered at: *astropixels.com/pubs/5MKSE2.html*

Printed in the United States of America

ISBN 978-1-941983-37-9

Astropixels Publication: AP038 (Version 1.0a)

Front Cover: The *Eclipsareon* is an 18th century instrument contrived by James Ferguson to show various solar eclipse phenomena including their time, duration, and quantity from all places on Earth (*Philos. Trans.*. Vol. 48, 1753–5, pp. 520–525).

Table of Contents

Preface – 1st Edition

Solar eclipse canons have traditionally been publications offering maps of past and future eclipse paths using the best ephemerides of their day for calculating the positions of the Sun and Moon. The first major work of this kind was Theodor von Oppolzer's 1887 *Canon der Finsternisse* (Translated as *Canon of Eclipses*, Gingerich, 1962). It stands as one of the greatest achievements in computational astronomy of the 19th century and contains the elements of all 8,000 solar eclipses (and 5,200 lunar eclipses) occurring between the years -1207 and +2161 (1208 BCE and 2161 CE, respectively), together with maps showing the approximate positions of the central lines. To accomplish this remarkable feat, a number of approximations were used in the calculations and maps. Consequently, the eclipse paths often differ by hundreds of miles compared to rigorous predictions generated with modern ephemerides. Furthermore, the 1887 canon took no account of the shifts imparted to ancient eclipse paths as a consequence of Earth's variable rotation rate and the secular acceleration of the Moon. Nevertheless, Oppolzer's canon remained the standard reference for nearly a century.

With the arrival of the electronic computer, the *Canon of Solar Eclipses* (Meeus, Grosjean, and Vanderleen, 1966) contains the Besselian elements of all solar eclipses from +1898 to +2510, together with central line tables and maps. The aim of this work was to provide data on future eclipses.

In comparison, the *Canon of Solar Eclipses, –2003 to +2526* (Mucke and Meeus, 1983) was intended primarily for historical research, serving as the modern day successor of Oppolzer's great canon. The Mucke-Meeus publication included Besselian elements and maps of all 10,774 solar eclipses during this time interval. Each orthographic map was oriented to show the day-side hemisphere of Earth. In this projection, the path of the Moon's penumbra and the central axis of the shadow cone could be approximated by straight lines.

Several other special canons have been produced. *Atlas of Historical Eclipse Maps, East Asia 1500 BC – AD 1900* (Stephenson and Houlden, 1986) provides the path maps of all total and annular eclipses visible from China. The *Fifty Year Canon of Solar Eclipses: 1986–2035* (Espenak, 1987) contains individual detailed maps and central path data for all solar eclipses from 1986 through 2035.

Without exception, all solar eclipse canons produced during the latter half of the 20th century were based on Newcomb's tables of the Sun (1895) and Brown's lunar theory (1905), subject to later modifications in the *Improved Lunar Ephemeris* (1954). These were the best ephemerides of their day, but they have all been superseded.

The recently published *Five Millennium Canon of Solar Eclipses: –1999 to +3000* (Espenak and Meeus, 2nd Edition 2021) contains individual maps for each of the 11,898 solar eclipses occurring over this period. The following points highlight the features and characteristics of this work.

- Based on modern theories of the Sun and the Moon constructed at the *Bureau des Longitudes* of Paris rather than the older Newcomb and Brown ephemerides.
- Ephemerides and eclipse predictions performed in Terrestrial Dynamical Time.
- Covers historical period of eclipses, as well as one millennium into the future.
- Global maps for each eclipse depict the actual northern and southern limits of the Moon's penumbral and umbral or antumbral shadows, as well as the sunrise and sunset curves.
- Maps include curve of eclipse magnitude 0.5.
- Maps include continental outlines with contemporary political boundaries and are large enough to identify geographic regions of eclipse visibility.
- Maps are based of the most current determination of the historical values of ΔT.
- Estimates of eclipse path accuracy based on the uncertainty in ΔT (i.e., standard error in ΔT)

A primary goal of this work is to assist historians and archeologists in the identification and dating of eclipses found in references and records from antiquity. This is no easy task because there are usually several possible candidates. Accurate maps using the best available values of ΔT coupled with estimates in the standard error of ΔT, are critical in discriminating among potential eclipse candidates. Ultimately, historical eclipse identification can lead to improved chronologies in the timeline of a particular culture.

The *Canon* is of value to educators, planetariums, and anyone interested in knowing when and where past and future eclipses occur. The general public is fascinated by eclipses. With each major eclipse, the question always arises as to when a particular location experienced its last and next eclipses. The maps presented in the *Canon* are ideally suited to addressing such queries.

To supplement the 11,898 eclipse maps in the *Five Millennium Canon of Solar Eclipses*, we offer the following catalog. It includes additional information for each eclipse that could not be included in the original 664-page publication because of size limits. The data tabulated for each eclipse include the catalog number, canon plate number, calendar date, Terrestrial Dynamical Time of greatest eclipse, ΔT, lunation number, Saros number, eclipse type, Quincena Lunar Eclipse parameter, gamma, eclipse magnitude, geographic coordinates of greatest eclipse (latitude and longitude), and the circumstances at greatest eclipse (i.e., Sun altitude and azimuth, path width, and central line duration).

The *Canon* and the *Catalog* both use the same solar and lunar ephemerides as well as the same values of ΔT. This 1-to-1 correspondence between them will enhance the value of each. The researcher may now search, evaluate, and compare eclipses graphically (*Canon*) or textually (*Catalog*).

— Fred Espenak and Jean Meeus, 2009 January

Preface – 2nd Edition

The *Five Millennium Catalog of Solar Eclipses: –1999 to +3000* was first published in January 2009 (NASA TP-2009-214174). As a NASA Technical Publication it had a limited publication run and distribution. The available supply of hard copies was depleted within 12 months of publication although the PDF version continues to be available,

A decade later, the *Five Millennium Catalog of Solar Eclipses* is available again in hard copy in this Second Edition through Astropixels Publishing. The *Catalog* serves as a supplement to the 2-volume *Five Millennium Canon of Solar Eclipses* which contains a map of every eclipse. The *Canon* and the *Catalog* both use the same solar and lunar ephemerides and the same value of ΔT.

A new section (Section 4: Eclipses and the Moon's Orbit) has also been added to this second edition.

— Fred Espenak and Jean Meeus, 2021 July

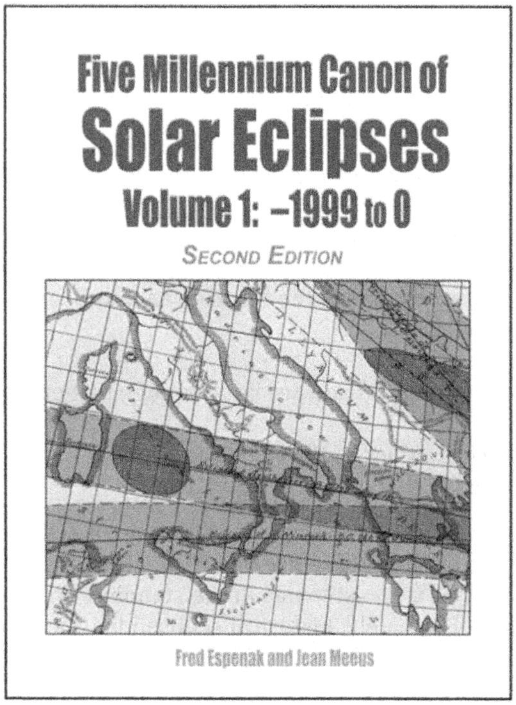

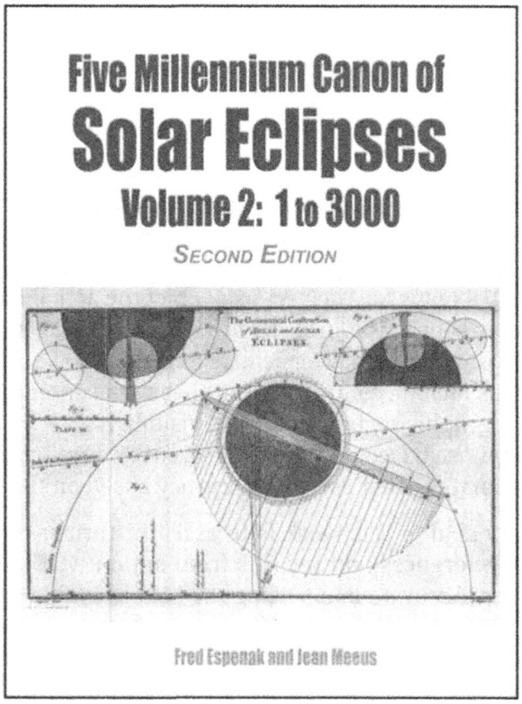

See: *http://astropixels.com/pubs/5MCSE2.html*

Section 1: Eclipse Catalog and Predictions

1.1 Introduction

Earth will experience 11,898 eclipses of the Sun during the 5000-year period from -1999 to +3000 (2000 BCE to 3000 CE). The catalog presented in the Appendix consists of a series of tables that summarize the principal characteristics of each solar eclipse over this time interval. As such, it serves to complement the previously published *Five Millennium Canon of Solar Eclipses*, which contains individual maps for each eclipse over the same period.

1.2 Explanation of Solar Eclipse Catalog

Each line in the catalog corresponds to one eclipse and provides concise parameters to characterize the eclipse. The calendar date and Dynamical Time of the instant of greatest eclipse are given along with the adopted value of delta T (ΔT). The lunation number (since 2000 Jan 06) and the Saros series are listed along with the eclipse type (P=Partial, A=Annular, T=Total, or H=Hybrid). Gamma is the distance of the shadow axis from Earth's center at greatest eclipse, while the eclipse magnitude is defined as the fraction of the Sun's diameter obscured at that instant. The geographic latitude and longitude of the umbra are given for greatest eclipse, along with the Sun's altitude and azimuth, the width of the path (kilometers), and the central line duration of totality or annularity. For both partial and non-central umbral/antumbral eclipses, the latitude and longitude correspond to the point closest to the shadow cone axis at greatest eclipse. The Sun's altitude is always 0° at this location. A more detailed description of each field in the catalog follows.

1.2.1 Catalog Number

The catalog number is the sequential number assigned to each eclipse in the catalog from 1 to 11,898.

1.2.2 Canon Plate

The *Five Millennium Canon of Solar Eclipses* consists of 595 plates with 20 eclipse maps per plate. The canon plate identifies the plate number in which each eclipse appears.

1.2.3 Calendar Date

All eclipse dates from 1582 Oct 15 onwards use the modern Gregorian calendar currently found throughout most of the world. The older Julian calendar is used for eclipse dates prior to 1582 Oct 04. Because of the Gregorian Calendar Reform, the day following 1582 Oct 04 (Julian calendar) is 1582 Oct 15 (Gregorian calendar).

The Gregorian calendar was decreed by Pope Gregory XIII in 1582 to correct a problem in a drift of the seasons. It adopts the convention of a year containing 365 days. Every fourth year is a leap year of 366 days if it is divisible by 4 (e.g., 2004, 2008, etc.). However, whole century years (e.g., 1700, 1800, 1900) are excluded from the leap year rule unless they are also divisible by 400 (e.g., 2000). This complicated dating scheme was designed to keep the vernal equinox on or within a day of March 21.

Prior to the Gregorian Calendar Reform in 1582, the Julian calendar was in wide use. It was simpler than the Gregorian in that all years divisible by 4 were counted as 366-day leap years. This simplicity came at a cost. After more than 16 centuries of use, the Julian calendar date of the vernal equnox had drifted 11 days from March 21. It was this failure of the Julian calendar that resulted in the Gregorian Calendar Reform.

The Julian calendar does not include the year 0, so the year 1 BCE is followed by the year 1 CE. This is awkward for arithmetic calculations. In this publication, dates are counted using the astronomical numbering system which recognizes the year 0. Historians should note the numerical difference of one year between astronomical dates and BCE dates. Thus, the year 0 corresponds to 1 BCE, and the year -100 corresponds to 101 BCE, etc.

There are a number of ways to write the calendar date through variations in the order of day, month, and year. The International Organization for Standardization's (ISO) 8601 advises a numeric date representation, which organizes the elements from the largest to the smallest. The exact format is YYYY-MM-DD where YYYY is the calendar year, MM is the month of the year between 01 (January) and 12 (December), and DD is the day of the month between 01 and 31. For example, the 27th day of April in the year 1943 would then be expressed as 1943-04-27. We follow the ISO convention, but have replaced the month number with the three-letter English abbreviation of the month name for additional clarity. From the previous example, we express the date as 1943 Apr 27.

1.2.4 TD of Greatest Eclipse

The instant of greatest eclipse occurs when the distance between the axis of the Moon's shadow cone and the center of Earth reaches a minimum. For partial eclipses, the instant of greatest eclipse differs slightly from the instant of greatest magnitude primarily because of Earth's flattening. For total eclipses, the instant of greatest eclipse differs slightly from the instant of greatest duration, although the differences are quite small.

Solar eclipses occur when the Moon is near one of the nodes of its orbit, and therefore, moving at an angle of about 5° to the ecliptic. Hence, unless the eclipse is perfectly central, the instant of greatest eclipse does not coincide with that of apparent ecliptic conjunction (i.e., New Moon), nor with the time of conjunction in Right Ascension.

Greatest eclipse is given in Terrestrial Dynamical Time (TD, Sect. 2.3), which is a time system based on International Atomic Time. As such, TD is the atomic time equivalent to its predecessor Ephemeris Time (Sect. 2.2) and is used in the theories of motion for bodies in the solar system. To determine the geographic visibility of an eclipse, TD is converted to Universal Time (Sect. 2.4) using the parameter ΔT (Sects. 2.6 and 2.7).

1.2.5 Delta T (ΔT)

Delta T (ΔT) is the arithmetic difference, in seconds, between Terrestrial Dynamical Time (Sect. 2.3) and Universal Time (Sect. 2.4). For more information on ΔT, see Section 2.6.

1.2.6 Lunation Number

The lunation number is the number of synodic months or lunations since New Moon on 2000 Jan 06. The Brown Lunation Number can be calculated from it by adding 953.

1.2.7 Saros Series Number

Each eclipse belongs to a Saros series (Sect. 5.2) using a numbering system first introduced by van den Bergh (1955). This system has been expanded to include negative values from the past, as well as additional series in the future. The eclipses with an odd Saros number take place at the ascending node of the Moon's orbit; those with an even Saros number take place at the descending node.

The Saros is a period of 223 synodic months, or approximately 18 years, 11 days, and 8 hours. Eclipses separated by this period belong to the same Saros series and share similar geometry and characteristics.

1.2.5 Delta T (ΔT)

Delta T (ΔT) is the arithmetic difference, in seconds, between Terrestrial Dynamical Time (Sect. 2.3) and Universal Time (Sect. 2.4). For more information on ΔT, see Section 2.6.

1.2.8 Eclipse Type

The first character in this 2-character parameter gives the eclipse type. The four basic types of solar eclipses are:

1) P = Partial Solar Eclipse (Moon's penumbral shadow traverses Earth; Moon's umbral and antumbral shadows completely misses Earth)
2) A = Annular Solar Eclipse (Moon's antumbral shadow traverses Earth; Moon is too far from Earth to completely cover the Sun)
3) T = Total Solar Eclipse (Moon's umbral shadow traverses Earth; Moon is close enough to Earth to completely cover the Sun)
4) Hybrid Solar Eclipse (Moon's umbral and antumbral shadows traverse different parts of Earth; eclipse appears either total or annular along different sections of its path. Hybrid eclipses are also known as annular-total eclipses.)

The second character of the eclipse type is a qualifier defined as follows.

1) m = Middle eclipse of Saros series.
2) n = Central eclipse with no northern limit.
3) s = Central eclipse with no southern limit.
4) + = Non-central eclipseb with no northern limit.
5) – = Non-central eclipse with no southern limit.
6) 2 = Hybrid eclipse path begins total and ends annular.
7) 3 = Hybrid eclipse path begins annular and ends total.
8) b = Saros series begins (first eclipse in a Saros series).
9) e = Saros series ends (last eclipse in a Saros series).

Qualifiers 1 through 5 are used with annular, total or hybrid eclipses but not partial eclipses. Qualifiers 6 and 7 apply only to special classes of hybrid eclipses while qualifiers 8 and 9 are used exclusively with partial eclipses.

1.2.9 Quincena Lunar Eclipse Parameter (QLE)

A lunar eclipse always occurs within ~15 days of a solar eclipse. The Quincena Lunar Eclipse parameter (QLE) identifies the type of the lunar eclipse and whether it precedes or succeeds a particular solar eclipse. There are three basic types of lunar eclipses:

1) n = penumbral lunar eclipse (Moon partly or completely within Earth's penumbral shadow)
2) p = partial lunar eclipse (Moon partly within Earth's umbral shadow)
3) t = total lunar eclipse (Moon completely within Earth's umbral shadow)

The QLE consists of a two-character string. The characters identify the type of lunar eclipse preceding and succeeding a solar eclipse, respectively. In most instances, one of the two characters in the QLE is "-" indicating a single lunar eclipse either precedes or succeeds the solar eclipse. On some occasions, a double quincena occurs in which a solar eclipse is both preceded and succeeded by lunar eclipses. The QLE then consists of two characters identifying the types of the two lunar eclipses.

1.2.10 Gamma

The quantity gamma is the minimum distance from the axis of the lunar shadow cone to the center of Earth, in units of Earth's equatorial radius. This distance is positive or negative, depending on whether the axis of the shadow cone passes north or south of Earth's center. If gamma is between +0.997 and –0.997, the eclipse is a central one (either total, annular, or hybrid). The limiting value 0.997 differs from unity because of the flattening of Earth.

The change in the value of gamma, after one Saros period, is larger when Earth is near its aphelion (June–July) than when it is near perihelion (December–January). Table 1-1 illustrates this point using eclipses from two different Saros series.

Table 1-1. Variation in Gamma at Aphelion vs. Perihelion

Date	Gamma		Date	Gamma
1955 Jun 20	-0.15278		1956 Dec 02	+1.09229
1973 Jun 30	-0.07853		1974 Dec 13	+1.07975
1991 Jul 11	-0.00412		1992 Dec 24	+1.07107
2009 Jul 22	+0.06977		2011 Jan 04	+1.06265
2027 Aug 02	+0.14209		2029 Jan 14	+1.05532

A similar situation exists in the case of lunar eclipses. The explanation can be found in van den Bergh [1955].

1.2.11 Eclipse Magnitude

The eclipse magnitude is defined as the fraction of the Sun's diameter occulted by the Moon. For partial eclipses, the eclipse magnitude at the instant of greatest eclipse is given for the geographic position closest to the axis of the Moon's shadow cone. For central eclipses (total, annular, and hybrid), the eclipse magnitude listed is actually the ratio of the topocentric apparent diameters of the Moon and Sun at greatest eclipse. The eclipse magnitude is always less than 1.0 for partial and annular eclipses, but equal to, or greater than, 1.0 for total and hybrid eclipses.

1.2.12 Latitude and Longitude

The geographic latitude and longitude corresponds to the position of greatest eclipse.

1.2.13 Altitude of Sun

The Sun's altitude at the geographic position intersected by the axis of the lunar shadow cone is given at the instant of greatest eclipse. For partial eclipses, the Sun's altitude is always 0° because the shadow axis misses Earth. In this case, the geographic position corresponds to the point closest to the shadow axis.

1.2. 14 Azimuth of Sun

The Sun's azimuth at the geographic position intersected by the axis of the lunar shadow cone is given at the instant of greatest eclipse. The values 0°, 90°, 180°, and 270° correspond to the cardinal directions north, east, south and west, respectively.

1.2.15 Path Width

For central eclipses (total, annular, or hybrid), the width of the path of totality or annularity (kilometers) is given at the geographic position intersected by the axis of the lunar shadow cone at the instant of greatest eclipse.

1.2.16 Central Line Duration

For central eclipses (total, annular, or hybrid), the central line duration of the total or annular phase (in minutes and seconds) is given at the geographic position intersected by the axis of the lunar shadow cone at the instant of greatest eclipse.

In the case of a total or hybrid eclipse, this duration is very nearly, but not exactly, the maximum duration of the total phase along the entire umbral path. For an annular eclipse, the duration at greatest eclipse may be near either the minimum or maximum duration of the annular phase along the path. If the annular phase duration exceeds approximately 2.3 min, then it corresponds to the near maximum duration along the central line track. If the annular phase duration is less, however, then it corresponds to a near minimum and the annular duration increases towards the ends of the central path.

1.3 Solar and Lunar Coordinates

The coordinates of the Sun used in these eclipse predictions have been calculated on the basis of the VSOP87 theory constructed by P. Bretagnon and G. Francou [1988] at the Bureau des Longitudes, Paris. This theory gives the ecliptic longitude and latitude of the planets, and their radius vector, as sums of periodic terms. The complete set of periodic terms of version D of VSOP87 (this version provides the positions referred to the mean equinox of the date) were used in the predictions.

For the Moon, use has been made of the theory ELP-2000/82 of M. Chapront-Touzé and J. Chapront [1983], again of the Bureau des Longitudes. This theory contains a total of 37862 periodic terms, namely 20560 for the Moon's longitude, 7684 for the latitude, and 9618 for the distance to Earth. But many of these terms are very small: some have an amplitude of only 0.00001 arcsecond for the longitude or the latitude, and of 2 centimeters for the distance. In our computer program, we neglected all periodic terms with coefficients smaller than 0.0005 arcsecond in longitude and latitude, and smaller than 1 meter in distance. Due to neglecting the very small periodic terms, the Moon's positions calculated in our program have a mean error (as compared to the full ELP theory) of about 0.0006 second of time in right ascension, and about 0.006 arcsecond in declination. The corresponding error in the calculated times of the phases of a solar eclipse is of the order of 1/40 second, which is much smaller than the uncertainties in predicted values of ΔT, and also much smaller than the error due to neglecting the irregularities (mountains and valleys) at the lunar limb

Improved expressions for the mean arguments L', D, M, M' and F have been taken from Chapront, Chapront-Touzé, and Francou [2002]. A major consequence of this work is to bring the secular acceleration of the Moon's longitude (–25.858 "/cy2, where arcsec/cy2 is arc seconds per Julian century squared) into good agreement with Lunar Laser Ranging (LLR) observations from 1972 to 2001 (Sect. 1.4).

The center of figure of the Moon does not coincide exactly with its center of mass. To compensate for this property in their eclipse predictions, many of the national institutes employ an empirical correction to the center of mass position of the Moon. This correction is typically +0.50" in longitude and –0.25" in latitude. Unfortunately, the large variation in lunar libration from one eclipse to the next minimizes the effectiveness of the empirical correction. We choose to ignore this convention and have performed all calculations using the Moon's center of mass position. In any case, it has no practical impact on the present work.

1.4 Secular Acceleration of the Moon

Ocean tides are caused by the gravitational pull of the Moon (and, to a lesser extent, the Sun). The resulting tidal bulge in Earth's oceans is dragged ahead of the Moon in its orbit due to the daily rotation of Earth. As a consequence, the ocean mass offset from the Earth-Moon line exerts a pull on the Moon and accelerates it in its orbit. Conversely, the Moon's gravitational tug on this mass exerts a torque that decelerates the rotation of Earth. The length of the day gradually increases as energy is transferred from Earth to the Moon, causing the lunar orbit and period of revolution about Earth to increase.

This secular acceleration of the Moon is small but it has a cumulative effect on the Moon's position when extrapolated over many centuries. Direct measurements of the acceleration have only been possible since 1969 using the Apollo retro-reflectors left on the Moon. The results from Lunar Laser Ranging show that the Moon's mean distance from Earth is increasing by 3.8 cm per year (Dickey, et al., 1994). The corresponding acceleration in the Moon's ecliptic longitude is –25.858 "/cy2 (Chapront, Chapront-Touzé, and Francou, 2002). This is the value we have adopted in our lunar ephemeris calculations.

There is a close correlation between the Moon's secular acceleration and changes in the length of the day. However, the relationship is not exact because the lunar orbit is inclined anywhere from about 18.5° to 28.5° to Earth's equator. The parameter ΔT (sections 2.6 and 2.7) is a measure of the accumulated difference in time between an ideal clock and one based on Earth's rotation as it gradually slows down. Published determinations of ΔT from historical eclipse records have assumed a secular acceleration of –26 "/cy^2 (Morrison and Stephenson, 2004). Since we have adopted a slightly different value for the secular acceleration, we must make a small correction "c" to the published values of ΔT as follows:

$$c = -\,0.91072\ (\,-25.858 + 26.0\,)\ u^2, \qquad (1\text{-}1)$$

where: u = (year–1955)/100

Then:

$$\Delta T\ (\text{corrected}) = \Delta T + c \qquad (1\text{-}2)$$

Evaluation of the correction at 1,000 year intervals over the period spanned by the *Catalog* is found in Table 1-2.

Table 1-2. Corrections to ΔT Due to Secular Acceleration

Year	Correction (seconds)
-2000	-202
-1000	-113
0	-49
+1000	-12
+2000	0
+3000	-14

Clearly the correction is only important for negative years although it is significantly smaller than the actual uncertainty in ΔT itself (Sect. 2.8).

The secular acceleration of the Moon is very poorly known and may not be constant. Careful records for its derivation only go back about a century. Before then, spurious and often incomplete eclipse and occultation observations from medieval and ancient manuscripts comprise the data base. In any case, the current value implies an increase in the length of day [LOD] of about 2.3 milliseconds per century. Such a small amount may seem insignificant, but it has very measurable cumulative effects. At this rate, time as measured through Earth's rotation is losing about 84 seconds per century^2 when compared to atomic time.

1.5 Mean Lunar Radius

A fundamental parameter used in eclipse predictions is the Moon's radius, k, expressed in units of Earth's equatorial radius. The Moon's actual radius varies as a function of position angle and libration due to irregularity in the limb profile. From 1968 through 1980, the Nautical Almanac Office used two separate values for k in their predictions. The larger value (k=0.2724880), representing a mean over topographic features, was used for all penumbral (exterior) contacts and for annular eclipses. A smaller value (k=0.272281), representing a mean minimum radius, was reserved exclusively for umbral (interior) contact calculations of total eclipses [Explanatory Supplement, 1974]. Unfortunately, the use of two different values of k for total and annular eclipses introduces a discontinuity in the case of hybrid eclipses.

In 1982, the *International Astronomical Union* (IAU) adopted a value of k=0.2725076 for the lunar radius, based on a mean including mountain peaks and valleys along the Moon's limb. This value is currently used by the Nautical Almanac Office for all solar eclipse predictions. The adoption of one single value for k eliminates the discontinuity in the case of hybrid eclipses and ends confusion arising from the use of two different values. However, the use of even the best mean value for the Moon's radius introduces a problem in predicting the true character and duration of umbral and antumbral eclipses, particularly total eclipses. A total eclipse can be defined as an eclipse in which the Sun's disk is completely occulted by the Moon. This cannot occur so long as any photospheric rays are visible through deep valleys along the Moon's limb [Meeus, Grosjean and Vanderleen, 1966]. But the use of the IAU's mean k guarantees that some annular or hybrid eclipses will be misidentified as total. A case in point is the eclipse of 1986 October 03. Using the IAU value for k, the Astronomical Almanac identified this event as a total eclipse of 3 seconds duration when it was, in fact, a beaded annular eclipse. Since a smaller value of k is more representative of the deeper lunar valleys and hence the minimum solid disk radius, it helps ensure the correct identification of an eclipse's actual type.

This publication adopts the two values for k used by the Nautical Almanac Office from 1968 through 1980. The larger value (k=0.2724880) is utilized for all partial (penumbral) eclipses. The magnitudes of these eclipses typically agree to within 0.0001 of the magnitudes calculated using the IAU value for k.

In order to avoid eclipse type misidentification and to predict central durations which are closer to the actual durations at total eclipses, the smaller value (k=0.272281) is used for all umbral and antumbral eclipses (total, annular and hybrid). This usage of the smaller k value is consistent with predictions in *Fifty Year Canon of Solar Eclipses: 1986 - 2035* [Espenak, 1987]. Consequently, the smaller k produces shorter central durations and narrower paths for total eclipses when compared with calculations using the IAU value for k. Similarly, predictions using the smaller k result in longer central durations and wider paths for annular eclipses than do predictions using the IAU's k.

1.6 Five Millennium Catalog of Solar Eclipses on EclipseWise.com

The tables in the *Catalog* are available via the website *EclipseWise.com* at:

http://eclipsewise.com/solar/SEcatalog/SEcatalog.html

Organized into 100-year intervals, the tables have individual links to eclipse maps and tables.

A ASCII text file listing basic details about each eclipse in the Catalog is also available on-line at:

http://eclipsewise.com/solar/SEpubs/5MKSE.html

All the information found on the maps is included (calendar date, Dynamical Time, eclipse type, Saros number, gamma, altitude and magnitude or duration). In addition, a number of other useful data are listed including ΔT, the geographic coordinates of greatest eclipse and the path width for umbral/antumbral eclipses. The eclipse type is augmented to indicate the first/last eclipse in a Saros series, umbral/antumbral eclipses with one limit, and non-central eclipses.

Section 2: Time

2.1 Greenwich Mean Time

For thousands of years, time has been measured using the length of the solar day. This is the interval between two successive returns of the Sun to an observer's local meridian. Unfortunately, the length of the apparent solar day can vary by tens of seconds over the course of a year. Earth's elliptical orbit around the Sun and the 23.5° inclination of Earth's axis of rotation are responsible for these variations. Apparent solar time was eventually replaced by mean solar time because it provides for a uniform time scale. The key to mean solar time is the mean solar day which has a constant length of 24 hours throughout the year.

Mean solar time on the 0° longitude meridian in Greenwich, England is known as Greenwich Mean Time (GMT). At the International Meridian Conference of 1884, GMT[1] was adopted as the reference time for all clocks around the world. It was also agreed that all longitudes would be measured east or west with respect to the Greenwich meridian. In 1972, Greenwich Mean Time was replaced by Coordinated Universal Time (UTC) as the international time reference. Nevertheless, UTC is colloquially referred to as Greenwich Mean Time although this is technically not correct.

2.2 Ephemeris Time

During the twentieth century, it was found that the rotational period of Earth (length of the day) was gradually slowing down. For the purposes of orbital calculations, time using Earth's rotation was abandoned for a more uniform time scale based on Earth's orbit about the Sun. In 1952, the International Astronomical Union (IAU) introduced Ephemeris Time to address this problem. The ephemeris second was defined as a fraction of the tropical year for 1900 Jan 01 as calculated from Newcomb's Tables of the Sun [1895]. Ephemeris Time was used for Solar System ephemeris calculations until it was replaced by Terrestrial Dynamical Time in 1979.

2.3 Terrestrial Dynamical Time

Terrestrial Dynamical Time (TD) was introduced by the International Astronomical Union (IAU) in 1979 as the coordinate time scale for an observer on the surface of Earth. It takes into account relativistic effects and is based on International Atomic Time (TAI) which is a high-precision standard using several hundred atomic clocks worldwide. As such Terrestrial Dynamical Time is the atomic time equivalent to its predecessor Ephemeris Time (ET) and is used in the theories of motion for bodies in the Solar System. To ensure continuity with Ephemeris Time, TD was defined to match ET for the date 1977 Jan 01. In 1991, the IAU refined the definition of TD to make it more precise. It was also renamed Terrestrial Time (TT) although we prefer and use the older name Terrestrial Dynamical Time.

2.4 Universal Time

For many centuries, the fundamental unit of time was the rotational period of Earth with respect to the Sun. Greenwich Mean Time (GMT) was the standard time reference based on the mean solar time on the 0° longitude meridian in Greenwich, England. Universal Time (UT) is the modern counterpart to GMT and is determined from Very Long Baseline Interferometry (VLBI) observations of the diurnal motion of quasars. Unfortunately, Universal Time is not a uniform time scale because Earth's rotational period is (on average) gradually increasing.

---

[1] GMT was originally reckoned from noon to noon. In 1925, some countries shifted GMT by 12 hours so that it would begin at Greenwich midnight. This new definition is the one in common usage for world time and in the navigational publications of English-speaking countries. The designation Greenwich Mean Astronomical Time (GMAT) is reserved for the reckoning of time from noon (and previously called GMT).

The change is primarily due to tidal friction between Earth's oceans and its rocky mantle through the gravitational attraction of the Moon and, to a lesser extent, the Sun. This secular acceleration (Sect. 1.4) gradually transfers angular momentum from Earth to the Moon. As Earth loses energy and slows down, the Moon gains this energy and its orbital period and distance from Earth increase. Shorter period fluctuations in terrestrial rotation also exist which can produce an accumulated clock error of ±20 seconds in one or more decades. These decade variations are attributed to several geophysical mechanisms including fluid interactions between the core and mantle of Earth. Climatological changes and variations in sea-level may also play significant roles since they alter Earth's moment of inertia.

The secular acceleration of the Moon implies an increase in the length of day [LOD] of about 2.3 milliseconds per century. Such a small amount may seem insignificant, but it has very measurable cumulative effects. At this rate, time as measured through Earth's rotation is losing about 84 seconds per century[2] when compared to atomic time.

2.5 Coordinated Universal Time

Coordinated Universal Time (UTC) is the present day basis of all civilian time throughout the world. Derived from International Atomic Time (TAI), the length of the UTC second is defined in terms an atomic transition of the element cesium and is accurate to approximately one nanosecond (billionth of a second) per day. Since most daily life is still organized around the solar day, UTC was defined to closely parallel Universal Time. However, the two time systems are intrinsically incompatible since UTC is uniform while UT is based on Earth's rotation which is gradually slowing. In order to keep the two times within 0.9 seconds of each other, a leap second is added to UTC about once every 12 to 18 months.

2.6 Delta T (ΔT)

The orbital positions of the Sun and Moon required by eclipse predictions, are calculated using Terrestrial Dynamical Time (TD) because it is a uniform time scale. World time zones and daily life, however, are based on Universal Time (UT)[2]. In order to convert eclipse predictions from TD to UT, the difference between these two time scales must be known. The parameter delta-T (ΔT) is the arithmetic difference, in seconds, between the two as:

$$\Delta T = TD - UT \qquad (2\text{-}1)$$

Past values of ΔT can be deduced from the historical records. In particular, hundreds of eclipse observations (both solar and lunar) were recorded in early European, Middle Eastern and Chinese annals, manuscripts, and canons. In spite of their relatively low precision, these data represent the only evidence for the value of ΔT prior to 1600 CE. In the centuries following the introduction of the telescope (circa 1609 CE), thousands of high quality observations have been made of lunar occultations of stars. The number and accuracy of these timings increase from the seventeenth through the twentieth century, affording valuable data in the determination of ΔT. A detailed analysis of these measurements fitted with cubic splines for ΔT from –500 to +1950 is presented in Table 2-1 and includes the standard error for each value (Morrison and Stephenson, 2004).

[2] World time zones are actually based on Coordinated Universal Time (UTC). It is an atomic time synchronized and adjusted to stay within 0.9 seconds of astronomically determined Universal Time (UT). Occasionally, a "leap second" is added to UTC to keep it in sync with UT (which changes due to variations in Earth's rotation rate).

Table 2-1. Values of ΔT Derived from Historical Records

Year	ΔT (seconds)	Standard Error (seconds)
-500	17190	430
-400	15530	390
-300	14080	360
-200	12790	330
-100	11640	290
0	10580	260
100	9600	240
200	8640	210
300	7680	180
400	6700	160
500	5710	140
600	4740	120
700	3810	100
800	2960	80
900	2200	70
1000	1570	55
1100	1090	40
1200	740	30
1300	490	20
1400	320	20
1500	200	20
1600	120	20
1700	9	5
1750	13	2
1800	14	1
1850	7	<1
1900	-3	<1
1950	29	<0.1

In modern times, the determination of ΔT is made using atomic clocks and radio observations of quasars, so it is completely independent of the lunar ephemeris. Table 2-2 gives the value of ΔT every five years from 1955 to 2005 (*Astronomical Almanac* for 2006, page K9).

Table 2-2. Recent Values of ΔT from Direct Observations

Year	ΔT (seconds)	5-Year Change (seconds)	Average 1-Year Change (seconds)
1955.0	+31.1	—	—
1960.0	+33.2	2.1	0.42
1965.0	+35.7	2.5	0.50
1970.0	+40.2	4.5	0.90
1975.0	+45.5	5.3	1.06
1980.0	+50.5	5.0	1.00
1985.0	+54.3	3.8	0.76
1990.0	+56.9	2.6	0.52
1995.0	+60.8	3.9	0.78
2000.0	+63.8	3.0	0.60
2005.0	+64.7	0.9	0.18

The average annual change of ΔT was 0.99 second from 1965 to 1980. However, the average annual increase was just 0.63 second from 1985 to 2000, and only 0.18 second from 2000 to 2005. Future changes and trends in ΔT cannot be predicted with certainty since theoretical models of the physical causes are not of high enough precision. Extrapolations from the table weighted by the long period trend from tidal braking of the Moon offer reasonable estimates of +67 seconds in 2010, +93 seconds in 2050, +203 seconds in 2100, and +442 seconds in the year 2200.

Outside the period of observations (500 BCE to 2005 CE), the value of ΔT can be extrapolated from measured values using the long-term mean parabolic trend:

$$\Delta T = -20 + 32\ u^2 \text{ seconds}, \qquad (2\text{-}2)$$

where: u = (year–1820)/100, and is defined as time measured in centuries.

2.7 Polynomial Expressions for ΔT

Using the ΔT values derived from the historical record and from direct observations (Tables 2-1 and 2-2, respectively), a series of polynomial expressions have been created to simplify the evaluation of ΔT for any time during the interval -1999 to +3000. We define the decimal year "y" as follows:

$$y = \text{year} + (\text{month} - 0.5)/12 \qquad (2\text{-}3)$$

This gives "y" for the middle of the month, which is accurate enough given the precision in the known values of ΔT. The following polynomial expressions can be used to calculate the value of ΔT (in seconds) over the interval of the *Catalog*.

Before the year –500, calculate:

$$\Delta T = -20 + 32\ u^2, \qquad (2\text{-}4)$$

where: u = (y–1820)/100.

Between years –500 and +500, we use the data from Table 2-1, except that for the year –500 we changed the value 17190 to 17203.7 in order to avoid a discontinuity with the previous formula at that epoch. The value for ΔT is given by a polynomial of the 6th degree, which reproduces the values in Table 2-1 with an error not larger than 4 seconds:

$$\Delta T = 10583.6 - 1014.41\ u + 33.78311\ u^2 - 5.952053\ u^3 - 0.1798452\ u^4 + 0.022174192\ u^5 + 0.0090316521\ u^6, \qquad (2\text{-}5)$$

where: u = y/100.

Between years +500 and +1600, we again use the data from Table 2-1. Calculate u = (y–1000)/100. The value for ΔT is given by the following polynomial of the 6th degree with a divergence from Table 2-1 not larger than 4 seconds:

$$\Delta T = 1574.2 - 556.01\ u + 71.23472\ u^2 + 0.319781\ u^3 - 0.8503463\ u^4 - 0.005050998\ u^5 + 0.0083572073\ u^6, \qquad (2\text{-}6)$$

where: u = (y–1000)/100.

Between years +1600 and +1700, calculate:

$$\Delta T = 120 - 0.9808\ t - 0.01532\ t^2 + t^3 / 7129, \qquad (2\text{-}7)$$

where: t = y – 1600.

Between years +1700 and +1800, calculate:

$$\Delta T = 8.83 + 0.1603\, t - 0.0059285\, t^2 + 0.00013336\, t^3 - t^4 / 1174000, \qquad (2\text{-}8)$$

where: t = y – 1700.

Between years +1800 and +1860, calculate:

$$\Delta T = 13.72 - 0.332447\, t + 0.0068612\, t^2 + 0.0041116\, t^3 - 0.00037436\, t^4 + 0.0000121272\, t^5 - 0.0000001699\, t^6 + 0.000000000875\, t^7, \qquad (2\text{-}9)$$

where: t = y – 1800.

Between years 1860 and 1900, calculate:

$$\Delta T = 7.62 + 0.5737\, t - 0.251754\, t^2 + 0.01680668\, t^3 - 0.0004473624\, t^4 + t^5 / 233174, \qquad (2\text{-}10)$$

where: t = y – 1860.

Between years 1900 and 1920, calculate:

$$\Delta T = -2.79 + 1.494119\, t - 0.0598939\, t^2 + 0.0061966\, t^3 - 0.000197\, t^4, \qquad (2\text{-}11)$$

where: t = y – 1900.

Between years 1920 and 1941, calculate:

$$\Delta T = 21.20 + 0.84493\, t - 0.076100\, t^2 + 0.0020936\, t^3, \qquad (2\text{-}12)$$

where: t = y – 1920.

Between years 1941 and 1961, calculate:

$$\Delta T = 29.07 + 0.407\, t - t^2 / 233 + t^3 / 2547, \qquad (2\text{-}13)$$

where: t = y – 1950.

Between years 1961 and 1986, calculate:

$$\Delta T = 45.45 + 1.067\, t - t^2 /260 - t^3 / 718, \qquad (2\text{-}14)$$

where: t = y – 1975.

Between years 1986 and 2005, calculate:

$$\Delta T = 63.86 + 0.3345\, t - 0.060374\, t^2 + 0.0017275\, t^3 + 0.000651814\, t^4 + 0.00002373599\, t^5, \qquad (2\text{-}15)$$

where: t = y – 2000.

Between years 2005 and 2050, calculate:

$$\Delta T = 62.92 + 0.32217\, t + 0.005589\, t^2, \qquad (2\text{-}16)$$

where: t = y – 2000.

This expression is derived from estimated values of ΔT in the years 2010 and 2050. The value for 2010 (66.9 seconds) is based on a linearly extrapolation from 2005 using 0.39 seconds/year (average from 1995 to 2005). The value for 2050 (93 seconds) is linearly extrapolated from 2010 using 0.66 seconds/year (average rate from 1901 to 2000).

Between years 2050 and 2150, calculate:

$$\Delta T = -20 + 32\, ((y-1820)/100)^2 - 0.5628\, (2150 - y), \qquad (2\text{-}17)$$

The last term is introduced to eliminate the discontinuity at 2050.

After 2150, calculate:

$$\Delta T = -20 + 32\,u^2, \qquad (2\text{-}18)$$

where: u = (y–1820)/100.

All values of ΔT based on Morrison and Stephenson [2004] assume a value for the Moon's secular acceleration of –26 arcsec/cy². However, the ELP-2000/82 lunar ephemeris employed in the *Catalog* uses a slightly different value of –25.858 arcsec/cy². Thus, a small correction "c" must be added to the values derived from the polynomial expressions for ΔT before they can be used in the *Catalog*:

$$c = -0.000012932\,(y - 1955)^2. \qquad (2\text{-}19)$$

Since the values of ΔT for the interval 1955 to 2005 were derived independent of any lunar ephemeris, no correction is needed for this period.

2.8 Uncertainty in ΔT

The uncertainty in the value of ΔT is of particular interest in the calculation of eclipse paths in the distant past and future. Unfortunately, estimating the standard error in ΔT prior to 1600 CE is a difficult problem. It depends on a number of factors which include the accuracy of determining ΔT from historical eclipse records and modeling the physical processes producing changes in Earth's rotation. Morrison and Stephenson [2004] propose a simple parabolic relation to estimate the standard error (σ) which is valid over the period 1000 BCE to 1200 CE:

$$\sigma = 0.8\,t^2 \text{ seconds}, \qquad (2\text{-}20)$$

where: t = (year–1820)/100.

Table 2-3 gives the errors in ΔT along with the corresponding uncertainties in the longitude of an eclipse path.

Table 2-3. Uncertainty of ΔT, Part I

Year	σ (seconds)	Longitude
-1000	636	2.65°
-500	431	1.79°
0	265	1.10°
+500	139	0.58°
+1000	54	0.22°
+1200	31	0.13°

The decade fluctuations in ΔT result in an uncertainty of ~20 seconds (0.08°) for the period 1300 CE to 1600 CE.

During the telescopic era (1600 CE to present), records of astronomical observations pin down the decade fluctuations with increasing reliability. The uncertainties in ΔT are presented in Table 2-4 [Stephenson and Houlden, 1986].

Table 2-4. Uncertainty of ΔT, Part II

Year	σ (seconds)	Longitude
+1700	5	0.021°
+1800	1	0.004°
+1900	0.1	0.0004°

The estimation in the uncertainty of ΔT prior to 1000 BCE must rely on a certain amount of modeling and theoretical arguments since no measurements of ΔT are available for this period. Huber (2000) has proposed a Brownian motion model including drift to estimate the standard error in ΔT for periods outside the epoch of measured values. The intrinsic variability in the LOD (length of day) during the 2500 years of observations (500 BCE to 2000 CE) is 1.780 ms/cy with a standard error of 0.56 ms/cy. This rate is not due entirely to tidal friction but includes a drift in LOD from imperfectly understood effects such as changes in sea level due to variations in polar ice caps. Presumably, the same mechanisms operating during the present era also operated prior to 1000 BCE as well as one millennium into the future.

Huber's derived estimate for the total standard error (fluctuations plus drift) in ΔT is as follows.

$$\sigma = 365.25\ N\ \mathrm{SQRT}\,[\,(N\ Q/3)\ (1 + N/M)\,]\,/\,1000, \qquad (2\text{-}21)$$

where:

N = Difference between target year and calibration year
M = 2500 years (–500 to +2000);
this covers period of observed ΔT measurements
Q = 0.058 ms^2 /yr

The calibration year is taken as –500 for target years before 500 BCE, while the calibration year is 2005 CE for target years in the future. Evaluation of this expression at 1,000 year intervals is found in Table 2-5. It shows estimates in the standard error of ΔT along with the equivalent shift in longitude

Table 2-5. Uncertainty of ΔT, Part III

Year	σ (seconds)	Longitude
-4500	20,717	86.3°
-4000	16,291	67.9°
-3500	12,378	51.6°
-3000	8,978	37.4°
-2500	6,094	25.4°
-2000	3,732	15.6°
-1500	1,900	7.9°
-1000	622	2.6°
—	—	—
+2500	612	2.6°
+3000	1,885	7.9°
+3500	3,711	15.6°
+4000	6,068	25.3°
+4500	8,946	37.3°
+5000	12,341	51.4°

Section 3: Solar Eclipse Statistics

3.1 Statistical Distribution of Eclipse Types

Eclipses of the Sun can only occur during the New Moon phase. It is then possible for the Moon's penumbral, umbral or antumbral shadows to sweep across Earth's surface thereby producing an eclipse. There are four types of solar eclipses:

1. **Partial** — Moon's penumbral shadow traverses Earth (umbral and antumbral shadows miss Earth)
2. **Annular** — Moon's antumbral shadow traverses Earth (Moon is too far from Earth to cover the entire Sun)
3. **Total** — Moon's umbral shadow traverses Earth (Moon is close enough to Earth to cover the entire Sun)
4. **Hybrid** — Moon's umbral and antumbral shadows traverse Earth (eclipse appears annular and total along different sections of its path). Hybrid eclipses are also known as annular-total eclipses.

During the 5000-year period from –1999 to +3000 (2000 BCE to 3000 CE), Earth will experience 11,898 eclipses of the Sun. The statistical distribution of the four basic eclipse types over this interval is show in Table 3-1.

Table 3-1. Distribution of Basic Eclipse Types

Eclipse Type	Abbreviation	Number	Percent
All Eclipses	—	11898	100.0%
Partial	P	4200	35.3%
Annular	A	3956	33.2%
Total	T	3173	26.7%
Hybrid	H	569	4.8%

All partial eclipses are events in which some portion of the Moon's penumbral shadow passes across Earth's surface. In comparison all annular, total and hybrid eclipses can be characterized as events in which some portion of the Moon's umbral and/or antumbral shadow crosses Earth.

In the case of umbral or antumbral eclipses (annular, total or hybrid), they can be further categorized as:

1. **Central (two limits)** — The central axis of the Moon's umbral or antumbral shadow traverses Earth thereby producing a central line in the eclipse track. The umbra or antumbra falls entirely upon Earth producing a ground track with both a northern and southern limit.
2. **Central (one limit)** — The central axis of the Moon's umbral or antumbral shadow traverses Earth. However, a portion of the umbra or antumbra misses Earth throughout the eclipse thereby producing a ground track with just one limit.
3. **Non-Central** — The central axis of the Moon's umbral or antumbral shadow misses Earth. However, one edge of the umbra or antumbra grazes Earth thereby producing a ground track with one limit and no central line.

Using the above categories, the distribution of the 3956 annular eclipses is shown in Table 3-2.

Table 3-2. Statistics of Annular Eclipses

Annular Eclipses	Number	Percent
All Annular Eclipses	3956	100.0%
Central (two limits)	3827	96.7%
Central (one limit)	61	1.5%
Non-Central (one limit)	68	1.7%

Examples of central annular eclipses with one limit include: 1874 Oct 10, 2003 May 31, 2044 Feb 28, and 2101 Feb 28. Some examples of non-central annular eclipses are: 1950 Mar 18, 1957 Apr 30, 2014 Apr 29, and 2043 Oct 03.

Similarly, the distribution of the 3173 total eclipses is shown in Table 3-3.

Table 3-3. Statistics of Total Eclipses

Total Eclipses	Number	Percent
All Total Eclipses	3173	100.0%
Central (two limits)	3121	98.4%
Central (one limit)	26	0.8%
Non-Central (one limit)	26	0.8%

Examples of central total eclipses with one limit include: 1494 Mar 07, 1523 Aug 11, 2185 Jul 26, and 2195 Aug 05. The most recent examples of non-central total eclipses are: 1957 Oct 23, 1967 Nov 02, 2043 Apr 09, and 2459 Jun 01.

All 569 hybrid eclipses are central with two limits. Hybrid eclipses with a single limit (both central and non-central) are exceedingly rare. An estimate of the mean frequency of non-central hybrid eclipses is one out of every 600 million eclipses or once every 250 million years [Meeus, 2002].

The central path of most hybrid eclipses begins annular, changes to total and finally reverts back to annular. This combination (ATA) occurs in 519 out of the 569 hybrid eclipses in the Catalog. However, there are two other possibilities. If the vertex of the Moon's umbral shadow passes through Earth's fundamental plane during the eclipse, then the hybrid eclipse can begin as total and end as annular (TA) or it can begin as annular and end as total (AT). Table 3.4 shows the distribution of the three different classes of hybrid eclipses.

Table 3-4. Statistics of Hybrid Eclipses

Hybrid Eclipses	Number	Percent
All Hybrid Eclipses	569	100.0%
Hybrid (ATA)	519	92.2%
Hybrid (TA)	24	4.2%
Hybrid (AT)	26	4.6%

Examples of ATA hybrid eclipses include: 1986 Oct 03, 1987 Mar 29, 2005 Apr 08, and 2023 Apr 20. Examples of the relatively rare TA hybrid eclipse are: 1564 Jun 08, 1703 Jan 17, 1825 Dec 09, and 2386 Apr 29. Finally, examples of the rare AT hybrid eclipse include: 1489 Jun 28, 1854 Nov 20, 2013 Nov 03, and 2172 Oct 17.

3.2 Distribution of Eclipse Types by Century

Eclipses Table 3-5 summarize 5000 years of eclipses by eclipse type in 100-year intervals. The number of central and non-central (in square brackets) events are given for annular and total eclipses. The number of eclipses in any one century ranges from 222 to 255 with an average of 238.0. Over the 1000-year interval 1501-2500 CE (centered on the present era), the average is 238.9 eclipses per century.

Some remarkable patterns are present in this table. There exists a cyclical variation in the number of eclipses per century with a length of a little under six centuries, giving alternating "rich" and "poor" periods [Meeus, 1997]. The 20th and 21st centuries (1901 to 2100) are poor periods, with only 228 and 224 eclipses. This cycle is also present when only central eclipses are considered.

The cycle appears to have a period of approximately 600 years with an amplitude of ~30 eclipses. This is close to a known eclipse period called the tetradia which has a period of 586.02 years. The tetradia governs the recurrence of tetrads or groups of four successive total lunar eclipses each separated by six lunations. The tetradia cycle for lunar eclipses tetrads appears to be 180 degrees out of phase with the cycle for solar eclipses. When there are many tetrads, there are fewer solar eclipses. We are currently in a tetrad rich period with recent tetrads in 2003-2004, 2014-2015 and 2032-2033.

The number of hybrid solar eclipses per century also varies cyclically with a period of approximately 17 centuries.

Table 3-5a. Eclipse Types by Century: –1999 to +0000 (2000 BCE to 0000 CE)

Century Interval	Number of Eclipses	Number of Partial Eclipses	Number of Annular Eclipses*	Number of Total Eclipses*	Number of Hybrid Eclipses
-1999 to -1900	239	84	70 [1]	62 [0]	22
-1899 to -1800	253	93	80 [0]	62 [1]	17
-1799 to -1700	254	95	73 [1]	63 [1]	21
-1699 to -1600	230	75	70 [1]	60 [0]	24
-1599 to -1500	225	78	65 [2]	59 [0]	21
-1499 to -1400	226	77	65 [4]	61 [1]	18
-1399 to -1300	234	76	83 [1]	68 [0]	6
-1299 to -1200	250	93	86 [0]	64 [0]	7
-1199 to -1100	252	93	89 [0]	63 [0]	7
-1099 to -1000	238	79	89 [2]	67 [1]	0
-0999 to -0900	226	84	74 [1]	58 [3]	6
-0899 to -0800	225	80	73 [2]	64 [2]	4
-0799 to -0700	234	79	88 [0]	64 [0]	3
-0699 to -0600	253	96	86 [1]	63 [0]	7
-0599 to -0500	255	96	85 [1]	65 [0]	8
-0499 to -0400	241	84	76 [2]	62 [0]	17
-0399 to -0300	225	83	62 [1]	56 [0]	23
-0299 to -0200	226	83	61 [1]	55 [2]	24
-0199 to -0100	237	80	71 [2]	62 [1]	21
-0099 to 0000	251	92	77 [0]	64 [1]	17

* The first quantity is the number of central eclipses, while the second quantity [in square brackets] is the number of non-central eclipses.

Table 3-5b. Eclipse Types by Century: 0001 to +3000 (0001 CE to 3000 CE)

Century Interval	Number of Eclipses	Number of Partial Eclipses	Number of Annular Eclipses*	Number of Total Eclipses*	Number of Hybrid Eclipses
0001 to 0100	248	90	74 [1]	58 [0]	25
0101 to 0200	237	80	75 [2]	63 [1]	16
0201 to 0300	227	79	70 [4]	69 [0]	5
0301 to 0400	222	73	74 [2]	65 [1]	7
0401 to 0500	233	80	83 [1]	67 [0]	2
0501 to 0600	251	93	86 [1]	65 [0]	6
0601 to 0700	251	90	89 [1]	67 [0]	4
0701 to 0800	233	77	86 [2]	66 [0]	2
0801 to 0900	222	78	72 [2]	62 [2]	6
0901 to 1000	227	76	83 [1]	65 [1]	1
1001 to 1100	241	84	90 [0]	61 [0]	6
1101 to 1200	250	92	82 [0]	61 [0]	15
1201 to 1300	246	87	80 [1]	60 [0]	18
1301 to 1400	229	76	72 [3]	54 [0]	24
1401 to 1500	222	77	62 [3]	60 [1]	19
1501 to 1600	228	75	69 [3]	62 [0]	19
1601 to 1700	248	89	74 [0]	60 [1]	24
1701 to 1800	251	92	78 [0]	62 [0]	19
1801 to 1900	242	87	77 [0]	63 [0]	15
1901 to 2000	228	78	71 [2]	68 [3]	6
2001 to 2100	224	77	70 [2]	67 [1]	7
2101 to 2200	235	79	82 [5]	65 [0]	4
2201 to 2300	248	92	86 [0]	67 [0]	3
2301 to 2400	248	88	86 [0]	66 [0]	8
2401 to 2500	237	81	87 [2]	65 [1]	1
2501 to 2600	225	83	71 [1]	63 [1]	6
2601 to 2700	227	77	78 [3]	64 [0]	5
2701 to 2800	242	84	92 [0]	63 [0]	3
2801 to 2900	254	95	86 [1]	63 [0]	9
2901 to 3000	248	91	80 [2]	64 [0]	11

* The first quantity is the number of central eclipses, while the second quantity [in square brackets] is the number of non-central eclipses.

3.3 Distribution of Eclipse Types by Month

Table 3-6 summarizes 5000 years of eclipses by eclipse type in each month of the year. The first value in each column is the number of eclipses of a given type for the corresponding month. The second number in square brackets [] is the number of eclipses divided by the number of days in that month. This normalization allows direct comparison of eclipse frequencies in different months.

A brief examination of the values in the column Number of All Eclipses shows that eclipses are equally distributed around the year. The same holds true for partial eclipses. However, the columns for annular and total eclipses reveal something interesting. Annular eclipses are 4/3 times more likely during the period November-December-January compared to the months May-June-July. This effect is attributed to Earth's elliptical orbit. Earth currently reaches perihelion in early January and aphelion in early July. Consequently the Sun's apparent diameter varies from 1952 to 1887 arc-seconds between perihelion and aphelion. The Sun's larger apparent diameter at perihelion makes annular eclipses more frequent at that time.

The opposite argument holds true for total eclipses which are nearly 3/2 times more likely during the period May-June-July compared to the months November-December-January. In this case the Sun's smaller apparent size around aphelion increases the frequency of total eclipses at that time. Total eclipses actually outnumber annular eclipses during the season May-June-July [Meeus, 2002].

Table 3-6. Eclipse Types by Month: -1999 to +3000 (2000 BCE to 3000 CE)

Month	Number of All Eclipses	Number of Partial Eclipses	Number of Annular Eclipses	Number of Total Eclipses	Number of Hybrid Eclipses
January	1010 [32.6]	357 [11.5]	380 [12.3]	222 [7.2]	51 [1.6]
February	919 [32.8]	317 [11.3]	334 [11.9]	225 [8.0]	43 [1.5]
March	1009 [32.5]	359 [11.6]	319 [10.3]	280 [9.0]	51 [1.6]
April	981 [32.7]	345 [11.5]	294 [9.8]	299 [10.0]	43 [1.4]
May	1009 [32.5]	353 [11.4]	294 [9.5]	313 [10.1]	49 [1.6]
June	973 [32.4]	348 [11.6]	279 [9.3]	310 [10.3]	36 [1.2]
July	1008 [32.5]	354 [11.4]	299 [9.6]	312 [10.1]	43 [1.4]
August	1008 [32.5]	358 [11.5]	308 [9.9]	303 [9.8]	39 [1.3]
September	982 [32.7]	354 [11.8]	333 [11.1]	248 [8.3]	47 [1.6]
October	1008 [32.5]	355 [11.5]	362 [11.7]	230 [7.4]	61 [2.0]
November	977 [32.6]	344 [11.5]	367 [12.2]	210 [7.0]	56 [1.9]
December	1014 [32.7]	356 [11.5]	387 [12.5]	221 [7.1]	50 [1.6]

(numbers in square brackets [] are number of eclipses divided by number of days in the month)

3.4 Eclipse Frequency and the Calendar Year

There are 2 to 5 solar eclipses in every calendar year. Table 3-7 shows the distribution in the number of eclipses per year for the 5000 years covered in the *Catalog*.

Table 3-7. Number of Eclipses per Year

Number of Eclipses per Year	Number of Years	Percent
2	3625	72.5%
3	877	17.5%
4	473	9.5%
5	25	0.5%

When two eclipses occur in one calendar year, they can be any combination of P, A, T or H (partial, annular, total or hybrid) with the one exception that they cannot both be T. Table 3-8 lists the frequency of each eclipse combination along with the five most recent years when the combination occurs. The table makes no distinction in the order of any two eclipses. For example, the eclipse combination PA includes all years where the order is either PA or AP.

Table 3-8. Two Eclipses in One Year

Eclipse Combinations[a]	Number of Years	Percent	Examples (Years) [b]
PP	177	4.9%	..., 2004, 2007, 2022, 2025, 2040, ...
PA	97	2.7%	..., 2014, 2032, 2101, 2102, 2119, ...
PH	19	0.5%	..., 0227, 0245, 1909, 1986, 2050]
PT	236	6.5%	..., 2015, 2033, 2037, 2055, 2068, ...
AA	292	8.1%	..., 1951, 1969, 2056, 2074, 2085, ...
AH	239	6.6%	..., 2005, 2013, 2023, 2031, 2049, ...
AT	2402	66.3%	..., 2006, 2008, 2009, 2010, 2012, ...
HH	84	2.3%	..., 1753, 1771, 1789, 1807, 1825]
HT	79	2.2%	..., 1843, 1894, 1912, 1930, 2910, ...

[a] P = Partial, A = Annular, T = Total and H = Hybrid.
[b] When years end with a square bracket], there are no other examples beyond the last year.

When three eclipses occur in one calendar year, there are fourteen possible combinations of P, A, T or H. Table 3-9 lists the frequency of each eclipse combination along with the five most recent years when each combination occurs. The table makes no distinction in the order of eclipses in any combination. For example, the eclipse combination PAT includes all years where the order is PAT, PTA, APT, ATP, TAP and TPA. The rarest combinations PHT and AAH (actually HTP and AHA, respectively) each occurred only twice in the five millennium span of this work.

Table 3-9. Three Eclipses in One Year

Eclipse Combinations[a]	Number of Years	Percent	Examples (Years) [b]
PPP	396	45.2%	..., 1971, 2018, 2036, 2054, 2058, ...
PPA	71	8.1%	..., 1722, 1740, 1899, 2224, 2242, ...
PPH	7	0.8%	-1906,-1888,-1794,-0224, 1544, 1609, 1703
PPT	74	8.4%	..., 1834, 1852, 1928, 2130, 2271, ...
PAA	18	2.1%	..., 0650, 0791, 1704, 2419, 2437, ...
PAH	5	0.6%	[-1907,-0457,-0316,-0101,-0055]
PAT	145	16.5%	..., 1992, 2019, 2084, 2149, 2225, ...
PHH	5	0.6%	[-1683,-0037,-0019,-0001, 1768]
PHT	2	0.2%	[-1488, 1786]
AAH	2	0.2%	[-1944, 1489]
AAT	102	11.6%	..., 1954, 1973, 2038, 2103, 2122, ...
AHH	8	0.9%	[-484,-0400,-0139, 1144, 1228, 1339, 1405, 1666]
AHT	13	1.5%	[-1833,-1702,-1507,-0660,-0465,-0419,-0074, 0121, 1163, 1386, 1731, 1908, 2950]
ATT	29	3.3%	..., 1554, 1712, 1889, 2057, 2252, ...

[a] P = Partial, A = Annular, T = Total and H = Hybrid.
[b] When years are enclosed in square brackets [], they include all examples in 5,000 years.

When four eclipses occur in one calendar year, there are seven possible combinations of eclipse types P, A, T and H. Table 3-10 lists the frequency of each eclipse combination along with the five most recent years when each combination occurs. The table makes no distinction in the order of eclipses in the seven combinations. The rarest combination PPAH (actually HAPP) occurred only once in year –1748 (1749 BCE).

Table 3-10. Four Eclipses in One Year

Eclipse Combinations[a]	Number of Years	Percent	Recent Examples (Years) [b]
PPPP	327	69.1%	..., 2000, 2011, 2029, 2047, 2065, ...
PPPA	79	16.7%	..., 1758, 1917, 2141, 2159, 2177, ...
PPPH	7	1.5%	[-1925,-1870,-0120, 1573, 1591, 1685, 1750]
PPPT	41	8.7%	..., 1693, 1870, 2076, 2094, 2112, ...
PPAA	3	0.6%	[-1209,-1032, 0596]
PPAH	1	0.2%	[-1748]
PPAT	15	3.2%	[-1795,-1162,-0688,-0641,-0576,-0511,-0446, 0010, 0075, 0661, 1182, 1880, 2195, 2782, 2912]

[a] P = Partial, A = Annular, T = Total and H = Hybrid.
[b] When years are enclosed in square brackets [], they include all examples in 5,000 years.

The maximum number of five solar eclipses in one calendar year is quite rare. Over the 5000-year span of the *Catalog* there are only 25 years containing five solar eclipses. They occur in three possible combinations of eclipse types where four out of the five eclipses are always of type P. The first eclipse of such a quintet always occurs in the first half of January while the last eclipse falls in the latter half of December. Table 3-11 lists all 25 years containing five eclipses along with their eclipse combinations and frequencies. The rarest combination PPPPH occurred only once in year –1852 (1853 BCE). Once again, the table makes no distinction in the order of eclipses in any combination.

Table 3-11. Five Eclipses in One Year

Eclipse Combinations[a]	Number of Years	Percent	All Examples (Years)
PPPPA	18	72.0%	-1805,-1787,-1675,-1089,-0568,-0503,-0373, 0018, 0148, 0604, 0734, 1255, 1805, 1935, 2206, 2709, 2839, 2904
PPPPH	1	4.0%	-1852
PPPPT	6	24.0%	-1740,-1154,-0438, 0083, 0669, 2774

[a] P = Partial, A = Annular, T = Total and H = Hybrid.

3.5 Extremes in Eclipse Magnitude – Partial Eclipses

Eclipse magnitude is the fraction of the Sun's diameter covered by the Moon. It reaches a maximum at greatest eclipse. The *Catalog* reveals some interesting cases involving extreme values of the eclipse magnitude.

Thirteen partial eclipses have a maximum magnitude less than 0.005 (Table 3-12). These events are all the first or last members in a Saros series. The smallest magnitude was the partial of –1838 Apr 04 with a magnitude of just 0.00002.

Table 3-12. Partial Eclipses with Magnitude 0.005 or Less

Date (Dynamical Time)	Saros	Gamma	Eclipse Magnitude
-1838 Apr 04	-10	1.5615	0.00002
-1512 Apr 29	43	1.5386	0.0041
-0756 Mar 12	66	-1.5417	0.0047
0662 Jun 21	115	1.5377	0.0030
0929 Jul 09	80	1.5267	0.0049
1175 Oct 16	91	-1.5690	0.0019
1512 Apr 16	140	-1.5289	0.0003
1639 Jan 04	145	1.5650	0.0009
1935 Jan 05	111	-1.5381	0.0013
2883 Aug 23	188	-1.5524	0.0010
2893 Dec 29	146	1.5706	0.0028
2904 Jun 05	142	1.5428	0.0040
2995 Aug 17	190	-1.5542	0.0036

Table 3-13 lists the eight partial eclipses having a maximum magnitude greater than 0.995. The greatest partial eclipse occurred on –1577 Mar 30 with a maximum magnitude of 0.9998.

Table 3-13. Partial Eclipses with Magnitude 0.995 or More

Date (Dynamical Time)	Saros	Gamma	Eclipse Magnitude
-1585 Mar 28	33	1.0137	0.9960
-1577 Mar 30	4	1.0109	0.9998
-0944 Sep 14	29	-1.0056	0.9987
-0927 Nov 04	57	1.0005	0.9990
-0018 Jun 10	56	1.0154	0.9954
0257 Aug 26	68	1.0060	0.9969
0654 May 22	106	-1.0131	0.9990
1750 Jul 03	142	-0.9985	0.9956

3.6 Extremes in Eclipse Magnitude – Annular Eclipses

Sixteen annular eclipses have a maximum magnitude (at greatest eclipse) less than or equal to 0.910 (Table 3-14). Eleven of these events are central, three are central with one limit and two are non-central (with one limit). The annular eclipses with the smallest magnitude (at greatest eclipse) occurred on -1682 Nov 12 and 1601 Dec 24 and had a magnitude of just 0.9078.

Table 3-14. Annular Eclipses with Magnitude 0.910 or Less

Date (Dynamical Time)	Saros	Gamma	Eclipse Magnitude	Central Duration
-1718 Oct 21	6	0.9195	0.9091	08m 18s
-1700 Oct 31	6	0.9254	0.9081	08m 44s
-1682 Nov 12	6	0.9295	0.9078	09m 08s
-1664 Nov 22	6	0.9323	0.9083	09m 26s
-1646 Dec 03	6	0.9353	0.9095	09m 36s
-0984 Nov 04 [a]	27	-1.0234	0.9099	-
0123 Nov 06 [b]	64	0.9783	0.9098	08m 20s
0141 Nov 16 [b]	64	0.9854	0.9089	08m 31s
0159 Nov 27 [b]	64	0.9908	0.9087	08m 34s
0177 Dec 08	64	0.9944	0.9093	08m 28s
1565 Nov 22	135	0.9564	0.9092	09m 37s
1583 Dec 14	135	0.9471	0.9083	10m 03s
1601 Dec 24	135	0.9402	0.9078	10m 14s
1620 Jan 04	135	0.9321	0.9081	10m 13s
1638 Jan 15	135	0.9242	0.9090	10m 00s
2485 Dec 07 [a]	140	1.0242	0.9100	-

[a] Non-central annular eclipse (with one limit).
[b] Central annular eclipse with one limit.

Seventeen annular eclipses have a maximum magnitude (at greatest eclipse) greater than or equal to 0.9995 (Table 3-15). All of these events have central durations lasting 3 seconds or less. The annular eclipse with the largest magnitude (at greatest eclipse) occurs on 2931 Dec 30 with a magnitude of 0.99998.

Table 3-15. Annular Eclipses with Magnitude 0.9995 or More

Date (Dynamical Time)	Saros	Gamma	Eclipse Magnitude	Central Duration
-1800 Apr 03	10	0.1778	0.9997	00m 02s
-1734 Sep 18	26	-0.5105	0.9995	00m 03s
-1725 Mar 17	2	0.0105	0.9997	00m 01s
-1624 Oct 02	8	0.9377	0.9995	00m 02s
-1590 Jun 20	21	-0.0376	0.9997	00m 02s
-1482 Feb 27	16	0.3992	0.9997	00m 02s
-1326 Apr 14	27	0.0409	0.9996	00m 02s
-0124 Sep 07	81	0.7642	0.9999	00m 00s
1087 Aug 01	111	0.1644	0.9996	00m 02s
1384 Aug 17	125	0.5354	0.9999	00m 01s
1704 Nov 27	118	0.6716	0.9999	00m 01s
1822 Feb 21	137	0.6914	0.9996	00m 02s
1858 Mar 15	137	0.6461	0.9996	00m 02s
1876 Mar 25	137	0.6142	0.9999	00m 01s
1948 May 09	137	0.4133	0.9999	00m 00s
2862 Sep 15	158	0.5956	0.9999	00m 01s
2931 Dec 30	166	0.1511	0.99998	00m 00s

3.7 Extremes in Eclipse Magnitude – Total Eclipses

Nineteen total eclipses have a maximum magnitude less than or equal to 1.0075 (Table 3-16). Six of these eclipses are central while the remaining thirteen are non-central. The smallest magnitude was the total eclipse of –0839 Jul 26 with a magnitude of just 1.0002.

Table 3-16. Total Eclipses with Magnitude 1.0075 or Less

Date (Dynamical Time)	Saros	Gamma	Eclipse Magnitude	Central Duration
-1038 Apr 09 [a]	22	1.0023	1.0034	–
-0915 Feb 28 [b]	25	-1.0012	1.0004	–
-0909 Nov 15 [a]	57	0.9976	1.0050	–
-0905 Mar 10 [b]	54	-1.0053	1.0072	–
-0839 Jul 26 [a]	32	1.0095	1.0002	–
-0829 Aug 05 [a]	61	0.9972	1.0064	–
-0159 Jul 08 [b]	53	-1.0096	1.0051	–
0854 Feb 01	83	-0.9582	1.0065	00m 22s
0861 Sep 08 [b]	87	-1.0032	1.0053	–
0865 Jan 01	84	0.9518	1.0073	00m 36s
0883 Jan 12	84	0.9609	1.0057	00m 27s
0890 Feb 23 [b]	83	-1.0005	1.0005	–
0901 Jan 23	84	0.9731	1.0042	00m 19s
0919 Feb 03	84	0.9909	1.0020	00m 09s
0994 Aug 09 [a]	119	0.9985	1.0017	–
1957 Oct 23 [b]	123	-1.0022	1.0013	–
2459 Jun 01 [b]	164	-1.0097	1.0038	–
2518 Mar 12	138	0.9200	1.0071	00m 31s
2542 Dec 08 [b]	170	-0.9975	1.0072	–

[a] Non-central total eclipse at high northern latitudes.
[b] Non-central total eclipse at high southern latitudes.

Sixteen total eclipses have a maximum magnitude greater than or equal to 1.080. Their central durations all exceed six minutes with nearly half exceeding seven minutes. All take place during the period of the year when Earth is near aphelion (May-July), resulting in a smaller than normal diameter of the solar disk. The total eclipse with the largest magnitude (1.0813) occurred on 0504 May 29. The total eclipse with the longest duration of totality occurs on 2186 Jul 16 with a magnitude of 1.0805. The sixteen eclipses in Table 3-17 belong to just five Saros series.

Table 3-17. Total Eclipses with Magnitude 1.080 or More

Date (Dynamical Time)	Saros	Gamma	Eclipse Magnitude	Central Duration
-1337 May 14	26	0.1487	1.0801	06m 51s
-1319 May 25	26	0.2236	1.0807	06m 41s
-1301 Jun 05	26	0.2982	1.0805	06m 25s
-1160 May 07	29	-0.2990	1.0806	06m 45s
-1142 May 18	29	-0.3742	1.0809	06m 56s
-1124 May 28	29	-0.4490	1.0804	07m 03s
0327 Jun 06	81	-0.0413	1.0810	07m 03s
0345 Jun 16	81	-0.1162	1.0811	07m 17s
0363 Jun 27	81	-0.1899	1.0804	07m 24s
0486 May 19	84	0.1193	1.0806	06m 54s
0504 May 29	84	0.1927	1.0813	06m 44s
0522 Jun 10	84	0.2675	1.0812	06m 28s
0540 Jun 20	84	0.3414	1.0801	06m 07s
2150 Jun 25	139	-0.0910	1.0802	07m 14s
2168 Jul 05	139	-0.1660	1.0807	07m 26s
2186 Jul 16	139	-0.2396	1.0805	07m 29s

3.8 Extremes in Eclipse Magnitude – Hybrid Eclipses

Fourteen hybrid eclipses have a maximum magnitude (at greatest eclipse) less than or equal to 1.00025. All of these events are central with a central duration of totality of 1 second or less.

Table 3-18. Hybrid Eclipses with Magnitude 1.00025 or Less

Date (Dynamical Time)	Saros	Gamma	Eclipse Magnitude	Central Duration
-1747 Nov 10	5	-0.7406	1.0001	00m 00s
-1716 Sep 28	26	-0.4927	1.0002	00m 01s
-1641 Mar 17	13	-0.2772	1.0002	00m 01s
-0819 Jan 18	47	0.3047	1.0001	00m 00s
-0097 Mar 17	57	-0.5539	1.0001	00m 00s
0121 Dec 27	82	-0.6196	1.0002	00m 01s
0403 Nov 01	88	-0.1968	1.0001	00m 01s
1339 Jul 07	106	0.6451	1.0002	00m 01s
1612 Nov 22	136	-0.7691	1.0002	00m 01s
1627 Aug 11	139	0.9401	1.0001	00m 00s
1702 Jul 24	131	0.3160	1.0001	00m 01s
1804 Feb 11	137	0.7053	1.0000	00m 00s
1894 Apr 06	137	0.5740	1.0001	00m 01s
1986 Oct 03	124	0.9931	1.0000	00m 00s

Seven hybrid eclipses have a maximum magnitude (at greatest eclipse) greater than or equal to 1.0170. All of these events are central with a duration of totality of 1 minute 34 seconds or more.

Table 3-19. Hybrid Eclipses with Magnitude 1.0170 or More

Date (Dynamical Time)	Saros	Gamma	Eclipse Magnitude	Central Duration
–0437 Dec 17	54	0.1286	1.0173	01m 45s
–0100 May 17	65	–0.1912	1.0170	01m 44s
0508 Sep 11	91	0.0826	1.0173	01m 45s
1199 Jan 28	108	0.0033	1.0174	01m 45s
1228 Jan 08	109	–0.0068	1.0176	01m 40s
1564 Jun 08	120	0.1253	1.0174	01m 44s
2172 Oct 17	146	–0.1484	1.0174	01m 34s

3.9 Greatest Central Duration – Annular Eclipses

Ten annular eclipses have a central duration (i.e., central line duration at greatest eclipse) of twelve minutes or more. There are no cases between the years 1974 and 3000.

Table 3-20. Annular Eclipses with Central Duration of 12 Minutes or More

Date (Dynamical Time)	Saros	Gamma	Eclipse Magnitude	Central Line Duration
-1655 Dec 12	25	0.6207	0.9147	12m 07s
-0195 Dec 11	58	0.4971	0.9153	12m 04s
-0177 Dec 22	58	0.5030	0.9165	12m 08s
0132 Nov 25	83	0.5691	0.9144	12m 16s
0150 Dec 07	83	0.5630	0.9147	12m 23s
0168 Dec 17	83	0.5579	0.9156	12m 14s
1628 Dec 25	116	0.6265	0.9153	12m 02s
1937 Dec 02	141	0.4389	0.9184	12m 00s
1955 Dec 14	141	0.4266	0.9176	12m 09s
1973 Dec 24	141	0.4171	0.9174	12m 02s

3.10 Greatest Central Duration – Total Eclipses

Forty-four total eclipses have a central duration (i.e., central line duration at greatest eclipse) of 7 minutes or more. These eclipses all take place during the period of the year when Earth is near the aphelion of its orbit (June-July), resulting in a smaller than normal diameter of the solar disk. The total eclipse with the longest duration of totality occurs on 2186 Jul 16. Its central duration of 7 minutes 29 seconds is very close to the theoretical maximum of 7 minutes 32.1 seconds during that epoch. All 44 eclipses belong to just 12 Saros series. The eclipses of 1937, 1955 and 1973 all belong to Saros 136. This is the same Saros producing the 6+ minute eclipses in 1991, 2009 and 2027.

Table 3-21. Total Eclipses with Central Duration of 7 Minutes or More

Date (Dynamical Time)	Saros	Gamma	Eclipse Magnitude	Central Duration
-1460 Jun 22	23	-0.226	1.078	07m 04s
-1442 Jul 03	23	-0.293	1.076	07m 05s
-1124 May 28	29	-0.449	1.080	07m 03s
-1106 Jun 09	29	-0.524	1.079	07m 04s
-0779 May 24	54	-0.548	1.079	07m 12s
-0761 Jun 05	54	-0.474	1.080	07m 25s
-0743 Jun 15	54	-0.400	1.079	07m 28s
-0725 Jun 26	54	-0.329	1.078	07m 18s
-0707 Jul 07	54	-0.261	1.075	07m 00s
-0443 Apr 30	60	-0.319	1.077	07m 01s
-0425 May 12	60	-0.247	1.078	07m 12s
-0407 May 22	60	-0.173	1.078	07m 13s
-0389 Jun 02	60	-0.098	1.077	07m 04s
0114 May 22	78	-0.268	1.075	07m 06s
0132 Jun 01	78	-0.193	1.077	07m 14s
0150 Jun 12	78	-0.119	1.079	07m 13s
0168 Jun 23	78	-0.044	1.079	07m 03s
0327 Jun 06	81	-0.041	1.081	07m 03s
0345 Jun 16	81	-0.116	1.081	07m 17s
0363 Jun 27	81	-0.190	1.080	07m 24s
0381 Jul 08	81	-0.261	1.079	07m 22s
0399 Jul 19	81	-0.329	1.076	07m 11s
0681 May 23	87	-0.354	1.080	07m 10s
0699 Jun 03	87	-0.429	1.079	07m 17s
0717 Jun 13	87	-0.503	1.078	07m 15s
0735 Jun 25	87	-0.578	1.076	07m 02s
1044 May 29	112	-0.553	1.077	07m 12s
1062 Jun 09	112	-0.479	1.078	07m 20s
1080 Jun 20	112	-0.405	1.078	07m 18s
1098 Jul 01	112	-0.332	1.077	07m 05s
1937 Jun 08	136	-0.225	1.075	07m 04s
1955 Jun 20	136	-0.153	1.078	07m 08s
1973 Jun 30	136	-0.079	1.079	07m 04s
2150 Jun 25	139	-0.091	1.080	07m 14s
2168 Jul 05	139	-0.166	1.081	07m 26s
2186 Jul 16	139	-0.240	1.080	07m 29s
2204 Jul 27	139	-0.313	1.079	07m 22s
2222 Aug 08	139	-0.384	1.077	07m 06s
2504 Jun 14	145	-0.428	1.077	07m 10s
2522 Jun 25	145	-0.499	1.077	07m 12s
2540 Jul 05	145	-0.572	1.076	07m 04s
2867 Jun 23	170	-0.462	1.077	07m 10s
2885 Jul 03	170	-0.391	1.078	07m 11s
2903 Jul 16	170	-0.318	1.078	07m 04s

3.11 Greatest Central Duration – Hybrid Eclipses

Ten hybrid eclipses have a central duration (i.e., central line duration at greatest eclipse) greater than or equal to 1 minute 40 seconds.

Table 3-22. Hybrid Eclipses with Central Duration 1m 40s or More

Date (Dynamical Time)	Saros	Gamma	Eclipse Magnitude	Central Duration
-1297 Sep 17	33	0.0674	1.0168	01m 40s
-0979 Aug 13	39	-0.2387	1.0168	01m 48s
-0437 Dec 17	54	0.1286	1.0173	01m 45s
-0100 May 17	65	-0.1912	1.0170	01m 44s
0508 Sep 11	91	0.0826	1.0173	01m 45s
1199 Jan 28	108	0.0033	1.0174	01m 45s
1228 Jan 08	109	-0.0068	1.0176	01m 40s
1350 Nov 30	112	0.2227	1.0166	01m 42s
1423 Jul 08	117	-0.1158	1.0161	01m 45s
1564 Jun 08	120	0.1253	1.0174	01m 44s

3.12 Theoretical Maximum Duration of Annularity

The theoretical maximum duration of an annular solar eclipse slowly varies due to long term secular changes in the eccentricity of Earth's orbit and the longitude of its perihelion. Although the maximum theoretical duration differs between the ascending and descending nodes, the durations are equal in the year +1246 because the Sun's perihelion then coincides with longitude 270°.

Table 3-23 lists the maximum duration theoretically possible over the period –2000 to +7000 [Meeus, 2007]. The values here are 0.2 seconds smaller than those in Meeus due to the use of a slightly larger value for the Moon's radius *k* (Sect. 1.5).

Table 3-23. Theoretical Maximum Duration of Annularity

Year	Duration at Ascending Node	Duration at Descending Node
-2000	12m 16.8s	11m 40.9s
-1000	12m 30.2s	12m 04.8s
0000	12m 35.5s	12m 21.3s
+1000	12m 32.3s	12m 29.5s
+2000	12m 20.7s	12m 29.2s
+3000	12m 01.4s	12m 20.6s
+4000	11m 35.6s	12m 04.6s
+5000	11m 04.9s	11m 42.4s
+6000	10m 31.0s	11m 15.9s
+7000	10m 33.1s	11m 15.7s

The absolute maximum of 12 minutes 35.6 seconds occurred at the Moon's ascending node about the year +125. An inflexion point occurs between years +6000 and +7000, when the maximum possible durations increase once again.

All calculations in the *Catalog* use the same mean lunar radius "k" for both annular and total eclipses (Sect. 1.5). Consequently, the annular durations are extended several seconds since they include the appearance of Baily's beads at the start and end of the antumbral phase.

3.13 Theoretical Maximum Duration of Totality

The theoretical maximum duration of a total solar eclipse for a point on Earth's surface slowly varies with time. This effect is due to long term secular changes in the eccentricity of Earth's orbit and the longitude of its perihelion. That eccentricity is now 0.01671, but at some epochs in the distant past or future the orbit was (will be) almost exactly circular, and at other times the eccentricity can be as large as 0.06.

Table 3-24 lists the maximum duration theoretically possible from -2000 to +7000 [Meeus 2003]. The values here are 0.1 to 0.2 seconds larger than those in Meeus due to the use of a slightly larger value for the Moon's radius k (Sect. 1.5).

Table 3-24. Theoretical Maximum Duration of Totality

Year	Duration at Ascending Node	Duration at Descending Node
-2000	7m 07.4s	7m 29.8s
-1000	7m 19.1s	7m 34.6s
0000	7m 27.4s	7m 36.0s
+1000	7m 31.9s	7m 33.6s
+2000	7m 32.3s	7m 27.1s
+3000	7m 28.8s	7m 17.1s
+4000	7m 22.1s	7m 04.0s
+5000	7m 12.9s	6m 48.7s
+6000	7m 03.3s	6m 32.5s
+7000	7m 01.9s	6m 32.8s

The absolute maximum of 7 minutes 36.1 seconds occurred at the Moon's descending node about the year -120. Prior to -2000 there must have been epochs when the maximum possible duration was even larger due to an even greater value of the eccentricity of Earth's orbit.

3.14 Eclipse Duos

A duo is a pair of eclipses separated by one lunation (synodic month). Of the 11,898 eclipses in the *Catalog*, 1361 eclipses (11.4%) belong to a duo. In most cases, both eclipses in a duo are partial eclipses. However there are 14 instances where one eclipse is partial and the other is total. The dates and combinations are listed in Table 3-25.

Table 3-25. Eclipse Duos of Two Types

Dates (Dynamical Time)	Eclipse Combinations
-1859 May-Jun	TP
-1718 Apr-May	TP
-1310 May-Jun	PT
-1169 Apr-May	PT
-1028 Mar-Apr	PT
-0575 May-Jun	TP
-0434 Apr-May	TP
-0159 Jul-Aug	TP
-0026 May-Jun	PT
1248 May-Jun	TP
1928 May-Jun	TP
2195 Jul-Aug	PT
2459 May-Jun	PT
2912 Jul-Aug	TP

3.15 Eclipse Duos in One Calendar Month

There are 43 instances where both members of an eclipse duo occur in one calendar month. In all cases, both eclipses in the duos are partial. The year and month of each occurrence appears in Table 3-26.

Table 3-26. Two Eclipses in One Calendar Month

-1957 Mar	-1035 Aug	-0416 May	0629 Mar	2206 Dec
-1805 Jan	-1024 Jul	0007 Aug	1063 May	2261 Jan
-1610 Jul	-1013 Jun	0018 Jul	1150 Mar	2282 Nov
-1534 Jun	-0688 Dec	0097 Apr	1215 Mar	2304 Sep
-1523 May	-0677 Nov	0463 Aug	1631 May	2380 Aug
-1447 Apr	-0601 Oct	0528 Aug	1696 May	2684 Oct
-1209 Dec	-0590 Sep	0539 Jul	1805 Jan	2785 May
-1122 Oct	-0514 Aug	0542 May	1880 Dec	
-1111 Sep	-0503 Jul	0618 Apr	2000 Jul	

3.16 January–March Eclipse Duos

The mean length of one synodic month is 29.5306 days (2000). Since this is longer than the month of February, it is possible to have one member of an eclipse duo in January followed by the second in March. There are four instances of such a rare January/March duo in the *Catalog*: -1881, -1295, 1291 and 1794. In all cases, both eclipses in the duos are partial.

3.17 Eclipses on February 29

There are nine instances of a solar eclipse occurring on February 29. Five eclipses are partial, two are annular and two are total. A list of eclipses on February 29 with physical parameters appears in Table 3-27.

Table 3-27. Eclipses on February 29

Date (Dynamical Time)	Type	Saros	Gamma	Eclipse Magnitude	Central Duration
-1436 Feb 29	P	7	-1.0586	0.9059	-
-0896 Feb 29	T	35	-0.3068	1.0652	05m 04s
-0356 Feb 29	T	63	0.4386	1.0628	05m 11s
0108 Feb 29	P	51	-1.5625	0.0082	-
0184 Feb 29	P	91	1.1684	0.6947	
0648 Feb 29	A	79	-0.7722	0.9257	06m 44s
1188 Feb 29	A	107	0.0292	0.9294	08m 14s
2416 Feb 29	P	127	-1.4865	0.1279	-
2872 Feb 29	P	144	1.3315	0.3864	-

3.18 Eclipse Seasons

The 5.1° inclination of the lunar orbit around Earth means that the Moon's orbit crosses the ecliptic at two points or nodes. If New Moon takes place within about 16° of a node[3], then a solar eclipse will be visible from some location on Earth.

[3] The actual value ranges from 15.2° to 16.8° because of the eccentricity of the Moon's and Earth's orbits. If an eclipse occurs within 9.5° to 10.9° of a node, the eclipse will be central (annular, hybrid, or total).

The Sun makes one complete circuit of the ecliptic in 365.24 days, so its average angular velocity is 0.99° per day. At this rate, it takes 34.5 days for the Sun to cross the 34° wide eclipse zone centered on each node. Because the Moon's orbit with respect to the Sun has a mean duration of 29.53 days, there will always be one and possibly two solar eclipses during each 34.5-day interval when the Sun passes through the nodal eclipse zones. These time periods are called eclipse seasons.

The mid-point of each eclipse season is separated by 173.3 days because this is the mean time for the Sun to travel from one node to the next. The period is a little less than half a calendar year because the lunar nodes slowly regress westward by 19.3° per year.

3.19 Quincena

The mean time interval between New Moon and Full Moon is 14.77 days. This is less than half the duration of an eclipse season. As a consequence, the same Sun–node alignment geometry responsible for producing a solar eclipse always results in a complementary lunar eclipse within a fortnight. The lunar eclipse may either precede or succeed the solar eclipse. In either case, the pair of eclipses are referred to here as a quincena. The QLE (Quincena Lunar Eclipse parameter) identifies the type of the lunar eclipse and whether it precedes or succeeds a particular solar eclipse. There are three basic types of lunar eclipses:

1) **n = penumbral lunar eclipse (Moon partly or completely within Earth's penumbral shadow)**
2) **p = partial lunar eclipse (Moon partly within Earth's umbral shadow)**
3) **t = total lunar eclipse (Moon completely within Earth's umbral shadow)**

The QLE is a two character string consisting of one or more of the above lunar eclipse types. The first character in the QLE identifies a lunar eclipse preceding a solar eclipse while the second character identifies a lunar eclipse succeeding a solar eclipse. In most instances, one of the two characters is "-" indicating a single lunar eclipse either precedes or succeeds the solar eclipse. On rare occasions, a double quincena occurs in which a solar eclipse is both preceded and succeeded by lunar eclipses.

3.20 Quincena Combinations with Total Solar Eclipses

A total solar eclipse can be preceded or succeeded by a total lunar eclipse (8.8%), a partial lunar eclipse (49.8%), or a penumbral lunar eclipse (28.2%). Double quincenas (a solar eclipse is both preceded and succeeded by a lunar eclipse) occur with a frequency of 13.1% and always consist of two penumbral lunar eclipses. A detailed list of total solar eclipse and quincena lunar eclipse combinations appears in Table 3-28.

Table 3-28. Quincena Combinations with Total Solar Eclipses

Quincena Lunar Eclipse	QLE	Number	Percent	Examples (Years)
- total	-t	147	4.6%	..., 1957, 1968, 2015, 2033, 2044,...
total -	t-	133	4.2%	..., 1985, 2003, 2043, 2061, 2072,...
- partial	-p	801	25.2%	..., 2001, 2008, 2019, 2026, 2037,...
partial -	p-	782	24.6%	..., 1992, 1999, 2010, 2017, 2021,...
- penumbral	-n	432	13.6%	..., 1994, 1998, 2012, 2016, 2030,...
penumbral -	n-	462	14.6%	..., 2002, 2006, 2020, 2024, 2038,...
penumbral - penumbral	nn	416	13.1%	..., 1973, 1991, 2009, 2027, 2096,...

3.21 Quincena Combinations with Annular Solar Eclipses

An annular solar eclipse can be preceded or succeeded by a total lunar eclipse (9.0%), a partial lunar eclipse (57.4%), or a penumbral lunar eclipse (8.5%). Double quincenas consisting of two penumbral lunar eclipses (23.8%) are common, but penumbral-partial combinations are rare (1.3%). A list of annular solar eclipse and quincena lunar eclipse combinations is found in Table 3-29.

Table 3-29. Quincena Combinations with Annular Solar Eclipses

Quincena Lunar Eclipse	QLE	Number	Percent	Examples (Years)
– total	–t	178	4.5%	..., 1990, 2008, 2026, 2044, 2102,...
total –	t–	178	4.5%	..., 1891, 2003, 2014, 2021, 2032,...
– partial	–p	1147	29.0%	..., 1994, 2005, 2012, 2023, 2030,...
partial –	p–	1122	28.4%	..., 1995, 2006, 2010, 2024, 2028,...
– penumbral	–n	160	4.0%	..., 1991, 2001, 2009, 2016, 2019,...
penumbral –	n–	179	4.5%	..., 1981, 1999, 2017, 2035, 2042,...
partial - penumbral	pn	27	0.7%	..., 1608, 1749, 2013, 2147, 2288,...
penumbral - partial	np	24	0.6%	..., 1694, 1835, 1958, 2819, 2960]
penumbral - penumbral	nn	941	23.8%	..., 1998, 2002, 2020, 2031, 2038,...

3.22 Quincena Combinations with Hybrid Solar Eclipses

A hybrid solar eclipse can be preceded or succeeded by a total lunar eclipse (3.0%), a partial lunar eclipse (51.1%), or a penumbral lunar eclipse (24.9%). Double quincenas consisting of two penumbral lunar eclipses (20.9%) are also fairly common. A complete list of hybrid solar eclipse and quincena lunar eclipse combinations appears in Table 3-30.

Table 3-30. Quincena Combinations with Hybrid Solar Eclipses

Quincena Lunar Eclipse	QLE	Number	Percent	Examples (Years)
– total	–t	5	0.9%	...,-1989, -1848, -1642, 163, 1986]
total –	t–	12	2.1%	..., 1627, 1645, 1768, 1909, 2050]
– partial	–p	124	21.8%	..., 1827, 1845, 2164, 2182, 2323,...
partial –	p–	167	29.3%	..., 1912, 1930, 2209, 2350, 2368,...
– penumbral	–n	85	14.9%	..., 1987, 2005, 2023, 2385, 2508,...
penumbral –	n–	57	10.0%	..., 1702, 1908, 2013, 2031, 2049,...
penumbral - penumbral	nn	119	20.9%	..., 1843, 1846, 2172, 2190, 2208,...

3.23 Quincena Combinations with Partial Solar Eclipses

A partial solar eclipse is almost always preceded or succeeded by a total lunar eclipse (99.6 %). On very rare occasions (0.3%), a partial lunar eclipse occurs before a partial solar eclipse. However, there are no instances of a partial lunar eclipse following a partial solar eclipse. No double quincenas occur with partial solar eclipses. A list of partial solar eclipse and quincena lunar eclipse combinations is found in Table 3-31.

Table 3-31. Quincena Combinations with Partial Solar Eclipses

Quincena Lunar Eclipse	QLE	Number	Percent	Examples (Years)
– total	–t	2102	50.0%	..., 2000, 2004, 2011, 2015, 2018,...
total –	t–	2085	49.6%	..., 2000, 2007, 2011, 2014, 2018,...
partial –	p–	13	0.3%	..., -753, -196, 2086, 2607, 2625]

Section 4: Eclipses and the Moon's Orbit

4.1 Introduction

The Moon revolves around Earth in an elliptical orbit with a mean eccentricity of 0.0549. Thus, the Moon's center-to-center distance from Earth varies with mean values of 363,396 km at perigee to 405,504 km at apogee. The lunar orbital period with respect to the stars (sidereal month) is 27.32166 days (27d 07h 43m 12s). However, there are three other orbital periods or months that are crucial to the understanding and prediction of eclipses. These three cycles and the harmonics between them determine when, where, and how solar (and lunar) eclipses occur.

The mutual gravitational force between the Sun and Moon is over twice as large as between the Moon and Earth. For this reason, the Sun plays a dominant role in perturbing the Moon's motion. The ever changing distances and relative positions between the Sun, Moon, and Earth, the inclination of the Moon's orbit, the oblateness of Earth, and (to a lesser extent) the gravitational attraction of the other planets all act to throw the Moon's orbital parameters into a constant state of change. Although the Moon's position and velocity can be described by the classic Keplerian orbital elements, such osculating elements are only valid for a single instant in time (Chapront-Touze' and Chapront, 1991). Nevertheless, these instantaneous parameters are of value in understanding the Moon's complex motions particularly with respect to the three major orbital cycles that govern eclipses.

4.2 Synodic Month

The most familiar lunar cycle is the synodic month because it governs the well-known cycle of the Moon's phases. The Moon has no light of its own but shines by reflected sunlight. As a consequence, the geometry of its orbital position relative to the Sun and Earth determines the Moon's apparent phase.

The mean length of the synodic month is 29.53059 days (29d 12h 44m 03s). This is nearly 2.21 days longer than the sidereal month. As the Moon revolves around Earth, both objects also progress in orbit around the Sun. After completing one revolution with respect to the stars, the Moon must continue a little farther along its orbit to catch up to the same position it started from relative to the Sun and Earth. This explains why the mean synodic month is longer than the sidereal month.

According to astronomical convention, New Moon is defined as the instant when the geocentric ecliptic longitudes of the Sun and Moon are equal. When the synodic month is measured from New Moon to New Moon, it is sometimes referred to as a lunation, and we will follow that usage here. Historically, the phases of the Moon have been used as the basis of lunar calendars by many cultures around the world. The major problem with such calendars is that the year, based on the solar calendar, is not evenly divisible by a whole number of lunations. Consequently, most lunar calendars are actually lunisolar calendars (e.g., Chinese, Hebrew, and Hindu) that include intercalary months to keep the seasons in step with the year.

The duration of the lunation actually varies from its mean value by up to seven hours. For instance, Table 4-1 contains details for all lunations in 2008. The first column lists the decimal date of every New Moon throughout the year (Terrestrial Dynamical Time), while the second column gives the duration of each lunation. The third column is the difference between the actual and mean lunation. The first lunation of the year (Jan 08) was 03h 23m longer than the mean. Continuing through 2008, the length of each lunation drops and reaches a minimum of 05h 48m shorter than the mean value (Jun 03). The duration now increases with each succeeding lunation until the maximum value of the year is reached of 06h 49m longer than the mean (Dec 27).

Table 4-1 New Moon and Lunation Length in 2008

Date of New Moon (Dynamical Time)	Length of Lunation	Difference From Mean Lunation	Moon's True Anomaly
2008 Jan 08.4849	29d 16h 07m	+03h 23m	242.4°
2008 Feb 07.1567	29d 13h 30m	+00h 46m	280.0°
2008 Mar 07.7190	29d 10h 41m	-02h 03m	310.8°
2008 Apr 06.1642	29d 08h 23m	-04h 21m	332.7°
2008 May 05.5134	29d 07h 04m	-05h 40m	349.4°
2008 Jun 03.8081	29d 06h 56m	-05h 48m	4.4°
2008 Jul 03.0970	29d 07h 54m	-04h 50m	20.1°
2008 Aug 01.4261	29d 09h 45m	-02h 59m	39.2°
2008 Aug 30.8327	29d 12h 14m	-00h 30m	64.9°
2008 Sep 29.3426	29d 15h 02m	+02h 18m	98.7°
2008 Oct 28.9687	29d 17h 41m	+04h 57m	133.4°
2008 Nov 27.7053	29d 19h 28m	+06h 44m	161.9°
2008 Dec 27.5163	29d 19h 33m	+06h 49m	186.6°

What is the cause of this odd behavior? The last column in Table 4-1 gives a clue; it contains the Moon's true anomaly at the instant of New Moon. The true anomaly is the angle between the Moon's position and the point of perigee along its orbit. In other words, it is the orbital longitude of the Moon with respect to perigee. Table 4-1 shows that when New Moon occurs near perigee (true anomaly = 0°), the length of the lunation is at a minimum (e.g., Jun 03). Similarly, when New Moon occurs near apogee (true anomaly = 180°), the length of the lunation reaches a maximum (e.g., Dec 27).

Figure 4–1: Length of Lunation for 2008 – 2010

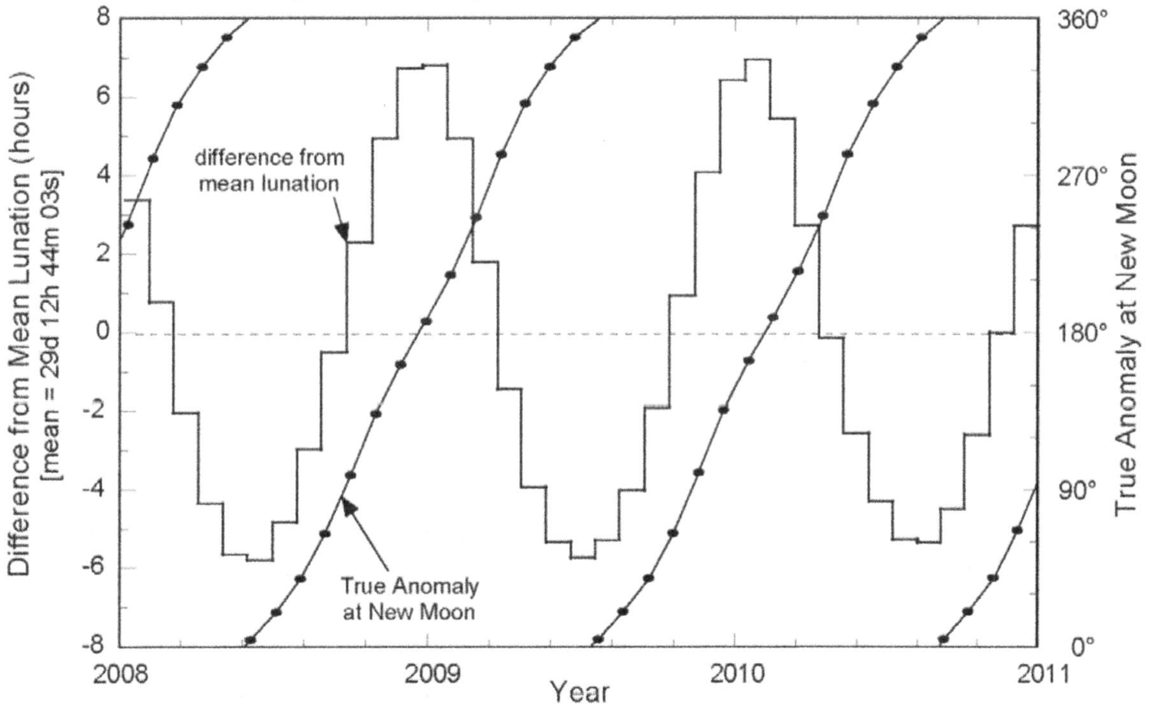

This relationship is quite apparent when viewed graphically. Figure 4-1 plots the difference from mean lunation (histogram) and the Moon's true anomaly (diagonal curves) for every New Moon from 2008 through 2010. The left-hand scale is for the difference from mean lunation, while the right-hand scale is for the true anomaly. The shortest lunations are clearly correlated with New Moon at perigee, while the longest lunations occur at apogee. From the figure, the length of this cycle appears to be about 412 days. The reason why must wait until the next section.

The Moon's orbital period with respect to perigee is the anomalistic month and has a duration of approximately 27.55 days. The lock-step rhythm between the lunation length and true anomaly can be explained with the help of the anomalistic month and Figure 4-2. It illustrates the Moon's orbit around Earth and Earth's orbit around the Sun. The relative sizes and distances of the Sun, Moon, and Earth as well as the eccentricity of the Moon's orbit are all exaggerated for clarity. The major axis of the Moon's orbit marks the positions of perigee and apogee.

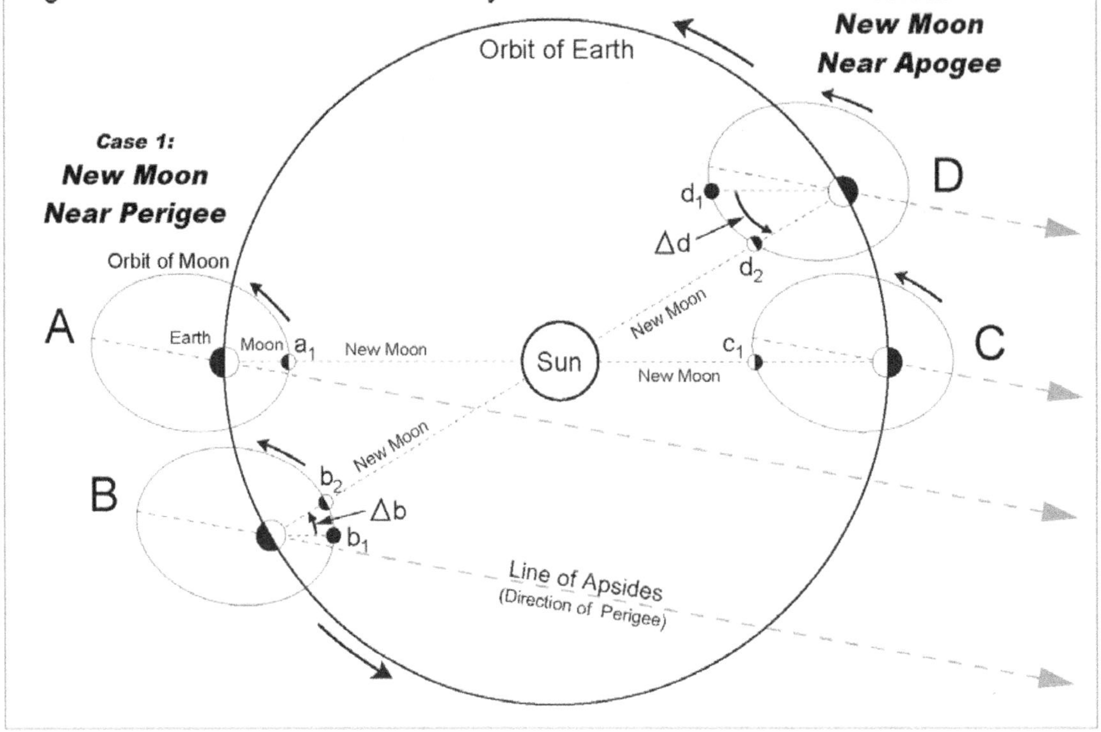

Figure 4–2: Moon's Orbit and the Synodic Month

Two distinct cases—each consisting of two revolutions of the Moon around Earth—are depicted in Figure 4-2. The first case covers the New Moon geometry around perigee. The orbit marked A shows New Moon taking place near perigee at position a_1. One anomalistic month later (orbit B), the Moon has returned to the same position relative to perigee (marked b_1). However, Earth has traveled about 30° around its orbit so the Sun's direction relative to the Moon's major axis has shifted. The Moon must travel an additional distance of Δb in its orbit before reaching the New Moon phase at b_2. This graphically demonstrates why the synodic month is longer (~1.98 days) than the anomalistic month.

The second case takes place about half a year later. New Moon then occurs near apogee (orbit C, position c_1). After one anomalistic month, the Moon has returned to the same location with respect to apogee (orbit D, position d_1). Once again, Earth has traveled about 30° around its orbit so the Moon must revolve an additional distance of Δd before reaching the New Moon phase at position d_2.

An inspection of orbits B and D reveals that the orbital arc Δd is longer that Δb. This means that the Moon must cover a greater orbital distance to reach New Moon near apogee as compared to perigee. Furthermore, the Moon's orbital velocity is slower at apogee so it takes longer to travel a given distance. Thus, the length of the lunation is shorter than average when New Moon occurs near perigee and longer than average when New Moon occurs near apogee.

Earth's elliptical orbit around the Sun also factors into the length of the lunation. With an eccentricity of 0.0167, Earth's orbit is about one third as elliptical as the Moon's orbit. Nevertheless, it affects the length of the lunation by producing shorter lunations near aphelion and longer lunations near perihelion.

During the 5000-year period covered in this catalog, there are 61841 complete lunations. The shortest lunation began on –1602 Jun 03 and lasted 29.26574 days (29d 06h 22m 40s; 6h 21m 23s shorter than the mean). The longest lunation began on –1868 Nov 27 and lasted 29.84089 days (29d 20h 10m 53s; 7h 26m 50s longer than the mean). Thus, the duration of the lunation varies over a range of 13h 48m 13s during this time interval.

Figure 4–3: Length of Lunation Over 5000 Years

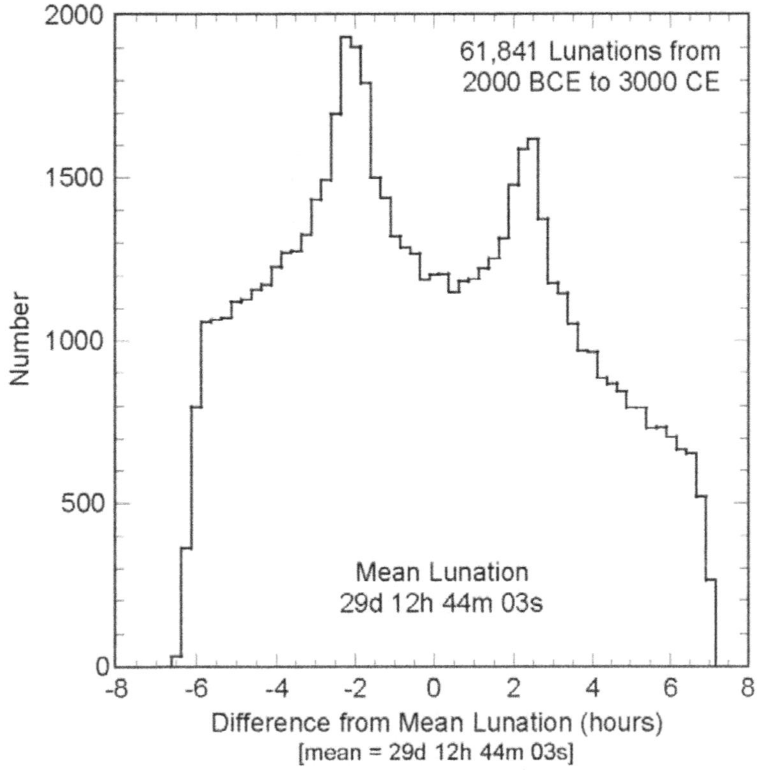

The histogram presented in Figure 4-3 shows the distribution in the length of the lunation over 5000 years. To create the histogram, the durations of individual lunations were binned into 30-minute groups. It might seem reasonable to expect a simple bell-shaped Gaussian curve. However, the results are surprising because the distribution in lunation length has two distinct peaks. This bifurcation can be understood if the lunation length, which depends primarily on the Moon's distance, is considered as a series of sine functions. The extremes of a sine function always occur more frequently than the mean, which is just what is seen in Figure 4-3. For a more detailed discussion, see Meeus (1997).

4.3 Anomalistic Month

The anomalistic month is defined as the revolution of the Moon around its elliptical orbit as measured from perigee to perigee. The length of this period can vary by several days from its mean value of 27.55455 days (27d 13h 18m 33s). Figure 4-4 plots the difference of the anomalistic month from the mean value for the 3-year interval 2008 through 2010. Also plotted is the difference between the mean longitudes of the Sun and perigee. This is just the angle between the Sun and the Moon's major axis in the direction of perigee. The left-hand scale is the length of the anomalistic month minus the mean value, while the right-hand scale is for the difference in longitude (Sun–perigee). For comparison, the lunation length minus its mean value is also plotted (light gray).

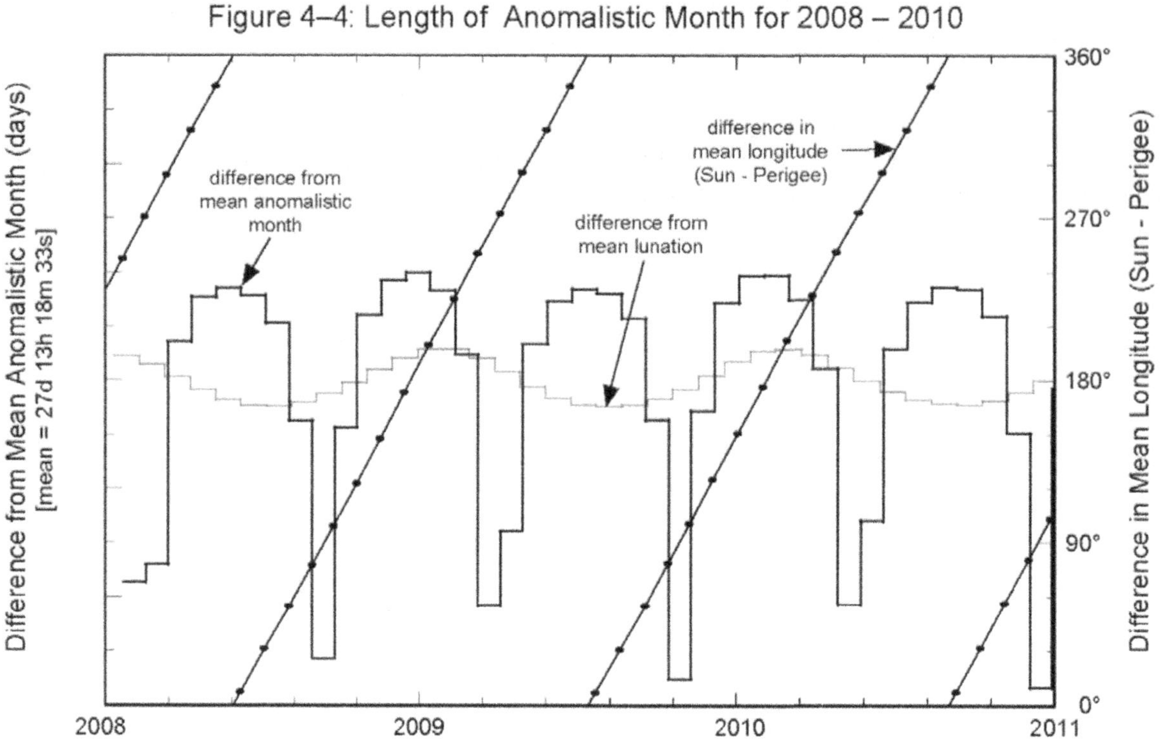

The variation in the length of the anomalistic month is much larger than that of the lunation. Figure 4-4 shows the anomalistic month is typically within 1 day of its mean value. But once or twice every 7 to 8 months, the anomalistic month is significantly shorter than the mean by 2 to nearly 3 days. The difference in longitude of the Sun and perigee show that the shortest anomalistic months are correlated with values of 90° and 270°, when the line of apsides is perpendicular to the Sun's direction.

In comparison, the longest anomalistic months take place when the difference in longitude passes through 0° or 180°. The line of apsides is then directed towards or away from the Sun. The maximum duration of the anomalistic month is then about 28.5 days (1.0 day longer than the mean). The Earth–Sun distance also influences the anomalistic month by causing greater extremes near perihelion. This currently occurs in early January each year.

In an earlier discussion on the synodic month, it was assumed that the lunar orbit's line of apsides has a fixed and permanent direction in space. In fact, the length of the mean anomalistic month (27.55 days) exceeds the mean sidereal month (27.32 days) by 0.23 days. Thus, the Moon's major axis slowly shifts with a mean rate of 0.11140° per day in the direct sense, that is, in the same direction as the Moon's orbital motion. This corresponds to an average of 40.7° per year, so it takes 8.85 years (3231.6 days) for the line of apsides to make one complete revolution with respect to the stars.

What impact do the varying length of the anomalistic month and the direct (eastward) rotation of the Moon's elliptical orbit have on the length of the lunation? To answer this, one must first consider Earth's elliptical orbit around the Sun, which has a mean eccentricity of 0.0167. The center-to-center distance between Earth and the Sun varies with mean values of 147,098,074 km at perihelion to 152,097,701 km at aphelion. The direction of Earth's orbital line of apsides also changes but at a rate far slower than the Moon's. Having a direct (eastward) shift with a mean value of 0.0172° per year, it takes about 20,500 years for Earth's major axis to make one complete revolution. This is only 0.0004 of the lunar rate, so it can be treated as fixed for the purpose of the following discussion.

At certain times, the perigee of the lunar orbit and the perihelion of Earth's orbit can have the same ecliptic longitude. Ignoring the 5.1° tilt of the Moon's obit, the major axes are then essentially parallel to each other and point in the same direction. As time passes, the major axis of the lunar orbit slowly rotates east with respect to Earth's major axis until it becomes perpendicular to it 2.21 years later. In another 2.21 years (4.42 years from the start), the major axes of the orbits are again parallel to each other, but the perigee and the perihelion are 180° apart as they point in opposite directions. After an additional period of 2.21 years, the axes are once more perpendicular. Finally, the Moon's perigee and Earth's perihelion again share the same ecliptic longitude after a total interval of 8.85 years.

Figure 4–5: Length of Lunation for 2008 – 2027

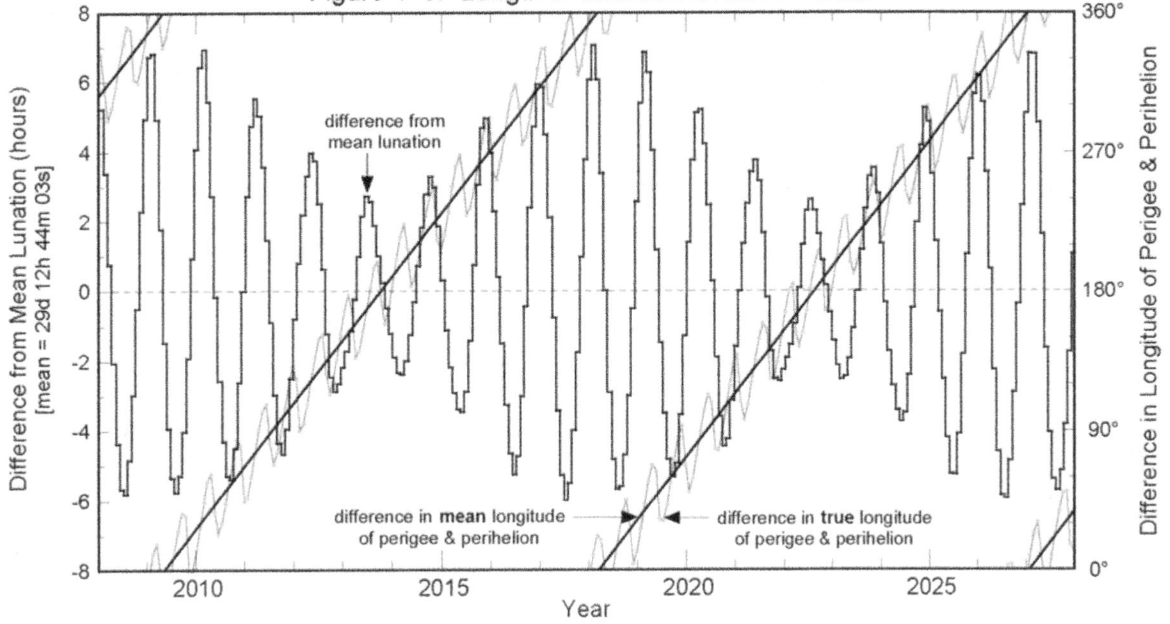

The length of each lunation minus the mean lunation is plotted in Figure 4-5 for the 20-year period from 2008 through 2027. The periodic rhythm between the lunation length and the true anomaly, as described earlier (via Figure 4-1), can now be seen over the course of two decades. The 412-day mean period of this cycle corresponds to the time between two consecutive alignments of the major axis in the direction of the Sun. It is slightly longer than a year because of the slow eastward shift of the Moon's major axis.

An interesting feature revealed in Figure 4-5 is how the extremes in the lunation length slowly vary over a period of nearly 9 years. The envelope defined by the minima and maxima appears to oscillate over a range of values from ±2 h to ±6 h. This behavior is evidence revealing the influence of the 8.85-year cycle in the alignment of the major axes of the orbits of the Moon and Earth.

The amplitude of the envelope is due to the eccentricity of Earth's orbit. When Earth is at perihelion, its orbital velocity is at its maximum value so Earth travels a larger distance around its orbit in a given time as compared to aphelion. Thus, the Moon must travel a greater distance to align with the Sun, which results in a longer lunation. Near aphelion, the opposite conditions produce a shorter lunation.

Using the axis scale on the right, the diagonal lines in Figure 4-5 plot the angle between the Moon's perigee and Earth's perihelion. This is the difference between the Moon's mean longitude of perigee and Earth's true longitude of perihelion. When the angle between the perigee and perihelion is 0°, the length of the lunation varies from a minimum of 29.273 days (-6.17 hours from mean) to a maximum of 29.820 days (+6.93 hours from mean). Similarly, when the angle between the perigee and perihelion is 180°, the length of the lunation varies from a minimum of 29.452 days (-1.88 hours from mean) to a maximum of 29.628 days (+2.33 hours from mean). To summarize, the greatest extremes in the length of the lunation occur when the longitudes of the Moon's perigee and Earth's perihelion are equal. The smallest extremes in the lunation length occur when their longitudes differ by 180°.

Although the Moon's major axis rotates eastward at a mean rate of 0.1114° per day, the true rate varies considerably. Figure 4-5 illustrates the variation by plotting the difference between the true longitudes of the Moon's perigee and Earth's perihelion. This quasi-sinusoidal oscillation about the difference in the mean longitudes shows peak departures of ±30° from average. Indeed, the Moon's major axis can swing both east and west of its mean value, taking on an actual retrograde shift west during some anomalistic months.

This dynamic behavior is due to the gravitational pull of the Sun on the Moon as it orbits Earth. Consequently, a continuous torque is applied to the lunar orbit in an unsuccessful effort to permanently align the major axis towards the Sun. The annual orbit of the Earth–Moon system around the Sun coupled with the Moon's synodic orbit around Earth mean that the conditions for such a permanent alignment are always changing. The overall effect is to twist and distort the shape and orientation of the Moon's elliptical orbit.

Figure 4–6: Moon's Eccentricity for 2008 – 2010

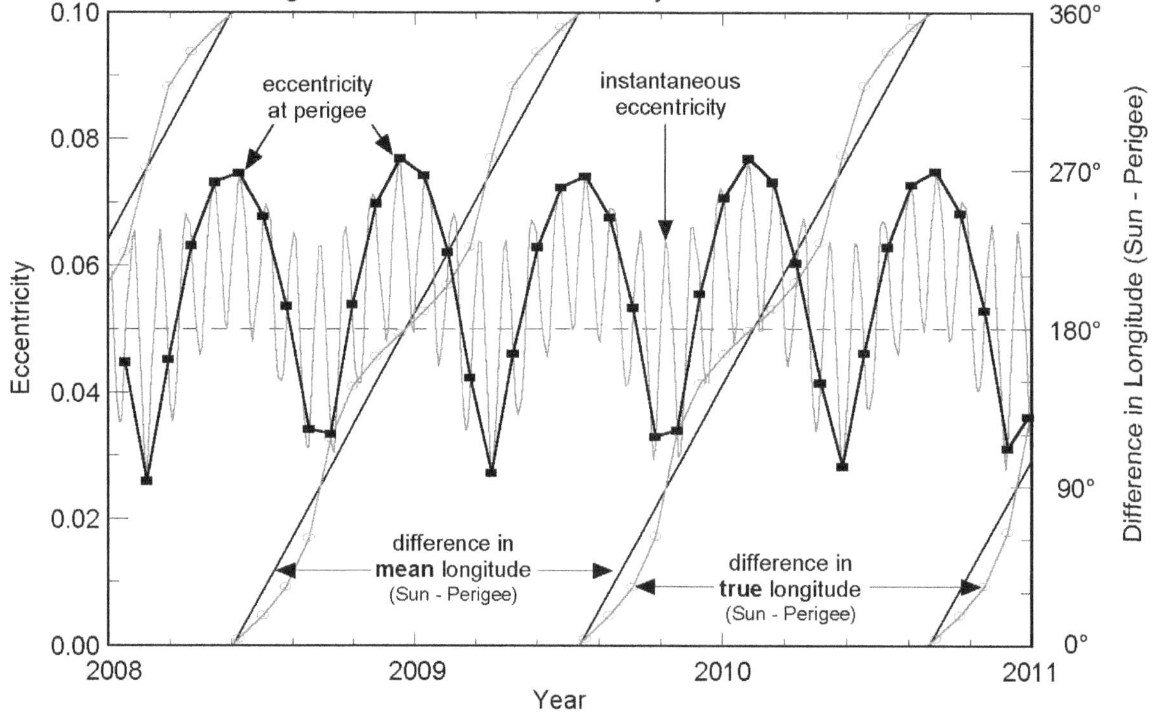

It was stated earlier that the Moon's mean orbital eccentricity is 0.0549, but this too is subject to large changes because of solar perturbations. Figure 4-6 plots the variation in the Moon's orbital eccentricity from 2008 through 2010. The instantaneous eccentricity (light gray curve) oscillates with a period tied to the synodic month and ranges from 0.0266 to 0.0762 over this 3-year interval. Superimposed on the instantaneous eccentricity is the eccentricity at the instant of perigee, which occurs at the beginning of each anomalistic month (heavy black curve). The straight diagonal lines represent the difference between the mean longitudes of the Sun and perigee. In other words, it is the angle between the Moon's perigee-directed major axis and the Sun. Oscillating about this line is the difference between the true longitudes of Sun and perigee. The scale for these angles appears along the right side of Figure 4-6. The extreme range of the Moon's orbital eccentricity at perigee during the 5000 years of the catalog is 0.0255 to 0.0775.

Figure 4-6 shows that the eccentricity reaches a maximum when the major axis of the lunar orbit is pointed directly towards or directly away from the Sun (angles of 0° and 180°, respectively). This occurs at a mean interval of 205.9 days, which is somewhat longer than half a year because of the eastward shift of the major axis. The eccentricity reaches a minimum when the major axis of the lunar orbit is perpendicular to the Sun (angles of 90° and 270°).

Figure 4–7: Perigee & Apogee for 2008 – 2010

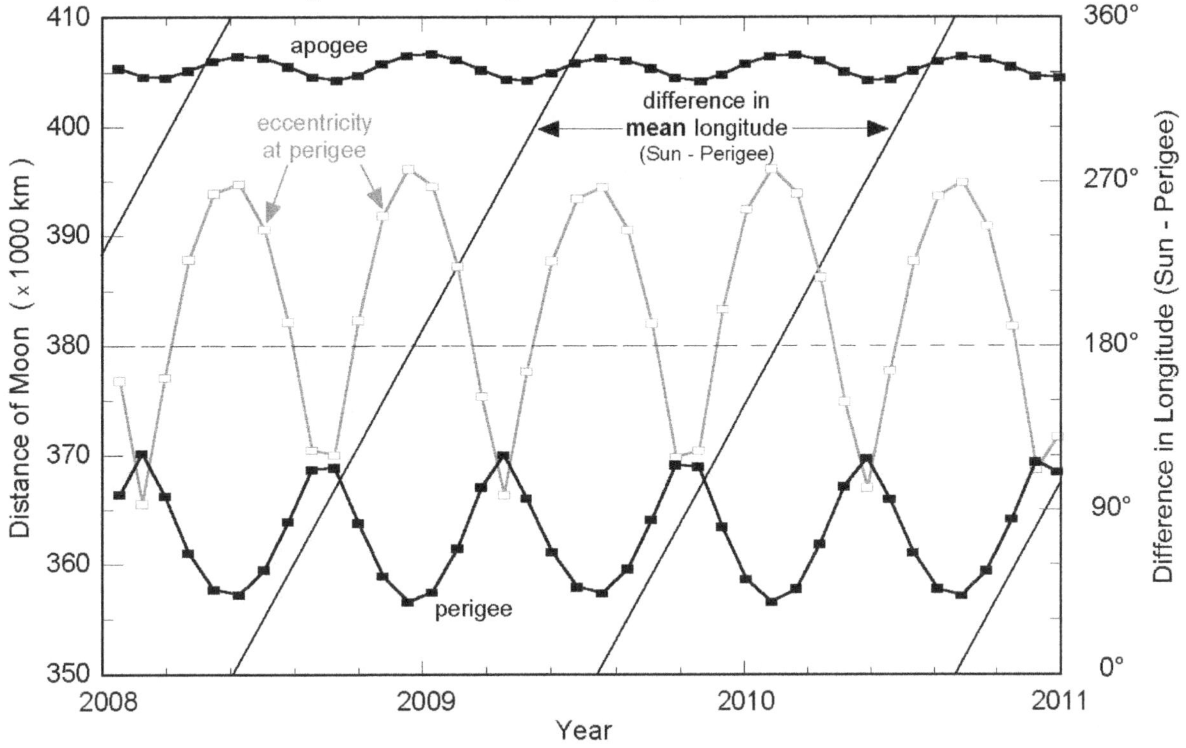

Such changes in orbital eccentricity produce significant variations in the Moon's distance at perigee and apogee. Figure 4-7 plots the Moon's distance for all perigees and apogees from 2008 through 2010. Also shown is the orbital eccentricity at perigee as well as the angle between the perigee directed major axis and the Sun. The closest perigee (minimum perigee distance) and farthest apogee (maximum apogee distance) occur when the eccentricity is at maximum. This corresponds to times when the Moon's major axis points directly towards or directly away from the Sun (angles of 0° and 180°, respectively). The farthest perigee (maximum perigee distance) and closest apogee (minimum apogee distance) occur when the eccentricity is at minimum. At such times, the major axis is oriented perpendicular to the Sun. During the 3-year interval covered in Figure 4-7, the Moon's perigee distance ranges from 356,568 to 370,216 km while the apogee distance ranges from 404,168 to 406,602 km.

Over the 5000-year period of the catalog, there are 66,276 perigees and apogees. During this epoch, the distance of the Moon's perigee varies from 356,355 to 370,399 km while the apogee varies from 404,042 to 406,725 km. The minimum and maximum extremes in orbital eccentricity are 0.0255 to 0.0775 and the extremes in the length of the anomalistic month are 24.629 days (2.925 days shorter than the mean) to 28.565 days (1.011 days longer than the mean). A histogram showing the distribution in the length of the anomalistic month is presented in Figure 4-8 where the durations of individual anomalistic months have been binned into 2-hour groups. The sharply asymmetric distribution shows that anomalistic months longer than the mean cluster over a much shorter range of values compared to anomalistic months shorter than the mean.

Figure 4–8: Length of Anomalistic Month Over 5000 Years

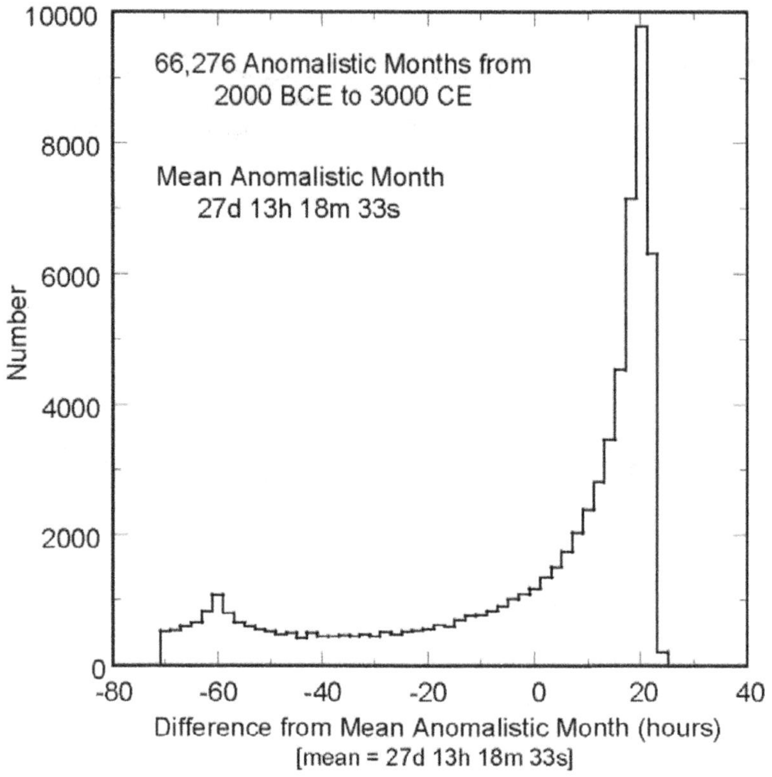

4.4 Draconic Month

The plane of the Moon's orbit is inclined at a mean angle of 5.145° to the plane of Earth's orbit around the Sun. The intersection of these planes defines two points or nodes on the celestial sphere. The node where the Moon's path crosses the ecliptic from south to north is the ascending node, while the node where the Moon's path crosses the ecliptic from north to south is the descending node.

Figure 4–9: Length of Draconic Month for 2008 – 2010

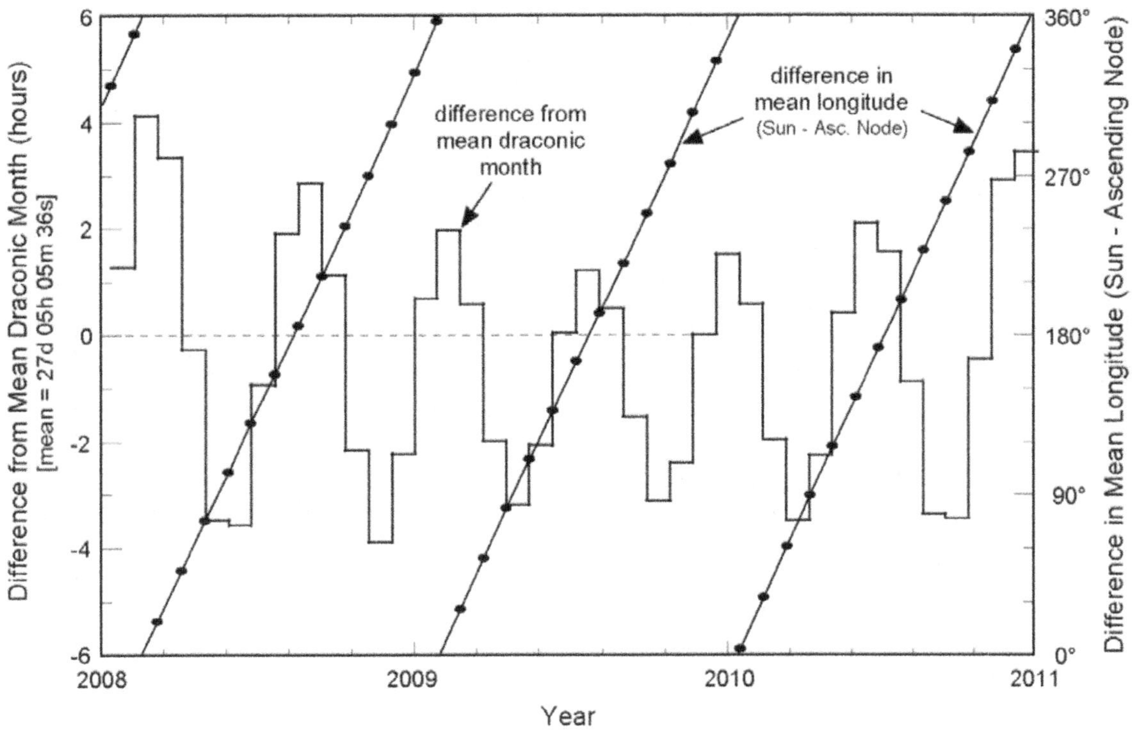

The draconic month is defined as one revolution of the Moon about its orbit with respect to the ascending node. The mean length of this nodical period is 27.21222 days (27d 05h 05m 36s). However, the actual duration can vary by over 6 h from the mean. Figure 4-9 plots the duration of the draconic month minus its mean value for 2008 through 2010. The shortest month over this 3-year period is 27.05115 days (27d 01h 14m), while the longest month is 27.38409 days (27d 09h 13m).

The most significant characteristic of this variation is that it is synchronized with the ascending node relative to the Sun's position along the ecliptic. The mean angle between the Sun and the ascending node (i.e., difference in mean longitude) is also plotted in Figure 4-9 (diagonal lines) to illustrate this relationship. The longitude difference at the start of each draconic month is plotted as a black dot. Longitude values can be read using the scale along the right side of the figure. The longest draconic months occur when the difference in the mean longitudes of the Sun and the ascending node is either 0° or 180°. In contrast, the shortest months occur when the angle between the Sun and the ascending node is either 90° or 270°.

The mean draconic month is 0.10944 day (2h 36m 36s) shorter than the sidereal month. Consequently, the lunar nodes slowly rotate west or retrograde (opposite the Moon's orbital motion) along the ecliptic at a rate of 0.05295° per day. One complete rotation of the ascending node about the ecliptic requires 18.6 years (6793.48 days) with respect to the fixed stars.

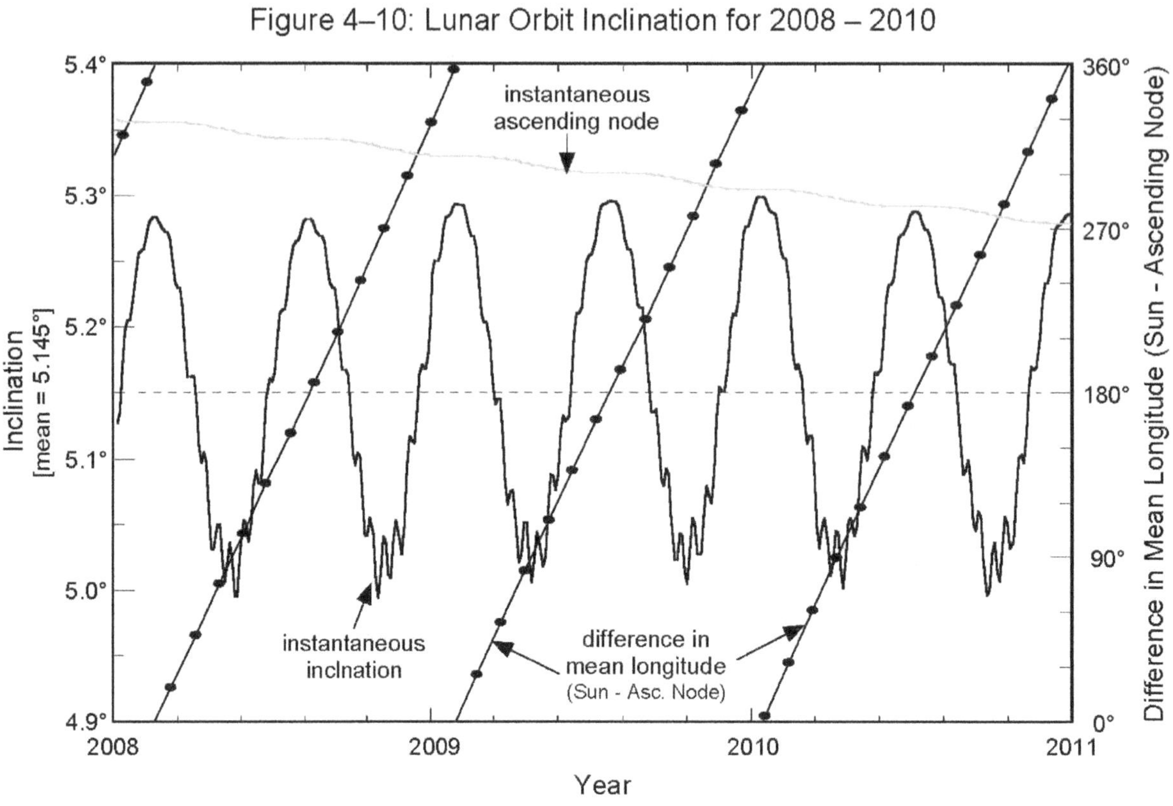

Figure 4-10 plots the instantaneous inclination of the lunar orbit over the 3-year period 2008–2010. The mean angle between the Sun and the ascending node (i.e., difference in mean longitude) is also plotted. The largest inclination of 5.30° occurs when the difference in longitude is either 0° or 180°. In other words, the inclination is always near its maximum value for both solar and lunar eclipses. The smallest inclination of 5.00° occurs when the difference in longitude is either 90° or 270°. Note the small monthly oscillations in the inclination when near its minimum. The figure also plots the longitude of the instantaneous ascending node. Its westward motion draws to a near standstill whenever the Sun aligns with either of the nodes. This corresponds to a difference in longitude of either 0° or 180°.

The mean interval in the periodic variation of both the draconic month and the orbital inclination is 173.3 days. This is the average time it takes for the Sun to travel from one node to the other. It is also equivalent to the interval between the midpoints of two eclipse seasons. The period is slightly less than half a year because of the retrograde motion of the nodes.

Figure 4–11: Draconic Month and Perigee for 2008 – 2017

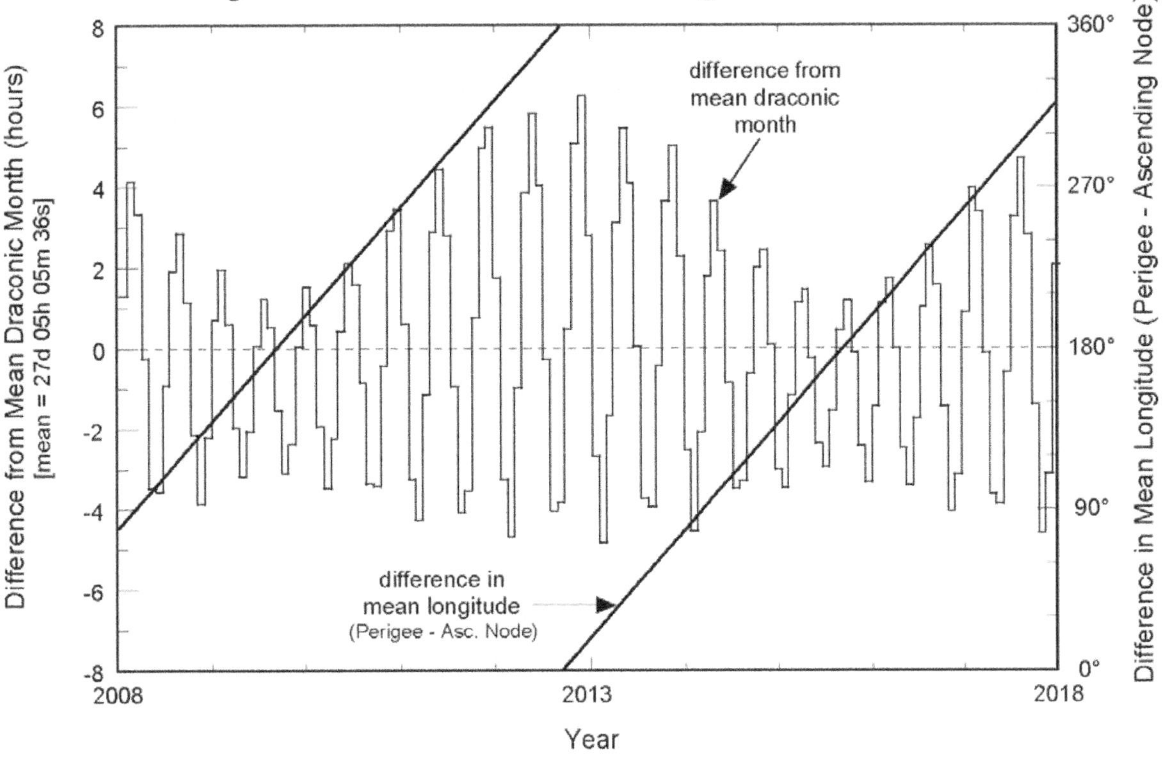

The length of the draconic month is strongly modulated by the position of the nodes with respect to the major axis of the Moon's orbit. The histogram in Figure 4-11 shows how the draconic month changes from 2008 through 2017. The 173-day alignment of the Sun with a node appears as the rapid oscillation in the month length. The quasi-sinusoidal envelopes surrounding the minima and maxima form two longer period oscillations. Over the 10-year period covered in this figure, the minimum month duration varies from 27.089 to 27.011 days (3.0 to 4.8 hours shorter than the mean). The maximum month duration ranges from 27.261 to 27.472 days (1.2 to 6.2 hours longer than the mean).

The difference in the mean longitudes of perigee and the ascending node appear as diagonal lines in Figure 4-11. This is the angle between these orbital parameters measured along the ecliptic. The greatest extremes in the draconic month occur when the angle between perigee and the ascending node is 0°. Likewise, the smallest extremes of the month take place when the difference in longitude is 180°. The mean rates of the major axis and the ascending node are 0.11140° east and 0.05295° west per day, respectively. Therefore, the mean period between alignments of the axis and node is 2190.4 days or 6.0 years. This period is clearly seen in Figure 4-11.

There are 67,111 draconic months during the 5000 years covered in this catalog. The shortest and longest months are 27.004 days (0.208 days or 5.0 hours shorter than the mean) and 27.487 days (0.275 days or 6.6 hours longer than the mean), respectively. A histogram of the distribution in the length of the draconic month over the five millennia appears in Figure 4-12 where the duration of individual draconic months have been binned into 30-min groups. The width and bifurcated symmetry of the distribution resemble the distribution for the lunation (synodic month) in Figure 4-4.

Figure 4–12: Length of Draconic Month Over 5000 Years

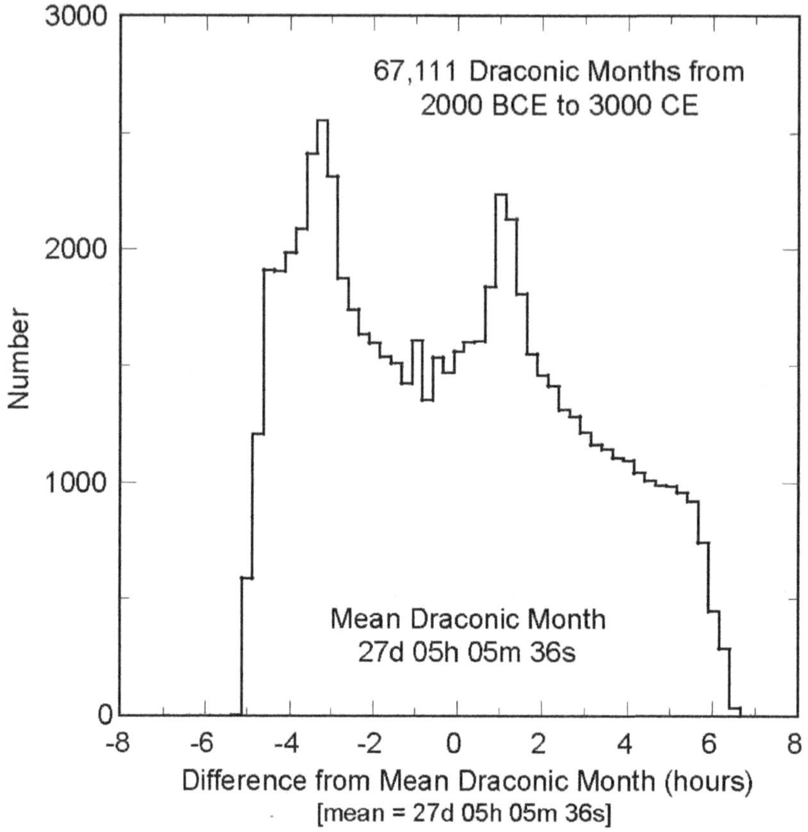

4.5 Eclipse Cycles

The interaction and harmonics of the synodic, anomalistic, and draconic months not only determine how frequently eclipses occur, but they also control the geometric characteristics and classification of each eclipse. The commensurability of these periods over long time scales results in several important eclipse cycles, which will be the subject of the next section.

Section 5: Solar Eclipse Periodicity

5.1 Interval Between Two Successive Solar Eclipses

The time interval between any two successive solar eclipses can be either 1, 5 or 6 lunations (synodic months). The distribution of these intervals among the 11,898 in the *Catalog* is found in Table 5-1.

Table 5-1. Interval Between Successive Eclipses

Number of Lunations	Number of Eclipses	Percent
1	1361	11.4%
5	2743	23.1%
6	7793	65.5%

5.2 Solar Eclipse Repetition

Eclipses separated by 1, 5 or 6 lunations are usually quite dissimilar. They are often of unlike types (i.e., partial, annular, total or hybrid) with diverse Sun-Moon-Earth alignment geometries, and with different lunar orbital characteristics (i.e., longitude of perigee and longitude of ascending node). More importantly, these short periods are of no value as predictors of future eclipses because they do not repeat in any simple pattern.

A useful eclipse repetition cycle can be found by requiring that certain orbital parameters be repeated. The Moon must be in the new phase with the same longitude of perigee and same longitude of the ascending node. These conditions are met by searching for an integral multiple in the Moon's three major periods—the synodic, anomalistic and draconic months. A fourth condition might require that an eclipse occur at approximately the same time of year to preserve the axial tilt of Earth and thus, the same season, as well as the distance from the Sun.

5.3 Saros Series

The Saros arises from a harmonic between three of the Moon's orbital cycles. All three periods are subject to slow variations over long time scales, but there current values (2000 CE) are:

Synodic Month (New Moon to New Moon) = 29.530589 days = 29d 12h 44m
Anomalistic Month (perigee to perigee)
= 27.554550 days = 27d 13h 19m
Draconic Month (node to node) = 27.212221 days = 27d 05h 06m

One Saros is equal to 223 synodic months. However, 239 anomalistic months, and 242 draconic months are also equal (within a few hours) to this same period:

223 Synodic Months	= 6585.3223 days	= 6585d 07h 43m
239 Anomalistic Months	= 6585.5375 days	= 6585d 12h 54m
242 Draconic Months	= 6585.3575 days	= 6585d 08h 35m

With a period of approximately 6,585.32 days (~18 years 11 days 8 hours), the Saros is valuable tool in investigating the periodicity and recurrence of eclipses. It was first known to the Chaldeans as an interval when lunar eclipses repeat, but the Saros is applicable to solar eclipses as well.

Any two eclipses separated by one Saros cycle share similar characteristics. They occur at the same node with the Moon at nearly the same distance from Earth and at the same time of year. Because the Saros period is not equal to a whole number of days, its biggest drawback as an eclipse predictor is that subsequent eclipses are visible from different parts of the globe. The extra 1/3 day displacement means that Earth must rotate an additional ~8 hours or ~120° with each cycle. For solar eclipses, this results in a shift of each succeeding eclipse path by ~120° west. Thus, a Saros series returns to approximately the same geographic region every three Saros periods (~54 years and 34 days). This triple Saros cycle is known as the Exeligmos. Figure 5-1 shows the path of totality for nine eclipses belonging to Saros 136. This series is of particular interest because it is producing the longest total eclipses of the 20th and 21st centuries. The westward migration of each eclipse path from 1901 through 2045 illustrates the consequences of the extra 1/3 day in the Saros period. The northward shift of each path is due to the progressive increase in gamma from -0.3626 (1901) to 0.2116 (2045).

Figure 5-1 — Eclipses from Saros 136: 1901 to 2045

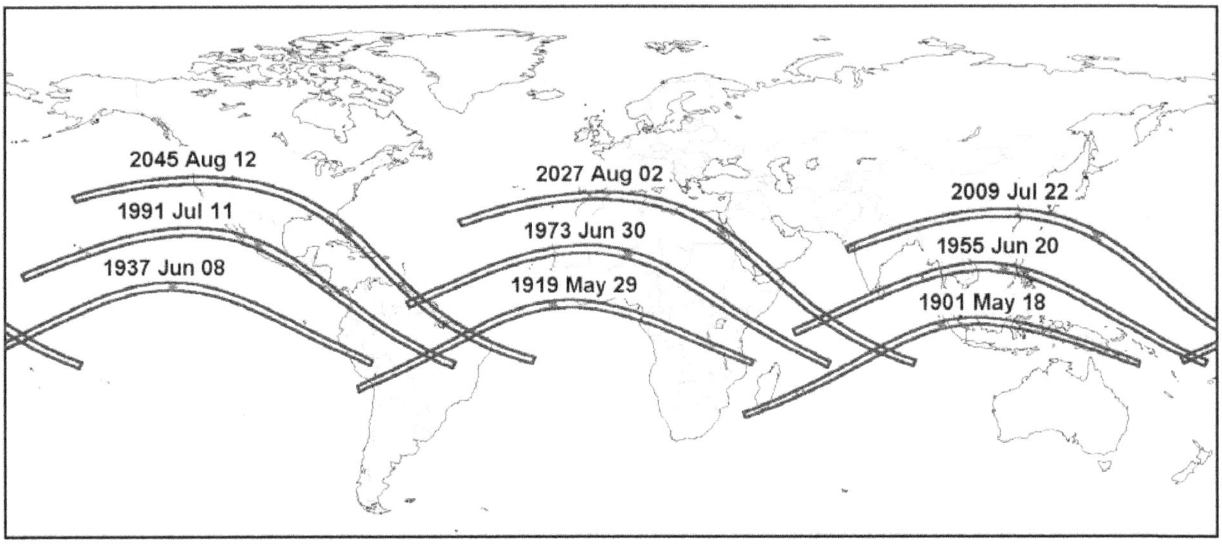

Saros series do not last indefinitely because the synodic, draconic and anomalistic months are not perfectly commensurate with one another. In particular, the Moon's node shifts westward by about 0.5° with each eclipse in a series. The following narrative describes the life cycle of a typical Saros series at the Moon's descending node. The series begins when New Moon occurs ~17° east of the node. The Moon's umbral/antumbral shadow passes about 3500 km south of Earth and a small partial eclipse will be visible from high southern latitudes. One Saros period later, the umbra/antumbra passes ~250 km closer to Earth (gamma increases) and a partial eclipse of slightly larger magnitude will result. After about ten Saros cycles (~200 years), the first umbral/antumbral eclipse occurs near the south pole of Earth. Over the course of the next seven to ten centuries, a central eclipse occurs every 18.031 years (= Saros) but will be displaced northward by about 250 km with respect to Earth's center. Halfway through this period, eclipses of long duration occur near the equator. The last central eclipse of the series takes place at high northern latitudes. Approximately ten more eclipses will be partial with successively smaller magnitudes. Finally, the Saros series ends a dozen or more centuries after it began at the opposite pole.

Based on the above description, the path of each umbral/antumbral eclipse should shift uniformly north in latitude after every Saros period. As Figure 5-2 shows, this is not always the case. Nine members from Saros 136 are plotted for the years 2117 through 2261. Although the paths of previous eclipses in this series were shifting progressively northward (Figure 5-1), the trend here is reversed and the paths shift south. This temporary effect is due to the tilt of Earth's axis combined with the passage of Saros 136 eclipses through northern hemisphere winter solstice. Note that the season for this group of eclipses runs from September through December. With each successive eclipse, Earth's northern hemisphere tips further and further away from the Sun. This motion shifts geographic features and circles of latitude northward with respect to the Sun-Earth line at a rate that is faster than the change in gamma. Consequently, the eclipse paths appear to shift south in latitude until the winter solstice when they again resume a northward trend.

Figure 5-2 — Eclipses from Saros 136: 2117 to 2261

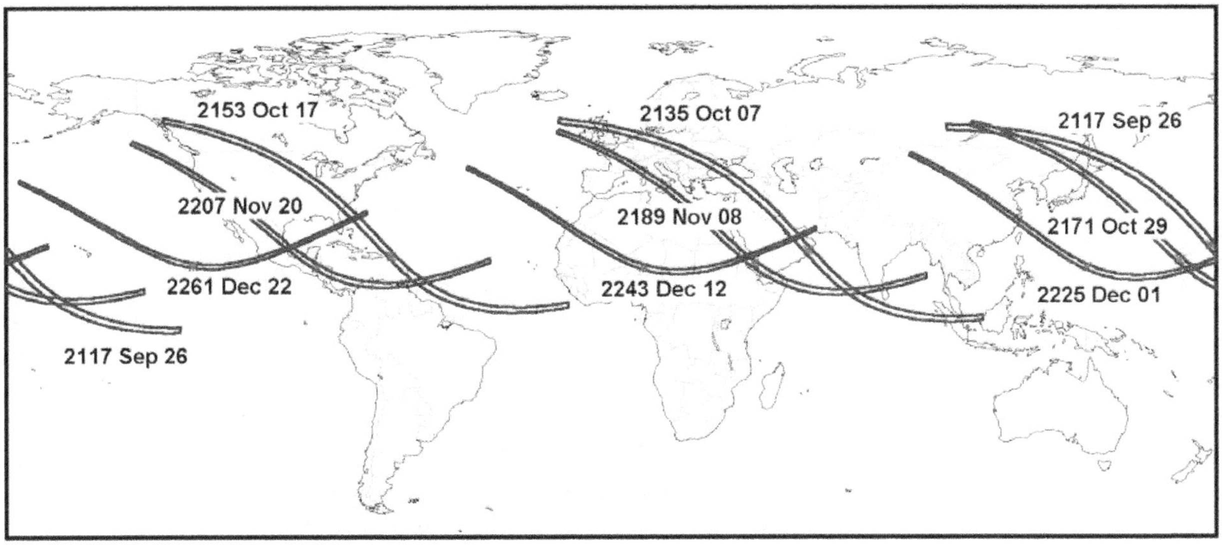

The scenario for a Saros series at the ascending node is similar except that gamma decreases as each successive eclipse shifts south of the previous one. The southern latitude trend in eclipse paths reverses to the north near the northern hemisphere summer solstice.

Due to the ellipticity of the orbits of Earth and the Moon, the exact duration and number of eclipses in a complete Saros series is not constant. A series may last 1226 to 1551 years and is comprised of 69 to 87 eclipses, of which 39 to 59 are umbral/antumbral (i.e., annular, total or hybrid). At present (2008), there are thirty-nine active Saros series numbered 117 to 155. The number of eclipses in each of these series ranges from 70 to 82. However, the majority of the series (84.6%) are composed of 70 to 73 eclipses.

5.4 Gamma And Saros Series

Gamma changes monotonically throughout any single Saros series. As mentioned previously (Sect. 1.2.10), the change in gamma is larger when Earth is near its aphelion (June-July) than when it is near perihelion (December-January). For odd numbered series (ascending node), gamma decreases while for even numbered series (descending node), gamma increases. This simple rule describes the current behavior of gamma, but this has not always been the case. The eccentricity of Earth's orbit is presently 0.0167, and is slowly decreasing. It was 0.0181 in the year –2000 and will be 0.0163 in +3000. In the past when the eccentricity was larger, there were Saros series in which the trend in gamma reversed for a few cycles before resuming its original direction. These instances occur near perihelion when the Sun's apparent motion is highest and may, in fact, overtake the eastward shift of the node. The resulting effect is a relative shift west of the node after one Saros cycle instead of the usual eastward shift. Consequently, gamma reverses direction.

The most unusual case of this occurs in Saros series 0. It began in –2955 with 11 partial eclipses, followed by 1 total, 1 hybrid and 4 annulars. Gamma increased with each eclipse until it reversed direction with the second annular. It continued to decrease and the series began to once again produce partial eclipses. With the third partial eclipse, gamma resumed its original northward shift. The series went on to produce 45 more annular eclipses before ending in the year –1675 after 7 partial eclipses.

Among several hundred Saros series examined (–34 to 247), there are many other examples of temporary shifts in the monotonic nature of gamma, although none as bizarre as Saros 0. Some series have two separate reversals in gamma (e.g., series 15, 34 and 52) or even three (e.g., series –5 and 13). The most recent eclipse with a gamma reversal was in 1674 (Saros 107). The next and last in the *Catalog* will occur in 2290 (Saros 165). In past millennia, the gamma reversals were more frequent since Earth's orbital eccentricity was larger.

5.5 Saros Series Statistics

Eclipses belonging to 204 different Saros series fall within the five millennium span of the *Catalog*. Two series (–13 and 190) have only one member represented while 53 have a larger but incomplete subset of their members included (–12 to –1, 145, 147 and 151 to 189). Finally, 149 complete Saros series are contained within the *Catalog* (0 to 144, 146, 148, 149 and 150).

The number of eclipses in each of these series ranges from 69 to 87. However, over a quarter (27.9%) of the series contain 72 eclipses while nearly three quarters (71.1%) of them have 70 to 73 eclipses. Table 5-2 presents the statistical distribution of the number of eclipses in each Saros series. The approximate duration (years) as a function of the number of eclipses is listed along with the first five Saros series containing the corresponding number of eclipses.

Table 5-2. Number of Eclipses in Saros Series

Number of Eclipses	Duration (years)	Number of Series	Saros Series Numbers
69	1226	4	156,171,174,177
70	1244	25	104,116,122,123,131,...
71	1262	40	22,25,61,62,64,...
72	1280	57	-11,0,1,3,4,...
73	1298	25	-13,-12,-3,2,5,...
74	1316	10	-8,-1,9,17,31,...
75	1334	8	-10,-9,-2,15,74,...
76	1352	3	11,108,146
77	1370	3	145,166,184
78	1388	1	69
79	1406	2	111,182
80	1424	4	-4,129,147,164
81	1442	1	109
82	1460	2	71,127
83	1478	4	30,72,88,90
84	1496	5	32,33,35,53,70
85	1514	4	13,14,16,51
86	1533	5	-7,-5,12,34,52
87	1551	1	-6

All Saros series begin and end with a number of partial eclipses. Among the 204 Saros series with members falling within the scope of this *Catalog*, the number of partial eclipses in the initial phase ranges from 6 to 25. Similarly, the number of partial eclipses in the final phase varies from 6 to 24. The middle life of a Saros series is composed of umbral/antumbral eclipses (i.e., annular, total or hybrid) which range in number from 39 to 59. Tables 5-3 presents the statistical distribution in the number of umbral/antumbral eclipses in the Saros series represented in the *Catalog*.

Saros 0 is an exception to the above scheme. After beginning with 11 partial eclipses, Saros 0 proceeds with a total, a hybrid, and an annular eclipse. The series then reverts back to three more partial eclipses. It finally resumes with a string of 45 annular eclipses before ending with 7 partial eclipses. This odd behavior is due to the higher orbital eccentricity of Earth in the past and fortuitous timing.

Table 5-3. Number of A/T/H Eclipses in Saros Series

Number of A/T/H Eclipses	Duration (years)	Number of Series	Saros Series Numbers
39	703	4	110,144,162,165
40	721	19	-6,31,34,37,...
41	739	21	-9,-3,12,13,...
42	757	17	10,15,16,28,...
43	775	30	-8,-7,-5,-4,...
44	793	18	-2,11,17,18,...
45	811	7	29,48,77,85,...
46	829	3	-10,114,151
47	847	1	140
48	865	5	-1,38,66,171,188
49	883	2	27,153
50	902	1	103
51	920	2	0,190
52	938	4	57,64,156,189
53	956	8	40,101,116,133,...
54	974	6	47,98,119,134,...
55	992	14	43,59,82,83,...
56	1010	17	-11,1,6,8,...
57	1028	13	3,4,7,20,...
58	1046	10	2,21,26,41,...
59	1064	2	5,23

A concise summary of all 204 Saros series (–13 to 190) is presented in Tables 5-4 to 5-8. The number of eclipses in each series is listed followed by the calendar dates of the first and last eclipses in the Saros. Finally, the chronological sequence of eclipse types in the series is tabulated. The number and type of eclipses varies from one Saros series to the next as reflected in the sequence diversity. Note that the tables make no distinction between central and non-central umbral/antumbral eclipses. The following abbreviations are used in the eclipse sequences.

P = Partial Eclipse
A = Annular Eclipse
T = Total Eclipse
H = Hybrid Eclipse

Table 5-4. Summary of Saros Series –13 to 31

Saros Series	Number of Eclipses	First Eclipse	Last Eclipse	Eclipse Sequence
–13	73	–3277 Mar 15	–1979 May 02	7P 39T 2H 17A 8P
–12	73	–3230 Mar 06	–1932 Apr 22	8P 1T 2H 42A 20P
–11	72	–3147 Mar 17	–1867 Apr 24	6P 24A 3H 29T 10P
–10	75	–3172 Jan 24	–1838 Apr 04	9P 40T 2H 4A 20P
–9	75	–3125 Jan 15	–1791 Mar 25	10P 1T 2H 38A 24P
–8	74	–3042 Jan 27	–1726 Mar 27	8P 25A 3H 15T 23P
–7	86	–3248 Aug 18	–1715 Feb 24	21P 41T 1H 1A 22P
–6	87	–3237 Jul 19	–1686 Feb 03	25P 3H 37A 22P
–5	86	–3136 Aug 10	–1603 Feb 16	21P 26A 3H 14T 22P
–4	80	–3143 Jun 29	–1719 Nov 01	23P 41T 1H 1A 14P
–3	73	–3096 Jun 20	–1798 Aug 07	24P 41A 8P
–2	75	–3013 Jul 03	–1679 Sep 10	21P 27A 4H 13T 10P
–1	74	–3002 Jun 01	–1686 Jul 31	18P 44T 3H 1A 8P
0	72	–2955 May 23	–1675 Jun 29	11P 1T 1H 4A 3P 45A 7P
1	72	–2872 Jun 04	–1592 Jul 11	9P 39A 5H 12T 7P
2	73	–2861 May 04	–1563 Jun 21	8P 43T 12H 3A 7P
3	72	–2814 Apr 24	–1534 Jun 01	8P 5T 2H 50A 7P
4	72	–2731 May 06	–1451 Jun 13	7P 29A 17H 11T 8P
5	73	–2720 Apr 04	–1422 May 24	7P 44T 4H 11A 7P
6	72	–2673 Mar 27	–1393 May 03	7P 7T 2H 47A 9P
7	72	–2590 Apr 08	–1310 May 16	6P 30A 6H 21T 9P
8	73	–2579 Mar 07	–1281 Apr 26	7P 45T 1H 10A 10P
9	74	–2568 Feb 06	–1252 Apr 04	9P 8T 3H 32A 22P
10	73	–2467 Feb 28	–1169 Apr 18	8P 30A 3H 9T 23P
11	76	–2492 Jan 06	–1140 Mar 28	10P 44T 22P
12	86	–2662 Aug 20	–1129 Feb 25	23P 8T 3H 30A 22P
13	85	–2543 Sep 23	–1028 Mar 19	20P 30A 3H 8T 24P
14	85	–2550 Aug 11	–1035 Feb 06	21P 43T 21P
15	75	–2557 Jul 01	–1223 Sep 08	24P 10T 3H 29A 9P
16	85	–2456 Jul 23	–0941 Jan 18	22P 33A 2H 7T 21P
17	74	–2427 Jul 03	–1111 Sep 01	21P 44T 9P
18	73	–2416 Jun 02	–1118 Jul 21	22P 13T 3H 28A 7P
19	73	–2333 Jun 15	–1035 Aug 01	21P 36A 2H 6T 8P
20	72	–2286 Jun 05	–1006 Jul 13	8P 12A 2H 43T 7P
21	72	–2275 May 05	–0995 Jun 11	8P 26T 4H 28A 6P
22	71	–2174 May 28	–0912 Jun 23	8P 49A 2H 5T 7P
23	72	–2145 May 07	–0865 Jun 15	6P 14A 3H 42T 7P
24	72	–2134 Apr 06	–0854 May 14	8P 15T 16H 26A 7P
25	71	–2033 Apr 30	–0771 May 26	7P 52A 1H 3T 8P
26	72	–2004 Apr 08	–0724 May 17	6P 10A 7H 41T 8P
27	72	–1993 Mar 09	–0713 Apr 16	8P 14T 15H 20A 15P
28	72	–1910 Mar 22	–0630 Apr 28	7P 42A 23P
29	73	–1881 Mar 01	–0583 Apr 19	7P 3A 14H 28T 21P
30	83	–2051 Oct 12	–0572 Mar 18	19P 14T 5H 24A 21P
31	74	–1805 Jan 31	–0489 Mar 31	10P 40A 24P

Table 5-5. Summary of Saros Series 32 to 76

Saros Series	Number of Eclipses	First Eclipse	Last Eclipse	Eclipse Sequence
32	84	-1957 Sep 24	-0460 Mar 10	19P 2A 3H 39T 21P
33	84	-1982 Aug 02	-0485 Jan 17	23P 15T 4H 23A 19P
34	86	-1917 Aug 04	-0384 Feb 09	23P 40A 23P
35	84	-1870 Jul 25	-0373 Jan 09	22P 3A 2H 38T 19P
36	73	-1859 Jun 23	-0561 Aug 11	22P 18T 3H 23A 7P
37	73	-1794 Jun 25	-0496 Aug 12	24P 40A 9P
38	73	-1729 Jun 26	-0431 Aug 14	17P 8A 2H 38T 8P
39	72	-1718 May 26	-0438 Jul 03	9P 32T 3H 22A 6P
40	72	-1653 May 28	-0373 Jul 04	11P 53A 8P
41	72	-1588 May 28	-0308 Jul 05	7P 19A 2H 37T 7P
42	72	-1577 Apr 28	-0297 Jun 05	8P 34T 3H 21A 6P
43	72	-1512 Apr 29	-0232 Jun 05	8P 55A 9P
44	72	-1447 Apr 30	-0167 Jun 07	6P 21A 2H 35T 8P
45	72	-1436 Mar 30	-0156 May 07	7P 36T 3H 18A 8P
46	72	-1371 Apr 01	-0091 May 08	8P 43A 21P
47	72	-1306 Apr 02	-0026 May 10	6P 21A 3H 30T 12P
48	74	-1331 Feb 08	-0015 Apr 09	9P 37T 2H 6A 20P
49	72	-1248 Feb 22	0032 Mar 29	9P 40A 23P
50	73	-1201 Feb 11	0097 Apr 01	8P 22A 3H 18T 22P
51	85	-1407 Sep 02	0108 Feb 29	21P 36T 4H 3A 21P
52	86	-1378 Aug 14	0155 Feb 19	24P 40A 22P
53	84	-1277 Sep 06	0220 Feb 21	20P 22A 4H 17T 21P
54	74	-1284 Jul 25	0032 Sep 23	21P 26T 15H 3A 9P
55	73	-1255 Jul 06	0043 Aug 23	24P 41A 8P
56	74	-1172 Jul 17	0144 Sep 15	21P 13A 15H 15T 10P
57	73	-1161 Jun 17	0137 Aug 04	14P 33T 13H 6A 7P
58	72	-1114 Jun 07	0166 Jul 14	21P 44A 7P
59	72	-1031 Jun 19	0249 Jul 27	9P 23A 16H 16T 8P
60	72	-1020 May 18	0260 Jun 26	8P 40T 4H 14A 6P
61	71	-0973 May 10	0289 Jun 05	8P 3T 1H 52A 7P
62	71	-0890 May 22	0372 Jun 17	7P 25A 5H 27T 7P
63	72	-0879 Apr 20	0401 May 29	7P 42T 2H 14A 7P
64	71	-0832 Apr 11	0430 May 08	8P 4T 2H 46A 11P
65	71	-0749 Apr 24	0513 May 20	6P 27A 4H 25T 9P
66	73	-0756 Mar 12	0542 May 01	8P 43T 1H 4A 17P
67	72	-0709 Mar 04	0571 Apr 10	9P 5T 2H 34A 22P
68	72	-0626 Mar 16	0654 Apr 22	7P 28A 3H 11T 23P
69	78	-0724 Dec 09	0665 Mar 22	14P 43T 21P
70	84	-0821 Sep 05	0676 Feb 19	23P 5T 3H 32A 21P
71	82	-0684 Oct 19	0777 Mar 14	18P 29A 3H 9T 23P
72	83	-0727 Aug 16	0752 Jan 21	22P 43T 18P
73	72	-0698 Jul 27	0582 Sep 03	23P 7T 3H 31A 8P
74	75	-0615 Aug 08	0719 Oct 18	22P 30A 3H 8T 12P
75	73	-0604 Jul 07	0694 Aug 26	21P 44T 8P
76	72	-0575 Jun 18	0705 Jul 25	22P 8T 5H 30A 7P

Table 5-6. Summary of Saros Series 77 to 121

Saros Series	Number of Eclipses	First Eclipse	Last Eclipse	Eclipse Sequence
77	71	-0474 Jul 11	0788 Aug 06	18P 36A 2H 7T 8P
78	72	-0463 Jun 09	0817 Jul 18	9P 9A 2H 45T 7P
79	71	-0434 May 21	0828 Jun 16	8P 11T 16H 30A 6P
80	71	-0333 Jun 13	0929 Jul 09	7P 48A 2H 6T 8P
81	72	-0322 May 12	0958 Jun 19	7P 5A 9H 44T 7P
82	71	-0293 Apr 22	0969 May 19	8P 11T 5H 39A 8P
83	71	-0210 May 05	1052 May 30	7P 51A 1H 3T 9P
84	72	-0181 Apr 14	1099 May 22	7P 1A 11H 43T 10P
85	72	-0170 Mar 14	1110 Apr 20	8P 12T 4H 29A 19P
86	71	-0069 Apr 06	1193 May 02	7P 41A 23P
87	73	-0076 Feb 23	1222 Apr 13	9P 2H 42T 20P
88	83	-0246 Oct 06	1233 Mar 12	20P 13T 4H 26A 20P
89	73	0018 Feb 04	1316 Mar 24	10P 40A 23P
90	83	-0134 Sep 28	1345 Mar 04	20P 2H 40T 21P
91	75	-0159 Aug 06	1175 Oct 16	23P 14T 3H 25A 10P
92	74	-0076 Aug 19	1240 Oct 16	23P 40A 11P
93	74	-0029 Aug 09	1287 Oct 08	20P 3A 1H 40T 10P
94	72	-0018 Jul 09	1262 Aug 16	21P 18T 2H 24A 7P
95	71	0047 Jul 11	1309 Aug 06	22P 41A 8P
96	72	0094 Jul 01	1374 Aug 08	10P 14A 2H 39T 7P
97	71	0123 Jun 11	1385 Jul 08	8P 32T 2H 23A 6P
98	71	0188 Jun 12	1450 Jul 09	9P 54A 8P
99	72	0235 Jun 03	1515 Jul 11	7P 18A 2H 37T 8P
100	71	0264 May 13	1526 Jun 10	7P 34T 2H 21A 7P
101	71	0329 May 15	1591 Jun 21	8P 53A 10P
102	71	0376 May 05	1638 Jun 12	7P 19A 3H 34T 8P
103	72	0387 Apr 04	1667 May 22	8P 34T 3H 13A 14P
104	70	0470 Apr 17	1714 May 13	7P 41A 22P
105	72	0499 Mar 27	1779 May 16	7P 20A 4H 21T 20P
106	75	0456 Jan 23	1790 Apr 14	12P 34T 4H 5A 20P
107	72	0557 Feb 15	1837 Apr 05	10P 40A 22P
108	76	0550 Jan 04	1902 Apr 08	12P 20A 5H 18T 21P
109	81	0416 Sep 07	1859 Feb 03	21P 24T 15H 4A 17P
110	72	0463 Aug 30	1743 Oct 17	23P 39A 10P
111	79	0528 Aug 30	1935 Jan 05	21P 11A 14H 17T 16P
112	72	0539 Jul 31	1819 Sep 19	21P 24T 14H 5A 8P
113	71	0586 Jul 22	1848 Aug 28	23P 40A 8P
114	72	0651 Jul 23	1931 Sep 12	18P 13A 16H 17T 8P
115	72	0662 Jun 21	1942 Aug 12	10P 37T 4H 14A 7P
116	70	0727 Jun 23	1971 Jul 22	10P 53A 7P
117	71	0792 Jun 24	2054 Aug 03	8P 23A 5H 28T 7P
118	72	0803 May 24	2083 Jul 15	8P 40T 2H 15A 7P
119	71	0850 May 15	2112 Jun 24	8P 2T 1H 51A 9P
120	71	0933 May 27	2195 Jul 07	7P 25A 4H 26T 9P
121	71	0944 Apr 25	2206 Jun 07	7P 42T 2H 11A 9P

Table 5-7. Summary of Saros Series 122 to 167

Saros Series	Number of Eclipses	First Eclipse	Last Eclipse	Eclipse Sequence
122	70	0991 Apr 17	2235 May 17	8P 3T 2H 37A 20P
123	70	1074 Apr 29	2318 May 31	6P 27A 3H 14T 20P
124	73	1049 Mar 06	2347 May 11	9P 43T 1H 20P
125	73	1060 Feb 04	2358 Apr 09	12P 4T 2H 34A 21P
126	72	1179 Mar 10	2459 May 03	8P 28A 3H 10T 23P
127	82	0991 Oct 10	2452 Mar 21	20P 42T 20P
128	73	0984 Aug 29	2282 Nov 01	24P 4T 4H 32A 9P
129	80	1103 Oct 03	2528 Feb 21	20P 29A 3H 9T 19P
130	73	1096 Aug 20	2394 Oct 25	21P 43T 9P
131	70	1125 Aug 01	2369 Sep 02	22P 6T 5H 30A 7P
132	71	1208 Aug 13	2470 Sep 25	20P 33A 2H 7T 9P
133	72	1219 Jul 13	2499 Sep 05	12P 6A 1H 46T 7P
134	71	1248 Jun 22	2510 Aug 06	10P 8T 16H 30A 7P
135	71	1331 Jul 05	2593 Aug 17	10P 45A 2H 6T 8P
136	71	1360 Jun 14	2622 Jul 30	8P 6A 6H 44T 7P
137	70	1389 May 25	2633 Jun 28	8P 10T 6H 4A 3H 32A 7P
138	70	1472 Jun 06	2716 Jul 11	7P 50A 1H 3T 9P
139	71	1501 May 17	2763 Jul 03	7P 12H 43T 9P
140	71	1512 Apr 16	2774 Jun 01	8P 11T 4H 32A 16P
141	70	1613 May 19	2857 Jun 13	7P 41A 22P
142	72	1624 Apr 17	2904 Jun 05	8P 1H 43T 20P
143	72	1617 Mar 07	2897 Apr 23	10P 12T 4H 26A 20P
144	70	1736 Apr 11	2980 May 05	8P 39A 23P
145	77	1639 Jan 04	3009 Apr 17	14P 1A 1H 41T 20P
146	76	1541 Sep 19	2893 Dec 29	22P 13T 4H 24A 13P
147	80	1624 Oct 12	3049 Feb 24	21P 40A 19P
148	75	1653 Sep 21	2987 Dec 12	20P 2A 1H 40T 12P
149	71	1664 Aug 21	2926 Sep 28	21P 17T 3H 23A 7P
150	71	1729 Aug 24	2991 Sep 29	22P 40A 9P
151	72	1776 Aug 14	3056 Oct 01	18P 6A 1H 39T 8P
152	70	1805 Jul 26	3049 Aug 20	9P 30T 3H 22A 6P
153	70	1870 Jul 28	3114 Aug 22	13P 49A 8P
154	71	1917 Jul 19	3179 Aug 25	7P 17A 3H 36T 8P
155	71	1928 Jun 17	3190 Jul 24	8P 33T 3H 20A 7P
156	69	2011 Jul 01	3237 Jul 14	8P 52A 9P
157	70	2058 Jun 21	3302 Jul 17	6P 19A 3H 34T 8P
158	70	2069 May 20	3313 Jun 16	7P 35T 2H 16A 10P
159	70	2134 May 23	3378 Jun 17	8P 41A 21P
160	71	2181 May 13	3443 Jun 20	7P 20A 3H 22T 19P
161	72	2174 Apr 01	3454 May 20	9P 35T 3H 5A 20P
162	70	2257 Apr 15	3501 May 10	9P 39A 22P
163	72	2286 Mar 25	3566 May 13	9P 20A 4H 18T 21P
164	80	2098 Oct 24	3523 Mar 10	20P 36T 4H 3A 17P
165	72	2145 Oct 16	3425 Dec 02	22P 39A 11P
166	77	2228 Oct 29	3599 Feb 08	19P 21A 5H 16T 16P
167	72	2203 Sep 06	3483 Oct 24	21P 26T 14H 3A 8P

Table 5-8. Summary of Saros Series 168 to 190

Saros Series	Number of Eclipses	First Eclipse	Last Eclipse	Eclipse Sequence
168	70	2250 Aug 28	3494 Sep 22	23P 40A 7P
169	71	2333 Sep 10	3595 Oct 16	19P 13A 16H 15T 8P
170	71	2344 Aug 09	3606 Sep 15	11P 36T 11H 6A 7P
171	69	2391 Aug 01	3617 Aug 14	14P 48A 7P
172	70	2474 Aug 13	3718 Sep 08	8P 23A 16H 15T 8P
173	70	2485 Jul 12	3729 Aug 08	7P 41T 3H 12A 7P
174	69	2532 Jul 04	3758 Jul 18	8P 1T 2H 50A 8P
175	70	2597 Jul 05	3841 Jul 31	7P 26A 5H 24T 8P
176	71	2608 Jun 04	3870 Jul 12	7P 43T 2H 10A 9P
177	69	2655 May 27	3881 Jun 10	8P 3T 3H 37A 18P
178	70	2738 Jun 09	3982 Jul 04	6P 28A 4H 11T 21P
179	71	2731 Apr 28	3993 Jun 03	8P 44T 19P
180	70	2760 Apr 08	4004 May 02	10P 5T 2H 33A 20P
181	71	2843 Apr 20	4105 May 27	8P 29A 3H 9T 22P
182	79	2691 Dec 11	4098 Apr 15	18P 42T 19P
183	72	2666 Oct 20	3946 Dec 06	22P 6T 4H 30A 10P
184	77	2785 Nov 24	4156 Mar 05	19P 30A 3H 7T 18P
185	73	2760 Oct 01	4058 Nov 29	21P 42T 10P
186	70	2789 Sep 11	4033 Oct 06	22P 8T 4H 29A 7P
187	70	2872 Sep 23	4116 Oct 19	20P 34A 2H 5T 9P
188	71	2883 Aug 23	4145 Sep 30	16P 3A 1H 44T 7P
189	70	2912 Aug 04	4156 Aug 29	11P 19T 6H 27A 7P
190	70	2995 Aug 17	4239 Sep 12	11P 46A 2H 3T 8P

5.6 Saros and Other Periods

The numbering system used for the Saros series was introduced by van den Bergh in his book *Periodicity and Variation of Solar (and Lunar) Eclipses* (1955). He assigned the number 1 to a pair of solar and lunar eclipse series that were in progress during the second millennium BCE based on an extrapolation from Oppolzer's *Canon der Finsternisse* [1887].

There is an interval of 1, 5 or 6 synodic months between any sequential pair of solar eclipses. Interestingly, the number of lunations between two eclipses permits the determination of the Saros series number of the second eclipse when the Saros series number of the first eclipse is known. Let the Saros series number of the first eclipse in a pair be "s". The Saros series number of the second eclipse can be found from the relationships in Table 5-9 [Meeus, Grosjean, and Vanderleen, 1966].

Table 5-9. Some Eclipse Periods and their Relationships to the Saros Number

Number of Synodic Months	Length of Time	Saros Series Number	Period Name
1	~1 month	s + 38	Lunation
5	~5 months	s – 33	Short Semester
6	~6 months	s + 5	Semester
135	~11 years - 1 month	s + 1	Tritos
223	~18 years + 11 days	s	Saros
235	~19 years	s + 10	Metonic Cycle
358	~29 years - 20 days	s + 1	Inex
669	~54 years + 33 days	s	Exeligmos (Triple Saros)

5.7 Saros and Inex

A number of different eclipse cycles were investigated by van den Bergh, but the most useful were the Saros and the Inex. The Inex is equal to 358 synodic months (~29 years - 20 days) which is very nearly 388.5 draconic months.

358 Synodic Months	= 10,571.9509 days	= 10,571d 22h 49m
388.5 Draconic Months	= 10,571.9479 days	= 10,571d 22h 55m

The extra 0.5 in the number of draconic months means that eclipses separated by one Inex period occur at opposite nodes. Consequently, an eclipse visible from the northern hemisphere will be followed one Inex later by an eclipse visible from the southern hemisphere, and vice versa. The Inex is is equal to ~383.67 anomalistic months, which is far from an integer number. Thus, eclipses separated by one Inex will very likely be of different types, especially if they are central (i.e., total or annular).

The mean time difference between 358 synodic months and 388.5 draconic months making up an Inex is only 6 minutes. In comparison, the mean difference between these two cycles in the Saros is 52 minutes. This means that after one Inex, the shift of the Moon with respect to the node (+0.04°) is much smaller than for the Saros (-0.48°). While a Saros series lasts 12 to 15 centuries, an Inex series typically lasts 225 centuries and contains about 780 eclipses. Although the Inex posesses a long lifespan, its mean duration is not easily characterized because the length of the synodic and draconic months are changing over long time scales. If the instantaneous mean durations of the synodic and draconic months for the years -2000, +2000 and +4000 are used to calculate the mean duration of the Inex, the resulting lengths are about 14,500, 26,600 and 51,000 years, respectively [Meeus, 2004].

5.8 Saros—Inex Panorama

Van den Bergh placed all 8000 solar eclipses in Oppolzer's *Canon der Finsternisse* [1887] into a large two dimensional matrix. Each Saros series was arranged as a separate column containing every eclipse in chronological order. The individual Saros columns were then staggered so that the horizontal rows each corresponded to different Inex series. This "Saros-Inex Panorama" proved useful in organizing eclipses. For instance, one step down in the panorama is a change of one Saros period (6585.32 days) later, while one step to the right is a change of one Inex period (10571.95 days) later. The rows and columns were then numbered with the Saros and Inex numbers.

The panorama also made it possible to predict the approximate circumstances of solar (and lunar) eclipses occurring before or after the period spanned by Oppolzer's canon. The time interval "t" between any two solar eclipses can be found through an integer combination of Saros and Inex periods via the following relationship.

$$t = a\,i + b\,s, \qquad (5\text{-}1)$$

where: t = interval in days,
i = Inex period of 10571.95 days (358 synodic months),
s = Saros period of 6585.32 days (223 synodic months), and
a, b = integers (negative, zero or positive).

From this equation, a number of useful combinations of Inex and Saros periods can be employed to extend Oppolzer's canon from -1207 back to -1600 using nothing more than simple arithmetic [van den Bergh, 1954]. The ultimate goal of the effort was to a produce an eclipse canon for dating historical events prior to -1207. Periods formed by various combinations of Inex and Saros were evaluated in order to satisfy one or more of the following conditions:

1) The deviation from a multiple of 0.5 draconic months should be small (i.e., Moon should be nearly the same distance from the node).
2) The deviation from an integral multiple of anomalistic months should be small (i.e., Moon should be nearly the same distance from Earth).
3) The deviation from an integral multiple of anomalistic years should be small (i.e., eclipse should occur on nearly the same calendar date).

No single Inex-Saros combination meets all three criteria, but there are periods which do a reasonably good job for any one of them. Note that secular changes in the Moon's elements cause a particular period to be of high accuracy for a limited number of centuries. The direct application of the Saros-Inex panorama allows for the determination of eclipse dates in the past (or future). However, the application of the longer Saros-Inex combinations permit the rapid estimation of a number of eclipse characteristics without lengthy calculations. Table 5-9 lists several of the most useful periods.

5-9. Some Useful Eclipse Periods

Period Name	Period (Inex + Saros)	Period (years)	Use
Heliotrope	58i + 6s	1787	Geographic longitude of central line
Accuratissima	58i + 9s	1841	Geographic latitude of central line
Horologia	110i + 7s	3310	Time of ecliptic conjunction

Modern digital computers using high precision solar and lunar ephemerides can predict the dates and circumstances of eclipses directly. Nevertheless, the Saros and Inex cycles are of great value in understanding the periodicity and frequency of solar eclipses.

5.9 Secular Variations in the Saros and Inex

Because of long secular variations in the average ellipticity of the Moon's and Earth's orbits, the mean lengths of the synodic, draconic, and anomalistic months are slowly changing. The mean synodic and draconic months are increasing by approximately 0.2 and 0.4 s per millennium, respectively. Meanwhile, the anomalistic month is decreasing by about 0.8 seconds per millennium.

Although small, the cumulative effects of such changes has an impact on both the Saros and Inex. Table 5-10 shows how the number of draconic and anomalistic months change with respect to 223 synodic months (Saros period) over an interval of 7000 years. Of particular interest is the last column, which shows the mean shift of the Moon's node after a period of 1 Saros. It is gradually increasing, which means that the average number of eclipses in a typical Saros series is decreasing.

Table 5-10: Number of Anomalistic and Draconic Months in 1 Saros

Year	Anomalistic Months (223 Lunations)	Draconic Months (223 Lunations)	Node Shift (after 1 Saros)
-3000	238.991679	241.998742	0.4529
-2000	238.991763	241.998730	0.4571
-1000	238.991854	241.998717	0.4618
0	238.991950	241.998703	0.4668
1000	238.992051	241.998688	0.4722
2000	238.992157	241.998673	0.4779
3000	238.992267	241.998656	0.4838
4000	238.992379	241.998639	0.4899

Table 5-11 shows how the number of draconic months is changing with respect to 358 synodic months (Inex period) over a 7000-year interval. The mean shift in the lunar node after 1 Inex is much smaller than the Saros and is gradually decreasing. This explains why the lifetime of the Inex is so much longer than the Saros and is still increasing.

Table 5-11: Number of Draconic Months in 1 Inex

Year	Draconic Months (358 Lunations)	Node Shift (after 1 Inex)
-3000	388.500223	-0.0801
-2000	388.500204	-0.0734
-1000	388.500183	-0.0659
0	388.500160	-0.0578
1000	388.500136	-0.0491
2000	388.500111	-0.0400
3000	388.500085	-0.0305
4000	388.500057	-0.0207

Although the Inex possesses a long lifespan, its mean duration is not easily characterized because of the decreasing nodal shift seen in Table 5-11. If the instantaneous mean durations of the synodic and draconic months for the years -2000, +2000, and +4000 are used to calculate the mean duration of the Inex, the resulting lengths are about 14,500, 26,600, and 51,000 years, respectively (Meeus, 2004a).

Abbreviations

arcsec	Arc second
AT	Hybrid eclipse that begins as annular, then changes to total.
ATA	Hybrid eclipse that begins as annular, changes to total, and then reverts back to annular.
BCE	Before the Common Era
CE	Common Era
cm	Centimeter
ET	Ephemeris Time
GMAT	Greenwich Mean Astronomical Time
GMT	Greenwich Mean Time
IAU	International Astronomical Union
ISO	International Standards Organization
LLR	Lunar Laser Ranging
LOD	Length of Day
m	Meter (or minutes in tables)
min	Minutes
ms	Millisecond
s	Second
arcsec/cy2	Arc seconds per Julian century squared
TA	Hybrid eclipse that begins as total and ends as annular.
TAI	International Atomic Time
TD	Terrestrial Dynamical Time
TT	Terrestrial Time
UT	Universal Time
UTC	Coordinated Universal Time
VLBI	Very Long Baseline Interferometry
yr	Year

References

Astronomical Almanac for 1986, Washington: US Government Printing Office; London: HM Stationery Office (1985).

Astronomical Almanac for 2006, Washington: US Government Printing Office; London: HM Stationery Office (2004).

Bretagnon, P., and Francou G., "Planetary theories in rectangular and spherical variables: VSOP87 solution," Astron. Astro- phys., 202(309) (1988).

Brown, E.W., "Theory of the Motion of the Moon," *Mem. Royal Astron. Soc.*, Vol. LVII, Part II, pp. 136–141, London (1905).

Chapront-Touzé, M., and Chapront, J., "The Lunar Ephemeris ELP 2000," *Astron. Astrophys.*, vol. 124, no. 1, pp. 50–62 (1983).

Chapront-Touzé, M., and Chapront, J., *Lunar Tables and Programs from 4000 B.C. to A.D. 8000*, Willmann-Bell, (1991).

Chapront, J., Chapront-Touzé, M., and Francou, G., "A new determination of lunar orbital parameters, precession constant and tidal acceleration from LLR measurements," *Astron. Astrophys.*, vol. 387, pp. 700–709 (2002).

Dickey, J.O., Bender, P.L., Faller, J.E., Newhall, X.X., Ricklefs, R.L., Ries,, J.G., Shelus, P.J., Veillet, C., Whipple, A.L., Wiant, J.R., Williams, J.G., and Yoder, C.F., "Lunar Laser Ranging: a Continuing Legacy of the Apollo Program," *Science*, **265**, pp. 482–490 (1994).

Espenak, F., *Fifty Year Canon of Solar Eclipses: 1986–2035*, Sky Publishing Corp., Cambridge, Massachusetts (1987).

Espenak, F., and Meeus, J., *Five Millennium Canon of Solar Eclipses: –1999 to +3000 (2000 BCE to 3000 CE)*, NASA Tech. Pub. 2006–214141, NASA Goddard Space Flight Center, Greenbelt, Maryland (2006).

Explanatory Supplement to the Ephemeris, H.M. Almanac Office, London (1974).

Gingerich, O., (Translator) *Canon of Eclipses*, Dover Publications, New York (1962) (from the original T.R. von Oppolzer, book, *Canon der Finsternisse*, Wien, [1887]).

Huber, P.J., "Modeling the Length of Day and Extrapolating the Rotation of the Earth," *Astronomical Amusements*, F. Bònoli, S. De Meis, and A. Panaino, Eds., Rome (2000).

Improved Lunar Ephemeris 1952–1959, Nautical Almanac Office, U.S. Naval Observatory, Washington, DC (1954).

Meeus, J., *Mathematical Astronomy Morsels*, Willmann-Bell, pp. 56–62 (1997).

——, *More Mathematical Astronomy Morsels*, Willmann-Bell, pp. 120–126 (2002a).

——, *More Mathematical Astronomy Morsels*, Willmann-Bell, pp. 70–72 (2002b).

——, "The maximum possible duration of a total solar eclipse," *J. Br. Astron. Assoc.*, **113**(6) (2003).

——, *Mathematical Astronomy Morsels III*, Willmann-Bell, pp. 109–111, (2004).

——, *Mathematical Astronomy Morsels IV*, Willmann-Bell, pp. 44–45, (2007).

——, Grosjean, C.C., and Vanderleen, W., *Canon of Solar Eclipses*, Pergamon Press, Oxford, United Kingdom (1966).

Morrison, L., and Stephenson, F.R., "Historical Values of the Earth's Clock Error ΔT and the Calculation of Eclipses," *J. Hist. Astron.*, Vol. 35 Part 3, August 2004, No. 120, pp, 327–336 (2004).

Mucke, H., and Meeus, J., *Canon of Solar Eclipses: -2003 to +2526*, Astronomisches Büro, Vienna (1983).

Newcomb, S., "Tables of the Motion of the Earth on its Axis Around the Sun," *Astron. Papers Amer. Eph.*, Vol. 6, Part I (1895).

Stephenson, F.R., *Historical Eclipses and Earth's Rotation*, Cambridge University Press, Cambridge (1997).

——, and Houlden, M.A., *Atlas of Historical Eclipse Maps, East Asia 1500BC—AD 1900*, Cambridge University Press, Cambridge/New York (1986).

van den Bergh, *Eclipses in the Second Millennium B.C. -1600 to -1207*, Tjeenk Willink, and Haarlem, Netherlands (1954).

——, *Periodicity and Variation of Solar (and Lunar) Eclipses*, Tjeenk Willink, and Haarlem, Netherlands (1955).

von Oppolzer, T.R., *Canon der Finsternisse*, Wien, (1887).

Five Millennium Catalog of Solar Eclipses: –1999 to +3000

Appendix

Key to Solar Eclipse Catalog

Cat Num — sequential Catalog Number assigned to each eclipse from 1 to 11,898

Calendar Date — Julian or Gregorian date of Greatest Eclipse

TD of Greatest Eclipse — Terrestrial Dynamical Time of Greatest Eclipse

ΔT — arithmetic difference between Terrestrial Dynamical Time Universal Time (seconds)

Luna Num — number of synodic months, or lunations, since New Moon on 2000 Jan 06

Saros Num — Saros Series Number of eclipse

Ecl Type — Solar Eclipse Type

P = Partial Solar Eclipse
A = Annular Solar Eclipse
T = Total Solar Eclipse
H = Hybrid Solar Eclipse
m = Middle eclipse of Saros series
n = Central eclipse of with no northern limit
s = Central eclipse of with no southern limit
+ = Non-central eclipse of with no northern limit
– = Non-central eclipse of with no southern limit
2 = Hybrid eclipse path begins total and ends annular
3 = Hybrid eclipse path begins annular and ends total
b = Saros series begins (first eclipse in a Saros series)
e = Saros series ends (last eclipse in a Saros series)

QLE — Quincena Lunar Eclipse Parameter

n = Penumbral Lunar Eclipse
p = Partial Lunar Eclipse
t = Total Lunar Eclipse

Gamma — minimum distance from the axis of the lunar shadow to the center of Earth

Ecl Mag — Eclipse Magnitude; fraction of the Sun's diameter obscured by the Moon

Lat & Lng — latitude and longitude where the Sun appears in zenith at greatest eclipse

Sun Alt & **Sun Azm** — altitude and azimuth of the Sun at greatest eclipse

Path Width — width of the central path (km) at greatest eclipse (total, annular & hybrid eclipses)

Central Line Dur — Central Line Duration (minutes. seconds) at greatest eclipse

Cat Num	Canon Plate	Calendar Date	TD of Greatest Eclipse	ΔT s	Luna Num	Saros Num	Ecl. Type	QLE	Gamma	Ecl. Mag.	Lat. °	Long. °	Sun Alt °	Sun Azm °	Path Width km	Central Line Dur.
1	001	-1999 Jun 12	03:14:51	46438	-49456	5	T	-n	-0.2701	1.0733	6.0N	33.3W	74	344	247	06m37s
2	001	-1999 Dec 05	23:45:23	46426	-49450	10	A	n-	-0.2317	0.9382	32.9S	10.8E	76	21	236	06m44s
3	001	-1998 Jun 01	18:09:16	46415	-49444	15	T	p-	0.4994	1.0284	46.2N	83.4E	60	151	111	02m15s
4	001	-1998 Nov 25	05:57:03	46403	-49438	20	A	p-	-0.9045	0.9806	67.8S	143.8W	25	74	162	01m14s
5	001	-1997 Apr 22	13:19:56	46393	-49433	-13	P	-t	-1.4670	0.1611	60.6S	106.4W	0	281		
6	001	-1997 May 22	02:45:35	46391	-49432	25	P	t-	1.3253	0.4035	61.7N	151.7W	0	55		
7	001	-1997 Oct 16	08:01:52	46381	-49427	-8	P	-t	1.1669	0.6954	60.6N	22.7W	0	265		
8	001	-1997 Nov 14	18:48:49	46379	-49426	30	P	t-	-1.5183	0.0377	61.5S	27.7W	0	120		
9	001	-1996 Apr 10	13:54:52	46369	-49421	-3	A	-p	-0.7231	0.9464	38.2S	167.2W	43	321	277	05m11s
10	001	-1996 Oct 04	23:23:37	46358	-49415	2	T	-p	0.5166	1.0257	28.8N	38.6E	59	214	101	02m04s
11	001	-1995 Mar 30	17:24:52	46346	-49409	7	A	nn	0.0609	0.9873	0.2S	112.7E	87	151	45	01m17s
12	001	-1995 Sep 24	10:31:54	46334	-49403	12	A	nn	-0.1863	0.9766	3.1S	150.0W	79	29	85	02m22s
13	001	-1994 Mar 20	03:59:50	46322	-49397	17	T	p-	0.8091	1.0333	39.4N	73.3W	36	142	186	02m35s
14	001	-1994 Sep 13	14:32:00	46310	-49391	22	A	p-	-0.9265	0.9249	48.2S	116.1E	22	42	733	06m57s
15	001	-1993 Feb 08	11:41:40	46301	-49386	-11	P	-t	-1.0699	0.8826	62.7S	23.1W	0	222		
16	001	-1993 Mar 09	19:48:09	46299	-49385	27	Pb	t-	1.4907	0.0754	61.2N	11.4E	0	114		
17	001	-1993 Aug 03	21:35:06	46289	-49380	-6	P	-t	1.3116	0.4292	63.5N	166.7W	0	327		
18	001	-1992 Jan 29	02:34:14	46277	-49374	-1	T	-p	-0.3875	1.0181	44.1S	13.3W	67	340	67	01m30s
19	001	-1992 Jul 23	04:03:17	46265	-49368	4	H	-p	0.5182	1.0059	54.0N	39.0W	59	197	24	00m29s
20	001	-1991 Jan 17	11:29:42	46253	-49362	9	A	n-	0.3460	0.9599	3.5S	160.1W	70	170	155	05m03s
21	002	-1991 Jul 12	17:34:12	46242	-49356	14	T	n-	-0.2421	1.0606	10.0N	106.3E	76	6	205	05m50s
22	002	-1990 Jan 06	13:10:01	46230	-49350	19	P	t-	1.0749	0.8256	65.4N	163.3E	0	167		
23	002	-1990 Jul 02	10:30:06	46218	-49344	24	T	t-	-0.9623	1.0609	51.2S	149.5W	15	2	756	04m30s
24	002	-1990 Nov 26	19:21:37	46208	-49339	-9	P	-t	-1.2517	0.5316	69.3S	67.4W	0	150		
25	002	-1989 May 23	16:56:38	46196	-49333	-4	H	-t	0.8616	1.0067	72.1N	79.4E	30	139	46	00m27s
26	002	-1989 Nov 16	03:39:33	46185	-49327	1	H	-n	-0.5096	1.0036	44.0S	54.1W	59	19	14	00m18s
27	002	-1988 May 11	23:28:58	46173	-49321	6	A	nn	0.0987	0.9652	18.3N	16.8E	84	165	126	04m11s
28	002	-1988 Nov 04	17:46:47	46161	-49315	11	T	n-	0.1768	1.0435	0.6S	106.5E	80	196	148	04m04s
29	002	-1987 May 01	00:29:51	46149	-49309	16	A	p-	-0.6733	0.9464	31.9S	17.1E	47	341	264	06m10s
30	002	-1987 Oct 25	09:21:08	46138	-49303	21	T	p-	0.8460	1.0260	48.3N	107.1W	32	207	165	01m58s
31	002	-1986 Mar 21	15:06:51	46128	-49298	-12	P	-t	1.3284	0.3949	71.2N	79.3E	0	111		
32	002	-1986 Apr 20	02:55:09	46126	-49297	26	P	t-	-1.4035	0.2645	71.5S	44.6E	0	285		
33	002	-1986 Sep 15	07:15:14	46116	-49292	-7	P	-t	-1.2711	0.4983	71.0S	155.5W	0	58		
34	002	-1985 Mar 11	03:26:36	46104	-49286	-2	T	-p	0.5312	1.0490	20.3N	44.2W	58	163	191	04m29s
35	002	-1985 Sep 04	09:29:03	46093	-49280	3	A	-p	-0.5977	0.9272	21.9S	138.3W	53	15	338	09m22s
36	002	-1984 Feb 28	19:51:38	46081	-49274	8	T	n-	-0.1975	1.0690	25.5S	81.9E	78	345	229	05m53s
37	002	-1984 Aug 23	09:09:06	46069	-49268	13	A	nn	0.1151	0.9402	24.0N	123.1W	83	191	223	07m27s
38	002	-1983 Feb 17	11:43:36	46057	-49262	18	T	p-	-0.9309	1.0259	79.1S	89.1W	21	286	246	01m30s
39	002	-1983 Aug 12	13:43:07	46046	-49256	23	A	p-	0.8332	0.9820	75.4N	168.1W	33	209	116	01m15s
40	002	-1982 Jan 08	07:26:45	46036	-49251	-10	P	-t	1.1451	0.7132	65.6N	85.9W	0	191		
41	003	-1982 Jul 03	17:34:50	46024	-49245	-5	P	-t	-1.0663	0.8933	64.8S	127.1E	0	341		
42	003	-1982 Aug 02	01:40:26	46022	-49244	33	Pb	t-	1.4996	0.0678	67.5N	157.3E	0	351		
43	003	-1982 Dec 28	07:38:06	46012	-49239	0	A	-p	0.4710	0.9191	4.2N	97.2W	62	189	346	11m38s
44	003	-1981 Jun 23	10:42:06	46000	-49233	5	T	-n	-0.3407	1.0693	3.3N	146.8W	70	348	240	06m28s
45	003	-1981 Dec 17	07:52:15	45989	-49227	10	A	n-	-0.2249	0.9420	35.0S	110.7W	77	16	220	06m21s
46	003	-1980 Jun 12	01:14:01	45977	-49221	15	T	p-	0.4248	1.0242	45.0N	19.9W	65	157	91	02m00s
47	003	-1980 Dec 05	14:32:25	45965	-49215	20	A	p-	-0.9018	0.9845	72.0S	82.1E	25	77	127	00m58s
48	003	-1979 May 02	19:42:02	45955	-49210	-13	Pe	-t	-1.5543	0.0110	60.9S	147.5E	0	290		
49	003	-1979 Jun 01	09:17:39	45953	-49209	25	P	t-	1.2453	0.5442	62.3N	99.3E	0	47		
50	003	-1979 Oct 26	16:52:42	45944	-49204	-8	P	-t	1.1691	0.6910	60.8N	165.9W	0	256		
51	003	-1979 Nov 25	03:41:17	45942	-49203	30	P	t-	-1.5170	0.0392	62.1S	171.7W	0	130		
52	003	-1978 Apr 21	20:18:29	45932	-49198	-3	A	-p	-0.8065	0.9476	40.6S	97.8E	36	321	315	05m00s
53	003	-1978 Oct 16	08:02:30	45920	-49192	2	T	-p	0.5237	1.0214	24.8N	94.0W	58	213	84	01m47s
54	003	-1977 Apr 11	00:17:50	45908	-49186	7	A	nn	-0.0139	0.9927	0.4N	8.5E	89	330	26	00m44s
55	003	-1977 Oct 05	18:43:53	45897	-49180	12	A	nn	0.1747	0.9713	6.9S	84.7E	80	30	104	02m53s
56	003	-1976 Mar 30	11:26:06	45885	-49174	17	T	p-	0.7436	1.0404	37.9N	174.7E	42	142	197	03m05s
57	003	-1976 Sep 23	22:12:57	45873	-49168	22	A	p-	-0.9053	0.9215	48.2S	2.8W	25	46	678	07m10s
58	003	-1975 Feb 18	19:47:20	45863	-49163	-11	P	-t	-1.1099	0.8070	62.0S	155.5W	0	231		
59	003	-1975 Mar 20	03:32:45	45862	-49162	27	P	t-	1.4354	0.1811	60.9N	115.5W	0	105		
60	003	-1975 Aug 14	04:48:27	45852	-49157	-6	P	-t	1.3569	0.3516	62.7N	73.8E	0	318		

Cat Num	Canon Plate	Calendar Date	TD of Greatest Eclipse	ΔT s	Luna Num	Saros Num	Ecl. Type	QLE	Gamma	Ecl. Mag.	Lat. °	Long. °	Sun Alt °	Sun Azm °	Path Width km	Central Line Dur.
61	004	-1974 Feb 08	10:34:06	45840	-49151	-1	T	-p	-0.4230	1.0179	43.3S	131.6W	65	335	68	01m28s
62	004	-1974 Aug 03	11:35:27	45828	-49145	4	H	-p	0.5677	1.0057	55.0N	147.6W	55	205	24	00m27s
63	004	-1973 Jan 28	19:13:23	45817	-49139	9	A	n-	0.3168	0.9602	4.2S	82.0E	72	166	152	04m54s
64	004	-1973 Jul 24	01:18:22	45805	-49133	14	T	n-	-0.1865	1.0601	12.9N	11.3W	79	10	201	05m38s
65	004	-1972 Jan 17	20:43:43	45793	-49127	19	P	t-	1.0502	0.8677	64.4N	38.0E	0	157		
66	004	-1972 Jul 12	18:09:32	45781	-49121	24	T	p-	-0.9033	1.0603	40.7S	89.6E	25	8	464	04m50s
67	004	-1972 Dec 07	03:38:57	45772	-49116	-9	P	-t	-1.2500	0.5347	68.4S	154.4E	0	162		
68	004	-1971 Jun 02	23:50:31	45760	-49110	-4	A	-t	0.9435	0.9992	81.4N	79.8W	19	87	8	00m03s
69	004	-1971 Nov 26	12:22:23	45748	-49104	1	H	-n	-0.5060	1.0069	47.7S	175.3E	59	15	28	00m34s
70	004	-1970 May 23	05:53:04	45736	-49098	6	A	nn	0.1869	0.9625	26.8N	82.7W	79	167	138	04m23s
71	004	-1970 Nov 16	02:41:09	45725	-49092	11	T	n-	0.1785	1.0442	4.4S	29.3W	80	193	150	04m10s
72	004	-1969 May 12	06:45:38	45713	-49086	16	A	p-	-0.5839	0.9485	22.1S	83.5W	54	344	231	06m24s
73	004	-1969 Nov 05	18:10:59	45701	-49080	21	T	p-	0.8487	1.0233	45.1N	115.1E	32	202	149	01m51s
74	004	-1968 Mar 31	22:13:17	45692	-49075	-12	P	-t	1.3890	0.2843	71.5N	42.9W	0	98		
75	004	-1968 Apr 30	09:34:02	45690	-49074	26	P	t-	-1.3203	0.4110	71.2S	70.6W	0	298		
76	004	-1968 Sep 25	15:13:22	45680	-49069	-7	P	-t	-1.2916	0.4628	71.5S	69.3E	0	72		
77	004	-1967 Mar 21	11:03:12	45668	-49063	-2	T	-p	0.5876	1.0537	27.8N	163.6W	54	160	218	04m38s
78	004	-1967 Sep 14	16:59:19	45656	-49057	3	A	-p	-0.6263	0.9232	27.4S	104.6E	51	18	368	09m27s
79	004	-1966 Mar 11	03:42:14	45645	-49051	8	T	n-	-0.1456	1.0719	18.7S	38.7W	82	344	236	06m14s
80	004	-1966 Sep 03	16:35:18	45633	-49045	13	A	nn	0.0794	0.9390	18.5N	123.3E	85	194	227	07m41s
81	005	-1965 Feb 28	19:33:01	45621	-49039	18	T	p-	-0.8890	1.0279	71.9S	125.0E	27	312	208	01m44s
82	005	-1965 Aug 23	21:25:13	45610	-49033	23	A	p-	0.7919	0.9831	67.9N	73.2E	37	209	98	01m14s
83	005	-1964 Jan 19	15:11:49	45600	-49028	-10	P	-t	1.1713	0.6686	66.7N	145.4E	0	180		
84	005	-1964 Jul 14	01:14:53	45588	-49022	-5	P	-t	-1.1238	0.7821	65.8S	0.1E	0	351		
85	005	-1964 Aug 12	09:35:44	45586	-49021	33	P	t-	1.4540	0.1540	68.5N	25.2E	0	340		
86	005	-1963 Jan 07	15:19:18	45577	-49016	0	A	-p	0.4901	0.9215	5.4N	144.8E	61	185	340	11m26s
87	005	-1963 Jul 03	18:12:59	45565	-49010	5	T	-p	-0.4077	1.0646	0.1S	98.3E	66	352	231	06m12s
88	005	-1963 Dec 27	15:54:04	45553	-49004	10	A	n-	-0.2134	0.9464	35.9S	129.6E	77	11	202	05m54s
89	005	-1962 Jun 23	08:20:59	45541	-48998	15	T	n-	0.3533	1.0193	43.0N	124.2W	69	164	71	01m41s
90	005	-1962 Dec 16	23:03:15	45530	-48992	20	A	p-	-0.8949	0.9893	76.3S	47.3W	26	77	85	00m39s
91	005	-1961 Jun 12	15:51:46	45518	-48986	25	P	t-	1.1676	0.6797	63.1N	10.3W	0	38		
92	005	-1961 Nov 07	01:46:11	45508	-48981	-8	P	-t	1.1690	0.6910	61.1N	50.2E	0	247		
93	005	-1961 Dec 06	12:31:33	45506	-48980	30	P	t-	-1.5144	0.0428	62.9S	44.7E	0	139		
94	005	-1960 May 02	02:39:40	45497	-48975	-3	A	-p	-0.8919	0.9479	45.2S	4.7E	27	319	410	04m49s
95	005	-1960 Oct 26	16:45:10	45485	-48969	2	T	-p	0.5271	1.0175	21.0N	132.2E	58	210	70	01m31s
96	005	-1959 Apr 21	07:07:29	45473	-48963	7	A	nn	-0.0923	0.9977	0.6N	94.7W	85	331	8	00m14s
97	005	-1959 Oct 16	03:01:44	45462	-48957	12	A	nn	-0.1688	0.9665	11.1S	42.0W	80	30	122	03m23s
98	005	-1958 Apr 10	18:45:52	45450	-48951	17	T	p-	0.6719	1.0468	37.0N	64.7E	48	142	206	03m32s
99	005	-1958 Oct 05	06:02:20	45438	-48945	22	A	p-	-0.8914	0.9183	49.8S	124.6W	27	49	664	07m16s
100	005	-1957 Mar 02	03:43:57	45428	-48940	-11	P	-t	-1.1573	0.7169	61.4S	74.5E	0	240		
101	006	-1957 Mar 31	11:10:33	45426	-48939	27	P	t-	1.3744	0.2986	60.7N	119.5E	0	96		
102	006	-1957 Aug 25	12:12:43	45417	-48934	-6	P	-t	1.3938	0.2884	62.0N	48.2W	0	308		
103	006	-1957 Sep 24	05:12:58	45415	-48933	32	Pb	t-	-1.5660	0.0061	60.7S	147.1W	0	77		
104	006	-1956 Feb 19	18:25:24	45405	-48928	-1	T	-p	-0.4649	1.0176	42.2S	111.9E	62	330	68	01m25s
105	006	-1956 Aug 13	19:17:10	45393	-48922	4	H	-p	0.6100	1.0054	54.4N	100.8E	52	212	23	00m25s
106	006	-1955 Feb 08	02:47:50	45382	-48916	9	A	nn	0.2811	0.9608	4.4S	33.4W	74	162	148	04m41s
107	006	-1955 Aug 03	09:08:54	45370	-48910	14	T	n-	-0.1359	1.0589	14.5N	130.1W	82	15	196	05m22s
108	006	-1954 Jan 28	04:09:29	45358	-48904	19	A+	t-	1.0192	0.9207	63.4N	84.9W	0	147	-	-
109	006	-1954 Jul 24	01:54:42	45347	-48898	24	T	p-	-0.8494	1.0580	34.1S	31.5W	32	12	361	04m51s
110	006	-1954 Dec 18	11:52:50	45337	-48893	-9	P	-t	-1.2518	0.5318	67.3S	17.7E	0	173		
111	006	-1953 Jun 14	06:46:26	45325	-48887	-4	P	-t	1.0223	0.9489	68.0N	99.1E	0	13		
112	006	-1953 Dec 07	21:03:10	45314	-48881	1	H	-n	-0.5050	1.0107	50.8S	46.4E	59	10	43	00m52s
113	006	-1952 Jun 02	12:18:03	45302	-48875	6	A	np	0.2734	0.9595	34.9N	178.5E	74	170	153	04m34s
114	006	-1952 Nov 26	11:34:24	45290	-48869	11	T	n-	0.1792	1.0452	7.7S	164.7W	80	190	153	04m18s
115	006	-1951 May 22	13:02:16	45279	-48863	16	A	p-	-0.4945	0.9502	13.1S	176.9E	60	348	209	06m34s
116	006	-1951 Nov 16	03:00:41	45267	-48857	21	T	p-	0.8520	1.0211	42.4N	22.8W	31	198	137	01m44s
117	006	-1950 Apr 12	05:14:40	45257	-48852	-12	P	-t	1.4542	0.1642	71.5N	164.0W	0	85		
118	006	-1950 May 11	16:12:21	45255	-48851	26	P	t-	-1.2352	0.5621	70.6S	174.6E	0	311		
119	006	-1950 Oct 06	23:18:01	45246	-48846	-7	P	-t	-1.3058	0.4387	71.7S	67.9W	0	86		
120	006	-1949 Apr 01	18:33:43	45234	-48840	-2	T	-p	0.6496	1.0576	36.0N	77.8E	49	157	248	04m39s

Cat Num	Canon Plate	Calendar Date	TD of Greatest Eclipse	ΔT s	Luna Num	Saros Num	Ecl. Type	QLE	Gamma	Ecl. Mag.	Lat. °	Long. °	Sun Alt °	Sun Azm °	Path Width km	Central Line Dur.
121	007	-1949 Sep 26	00:37:49	45222	-48834	3	A	-p	-0.6477	0.9197	32.8S	14.4W	49	21	394	09m25s
122	007	-1948 Mar 21	11:25:32	45211	-48828	8	T	nn	-0.0878	1.0742	11.4S	158.1W	85	343	242	06m31s
123	007	-1948 Sep 14	00:12:27	45199	-48822	13	A	nn	0.0521	0.9380	13.1N	6.6E	87	196	231	07m50s
124	007	-1947 Mar 11	03:11:51	45187	-48816	18	T	p-	-0.8390	1.0294	63.3S	4.3W	33	323	183	01m59s
125	007	-1947 Sep 03	05:17:59	45176	-48810	23	A	p-	0.7585	0.9839	61.0N	48.4W	40	209	88	01m15s
126	007	-1946 Jan 29	22:46:25	45166	-48805	-10	P	-t	1.2051	0.6110	67.8N	18.9E	0	169		
127	007	-1946 Jul 25	09:01:49	45154	-48799	-5	P	-t	-1.1756	0.6819	66.8S	129.1W	0	1		
128	007	-1946 Aug 23	17:40:35	45153	-48798	33	P	t-	1.4157	0.2262	69.4N	109.9W	0	328		
129	007	-1945 Jan 18	22:51:51	45143	-48793	0	A	-p	0.5161	0.9243	7.8N	28.7E	59	180	333	10m58s
130	007	-1945 Jul 15	01:49:25	45131	-48787	5	T	-p	-0.4692	1.0593	4.1S	18.5W	62	356	221	05m48s
131	007	-1944 Jan 07	23:51:29	45120	-48781	10	A	n-	-0.1977	0.9514	35.5S	11.1E	78	6	182	05m24s
132	007	-1944 Jul 03	15:30:05	45108	-48775	15	H	n-	0.2846	1.0139	40.3N	130.4E	73	170	50	01m16s
133	007	-1944 Dec 27	07:31:39	45096	-48769	20	A	p-	-0.8854	0.9946	80.4S	169.0W	27	69	41	00m20s
134	007	-1943 Jun 22	22:27:06	45085	-48763	25	P	t-	1.0914	0.8113	63.9N	120.4W	0	28		
135	007	-1943 Nov 17	10:40:46	45075	-48758	-8	P	-t	1.1685	0.6919	61.6N	94.1W	0	237		
136	007	-1943 Dec 16	21:18:48	45073	-48757	30	P	t-	-1.5092	0.0512	63.7S	98.4W	0	149		
137	007	-1942 May 13	08:59:51	45063	-48752	-3	A	-t	-0.9784	0.9462	54.8S	81.5W	11	313	978	04m34s
138	007	-1942 Nov 07	01:29:55	45052	-48746	2	H3	-p	0.5284	1.0143	17.4N	2.2W	58	208	57	01m17s
139	007	-1941 May 02	13:57:20	45040	-48740	7	Hm	nn	-0.1717	1.0021	0.5N	161.9E	80	333	7	00m12s
140	007	-1941 Oct 27	11:21:18	45028	-48734	12	A	nn	-0.1649	0.9623	15.5S	169.1W	80	29	138	03m49s
141	008	-1940 Apr 21	02:04:36	45017	-48728	17	T	p-	0.5983	1.0524	36.8N	44.7W	53	143	213	03m56s
142	008	-1940 Oct 15	13:56:25	45005	-48722	22	A	p-	-0.8816	0.9156	52.3S	112.2E	28	53	662	07m17s
143	008	-1939 Mar 12	11:31:58	44996	-48717	-11	P	-t	-1.2114	0.6133	61.0S	53.1W	0	249		
144	008	-1939 Apr 10	18:42:36	44994	-48716	27	P	t-	1.3086	0.4259	60.6N	4.1W	0	87		
145	008	-1939 Sep 04	19:47:20	44984	-48711	-6	P	-t	1.4230	0.2384	61.4N	172.6W	0	299		
146	008	-1939 Oct 04	13:00:58	44982	-48710	32	P	t-	-1.5481	0.0367	60.5S	85.4E	0	86		
147	008	-1938 Mar 02	02:05:12	44972	-48705	-1	T	-p	-0.5156	1.0169	41.0S	2.1W	59	327	67	01m21s
148	008	-1938 Aug 25	03:09:40	44961	-48699	4	H	-p	0.6444	1.0050	52.6N	15.0W	50	217	22	00m23s
149	008	-1937 Feb 19	10:09:18	44949	-48693	9	A	nn	0.2352	0.9613	4.4S	145.3W	76	158	144	04m29s
150	008	-1937 Aug 14	17:09:41	44937	-48687	14	T	n-	-0.0934	1.0574	14.8N	108.7E	85	19	190	05m05s
151	008	-1936 Feb 08	11:24:38	44926	-48681	19	An	t-	0.9793	0.9202	53.9N	167.7E	11	148	-	07m38s
152	008	-1936 Aug 03	09:46:19	44914	-48675	24	T	p-	-0.8011	1.0548	29.8S	153.4W	37	17	300	04m40s
153	008	-1936 Dec 28	20:02:32	44905	-48670	-9	P	-t	-1.2579	0.5216	66.2S	117.4W	0	185		
154	008	-1935 Jun 24	13:43:01	44893	-48664	-4	P	-t	1.0997	0.8073	67.0N	18.0W	0	3		
155	008	-1935 Dec 18	05:41:01	44881	-48658	1	H2	-p	-0.5070	1.0150	53.2S	80.5W	59	4	60	01m11s
156	008	-1934 Jun 13	18:42:52	44870	-48652	6	A	-p	0.3594	0.9558	42.5N	81.0E	69	173	173	04m44s
157	008	-1934 Dec 07	20:26:30	44858	-48646	11	T	n-	0.1788	1.0468	10.4S	60.5E	80	186	158	04m28s
158	008	-1933 Jun 02	19:20:26	44847	-48640	16	A	p-	-0.4060	0.9512	4.8S	77.7E	66	351	195	06m39s
159	008	-1933 Nov 27	11:49:40	44835	-48634	21	T	p-	0.8553	1.0194	40.1N	160.5W	31	193	128	01m39s
160	008	-1932 Apr 22	12:14:51	44825	-48629	-12	Pe	-t	1.5210	0.0401	71.2N	75.4E	0	71		
161	009	-1932 May 21	22:52:55	44823	-48628	26	P	t-	-1.1505	0.7135	69.9S	50.9E	0	323		
162	009	-1932 Oct 17	07:27:29	44814	-48623	-7	P	-t	-1.3152	0.4228	71.6S	153.6E	0	101		
163	009	-1931 Apr 12	01:59:44	44802	-48617	-2	T	-p	0.7158	1.0608	44.8N	40.6W	44	154	284	04m33s
164	009	-1931 Oct 06	08:23:27	44790	-48611	3	A	-p	-0.6630	0.9167	38.0S	135.1W	48	23	418	09m19s
165	009	-1930 Apr 01	19:02:33	44779	-48605	8	Tm	nn	-0.0247	1.0760	3.7S	83.6E	89	342	246	06m43s
166	009	-1930 Sep 25	07:58:25	44767	-48599	13	A	nn	0.0317	0.9374	7.7N	112.5W	88	197	233	07m55s
167	009	-1929 Mar 22	10:43:01	44756	-48593	18	T	p-	-0.7834	1.0303	54.5S	127.0W	38	329	164	02m13s
168	009	-1929 Sep 14	13:20:52	44744	-48587	23	A	p-	0.7326	0.9846	54.7N	172.8W	43	208	80	01m14s
169	009	-1928 Feb 10	06:11:20	44734	-48582	-10	P	-t	1.2457	0.5418	68.8N	105.8W	0	157		
170	009	-1928 Aug 04	16:56:04	44723	-48576	-5	P	-t	-1.2216	0.5932	67.8S	99.4E	0	12		
171	009	-1928 Sep 03	01:53:33	44721	-48575	33	P	t-	1.3838	0.2863	70.2N	112.4E	0	315		
172	009	-1927 Jan 29	06:15:48	44711	-48570	0	A	-p	0.5491	0.9274	11.5N	85.7W	57	176	326	10m18s
173	009	-1927 Jul 25	09:31:57	44700	-48564	5	T	-p	-0.5249	1.0535	8.6S	137.3W	58	1	208	05m16s
174	009	-1926 Jan 18	07:40:04	44688	-48558	10	A	nn	-0.1741	0.9570	33.5S	105.5W	80	0	159	04m50s
175	009	-1926 Jul 14	22:44:43	44677	-48552	15	H	nn	0.2216	1.0079	36.8N	22.8E	77	176	28	00m46s
176	009	-1925 Jan 07	15:52:48	44665	-48546	20	H	p-	-0.8696	1.0006	83.1S	91.4E	29	41	4	00m02s
177	009	-1925 Jul 04	05:07:17	44653	-48540	25	P	t-	1.0195	0.9342	64.8N	128.0E	0	19		
178	009	-1925 Nov 28	19:33:50	44644	-48535	-8	P	-t	1.1693	0.6904	62.3N	121.8E	0	227		
179	009	-1925 Dec 28	06:00:28	44642	-48534	30	P	t-	-1.4998	0.0675	64.7S	119.6E	0	159		
180	009	-1924 May 23	15:21:27	44632	-48529	-3	P	-t	-1.0637	0.8577	61.8S	170.4W	0	307		

Cat Num	Canon Plate	Calendar Date	TD of Greatest Eclipse	ΔT s	Luna Num	Saros Num	Ecl. Type	QLE	Gamma	Ecl. Mag.	Lat. °	Long. °	Sun Alt °	Sun Azm °	Path Width km	Central Line Dur.
181	010	-1924 Nov 17	10:15:36	44620	-48523	2	H	-p	0.5283	1.0116	14.2N	137.0W	58	204	47	01m06s
182	010	-1923 May 12	20:45:04	44609	-48517	7	H	nn	-0.2538	1.0059	0.2S	59.1E	75	335	21	00m36s
183	010	-1923 Nov 06	19:44:18	44597	-48511	12	A	nn	-0.1649	0.9588	19.8S	63.2E	80	28	152	04m14s
184	010	-1922 May 02	09:19:50	44586	-48505	17	T	p-	0.5213	1.0573	36.7N	152.8W	58	145	219	04m18s
185	010	-1922 Oct 26	21:55:31	44574	-48499	22	A	p-	-0.8763	0.9134	55.6S	12.7W	28	57	671	07m15s
186	010	-1921 Mar 23	19:12:17	44565	-48494	-11	P	-t	-1.2716	0.4978	60.7S	178.8W	0	258		
187	010	-1921 Apr 22	02:10:29	44563	-48493	27	P	t-	1.2395	0.5603	60.7N	126.7W	0	79		
188	010	-1921 Sep 16	03:32:55	44553	-48488	-6	P	-t	1.4440	0.2026	61.0N	60.3E	0	290		
189	010	-1921 Oct 15	20:56:53	44551	-48487	32	P	t-	-1.5362	0.0567	60.5S	44.0W	0	95		
190	010	-1920 Mar 12	09:36:57	44541	-48482	-1	T	-p	-0.5721	1.0158	39.9S	114.4W	55	324	65	01m16s
191	010	-1920 Sep 04	11:10:54	44530	-48476	4	H	-p	0.6724	1.0046	49.9N	134.7W	48	220	21	00m21s
192	010	-1919 Mar 01	17:22:05	44518	-48470	9	A	nn	0.1830	0.9619	4.1S	105.2E	79	155	140	04m17s
193	010	-1919 Aug 25	01:18:07	44507	-48464	14	T	nn	-0.0570	1.0556	13.9N	14.5W	87	23	184	04m49s
194	010	-1918 Feb 18	18:29:52	44495	-48458	19	A	p-	0.9312	0.9255	46.0N	61.8E	21	148	755	07m30s
195	010	-1918 Aug 14	17:45:39	44484	-48452	24	T	p-	-0.7596	1.0510	27.3S	83.1E	40	21	256	04m21s
196	010	-1917 Jan 09	04:06:28	44474	-48447	-9	P	-t	-1.2694	0.5017	65.2S	109.5E	0	195		
197	010	-1917 Jul 05	20:44:02	44463	-48441	-4	P	-t	1.1718	0.6764	66.0N	135.7W	0	352		
198	010	-1917 Aug 04	07:51:37	44461	-48440	34	Pb	t-	-1.5248	0.0346	63.5S	148.9W	0	34		
199	010	-1917 Dec 29	14:13:35	44451	-48435	1	T	-p	-0.5141	1.0197	54.8S	154.9E	59	357	79	01m31s
200	010	-1916 Jun 24	01:12:09	44439	-48429	6	A	-p	0.4409	0.9519	49.4N	15.9W	64	178	197	04m53s
201	011	-1916 Dec 18	05:14:10	44428	-48423	11	T	n-	0.1747	1.0488	12.6S	73.1W	80	182	165	04m39s
202	011	-1915 Jun 13	01:41:44	44416	-48417	16	A	p-	-0.3194	0.9518	2.8N	21.3W	71	355	186	06m40s
203	011	-1915 Dec 07	20:35:33	44405	-48411	21	T	p-	0.8567	1.0182	38.2N	62.4E	31	188	121	01m36s
204	011	-1914 Jun 02	05:36:42	44393	-48405	26	P	t-	-1.0668	0.8639	69.0S	55.2W	0	335		
205	011	-1914 Oct 28	15:39:52	44384	-48400	-7	P	-t	-1.3215	0.4125	71.3S	14.6E	0	115		
206	011	-1913 Apr 23	09:22:31	44372	-48394	-2	T	-p	0.7853	1.0629	54.3N	159.8W	38	149	332	04m19s
207	011	-1913 Oct 17	16:15:51	44361	-48388	3	A	-p	-0.6726	0.9145	43.1S	103.1E	47	25	436	09m09s
208	011	-1912 Apr 12	02:33:41	44349	-48382	8	T	nn	0.0431	1.0769	4.3N	33.4W	88	162	249	06m48s
209	011	-1912 Oct 05	15:53:35	44337	-48376	13	A	nn	0.0182	0.9371	2.4N	126.0E	89	198	234	07m55s
210	011	-1911 Apr 01	18:05:15	44326	-48370	18	T	p-	-0.7212	1.0307	45.5S	114.4E	44	334	149	02m26s
211	011	-1911 Sep 24	21:33:45	44314	-48364	23	A	p-	0.7144	0.9854	48.8N	60.2E	44	207	73	01m14s
212	011	-1910 Feb 20	13:23:40	44305	-48359	-10	P	-t	1.2954	0.4569	69.7N	132.1E	0	145		
213	011	-1910 Mar 22	03:40:50	44303	-48358	28	Pb	t-	-1.5388	0.0265	71.2S	61.6E	0	250		
214	011	-1910 Aug 16	00:59:14	44293	-48353	-5	P	-t	-1.2603	0.5188	68.8S	34.8W	0	24		
215	011	-1910 Sep 14	10:16:33	44291	-48352	33	P	t-	1.3600	0.3312	70.9N	28.4W	0	302		
216	011	-1909 Feb 09	13:30:39	44282	-48347	0	A	-p	0.5892	0.9307	16.3N	161.6E	54	172	320	09m28s
217	011	-1909 Aug 05	17:20:50	44270	-48341	5	T	-p	-0.5746	1.0475	13.5S	101.7E	55	5	193	04m38s
218	011	-1908 Jan 29	15:22:41	44259	-48335	10	A	nn	-0.1448	0.9628	30.3S	138.8E	81	356	136	04m13s
219	011	-1908 Jul 25	06:03:48	44247	-48329	15	H	nn	0.1634	1.0017	32.7N	86.7W	80	181	6	00m10s
220	011	-1907 Jan 18	00:08:49	44236	-48323	20	H	p-	-0.8487	1.0071	81.7S	4.7E	32	1	47	00m27s
221	012	-1907 Jul 14	11:50:45	44224	-48317	25	A	t-	0.9506	0.9452	82.7N	36.3E	18	29	673	03m30s
222	012	-1907 Dec 09	04:25:19	44214	-48312	-8	P	-t	1.1714	0.6864	63.1N	22.1W	0	218		
223	012	-1906 Jan 07	14:37:00	44213	-48311	30	P	t-	-1.4864	0.0912	65.7S	21.5W	0	170		
224	012	-1906 Jun 03	21:45:37	44203	-48306	-3	P	-t	-1.1470	0.7138	62.5S	82.6E	0	316		
225	012	-1906 Nov 28	18:58:47	44191	-48300	2	H	-p	0.5296	1.0095	11.6N	89.1E	58	200	38	00m56s
226	012	-1905 May 24	03:36:37	44180	-48294	7	H	-n	-0.3336	1.0090	1.6S	45.0W	71	338	33	00m56s
227	012	-1905 Nov 18	04:06:24	44168	-48288	12	A	nn	-0.1653	0.9559	23.9S	63.9W	80	26	163	04m36s
228	012	-1904 May 12	16:35:44	44157	-48282	17	T	p-	0.4441	1.0613	36.6N	99.2E	63	148	223	04m39s
229	012	-1904 Nov 06	05:56:20	44145	-48276	22	A	p-	-0.8727	0.9118	59.4S	138.0W	29	60	680	07m10s
230	012	-1903 Apr 03	02:45:51	44136	-48271	-11	P	-t	-1.3369	0.3724	60.6S	57.2E	0	267		
231	012	-1903 May 02	09:35:23	44134	-48270	27	P	t-	1.1681	0.6993	61.0N	111.5E	0	70		
232	012	-1903 Sep 26	11:27:50	44124	-48265	-6	P	-t	1.4585	0.1777	60.7N	68.9W	0	281		
233	012	-1903 Oct 26	04:57:57	44122	-48264	32	P	t-	-1.5275	0.0711	60.7S	174.7W	0	104		
234	012	-1902 Mar 23	16:57:33	44113	-48259	-1	T	-p	-0.6371	1.0140	39.4S	136.0E	50	322	61	01m07s
235	012	-1902 Sep 15	19:22:56	44101	-48253	4	H	-p	0.6925	1.0044	46.6N	101.3E	46	222	21	00m20s
236	012	-1901 Mar 13	00:23:10	44090	-48247	9	Am	nn	0.1217	0.9623	3.6S	1.2W	83	153	137	04m09s
237	012	-1901 Sep 05	09:35:37	44078	-48241	14	T	nn	-0.0279	1.0537	12.0N	140.2W	88	27	178	04m33s
238	012	-1900 Mar 01	01:25:12	44067	-48235	19	A	p-	0.8748	0.9302	41.1N	42.9W	29	146	523	07m09s
239	012	-1900 Aug 25	01:52:52	44055	-48229	24	T	p-	-0.7250	1.0467	26.4S	42.2W	43	25	222	03m56s
240	012	-1899 Jan 19	12:03:25	44046	-48224	-9	P	-t	-1.2871	0.4706	64.2S	21.5W	0	205		

Cat Num	Canon Plate	Calendar Date	TD of Greatest Eclipse	ΔT s	Luna Num	Saros Num	Ecl. Type	QLE	Gamma	Ecl. Mag.	Lat. °	Long. °	Sun Alt °	Sun Azm °	Path Width km	Central Line Dur.
241	013	-1899 Jul 16	03:48:10	44034	-48218	-4	P	-t	1.2402	0.5537	65.0N	106.2E	0	342		
242	013	-1899 Aug 14	15:31:53	44032	-48217	34	P	t-	-1.4863	0.1079	62.7S	84.8E	0	43		
243	013	-1898 Jan 08	22:40:58	44023	-48212	1	T	-p	-0.5261	1.0248	55.5S	32.1E	58	349	99	01m53s
244	013	-1898 Jul 05	07:45:10	44011	-48206	6	A	-p	0.5188	0.9476	55.2N	111.5W	58	185	226	05m02s
245	013	-1898 Dec 29	13:56:59	44000	-48200	11	T	n-	0.1665	1.0512	14.2S	154.8E	81	178	172	04m51s
246	013	-1897 Jun 24	08:08:49	43988	-48194	16	A	nn	-0.2370	0.9519	9.3N	121.0W	76	359	181	06m36s
247	013	-1897 Dec 19	05:18:24	43977	-48188	21	T	p-	0.8559	1.0176	36.5N	73.9W	31	183	117	01m35s
248	013	-1896 Jun 12	12:24:24	43965	-48182	26	A	t-	-0.9851	0.9821	59.4S	174.8W	9	350	402	01m29s
249	013	-1896 Nov 07	23:54:29	43956	-48177	-7	P	-t	-1.3252	0.4066	70.7S	124.5W	0	129		
250	013	-1895 May 03	16:43:42	43944	-48171	-2	T	-p	0.8566	1.0640	64.4N	77.3E	31	141	407	04m00s
251	013	-1895 Oct 28	00:12:42	43933	-48165	3	A	-p	-0.6781	0.9130	48.0S	19.1W	47	25	449	08m57s
252	013	-1894 Apr 23	10:00:21	43921	-48159	8	T	nn	0.1144	1.0770	12.5N	149.2W	83	162	251	06m45s
253	013	-1894 Oct 16	23:55:45	43910	-48153	13	A	nn	0.0099	0.9375	2.5S	2.9E	89	198	232	07m51s
254	013	-1893 Apr 13	01:21:49	43898	-48147	18	T	p-	-0.6552	1.0302	36.7S	1.7W	49	337	134	02m36s
255	013	-1893 Oct 06	05:53:28	43887	-48141	23	A	p-	0.7011	0.9865	43.4N	68.7W	45	205	67	01m11s
256	013	-1892 Mar 02	20:26:50	43877	-48136	-10	P	-t	1.3514	0.3609	70.5N	11.7E	0	133		
257	013	-1892 Apr 01	10:34:53	43875	-48135	28	P	t-	-1.4763	0.1369	71.5S	57.4W	0	263		
258	013	-1892 Aug 26	09:10:32	43866	-48130	-5	P	-t	-1.2926	0.4570	69.7S	171.7W	0	36		
259	013	-1892 Sep 24	18:46:57	43864	-48129	33	P	t-	1.3420	0.3652	71.4N	171.6W	0	288		
260	013	-1891 Feb 19	20:35:10	43854	-48124	0	A	-p	0.6381	0.9340	22.4N	50.9E	50	168	318	08m31s
261	014	-1891 Aug 16	01:16:58	43843	-48118	5	T	-p	-0.6177	1.0412	18.6S	21.6W	52	9	175	03m57s
262	014	-1890 Feb 08	22:55:57	43831	-48112	10	A	nn	-0.1069	0.9690	25.8S	24.5E	84	352	112	03m32s
263	014	-1890 Aug 05	13:30:29	43820	-48106	15	Am	nn	0.1125	0.9952	28.1N	161.0E	83	185	17	00m30s
264	014	-1889 Jan 29	08:15:48	43808	-48100	20	T	p-	-0.8198	1.0140	76.7S	101.4W	35	342	85	00m54s
265	014	-1889 Jul 25	18:42:25	43797	-48094	25	A	p-	0.8889	0.9427	86.2N	90.3E	27	189	470	04m01s
266	014	-1889 Dec 20	13:12:37	43787	-48089	-8	P	-t	1.1769	0.6761	64.0N	165.3W	0	208		
267	014	-1888 Jan 18	23:06:19	43785	-48088	30	P	t-	-1.4674	0.1258	66.7S	161.3W	0	181		
268	014	-1888 Jun 14	04:13:21	43776	-48083	-3	P	-t	-1.2276	0.5742	63.3S	25.4W	0	325		
269	014	-1888 Dec 09	03:39:58	43764	-48077	2	H	-p	0.5320	1.0079	9.8N	44.3W	58	196	32	00m48s
270	014	-1887 Jun 03	10:29:53	43753	-48071	7	H	-p	-0.4126	1.0115	3.7S	149.8W	66	341	43	01m13s
271	014	-1887 Nov 28	12:27:35	43742	-48065	12	Am	nn	-0.1654	0.9537	27.4S	169.8E	80	22	171	04m54s
272	014	-1886 May 23	23:51:15	43730	-48059	17	T	p-	0.3658	1.0644	36.1N	8.4W	68	152	226	04m59s
273	014	-1886 Nov 17	13:59:10	43719	-48053	22	A	p-	-0.8709	0.9109	63.5S	96.4E	29	63	688	07m03s
274	014	-1885 Apr 14	10:13:55	43709	-48048	-11	P	-t	-1.4065	0.2387	60.7S	65.3W	0	275		
275	014	-1885 May 13	16:58:46	43707	-48047	27	P	t-	1.0951	0.8412	61.4N	10.0W	0	61		
276	014	-1885 Oct 07	19:30:51	43698	-48042	-6	P	-t	1.4674	0.1624	60.6N	159.9E	0	272		
277	014	-1885 Nov 06	13:03:01	43696	-48041	32	P	t-	-1.5214	0.0810	61.1S	53.6E	0	113		
278	014	-1884 Apr 03	00:11:40	43686	-48036	-1	H3	-p	-0.7065	1.0117	39.5S	27.9E	45	321	56	00m56s
279	014	-1884 Sep 26	03:43:13	43675	-48030	4	H	-p	0.7067	1.0045	43.1N	26.0W	45	221	22	00m21s
280	014	-1883 Mar 23	07:14:53	43663	-48024	9	A	nn	0.0534	0.9625	3.0S	105.1W	87	151	136	04m04s
281	015	-1883 Sep 15	18:00:59	43652	-48018	14	T	nn	-0.0050	1.0518	9.2N	91.9E	90	35	172	04m20s
282	015	-1882 Mar 12	08:11:58	43640	-48012	19	A	p-	0.8112	0.9347	37.7N	145.4W	36	145	402	06m41s
283	015	-1882 Sep 05	10:07:48	43629	-48006	24	T	p-	-0.6972	1.0423	26.8S	169.2W	46	28	193	03m30s
284	015	-1881 Jan 30	19:52:53	43619	-48001	-9	P	-t	-1.3118	0.4269	63.3S	150.3W	0	215		
285	015	-1881 Mar 01	09:41:45	43618	-48000	29	Pb	t-	1.5127	0.0793	61.4N	157.8E	0	120		
286	015	-1881 Jul 27	10:59:30	43608	-47995	-4	P	-t	1.3015	0.4449	64.0N	13.3W	0	333		
287	015	-1881 Aug 25	23:20:13	43606	-47994	34	P	t-	-1.4549	0.1672	62.0S	43.2W	0	52		
288	015	-1880 Jan 20	07:01:48	43597	-47989	1	T	-p	-0.5444	1.0301	55.3S	88.9W	57	341	121	02m13s
289	015	-1880 Jul 15	14:24:03	43585	-47983	6	A	-p	0.5912	0.9430	59.7N	154.2E	53	195	262	05m12s
290	015	-1879 Jan 08	22:33:28	43574	-47977	11	T	n-	0.1528	1.0539	15.2S	24.5E	81	173	180	05m01s
291	015	-1879 Jul 04	14:42:43	43562	-47971	16	A	nn	-0.1592	0.9516	14.7N	138.6E	81	3	179	06m28s
292	015	-1879 Dec 29	13:53:36	43551	-47965	21	T	p-	0.8495	1.0175	34.7N	151.7E	32	177	114	01m36s
293	015	-1878 Jun 23	19:18:52	43539	-47959	26	A	t-	-0.9074	0.9881	42.6S	71.0E	24	358	101	01m09s
294	015	-1878 Nov 19	08:08:52	43530	-47954	-7	P	-t	-1.3285	0.4012	70.0S	97.0E	0	142		
295	015	-1877 May 15	00:04:14	43518	-47948	-2	T	-t	0.9288	1.0633	74.5N	60.5W	21	117	570	03m33s
296	015	-1877 Nov 08	08:12:04	43507	-47942	3	A	-p	-0.6816	0.9122	52.8S	141.0W	47	25	457	08m43s
297	015	-1876 May 03	17:23:23	43496	-47936	8	T	-n	0.1886	1.0763	20.7N	96.0E	79	163	251	06m34s
298	015	-1876 Oct 27	08:04:34	43484	-47930	13	A	nn	0.0064	0.9384	7.2S	121.6W	90	199	229	07m43s
299	015	-1875 Apr 23	08:29:55	43473	-47924	18	T	p-	-0.5830	1.0291	27.8S	115.1W	54	340	120	02m42s
300	015	-1875 Oct 16	14:21:40	43461	-47918	23	A	p-	0.6941	0.9877	38.6N	160.3E	46	203	60	01m07s

Cat Num	Canon Plate	Calendar Date	TD of Greatest Eclipse	ΔT s	Luna Num	Saros Num	Ecl. Type	QLE	Gamma	Ecl. Mag.	Lat. °	Long. °	Sun Alt °	Sun Azm °	Path Width km	Central Line Dur.
301	016	-1874 Mar 14	03:18:38	43452	-47913	-10	P	-t	1.4157	0.2506	71.1N	106.4W	0	120		
302	016	-1874 Apr 12	17:19:20	43450	-47912	28	P	t-	-1.4066	0.2599	71.5S	174.2W	0	277		
303	016	-1874 Sep 06	17:30:25	43440	-47907	-5	P	-t	-1.3183	0.4083	70.5S	48.6E	0	49		
304	016	-1874 Oct 06	03:24:58	43438	-47906	33	P	t-	1.3299	0.3878	71.6N	43.0E	0	274		
305	016	-1873 Mar 03	03:31:18	43429	-47901	0	A	-p	0.6938	0.9372	29.6N	58.7W	46	164	321	07m32s
306	016	-1873 Aug 27	09:20:55	43418	-47895	5	T	-p	-0.6537	1.0348	23.9S	147.1W	49	13	154	03m15s
307	016	-1872 Feb 20	06:23:37	43406	-47889	10	A	nn	-0.0635	0.9753	20.3S	89.3W	86	348	88	02m49s
308	016	-1872 Aug 15	21:02:20	43395	-47883	15	A	nn	0.0669	0.9887	23.1N	46.6E	86	189	40	01m14s
309	016	-1871 Feb 08	16:17:25	43383	-47877	20	T	p-	-0.7858	1.0210	70.3S	141.1E	38	336	117	01m24s
310	016	-1871 Aug 05	01:40:13	43372	-47871	25	A	p-	0.8327	0.9393	77.5N	4.6W	33	202	410	04m37s
311	016	-1871 Dec 30	21:54:24	43362	-47866	-8	P	-t	1.1869	0.6574	65.0N	52.7E	0	198		
312	016	-1870 Jan 29	07:27:56	43360	-47865	30	P	t-	-1.4425	0.1718	67.8S	60.4E	0	192		
313	016	-1870 Jun 25	10:47:42	43351	-47860	-3	P	-t	-1.3031	0.4433	64.2S	135.4W	0	334		
314	016	-1870 Jul 25	01:47:41	43349	-47859	35	Pb	t-	1.5600	0.0081	66.8N	151.8E	0	359		
315	016	-1870 Dec 20	12:15:47	43340	-47854	2	H	-p	0.5379	1.0066	9.0N	176.3W	57	192	27	00m42s
316	016	-1869 Jun 14	17:29:24	43328	-47848	7	H	-p	-0.4875	1.0133	6.6S	103.4E	61	345	52	01m26s
317	016	-1869 Dec 09	20:43:48	43317	-47842	12	A	nn	-0.1623	0.9521	30.1S	45.4E	80	18	178	05m11s
318	016	-1868 Jun 03	07:10:24	43305	-47836	17	T	n-	0.2900	1.0667	35.3N	116.9W	73	157	227	05m17s
319	016	-1868 Nov 27	22:00:10	43294	-47830	22	A	p-	-0.8678	0.9107	67.8S	27.5W	29	65	686	06m55s
320	016	-1867 Apr 24	17:36:03	43284	-47825	-11	Pe	-t	-1.4802	0.0973	60.8S	173.6E	0	284		
321	017	-1867 May 24	00:20:47	43283	-47824	27	P	t-	1.0213	0.9844	61.9N	131.3W	0	53		
322	017	-1867 Oct 18	03:41:29	43273	-47819	-6	P	-t	1.4709	0.1561	60.7N	26.8E	0	262		
323	017	-1867 Nov 16	21:10:19	43271	-47818	32	P	t-	-1.5165	0.0884	61.6S	78.8W	0	123		
324	017	-1866 Apr 14	07:17:07	43262	-47813	-1	H	-p	-0.7818	1.0084	40.8S	77.7W	38	320	45	00m41s
325	017	-1866 Oct 07	12:11:37	43250	-47807	4	H	-p	0.7148	1.0050	39.4N	156.4W	44	219	24	00m24s
326	017	-1865 Apr 03	13:57:33	43239	-47801	9	A	nn	-0.0214	0.9623	2.6S	153.4E	89	331	137	04m04s
327	017	-1865 Sep 27	02:35:16	43227	-47795	14	T	nn	0.0102	1.0500	5.6N	38.6W	89	207	166	04m09s
328	017	-1864 Mar 22	14:50:16	43216	-47789	19	A	p-	0.7404	0.9389	35.4N	114.6E	42	144	326	06m12s
329	017	-1864 Sep 15	18:30:18	43205	-47783	24	T	p-	-0.6763	1.0377	28.4S	61.8E	47	31	168	03m03s
330	017	-1863 Feb 10	03:35:11	43195	-47778	-9	P	-t	-1.3429	0.3711	62.6S	83.1E	0	225		
331	017	-1863 Mar 11	16:45:13	43193	-47777	29	P	t-	1.4511	0.1845	61.0N	41.3E	0	111		
332	017	-1863 Aug 06	18:16:57	43184	-47772	-4	P	-t	1.3569	0.3480	63.1N	133.9W	0	323		
333	017	-1863 Sep 05	07:16:15	43182	-47771	34	P	t-	-1.4302	0.2136	61.4S	173.0W	0	62		
334	017	-1862 Jan 30	15:14:50	43172	-47766	1	T	-p	-0.5698	1.0355	54.4S	151.5E	55	334	145	02m34s
335	017	-1862 Jul 26	21:10:21	43161	-47760	6	A	-p	0.6570	0.9383	62.4N	60.2E	49	206	304	05m24s
336	017	-1861 Jan 20	07:03:23	43150	-47754	11	T	n-	0.1335	1.0568	15.7S	104.1W	82	168	189	05m11s
337	017	-1861 Jul 15	21:24:00	43138	-47748	16	A	nn	-0.0870	0.9510	19.0N	37.1E	85	8	180	06m18s
338	017	-1860 Jan 09	22:23:09	43127	-47742	21	T	p-	0.8391	1.0178	33.0N	19.0E	33	172	112	01m39s
339	017	-1860 Jul 04	02:20:00	43116	-47736	26	A	t-	-0.8339	0.9917	32.9S	40.2W	33	3	53	00m52s
340	017	-1860 Nov 29	16:22:01	43106	-47731	-7	P	-t	-1.3320	0.3957	69.0S	40.5W	0	155		
341	018	-1859 May 25	07:24:25	43095	-47725	-2	T+	-t	1.0015	1.0203	69.5N	100.9E	0	33	-	-
342	018	-1859 Jun 23	14:36:03	43093	-47724	36	Pb	t-	-1.5239	0.0181	67.0S	142.0E	0	357		
343	018	-1859 Nov 18	16:12:53	43083	-47719	3	A	-p	-0.6838	0.9121	57.3S	98.3E	47	23	460	08m27s
344	018	-1858 May 15	00:44:54	43072	-47713	8	T	-n	0.2636	1.0746	28.8N	18.0W	75	165	251	06m15s
345	018	-1858 Nov 07	16:16:01	43061	-47707	13	A	nn	0.0044	0.9400	11.5S	113.4E	90	201	222	07m30s
346	018	-1857 May 04	15:35:26	43049	-47701	18	T	p-	-0.5094	1.0272	19.3S	132.9E	59	343	107	02m41s
347	018	-1857 Oct 27	22:54:36	43038	-47695	23	A	p-	0.6901	0.9895	34.3N	28.1E	46	201	51	00m59s
348	018	-1856 Mar 24	10:02:17	43028	-47690	-10	P	-t	1.4854	0.1309	71.5N	137.2E	0	106		
349	018	-1856 Apr 22	23:58:22	43026	-47689	28	P	t-	-1.3330	0.3895	71.3S	70.4E	0	290		
350	018	-1856 Sep 17	01:57:42	43017	-47684	-5	P	-t	-1.3381	0.3709	71.1S	93.4W	0	62		
351	018	-1856 Oct 16	12:08:28	43015	-47683	33	P	t-	1.3223	0.4022	71.5N	103.9W	0	260		
352	018	-1855 Mar 13	10:18:56	43006	-47678	0	A	-p	0.7566	0.9402	37.9N	167.3W	41	159	335	06m34s
353	018	-1855 Sep 06	17:32:50	42994	-47672	5	T	-p	-0.6822	1.0286	29.2S	85.1E	47	17	132	02m35s
354	018	-1854 Mar 02	13:41:30	42983	-47666	10	A	nn	-0.0113	0.9817	13.9S	158.6E	89	347	65	02m05s
355	018	-1854 Aug 27	04:43:27	42972	-47660	15	A	nn	0.0302	0.9822	17.9N	70.6W	88	192	63	01m59s
356	018	-1853 Feb 20	00:10:17	42960	-47654	20	T	p-	-0.7440	1.0282	63.0S	21.0E	42	335	143	01m57s
357	018	-1853 Aug 16	08:46:38	42949	-47648	25	A	p-	0.7840	0.9356	69.7N	111.7W	38	205	388	05m17s
358	018	-1852 Jan 11	06:29:27	42939	-47643	-8	P	-t	1.2024	0.6283	66.0N	88.2W	0	187		
359	018	-1852 Feb 09	15:41:30	42937	-47642	30	P	t-	-1.4111	0.2304	68.7S	76.4W	0	203		
360	018	-1852 Jul 05	17:28:58	42928	-47637	-3	P	-t	-1.3730	0.3218	65.1S	112.5E	0	344		

Cat Num	Canon Plate	Calendar Date	TD of Greatest Eclipse	ΔT s	Luna Num	Saros Num	Ecl. Type	QLE	Gamma	Ecl. Mag.	Lat. °	Long. °	Sun Alt °	Sun Azm °	Path Width km	Central Line Dur.
361	019	-1852 Aug 04	08:41:15	42926	-47636	35	P	t-	1.5009	0.1091	67.9N	35.6E	0	348		
362	019	-1852 Dec 30	20:45:47	42917	-47631	2	H	-p	0.5483	1.0059	9.2N	53.2E	57	187	24	00m38s
363	019	-1851 Jun 25	00:33:02	42905	-47625	7	H2	-p	-0.5598	1.0145	10.4S	5.0W	56	349	60	01m34s
364	019	-1851 Dec 20	04:56:02	42894	-47619	12	A	nn	-0.1566	0.9511	31.8S	77.5W	81	14	181	05m24s
365	019	-1850 Jun 14	14:32:21	42883	-47613	17	T	n-	0.2160	1.0680	33.7N	133.8E	77	162	227	05m33s
366	019	-1850 Dec 09	05:58:42	42871	-47607	22	A	p-	-0.8628	0.9113	72.0S	148.4W	30	64	673	06m47s
367	019	-1849 Jun 04	07:44:24	42860	-47601	27	T	t-	0.9488	1.0660	71.4N	148.2E	18	82	698	03m25s
368	019	-1849 Oct 29	11:58:23	42851	-47596	-6	P	-t	1.4700	0.1571	60.9N	107.9W	0	253		
369	019	-1849 Nov 28	05:19:01	42849	-47595	32	P	t-	-1.5121	0.0947	62.2S	148.2E	0	132		
370	019	-1848 Apr 24	14:17:16	42839	-47590	-1	H	-t	-0.8605	1.0042	43.8S	178.6E	30	319	28	00m20s
371	019	-1848 Oct 17	20:46:35	42828	-47584	4	H	-p	0.7184	1.0059	35.7N	71.0E	44	217	29	00m29s
372	019	-1847 Apr 13	20:33:27	42817	-47578	9	A	nn	-0.1011	0.9617	2.3S	53.7E	84	331	139	04m09s
373	019	-1847 Oct 07	11:15:47	42805	-47572	14	T	nn	0.0209	1.0485	1.7N	170.8W	89	209	161	04m01s
374	019	-1846 Apr 02	21:21:37	42794	-47566	19	A	p-	0.6631	0.9428	34.0N	16.8E	48	144	274	05m46s
375	019	-1846 Sep 27	02:59:55	42782	-47560	24	T	p-	-0.6615	1.0333	31.0S	68.9W	48	34	147	02m38s
376	019	-1845 Feb 21	11:10:37	42773	-47555	-9	P	-t	-1.3808	0.3024	61.9S	41.7W	0	234		
377	019	-1845 Mar 22	23:43:44	42771	-47554	29	P	t-	1.3838	0.3009	60.7N	73.8W	0	102		
378	019	-1845 Aug 18	01:41:20	42762	-47549	-4	P	-t	1.4057	0.2638	62.4N	103.9E	0	314		
379	019	-1845 Sep 16	15:18:31	42760	-47548	34	P	t-	-1.4111	0.2495	61.0S	55.8E	0	71		
380	019	-1844 Feb 10	23:20:42	42750	-47543	1	T	-p	-0.6022	1.0408	52.9S	33.0E	53	327	170	02m53s
381	020	-1844 Aug 06	04:04:34	42739	-47537	6	A	-p	0.7161	0.9335	63.3N	35.0W	44	217	355	05m37s
382	020	-1843 Jan 30	15:24:14	42728	-47531	11	T	n-	0.1064	1.0598	15.7S	129.6E	84	164	198	05m19s
383	020	-1843 Jul 26	04:14:32	42716	-47525	16	A	nn	-0.0216	0.9502	21.9N	66.1W	89	14	183	06m09s
384	020	-1842 Jan 20	06:43:06	42705	-47519	21	T	p-	0.8212	1.0185	31.3N	110.9W	35	167	110	01m43s
385	020	-1842 Jul 15	09:30:31	42694	-47513	26	A	p-	-0.7668	0.9941	26.0S	152.4W	40	8	32	00m38s
386	020	-1842 Dec 11	00:30:01	42684	-47508	-7	P	-t	-1.3391	0.3838	68.0S	176.1W	0	167		
387	020	-1841 Jun 05	14:47:29	42673	-47502	-2	P	-t	1.0723	0.8834	68.7N	23.7W	0	22		
388	020	-1841 Jul 04	22:03:49	42671	-47501	36	P	t-	-1.4564	0.1461	66.0S	17.6E	0	7		
389	020	-1841 Nov 30	00:13:30	42662	-47496	3	A	-p	-0.6864	0.9127	61.4S	20.6W	46	18	459	08m08s
390	020	-1840 May 25	08:04:26	42650	-47490	8	T	-n	0.3401	1.0720	36.8N	130.7W	70	167	249	05m50s
391	020	-1840 Nov 18	00:31:24	42639	-47484	13	A	nn	0.0048	0.9422	15.3S	12.2W	90	199	214	07m13s
392	020	-1839 May 14	22:35:50	42628	-47478	18	T	p-	-0.4321	1.0246	11.0S	22.7E	64	346	92	02m34s
393	020	-1839 Nov 07	07:32:22	42616	-47472	23	A	p-	0.6895	0.9917	30.5N	105.4W	46	197	40	00m48s
394	020	-1838 Apr 04	16:36:35	42607	-47467	-10	Pe	-t	1.5615	0.0000	71.6N	23.0E	0	93		
395	020	-1838 May 04	06:31:11	42605	-47466	28	P	t-	-1.2547	0.5269	70.9S	43.1W	0	303		
396	020	-1838 Sep 28	10:33:15	42596	-47461	-5	P	-t	-1.3514	0.3459	71.4S	122.1E	0	76		
397	020	-1838 Oct 27	20:57:36	42594	-47460	33	P	t-	1.3189	0.4084	71.2N	108.1E	0	246		
398	020	-1837 Mar 24	17:00:38	42584	-47455	0	A	-p	0.8244	0.9428	47.3N	83.7E	34	154	369	05m39s
399	020	-1837 Sep 18	01:51:07	42573	-47449	5	T	-p	-0.7050	1.0226	34.4S	44.4W	45	20	108	01m58s
400	020	-1836 Mar 12	20:55:13	42562	-47443	10	A	nn	0.0456	0.9880	6.9S	46.8E	87	163	43	01m22s
401	021	-1836 Sep 06	12:30:48	42550	-47437	15	A	nn	-0.0004	0.9759	12.6N	170.3E	90	174	86	02m45s
402	021	-1835 Mar 02	07:56:34	42539	-47431	20	T	p-	-0.6959	1.0353	55.3S	99.4W	46	335	165	02m34s
403	021	-1835 Aug 26	16:01:09	42528	-47425	25	A	p-	0.7422	0.9317	62.4N	137.4E	42	206	381	06m00s
404	021	-1834 Jan 21	14:57:13	42518	-47420	-8	P	-t	1.2239	0.5877	67.1N	132.4E	0	176		
405	021	-1834 Feb 19	23:47:18	42517	-47419	30	P	t-	-1.3735	0.3016	69.6S	148.2E	0	215		
406	021	-1834 Jul 17	00:19:47	42507	-47414	-3	P	-t	-1.4360	0.2124	66.1S	2.3W	0	354		
407	021	-1834 Aug 15	15:45:30	42505	-47413	35	P	t-	1.4493	0.1971	68.9N	83.8W	0	336		
408	021	-1833 Jan 11	05:07:17	42496	-47408	2	H	-p	0.5652	1.0052	10.7N	75.3W	56	183	22	00m34s
409	021	-1833 Jul 06	07:46:09	42485	-47402	7	T	-p	-0.6253	1.0151	14.7S	116.4W	51	353	66	01m37s
410	021	-1833 Dec 31	13:00:55	42473	-47396	12	A	nn	-0.1456	0.9506	32.3S	161.7E	81	9	183	05m35s
411	021	-1832 Jun 24	21:59:15	42462	-47390	17	T	n-	0.1456	1.0685	31.5N	22.9E	81	167	226	05m46s
412	021	-1832 Dec 19	13:52:26	42451	-47384	22	A	p-	-0.8539	0.9126	76.0S	97.1E	31	58	646	06m40s
413	021	-1831 Jun 14	15:09:59	42439	-47378	27	T	p-	0.8784	1.0667	74.3N	64.4E	28	111	459	03m40s
414	021	-1831 Nov 08	20:18:55	42430	-47373	-6	P	-t	1.4672	0.1612	61.3N	116.5E	0	244		
415	021	-1831 Dec 08	13:25:07	42428	-47372	32	P	t-	-1.5050	0.1053	63.0S	15.7E	0	142		
416	021	-1830 May 05	21:11:40	42419	-47367	-1	A	-t	-0.9428	0.9984	50.0S	79.3E	19	316	17	00m08s
417	021	-1830 Oct 29	05:27:09	42408	-47361	4	H	-p	0.7181	1.0074	32.1N	63.4W	44	213	36	00m38s
418	021	-1829 Apr 25	03:02:58	42396	-47355	9	A	nn	-0.1855	0.9607	2.5S	44.3W	79	332	145	04m21s
419	021	-1829 Oct 18	20:01:55	42385	-47349	14	T	nn	0.0269	1.0472	2.6S	55.4E	88	210	157	03m56s
420	021	-1828 Apr 13	03:47:21	42374	-47343	19	A	p-	0.5808	0.9464	33.1N	79.3W	54	144	237	05m23s

Cat Num	Canon Plate	Calendar Date	TD of Greatest Eclipse	ΔT s	Luna Num	Saros Num	Ecl. Type	QLE	Gamma	Ecl. Mag.	Lat. °	Long. °	Sun Alt °	Sun Azm °	Path Width km	Central Line Dur.
421	022	-1828 Oct 07	11:36:07	42362	-47337	24	T	p-	-0.6527	1.0290	34.4S	158.7E	49	36	127	02m15s
422	022	-1827 Mar 03	18:37:30	42353	-47332	-9	P	-t	-1.4261	0.2193	61.4S	164.1W	0	243		
423	022	-1827 Apr 02	06:34:52	42351	-47331	29	P	t-	1.3094	0.4314	60.5N	172.9E	0	93		
424	022	-1827 Aug 28	09:13:31	42342	-47326	-4	P	-t	1.4472	0.1933	61.7N	19.9W	0	305		
425	022	-1827 Sep 26	23:28:24	42340	-47325	34	P	t-	-1.3988	0.2727	60.7S	77.2W	0	80		
426	022	-1826 Feb 21	07:18:40	42330	-47320	1	T	-p	-0.6416	1.0459	51.3S	84.2W	50	322	198	03m11s
427	022	-1826 Aug 17	11:07:52	42319	-47314	6	A	-p	0.7672	0.9289	62.6N	133.7W	40	227	414	05m53s
428	022	-1825 Feb 10	23:37:36	42308	-47308	11	T	nn	0.0732	1.0627	15.2S	5.3E	86	160	206	05m25s
429	022	-1825 Aug 06	11:15:15	42297	-47302	16	A	nn	0.0361	0.9491	23.4N	171.5W	88	196	187	06m01s
430	022	-1824 Jan 31	14:55:54	42285	-47296	21	T	p-	0.7981	1.0194	29.8N	121.5E	37	162	109	01m47s
431	022	-1824 Jul 25	16:48:41	42274	-47290	26	A	p-	-0.7049	0.9959	21.1S	94.4E	45	12	20	00m26s
432	022	-1824 Dec 21	08:34:10	42265	-47285	-7	P	-t	-1.3486	0.3679	66.9S	49.9E	0	178		
433	022	-1823 Jun 15	22:13:19	42254	-47279	-2	P	-t	1.1413	0.7498	67.7N	148.4W	0	11		
434	022	-1823 Jul 15	05:38:07	42252	-47278	36	P	t-	-1.3935	0.2659	65.0S	108.1W	0	17		
435	022	-1823 Dec 10	08:10:48	42242	-47273	3	A	-p	-0.6918	0.9140	65.0S	136.2W	46	12	456	07m48s
436	022	-1822 Jun 05	15:25:30	42231	-47267	8	T	-p	0.4148	1.0686	44.3N	117.3E	65	170	246	05m20s
437	022	-1822 Nov 29	08:46:06	42220	-47261	13	A	nn	0.0036	0.9451	18.6S	137.2W	90	201	202	06m49s
438	022	-1821 May 26	05:35:56	42208	-47255	18	T	n-	-0.3555	1.0212	3.3S	86.7W	69	349	77	02m18s
439	022	-1821 Nov 18	16:10:47	42197	-47249	23	A	p-	0.6886	0.9945	27.2N	120.9E	46	194	26	00m33s
440	022	-1820 May 14	13:01:43	42186	-47243	28	P	t-	-1.1748	0.6667	70.3S	155.6W	0	316		
441	023	-1820 Oct 08	19:14:50	42177	-47238	-5	P	-t	-1.3602	0.3298	71.6S	24.2W	0	90		
442	023	-1820 Nov 07	05:49:12	42175	-47237	33	P	t-	1.3177	0.4108	70.7N	40.2W	0	232		
443	023	-1819 Apr 03	23:34:14	42165	-47232	0	A	-p	0.8990	0.9446	58.2N	28.1W	26	144	465	04m48s
444	023	-1819 Sep 28	10:17:02	42154	-47226	5	T	-p	-0.7208	1.0171	39.5S	175.6W	44	23	84	01m26s
445	023	-1818 Mar 24	04:01:15	42143	-47220	10	Am	nn	0.1098	0.9939	0.8N	63.6W	84	162	21	00m40s
446	023	-1818 Sep 17	20:26:38	42132	-47214	15	A	nn	-0.0232	0.9700	7.3N	48.8E	89	17	108	03m28s
447	023	-1817 Mar 13	15:35:24	42120	-47208	20	T	p-	-0.6407	1.0421	47.2S	141.1E	50	336	183	03m14s
448	023	-1817 Sep 06	23:25:42	42109	-47202	25	A	p-	0.7087	0.9279	55.6N	23.3E	45	206	382	06m45s
449	023	-1816 Feb 01	23:16:53	42100	-47197	-8	P	-t	1.2520	0.5345	68.1N	5.6W	0	165		
450	023	-1816 Mar 02	07:44:54	42098	-47196	30	P	t-	-1.3292	0.3861	70.4S	14.3E	0	228		
451	023	-1816 Jul 27	07:19:00	42088	-47191	-3	P	-t	-1.4927	0.1141	67.1S	119.6W	0	5		
452	023	-1816 Aug 25	22:59:47	42087	-47190	35	P	t-	1.4050	0.2724	69.8N	153.6E	0	324		
453	023	-1815 Jan 21	13:21:20	42077	-47185	2	H	-p	0.5875	1.0048	13.2N	157.8E	54	178	21	00m31s
454	023	-1815 Jul 16	15:06:20	42066	-47179	7	T	-p	-0.6856	1.0151	19.6S	129.8E	47	357	71	01m35s
455	023	-1814 Jan 10	20:57:33	42055	-47173	12	A	nn	-0.1287	0.9506	31.3S	43.0E	82	3	183	05m43s
456	023	-1814 Jul 06	05:31:47	42043	-47167	17	T	nn	0.0796	1.0682	28.5N	90.0W	85	173	223	05m56s
457	023	-1814 Dec 30	21:41:13	42032	-47161	22	A	p-	-0.8414	0.9146	78.8S	6.8W	32	42	608	06m34s
458	023	-1813 Jun 25	22:38:32	42021	-47155	27	T	p-	0.8105	1.0656	74.1N	23.1W	36	138	369	03m49s
459	023	-1813 Nov 20	04:41:44	42012	-47150	-6	P	-t	1.4632	0.1670	61.9N	19.9W	0	234		
460	023	-1813 Dec 19	21:29:10	42010	-47149	32	P	t-	-1.4956	0.1196	63.9S	116.5W	0	152		
461	024	-1812 May 16	04:03:20	42000	-47144	-1	P	-t	-1.0261	0.9429	61.5S	6.4W	0	302		
462	024	-1812 Nov 08	14:11:00	41989	-47138	4	H	-p	0.7160	1.0094	28.8N	161.2E	44	209	45	00m49s
463	024	-1811 May 05	09:27:26	41978	-47132	9	A	np	-0.2735	0.9592	3.4S	141.1W	74	333	153	04m39s
464	024	-1811 Oct 29	04:52:15	41967	-47126	14	T	nn	0.0298	1.0464	6.8S	79.3W	88	209	155	03m53s
465	024	-1810 Apr 24	10:09:51	41956	-47120	19	A	p-	0.4952	0.9495	32.5N	174.3W	60	146	210	05m06s
466	024	-1810 Oct 18	20:16:20	41944	-47114	24	T	p-	-0.6476	1.0251	38.4S	25.5E	49	38	110	01m54s
467	024	-1809 Mar 15	01:59:09	41935	-47109	-9	P	-t	-1.4765	0.1261	61.0S	74.9E	0	252		
468	024	-1809 Apr 13	13:24:19	41933	-47108	29	P	t-	1.2320	0.5684	60.5N	60.1E	0	85		
469	024	-1809 Sep 08	16:53:49	41924	-47103	-4	P	-t	1.4811	0.1367	61.2N	145.6W	0	295		
470	024	-1809 Oct 08	07:43:56	41922	-47102	34	P	t-	-1.3919	0.2862	60.6S	148.5E	0	89		
471	024	-1808 Mar 03	15:09:43	41913	-47097	1	T	-p	-0.6876	1.0506	49.8S	159.7E	46	319	229	03m28s
472	024	-1808 Aug 27	18:20:00	41901	-47091	6	A	-p	0.8110	0.9244	60.9N	123.0E	36	233	482	06m12s
473	024	-1807 Feb 21	07:42:02	41890	-47085	11	Tm	nn	0.0323	1.0654	14.4S	116.8W	88	155	214	05m31s
474	024	-1807 Aug 16	18:27:18	41879	-47079	16	A	nn	0.0854	0.9481	23.6N	80.3E	85	201	192	05m55s
475	024	-1806 Feb 10	22:57:03	41868	-47073	21	T	p-	0.7659	1.0204	28.3N	2.6W	40	158	106	01m51s
476	024	-1806 Aug 06	00:18:42	41856	-47067	26	A	p-	-0.6514	0.9971	17.9S	21.3W	49	16	13	00m18s
477	024	-1805 Jan 01	16:30:32	41847	-47062	-7	P	-t	-1.3642	0.3415	65.8S	81.6W	0	189		
478	024	-1805 Jan 31	08:30:57	41845	-47061	31	Pb	t-	1.5526	0.0081	63.3N	167.9W	0	144		
479	024	-1805 Jun 27	05:43:43	41836	-47056	-2	P	-t	1.2069	0.6227	66.7N	86.3E	0	0		
480	024	-1805 Jul 26	13:20:27	41834	-47055	36	P	t-	-1.3363	0.3747	64.1S	124.7E	0	27		

Cat Num	Canon Plate	Calendar Date	TD of Greatest Eclipse	ΔT s	Luna Num	Saros Num	Ecl. Type	QLE	Gamma	Ecl. Mag.	Lat. °	Long. °	Sun Alt °	Sun Azm °	Path Width km	Central Line Dur.
481	025	-1805 Dec 21	16:04:21	41825	-47050	3	A	-p	-0.7005	0.9159	67.8S	111.9E	45	2	451	07m26s
482	025	-1804 Jun 15	22:47:41	41814	-47044	8	T	-p	0.4884	1.0643	51.4N	6.6E	61	174	242	04m47s
483	025	-1804 Dec 09	17:00:35	41802	-47038	13	A	nn	0.0015	0.9487	21.3S	98.2E	90	349	188	06m18s
484	025	-1803 Jun 05	12:33:29	41791	-47032	18	T	nn	-0.2775	1.0171	4.0N	165.2E	74	353	61	01m54s
485	025	-1803 Nov 29	00:51:05	41780	-47026	23	A	p-	0.6885	0.9979	24.4N	13.2W	46	190	10	00m13s
486	025	-1802 May 25	19:29:58	41769	-47020	28	P	t-	-1.0933	0.8083	69.5S	93.0E	0	328		
487	025	-1802 Oct 20	04:01:48	41759	-47015	-5	P	-t	-1.3649	0.3213	71.4S	171.9W	0	105		
488	025	-1802 Nov 18	14:42:23	41758	-47014	33	P	t-	1.3175	0.4110	69.9N	171.6E	0	219		
489	025	-1801 Apr 15	06:05:13	41748	-47009	0	A	-t	0.9760	0.9447	70.3N	160.1W	12	114	988	04m02s
490	025	-1801 Oct 09	18:49:02	41737	-47003	5	H3	-p	-0.7313	1.0119	44.5S	52.0E	43	26	60	00m58s
491	025	-1800 Apr 03	11:04:05	41726	-46997	10	A	nn	0.1778	0.9997	8.7N	173.4W	80	162	1	00m02s
492	025	-1800 Sep 28	04:28:25	41715	-46991	15	A	nn	-0.0402	0.9644	2.0N	74.3W	88	18	129	04m09s
493	025	-1799 Mar 23	23:09:16	41703	-46985	20	T	p-	-0.5805	1.0485	38.9S	22.6E	54	337	197	03m55s
494	025	-1799 Sep 17	06:58:25	41692	-46979	25	A	p-	0.6823	0.9244	49.3N	93.2W	47	206	388	07m32s
495	025	-1798 Feb 12	07:27:30	41683	-46974	-8	P	-t	1.2872	0.4675	69.1N	141.8W	0	153		
496	025	-1798 Mar 13	15:34:58	41681	-46973	30	P	t-	-1.2789	0.4828	71.0S	118.3W	0	241		
497	025	-1798 Aug 07	14:29:58	41672	-46968	-3	Pe	-t	-1.5404	0.0313	68.1S	119.6E	0	16		
498	025	-1798 Sep 06	06:25:28	41670	-46967	35	P	t-	1.3692	0.3331	70.6N	27.6E	0	311		
499	025	-1797 Feb 01	21:25:34	41661	-46962	2	H	-p	0.6170	1.0043	17.0N	32.9E	52	174	19	00m28s
500	025	-1797 Jul 27	22:35:52	41649	-46956	7	T	-p	-0.7391	1.0147	25.0S	12.9E	42	2	75	01m30s
501	026	-1796 Jan 22	04:44:37	41638	-46950	12	A	nn	-0.1047	0.9509	29.0S	73.7W	84	358	181	05m49s
502	026	-1796 Jul 16	13:11:22	41627	-46944	17	T	nn	0.0193	1.0673	25.0N	154.7E	89	178	220	06m01s
503	026	-1795 Jan 10	05:21:00	41616	-46938	22	A	p-	-0.8213	0.9173	79.0S	98.4W	34	16	556	06m29s
504	026	-1795 Jul 06	06:11:22	41605	-46932	27	T	p-	0.7465	1.0633	71.5N	117.7W	41	160	313	03m53s
505	026	-1795 Nov 30	13:04:30	41595	-46927	-6	P	-t	1.4602	0.1709	62.6N	156.4W	0	225		
506	026	-1795 Dec 30	05:27:20	41594	-46926	32	P	t-	-1.4807	0.1432	64.9S	112.3E	0	162		
507	026	-1794 May 27	10:52:48	41584	-46921	-1	P	-t	-1.1099	0.7897	62.1S	119.5W	0	310		
508	026	-1794 Jun 25	20:54:42	41582	-46920	37	Pb	t-	1.5328	0.0144	64.4N	115.8W	0	24		
509	026	-1794 Nov 19	22:56:01	41573	-46915	4	H2	-p	0.7137	1.0120	26.0N	25.7E	44	205	58	01m05s
510	026	-1793 May 16	15:49:33	41562	-46909	9	A	-p	-0.3627	0.9571	5.0S	122.5E	69	336	166	05m03s
511	026	-1793 Nov 09	13:46:13	41551	-46903	14	Tm	nn	0.0296	1.0461	11.0S	145.2E	88	207	153	03m54s
512	026	-1792 May 04	16:28:25	41539	-46897	19	A	p-	0.4057	0.9522	31.8N	92.1E	66	148	190	04m53s
513	026	-1792 Oct 29	05:01:32	41528	-46891	24	T	p-	-0.6471	1.0216	42.9S	108.8W	49	39	95	01m36s
514	026	-1791 Mar 25	09:14:19	41519	-46886	-9	Pe	-t	-1.5327	0.0211	60.8S	44.4W	0	260		
515	026	-1791 Apr 23	20:09:41	41517	-46885	29	P	t-	1.1501	0.7149	60.7N	51.7W	0	76		
516	026	-1791 Sep 19	00:42:05	41508	-46880	-4	P	-t	1.5081	0.0925	60.8N	86.8E	0	286		
517	026	-1791 Oct 18	16:05:01	41506	-46879	34	P	t-	-1.3898	0.2908	60.7S	12.8E	0	99		
518	026	-1790 Mar 14	22:53:21	41497	-46874	1	T	-p	-0.7405	1.0548	48.7S	45.1E	42	316	266	03m42s
519	026	-1790 Sep 08	01:42:19	41486	-46868	6	A	-p	0.8463	0.9202	58.7N	14.4E	32	237	558	06m33s
520	026	-1789 Mar 04	15:39:33	41474	-46862	11	T	nn	-0.0146	1.0677	13.2S	122.9E	89	336	221	05m36s
521	027	-1789 Aug 28	01:48:41	41463	-46856	16	A	nn	0.1279	0.9470	22.6N	30.3W	83	205	196	05m52s
522	027	-1788 Feb 22	06:50:44	41452	-46850	21	T	p-	0.7280	1.0213	27.2N	124.2W	43	154	104	01m55s
523	027	-1788 Aug 16	07:57:40	41441	-46844	26	A	p-	-0.6041	0.9980	16.2S	138.8W	53	20	9	00m12s
524	027	-1787 Jan 12	00:19:01	41432	-46839	-7	P	-t	-1.3854	0.3057	64.8S	149.4E	0	199		
525	027	-1787 Feb 10	16:07:23	41430	-46838	31	P	t-	1.5213	0.0626	62.5N	66.9E	0	135		
526	027	-1787 Jul 07	13:19:08	41421	-46833	-2	P	-t	1.2685	0.5034	65.7N	39.9W	0	350		
527	027	-1787 Aug 05	21:11:03	41419	-46832	36	P	t-	-1.2853	0.4718	63.2S	4.3W	0	36		
528	027	-1787 Dec 31	23:51:46	41409	-46827	3	A	-p	-0.7145	0.9183	69.5S	4.2E	44	350	446	07m03s
529	027	-1786 Jun 27	06:14:07	41398	-46821	8	T	-p	0.5578	1.0592	57.5N	102.8W	56	181	235	04m13s
530	027	-1786 Dec 21	01:10:23	41387	-46815	13	Am	nn	-0.0054	0.9529	23.5S	25.0W	90	1	172	05m41s
531	027	-1785 Jun 16	19:34:08	41376	-46809	18	H	nn	-0.2030	1.0124	10.4N	57.1E	78	356	44	01m23s
532	027	-1785 Dec 10	09:29:08	41365	-46803	23	H	p-	0.6857	1.0019	22.0N	146.8W	47	185	9	00m12s
533	027	-1784 Jun 05	01:57:48	41354	-46797	28	A-	t-	-1.0116	0.9493	68.6S	17.8W	0	339	-	-
534	027	-1784 Oct 30	12:52:51	41344	-46792	-5	P	-t	-1.3666	0.3184	71.0S	39.7E	0	119		
535	027	-1784 Nov 28	23:35:25	41342	-46791	33	P	t-	1.3172	0.4116	69.0N	24.2E	0	206		
536	027	-1783 Apr 25	12:31:39	41333	-46786	0	P	-t	1.0571	0.8684	71.3N	51.5E	0	66		
537	027	-1783 Oct 20	03:26:05	41322	-46780	5	H	-p	-0.7370	1.0075	49.4S	81.1W	42	28	38	00m35s
538	027	-1782 Apr 14	18:01:43	41311	-46774	10	H	nn	0.2512	1.0049	17.1N	77.8E	75	161	17	00m30s
539	027	-1782 Oct 09	12:37:27	41300	-46768	15	A	nn	-0.0506	0.9593	3.1S	160.8E	87	18	148	04m46s
540	027	-1781 Apr 04	06:37:38	41289	-46762	20	T	p-	-0.5149	1.0545	30.4S	94.7W	59	339	209	04m36s

Cat Num	Canon Plate	Calendar Date	TD of Greatest Eclipse	ΔT s	Luna Num	Saros Num	Ecl. Type	QLE	Gamma	Ecl. Mag.	Lat. °	Long. °	Sun Alt °	Sun Azm °	Path Width km	Central Line Dur.
541	028	-1781 Sep 28	14:39:03	41278	-46756	25	A	p-	0.6625	0.9211	43.4N	148.1E	48	205	395	08m19s
542	028	-1780 Feb 23	15:29:49	41268	-46751	-8	P	-t	1.3291	0.3875	69.9N	83.4E	0	141		
543	028	-1780 Mar 23	23:18:05	41266	-46750	30	P	t-	-1.2229	0.5913	71.4S	110.5E	0	254		
544	028	-1780 Sep 16	14:01:07	41255	-46744	35	P	t-	1.3406	0.3814	71.2N	101.6W	0	298		
545	028	-1779 Feb 12	05:19:31	41246	-46739	2	H	-p	0.6544	1.0038	22.0N	90.0W	49	170	17	00m24s
546	028	-1779 Aug 07	06:14:34	41235	-46733	7	T	-p	-0.7859	1.0139	30.8S	107.0W	38	7	77	01m21s
547	028	-1778 Feb 01	12:22:05	41224	-46727	12	A	nn	-0.0733	0.9516	25.4S	171.3E	86	354	178	05m53s
548	028	-1778 Jul 27	20:58:21	41213	-46721	17	T	nn	-0.0353	1.0658	20.8N	37.0E	88	2	215	06m01s
549	028	-1777 Jan 21	12:53:29	41201	-46715	22	A	p-	-0.7951	0.9205	75.8S	166.7E	37	355	501	06m26s
550	028	-1777 Jul 17	13:49:57	41190	-46709	27	T	p-	0.6876	1.0601	67.3N	139.2E	46	174	273	03m53s
551	028	-1777 Dec 11	21:26:59	41181	-46704	-6	P	-t	1.4583	0.1729	63.4N	66.9E	0	215		
552	028	-1776 Jan 10	13:21:28	41179	-46703	32	P	t-	-1.4616	0.1738	66.0S	18.2W	0	173		
553	028	-1776 Jun 06	17:41:15	41170	-46698	-1	P	-t	-1.1934	0.6385	62.8S	127.4E	0	319		
554	028	-1776 Jul 06	04:07:49	41168	-46697	37	P	t-	1.4717	0.1295	65.3N	124.2E	0	15		
555	028	-1776 Nov 30	07:41:49	41159	-46692	4	T	-p	0.7115	1.0151	23.7N	110.0W	44	201	73	01m24s
556	028	-1775 May 26	22:10:28	41148	-46686	9	A	-p	-0.4519	0.9546	7.5S	26.2E	63	339	184	05m34s
557	028	-1775 Nov 19	22:40:15	41137	-46680	14	T	nn	0.0296	1.0462	14.7S	10.0E	88	205	154	03m58s
558	028	-1774 May 15	22:47:48	41126	-46674	19	A	pn	0.3163	0.9543	31.0N	1.7W	71	152	175	04m47s
559	028	-1774 Nov 09	13:48:09	41114	-46668	24	T	p-	-0.6486	1.0186	47.5S	117.2E	49	39	83	01m22s
560	028	-1773 May 05	02:55:56	41103	-46662	29	P	t-	1.0674	0.8642	61.0N	163.7W	0	68		
561	029	-1773 Sep 30	08:36:13	41094	-46657	-4	P	-t	1.5296	0.0580	60.6N	42.1W	0	277		
562	029	-1773 Oct 30	00:28:20	41092	-46656	34	P	t-	-1.3904	0.2907	61.0S	123.4W	0	108		
563	029	-1772 Mar 25	06:31:35	41083	-46651	1	T	-p	-0.7985	1.0583	48.3S	68.3W	37	314	313	03m53s
564	029	-1772 Sep 18	09:13:28	41072	-46645	6	A	-p	0.8744	0.9165	56.3N	98.7W	29	238	641	06m57s
565	029	-1771 Mar 14	23:27:17	41061	-46639	11	T	nn	-0.0696	1.0696	12.1S	5.0E	86	332	227	05m39s
566	029	-1771 Sep 07	09:22:22	41050	-46633	16	A	nn	0.1610	0.9462	20.5N	144.5W	81	207	201	05m50s
567	029	-1770 Mar 04	14:33:02	41039	-46627	21	T	p-	0.6815	1.0220	26.4N	117.6E	47	151	100	01m57s
568	029	-1770 Aug 27	15:48:22	41028	-46621	26	A	p-	-0.5657	0.9986	16.0S	100.9E	55	24	6	00m08s
569	029	-1769 Jan 23	07:57:06	41018	-46616	-7	P	-t	-1.4143	0.2567	63.8S	23.4E	0	209		
570	029	-1769 Feb 21	23:32:24	41016	-46615	31	P	t-	1.4816	0.1318	61.8N	55.2W	0	125		
571	029	-1769 Jul 18	21:01:54	41007	-46610	-2	P	-t	1.3244	0.3955	64.7N	167.4W	0	340		
572	029	-1769 Aug 17	05:11:34	41005	-46609	36	P	t-	-1.2420	0.5539	62.4S	135.5W	0	46		
573	029	-1768 Jan 12	07:33:02	40996	-46604	3	A	-p	-0.7337	0.9212	69.8S	100.6W	43	336	441	06m38s
574	029	-1768 Jul 07	13:43:11	40985	-46598	8	T	-p	0.6244	1.0535	62.5N	150.3E	51	190	227	03m39s
575	029	-1768 Dec 31	09:17:22	40974	-46592	13	A	nn	-0.0155	0.9578	24.9S	147.1W	89	357	154	04m59s
576	029	-1767 Jun 27	02:35:28	40963	-46586	18	Hm	nn	-0.1298	1.0070	16.0N	50.3W	83	0	25	00m47s
577	029	-1767 Dec 20	18:04:43	40952	-46580	23	H	p-	0.6801	1.0065	20.0N	80.3E	47	181	31	00m41s
578	029	-1766 Jun 16	08:27:00	40941	-46574	28	A	t-	-0.9312	0.9509	46.9S	133.4W	21	355	499	05m03s
579	029	-1766 Nov 10	21:46:53	40931	-46569	-5	P	-t	-1.3664	0.3191	70.4S	109.0W	0	132		
580	029	-1766 Dec 10	08:27:16	40929	-46568	33	P	t-	1.3157	0.4144	68.0N	122.3W	0	194		
581	030	-1765 May 06	18:57:36	40920	-46563	0	P	-t	1.1387	0.7279	70.8N	60.2W	0	54		
582	030	-1765 Oct 31	12:05:53	40909	-46557	5	H	-p	-0.7400	1.0034	54.1S	146.0E	42	28	18	00m16s
583	030	-1764 Apr 25	00:58:42	40898	-46551	10	H	-n	0.3262	1.0096	25.6N	30.8W	71	161	35	00m57s
584	030	-1764 Oct 19	20:50:22	40887	-46545	15	A	nn	-0.0567	0.9548	8.0S	35.2E	87	18	165	05m20s
585	030	-1763 Apr 14	14:01:43	40876	-46539	20	T	p-	-0.4451	1.0598	21.9S	149.1E	63	341	218	05m15s
586	030	-1763 Oct 08	22:26:39	40865	-46533	25	A	p-	0.6488	0.9183	38.1N	27.5E	49	203	403	09m05s
587	030	-1762 Mar 05	23:23:27	40856	-46528	-8	P	-t	1.3777	0.2947	70.6N	49.7W	0	128		
588	030	-1762 Apr 04	06:55:36	40854	-46527	30	P	t-	-1.1623	0.7087	71.6S	19.5W	0	268		
589	030	-1762 Sep 27	21:45:47	40843	-46521	35	P	t-	1.3186	0.4185	71.6N	126.6E	0	284		
590	030	-1761 Feb 23	13:03:27	40833	-46516	2	H	-p	0.6992	1.0028	28.1N	149.0E	45	166	14	00m17s
591	030	-1761 Aug 18	14:04:01	40822	-46510	7	T	-p	-0.8246	1.0130	36.6S	129.7E	34	12	79	01m12s
592	030	-1760 Feb 12	19:48:27	40811	-46504	12	A	nn	-0.0336	0.9523	20.5S	58.2E	88	351	175	05m54s
593	030	-1760 Aug 07	04:53:14	40800	-46498	17	Tm	nn	-0.0834	1.0638	16.3N	83.4W	85	6	210	05m54s
594	030	-1759 Jan 31	20:15:58	40789	-46492	22	A	p-	-0.7606	0.9242	70.3S	63.6E	40	344	443	06m25s
595	030	-1759 Jul 27	21:35:14	40778	-46486	27	T	p-	0.6349	1.0561	62.3N	29.0E	50	184	240	03m49s
596	030	-1759 Dec 22	05:44:51	40769	-46481	-6	P	-t	1.4605	0.1673	64.3N	68.9W	0	205		
597	030	-1758 Jan 20	21:06:35	40767	-46480	32	P	t-	-1.4347	0.2181	67.1S	146.9W	0	183		
598	030	-1758 Jun 18	00:31:08	40758	-46475	-1	P	-t	-1.2740	0.4936	63.6S	13.8E	0	329		
599	030	-1758 Jul 17	11:26:42	40756	-46474	37	P	t-	1.4162	0.2332	66.2N	2.5E	0	5		
600	030	-1758 Dec 11	16:25:49	40747	-46469	4	T	-p	0.7109	1.0187	22.2N	114.8E	45	196	90	01m47s

Cat Num	Canon Plate	Calendar Date	TD of Greatest Eclipse	ΔT s	Luna Num	Saros Num	Ecl. Type	QLE	Gamma	Ecl. Mag.	Lat. °	Long. °	Sun Alt °	Sun Azm °	Path Width km	Central Line Dur.
601	031	-1757 Jun 07	04:31:26	40736	-46463	9	A	-p	-0.5404	0.9515	11.0S	70.7W	57	342	208	06m08s
602	031	-1757 Dec 01	07:34:28	40725	-46457	14	T	nn	0.0292	1.0468	17.8S	124.9W	88	201	156	04m05s
603	031	-1756 May 26	05:07:02	40714	-46451	19	A	nn	0.2263	0.9560	29.7N	95.3W	77	156	164	04m46s
604	031	-1756 Nov 19	22:36:12	40703	-46445	24	T	p-	-0.6516	1.0162	52.1S	16.4W	49	38	73	01m10s
605	031	-1755 May 15	09:40:36	40692	-46439	29	A	t-	0.9819	0.9858	65.4N	105.4E	10	78	284	00m51s
606	031	-1755 Oct 10	16:37:28	40682	-46434	-4	P	-t	1.5447	0.0345	60.5N	172.8W	0	268		
607	031	-1755 Nov 09	08:54:37	40680	-46433	34	P	t-	-1.3940	0.2852	61.4S	99.5E	0	117		
608	031	-1754 Apr 05	14:04:42	40671	-46428	1	T	-p	-0.8613	1.0607	49.1S	179.8W	30	312	385	03m59s
609	031	-1754 Sep 29	16:52:56	40660	-46422	6	A	-p	0.8956	0.9133	54.1N	144.1E	26	236	725	07m23s
610	031	-1753 Mar 26	07:09:38	40649	-46416	11	T	-n	-0.1289	1.0709	10.9S	111.5W	83	331	232	05m43s
611	031	-1753 Sep 18	17:05:43	40638	-46410	16	A	nn	0.1868	0.9456	17.6N	98.4E	79	209	204	05m50s
612	031	-1752 Mar 14	22:07:20	40627	-46404	21	T	p-	0.6290	1.0224	26.0N	2.0E	51	149	96	01m57s
613	031	-1752 Sep 06	23:48:10	40616	-46398	26	A	p-	-0.5338	0.9991	17.0S	21.6W	58	27	4	00m05s
614	031	-1751 Feb 02	15:26:22	40607	-46393	-7	P	-t	-1.4497	0.1966	62.9S	100.1W	0	219		
615	031	-1751 Mar 04	06:47:43	40605	-46392	31	P	t-	1.4347	0.2136	61.3N	174.7W	0	116		
616	031	-1751 Jul 29	04:51:39	40596	-46387	-2	P	-t	1.3749	0.2987	63.8N	63.6E	0	330		
617	031	-1751 Aug 27	13:20:51	40594	-46386	36	P	t-	-1.2053	0.6232	61.8S	91.3E	0	55		
618	031	-1750 Jan 22	15:05:02	40585	-46381	3	A	-p	-0.7609	0.9243	69.1S	156.6E	40	323	440	06m12s
619	031	-1750 Jul 18	21:18:48	40574	-46375	8	T	-p	0.6850	1.0472	65.9N	44.9E	46	202	216	03m07s
620	031	-1749 Jan 11	17:17:46	40563	-46369	13	A	nn	-0.0323	0.9631	25.7S	92.6E	88	352	133	04m13s
621	032	-1749 Jul 08	09:40:54	40552	-46363	18	H	nn	-0.0609	1.0013	20.5N	158.0W	87	5	5	00m09s
622	032	-1748 Jan 01	02:35:12	40541	-46357	23	H	p-	0.6695	1.0117	18.2N	51.2W	48	176	54	01m12s
623	032	-1748 Jun 26	14:59:14	40530	-46351	28	A	p-	-0.8533	0.9496	35.5S	121.5E	31	0	356	05m50s
624	032	-1748 Nov 21	06:40:48	40520	-46346	-5	P	-t	-1.3662	0.3196	69.6S	102.9E	0	146		
625	032	-1748 Dec 20	17:14:20	40519	-46345	33	P	t-	1.3104	0.4242	66.9N	93.0E	0	182		
626	032	-1747 May 17	01:22:07	40509	-46340	0	P	-t	1.2218	0.5842	70.1N	170.9W	0	41		
627	032	-1747 Nov 10	20:47:50	40498	-46334	5	H	-p	-0.7406	1.0001	58.7S	13.8E	42	28	1	00m00s
628	032	-1746 May 06	07:54:16	40487	-46328	10	H	-p	0.4033	1.0136	34.2N	138.8W	66	162	51	01m16s
629	032	-1746 Oct 31	05:06:38	40476	-46322	15	A	nn	-0.0594	0.9510	12.5S	91.0W	87	16	180	05m50s
630	032	-1745 Apr 25	21:23:14	40465	-46316	20	T	p-	-0.3724	1.0644	13.6S	33.8E	68	343	226	05m50s
631	032	-1745 Oct 20	06:21:00	40454	-46310	25	A	p-	0.6405	0.9160	33.2N	94.8W	50	201	411	09m50s
632	032	-1744 Mar 16	07:08:25	40445	-46305	-8	P	-t	1.4332	0.1887	71.2N	178.9E	0	115		
633	032	-1744 Apr 14	14:26:42	40443	-46304	30	P	t-	-1.0965	0.8368	71.5S	147.9W	0	281		
634	032	-1744 Oct 08	05:39:41	40432	-46298	35	P	t-	1.3033	0.4443	71.7N	7.8W	0	270		
635	032	-1743 Mar 05	20:37:21	40423	-46293	2	H	-p	0.7516	1.0016	35.4N	29.4E	41	162	8	00m09s
636	032	-1743 Aug 28	22:03:23	40412	-46287	7	T	-p	-0.8559	1.0120	42.3S	3.2E	31	17	80	01m02s
637	032	-1742 Feb 23	03:04:13	40401	-46281	12	A	nn	0.0142	0.9531	14.6S	53.0W	89	166	171	05m53s
638	032	-1742 Aug 18	12:56:59	40390	-46275	17	T	-n	-0.1242	1.0614	11.4N	153.6E	83	10	203	05m43s
639	032	-1741 Feb 12	03:30:55	40379	-46269	22	A	p-	-0.7199	0.9282	63.6S	43.9W	44	339	390	06m24s
640	032	-1741 Aug 08	05:26:17	40368	-46263	27	T	p-	0.5875	1.0516	56.7N	86.0W	54	191	212	03m41s
641	033	-1740 Jan 02	13:59:35	40359	-46258	-6	P	t	1.4660	0.1558	65.3N	155.7E	0	194		
642	033	-1740 Feb 01	04:46:24	40357	-46257	32	P	t-	-1.4026	0.2717	68.1S	85.2E	0	195		
643	033	-1740 Jun 28	07:23:09	40348	-46252	-1	P	-t	-1.3513	0.3560	64.5S	100.6W	0	338		
644	033	-1740 Jul 27	18:50:24	40346	-46251	37	P	t-	1.3653	0.3273	67.2N	120.9W	0	354		
645	033	-1740 Dec 22	01:06:10	40337	-46246	4	T	-p	0.7143	1.0228	21.7N	19.3W	44	191	110	02m11s
646	033	-1739 Jun 17	10:54:45	40326	-46240	9	A	-p	-0.6264	0.9480	15.5S	168.7W	51	346	243	06m43s
647	033	-1739 Dec 11	16:25:31	40315	-46234	14	T	nn	0.0317	1.0479	20.1S	101.4E	88	197	159	04m15s
648	033	-1738 Jun 06	11:30:50	40304	-46228	19	A	nn	0.1392	0.9572	27.9N	169.6E	82	160	157	04m50s
649	033	-1738 Dec 01	07:21:07	40293	-46222	24	T	p-	-0.6526	1.0142	56.4S	147.5W	49	35	64	01m01s
650	033	-1737 May 26	16:29:39	40282	-46216	29	A	t-	0.8985	0.9933	68.0N	33.4E	26	107	54	00m26s
651	033	-1737 Oct 22	00:43:06	40273	-46211	-4	P	-t	1.5554	0.0182	60.7N	55.5E	0	259		
652	033	-1737 Nov 20	17:20:19	40271	-46210	34	P	t-	-1.3979	0.2791	61.9S	37.6W	0	127		
653	033	-1736 Apr 15	21:32:48	40262	-46205	1	T	-t	-0.9288	1.0617	52.0S	72.2E	21	308	538	03m57s
654	033	-1736 Oct 10	00:40:09	40251	-46199	6	A	-p	0.9103	0.9108	52.1N	23.5E	24	233	802	07m50s
655	033	-1735 Apr 05	14:43:41	40240	-46193	11	T	-n	-0.1950	1.0715	9.9S	134.0E	79	330	236	05m46s
656	033	-1735 Sep 29	00:59:34	40229	-46187	16	A	nn	0.2048	0.9454	14.1N	21.8W	78	210	205	05m50s
657	033	-1734 Mar 26	05:31:07	40218	-46181	21	T	p-	0.5685	1.0223	25.9N	110.3W	55	147	91	01m56s
658	033	-1734 Sep 18	07:59:14	40207	-46175	26	A	p-	-0.5105	0.9995	19.1S	146.9W	59	30	2	00m03s
659	033	-1733 Feb 13	22:44:34	40198	-46170	-7	P	-t	-1.4933	0.1221	62.2S	139.4E	0	228		
660	033	-1733 Mar 15	13:52:14	40196	-46169	31	P	t-	1.3796	0.3097	60.9N	68.6E	0	107		

Cat Num	Canon Plate	Calendar Date	TD of Greatest Eclipse	ΔT s	Luna Num	Saros Num	Ecl. Type	QLE	Gamma	Ecl. Mag.	Lat. °	Long. °	Sun Alt °	Sun Azm °	Path Width km	Central Line Dur.
661	034	-1733 Aug 09	12:48:19	40187	-46164	-2	P	-t	1.4200	0.2127	63.0N	66.8W	0	321		
662	034	-1733 Sep 07	21:38:26	40185	-46163	36	P	t-	-1.1752	0.6800	61.2S	43.7W	0	64		
663	034	-1732 Feb 02	22:29:32	40176	-46158	3	A	-p	-0.7944	0.9278	67.6S	54.0E	37	312	446	05m46s
664	034	-1732 Jul 29	04:59:19	40165	-46152	8	T	-p	0.7411	1.0404	67.3N	59.6W	42	215	202	02m36s
665	034	-1731 Jan 22	01:11:38	40154	-46146	13	A	nn	-0.0552	0.9690	25.9S	26.0W	87	347	112	03m25s
666	034	-1731 Jul 18	16:50:06	40143	-46140	18	A	nn	0.0039	0.9951	23.9N	94.2E	90	186	17	00m32s
667	034	-1730 Jan 11	11:00:39	40132	-46134	23	T	p-	0.6541	1.0173	16.8N	178.8E	49	172	78	01m46s
668	034	-1730 Jul 07	21:35:47	40121	-46128	28	A	p-	-0.7792	0.9472	27.3S	17.4E	39	5	310	06m33s
669	034	-1730 Dec 02	15:33:38	40112	-46123	-5	P	-t	-1.3670	0.3181	68.7S	44.3W	0	158		
670	034	-1729 Jan 01	01:57:02	40110	-46122	33	P	t-	1.3017	0.4405	65.9N	50.1W	0	171		
671	034	-1729 May 28	07:49:57	40101	-46117	0	P	-t	1.3024	0.4438	69.3N	78.1E	0	29		
672	034	-1729 Jun 26	22:02:21	40099	-46116	38	Pb	t-	-1.5245	0.0649	66.7S	13.9E	0	0		
673	034	-1729 Nov 22	05:30:00	40090	-46111	5	A	-p	-0.7405	0.9973	63.0S	116.5W	42	25	14	00m12s
674	034	-1728 May 16	14:50:12	40079	-46105	10	T	-p	0.4811	1.0170	42.8N	113.6E	61	162	66	01m30s
675	034	-1728 Nov 10	13:24:24	40068	-46099	15	A	nn	-0.0600	0.9478	16.6S	142.8E	86	14	192	06m17s
676	034	-1727 May 06	04:43:17	40057	-46093	20	T	n-	-0.2975	1.0681	5.4S	80.7W	73	345	232	06m18s
677	034	-1727 Oct 30	14:18:33	40046	-46087	25	A	p-	0.6350	0.9144	28.9N	142.0E	50	199	418	10m32s
678	034	-1726 Mar 27	14:45:53	40037	-46082	-8	Pe	-t	1.4941	0.0722	71.5N	49.0E	0	102		
679	034	-1726 Apr 25	21:54:40	40035	-46081	30	P	t-	-1.0281	0.9698	71.2S	84.6E	0	294		
680	034	-1726 Oct 19	13:40:25	40024	-46075	35	P	t-	1.2925	0.4624	71.6N	144.0W	0	255		
681	035	-1725 Mar 17	04:02:01	40015	-46070	2	A	-p	0.8105	0.9997	43.9N	89.3W	36	156	2	00m01s
682	035	-1725 Sep 09	06:11:45	40004	-46064	7	T	-p	-0.8805	1.0111	47.8S	126.2W	28	23	81	00m54s
683	035	-1724 Mar 05	10:09:29	39993	-46058	12	A	nn	0.0698	0.9538	7.9S	162.3W	86	165	169	05m50s
684	035	-1724 Aug 28	21:09:28	39982	-46052	17	T	-n	-0.1577	1.0589	6.3N	28.0E	81	13	196	05m28s
685	035	-1723 Feb 22	10:33:54	39971	-46046	22	A	p-	-0.6693	0.9324	56.1S	151.7W	48	338	340	06m23s
686	035	-1723 Aug 18	13:25:56	39960	-46040	27	T	p-	0.5480	1.0467	51.0N	154.6E	57	195	186	03m30s
687	035	-1722 Jan 12	22:07:13	39951	-46035	-6	P	-t	1.4777	0.1330	66.3N	21.7E	0	184		
688	035	-1722 Feb 11	12:16:43	39949	-46034	32	P	t-	-1.3621	0.3405	69.1S	40.9W	0	206		
689	035	-1722 Jul 09	14:19:32	39940	-46029	-1	P	-t	-1.4238	0.2284	65.5S	143.5E	0	348		
690	035	-1722 Aug 08	02:21:22	39938	-46028	37	P	t-	1.3210	0.4084	68.2N	113.3E	0	343		
691	035	-1721 Jan 02	09:41:25	39929	-46023	4	T	-p	0.7224	1.0271	22.3N	152.2W	44	187	133	02m37s
692	035	-1721 Jun 28	17:22:09	39918	-46017	9	A	-p	-0.7083	0.9440	21.1S	91.6E	45	350	291	07m15s
693	035	-1721 Dec 23	01:13:39	39907	-46011	14	T	nn	0.0364	1.0494	21.3S	31.3W	88	192	164	04m27s
694	035	-1720 Jun 16	17:56:39	39896	-46005	19	A	nn	0.0531	0.9579	25.3N	73.7E	87	166	153	04m58s
695	035	-1720 Dec 11	16:04:48	39885	-45999	24	H3	p-	-0.6533	1.0129	60.1S	83.5E	49	30	58	00m55s
696	035	-1719 Jun 05	23:20:36	39874	-45993	29	A	t-	0.8156	0.9985	67.9N	50.6W	35	127	9	00m06s
697	035	-1719 Nov 01	08:52:28	39865	-45988	-4	Pe	-t	1.5624	0.0080	61.0N	77.3W	0	250		
698	035	-1719 Dec 01	01:44:30	39863	-45987	34	P	t-	-1.4014	0.2734	62.6S	174.4W	0	136		
699	035	-1718 Apr 27	04:58:09	39854	-45982	1	T-	-t	-0.9988	1.0262	60.9S	13.9W	0	287	-	-
700	035	-1718 May 26	11:58:02	39852	-45981	39	Pb	t-	1.5157	0.0319	62.2N	37.2E	0	50		
701	036	-1718 Oct 21	08:33:49	39843	-45976	6	A	-p	0.9195	0.9091	50.2N	100.1W	23	229	861	08m18s
702	036	-1717 Apr 16	22:13:43	39832	-45970	11	T	-n	-0.2642	1.0714	9.3S	20.5E	75	331	240	05m49s
703	036	-1717 Oct 10	09:00:26	39821	-45964	16	A	nn	0.2176	0.9456	10.2N	144.0W	77	210	205	05m51s
704	036	-1716 Apr 05	12:48:28	39810	-45958	21	T	p-	0.5035	1.0218	26.2N	139.4E	60	147	85	01m52s
705	036	-1716 Sep 28	16:18:00	39799	-45952	26	H	p-	-0.4927	1.0002	22.0S	85.8E	60	32	1	00m01s
706	036	-1715 Feb 24	05:51:54	39790	-45947	-7	Pe	-t	-1.5449	0.0340	61.6S	21.9E	0	237		
707	036	-1715 Mar 25	20:46:57	39789	-45946	31	P	t-	1.3171	0.4188	60.7N	45.6W	0	99		
708	036	-1715 Aug 19	20:53:33	39779	-45941	-2	P	-t	1.4584	0.1402	62.3N	160.9E	0	311		
709	036	-1715 Sep 18	06:04:44	39778	-45940	36	P	t-	-1.1521	0.7235	60.8S	179.2E	0	74		
710	036	-1714 Feb 13	05:44:32	39768	-45935	3	A	-p	-0.8361	0.9311	65.8S	47.3W	33	302	468	05m21s
711	036	-1714 Aug 09	12:47:18	39758	-45929	8	T	-p	0.7902	1.0335	66.8N	166.8W	37	228	184	02m07s
712	036	-1713 Feb 02	08:57:33	39747	-45923	13	A	nn	-0.0858	0.9750	25.6S	142.5W	85	342	90	02m39s
713	036	-1713 Jul 30	00:06:08	39736	-45917	18	A	nn	0.0620	0.9888	26.0N	14.9W	86	194	40	01m11s
714	036	-1712 Jan 22	19:18:39	39725	-45911	23	T	p-	0.6319	1.0233	15.7N	50.9E	51	167	102	02m19s
715	036	-1712 Jul 18	04:17:46	39714	-45905	28	A	p-	-0.7099	0.9441	21.2S	87.0W	45	10	291	07m11s
716	036	-1712 Dec 13	00:23:31	39705	-45900	-5	P	-t	-1.3704	0.3119	67.6S	169.8E	0	170		
717	036	-1711 Jan 11	10:32:45	39703	-45899	33	P	t-	1.2875	0.4674	64.8N	169.1E	0	161		
718	036	-1711 Jun 07	14:20:18	39694	-45894	0	P	-t	1.3815	0.3056	68.4N	33.0W	0	18		
719	036	-1711 Jul 07	04:35:37	39692	-45893	38	P	t-	-1.4482	0.1959	65.7S	96.7W	0	11		
720	036	-1711 Dec 02	14:09:37	39683	-45888	5	A	-p	-0.7421	0.9951	67.0S	116.2E	42	19	26	00m21s

Cat Num	Canon Plate	Calendar Date	TD of Greatest Eclipse	ΔT s	Luna Num	Saros Num	Ecl. Type	QLE	Gamma	Ecl. Mag.	Lat. °	Long. °	Sun Alt °	Sun Azm °	Path Width km	Central Line Dur.
721	037	-1710 May 27	21:48:23	39672	-45882	10	T	-p	0.5580	1.0196	51.3N	6.4E	56	164	81	01m36s
722	037	-1710 Nov 21	21:42:44	39661	-45876	15	A	nn	-0.0598	0.9453	20.3S	16.8E	86	11	202	06m38s
723	037	-1709 May 17	12:02:25	39650	-45870	20	T	n-	-0.2214	1.0711	2.5N	165.4E	77	347	237	06m38s
724	037	-1709 Nov 10	22:19:39	39639	-45864	25	A	p-	0.6325	0.9133	25.1N	18.0E	51	195	423	11m10s
725	037	-1708 May 06	05:18:38	39628	-45858	30	T	t-	-0.9564	1.0657	58.2S	66.9W	16	330	750	04m22s
726	037	-1708 Oct 29	21:47:25	39618	-45852	35	P	t-	1.2865	0.4725	71.2N	78.5E	0	241		
727	037	-1707 Mar 27	11:16:25	39608	-45847	2	A	-t	0.8766	0.9970	53.7N	151.4E	28	149	22	00m14s
728	037	-1707 Sep 19	14:30:13	39598	-45841	7	T	-p	-0.8974	1.0105	52.9S	101.4E	26	29	82	00m48s
729	037	-1706 Mar 16	17:05:16	39587	-45835	12	A	nn	0.1323	0.9544	0.5S	90.1E	82	163	168	05m44s
730	037	-1706 Sep 09	05:29:48	39576	-45829	17	T	-n	-0.1846	1.0563	1.1N	99.8W	79	15	189	05m11s
731	037	-1705 Mar 05	17:29:56	39565	-45823	22	A	p-	-0.6127	0.9367	48.0S	100.7E	52	338	297	06m22s
732	037	-1705 Aug 29	21:32:37	39554	-45817	27	T	p-	0.5150	1.0416	45.2N	32.0E	59	198	162	03m16s
733	037	-1704 Jan 24	06:09:21	39545	-45812	-6	P	-t	1.4945	0.1007	67.4N	111.5W	0	173		
734	037	-1704 Feb 22	19:40:22	39543	-45811	32	P	t-	-1.3152	0.4213	70.0S	165.9W	0	219		
735	037	-1704 Jul 19	21:19:49	39534	-45806	-1	P	-t	-1.4918	0.1103	66.5S	26.2E	0	358		
736	037	-1704 Aug 18	09:58:25	39532	-45805	37	P	t-	1.2823	0.4782	69.2N	14.5W	0	331		
737	037	-1703 Jan 12	18:10:50	39523	-45800	4	T	-p	0.7358	1.0318	23.9N	76.3E	42	182	158	03m01s
738	037	-1703 Jul 08	23:54:56	39512	-45794	9	A	-p	-0.7850	0.9397	27.8S	10.3W	38	354	361	07m38s
739	037	-1702 Jan 02	09:54:44	39501	-45788	14	T	nn	0.0469	1.0512	21.3S	162.1W	87	187	170	04m40s
740	037	-1702 Jun 28	00:30:18	39490	-45782	19	A	nn	-0.0270	0.9582	22.1N	24.7W	89	349	152	05m09s
741	038	-1702 Dec 23	00:42:38	39479	-45776	24	H	p-	-0.6502	1.0120	62.7S	41.6W	49	23	54	00m51s
742	038	-1701 Jun 17	06:16:54	39469	-45770	29	H	p-	0.7360	1.0025	66.4N	140.2W	42	144	13	00m11s
743	038	-1701 Dec 12	10:04:20	39458	-45764	34	P	t-	-1.4024	0.2720	63.5S	49.6E	0	146		
744	038	-1700 May 07	12:21:06	39449	-45759	1	P	-t	-1.0709	0.8874	61.3S	135.2W	0	295		
745	038	-1700 Jun 05	19:15:14	39447	-45758	39	P	t-	1.4394	0.1769	62.9N	83.1W	0	41		
746	038	-1700 Oct 31	16:31:53	39438	-45753	6	A	-p	0.9254	0.9081	48.5N	134.6E	22	223	905	08m44s
747	038	-1699 Apr 27	05:37:08	39427	-45747	11	T	-n	-0.3386	1.0703	9.3S	91.3W	70	332	242	05m50s
748	038	-1699 Oct 20	17:09:38	39416	-45741	16	Am	nn	0.2245	0.9464	6.1N	91.4E	77	209	202	05m50s
749	038	-1698 Apr 16	19:57:41	39405	-45735	21	T	p-	0.4325	1.0206	26.5N	31.6E	64	147	77	01m47s
750	038	-1698 Oct 10	00:44:55	39394	-45729	26	H	p-	-0.4809	1.0011	25.6S	43.4W	61	33	4	00m06s
751	038	-1697 Apr 06	03:33:07	39383	-45723	31	P	t-	1.2482	0.5389	60.6N	157.5W	0	90		
752	038	-1697 Aug 31	05:06:54	39374	-45718	-2	P	-t	1.4904	0.0804	61.7N	26.8E	0	302		
753	038	-1697 Sep 29	14:39:00	39373	-45717	36	P	t-	-1.1353	0.7547	60.6S	40.3E	0	83		
754	038	-1696 Feb 24	12:50:43	39363	-45712	3	A	-p	-0.8850	0.9344	64.1S	146.6W	27	294	522	04m57s
755	038	-1696 Aug 19	20:41:34	39353	-45706	8	T	-p	0.8335	1.0263	65.1N	81.8E	33	237	161	01m39s
756	038	-1695 Feb 12	16:36:30	39342	-45700	13	A	nn	-0.1229	0.9814	24.9S	102.6E	83	338	67	01m54s
757	038	-1695 Aug 09	07:28:10	39331	-45694	18	A	nn	0.1143	0.9822	26.8N	125.3W	83	199	63	01m50s
758	038	-1694 Feb 02	03:29:44	39320	-45688	23	T	p-	0.6032	1.0296	14.8N	75.0W	53	163	125	02m51s
759	038	-1694 Jul 29	11:07:04	39309	-45682	28	A	p-	-0.6466	0.9407	16.9S	167.5E	50	14	285	07m41s
760	038	-1694 Dec 24	09:09:54	39300	-45677	-5	P	-t	-1.3768	0.2997	66.6S	25.5E	0	181		
761	039	-1693 Jan 22	19:02:44	39298	-45676	33	P	t	1.2606	0.5034	63.9N	30.1E	0	151		
762	039	-1693 Jun 18	20:55:31	39289	-45671	0	P	-t	1.4570	0.1734	67.4N	144.8W	0	7		
763	039	-1693 Jul 18	11:15:43	39287	-45670	38	P	t-	-1.3762	0.3192	64.7S	151.5E	0	21		
764	039	-1693 Dec 13	22:46:18	39278	-45665	5	A	-p	-0.7455	0.9933	70.3S	6.8W	41	10	36	00m28s
765	039	-1692 Jun 07	04:50:08	39268	-45659	10	T	-p	0.6331	1.0216	59.6N	100.2W	50	167	95	01m38s
766	039	-1692 Dec 02	05:58:03	39257	-45653	15	A	nn	-0.0611	0.9435	23.4S	107.9W	86	8	209	06m54s
767	039	-1691 May 27	19:22:45	39246	-45647	20	T	n-	-0.1459	1.0731	9.8N	51.8E	82	350	240	06m48s
768	039	-1691 Nov 21	06:20:30	39235	-45641	25	A	p-	0.6297	0.9131	21.7N	105.9W	51	192	424	11m40s
769	039	-1690 May 17	12:42:18	39224	-45635	30	T	p-	-0.8849	1.0681	45.7S	167.5E	27	342	477	05m07s
770	039	-1690 Nov 10	05:56:47	39213	-45629	35	P	t-	1.2815	0.4809	70.6N	59.0W	0	227		
771	039	-1689 Apr 07	18:23:32	39204	-45624	2	A	-t	0.9476	0.9929	65.1N	24.3E	18	132	80	00m30s
772	039	-1689 Sep 30	22:56:52	39193	-45618	7	T	-p	-0.9084	1.0102	57.6S	33.2W	24	35	84	00m44s
773	039	-1688 Mar 26	23:50:07	39183	-45612	12	A	nn	0.2032	0.9547	7.6N	15.2W	78	162	169	05m37s
774	039	-1688 Sep 19	13:58:44	39172	-45606	17	T	-n	-0.2045	1.0537	4.1S	130.1E	78	17	181	04m53s
775	039	-1687 Mar 16	00:15:31	39161	45600	22	A	p-	-0.5470	0.9410	39.5S	5.1W	57	338	260	06m20s
776	039	-1687 Sep 09	05:47:51	39150	-45594	27	T	p-	0.4896	1.0364	39.5N	93.5W	60	200	140	02m59s
777	039	-1686 Feb 03	14:02:23	39141	-45589	-6	Pe	-t	1.5192	0.0538	68.4N	117.1E	0	161		
778	039	-1686 Mar 05	02:54:52	39139	-45588	32	P	t-	-1.2597	0.5183	70.7S	70.8E	0	231		
779	039	-1686 Jul 31	04:27:07	39130	-45583	-1	Pe	-t	-1.5528	0.0060	67.6S	93.2W	0	9		
780	039	-1686 Aug 29	17:43:54	39128	-45582	37	P	t-	1.2512	0.5337	70.1N	145.0W	0	319		

Cat Num	Canon Plate	Calendar Date	TD of Greatest Eclipse	ΔT s	Luna Num	Saros Num	Ecl. Type	QLE	Gamma	Ecl. Mag.	Lat. °	Long. °	Sun Alt °	Sun Azm °	Path Width km	Central Line Dur.
781	040	-1685 Jan 24	02:33:47	39119	-45577	4	T	-p	0.7549	1.0365	26.8N	54.0W	41	177	187	03m23s
782	040	-1685 Jul 20	06:33:02	39108	-45571	9	A	-p	-0.8568	0.9349	35.8S	114.7W	31	359	473	07m48s
783	040	-1684 Jan 13	18:30:49	39098	-45565	14	T	-n	0.0613	1.0533	20.2S	68.2E	87	182	177	04m55s
784	040	-1684 Jul 08	07:09:27	39087	-45559	19	Am	nn	-0.1029	0.9580	18.1N	125.2W	84	354	153	05m22s
785	040	-1683 Jan 02	09:14:40	39076	-45553	24	H	p-	-0.6429	1.0116	63.7S	163.3W	50	13	52	00m50s
786	040	-1683 Jun 27	13:18:34	39065	-45547	29	H	p-	0.6596	1.0056	63.7N	125.6E	48	158	26	00m25s
787	040	-1683 Dec 22	18:19:28	39054	-45541	34	P	t-	-1.4005	0.2754	64.4S	85.5W	0	156		
788	040	-1682 May 18	19:44:10	39045	-45536	1	P	-t	-1.1432	0.7472	61.8S	103.4E	0	304		
789	040	-1682 Jun 17	02:36:40	39043	-45535	39	P	t-	1.3656	0.3177	63.7N	155.4E	0	31		
790	040	-1682 Nov 12	00:32:11	39034	-45530	6	A	-p	0.9295	0.9078	47.1N	8.3E	21	218	936	09m08s
791	040	-1681 May 08	12:59:02	39024	-45524	11	T	-p	-0.4138	1.0685	10.0S	157.0E	65	334	243	05m49s
792	040	-1681 Nov 01	01:23:34	39013	-45518	16	A	nn	0.2281	0.9477	2.2N	34.3W	77	207	197	05m47s
793	040	-1680 Apr 27	03:00:58	39002	-45512	21	T	n-	0.3575	1.0188	26.7N	74.2W	69	149	68	01m39s
794	040	-1680 Oct 20	09:17:56	38991	-45506	26	H	p-	-0.4736	1.0023	29.7S	174.1W	62	34	9	00m12s
795	040	-1679 Apr 16	10:11:56	38980	-45500	31	P	t-	1.1740	0.6681	60.6N	92.4E	0	81		
796	040	-1679 Sep 10	13:28:28	38971	-45495	-2	Pe	-t	1.5160	0.0331	61.2N	109.2W	0	293		
797	040	-1679 Oct 09	23:19:31	38970	-45494	36	P	t-	-1.1237	0.7763	60.6S	100.2W	0	92		
798	040	-1678 Mar 06	19:47:59	38961	-45489	3	A	-t	-0.9415	0.9369	63.1S	119.2E	19	283	696	04m34s
799	040	-1678 Aug 31	04:44:18	38950	-45483	8	T	-p	0.8690	1.0192	62.7N	35.2W	29	242	132	01m13s
800	040	-1677 Feb 24	00:07:48	38939	-45477	13	A	nn	-0.1674	0.9877	23.8S	10.5W	80	334	44	01m12s
801	041	-1677 Aug 20	14:57:08	38928	-45471	18	A	nn	0.1598	0.9757	26.3N	122.6E	81	203	88	02m28s
802	041	-1676 Feb 13	11:32:52	38917	-45465	23	T	p-	0.5673	1.0361	14.3N	161.5E	55	159	146	03m21s
803	041	-1676 Aug 08	18:04:23	38906	-45459	28	A	p-	-0.5900	0.9371	14.1S	60.5E	54	18	286	08m04s
804	041	-1675 Jan 03	17:48:56	38897	-45454	-5	P	-t	-1.3891	0.2764	65.5S	116.5W	0	192		
805	041	-1675 Feb 02	03:22:41	38896	-45453	33	P	t-	1.2417	0.5548	63.0N	106.1W	0	141		
806	041	-1675 Jun 29	03:37:03	38887	-45448	0	Pe	-t	1.5277	0.0493	66.4N	102.4E	0	357		
807	041	-1675 Jul 28	18:05:50	38885	-45447	38	P	t-	-1.3114	0.4301	63.7S	37.5E	0	30		
808	041	-1675 Dec 24	07:17:15	38876	-45442	5	A	-p	-0.7531	0.9920	72.6S	124.5W	41	357	43	00m34s
809	041	-1674 Jun 18	11:56:44	38865	-45436	10	T	-p	0.7047	1.0227	67.4N	155.1E	45	173	110	01m36s
810	041	-1674 Dec 13	14:09:35	38854	-45430	15	A	nn	-0.0650	0.9423	25.9S	128.7E	86	4	214	07m04s
811	041	-1673 Jun 08	02:45:01	38843	-45424	20	T	nn	-0.0714	1.0742	16.6N	61.6W	86	353	241	06m49s
812	041	-1673 Dec 02	14:21:47	38833	-45418	25	A	p-	0.6274	0.9135	18.9N	130.1E	51	188	423	12m00s
813	041	-1672 May 27	20:03:02	38822	-45412	30	T	p-	-0.8111	1.0687	35.5S	48.6E	36	348	382	05m39s
814	041	-1672 Nov 20	14:09:47	38811	-45406	35	P	t-	1.2789	0.4852	69.7N	163.1E	0	214		
815	041	-1671 Apr 18	01:22:18	38802	-45401	2	P	-t	1.0243	0.9436	71.6N	143.4W	0	74		
816	041	-1671 Oct 11	07:30:53	38791	-45395	7	T	-p	-0.9140	1.0104	61.8S	169.6W	23	41	89	00m43s
817	041	-1670 Apr 07	06:27:44	38780	-45389	12	A	np	0.2792	0.9546	16.1N	119.0W	74	161	172	05m29s
818	041	-1670 Sep 30	22:34:46	38770	-45383	17	T	-n	-0.2186	1.0513	9.2S	1.7W	77	18	174	04m37s
819	041	-1669 Mar 27	06:55:32	38759	-45377	22	A	p-	-0.4766	0.9451	30.8S	109.9W	61	339	229	06m14s
820	041	-1669 Sep 20	14:08:41	38748	-45371	27	T	n-	0.4697	1.0313	33.9N	138.9E	62	201	119	02m40s
821	042	-1668 Mar 15	10:04:15	38737	-45365	32	P	t-	-1.1992	0.6254	71.3S	51.7W	0	245		
822	042	-1668 Sep 09	01:35:56	38727	-45359	37	P	t-	1.2266	0.5771	70.8N	82.3E	0	306		
823	042	-1667 Feb 03	10:48:21	38718	-45354	4	T	-p	0.7813	1.0412	30.8N	177.4E	38	172	221	03m40s
824	042	-1667 Jul 30	13:19:42	38707	-45348	9	A	-t	-0.9208	0.9297	45.2S	137.0E	23	4	682	07m44s
825	042	-1666 Jan 24	02:58:13	38696	-45342	14	T	-n	0.0825	1.0554	17.7S	59.6W	85	177	184	05m10s
826	042	-1666 Jul 19	13:58:11	38685	-45336	19	A	nn	-0.1715	0.9577	13.7N	131.3E	80	359	156	05m33s
827	042	-1665 Jan 13	17:38:08	38674	-45330	24	H	p-	-0.6296	1.0115	62.8S	77.6E	51	3	51	00m51s
828	042	-1665 Jul 08	20:28:30	38664	-45324	29	H	p-	0.5893	1.0079	60.0N	25.9E	54	170	34	00m37s
829	042	-1664 Jan 03	02:26:41	38653	-45318	34	P	t-	-1.3932	0.2881	65.4S	141.0E	0	166		
830	042	-1664 May 29	03:06:18	38644	-45313	1	P	-t	-1.2163	0.6052	62.4S	18.0W	0	313		
831	042	-1664 Jun 27	10:01:58	38642	-45312	39	P	t-	1.2944	0.4538	64.6N	32.7E	0	22		
832	042	-1664 Nov 22	08:33:54	38633	-45307	6	A	-p	0.9323	0.9083	46.1N	118.5W	21	212	955	09m26s
833	042	-1663 May 18	20:17:26	38622	-45301	11	T	-p	-0.4912	1.0656	11.5S	46.1E	61	337	244	05m44s
834	042	-1663 Nov 11	09:41:25	38612	-45295	16	A	nn	0.2290	0.9497	1.6S	161.0W	77	205	189	05m40s
835	042	-1662 May 08	09:59:19	38601	-45289	21	H3	nn	0.2795	1.0162	26.7N	178.6W	74	151	58	01m28s
836	042	-1662 Oct 31	17:56:59	38590	-45283	26	H	p-	-0.4710	1.0040	34.1S	54.1E	62	33	15	00m20s
837	042	-1661 Apr 27	16:44:02	38579	-45277	31	P	t-	1.0949	0.8056	60.8N	16.0W	0	73		
838	042	-1661 Oct 21	08:05:55	38569	-45271	36	P	t-	-1.1170	0.7887	60.7S	117.9E	0	101		
839	042	-1660 Mar 17	02:38:17	38560	-45266	3	A-	-t	-1.0038	0.9574	60.7S	47.7E	0	255	-	-
840	042	-1660 Sep 10	12:54:09	38549	-45260	8	T	-t	0.8980	1.0123	60.3N	156.8W	26	244	95	00m47s

Cat Num	Canon Plate	Calendar Date	TD of Greatest Eclipse	ΔT s	Luna Num	Saros Num	Ecl. Type	QLE	Gamma	Ecl. Mag.	Lat. °	Long. °	Sun Alt °	Sun Azm °	Path Width km	Central Line Dur.
841	043	-1659 Mar 06	07:31:35	38538	-45254	13	A	nn	-0.2191	0.9941	22.6S	121.8W	77	332	21	00m34s
842	043	-1659 Aug 30	22:33:53	38527	-45248	18	A	-n	0.1980	0.9694	24.8N	8.1E	78	206	112	03m06s
843	043	-1658 Feb 23	19:29:26	38517	-45242	23	T	p-	0.5250	1.0425	14.1N	40.0E	58	156	165	03m48s
844	043	-1658 Aug 20	01:09:31	38506	-45236	28	A	p-	-0.5401	0.9334	12.8S	48.1W	57	21	291	08m21s
845	043	-1657 Jan 15	02:22:05	38497	-45231	-5	P	-t	-1.4062	0.2443	64.5S	103.4E	0	202		
846	043	-1657 Feb 13	11:36:16	38495	-45230	33	P	t-	1.2097	0.6162	62.3N	119.5E	0	132		
847	043	-1657 Aug 09	01:05:08	38484	-45224	38	P	t-	-1.2530	0.5298	62.9S	78.5W	0	40		
848	043	-1656 Jan 04	15:41:43	38475	-45219	5	A	-p	-0.7653	0.9909	73.4S	122.3E	40	341	50	00m37s
849	043	-1656 Jun 28	19:08:31	38465	-45213	10	T	-p	0.7729	1.0231	74.4N	55.8E	39	185	125	01m31s
850	043	-1656 Dec 23	22:15:04	38454	-45207	15	A	nn	-0.0731	0.9417	27.8S	7.1E	86	359	216	07m06s
851	043	-1655 Jun 18	10:11:15	38443	-45201	20	T	nn	0.0000	1.0745	22.6N	175.0W	90	180	242	06m41s
852	043	-1655 Dec 12	22:17:54	38432	-45195	25	A	p-	0.6207	0.9147	16.4N	7.5E	52	184	414	12m07s
853	043	-1654 Jun 08	03:26:19	38422	-45189	30	T	p-	-0.7398	1.0678	26.9S	68.6W	42	353	329	05m58s
854	043	-1654 Dec 01	22:21:49	38411	-45183	35	P	t-	1.2745	0.4928	68.8N	26.2E	0	202		
855	043	-1653 Apr 29	08:15:17	38402	-45178	2	P	-t	1.1044	0.7984	71.2N	97.9E	0	61		
856	043	-1653 May 28	18:12:41	38400	-45177	40	Pb	t-	-1.5359	0.0073	69.1S	94.5E	0	331		
857	043	-1653 Oct 22	16:10:27	38391	-45172	7	T	-p	-0.9157	1.0111	65.9S	53.0E	23	47	96	00m44s
858	043	-1652 Apr 17	12:57:24	38381	-45166	12	A	-p	0.3613	0.9542	25.1N	138.9E	69	161	179	05m19s
859	043	-1652 Oct 11	07:17:32	38370	-45160	17	T	-n	-0.2268	1.0492	14.2S	134.9W	77	19	167	04m22s
860	043	-1651 Apr 06	13:26:49	38359	-45154	22	A	p-	-0.3984	0.9490	21.9S	147.3E	66	341	203	06m06s
861	044	-1651 Sep 30	22:37:42	38348	-45148	27	T	n-	0.4574	1.0264	28.6N	9.0E	63	201	100	02m20s
862	044	-1650 Mar 26	17:06:50	38338	-45142	32	P	t-	-1.1319	0.7457	71.6S	172.9W	0	258		
863	044	-1650 Sep 20	09:35:05	38327	-45136	37	P	t-	1.2085	0.6083	71.3N	52.8W	0	292		
864	044	-1649 Feb 14	18:55:44	38318	-45131	4	T	-p	0.8141	1.0456	36.1N	49.8E	35	167	261	03m51s
865	044	-1649 Aug 10	20:14:27	38307	-45125	9	As	-t	-0.9777	0.9237	57.8S	22.8E	11	13	-	07m21s
866	044	-1648 Feb 04	11:17:57	38296	-45119	14	T	-n	0.1102	1.0577	14.1S	173.9E	84	173	191	05m24s
867	044	-1648 Jul 29	20:54:51	38286	-45113	19	A	nn	-0.2343	0.9570	8.7N	25.2E	77	3	161	05m42s
868	044	-1647 Jan 24	01:53:47	38275	-45107	24	H	p-	-0.6104	1.0117	60.1S	41.1W	52	354	51	00m53s
869	044	-1647 Jul 19	03:46:15	38264	-45101	29	H	p-	0.5243	1.0096	55.3N	78.7W	58	178	39	00m47s
870	044	-1646 Jan 13	10:25:39	38254	-45095	34	P	t-	-1.3800	0.3109	66.4S	9.1E	0	177		
871	044	-1646 Jun 09	10:31:31	38245	-45090	1	P	-t	-1.2869	0.4678	63.1S	140.3W	0	322		
872	044	-1646 Jul 08	17:34:05	38243	-45089	39	P	t-	1.2284	0.5802	65.5N	92.0W	0	12		
873	044	-1646 Dec 03	16:35:07	38234	-45084	6	A	-p	0.9353	0.9095	45.6N	114.7E	20	206	971	09m36s
874	044	-1645 May 30	03:35:26	38223	-45078	11	T	-p	-0.5683	1.0619	14.0S	65.1W	55	340	244	05m33s
875	044	-1645 Nov 22	18:00:57	38213	-45072	16	A	nn	0.2292	0.9523	4.9S	72.0E	77	202	179	05m30s
876	044	-1644 May 18	16:54:40	38202	-45066	21	H	nn	0.2002	1.0131	26.3N	78.1E	78	155	46	01m13s
877	044	-1644 Nov 11	02:38:09	38191	-45060	26	H	p-	-0.4694	1.0062	38.5S	77.6W	62	32	24	00m30s
878	044	-1643 May 07	23:11:49	38180	-45054	31	P	t-	1.0124	0.9483	61.2N	123.4W	0	64		
879	044	-1643 Oct 31	16:56:07	38170	-45048	36	P	t-	-1.1135	0.7952	61.0S	25.0W	0	110		
880	044	-1642 Mar 28	09:22:04	38161	-45043	3	P	-t	-1.0719	0.8423	60.5S	63.6W	0	264		
881	045	-1642 Sep 21	21:10:34	38150	-45037	8	H	-t	0.9211	1.0057	58.1N	77.9E	23	244	50	00m22s
882	045	-1641 Mar 17	14:48:59	38139	-45031	13	H	nn	-0.2772	1.0002	21.5S	128.4E	74	330	1	00m01s
883	045	-1641 Sep 11	06:18:17	38129	-45025	18	A	-n	0.2290	0.9633	22.4N	108.8W	77	209	136	03m43s
884	045	-1640 Mar 06	03:16:50	38118	-45019	23	T	p-	0.4746	1.0488	14.2N	78.8W	62	153	182	04m13s
885	045	-1640 Aug 30	08:24:11	38107	-45013	28	A	p-	-0.4982	0.9298	12.9S	159.0W	60	25	299	08m34s
886	045	-1639 Jan 25	10:45:55	38098	-45008	-5	P	-t	-1.4306	0.1979	63.6S	33.9W	0	212		
887	045	-1639 Feb 23	19:39:29	38097	-45007	33	P	t-	1.1695	0.6937	61.6N	12.0W	0	122		
888	045	-1639 Aug 19	08:15:30	38086	-45001	38	P	t-	-1.2026	0.6154	62.1S	163.1E	0	49		
889	045	-1638 Jan 14	23:57:02	38077	-44996	5	A	-p	-0.7844	0.9901	72.8S	11.7E	38	325	56	00m41s
890	045	-1638 Jul 10	02:27:59	38066	-44990	10	T	-p	0.8355	1.0227	79.4N	30.3W	33	212	142	01m24s
891	045	-1637 Jan 04	06:14:24	38056	-44984	15	A	nn	-0.0858	0.9417	28.9S	112.6W	85	354	217	07m01s
892	045	-1637 Jun 29	17:40:22	38045	-44978	20	Tm	nn	0.0694	1.0740	27.7N	71.7E	86	182	241	06m27s
893	045	-1637 Dec 24	06:11:25	38034	-44972	25	A	p-	0.6119	0.9165	14.4N	114.4W	52	179	401	11m58s
894	045	-1636 Jun 18	10:49:52	38024	-44966	30	T	p-	-0.6688	1.0659	19.5S	175.4E	48	357	290	06m06s
895	045	-1636 Dec 12	06:32:48	38013	-44960	35	P	t-	1.2687	0.5027	67.7N	109.8W	0	190		
896	045	-1635 May 09	15:02:53	38004	-44955	2	P	-t	1.1875	0.6484	70.6N	18.9W	0	48		
897	045	-1635 Jun 08	01:14:28	38002	-44954	40	P	t-	-1.4667	0.1372	68.2S	24.5W	0	343		
898	045	-1635 Nov 02	00:55:01	37993	-44949	7	T	-p	-0.9140	1.0124	69.8S	84.7W	23	52	106	00m47s
899	045	-1634 Apr 28	19:23:02	37983	-44943	12	A	-p	0.4456	0.9533	34.3N	37.9E	63	160	190	05m08s
900	045	-1634 Oct 22	16:04:12	37972	-44937	17	T	-n	-0.2317	1.0474	19.0S	91.1E	76	18	162	04m10s

Cat Num	Canon Plate	Calendar Date	TD of Greatest Eclipse	ΔT s	Luna Num	Saros Num	Ecl. Type	QLE	Gamma	Ecl. Mag.	Lat. °	Long. °	Sun Alt °	Sun Azm °	Path Width km	Central Line Dur.
901	046	-1633 Apr 17	19:55:43	37961	-44931	22	A	p-	-0.3178	0.9527	13.0S	45.3E	71	342	182	05m53s
902	046	-1633 Oct 12	07:11:21	37951	-44925	27	T	n-	0.4495	1.0217	23.7N	122.2W	63	200	83	02m00s
903	046	-1632 Apr 06	00:04:42	37940	-44919	32	P	t-	-1.0596	0.8763	71.7S	67.0E	0	272		
904	046	-1632 Sep 30	17:40:36	37929	-44913	37	P	t-	1.1964	0.6288	71.6N	170.1E	0	278		
905	046	-1631 Feb 25	02:54:56	37920	-44908	4	T	-p	0.8541	1.0495	42.8N	77.0W	31	162	315	03m54s
906	046	-1631 Aug 21	03:19:08	37910	-44902	9	P	-t	-1.0262	0.9099	69.6S	104.2W	0	32		
907	046	-1630 Feb 14	19:28:09	37899	-44896	14	T	-n	0.1455	1.0597	9.4S	49.2E	82	169	198	05m35s
908	046	-1630 Aug 10	04:03:27	37888	-44890	19	A	nn	-0.2879	0.9562	3.6N	84.5W	73	7	166	05m46s
909	046	-1629 Feb 04	09:59:41	37878	-44884	24	H	p-	-0.5842	1.0121	55.8S	159.8W	54	348	51	00m57s
910	046	-1629 Jul 30	11:12:58	37867	-44878	29	H	p-	0.4660	1.0108	50.2N	172.1E	62	185	42	00m55s
911	046	-1628 Jan 24	18:14:58	37856	-44872	34	P	t-	-1.3600	0.3452	67.5S	120.8W	0	188		
912	046	-1628 Jun 19	17:59:12	37848	-44867	1	P	-t	-1.3552	0.3351	64.0S	96.5E	0	332		
913	046	-1628 Jul 19	01:13:01	37846	-44866	39	P	t-	1.1677	0.6962	66.5N	141.1E	0	2		
914	046	-1628 Dec 14	00:32:53	37837	-44861	6	A	-p	0.9413	0.9113	46.1N	11.0W	19	200	1006	09m33s
915	046	-1627 Jun 09	10:53:30	37826	-44855	11	T	-p	-0.6445	1.0573	17.6S	176.8W	50	343	244	05m14s
916	046	-1627 Dec 03	02:20:57	37816	-44849	16	A	nn	0.2295	0.9557	7.5S	54.8W	77	198	166	05m13s
917	046	-1626 May 29	23:48:06	37805	-44843	21	Hm	nn	0.1201	1.0092	25.2N	24.8W	83	159	32	00m54s
918	046	-1626 Nov 22	11:21:31	37794	-44837	26	H	p-	-0.4691	1.0089	42.6S	150.9E	62	30	35	00m43s
919	046	-1625 May 19	05:36:22	37784	-44831	31	A	t-	0.9278	0.9523	66.7N	179.3E	21	100	469	03m19s
920	046	-1625 Nov 12	01:49:44	37773	-44825	36	P	t-	-1.1128	0.7965	61.4S	168.9W	0	120		
921	047	-1624 Apr 07	15:58:59	37764	-44820	3	P	-t	-1.1456	0.7162	60.5S	173.1W	0	272		
922	047	-1624 Oct 02	05:33:58	37753	-44814	8	A	-t	0.9377	0.9995	56.3N	51.0W	20	242	5	00m02s
923	047	-1623 Mar 27	22:00:44	37743	-44808	13	H	-n	-0.3407	1.0060	20.5S	19.9E	70	329	22	00m33s
924	047	-1623 Sep 21	14:09:41	37732	-44802	18	A	-n	0.2531	0.9575	19.3N	132.1E	75	210	159	04m20s
925	047	-1622 Mar 17	10:58:47	37722	-44796	23	T	p-	0.4190	1.0548	14.7N	164.1E	65	151	197	04m35s
926	047	-1622 Sep 10	15:47:27	37711	-44790	28	A	p-	-0.4637	0.9263	14.1S	88.0E	62	28	308	08m44s
927	047	-1621 Feb 05	19:02:47	37702	-44785	-5	P	-t	-1.4606	0.1410	62.8S	169.2W	0	222		
928	047	-1621 Mar 07	03:36:28	37700	-44784	33	P	t-	1.1243	0.7812	61.2N	141.8W	0	113		
929	047	-1621 Aug 30	15:35:02	37690	-44778	38	P	t-	-1.1588	0.6897	61.5S	42.6E	0	58		
930	047	-1620 Jan 26	08:04:19	37681	-44773	5	A	-p	-0.8094	0.9893	71.1S	99.3W	36	312	64	00m43s
931	047	-1620 Jul 20	09:55:17	37670	-44767	10	T	-p	0.8927	1.0217	79.8N	101.9W	26	255	166	01m16s
932	047	-1619 Jan 14	14:03:27	37659	-44761	15	A	nn	-0.1063	0.9420	29.6S	130.5E	84	349	216	06m51s
933	047	-1619 Jul 10	01:16:28	37649	-44755	20	T	nn	0.1332	1.0727	31.7N	42.4W	82	187	238	06m08s
934	047	-1618 Jan 03	13:57:11	37638	-44749	25	A	p-	0.5961	0.9190	12.6N	125.9E	53	175	382	11m34s
935	047	-1618 Jun 29	18:17:57	37628	-44743	30	T	p-	-0.6021	1.0630	13.5S	59.3E	53	2	259	06m01s
936	047	-1618 Dec 23	14:39:12	37617	-44737	35	P	t-	1.2582	0.5207	66.6N	115.9E	0	179		
937	047	-1617 May 20	21:47:47	37608	-44732	2	P	-t	1.2714	0.4977	69.8N	134.6W	0	36		
938	047	-1617 Jun 19	08:17:47	37606	-44731	40	P	t-	-1.3992	0.2632	67.3S	143.3W	0	354		
939	047	-1617 Nov 13	09:42:09	37597	-44726	7	T	-p	-0.9108	1.0143	73.8S	137.8E	24	56	120	00m52s
940	047	-1616 May 09	01:42:23	37587	-44720	12	A	-p	0.5345	0.9518	43.9N	61.5W	57	160	208	04m57s
941	048	-1616 Nov 02	00:55:43	37576	-44714	17	T	-n	-0.2327	1.0461	23.4S	43.6W	76	17	158	04m01s
942	048	-1615 Apr 28	02:19:23	37566	-44708	22	A	nn	-0.2322	0.9558	4.1S	55.3W	77	344	165	05m37s
943	048	-1615 Oct 22	15:50:19	37555	-44702	27	T	n-	0.4467	1.0176	19.2N	105.2E	63	198	67	01m40s
944	048	-1614 Apr 17	06:58:45	37544	-44696	32	A	t-	-0.9828	0.9847	67.3S	77.0W	10	308	317	01m04s
945	048	-1614 Oct 12	01:52:40	37534	-44690	37	P	t-	1.1902	0.6389	71.6N	31.2E	0	264		
946	048	-1613 Mar 08	10:47:14	37525	-44685	4	T	-p	0.9003	1.0528	50.8N	155.7E	25	155	401	03m49s
947	048	-1613 Sep 01	10:32:31	37514	-44679	9	P	-t	-1.0673	0.8396	70.4S	133.1E	0	44		
948	048	-1612 Feb 26	03:30:45	37504	-44673	14	T	-n	0.1870	1.0616	3.7S	74.2W	79	167	206	05m44s
949	048	-1612 Aug 20	11:21:53	37493	-44667	19	A	-n	-0.3338	0.9553	1.8S	162.9E	70	10	172	05m46s
950	048	-1611 Feb 14	17:55:34	37482	-44661	24	H	p-	-0.5506	1.0125	50.2S	81.6E	56	344	52	01m01s
951	048	-1611 Aug 09	18:49:44	37472	-44655	29	H	p-	0.4152	1.0115	44.7N	58.6E	65	190	44	01m01s
952	048	-1610 Feb 04	01:55:01	37461	-44649	34	P	t-	-1.3337	0.3905	68.5S	111.1E	0	200		
953	048	-1610 Jul 01	01:31:29	37452	-44644	1	P	-t	-1.4198	0.2100	64.9S	28.1W	0	341		
954	048	-1610 Jul 30	08:59:13	37451	-44643	39	P	t-	1.1127	0.8009	67.6N	12.0E	0	351		
955	048	-1610 Dec 25	08:26:41	37442	-44638	6	A	-p	0.9505	0.9134	47.8N	135.9W	18	194	1076	09m18s
956	048	-1609 Jun 20	18:13:40	37431	-44632	11	T	-p	-0.7181	1.0518	22.2S	70.4E	44	347	245	04m47s
957	048	-1609 Dec 14	10:38:54	37421	-44626	16	A	nn	0.2324	0.9596	9.2S	179.1E	77	194	151	04m50s
958	048	-1608 Jun 09	06:40:23	37410	-44620	21	H	nn	0.0401	1.0048	23.4N	127.6W	88	164	16	00m29s
959	048	-1608 Dec 02	20:03:58	37399	-44614	26	H	p-	-0.4676	1.0122	46.2S	20.8E	62	26	48	00m59s
960	048	-1607 May 29	12:00:07	37389	-44608	31	A	p-	0.8427	0.9525	66.9N	103.4E	32	121	322	03m32s

Cat Num	Canon Plate	Calendar Date	TD of Greatest Eclipse	ΔT s	Luna Num	Saros Num	Ecl. Type	QLE	Gamma	Ecl. Mag.	Lat. °	Long. °	Sun Alt °	Sun Azm °	Path Width km	Central Line Dur.
961	049	-1607 Nov 22	10:42:35	37378	-44602	36	P	t-	-1.1117	0.7986	62.1S	47.2E	0	129		
962	049	-1606 Apr 18	22:32:52	37369	-44597	3	P	-t	-1.2220	0.5843	60.7S	78.0E	0	281		
963	049	-1606 Oct 13	14:02:49	37359	-44591	8	A	-t	0.9490	0.9940	54.7N	177.1E	18	238	66	00m25s
964	049	-1605 Apr 08	05:07:38	37348	-44585	13	H	-p	-0.4092	1.0114	19.9S	87.5W	66	329	43	01m02s
965	049	-1605 Oct 02	22:07:42	37338	-44579	18	A	-n	0.2711	0.9523	15.7N	10.9E	74	211	180	04m58s
966	049	-1604 Mar 27	18:33:35	37327	-44573	23	T	n-	0.3569	1.0603	15.3N	49.2E	69	150	211	04m56s
967	049	-1604 Sep 20	23:20:23	37317	-44567	28	A	p-	-0.4370	0.9232	16.3S	27.4W	64	30	318	08m52s
968	049	-1603 Feb 16	03:08:36	37308	-44562	-5	Pe	-t	-1.4993	0.0673	62.1S	58.5E	0	231		
969	049	-1603 Mar 17	11:23:31	37306	-44561	33	P	t-	1.0711	0.8846	60.8N	91.0E	0	105		
970	049	-1603 Sep 09	23:06:46	37295	-44555	38	P	t-	-1.1239	0.7488	61.0S	80.8W	0	67		
971	049	-1602 Feb 05	16:01:25	37286	-44550	5	A	-p	-0.8424	0.9884	68.8S	149.9E	32	301	76	00m47s
972	049	-1602 Jul 31	17:30:59	37276	-44544	10	T	-p	0.9438	1.0198	75.5N	172.5E	19	285	209	01m06s
973	049	-1601 Jan 25	21:44:24	37265	-44538	15	A	nn	-0.1330	0.9427	29.6S	15.6E	82	344	214	06m36s
974	049	-1601 Jul 21	08:58:04	37255	-44532	20	T	-n	0.1927	1.0707	34.4N	157.1W	79	192	235	05m47s
975	049	-1600 Jan 14	21:36:47	37244	-44526	25	A	p-	0.5753	0.9221	11.3N	7.9E	55	170	358	10m55s
976	049	-1600 Jul 10	01:48:41	37234	-44520	30	T	p-	-0.5381	1.0593	8.6S	56.7W	57	6	231	05m44s
977	049	-1599 Jan 02	22:41:56	37223	-44514	35	P	t-	1.2439	0.5453	65.5N	16.9W	0	168		
978	049	-1599 May 31	04:30:55	37214	-44509	2	P	-t	1.3554	0.3481	69.0N	110.8E	0	25		
979	049	-1599 Jun 29	15:23:21	37213	-44508	40	P	t-	-1.3341	0.3833	66.3S	97.8E	0	4		
980	049	-1599 Nov 23	18:30:08	37204	-44503	7	T	-p	-0.9075	1.0167	78.0S	1.3E	24	58	138	00m59s
981	050	-1598 May 20	08:01:43	37193	-44497	12	A	-p	0.6228	0.9498	53.6N	160.5W	51	160	235	04m46s
982	050	-1598 Nov 13	09:49:09	37183	-44491	17	T	-n	-0.2322	1.0452	27.5S	178.2W	76	15	155	03m55s
983	050	-1597 May 09	08:42:30	37172	-44485	22	A	nn	-0.1455	0.9587	4.5N	155.3W	82	346	152	05m18s
984	050	-1597 Nov 03	00:31:48	37161	-44479	27	H	n-	0.4466	1.0138	15.1N	28.1W	63	196	53	01m22s
985	050	-1596 Apr 27	13:51:25	37151	-44473	32	A	t-	-0.9034	0.9937	52.6S	151.4E	25	332	51	00m31s
986	050	-1596 Oct 22	10:08:45	37140	-44467	37	P	t-	1.1881	0.6419	71.4N	108.6W	0	250		
987	050	-1595 Mar 18	18:31:52	37132	-44462	4	T	-t	0.9531	1.0547	60.8N	24.3E	17	142	607	03m33s
988	050	-1595 Sep 11	17:56:44	37121	-44456	9	P	-t	-1.0994	0.7850	71.1S	7.1E	0	57		
989	050	-1594 Mar 08	11:24:32	37110	-44450	14	T	-n	0.2355	1.0630	2.7N	164.0E	76	164	212	05m48s
990	050	-1594 Aug 31	18:51:18	37100	-44444	19	A	-n	-0.3714	0.9546	7.2S	47.3E	68	13	178	05m41s
991	050	-1593 Feb 26	01:41:29	37089	-44438	24	H	p-	-0.5096	1.0128	43.8S	36.3W	59	341	51	01m06s
992	050	-1593 Aug 21	02:36:50	37079	-44432	29	H	p-	0.3723	1.0120	39.0N	58.7W	68	193	44	01m05s
993	050	-1592 Feb 15	09:22:30	37068	-44426	34	P	t-	-1.2977	0.4525	69.5S	14.4W	0	211		
994	050	-1592 Jul 11	09:08:18	37059	-44421	1	Pe	-t	-1.4808	0.0925	65.9S	154.2W	0	351		
995	050	-1592 Aug 09	16:53:53	37058	-44420	39	P	t-	1.0643	0.8927	68.6N	119.8W	0	340		
996	050	-1591 Jan 04	16:14:07	37049	-44415	6	A	-p	0.9652	0.9160	51.3N	100.6E	15	187	1258	08m45s
997	050	-1591 Jul 01	01:36:49	37038	-44409	11	T	-p	-0.7886	1.0455	28.0S	43.9W	38	351	246	04m10s
998	050	-1591 Dec 24	18:53:12	37028	-44403	16	A	nn	0.2392	0.9642	9.9S	54.1E	76	189	133	04m20s
999	050	-1590 Jun 20	13:34:12	37017	-44397	21	A	nn	-0.0376	0.9997	20.8N	128.9E	88	346	1	00m02s
1000	050	-1590 Dec 14	04:45:51	37007	-44391	26	T	p-	-0.4650	1.0161	49.0S	108.1W	62	21	62	01m16s
1001	051	-1589 Jun 09	18:22:35	36996	-44385	31	A	p-	0.7568	0.9516	65.7N	23.0E	41	138	271	03m51s
1002	051	-1589 Dec 03	19:36:25	36986	-44379	36	P	t-	-1.1114	0.7994	62.8S	97.0W	0	139		
1003	051	-1588 Apr 29	05:03:27	36977	-44374	3	P	-t	-1.3012	0.4465	61.0S	30.0W	0	290		
1004	051	-1588 May 28	19:00:17	36975	-44373	41	Pb	t-	1.5121	0.0847	62.4N	83.4W	0	47		
1005	051	-1588 Oct 23	22:36:30	36966	-44368	8	A	-t	0.9560	0.9892	53.4N	42.8E	17	232	129	00m47s
1006	051	-1587 Apr 18	12:11:17	36956	-44362	13	T	-p	-0.4816	1.0163	19.9S	165.7E	61	329	63	01m29s
1007	051	-1587 Oct 13	06:11:49	36945	-44356	18	A	-n	0.2834	0.9476	11.9N	112.2W	73	210	200	05m35s
1008	051	-1586 Apr 08	02:05:05	36935	-44350	23	T	n-	0.2911	1.0653	16.0N	64.6W	73	149	222	05m16s
1009	051	-1586 Oct 02	06:59:43	36924	-44344	28	A	p-	-0.4154	0.9204	19.3S	144.4W	65	31	327	08m59s
1010	051	-1585 Mar 28	19:05:42	36914	-44338	33	P	t-	1.0137	0.9960	60.6N	35.0W	0	96		
1011	051	-1585 Sep 21	06:47:30	36903	-44332	38	P	t-	-1.0953	0.7971	60.6S	153.7E	0	76		
1012	051	-1584 Feb 16	23:47:55	36894	-44327	5	A	-t	-0.8833	0.9871	66.6S	40.7E	28	292	97	00m51s
1013	051	-1584 Aug 11	01:16:09	36884	-44321	10	T	-t	0.9879	1.0164	68.0N	75.4E	8	305	414	00m51s
1014	051	-1583 Feb 05	05:13:52	36873	-44315	15	A	nn	-0.1687	0.9436	29.1S	96.4W	80	339	211	06m20s
1015	051	-1583 Jul 31	16:48:13	36863	-44309	20	T	-n	0.2452	1.0683	35.7N	86.4E	76	197	230	05m26s
1016	051	-1582 Jan 25	05:06:22	36852	-44303	25	A	p-	0.5459	0.9257	10.1N	107.3W	57	166	331	10m07s
1017	051	-1582 Jul 21	09:26:34	36842	-44297	30	T	p-	-0.4804	1.0550	5.0S	173.9W	61	10	206	05m19s
1018	051	-1581 Jan 14	06:38:15	36831	-44291	35	P	t-	1.2234	0.5810	64.5N	147.6W	0	158		
1019	051	-1581 Jun 11	11:13:09	36822	-44286	2	P	-t	1.4387	0.2008	68.0N	3.0W	0	13		
1020	051	-1581 Jul 10	22:32:03	36821	-44285	40	P	t-	-1.2721	0.4967	65.3S	21.5W	0	14		

Cat Num	Canon Plate	Calendar Date	TD of Greatest Eclipse	ΔT s	Luna Num	Saros Num	Ecl. Type	QLE	Gamma	Ecl. Mag.	Lat. °	Long. °	Sun Alt °	Sun Azm °	Path Width km	Central Line Dur.
1021	052	-1581 Dec 05	03:18:02	36812	-44280	7	T	-p	-0.9051	1.0196	82.3S	132.7W	25	58	160	01m08s
1022	052	-1580 May 30	14:19:02	36801	-44274	12	A	-p	0.7124	0.9472	63.7N	101.5E	44	161	278	04m35s
1023	052	-1580 Nov 23	18:43:19	36791	-44268	17	T	-n	-0.2312	1.0450	31.1S	47.6E	76	12	154	03m51s
1024	052	-1579 May 19	15:04:38	36780	-44262	22	A	nn	-0.0570	0.9609	12.9N	105.5E	87	348	142	04m58s
1025	052	-1579 Nov 13	09:16:15	36770	-44256	27	H	n-	0.4492	1.0107	11.6N	162.0W	63	193	41	01m05s
1026	052	-1578 May 08	20:43:09	36759	-44250	32	H	t-	-0.8219	1.0005	41.0S	36.9E	34	341	3	00m03s
1027	052	-1578 Nov 02	18:28:07	36749	-44244	37	P	t-	1.1893	0.6394	70.9N	111.2E	0	236		
1028	052	-1577 Mar 30	02:11:02	36740	-44239	4	P	-t	1.0109	0.9998	71.6N	140.3W	0	97		
1029	052	-1577 Apr 28	09:28:38	36738	-44238	42	Pb	t-	-1.5198	0.0233	71.1S	106.8W	0	298		
1030	052	-1577 Sep 23	01:29:44	36730	-44233	9	P	-t	-1.1243	0.7428	71.6S	121.6W	0	71		
1031	052	-1576 Mar 18	19:09:49	36719	-44227	14	T	-n	0.2909	1.0640	9.8N	43.8E	73	163	218	05m47s
1032	052	-1576 Sep 11	02:31:18	36709	-44221	19	A	-n	-0.4011	0.9540	12.7S	71.0W	66	16	182	05m34s
1033	052	-1575 Mar 08	09:17:47	36698	-44215	24	H	p-	-0.4614	1.0130	36.7S	153.2W	62	340	50	01m10s
1034	052	-1575 Aug 31	10:33:31	36688	-44209	29	H	n-	0.3364	1.0123	33.2N	179.3W	70	196	45	01m08s
1035	052	-1574 Feb 25	16:40:43	36677	-44203	34	P	t-	-1.2549	0.5262	70.3S	138.2W	0	224		
1036	052	-1574 Aug 21	00:57:11	36667	-44197	39	P	t-	1.0228	0.9711	69.5N	105.6E	0	328		
1037	052	-1573 Jan 15	23:55:57	36658	-44192	6	An	-t	0.9847	0.9183	57.7N	22.6W	9	180	-	07m57s
1038	052	-1573 Jul 12	09:03:23	36647	-44186	11	T	-t	-0.8554	1.0385	35.1S	160.0W	31	356	250	03m26s
1039	052	-1572 Jan 05	03:03:16	36637	-44180	16	A	-n	0.2502	0.9693	9.5S	69.9W	76	185	114	03m45s
1040	052	-1572 Jun 30	20:30:31	36626	-44174	21	A	nn	-0.1120	0.9941	17.5N	24.3E	84	352	21	00m40s
1041	053	-1572 Dec 24	13:22:36	36616	-44168	26	T	p-	-0.4577	1.0206	50.4S	125.5E	63	14	79	01m37s
1042	053	-1571 Jun 20	00:48:42	36606	-44162	31	A	p-	0.6743	0.9499	63.3N	61.8W	47	152	249	04m15s
1043	053	-1571 Dec 14	04:26:39	36595	-44156	36	P	t-	-1.1088	0.8046	63.7S	119.3E	0	149		
1044	053	-1570 May 10	11:34:14	36586	-44151	3	P	-t	-1.3807	0.3072	61.4S	138.2W	0	298		
1045	053	-1570 Jun 09	01:21:29	36585	-44150	41	P	t-	1.4258	0.2328	63.1N	170.4E	0	38		
1046	053	-1570 Nov 04	07:12:05	36576	-44145	8	A	-t	0.9608	0.9850	52.4N	92.6W	16	226	191	01m08s
1047	053	-1569 Apr 29	19:13:12	36565	-44139	13	T	-p	-0.5562	1.0205	20.7S	59.2E	56	331	83	01m52s
1048	053	-1569 Oct 24	14:20:38	36555	-44133	18	A	-n	0.2910	0.9435	8.0N	123.3E	73	209	217	06m13s
1049	053	-1568 Apr 18	09:30:17	36544	-44127	23	T	n-	0.2195	1.0696	16.6N	176.5W	77	150	232	05m35s
1050	053	-1568 Oct 12	14:47:54	36534	-44121	28	A	p-	-0.4012	0.9181	22.9S	96.4E	66	32	335	09m05s
1051	053	-1567 Apr 08	02:40:01	36523	-44115	33	T	t-	0.9502	1.0617	57.0N	125.2W	18	116	649	03m40s
1052	053	-1567 Oct 01	14:38:25	36513	-44109	38	P	t-	-1.0743	0.8325	60.5S	25.7E	0	86		
1053	053	-1566 Feb 27	07:24:03	36504	-44104	5	A	-t	-0.9317	0.9851	64.7S	64.6W	21	283	145	00m57s
1054	053	-1566 Aug 22	09:11:28	36494	-44098	10	P	-t	1.0244	0.9567	62.0N	40.2W	0	308		
1055	053	-1565 Feb 16	12:34:13	36483	-44092	15	A	nn	-0.2112	0.9446	28.3S	153.7E	78	335	209	06m04s
1056	053	-1565 Aug 12	00:44:52	36473	-44086	20	T	-n	0.2924	1.0653	35.8N	31.7W	73	202	223	05m05s
1057	053	-1564 Feb 05	12:29:02	36462	-44080	25	A	p-	0.5104	0.9297	9.3N	139.6E	59	162	303	09m13s
1058	053	-1564 Jul 31	17:09:48	36452	-44074	30	T	n-	-0.4277	1.0500	2.7S	68.1E	65	14	183	04m48s
1059	053	-1563 Jan 24	14:27:55	36442	-44068	35	P	t-	1.1965	0.6282	63.5N	83.6E	0	148		
1060	053	-1563 Jun 21	17:57:26	36433	-44063	2	Pe	-t	1.5190	0.0603	67.0N	116.8W	0	3		
1061	054	-1563 Jul 21	05:46:11	36431	-44062	40	P	t-	-1.2151	0.5997	64.4S	141.7W	0	24		
1062	054	-1563 Dec 15	12:03:55	36422	-44057	7	T	-p	-0.9053	1.0228	86.7S	103.4E	25	48	186	01m18s
1063	054	-1562 Jun 10	20:38:45	36412	-44051	12	A	-p	0.7993	0.9440	74.1N	4.0E	37	162	347	04m26s
1064	054	-1562 Dec 05	03:35:56	36401	-44045	17	T	-n	-0.2314	1.0450	34.0S	85.5W	76	8	154	03m50s
1065	054	-1561 May 30	21:29:33	36391	-44039	22	A	nn	0.0299	0.9628	20.8N	6.4E	88	171	135	04m37s
1066	054	-1561 Nov 24	17:59:42	36380	-44033	27	H	n-	0.4519	1.0080	8.6N	64.4E	63	190	31	00m51s
1067	054	-1560 May 19	03:35:10	36370	-44027	32	H	p-	-0.7395	1.0062	30.8S	73.9W	42	346	32	00m37s
1068	054	-1560 Nov 13	02:48:31	36360	-44021	37	P	t-	1.1921	0.6342	70.2N	28.9W	0	222		
1069	054	-1559 Apr 09	09:44:40	36351	-44016	4	P	-t	1.0736	0.8801	71.6N	90.7E	0	84		
1070	054	-1559 May 08	16:46:41	36349	-44015	42	P	t-	-1.4450	0.1650	70.5S	128.8E	0	311		
1071	054	-1559 Oct 03	09:11:03	36340	-44010	9	P	-t	-1.1425	0.7121	71.8S	107.2E	0	85		
1072	054	-1558 Mar 30	02:47:55	36330	-44004	14	T	-n	0.3520	1.0642	17.5N	74.9W	69	161	224	05m39s
1073	054	-1558 Sep 22	10:21:59	36320	-43998	19	A	-n	-0.4227	0.9538	18.0S	168.0E	65	18	185	05m23s
1074	054	-1557 Mar 19	16:43:59	36309	-43992	24	H	n-	-0.4056	1.0128	29.2S	91.6E	66	340	48	01m13s
1075	054	-1557 Sep 11	18:40:22	36299	-43986	29	H	n-	0.3083	1.0124	27.5N	57.0E	72	198	45	01m10s
1076	054	-1556 Mar 07	23:47:10	36288	-43980	34	P	t-	-1.2031	0.6157	71.0S	100.4E	0	237		
1077	054	-1556 Aug 31	09:09:55	36278	-43974	39	Tn	t-	0.9891	1.0404	74.7N	51.1W	7	297	-	02m05s
1078	054	-1555 Jan 26	07:28:19	36269	-43969	6	A+	-t	1.0120	0.9351	67.8N	148.3W	0	169	-	-
1079	054	-1555 Jul 22	16:35:55	36259	-43963	11	T	-t	-0.9162	1.0306	43.7S	81.0E	23	1	260	02m35s
1080	054	-1554 Jan 15	11:07:55	36248	-43957	16	A	-n	0.2661	0.9749	7.9S	167.3E	75	180	93	03m03s

Cat Num	Canon Plate	Calendar Date	TD of Greatest Eclipse	ΔT s	Luna Num	Saros Num	Ecl. Type	QLE	Gamma	Ecl. Mag.	Lat. °	Long. °	Sun Alt °	Sun Azm °	Path Width km	Central Line Dur.
1081	055	-1554 Jul 12	03:30:02	36238	-43951	21	A	nn	-0.1826	0.9882	13.4N	81.7W	80	356	42	01m24s
1082	055	-1553 Jan 04	21:56:12	36227	-43945	26	T	p-	-0.4474	1.0254	50.5S	0.3E	63	7	97	02m01s
1083	055	-1553 Jul 01	07:17:10	36217	-43939	31	A	p-	0.5941	0.9476	59.7N	150.2W	53	164	240	04m44s
1084	055	-1553 Dec 25	13:14:00	36207	-43933	36	P	t-	-1.1037	0.8148	64.7S	23.9W	0	159		
1085	055	-1552 May 20	18:04:09	36198	-43928	3	P	-t	-1.4612	0.1652	62.0S	113.6E	0	307		
1086	055	-1552 Jun 19	07:45:20	36196	-43927	41	P	t-	1.3403	0.3798	63.9N	63.4E	0	28		
1087	055	-1552 Nov 14	15:50:05	36187	-43922	8	A	-t	0.9631	0.9815	51.5N	130.7E	15	220	245	01m27s
1088	055	-1551 May 10	02:15:08	36177	-43916	13	T	-p	-0.6318	1.0240	22.4S	47.5W	51	333	103	02m12s
1089	055	-1551 Nov 03	22:31:46	36167	-43910	18	A	-n	0.2961	0.9401	4.3N	1.8W	73	206	231	06m50s
1090	055	-1550 Apr 29	16:54:59	36156	-43904	23	T	n-	0.1468	1.0731	17.0N	71.9E	81	152	240	05m54s
1091	055	-1550 Oct 23	22:41:00	36146	-43898	28	A	p-	-0.3911	0.9164	26.9S	23.8W	67	32	342	09m09s
1092	055	-1549 Apr 19	10:10:16	36135	-43892	33	T	p-	0.8833	1.0653	55.0N	130.4E	28	124	449	04m00s
1093	055	-1549 Oct 12	22:37:05	36125	-43886	38	P	t-	-1.0587	0.8589	60.5S	104.1W	0	95		
1094	055	-1548 Mar 09	14:50:24	36116	-43881	5	A	-t	-0.9874	0.9813	62.9S	156.9W	8	265	462	01m08s
1095	055	-1548 Sep 01	17:16:45	36106	-43875	10	P	-t	1.0539	0.9018	61.4N	172.2W	0	299		
1096	055	-1547 Feb 26	19:42:22	36095	-43869	15	A	-n	-0.2636	0.9456	27.4S	46.9E	75	332	207	05m49s
1097	055	-1547 Aug 22	08:51:29	36085	-43863	20	T	-n	0.3316	1.0622	34.6N	152.7W	70	207	216	04m46s
1098	055	-1546 Feb 15	19:41:55	36075	-43857	25	A	p-	0.4660	0.9340	8.8N	29.3E	62	158	275	08m17s
1099	055	-1546 Aug 12	01:00:17	36064	-43851	30	T	n-	-0.3814	1.0447	1.7S	51.5W	68	18	160	04m13s
1100	055	-1545 Feb 04	22:09:58	36054	-43845	35	P	t-	1.1621	0.6891	62.7N	42.8W	0	138		
1101	056	-1545 Aug 01	13:05:55	36043	-43839	40	P	t-	-1.1630	0.6926	63.5S	97.1E	0	34		
1102	056	-1545 Dec 26	20:45:32	36035	-43834	7	T	-p	-0.9094	1.0265	88.0S	94.2E	24	285	220	01m28s
1103	056	-1544 Jun 21	02:59:53	36024	-43828	12	A	-t	0.8846	0.9399	85.4N	89.2W	27	167	485	04m19s
1104	056	-1544 Dec 15	12:26:09	36014	-43822	17	T	-n	-0.2338	1.0457	36.3S	142.5E	76	3	157	03m51s
1105	056	-1543 Jun 10	03:57:24	36004	-43816	22	Am	nn	0.1152	0.9640	28.0N	92.6W	83	174	131	04m16s
1106	056	-1543 Dec 05	02:41:32	35993	-43810	27	H	n-	0.4537	1.0060	6.2N	68.7W	63	186	23	00m39s
1107	056	-1542 May 30	10:30:04	35983	-43804	32	H	p-	-0.6579	1.0109	21.9S	176.3E	49	351	50	01m09s
1108	056	-1542 Nov 24	11:09:45	35972	-43798	37	P	t-	1.1960	0.6272	69.3N	168.5W	0	209		
1109	056	-1541 Apr 20	17:13:51	35964	-43793	4	P	-t	1.1403	0.7517	71.4N	37.0W	0	71		
1110	056	-1541 May 20	00:03:34	35962	-43792	42	P	t-	-1.3684	0.3112	69.8S	5.2E	0	323		
1111	056	-1541 Oct 14	17:00:21	35953	-43787	9	P	-t	-1.1543	0.6921	71.7S	26.0W	0	100		
1112	056	-1540 Apr 09	10:19:32	35943	-43781	14	T	-p	0.4183	1.0638	25.7N	167.7E	65	160	229	05m25s
1113	056	-1540 Oct 02	18:22:02	35933	-43775	19	A	-n	-0.4375	0.9538	23.2S	44.8E	64	20	186	05m10s
1114	056	-1539 Mar 30	00:01:51	35922	-43769	24	H	n-	-0.3438	1.0122	21.2S	21.9W	70	341	45	01m13s
1115	056	-1539 Sep 22	02:56:41	35912	-43763	29	H	n-	0.2873	1.0127	21.9N	69.4W	73	199	45	01m13s
1116	056	-1538 Mar 19	06:45:45	35901	-43757	34	P	t-	-1.1456	0.7152	71.4S	19.5W	0	250		
1117	056	-1538 Sep 11	17:29:52	35891	-43751	39	T	t-	0.9613	1.0409	73.4N	137.0E	15	253	514	02m18s
1118	056	-1537 Feb 06	14:54:09	35882	-43746	6	P	-t	1.0448	0.8815	68.8N	87.0E	0	158		
1119	056	-1537 Aug 03	00:14:13	35872	-43740	11	T	-t	-0.9713	1.0216	55.1S	42.3W	13	8	322	01m40s
1120	056	-1536 Jan 26	19:04:59	35862	-43734	16	A	-n	0.2895	0.9809	5.1S	46.1E	73	176	71	02m18s
1121	057	-1536 Jul 22	10:35:03	35851	-43728	21	A	-n	-0.2477	0.9820	8.8N	170.2E	76	1	66	02m13s
1122	057	-1535 Jan 15	06:22:23	35841	-43722	26	T	p-	-0.4303	1.0308	49.0S	123.2W	64	0	115	02m27s
1123	057	-1535 Jul 11	13:52:31	35831	-43716	31	A	p-	0.5194	0.9449	55.3N	116.6E	58	174	238	05m19s
1124	057	-1534 Jan 04	21:54:23	35820	-43710	36	P	t-	-1.0931	0.8353	65.7S	165.8W	0	169		
1125	057	-1534 Jun 01	00:38:21	35811	-43705	3	Pe	-t	-1.5387	0.0279	62.7S	4.3E	0	316		
1126	057	-1534 Jun 30	14:17:09	35810	-43704	41	P	t-	1.2601	0.5177	64.9N	46.0W	0	19		
1127	057	-1534 Nov 26	00:27:19	35801	-43699	8	A	-t	0.9654	0.9786	51.1N	5.9W	15	213	295	01m44s
1128	057	-1533 May 21	09:16:51	35791	-43693	13	T	-p	-0.7085	1.0266	25.3S	154.5W	45	335	125	02m27s
1129	057	-1533 Nov 15	06:45:08	35780	-43687	18	A	-n	0.2987	0.9374	0.9N	127.4W	73	204	243	07m26s
1130	057	-1532 May 10	00:16:30	35770	-43681	23	T	nn	0.0711	1.0758	17.0N	38.7W	86	154	246	06m12s
1131	057	-1532 Nov 03	06:39:37	35760	-43675	28	A	p-	-0.3854	0.9154	31.1S	145.0W	67	31	346	09m11s
1132	057	-1531 Apr 29	17:35:10	35749	-43669	33	T	p-	0.8123	1.0672	54.2N	24.5E	35	129	371	04m13s
1133	057	-1531 Oct 23	06:44:07	35739	-43663	38	P	t-	-1.0493	0.8751	60.6S	123.9E	0	104		
1134	057	-1530 Mar 20	22:07:07	35730	-43658	5	P	-t	-1.0501	0.8949	60.6S	100.3E	0	258		
1135	057	-1530 Sep 13	01:30:42	35720	-43652	10	P	-t	1.0771	0.8587	61.0N	53.9E	0	290		
1136	057	-1529 Mar 10	02:41:53	35710	-43646	15	A	-p	-0.3224	0.9466	26.4S	58.0W	71	330	206	05m38s
1137	057	-1529 Sep 02	17:05:29	35699	-43640	20	T	-n	0.3648	1.0588	32.5N	83.8E	68	210	207	04m29s
1138	057	-1528 Feb 27	02:45:44	35689	-43634	25	A	p-	0.4139	0.9385	8.6N	78.4W	65	155	248	07m24s
1139	057	-1528 Aug 22	08:57:37	35679	-43628	30	T	n-	-0.3414	1.0390	1.8S	172.6W	70	22	139	03m37s
1140	057	-1527 Feb 15	05:44:59	35668	-43622	35	P	t-	1.1210	0.7626	62.0N	167.3W	0	12		

Cat Num	Canon Plate	Calendar Date	TD of Greatest Eclipse	ΔT s	Luna Num	Saros Num	Ecl. Type	QLE	Gamma	Ecl. Mag.	Lat. °	Long. °	Sun Alt °	Sun Azm °	Path Width km	Central Line Dur.
1141	058	-1527 Aug 11	20:32:13	35658	-43616	40	P	t-	-1.1170	0.7734	62.6S	25.5W	0	43		
1142	058	-1526 Jan 06	05:22:05	35649	-43611	7	T	-p	-0.9184	1.0302	83.6S	12.6W	23	260	263	01m39s
1143	058	-1526 Jul 02	09:27:03	35639	-43605	12	A	-t	0.9646	0.9346	80.3N	14.8W	15	341	971	04m11s
1144	058	-1526 Dec 26	21:12:07	35629	-43599	17	T	-n	-0.2397	1.0466	37.7S	12.1E	76	357	160	03m54s
1145	058	-1525 Jun 21	10:29:50	35618	-43593	22	A	nn	0.1975	0.9649	34.5N	168.5E	78	179	129	03m58s
1146	058	-1525 Dec 16	11:19:44	35608	-43587	27	H	n-	0.4531	1.0044	4.4N	159.2E	63	182	17	00m29s
1147	058	-1524 Jun 09	17:28:51	35598	-43581	32	T	p-	-0.5781	1.0148	14.0S	66.8E	55	355	62	01m37s
1148	058	-1524 Dec 04	19:27:19	35587	-43575	37	P	t-	1.1975	0.6244	68.3N	53.5E	0	197		
1149	058	-1523 May 01	00:40:11	35579	-43570	4	P	-t	1.2096	0.6174	70.9N	163.8W	0	58		
1150	058	-1523 May 30	07:22:51	35577	-43569	42	P	t-	-1.2928	0.4563	68.9S	118.5W	0	335		
1151	058	-1523 Oct 25	00:55:50	35568	-43564	9	P	-t	-1.1616	0.6799	71.4S	160.6W	0	114		
1152	058	-1522 Apr 20	17:45:50	35558	-43558	14	T	-p	0.4886	1.0625	34.2N	51.5E	61	159	234	05m05s
1153	058	-1522 Oct 14	02:29:48	35548	-43552	19	A	-n	-0.4467	0.9544	28.2S	79.9W	63	20	185	04m55s
1154	058	-1521 Apr 10	07:11:26	35537	-43546	24	H	nn	-0.2758	1.0112	13.1S	133.6W	74	342	40	01m09s
1155	058	-1521 Oct 03	11:22:04	35527	-43540	29	H	n-	0.2732	1.0131	16.6N	161.7E	74	199	47	01m16s
1156	058	-1520 Mar 29	13:32:25	35517	-43534	34	P	t-	-1.0788	0.8307	71.7S	136.6W	0	263		
1157	058	-1520 Sep 22	01:59:24	35506	-43528	39	T	t-	0.9413	1.0400	68.3N	11.9W	19	234	403	02m24s
1158	058	-1519 Feb 16	22:10:22	35498	-43523	6	P	-t	1.0857	0.8139	69.7N	35.9W	0	146		
1159	058	-1519 Aug 13	07:59:47	35487	-43517	11	P	-t	-1.0198	0.9636	68.9S	176.9W	0	24		
1160	058	-1518 Feb 06	02:55:29	35477	-43511	16	A	-n	0.3191	0.9871	1.1S	73.9W	71	172	48	01m31s
1161	059	-1518 Aug 02	17:46:16	35467	-43505	21	A	-p	-0.3067	0.9756	3.7N	60.0E	72	5	92	03m03s
1162	059	-1517 Jan 26	14:43:34	35456	-43499	26	T	n-	-0.4084	1.0364	46.0S	113.6E	66	354	134	02m56s
1163	059	-1517 Jul 22	20:32:30	35446	-43493	31	A	p-	0.4487	0.9418	50.0N	19.7E	63	181	241	05m59s
1164	059	-1516 Jan 16	06:30:12	35436	-43487	36	P	t-	-1.0790	0.8627	66.8S	53.0E	0	180		
1165	059	-1516 Jul 10	20:55:20	35425	-43481	41	P	t-	1.1840	0.6484	65.9N	157.2W	0	9		
1166	059	-1516 Dec 06	09:02:31	35417	-43476	8	A	-t	0.9687	0.9762	51.4N	141.8W	14	206	348	01m59s
1167	059	-1515 May 31	16:22:02	35407	-43470	13	T	-p	-0.7831	1.0283	29.5S	97.2E	38	338	152	02m35s
1168	059	-1515 Nov 25	14:58:10	35396	-43464	18	A	-n	0.3005	0.9353	2.0S	107.3E	73	200	251	08m00s
1169	059	-1514 May 21	07:39:13	35386	-43458	23	T	nn	-0.0042	1.0776	16.5N	149.6W	90	321	251	06m29s
1170	059	-1514 Nov 14	14:39:37	35376	-43452	28	A	p-	-0.3807	0.9151	35.1S	94.0E	67	29	348	09m12s
1171	059	-1513 May 11	00:58:13	35365	-43446	33	T	p-	0.7399	1.0680	53.9N	81.7W	42	135	326	04m23s
1172	059	-1513 Nov 03	14:55:34	35355	-43440	38	P	t-	-1.0424	0.8872	61.0S	9.3W	0	113		
1173	059	-1512 Mar 31	05:13:52	35346	-43435	5	P	-t	-1.1200	0.7695	60.5S	16.7W	0	267		
1174	059	-1512 Apr 29	15:38:06	35345	-43434	43	Pb	t-	1.5386	0.0041	61.2N	16.3W	0	69		
1175	059	-1512 Sep 23	09:54:09	35336	-43429	10	P	-t	1.0935	0.8282	60.7N	82.3W	0	281		
1176	059	-1511 Mar 20	09:30:47	35326	-43423	15	A	-p	-0.3896	0.9473	25.6S	160.2W	67	328	208	05m31s
1177	059	-1511 Sep 13	01:28:05	35316	-43417	20	T	-n	0.3910	1.0555	29.6N	42.6W	67	212	198	04m15s
1178	059	-1510 Mar 09	09:40:57	35305	-43411	25	A	p-	0.3540	0.9430	8.6N	176.4E	69	153	223	06m36s
1179	059	-1510 Sep 02	17:03:29	35295	-43405	30	T	n-	-0.3092	1.0334	3.2S	64.1E	72	25	118	03m02s
1180	059	-1509 Feb 26	13:11:48	35285	-43399	35	P	t-	1.0720	0.8508	61.4N	70.4E	0	120		
1181	060	-1509 Aug 23	04:05:42	35274	-43393	40	P	t-	-1.0774	0.8417	61.9S	149.6W	0	53		
1182	060	-1508 Jan 17	13:52:29	35266	-43388	7	T	-p	-0.9330	1.0340	78.9S	137.2W	21	255	326	01m49s
1183	060	-1508 Jul 12	15:59:43	35256	-43382	12	P	-t	1.0399	0.8917	65.0N	113.2W	0	342		
1184	060	-1507 Jan 06	05:51:53	35245	-43376	17	T	-n	-0.2508	1.0479	38.4S	116.6W	75	352	165	03m57s
1185	060	-1507 Jul 01	17:09:19	35235	-43370	22	A	nn	0.2750	0.9653	39.9N	69.1E	74	184	131	03m43s
1186	060	-1507 Dec 26	19:53:46	35225	-43364	27	H	n-	0.4493	1.0034	3.0N	28.1E	63	178	13	00m22s
1187	060	-1506 Jun 21	00:32:06	35214	-43358	32	T	p-	-0.5007	1.0180	7.2S	42.9W	60	359	71	02m00s
1188	060	-1506 Dec 16	03:42:20	35204	-43352	37	P	t-	1.1975	0.6243	67.3N	83.2W	0	186		
1189	060	-1505 May 12	08:04:36	35196	-43347	4	P	-t	1.2809	0.4789	70.3N	70.4E	0	45		
1190	060	-1505 Jun 10	14:43:50	35194	-43346	42	P	t-	-1.2178	0.6009	68.0S	117.9E	0	346		
1191	060	-1505 Nov 05	08:56:27	35185	-43341	9	P	-t	-1.1646	0.6749	70.8S	63.9E	0	128		
1192	060	-1504 May 01	01:06:51	35175	-43335	14	T	-p	0.5627	1.0604	43.0N	63.5W	56	158	239	04m40s
1193	060	-1504 Oct 24	10:45:01	35165	-43329	19	A	-n	-0.4506	0.9555	33.0S	154.2E	63	20	181	04m38s
1194	060	-1503 Apr 20	14:15:00	35154	-43323	24	H	nn	-0.2038	1.0095	4.9S	116.2E	78	343	33	01m01s
1195	060	-1503 Oct 13	19:53:42	35144	-43317	29	H	n-	0.2637	1.0138	11.6N	31.2E	75	198	49	01m22s
1196	060	-1502 Apr 09	20:13:36	35134	-43311	34	A-	t-	-1.0080	0.9531	71.7S	107.4E	0	277	-	-
1197	060	-1502 Oct 03	10:35:41	35124	-43305	39	T	p-	0.9268	1.0389	63.1N	153.6W	22	224	350	02m28s
1198	060	-1501 Feb 28	05:19:15	35115	-43300	6	P	-t	1.1329	0.7347	70.5N	157.5W	0	133		
1199	060	-1501 Aug 24	15:52:09	35105	-43294	11	P	-t	-1.0619	0.8833	69.9S	51.1E	0	36		
1200	060	-1500 Feb 17	10:38:07	35095	-43288	16	A	-n	0.3562	0.9934	3.9N	167.5E	69	169	25	00m45s

Cat Num	Canon Plate	Calendar Date	TD of Greatest Eclipse	ΔT s	Luna Num	Saros Num	Ecl. Type	QLE	Gamma	Ecl. Mag.	Lat. °	Long. °	Sun Alt °	Sun Azm °	Path Width km	Central Line Dur.
1201	061	-1500 Aug 13	01:05:13	35084	-43282	21	A	-p	-0.3582	0.9692	1.6S	52.6W	69	8	119	03m52s
1202	061	-1499 Feb 05	22:55:25	35074	-43276	26	T	n-	-0.3784	1.0421	41.7S	8.7W	68	350	152	03m28s
1203	061	-1499 Aug 02	03:22:20	35064	-43270	31	A	p-	0.3864	0.9386	44.4N	81.5W	67	186	248	06m42s
1204	061	-1498 Jan 26	14:57:28	35053	-43264	36	P	t-	-1.0584	0.9028	67.8S	86.5W	0	191		
1205	061	-1498 Jul 22	03:42:06	35043	-43258	41	P	t-	1.1138	0.7687	66.9N	88.9E	0	359		
1206	061	-1498 Dec 17	17:33:27	35035	-43253	8	A	-t	0.9748	0.9742	52.8N	83.4E	12	200	429	02m11s
1207	061	-1497 Jun 11	23:29:47	35024	-43247	13	T	-p	-0.8561	1.0290	35.4S	12.2W	31	342	189	02m35s
1208	061	-1497 Dec 06	23:08:53	35014	-43241	18	A	-n	0.3041	0.9340	4.1S	17.3W	72	196	258	08m30s
1209	061	-1496 May 31	15:00:53	35004	-43235	23	Tm	nn	-0.0807	1.0785	15.2N	99.7E	85	340	254	06m44s
1210	061	-1496 Nov 24	22:41:35	34994	-43229	28	A	p-	-0.3774	0.9154	38.8S	26.8W	68	26	346	09m09s
1211	061	-1495 May 21	08:18:46	34983	-43223	33	T	p-	0.6657	1.0676	53.6N	172.9E	48	141	294	04m29s
1212	061	-1495 Nov 13	23:10:56	34973	-43217	38	P	t-	-1.0379	0.8956	61.5S	143.5W	0	122		
1213	061	-1494 Apr 11	12:13:12	34965	-43212	5	P	-t	-1.1948	0.6354	60.6S	131.8W	0	276		
1214	061	-1494 May 10	22:40:29	34963	-43211	43	P	t-	1.4688	0.1342	61.6N	132.4W	0	61		
1215	061	-1494 Oct 04	18:25:40	34954	-43206	10	P	-t	1.1040	0.8087	60.6N	139.5E	0	272		
1216	061	-1493 Mar 31	16:11:09	34944	-43200	15	A	-p	-0.4632	0.9478	25.3S	99.6E	62	327	213	05m27s
1217	061	-1493 Sep 24	09:58:07	34934	-43194	20	T	-n	0.4111	1.0522	26.2N	171.5W	66	212	188	04m03s
1218	061	-1492 Mar 19	16:28:54	34924	-43188	25	A	pn	0.2876	0.9476	8.9N	73.4E	73	151	200	05m54s
1219	061	-1492 Sep 13	01:16:24	34913	-43182	30	T	n-	-0.2831	1.0277	5.4S	61.0W	74	27	97	02m29s
1220	061	-1491 Mar 08	20:32:07	34903	-43176	35	P	t-	1.0164	0.9520	60.9N	50.1W	0	111		
1221	062	-1491 Sep 02	11:46:56	34893	-43170	40	P	t-	-1.0446	0.8972	61.4S	84.5E	0	62		
1222	062	-1490 Jan 27	22:16:48	34884	-43165	7	T	-p	-0.9536	1.0375	74.1S	99.0E	17	251	431	01m58s
1223	062	-1490 Jul 23	22:39:19	34874	-43159	12	P	-t	1.1094	0.7729	64.0N	135.5E	0	333		
1224	062	-1489 Jan 17	14:25:23	34864	-43153	17	T	-n	-0.2673	1.0493	38.4S	116.4E	74	346	170	04m00s
1225	062	-1489 Jul 12	23:56:21	34854	-43147	22	A	-n	0.3473	0.9653	44.1N	30.9W	69	190	134	03m31s
1226	062	-1488 Jan 07	04:19:45	34843	-43141	27	H	n-	0.4395	1.0028	2.1N	100.8W	64	173	11	00m18s
1227	062	-1488 Jul 01	07:42:06	34833	-43135	32	T	p-	-0.4278	1.0204	1.5S	153.4W	65	3	77	02m15s
1228	062	-1488 Dec 26	11:50:17	34823	-43129	37	P	t-	1.1921	0.6335	66.2N	142.4E	0	174		
1229	062	-1487 May 22	15:29:11	34815	-43124	4	P	-t	1.3520	0.3405	69.5N	54.9W	0	33		
1230	062	-1487 Jun 20	22:09:54	34813	-43123	42	P	t-	-1.1463	0.7387	67.0S	6.4W	0	357		
1231	062	-1487 Nov 15	16:58:34	34804	-43118	9	P	-t	-1.1669	0.6713	70.1S	71.4W	0	141		
1232	062	-1486 May 12	08:25:29	34794	-43112	14	T	-p	0.6384	1.0573	52.2N	177.8W	50	158	245	04m10s
1233	062	-1486 Nov 04	19:05:27	34784	-43106	19	A	-n	-0.4511	0.9572	37.5S	27.6E	63	19	174	04m19s
1234	062	-1485 May 01	21:11:15	34774	-43100	24	Hm	nn	-0.1265	1.0072	3.4N	8.0E	83	344	25	00m47s
1235	062	-1485 Oct 25	04:32:54	34763	-43094	29	H	n-	0.2597	1.0149	7.0N	101.0W	75	197	53	01m29s
1236	062	-1484 Apr 20	02:45:50	34753	-43088	34	A	t-	-0.9299	0.9493	57.7S	45.3W	21	326	506	04m24s
1237	062	-1484 Oct 13	19:19:27	34743	-43082	39	T	p-	0.9184	1.0376	58.6N	66.5E	23	216	320	02m30s
1238	062	-1483 Mar 10	12:19:02	34735	-43077	6	P	-t	1.1879	0.6414	71.1N	82.6E	0	120		
1239	062	-1483 Sep 03	23:53:00	34724	-43071	11	P	-t	-1.0963	0.8179	70.6S	83.7W	0	49		
1240	062	-1482 Feb 27	18:14:42	34714	-43065	16	A	-n	0.3992	0.9997	9.7N	49.8E	66	166	1	00m02s
1241	063	-1482 Aug 24	08:30:20	34704	-43059	21	A	-p	-0.4036	0.9628	7.3S	167.1W	66	12	147	04m37s
1242	063	-1481 Feb 17	07:01:44	34694	-43053	26	T	n-	-0.3430	1.0480	36.3S	131.0W	70	346	170	04m01s
1243	063	-1481 Aug 13	10:18:57	34683	-43047	31	A	p-	0.3298	0.9351	38.4N	174.1E	71	190	256	07m27s
1244	063	-1480 Feb 06	23:17:12	34673	-43041	36	P	t-	-1.0317	0.9547	68.8S	135.3E	0	203		
1245	063	-1480 Aug 01	10:37:55	34663	-43035	41	P	t-	1.0501	0.8778	67.9N	27.7W	0	348		
1246	063	-1480 Dec 28	01:59:17	34655	-43030	8	A	-t	0.9844	0.9722	56.1N	50.0W	9	193	611	02m20s
1247	063	-1479 Jun 22	06:43:32	34644	-43024	13	T	-p	-0.9253	1.0285	43.6S	123.6W	22	345	256	02m24s
1248	063	-1479 Dec 17	07:14:46	34634	-43018	18	A	-n	0.3109	0.9332	5.2S	140.5W	72	192	262	08m55s
1249	063	-1478 Jun 11	22:26:44	34624	-43012	23	T	nn	-0.1541	1.0785	13.2N	12.4W	81	345	257	06m57s
1250	063	-1478 Dec 06	06:41:47	34614	-43006	28	A	p-	-0.3725	0.9165	41.8S	146.1W	68	22	341	09m04s
1251	063	-1477 Jun 01	15:38:49	34604	-43000	33	T	p-	0.5914	1.0662	52.9N	67.8E	53	149	267	04m33s
1252	063	-1477 Nov 25	07:27:52	34593	-42994	38	P	t-	-1.0340	0.9033	62.2S	81.7E	0	132		
1253	063	-1476 Apr 21	19:05:07	34585	-42989	5	P	-t	-1.2741	0.4936	60.8S	114.9E	0	284		
1254	063	-1476 May 21	05:39:55	34583	-42988	43	P	t-	1.3975	0.2665	62.1N	112.1E	0	52		
1255	063	-1476 Oct 15	03:04:02	34575	-42983	10	P	-t	1.1097	0.7982	60.7N	0.3W	0	262		
1256	063	-1475 Apr 10	22:43:34	34565	-42977	15	A	-p	-0.5427	0.9478	25.7S	1.3E	57	327	223	05m30s
1257	063	-1475 Oct 04	18:35:57	34554	-42971	20	T	-n	0.4246	1.0492	22.4N	57.0E	65	212	179	03m53s
1258	063	-1474 Mar 30	23:10:04	34544	-42965	25	A	nn	0.2149	0.9519	9.3N	27.6W	78	150	180	05m17s
1259	063	-1474 Sep 24	09:36:28	34534	-42959	30	T	n-	-0.2638	1.0222	8.5S	171.9E	75	29	78	01m58s
1260	063	-1473 Mar 20	03:45:39	34524	-42953	35	A	t-	0.9541	0.9823	53.6N	140.1W	17	126	207	01m19s

Cat Num	Canon Plate	Calendar Date	TD of Greatest Eclipse	ΔT s	Luna Num	Saros Num	Ecl. Type	QLE	Gamma	Ecl. Mag.	Lat. °	Long. °	Sun Alt °	Sun Azm °	Path Width km	Central Line Dur.
1261	064	-1473 Sep 13	19:36:05	34514	-42947	40	P	t-	-1.0190	0.9395	60.9S	43.1W	0	71		
1262	064	-1472 Feb 08	06:32:03	34505	-42942	7	T	-t	-0.9820	1.0401	68.6S	17.3W	10	243	771	02m01s
1263	064	-1472 Aug 03	05:27:42	34495	-42936	12	P	-t	1.1715	0.6673	63.2N	22.3E	0	323		
1264	064	-1471 Jan 27	22:50:49	34485	-42930	17	T	-n	-0.2904	1.0508	37.7S	8.7W	73	341	176	04m04s
1265	064	-1471 Jul 23	06:52:23	34475	-42924	22	A	-p	0.4129	0.9650	46.8N	132.1W	65	197	139	03m23s
1266	064	-1470 Jan 17	12:38:40	34464	-42918	27	H	n-	0.4246	1.0025	1.6N	132.2E	65	169	9	00m16s
1267	064	-1470 Jul 12	14:59:27	34454	-42912	32	T	p-	-0.3600	1.0221	2.9N	95.0E	69	7	81	02m24s
1268	064	-1469 Jan 06	19:52:36	34444	-42906	37	P	t-	1.1831	0.6489	65.1N	9.9E	0	164		
1269	064	-1469 Jun 02	22:52:59	34436	-42901	4	P	-t	1.4237	0.2010	68.6N	179.4W	0	21		
1270	064	-1469 Jul 02	05:39:07	34434	-42900	42	P	t-	-1.0770	0.8722	66.0S	131.0W	0	7		
1271	064	-1469 Nov 27	01:03:08	34425	-42895	9	P	-t	-1.1673	0.6708	69.1S	153.4E	0	154		
1272	064	-1468 May 22	15:41:36	34415	-42889	14	T	-p	0.7155	1.0533	61.6N	68.7E	44	157	253	03m37s
1273	064	-1468 Nov 15	03:29:14	34405	-42883	19	A	-n	-0.4497	0.9596	41.5S	98.9W	63	16	164	03m59s
1274	064	-1467 May 12	04:04:18	34395	-42877	24	H	nn	-0.0476	1.0043	11.4N	99.0W	87	346	15	00m29s
1275	064	-1467 Nov 04	13:16:02	34385	-42871	29	H2	n-	0.2581	1.0164	2.9N	125.8E	75	195	58	01m39s
1276	064	-1466 May 01	09:14:05	34375	-42865	34	A	p-	-0.8490	0.9511	45.4S	156.9W	32	337	336	04m54s
1277	064	-1466 Oct 25	04:07:02	34365	-42859	39	T	p-	0.9133	1.0365	54.7N	73.0W	24	210	302	02m33s
1278	064	-1465 Mar 21	19:13:02	34356	-42854	6	P	-t	1.2478	0.5385	71.5N	36.2W	0	107		
1279	064	-1465 Sep 15	08:00:34	34346	-42848	11	P	-t	-1.1241	0.7653	71.2S	139.4E	0	63		
1280	064	-1464 Mar 10	01:42:32	34336	-42842	16	H	-p	0.4502	1.0058	16.5N	66.2W	63	163	22	00m36s
1281	065	-1464 Sep 03	16:04:01	34326	-42836	21	A	-p	-0.4408	0.9566	12.9S	76.0E	64	15	175	05m17s
1282	065	-1463 Feb 27	14:59:01	34315	-42830	26	T	n-	-0.2999	1.0537	30.1S	107.8E	72	344	186	04m35s
1283	065	-1463 Aug 23	17:25:29	34305	-42824	31	A	nn	0.2817	0.9318	32.3N	66.3E	73	194	266	08m10s
1284	065	-1462 Feb 17	07:27:51	34295	-42818	36	T-	t-	-0.9981	1.0200	69.7S	1.2W	0	215	-	-
1285	065	-1462 Aug 12	17:44:37	34285	-42812	41	An	t-	0.9943	0.9375	72.8N	153.1W	4	331	-	03m45s
1286	065	-1461 Jan 08	10:17:29	34277	-42807	8	A+	-t	0.9994	0.9805	66.1N	178.6E	0	187	-	-
1287	065	-1461 Jul 03	14:01:38	34266	-42801	13	Ts	-t	-0.9914	1.0252	59.2S	124.8E	6	347	-	01m52s
1288	065	-1461 Dec 28	15:15:29	34256	-42795	18	A	-n	0.3213	0.9331	5.2S	97.6E	71	188	263	09m12s
1289	065	-1460 Jun 22	05:54:44	34246	-42789	23	T	-n	-0.2256	1.0776	10.4N	125.4W	77	349	257	07m04s
1290	065	-1460 Dec 16	14:39:12	34236	-42783	28	A	p-	-0.3653	0.9183	43.7S	96.1E	68	17	333	08m56s
1291	065	-1459 Jun 11	22:59:34	34226	-42777	33	T	p-	0.5183	1.0638	51.5N	37.8W	59	156	244	04m34s
1292	065	-1459 Dec 05	15:45:47	34216	-42771	38	P	t-	-1.0304	0.9106	63.0S	53.5W	0	142		
1293	065	-1458 May 03	01:51:20	34207	-42766	5	P	-t	-1.3569	0.3463	61.1S	3.0E	0	293		
1294	065	-1458 Jun 01	12:36:59	34206	-42765	43	P	t-	1.3249	0.4003	62.7N	2.9W	0	43		
1295	065	-1458 Oct 26	11:47:51	34197	-42760	10	P	-t	1.1115	0.7949	60.9N	141.5W	0	253		
1296	065	-1457 Apr 22	05:10:13	34187	-42754	15	A	-p	-0.6262	0.9475	27.0S	95.6W	51	328	241	05m36s
1297	065	-1457 Oct 16	03:19:31	34177	-42748	20	T	-n	0.4335	1.0465	18.6N	76.3W	64	211	170	03m46s
1298	065	-1456 Apr 10	05:45:20	34167	-42742	25	A	nn	0.1365	0.9560	9.5N	126.9W	82	150	162	04m47s
1299	065	-1456 Oct 04	18:02:52	34157	-42736	30	H3	n-	-0.2502	1.0169	12.2S	43.3E	75	30	60	01m30s
1300	065	-1455 Mar 30	10:54:58	34147	-42730	35	A	t-	0.8869	0.9909	49.7N	118.4E	27	131	67	00m42s
1301	066	-1455 Sep 24	03:31:13	34136	-42724	40	A-	p-	-0.9989	0.9719	60.6S	172.2W	0	80	-	-
1302	066	-1454 Feb 18	14:40:46	34128	-42719	7	P	-t	-1.0164	0.9845	61.9S	130.7W	0	234		
1303	066	-1454 Aug 14	12:25:20	34118	-42713	12	P	-t	1.2259	0.5752	62.4N	93.0W	0	314		
1304	066	-1453 Feb 08	07:07:43	34108	-42707	17	T	-n	-0.3204	1.0523	36.5S	132.0W	71	336	183	04m07s
1305	066	-1453 Aug 03	13:57:48	34098	-42701	22	A	-p	0.4715	0.9644	48.0N	124.7E	62	203	145	03m18s
1306	066	-1452 Jan 28	20:47:42	34087	-42695	27	H	n-	0.4021	1.0025	1.5N	7.8E	66	165	9	00m16s
1307	066	-1452 Jul 22	22:25:55	34077	-42689	32	T	n-	-0.2987	1.0233	6.0N	18.4W	73	12	83	02m27s
1308	066	-1451 Jan 17	03:43:57	34067	-42683	37	P	t-	1.1656	0.6789	64.1N	119.4W	0	154		
1309	066	-1451 Jun 13	06:19:54	34059	-42678	4	Pe	-t	1.4929	0.0667	67.7N	55.8E	0	10		
1310	066	-1451 Jul 12	13:15:57	34057	-42677	42	P	t-	-1.0135	0.9944	65.0S	102.9E	0	17		
1311	066	-1451 Dec 07	09:06:01	34049	-42672	9	P	-t	-1.1698	0.6671	68.1S	19.2E	0	166		
1312	066	-1450 Jun 02	22:56:37	34039	-42666	14	T	-p	0.7928	1.0482	71.4N	44.6W	37	157	265	03m03s
1313	066	-1450 Nov 26	11:54:35	34029	-42660	19	A	-n	-0.4479	0.9626	45.0S	135.2E	63	13	151	03m36s
1314	066	-1449 May 23	10:52:54	34018	-42654	24	H	nn	0.0344	1.0008	19.3N	155.7E	88	169	3	00m05s
1315	066	-1449 Nov 15	22:02:29	34008	-42648	29	T	n-	0.2591	1.0185	0.8S	8.0W	75	192	65	01m53s
1316	066	-1448 May 11	15:36:15	33998	-42642	34	A	p-	-0.7633	0.9518	34.5S	98.5E	40	344	271	05m26s
1317	066	-1448 Nov 04	12:59:34	33988	-42636	39	T	p-	0.9124	1.0357	51.5N	147.1E	24	204	294	02m36s
1318	066	-1447 Apr 01	02:00:26	33980	-42631	6	P	-t	1.3133	0.4249	71.7N	153.6W	0	93		
1319	066	-1447 Apr 30	16:14:13	33978	-42630	44	Pb	t-	-1.5170	0.0757	71.0S	135.5E	0	303		
1320	066	-1447 Sep 25	16:14:53	33970	-42625	11	P	-t	-1.1457	0.7247	71.6S	0.2E	0	76		

Cat Num	Canon Plate	Calendar Date	TD of Greatest Eclipse	ΔT s	Luna Num	Saros Num	Ecl. Type	QLE	Gamma	Ecl. Mag.	Lat. °	Long. °	Sun Alt °	Sun Azm °	Path Width km	Central Line Dur.
1321	067	-1446 Mar 21	09:05:39	33959	-42619	16	H	-p	0.5060	1.0117	23.9N	178.5E	59	161	46	01m09s
1322	067	-1446 Sep 14	23:45:00	33949	-42613	21	A	-p	-0.4711	0.9508	18.5S	42.9W	62	17	203	05m50s
1323	067	-1445 Mar 10	22:49:45	33939	-42607	26	T	n-	-0.2507	1.0593	23.2S	12.6W	75	342	201	05m10s
1324	067	-1445 Sep 04	00:40:23	33929	-42601	31	A	nn	0.2404	0.9286	26.2N	44.3W	76	196	276	08m51s
1325	067	-1444 Feb 28	15:30:44	33919	-42595	36	T	t-	-0.9581	1.0513	75.5S	169.1E	16	280	614	02m48s
1326	067	-1444 Aug 23	01:01:34	33909	-42589	41	A	t-	0.9460	0.9403	78.4N	24.5E	18	261	701	04m06s
1327	067	-1443 Jan 18	18:27:07	33901	-42584	8	P	-t	1.0207	0.9425	67.1N	43.9E	0	176		
1328	067	-1443 Jul 13	21:28:21	33891	-42578	13	P	-t	-1.0510	0.9110	66.2S	4.6E	0	354		
1329	067	-1442 Jan 07	23:08:57	33880	-42572	18	A	-p	0.3369	0.9334	4.1S	22.6W	70	183	264	09m20s
1330	067	-1442 Jul 03	13:27:32	33870	-42566	23	T	-n	-0.2933	1.0760	6.9N	119.9E	73	353	257	07m05s
1331	067	-1442 Dec 27	22:31:42	33860	-42560	28	A	p-	-0.3539	0.9207	44.4S	19.9W	69	11	321	08m46s
1332	067	-1441 Jun 23	06:22:32	33850	-42554	33	T	p-	0.4478	1.0606	49.3N	144.5W	63	164	223	04m32s
1333	067	-1441 Dec 17	00:00:13	33840	-42548	38	P	t-	-1.0231	0.9247	63.9S	171.8E	0	152		
1334	067	-1440 May 13	08:32:40	33832	-42543	5	P	-t	-1.4422	0.1952	61.6S	107.8W	0	302		
1335	067	-1440 Jun 11	19:34:14	33830	-42542	43	P	t-	1.2530	0.5316	63.5N	118.1W	0	34		
1336	067	-1440 Nov 05	20:35:39	33822	-42537	10	P	-t	1.1108	0.7964	61.3N	76.2E	0	244		
1337	067	-1439 May 02	11:32:15	33812	-42531	15	A	-p	-0.7131	0.9465	29.6S	168.6E	44	329	273	05m46s
1338	067	-1439 Oct 26	12:07:11	33801	-42525	20	T	-n	0.4389	1.0443	14.8N	149.2E	64	209	163	03m41s
1339	067	-1438 Apr 21	12:16:55	33791	-42519	25	A	nn	0.0544	0.9598	9.6N	134.9E	87	152	146	04m24s
1340	067	-1438 Oct 16	02:35:27	33781	-42513	30	H	n-	-0.2425	1.0121	16.3S	86.9W	76	30	43	01m04s
1341	068	-1437 Apr 10	17:57:46	33771	-42507	35	A	p-	0.8132	0.9987	47.6N	16.1E	35	134	8	00m06s
1342	068	-1437 Oct 05	11:34:09	33761	-42501	40	As	p-	-0.9860	0.9460	59.4S	74.1E	9	74	-	03m49s
1343	068	-1436 Feb 29	22:40:19	33753	-42496	7	P	-t	-1.0586	0.9059	61.3S	98.9E	0	243		
1344	068	-1436 Mar 30	06:41:36	33751	-42495	45	Pb	t-	1.5035	0.0569	60.7N	135.2E	0	93		
1345	068	-1436 Aug 24	19:33:14	33743	-42490	12	P	-t	1.2719	0.4976	61.7N	149.5E	0	305		
1346	068	-1435 Feb 18	15:15:25	33733	-42484	17	T	-n	-0.3579	1.0536	35.0S	106.7E	69	332	189	04m10s
1347	068	-1435 Aug 13	21:14:14	33723	-42478	22	A	-p	0.5219	0.9637	47.8N	18.5E	58	209	153	03m16s
1348	068	-1434 Feb 08	04:48:24	33713	-42472	27	H	n-	0.3736	1.0026	1.8N	114.4W	68	161	10	00m16s
1349	068	-1434 Aug 03	05:59:35	33703	-42466	32	T	n-	-0.2425	1.0240	7.8N	133.1W	76	16	84	02m26s
1350	068	-1433 Jan 28	11:27:55	33692	-42460	37	P	t-	1.1428	0.7181	63.2N	113.5E	0	144		
1351	068	-1433 Jul 23	20:58:00	33682	-42454	42	T	t-	-0.9538	1.0553	48.3S	12.4W	17	17	615	04m08s
1352	068	-1433 Dec 18	17:06:34	33674	-42449	9	P	-t	-1.1742	0.6603	67.0S	113.7W	0	178		
1353	068	-1432 Jun 13	06:11:50	33664	-42443	14	T	-t	0.8691	1.0420	81.9N	159.8W	29	154	288	02m28s
1354	068	-1432 Dec 06	20:19:34	33654	-42437	19	A	-n	-0.4473	0.9662	47.8S	10.5E	63	8	136	03m10s
1355	068	-1431 Jun 02	17:41:14	33644	-42431	24	A	nn	0.1153	0.9967	26.7N	51.3E	83	172	12	00m22s
1356	068	-1431 Nov 26	06:48:35	33634	-42425	29	T	n-	0.2592	1.0210	3.9S	141.5W	75	189	74	02m09s
1357	068	-1430 May 22	21:58:08	33624	-42419	34	A	p-	-0.6776	0.9517	24.9S	3.5W	47	348	239	05m58s
1358	068	-1430 Nov 15	21:53:14	33614	-42413	39	T	p-	0.9127	1.0353	48.9N	7.2E	24	199	292	02m39s
1359	068	-1429 Apr 12	08:42:37	33605	-42408	6	P	-t	1.3832	0.3025	71.6N	90.4E	0	80		
1360	068	-1429 May 11	22:34:56	33604	-42407	44	P	t-	-1.4306	0.2237	70.4S	25.6E	0	316		
1361	069	-1429 Oct 07	00:35:25	33595	-42402	11	P	-t	-1.1614	0.6953	71.7S	140.7W	0	91		
1362	069	-1428 Mar 31	16:22:18	33585	-42396	16	T	-p	0.5682	1.0171	31.9N	64.3E	55	159	70	01m34s
1363	069	-1428 Sep 25	07:33:38	33575	-42390	21	A	-p	-0.4941	0.9454	24.1S	163.5W	60	19	230	06m17s
1364	069	-1427 Mar 21	06:32:50	33565	-42384	26	T	n-	-0.1946	1.0643	15.9S	131.8W	79	342	214	05m41s
1365	069	-1427 Sep 14	08:05:58	33555	-42378	31	A	nn	0.2080	0.9256	20.2N	158.0W	78	197	286	09m27s
1366	069	-1426 Mar 10	23:24:56	33545	-42372	36	T	p-	-0.9114	1.0557	68.4S	17.4E	24	310	451	03m20s
1367	069	-1426 Sep 03	08:29:04	33535	-42366	41	A	t-	0.9055	0.9410	71.8N	120.0W	25	232	519	04m24s
1368	069	-1425 Jan 30	02:27:26	33527	-42361	8	P	-t	1.0486	0.8929	68.2N	89.0W	0	165		
1369	069	-1425 Jul 25	05:02:03	33517	-42355	13	P	-t	-1.1054	0.8092	67.2S	121.2W	0	5		
1370	069	-1424 Jan 19	06:52:57	33507	-42349	18	A	-p	0.3595	0.9342	1.8S	140.6W	69	179	263	09m17s
1371	069	-1424 Jul 13	21:05:32	33497	-42343	23	T	-n	-0.3564	1.0736	2.8N	3.4E	69	358	256	06m58s
1372	069	-1423 Jan 07	06:18:43	33487	-42337	28	A	n-	-0.3381	0.9238	43.7S	134.4W	70	4	306	08m33s
1373	069	-1423 Jul 03	13:48:32	33477	-42331	33	T	n-	0.3805	1.0566	46.2N	107.2E	67	171	202	04m27s
1374	069	-1423 Dec 27	08:12:11	33467	-42325	38	A-	t-	-1.0132	0.9438	64.9S	37.4E	0	162	-	-
1375	069	-1422 May 24	15:11:38	33458	-42320	5	Pe	-t	-1.5278	0.0446	62.2S	141.8E	0	311		
1376	069	-1422 Jun 23	02:32:27	33456	-42319	43	P	t-	1.1828	0.6588	64.4N	126.1E	0	25		
1377	069	-1422 Nov 17	05:26:18	33448	-42314	10	P	-t	1.1083	0.8014	61.9N	66.9W	0	234		
1378	069	-1421 May 13	17:50:05	33438	-42308	15	A	-p	-0.8027	0.9448	34.0S	73.9E	36	331	331	05m56s
1379	069	-1421 Nov 06	20:58:56	33428	-42302	20	T	-n	0.4408	1.0425	11.2N	13.6E	64	206	157	03m39s
1380	069	-1420 May 01	18:45:54	33418	-42296	25	A	nn	-0.0302	0.9631	9.3N	37.4E	88	332	134	04m05s

Cat Num	Canon Plate	Calendar Date	TD of Greatest Eclipse	ΔT s	Luna Num	Saros Num	Ecl. Type	QLE	Gamma	Ecl. Mag.	Lat. °	Long. °	Sun Alt °	Sun Azm °	Path Width km	Central Line Dur.
1381	070	-1420 Oct 26	11:11:24	33408	-42290	30	H	n-	-0.2384	1.0077	20.6S	142.3E	76	30	27	00m41s
1382	070	-1419 Apr 21	00:58:56	33398	-42284	35	H	p-	0.7371	1.0055	46.5N	86.5W	42	137	28	00m26s
1383	070	-1419 Oct 15	19:41:51	33388	-42278	40	A	p-	-0.9779	0.9420	60.4S	52.2W	11	78	1058	04m02s
1384	070	-1418 Mar 12	06:33:44	33380	-42273	7	P	-t	-1.1062	0.8159	61.0S	29.9W	0	252		
1385	070	-1418 Apr 10	14:11:01	33378	-42272	45	P	t-	1.4371	0.1817	60.7N	12.5E	0	84		
1386	070	-1418 Sep 05	02:49:50	33370	-42267	12	P	-t	1.3109	0.4323	61.2N	29.9E	0	296		
1387	070	-1417 Mar 01	23:14:30	33360	-42261	17	T	-p	-0.4023	1.0546	33.4S	12.8W	66	329	196	04m12s
1388	070	-1417 Aug 25	04:41:00	33350	-42255	22	A	-p	0.5647	0.9630	46.4N	91.3W	55	214	161	03m16s
1389	070	-1416 Feb 19	12:36:46	33340	-42249	27	H	n-	0.3355	1.0027	2.2N	126.8E	70	157	10	00m16s
1390	070	-1416 Aug 13	13:44:07	33330	-42243	32	T	n-	-0.1945	1.0243	8.3N	109.6E	79	20	84	02m23s
1391	070	-1415 Feb 07	18:59:49	33320	-42237	37	P	t-	1.1105	0.7734	62.4N	10.3W	0	134		
1392	070	-1415 Aug 03	04:48:22	33310	-42231	42	T	t-	-0.9008	1.0560	40.7S	133.0W	25	21	423	04m22s
1393	070	-1415 Dec 29	01:01:31	33301	-42226	9	P	-t	-1.1836	0.6452	65.9S	115.4E	0	188		
1394	070	-1414 Jun 24	13:28:59	33291	-42220	14	T	-t	0.9429	1.0344	85.6N	71.0W	19	355	358	01m52s
1395	070	-1414 Dec 18	04:42:29	33281	-42214	19	A	-n	-0.4490	0.9705	49.9S	112.8W	63	1	119	02m42s
1396	070	-1413 Jun 14	00:27:17	33271	-42208	24	A	nn	0.1970	0.9919	33.5N	51.5W	78	176	29	00m51s
1397	070	-1413 Dec 07	15:35:10	33261	-42202	29	T	n-	0.2593	1.0242	6.3S	85.1E	75	185	85	02m29s
1398	070	-1412 Jun 02	04:17:50	33251	-42196	34	A	p-	-0.5902	0.9509	16.1S	103.7W	54	352	222	06m30s
1399	070	-1412 Nov 26	06:47:37	33241	-42190	39	T	p-	0.9137	1.0353	46.7N	132.8W	24	193	294	02m44s
1400	070	-1411 Apr 22	15:21:18	33233	-42185	6	P	-t	1.4562	0.1736	71.2N	24.6W	0	66		
1401	071	-1411 May 22	04:55:43	33231	-42184	44	P	t-	-1.3429	0.3744	69.6S	83.8W	0	328		
1402	071	-1411 Oct 17	09:01:51	33223	-42179	11	P	-t	-1.1718	0.6758	71.6S	76.9E	0	105		
1403	071	-1410 Apr 11	23:35:20	33213	-42173	16	T	-p	0.6340	1.0220	40.5N	49.6W	50	156	96	01m52s
1404	071	-1410 Oct 06	15:28:22	33203	-42167	21	A	-p	-0.5113	0.9405	29.5S	74.5E	59	21	255	06m39s
1405	071	-1409 Apr 01	14:10:45	33193	-42161	26	T	n-	-0.1337	1.0690	8.2S	110.0E	82	342	227	06m09s
1406	071	-1409 Sep 25	15:39:41	33183	-42155	31	A	nn	0.1825	0.9229	14.4N	86.1E	79	198	296	10m00s
1407	071	-1408 Mar 21	07:11:08	33173	-42149	36	T	p-	-0.8586	1.0592	59.5S	115.4W	30	323	379	03m51s
1408	071	-1408 Sep 13	16:07:15	33163	-42143	41	A	p-	0.8732	0.9414	64.7N	112.0E	29	221	445	04m41s
1409	071	-1407 Feb 09	10:18:04	33155	-42138	8	P	-t	1.0831	0.8315	69.2N	139.9E	0	153		
1410	071	-1407 Aug 04	12:44:32	33145	-42132	13	P	-t	-1.1531	0.7196	68.2S	110.3E	0	16		
1411	071	-1407 Sep 02	23:13:08	33143	-42131	51	Pb	t-	1.5266	0.0355	70.5N	100.0E	0	311		
1412	071	-1406 Jan 29	14:27:44	33135	-42126	18	A	-p	0.3890	0.9352	1.8N	103.3E	67	175	262	09m06s
1413	071	-1406 Jul 25	04:50:26	33125	-42120	23	T	-n	-0.4136	1.0707	1.8S	115.5W	66	2	252	06m42s
1414	071	-1405 Jan 18	13:57:34	33115	-42114	28	A	n-	-0.3152	0.9274	41.4S	112.6E	71	358	288	08m17s
1415	071	-1405 Jul 14	21:18:35	33105	-42108	33	T	n-	0.3172	1.0519	42.3N	3.4W	71	177	182	04m17s
1416	071	-1404 Jan 07	16:17:53	33095	-42102	38	A-	t-	-0.9974	0.9734	66.0S	95.8W	0	172	-	-
1417	071	-1404 Jul 03	09:33:53	33085	-42096	43	P	t-	1.1160	0.7784	65.3N	9.3E	0	15		
1418	071	-1404 Nov 27	14:16:25	33077	-42091	10	P	-t	1.1066	0.8051	62.6N	149.9E	0	225		
1419	071	-1403 May 24	00:07:06	33067	-42085	15	A	-t	-0.8922	0.9421	41.0S	20.1W	27	332	463	06m02s
1420	071	-1403 Nov 17	05:52:19	33057	-42079	20	T	-n	0.4410	1.0412	8.1N	122.4W	64	203	152	03m40s
1421	072	-1402 May 13	01:12:48	33047	-42073	25	Am	nn	-0.1173	0.9660	8.3N	59.6W	83	335	123	03m51s
1422	072	-1402 Nov 06	19:50:47	33037	-42067	30	H	n-	-0.2378	1.0039	24.9S	10.9E	76	28	14	00m21s
1423	072	-1401 May 02	07:56:36	33027	-42061	35	H	p-	0.6571	1.0118	45.9N	171.9E	49	140	53	00m55s
1424	072	-1401 Oct 27	03:54:45	33017	-42055	40	A	p-	-0.9744	0.9382	62.3S	177.1E	12	85	1052	04m13s
1425	072	-1400 Mar 22	14:18:16	33009	-42050	7	P	-t	-1.1615	0.7103	60.7S	156.3W	0	261		
1426	072	-1400 Apr 20	21:34:37	33007	-42049	45	P	t-	1.3652	0.3183	60.9N	108.7W	0	76		
1427	072	-1400 Sep 15	10:17:19	32999	-42044	12	P	-t	1.3414	0.3816	60.8N	92.3W	0	287		
1428	072	-1399 Mar 12	07:05:11	32989	-42038	17	T	-p	-0.4535	1.0551	31.9S	130.5W	63	327	203	04m14s
1429	072	-1399 Sep 04	12:17:50	32979	-42032	22	A	-p	0.6000	0.9624	44.2N	155.2E	53	217	168	03m18s
1430	072	-1398 Mar 01	20:16:49	32969	-42026	27	H	n-	0.2913	1.0027	3.0N	10.3E	73	154	10	00m16s
1431	072	-1398 Aug 24	21:37:16	32959	-42020	32	T	n-	-0.1528	1.0243	7.7N	9.8W	81	23	83	02m18s
1432	072	-1397 Feb 19	02:22:00	32949	-42014	37	P	t-	1.0711	0.8412	61.7N	131.5W	0	125		
1433	072	-1397 Aug 14	12:45:34	32939	-42008	42	T	p-	-0.8534	1.0554	36.4S	104.7E	31	24	346	04m23s
1434	072	-1396 Jan 09	08:51:43	32931	-42003	9	P	-t	-1.1971	0.6232	64.9S	13.9W	0	199		
1435	072	-1396 Jul 04	20:49:24	32921	-41997	14	P	-t	1.0130	0.9822	65.7N	171.2E	0	349		
1436	072	-1396 Dec 28	13:01:06	32911	-41991	19	A	-n	-0.4552	0.9753	51.0S	125.8E	63	355	99	02m13s
1437	072	-1395 Jun 24	07:16:36	32901	-41985	24	A	-p	0.2749	0.9867	39.5N	153.8W	74	181	49	01m21s
1438	072	-1395 Dec 18	00:18:52	32891	-41979	29	T	n-	0.2562	1.0278	8.2S	47.5W	75	181	97	02m50s
1439	072	-1394 Jun 13	10:39:05	32881	-41973	34	A	p-	-0.5042	0.9497	8.3S	156.8E	60	356	214	06m58s
1440	072	-1394 Dec 07	15:40:19	32871	-41967	39	T	p-	0.9135	1.0359	44.8N	87.7E	24	187	299	02m51s

Cat Num	Canon Plate	Calendar Date	TD of Greatest Eclipse	ΔT s	Luna Num	Saros Num	Ecl. Type	QLE	Gamma	Ecl. Mag.	Lat. °	Long. °	Sun Alt °	Sun Azm °	Path Width km	Central Line Dur.
1441	073	-1393 May 03	21:58:15	32863	-41962	6	Pe	-t	1.5310	0.0404	70.7N	138.7W	0	54		
1442	073	-1393 Jun 02	11:18:15	32861	-41961	44	P	t-	-1.2553	0.5257	68.7S	167.0E	0	339		
1443	073	-1393 Oct 28	17:31:46	32853	-41956	11	P	-t	-1.1784	0.6633	71.2S	66.2W	0	119		
1444	073	-1392 Apr 22	06:44:14	32843	-41950	16	T	-p	0.7040	1.0261	49.6N	163.2W	45	153	124	02m03s
1445	073	-1392 Oct 16	23:29:12	32833	-41944	21	A	-p	-0.5228	0.9363	34.7S	48.4W	58	21	276	06m55s
1446	073	-1391 Apr 11	21:43:28	32823	-41938	26	T	nn	-0.0681	1.0729	0.3S	7.1W	86	342	237	06m31s
1447	073	-1391 Oct 05	23:21:33	32813	-41932	31	A	nn	0.1637	0.9207	8.9N	31.9W	81	198	304	10m27s
1448	073	-1390 Apr 01	14:50:14	32803	-41926	36	T	p-	-0.8002	1.0619	50.4S	119.2E	37	331	337	04m22s
1449	073	-1390 Sep 24	23:55:59	32793	-41920	41	A	p-	0.8486	0.9417	58.2N	13.5W	32	215	405	04m58s
1450	073	-1389 Feb 20	17:56:51	32785	-41915	8	P	-t	1.1266	0.7544	70.1N	11.2E	0	141		
1451	073	-1389 Aug 15	20:35:54	32775	-41909	13	P	-t	-1.1942	0.6423	69.2S	21.0W	0	28		
1452	073	-1389 Sep 14	07:18:31	32774	-41908	51	P	t-	1.4962	0.0905	71.2N	36.4W	0	298		
1453	073	-1388 Feb 09	21:51:55	32765	-41903	18	A	-p	0.4267	0.9365	6.5N	10.7W	65	171	261	08m44s
1454	073	-1388 Aug 04	12:42:32	32755	-41897	23	T	-p	-0.4648	1.0672	6.8S	123.4E	62	6	247	06m19s
1455	073	-1387 Jan 28	21:28:28	32746	-41891	28	A	n-	-0.2853	0.9315	37.8S	0.7E	73	353	268	07m59s
1456	073	-1387 Jul 25	04:54:11	32736	-41885	33	T	n-	0.2592	1.0467	37.8N	116.4W	75	183	161	04m02s
1457	073	-1386 Jan 18	00:19:23	32726	-41879	38	A	t-	-0.9773	0.9560	78.5S	128.5E	11	186	817	02m35s
1458	073	-1386 Jul 14	16:37:21	32716	-41873	43	P	t-	1.0515	0.8922	66.3N	108.4W	0	5		
1459	073	-1386 Dec 08	23:06:43	32708	-41868	10	P	-t	1.1054	0.8081	63.4N	6.5E	0	215		
1460	073	-1385 Jun 04	06:23:36	32698	-41862	15	As	-t	-0.9813	0.9373	54.5S	109.8W	10	329	-	05m51s
1461	074	-1385 Nov 28	14:45:32	32688	-41856	20	T	-n	0.4414	1.0404	5.5N	101.7E	64	199	150	03m43s
1462	074	-1384 May 23	07:40:49	32678	-41850	25	A	nn	-0.2042	0.9684	6.6N	157.0W	78	339	116	03m41s
1463	074	-1384 Nov 17	04:30:43	32668	-41844	30	H	n-	-0.2385	1.0006	28.9S	120.1W	76	26	2	00m03s
1464	074	-1383 May 12	14:55:38	32658	-41838	35	T	p-	0.5768	1.0173	45.4N	70.0E	55	145	72	01m22s
1465	074	-1383 Nov 06	12:08:33	32648	-41832	40	A	p-	-0.9724	0.9348	64.7S	45.0E	13	92	1076	04m20s
1466	074	-1382 Apr 02	21:58:15	32640	-41827	7	P	-t	-1.2208	0.5962	60.7S	78.4E	0	270		
1467	074	-1382 May 02	04:57:24	32638	-41826	45	P	t-	1.2920	0.4585	61.2N	130.2E	0	67		
1468	074	-1382 Sep 26	17:53:21	32630	-41821	12	P	-t	1.3651	0.3422	60.6N	143.5E	0	278		
1469	074	-1381 Mar 23	14:46:27	32620	-41815	17	T	-p	-0.5121	1.0551	30.8S	114.0E	59	326	209	04m14s
1470	074	-1381 Sep 15	20:05:29	32610	-41809	22	A	-p	0.6273	0.9620	41.3N	37.7E	51	218	174	03m21s
1471	074	-1380 Mar 12	03:44:58	32600	-41803	27	H	nn	0.2381	1.0025	3.9N	103.1W	76	152	9	00m14s
1472	074	-1380 Sep 04	05:41:07	32590	-41797	32	T	n-	-0.1195	1.0242	6.0N	132.1W	83	26	83	02m14s
1473	074	-1379 Mar 01	09:31:59	32580	-41791	37	P	t-	1.0223	0.9252	61.2N	110.6E	0	116		
1474	074	-1379 Aug 24	20:52:28	32571	-41785	42	T	p-	-0.8141	1.0540	34.4S	20.1W	35	28	302	04m14s
1475	074	-1378 Jan 19	16:34:35	32562	-41780	9	P	-t	-1.2170	0.5903	63.9S	141.0W	0	209		
1476	074	-1378 Jul 16	04:12:53	32552	-41774	14	P	-t	1.0796	0.8554	64.7N	48.7E	0	339		
1477	074	-1378 Aug 14	13:12:41	32551	-41773	52	Pb	t-	-1.5201	0.0278	62.5S	68.7E	0	46		
1478	074	-1377 Jan 08	21:15:24	32543	-41768	19	A	-p	-0.4658	0.9806	51.2S	5.6E	62	347	78	01m42s
1479	074	-1377 Jul 05	14:07:10	32533	-41762	24	A	-p	0.3510	0.9810	44.5N	105.1E	69	187	72	01m51s
1480	074	-1377 Dec 29	08:58:42	32523	-41756	29	T	n-	0.2495	1.0320	9.4S	178.9W	76	176	112	03m13s
1481	075	-1376 Jun 23	17:02:24	32513	-41750	34	A	p-	-0.4201	0.9478	1.5S	57.7E	65	1	211	07m22s
1482	075	-1376 Dec 18	00:30:56	32503	-41744	39	T	p-	0.9116	1.0369	43.1N	51.2W	24	181	304	02m59s
1483	075	-1375 Jun 12	17:43:49	32493	-41738	44	P	t-	-1.1691	0.6748	67.7S	57.6E	0	350		
1484	075	-1375 Nov 08	02:03:59	32485	-41733	11	P	-t	-1.1825	0.6554	70.6S	150.7E	0	133		
1485	075	-1374 May 03	13:52:19	32475	-41727	16	T	-p	0.7755	1.0295	59.2N	82.0E	39	149	158	02m07s
1486	075	-1374 Oct 28	07:34:31	32465	-41721	21	A	-p	-0.5297	0.9326	39.7S	171.8W	58	21	295	07m08s
1487	075	-1373 Apr 23	05:11:28	32455	-41715	26	T	nn	0.0018	1.0762	7.7N	123.0W	90	169	247	06m47s
1488	075	-1373 Oct 17	07:10:35	32445	-41709	31	A	nn	0.1509	0.9190	3.7N	151.7W	81	198	310	10m51s
1489	075	-1372 Apr 11	22:23:02	32436	-41703	36	T	p-	-0.7369	1.0637	41.4S	2.2W	42	336	306	04m52s
1490	075	-1372 Oct 05	07:52:53	32426	-41697	41	A	p-	0.8301	0.9423	52.4N	139.3W	34	210	380	05m12s
1491	075	-1371 Mar 03	01:25:53	32417	-41692	8	P	-t	1.1769	0.6651	70.8N	115.6W	0	128		
1492	075	-1371 Apr 01	12:44:43	32416	-41691	46	Pb	t-	-1.5280	0.0258	71.5S	142.6W	0	268		
1493	075	-1371 Aug 26	04:36:55	32408	-41686	13	P	-t	-1.2282	0.5786	70.0S	155.2W	0	40		
1494	075	-1371 Sep 24	15:32:49	32406	-41685	51	P	t	1.4723	0.1337	71.5N	175.4W	0	284		
1495	075	-1370 Feb 20	05:06:20	32398	-41680	18	A	-p	0.4717	0.9378	12.2N	122.9W	62	167	261	08m17s
1496	075	-1370 Aug 15	20:41:22	32388	-41674	23	T	-p	-0.5101	1.0633	12.1S	0.2E	59	10	241	05m51s
1497	075	-1369 Feb 09	04:50:40	32378	-41668	28	A	nn	-0.2481	0.9359	32.9S	110.2W	75	349	246	07m38s
1498	075	-1369 Aug 05	12:35:59	32368	-41662	33	T	n-	0.2075	1.0410	32.8N	128.0E	78	187	140	03m41s
1499	075	-1368 Jan 29	08:11:40	32358	-41656	38	A	t-	-0.9492	0.9636	83.7S	42.8W	18	237	432	02m15s
1500	075	-1368 Jul 24	23:47:15	32348	-41650	43	An	t-	0.9934	0.9744	72.3N	130.2E	5	353	-	01m24s

Cat Num	Canon Plate	Calendar Date	TD of Greatest Eclipse	ΔT	Luna Num	Saros Num	Ecl. Type	QLE	Gamma	Ecl. Mag.	Lat.	Long.	Sun Alt	Sun Azm	Path Width	Central Line Dur.
				s							°	°	°	°	km	
1501	076	-1368 Dec 19	07:53:14	32340	-41645	10	P	-t	1.1072	0.8055	64.3N	136.3W	0	205		
1502	076	-1367 Jun 14	12:42:35	32330	-41639	15	P	-t	-1.0678	0.8463	63.6S	156.2E	0	328		
1503	076	-1367 Dec 08	23:36:35	32321	-41633	20	T	-n	0.4432	1.0401	3.8N	33.5W	64	195	149	03m48s
1504	076	-1366 Jun 03	14:10:48	32311	-41627	25	A	nn	-0.2899	0.9703	4.2N	104.7E	73	342	111	03m35s
1505	076	-1366 Nov 28	13:10:48	32301	-41621	30	A	n-	-0.2400	0.9979	32.4S	109.5E	76	22	7	00m11s
1506	076	-1365 May 23	21:53:13	32291	-41615	35	T	p-	0.4943	1.0220	44.6N	31.2W	60	150	86	01m47s
1507	076	-1365 Nov 17	20:25:06	32281	-41609	40	A	p-	-0.9735	0.9320	67.2S	90.3W	13	102	1159	04m25s
1508	076	-1364 Apr 13	05:31:40	32273	-41604	7	P	-t	-1.2855	0.4709	60.8S	45.3W	0	278		
1509	076	-1364 May 12	12:17:35	32271	-41603	45	P	t-	1.2162	0.6046	61.6N	9.7E	0	58		
1510	076	-1364 Oct 07	01:38:20	32263	-41598	12	P	-t	1.3816	0.3149	60.6N	17.1E	0	268		
1511	076	-1363 Apr 02	22:20:58	32253	-41592	17	T	-p	-0.5757	1.0543	30.2S	0.0E	55	325	216	04m13s
1512	076	-1363 Sep 26	04:03:15	32243	-41586	22	A	-p	0.6468	0.9619	38.1N	83.5W	49	218	177	03m25s
1513	076	-1362 Mar 23	11:05:00	32234	-41580	27	H	nn	0.1792	1.0020	4.9N	145.9E	80	151	7	00m12s
1514	076	-1362 Sep 15	13:53:07	32224	-41574	32	T	n-	-0.0924	1.0240	3.4N	103.5E	85	28	82	02m10s
1515	076	-1361 Mar 12	16:33:12	32214	-41568	37	A	t-	0.9669	0.9403	54.3N	18.6E	14	127	864	04m57s
1516	076	-1361 Sep 05	05:06:58	32204	-41562	42	T	p-	-0.7810	1.0521	33.9S	146.8W	38	31	270	04m01s
1517	076	-1360 Jan 31	00:09:43	32196	-41557	9	P	-t	-1.2435	0.5457	63.0S	94.3E	0	219		
1518	076	-1360 Jul 26	11:42:29	32186	-41551	14	P	-t	1.1403	0.7409	63.7N	75.0W	0	329		
1519	076	-1360 Aug 24	21:13:05	32184	-41550	52	P	t-	-1.4850	0.0962	61.9S	62.1W	0	55		
1520	076	-1359 Jan 19	05:23:36	32176	-41545	19	A	-p	-0.4825	0.9862	50.7S	113.1W	61	341	56	01m10s
1521	077	-1359 Jul 15	21:02:47	32166	-41539	24	A	-p	0.4217	0.9751	48.2N	4.0E	65	193	98	02m20s
1522	077	-1358 Jan 08	17:33:00	32157	-41533	29	T	n-	0.2376	1.0365	10.1S	51.1E	76	172	126	03m36s
1523	077	-1358 Jul 04	23:30:55	32147	-41527	34	A	pn	-0.3401	0.9456	4.1N	41.8W	70	5	213	07m38s
1524	077	-1358 Dec 29	09:15:55	32137	-41521	39	T	p-	0.9054	1.0386	41.3N	171.4E	25	175	307	03m09s
1525	077	-1357 Jun 24	00:13:59	32127	-41515	44	P	t-	-1.0855	0.8198	66.7S	52.5W	0	1		
1526	077	-1357 Nov 19	10:37:00	32119	-41510	11	P	-t	-1.1851	0.6502	69.7S	8.0E	0	146		
1527	077	-1356 May 13	20:59:28	32109	-41504	16	T	-p	0.8487	1.0317	69.4N	36.7W	32	141	203	02m03s
1528	077	-1356 Nov 07	15:41:51	32099	-41498	21	A	-p	-0.5343	0.9298	44.3S	65.2E	57	20	311	07m16s
1529	077	-1355 May 03	12:36:48	32089	-41492	26	Tm	nn	0.0743	1.0785	15.7N	122.0E	86	165	254	06m53s
1530	077	-1355 Oct 27	15:05:56	32080	-41486	31	A	nn	0.1429	0.9180	1.0S	87.2E	82	196	314	11m10s
1531	077	-1354 Apr 23	05:49:30	32070	-41480	36	T	p-	-0.6687	1.0648	32.5S	120.7W	48	340	283	05m19s
1532	077	-1354 Oct 16	15:58:40	32060	-41474	41	A	p-	0.8181	0.9430	47.4N	93.6E	35	206	362	05m24s
1533	077	-1353 Mar 14	08:43:42	32052	-41469	8	P	-t	1.2353	0.5616	71.4N	119.9E	0	115		
1534	077	-1353 Apr 12	19:56:23	32050	-41468	46	P	t-	-1.4667	0.1390	71.4S	94.0E	0	282		
1535	077	-1353 Sep 06	12:47:14	32042	-41463	13	P	-t	-1.2549	0.5283	70.7S	67.6E	0	53		
1536	077	-1353 Oct 05	23:56:17	32040	-41462	51	P	t-	1.4553	0.1643	71.7N	43.1E	0	270		
1537	077	-1352 Mar 02	12:09:04	32032	-41457	18	A	-p	0.5256	0.9391	19.1N	127.3E	58	164	264	07m44s
1538	077	-1352 Aug 26	04:48:59	32022	-41451	23	T	-p	-0.5478	1.0592	17.5S	125.6W	57	13	232	05m20s
1539	077	-1351 Feb 19	12:05:04	32013	-41445	28	A	nn	-0.2042	0.9406	27.1S	139.8E	78	346	225	07m12s
1540	077	-1351 Aug 15	20:23:41	32003	-41439	33	T	nn	0.1616	1.0350	27.5N	10.0E	81	191	119	03m16s
1541	078	-1350 Feb 08	15:58:43	31993	-41433	38	A	t-	-0.9158	0.9712	80.0S	139.2E	23	296	262	01m53s
1542	078	-1350 Aug 05	07:01:49	31983	-41427	43	A	p-	0.9399	0.9722	83.8N	51.1W	19	283	300	01m44s
1543	078	-1350 Dec 30	16:36:22	31975	-41422	10	P	-t	1.1125	0.7964	65.3N	81.4E	0	194		
1544	078	-1349 Jun 25	19:04:07	31965	-41416	15	P	-t	-1.1519	0.7023	64.6S	49.5E	0	338		
1545	078	-1349 Dec 20	08:24:42	31955	-41410	20	T	-n	0.4473	1.0402	2.9N	167.8W	63	190	150	03m54s
1546	078	-1348 Jun 13	20:45:31	31946	-41404	25	A	-p	-0.3727	0.9716	1.0N	4.8E	68	346	109	03m31s
1547	078	-1348 Dec 08	21:47:05	31936	-41398	30	A	n-	-0.2392	0.9958	35.1S	19.3W	76	18	15	00m24s
1548	078	-1347 Jun 03	04:55:30	31926	-41392	35	T	p-	0.4146	1.0260	43.3N	133.8W	65	156	97	02m09s
1549	078	-1347 Nov 28	04:39:56	31916	-41386	40	A	p-	-0.9739	0.9298	69.8S	134.2E	12	112	1219	04m28s
1550	078	-1346 Apr 24	13:01:27	31908	-41381	7	P	-t	-1.3532	0.3390	61.0S	168.0W	0	287		
1551	078	-1346 May 23	19:38:06	31907	-41380	45	P	t-	1.1401	0.7519	62.2N	111.0W	0	49		
1552	078	-1346 Oct 18	09:30:13	31898	-41375	12	P	-t	1.3929	0.2963	60.7N	111.0W	0	259		
1553	078	-1345 Apr 14	05:48:09	31889	-41369	17	T	-p	-0.6448	1.0528	30.5S	112.2W	50	326	224	04m08s
1554	078	-1345 Oct 07	12:09:59	31879	-41363	22	A	-p	0.6600	0.9623	34.7N	152.3E	49	217	178	03m27s
1555	078	-1344 Apr 02	18:14:26	31869	-41357	27	Hm	nn	0.1122	1.0010	5.8N	37.7E	84	150	4	00m06s
1556	078	-1344 Sep 25	22:15:23	31859	-41351	32	T	n-	-0.0731	1.0240	0.1N	23.8W	86	29	82	02m08s
1557	078	-1343 Mar 22	23:24:02	31849	-41345	37	A	p-	0.9031	0.9437	49.4N	77.9W	25	132	470	04m53s
1558	078	-1343 Sep 15	13:29:48	31840	-41339	42	T	p-	-0.7547	1.0500	34.9S	84.5E	41	35	247	03m46s
1559	078	-1342 Feb 10	07:36:39	31832	-41334	9	P	-t	-1.2772	0.4885	62.3S	28.2W	0	228		
1560	078	-1342 Aug 06	19:17:49	31822	-41328	14	P	-t	1.1956	0.6377	62.9N	160.2E	0	320		

Cat Num	Canon Plate	Calendar Date	TD of Greatest Eclipse	ΔT s	Luna Num	Saros Num	Ecl. Type	QLE	Gamma	Ecl. Mag.	Lat. °	Long. °	Sun Alt °	Sun Azm °	Path Width km	Central Line Dur.
1561	079	-1342 Sep 05	05:21:06	31820	-41327	52	P	t-	-1.4566	0.1517	61.4S	165.4E	0	65		
1562	079	-1341 Jan 30	13:24:26	31812	-41322	19	A	-p	-0.5061	0.9923	49.4S	129.5E	59	334	31	00m39s
1563	079	-1341 Jul 27	04:02:45	31802	-41316	24	A	-p	0.4878	0.9688	50.5N	97.2W	61	201	129	02m50s
1564	079	-1340 Jan 20	02:01:16	31792	-41310	29	T	n-	0.2203	1.0414	10.3S	77.2W	77	167	142	03m58s
1565	079	-1340 Jul 15	06:04:48	31783	-41304	34	A	nn	-0.2647	0.9430	8.4N	141.9W	75	9	218	07m50s
1566	079	-1339 Jan 08	17:55:37	31773	-41298	39	T	p-	0.8952	1.0406	39.5N	35.8E	26	170	306	03m20s
1567	079	-1339 Jul 04	06:50:46	31763	-41292	44	A-	t-	-1.0061	0.9574	65.7S	163.7W	0	11	-	-
1568	079	-1339 Nov 29	19:09:57	31755	-41287	11	P	-t	-1.1873	0.6460	68.8S	134.1W	0	159		
1569	079	-1338 May 25	04:06:56	31745	-41281	16	T	-t	0.9222	1.0327	79.4N	176.3W	22	112	291	01m54s
1570	079	-1338 Nov 18	23:51:03	31736	-41275	21	A	-p	-0.5367	0.9275	48.5S	57.1W	57	17	323	07m22s
1571	079	-1337 May 14	20:00:27	31726	-41269	26	T	nn	0.1487	1.0801	23.5N	8.0E	81	166	261	06m51s
1572	079	-1337 Nov 07	23:04:53	31716	-41263	31	A	nn	0.1379	0.9176	5.3S	34.6W	82	194	316	11m24s
1573	079	-1336 May 03	13:12:26	31706	-41257	36	T	p-	-0.5980	1.0649	24.0S	122.7E	53	344	263	05m41s
1574	079	-1336 Oct 27	00:10:24	31696	-41251	41	A	p-	0.8100	0.9442	42.9N	34.5W	36	202	347	05m33s
1575	079	-1335 Mar 24	15:53:17	31688	-41246	8	P	-t	1.2990	0.4487	71.7N	2.9W	0	101		
1576	079	-1335 Apr 23	03:03:14	31687	-41245	46	P	t-	-1.4024	0.2575	71.2S	27.9W	0	295		
1577	079	-1335 Sep 16	21:05:33	31679	-41240	13	P	-t	-1.2758	0.4892	71.2S	72.0W	0	66		
1578	079	-1335 Oct 16	08:25:36	31677	-41239	51	P	t-	1.4425	0.1872	71.5N	100.0W	0	256		
1579	079	-1334 Mar 13	19:03:14	31669	-41234	18	A	-p	0.5859	0.9402	26.7N	18.9E	54	161	271	07m09s
1580	079	-1334 Sep 06	13:04:19	31659	-41228	23	T	-p	-0.5790	1.0551	22.9S	106.6E	54	16	222	04m48s
1581	080	-1333 Mar 02	19:09:14	31649	-41222	28	A	nn	-0.1513	0.9454	20.3S	31.4E	81	344	203	06m43s
1582	080	-1333 Aug 27	04:19:26	31640	-41216	33	T	nn	0.1231	1.0289	22.0N	110.6W	83	193	99	02m46s
1583	080	-1332 Feb 19	23:36:23	31630	-41210	38	A	t-	-0.8738	0.9789	72.3S	1.8E	29	316	155	01m28s
1584	080	-1332 Aug 15	14:24:39	31620	-41204	43	A	p-	0.8938	0.9677	77.6N	143.6E	26	231	262	02m12s
1585	080	-1331 Jan 10	01:12:33	31612	-41199	10	P	-t	1.1238	0.7761	66.4N	59.6W	0	184		
1586	080	-1331 Feb 08	11:09:17	31610	-41198	48	Pb	t-	-1.5290	0.0163	69.1S	59.4W	0	207		
1587	080	-1331 Jul 06	01:32:17	31602	-41193	15	P	-t	-1.2301	0.5686	65.5S	59.1W	0	348		
1588	080	-1331 Dec 30	17:08:09	31593	-41187	20	T	-p	0.4551	1.0406	3.1N	59.0E	63	186	153	04m00s
1589	080	-1330 Jun 25	03:24:08	31583	-41181	25	A	-p	-0.4528	0.9724	3.1S	96.6W	63	351	110	03m29s
1590	080	-1330 Dec 20	06:20:46	31573	-41175	30	A	n-	-0.2369	0.9942	36.8S	146.8W	76	13	21	00m33s
1591	080	-1329 Jun 14	12:00:01	31563	-41169	35	T	p-	0.3360	1.0292	41.3N	122.7E	70	162	105	02m30s
1592	080	-1329 Dec 09	12:52:56	31554	-41163	40	A	p-	-0.9735	0.9284	72.4S	1.7W	13	124	1243	04m30s
1593	080	-1328 May 04	20:27:22	31545	-41158	7	P	-t	-1.4242	0.2007	61.4S	70.1E	0	296		
1594	080	-1328 Jun 03	02:59:25	31544	-41157	45	P	t-	1.0642	0.8990	62.9N	127.9E	0	41		
1595	080	-1328 Oct 28	17:29:06	31536	-41152	12	P	-t	1.3987	0.2867	61.0N	119.1E	0	250		
1596	080	-1327 Apr 24	13:10:37	31526	-41146	17	T	-p	-0.7174	1.0504	31.9S	136.6E	44	327	235	04m00s
1597	080	-1327 Oct 17	20:24:07	31516	-41140	22	A	-p	0.6680	0.9631	31.2N	25.6E	48	214	175	03m29s
1598	080	-1326 Apr 14	01:17:31	31506	-41134	27	A	nn	0.0409	0.9996	6.4N	68.7W	88	151	1	00m02s
1599	080	-1326 Oct 07	06:45:01	31497	-41128	32	T	n-	-0.0592	1.0241	3.7S	153.0W	87	29	82	02m07s
1600	080	-1325 Apr 03	06:05:37	31487	-41122	37	A	p-	0.8322	0.9462	46.8N	175.2W	33	134	347	04m45s
1601	081	-1325 Sep 26	22:00:17	31477	-41116	42	T	p-	-0.7348	1.0478	37.0S	46.2W	42	37	229	03m30s
1602	081	-1324 Feb 21	14:55:41	31469	-41111	9	P	-t	-1.3176	0.4192	61.7S	148.4W	0	237		
1603	081	-1324 Aug 17	03:00:16	31459	-41105	14	P	-t	1.2441	0.5482	62.2N	33.9E	0	311		
1604	081	-1324 Sep 15	13:36:18	31458	-41104	52	P	t-	-1.4346	0.1945	61.0S	31.3E	0	74		
1605	081	-1323 Feb 09	21:17:59	31450	-41099	19	A	-p	-0.5366	0.9983	47.8S	13.3E	57	329	7	00m08s
1606	081	-1323 Aug 06	11:10:03	31440	-41093	24	A	-p	0.5468	0.9625	51.3N	160.1E	57	208	162	03m20s
1607	081	-1322 Jan 30	10:22:25	31430	-41087	29	T	n-	0.1965	1.0465	10.0S	156.3E	79	163	158	04m19s
1608	081	-1322 Jul 26	12:45:15	31421	-41081	34	A	nn	-0.1947	0.9402	11.5N	116.9E	79	14	226	07m57s
1609	081	-1321 Jan 20	02:27:58	31411	-41075	39	T	p-	0.8791	1.0432	37.6N	97.5W	28	164	301	03m32s
1610	081	-1321 Jul 15	13:35:28	31401	-41069	44	A	t-	-0.9318	0.9509	44.6S	93.8E	21	12	498	05m04s
1611	081	-1321 Dec 11	03:38:45	31393	-41064	11	P	-t	-1.1923	0.6371	67.7S	85.6E	0	170		
1612	081	-1320 Jun 04	11:17:01	31383	-41058	16	Tn	-t	0.9944	1.0300	72.4N	18.0W	4	22	-	01m30s
1613	081	-1320 Nov 29	07:59:09	31374	-41052	21	A	-p	-0.5398	0.9260	52.2S	177.8W	57	12	331	07m23s
1614	081	-1319 May 25	03:23:45	31364	-41046	26	T	-n	0.2236	1.0807	31.0N	105.3W	77	169	267	06m41s
1615	081	-1319 Nov 18	07:06:24	31354	-41040	31	A	nn	0.1346	0.9178	9.1S	156.7W	82	191	315	11m29s
1616	081	-1318 May 14	20:31:36	31345	-41034	36	T	p-	-0.5244	1.0641	15.8S	7.8E	58	347	245	05m55s
1617	081	-1318 Nov 07	08:28:14	31335	-41028	41	A	p-	0.8063	0.9459	39.2N	163.9W	36	198	334	05m36s
1618	081	-1317 Apr 04	22:51:30	31327	-41023	8	P	-t	1.3708	0.3218	71.7N	123.0W	0	88		
1619	081	-1317 May 04	10:01:54	31325	-41022	46	P	t-	-1.3322	0.3865	70.7S	147.4W	0	308		
1620	081	-1317 Sep 28	05:33:18	31317	-41017	13	P	-t	-1.2897	0.4632	71.5S	145.6E	0	80		

Cat Num	Canon Plate	Calendar Date	TD of Greatest Eclipse	ΔT s	Luna Num	Saros Num	Ecl. Type	QLE	Gamma	Ecl. Mag.	Lat. °	Long. °	Sun Alt °	Sun Azm °	Path Width km	Central Line Dur.
1621	082	-1317 Oct 27	17:02:30	31316	-41016	51	P	t-	1.4354	0.1999	71.1N	115.3E	0	242		
1622	082	-1316 Mar 24	01:47:24	31307	-41011	18	A	-p	0.6539	0.9410	35.3N	87.9W	49	158	285	06m32s
1623	082	-1316 Sep 16	21:27:43	31298	-41005	23	T	-p	-0.6034	1.0510	28.3S	23.3W	53	19	211	04m18s
1624	082	-1315 Mar 13	02:06:52	31288	-40999	28	A	nn	-0.0924	0.9502	13.0S	76.1W	85	343	183	06m11s
1625	082	-1315 Sep 06	12:22:22	31278	-40993	33	T	nn	0.0914	1.0228	16.4N	126.6E	85	196	78	02m14s
1626	082	-1314 Mar 02	07:08:37	31269	-40987	38	A	p-	-0.8263	0.9867	64.0S	122.9W	34	325	84	00m59s
1627	082	-1314 Aug 26	21:53:09	31259	-40981	43	A	p-	0.8532	0.9627	69.9N	17.0E	31	220	260	02m46s
1628	082	-1313 Jan 21	09:43:54	31251	-40976	10	P	-t	1.1394	0.7475	67.4N	160.1E	0	173		
1629	082	-1313 Feb 19	19:12:20	31249	-40975	48	P	t-	-1.4916	0.0838	70.0S	165.9E	0	219		
1630	082	-1313 Jul 17	08:06:34	31241	-40970	15	P	-t	-1.3027	0.4449	66.6S	169.6W	0	358		
1631	082	-1312 Jan 11	01:44:54	31232	-40964	20	T	-p	0.4680	1.0413	4.4N	72.6W	62	182	156	04m06s
1632	082	-1312 Jul 05	10:11:12	31222	-40958	25	A	-p	-0.5266	0.9728	7.8S	159.1E	58	355	115	03m27s
1633	082	-1312 Dec 30	14:48:09	31212	-40952	30	Am	n-	-0.2307	0.9931	37.3S	87.6E	76	7	25	00m40s
1634	082	-1311 Jun 24	19:10:53	31203	-40946	35	T	n-	0.2617	1.0317	38.5N	17.0E	75	168	111	02m48s
1635	082	-1311 Dec 19	21:00:36	31193	-40940	40	A	p-	-0.9696	0.9277	75.6S	135.7W	13	134	1169	04m33s
1636	082	-1310 May 16	03:52:19	31185	-40935	7	Pe	-t	-1.4957	0.0613	61.9S	51.7W	0	305		
1637	082	-1310 Jun 14	10:23:54	31183	-40934	45	Tn	t-	0.9907	1.0593	69.1N	15.3E	7	40	-	02m48s
1638	082	-1310 Nov 09	01:31:57	31175	-40929	12	P	-t	1.4018	0.2815	61.4N	11.9W	0	241		
1639	082	-1309 May 05	20:27:00	31165	-40923	17	T	-p	-0.7942	1.0469	34.8S	27.0E	37	328	251	03m45s
1640	082	-1309 Oct 29	04:45:07	31156	-40917	22	A	-p	0.6711	0.9645	27.9N	103.2W	48	211	169	03m28s
1641	083	-1308 Apr 24	08:12:33	31146	-40911	27	A	nn	-0.0361	0.9976	6.6N	172.9W	88	331	8	00m14s
1642	083	-1308 Oct 17	15:21:35	31136	-40905	32	T	n-	-0.0506	1.0246	7.8S	76.0E	87	29	84	02m09s
1643	083	-1307 Apr 13	12:39:37	31127	-40899	37	A	p-	0.7554	0.9480	45.2N	88.9E	41	137	284	04m40s
1644	083	-1307 Oct 07	06:38:34	31117	-40893	42	T	p-	-0.7216	1.0456	40.0S	178.9W	44	40	215	03m16s
1645	083	-1306 Mar 03	22:06:21	31109	-40888	9	P	-t	-1.3651	0.3367	61.2S	93.5E	0	246		
1646	083	-1306 Apr 02	13:04:59	31107	-40887	47	Pb	t-	1.4985	0.1100	60.6N	25.3E	0	90		
1647	083	-1306 Aug 28	10:49:16	31099	-40882	14	P	-t	1.2864	0.4711	61.6N	93.9W	0	301		
1648	083	-1306 Sep 26	21:58:51	31098	-40881	52	P	t-	-1.4191	0.2249	60.8S	104.6W	0	83		
1649	083	-1305 Feb 21	05:04:15	31090	-40876	19	H	-p	-0.5739	1.0045	46.0S	101.8W	55	325	19	00m22s
1650	083	-1305 Aug 17	18:24:18	31080	-40870	24	A	-p	0.5994	0.9561	50.8N	55.0E	53	214	198	03m51s
1651	083	-1304 Feb 10	18:35:48	31070	-40864	29	T	n-	0.1657	1.0518	9.2S	31.8E	81	159	174	04m38s
1652	083	-1304 Aug 05	19:34:32	31061	-40858	34	A	nn	-0.1318	0.9373	13.3N	13.9E	83	18	235	08m02s
1653	083	-1303 Jan 30	10:53:53	31051	-40852	39	T	p-	0.8578	1.0459	36.0N	131.5E	31	159	294	03m44s
1654	083	-1303 Jul 25	20:28:00	31042	-40846	44	A	t-	-0.8627	0.9530	35.8S	12.6W	30	16	337	05m09s
1655	083	-1303 Dec 21	12:04:22	31033	-40841	11	P	-t	-1.1992	0.6248	66.6S	53.4W	0	182		
1656	083	-1302 Jun 15	18:30:34	31024	-40835	16	P	-t	1.0646	0.8882	67.4N	143.3W	0	7		
1657	083	-1302 Dec 10	16:05:12	31014	-40829	21	A	-p	-0.5439	0.9251	55.1S	63.5E	57	6	337	07m21s
1658	083	-1301 Jun 05	10:47:32	31004	-40823	26	T	-n	0.2982	1.0805	38.1N	142.3E	72	172	272	06m25s
1659	083	-1301 Nov 29	15:08:25	30995	-40817	31	A	nn	0.1315	0.9188	12.2S	81.3E	83	188	311	11m26s
1660	083	-1300 May 25	03:50:16	30985	-40811	36	T	p-	-0.4509	1.0624	8.3S	106.3W	63	350	229	06m00s
1661	084	-1300 Nov 17	16:47:32	30976	-40805	41	A	p-	0.8029	0.9482	36.0N	66.4E	36	194	317	05m33s
1662	084	-1299 Apr 15	05:43:56	30967	-40800	8	P	-t	1.4458	0.1893	71.5N	118.4E	0	74		
1663	084	-1299 May 14	16:58:50	30966	-40799	46	P	t-	-1.2611	0.5164	70.0S	93.9E	0	320		
1664	084	-1299 Oct 08	14:08:14	30958	-40794	13	P	-t	-1.2986	0.4466	71.6S	1.2E	0	95		
1665	084	-1299 Nov 07	01:43:03	30956	-40793	51	P	t-	1.4304	0.2084	70.5N	29.8W	0	228		
1666	084	-1298 Apr 04	08:23:37	30948	-40788	18	A	-p	0.7279	0.9415	44.7N	166.1E	43	154	312	05m56s
1667	084	-1298 Sep 28	05:58:24	30939	-40782	23	T	-p	-0.6217	1.0471	33.7S	154.9W	51	22	199	03m50s
1668	084	-1297 Mar 24	08:56:20	30929	-40776	28	A	nn	-0.0260	0.9550	5.1S	178.0E	88	343	164	05m35s
1669	084	-1297 Sep 17	20:33:36	30919	-40770	33	H3	nn	0.0674	1.0168	10.8N	1.4E	86	197	58	01m40s
1670	084	-1296 Mar 12	14:32:02	30910	-40764	38	A	p-	-0.7705	0.9943	55.2S	117.5E	39	330	31	00m27s
1671	084	-1296 Sep 06	05:31:22	30900	-40758	43	A	p-	0.8213	0.9575	62.8N	105.0W	34	215	271	03m23s
1672	084	-1295 Jan 31	18:07:10	30892	-40753	10	P	-t	1.1619	0.7057	68.5N	21.4E	0	161		
1673	084	-1295 Mar 02	03:07:17	30890	-40752	48	P	t-	-1.4471	0.1655	70.7S	32.7E	0	232		
1674	084	-1295 Jul 27	14:48:52	30882	-40747	15	P	-t	-1.3687	0.3330	67.6S	77.4E	0	9		
1675	084	-1294 Jan 21	10:14:58	30873	-40741	20	T	-p	0.4862	1.0422	6.8N	157.2E	61	177	161	04m11s
1676	084	-1294 Jul 16	17:05:20	30863	-40735	25	A	-p	-0.5951	0.9727	13.2S	52.5E	53	359	122	03m24s
1677	084	-1293 Jan 10	23:08:36	30853	-40729	30	A	n-	-0.2190	0.9924	36.4S	36.3W	77	2	27	00m45s
1678	084	-1293 Jul 06	02:26:36	30844	-40723	35	T	n-	0.1903	1.0335	35.0N	90.6W	79	174	115	03m04s
1679	084	-1293 Dec 31	05:03:18	30834	-40717	40	A	p-	-0.9620	0.9279	79.5S	90.0E	15	146	1036	04m35s
1680	084	-1292 Jun 24	17:51:46	30825	-40711	45	T	t-	0.9192	1.0642	80.8N	53.7W	23	84	544	03m22s

Cat Num	Canon Plate	Calendar Date	TD of Greatest Eclipse	ΔT s	Luna Num	Saros Num	Ecl. Type	QLE	Gamma	Ecl. Mag.	Lat. °	Long. °	Sun Alt °	Sun Azm °	Path Width km	Central Line Dur.
1681	085	-1292 Nov 19	09:37:28	30817	-40706	12	P	-t	1.4032	0.2788	62.0N	143.7W	0	231		
1682	085	-1291 May 16	03:41:12	30807	-40700	17	T	-t	-0.8723	1.0422	39.8S	81.8W	29	330	283	03m21s
1683	085	-1291 Nov 08	13:11:12	30797	-40694	22	A	-p	0.6710	0.9665	24.8N	126.5E	48	207	159	03m23s
1684	085	-1290 May 05	15:02:06	30788	-40688	27	A	nn	-0.1168	0.9951	6.3N	84.3E	83	334	17	00m30s
1685	085	-1290 Oct 29	00:03:50	30778	-40682	32	T	n-	-0.0460	1.0254	11.9S	56.4W	87	27	86	02m14s
1686	085	-1289 Apr 24	19:07:29	30769	-40676	37	A	p-	0.6738	0.9494	44.3N	5.5W	47	140	246	04m38s
1687	085	-1289 Oct 18	15:22:14	30759	-40670	42	T	p-	-0.7127	1.0436	43.6S	47.3E	44	41	204	03m03s
1688	085	-1288 Mar 14	05:09:51	30751	-40665	9	P	-t	-1.4191	0.2421	60.9S	22.6W	0	255		
1689	085	-1288 Apr 12	19:37:54	30749	-40664	47	P	t-	1.4205	0.2425	60.6N	83.1W	0	81		
1690	085	-1288 Sep 07	18:46:24	30741	-40659	14	P	-t	1.3215	0.4079	61.1N	136.5E	0	292		
1691	085	-1288 Oct 07	06:27:50	30740	-40658	52	P	t-	-1.4092	0.2444	60.8S	117.9E	0	92		
1692	085	-1287 Mar 03	12:44:13	30732	-40653	19	H	-p	-0.6178	1.0105	44.2S	144.2E	52	322	45	00m49s
1693	085	-1287 Aug 28	01:45:37	30722	-40647	24	A	-p	0.6455	0.9499	49.4N	53.0W	50	219	238	04m23s
1694	085	-1286 Feb 21	02:41:58	30712	-40641	29	T	n-	0.1281	1.0569	8.2S	90.8W	83	156	189	04m56s
1695	085	-1286 Aug 17	02:32:37	30703	-40635	34	A	nn	-0.0758	0.9343	13.8N	91.1W	86	22	246	08m08s
1696	085	-1285 Feb 10	19:09:44	30693	-40629	39	T	p-	0.8283	1.0488	34.3N	3.7E	34	155	285	03m56s
1697	085	-1285 Aug 06	03:31:11	30684	-40623	44	A	p-	-0.8011	0.9540	30.5S	121.3W	37	20	276	05m06s
1698	085	-1284 Jan 01	20:22:42	30676	-40618	11	P	-t	-1.2116	0.6030	65.5S	170.0E	0	193		
1699	085	-1284 Jun 26	01:49:39	30666	-40612	16	P	-t	1.1309	0.7633	66.4N	94.6E	0	356		
1700	085	-1284 Jul 25	11:02:31	30665	-40611	54	Pb	t-	-1.4893	0.0993	63.8S	109.3E	0	30		
1701	086	-1284 Dec 21	00:05:44	30656	-40606	21	A	-p	-0.5523	0.9249	57.1S	52.5W	56	359	340	07m16s
1702	086	-1283 Jun 15	18:14:09	30647	-40600	26	T	-n	0.3708	1.0793	44.5N	30.6E	68	177	276	06m03s
1703	086	-1283 Dec 09	23:10:06	30637	-40594	31	A	nn	0.1275	0.9205	14.7S	40.3W	83	184	303	11m11s
1704	086	-1282 Jun 05	11:06:36	30628	-40588	36	T	n-	-0.3759	1.0598	1.2S	140.9E	68	354	212	05m54s
1705	086	-1282 Nov 29	01:10:19	30618	-40582	41	A	p-	0.8014	0.9512	33.4N	64.0W	36	189	299	05m22s
1706	086	-1281 Apr 26	12:27:53	30610	-40577	8	Pe	-t	1.5267	0.0471	71.1N	2.3E	0	61		
1707	086	-1281 May 25	23:51:04	30609	-40576	46	P	t-	-1.1866	0.6517	69.2S	23.1W	0	332		
1708	086	-1281 Oct 19	22:50:18	30601	-40571	13	P	-t	-1.3024	0.4395	71.4S	145.0W	0	109		
1709	086	-1281 Nov 18	10:27:10	30599	-40570	51	P	t-	1.4278	0.2125	69.7N	175.3W	0	215		
1710	086	-1280 Apr 14	14:52:43	30591	-40565	18	A	-p	0.8073	0.9413	55.1N	59.4E	36	148	365	05m22s
1711	086	-1280 Oct 08	14:36:40	30581	-40559	23	T	-p	-0.6338	1.0435	38.9S	72.1E	50	23	187	03m26s
1712	086	-1279 Apr 03	15:41:31	30572	-40553	28	A	nn	0.0446	0.9596	3.0N	72.8E	87	162	147	04m58s
1713	086	-1279 Sep 28	04:50:38	30562	-40547	33	H	nn	0.0489	1.0110	5.4N	125.3W	87	198	38	01m07s
1714	086	-1278 Mar 23	21:51:30	30553	-40541	38	H	p-	-0.7102	1.0018	46.3S	0.6E	44	334	9	00m09s
1715	086	-1278 Sep 17	13:15:57	30543	-40535	43	A	p-	0.7954	0.9523	56.3N	133.1E	37	212	286	04m04s
1716	086	-1277 Feb 12	02:23:05	30535	-40530	10	P	-t	1.1908	0.6513	69.4N	116.1W	0	149		
1717	086	-1277 Mar 13	10:55:17	30534	-40529	48	P	t-	-1.3965	0.2597	71.2S	99.2W	0	245		
1718	086	-1277 Aug 07	21:40:06	30526	-40524	15	P	-t	-1.4274	0.2340	68.6S	38.4W	0	20		
1719	086	-1277 Sep 06	14:49:01	30524	-40523	53	Pb	t-	1.5325	0.0604	70.8N	149.9W	0	306		
1720	086	-1276 Feb 01	18:36:42	30516	-40518	20	T	-p	0.5111	1.0430	10.3N	28.8E	59	173	167	04m13s
1721	087	-1276 Jul 27	00:09:29	30506	-40512	25	A	-p	-0.6565	0.9724	18.9S	57.4W	49	3	131	03m20s
1722	087	-1275 Jan 21	07:20:02	30497	-40506	30	A	nn	-0.2007	0.9920	34.0S	158.4W	78	357	29	00m48s
1723	087	-1275 Jul 16	09:51:11	30487	-40500	35	T	nn	0.1255	1.0347	30.8N	158.6E	83	179	118	03m16s
1724	087	-1274 Jan 10	12:58:00	30478	-40494	40	A	p-	-0.9485	0.9288	84.2S	50.3W	18	166	873	04m39s
1725	087	-1274 Jul 06	01:24:05	30468	-40488	45	T	p-	0.8514	1.0658	81.2N	106.7W	31	146	415	03m41s
1726	087	-1274 Nov 30	17:43:47	30460	-40483	12	P	-t	1.4043	0.2764	62.8N	84.2E	0	222		
1727	087	-1273 May 27	10:52:41	30451	-40477	17	T	-t	-0.9513	1.0356	48.6S	171.8E	17	330	389	02m43s
1728	087	-1273 Nov 19	21:39:42	30441	-40471	22	A	-p	0.6696	0.9691	22.1N	4.3W	48	203	147	03m13s
1729	087	-1272 May 15	21:47:18	30431	-40465	27	A	nn	-0.2000	0.9919	5.2N	17.5W	78	337	29	00m52s
1730	087	-1272 Nov 08	08:50:37	30422	-40459	32	T	n-	-0.0451	1.0267	16.0S	170.4E	87	25	91	02m21s
1731	087	-1271 May 05	01:30:28	30412	-40453	37	A	p-	0.5883	0.9502	43.6N	98.4W	54	143	223	04m41s
1732	087	-1271 Oct 29	00:11:01	30403	-40447	42	T	p-	-0.7083	1.0418	47.8S	87.6W	45	42	196	02m51s
1733	087	-1270 Mar 25	12:06:54	30395	-40442	9	P	-t	-1.4786	0.1367	60.7S	137.1W	0	264		
1734	087	-1270 Apr 24	02:06:07	30393	-40441	47	P	t	1.3379	0.3840	60.9N	169.7E	0	73		
1735	087	-1270 Sep 19	02:50:50	30385	-40436	14	P	-t	1.3499	0.3575	60.8N	5.1E	0	283		
1736	087	-1270 Oct 18	15:02:56	30384	-40435	52	P	t-	-1.4049	0.2535	60.9S	21.0W	0	101		
1737	087	-1269 Mar 14	20:16:03	30376	-40430	19	T	-p	-0.6692	1.0162	42.9S	31.8E	48	320	74	01m15s
1738	087	-1269 Sep 08	09:15:41	30366	-40424	24	A	-p	0.6835	0.9439	47.3N	164.7W	47	221	279	04m58s
1739	087	-1268 Mar 03	10:40:37	30357	-40418	29	T	nn	0.0837	1.0619	6.9S	148.5E	85	153	204	05m13s
1740	087	-1268 Aug 27	09:40:13	30347	-40412	34	A	nn	-0.0277	0.9313	13.1N	161.4E	88	26	257	08m15s

Cat Num	Canon Plate	Calendar Date	TD of Greatest Eclipse	ΔT s	Luna Num	Saros Num	Ecl. Type	QLE	Gamma	Ecl. Mag.	Lat. °	Long. °	Sun Alt °	Sun Azm °	Path Width km	Central Line Dur.
1741	088	-1267 Feb 21	03:18:32	30338	-40406	39	T	p-	0.7933	1.0515	33.2N	121.6W	37	151	275	04m05s
1742	088	-1267 Aug 16	10:44:25	30328	-40400	44	A	p-	-0.7469	0.9546	27.4S	127.8E	41	23	244	04m59s
1743	088	-1266 Jan 12	04:35:06	30320	-40395	11	P	-t	-1.2280	0.5743	64.5S	35.4E	0	203		
1744	088	-1266 Jul 07	09:13:30	30311	-40389	16	P	-t	1.1940	0.6444	65.4N	28.2W	0	346		
1745	088	-1266 Aug 05	18:33:52	30309	-40388	54	P	t-	-1.4312	0.2056	63.0S	14.5W	0	40		
1746	088	-1265 Jan 01	08:01:37	30301	-40383	21	A	-p	-0.5640	0.9252	58.2S	166.4W	55	350	342	07m07s
1747	088	-1265 Jun 27	01:44:15	30291	-40377	26	T	-p	0.4405	1.0773	49.9N	80.2W	64	183	279	05m39s
1748	088	-1265 Dec 21	07:07:40	30282	-40371	31	A	nn	0.1197	0.9229	16.6S	160.6W	83	179	293	10m44s
1749	088	-1264 Jun 15	18:25:38	30272	-40365	36	T	n-	-0.3036	1.0564	4.9N	28.3E	72	358	195	05m38s
1750	088	-1264 Dec 09	09:31:31	30263	-40359	41	A	p-	0.7974	0.9548	31.1N	166.0E	37	184	274	05m03s
1751	088	-1263 Jun 05	06:43:23	30253	-40353	46	P	t-	-1.1127	0.7847	68.3S	139.5W	0	343		
1752	088	-1263 Oct 30	07:36:36	30245	-40348	13	P	-t	-1.3035	0.4376	70.9S	68.2E	0	123		
1753	088	-1263 Nov 28	19:11:33	30244	-40347	51	P	t-	1.4249	0.2172	68.8N	39.9E	0	202		
1754	088	-1262 Apr 25	21:16:42	30236	-40342	18	A	-t	0.8905	0.9403	66.4N	53.0W	27	136	487	04m49s
1755	088	-1262 Oct 19	23:20:22	30226	-40336	23	T	-p	-0.6412	1.0404	43.8S	61.7W	50	24	175	03m05s
1756	088	-1261 Apr 14	22:19:22	30217	-40330	28	Am	nn	0.1220	0.9638	11.5N	30.6W	83	162	132	04m21s
1757	088	-1261 Oct 09	13:15:03	30207	-40324	33	H	nn	0.0373	1.0056	0.3N	106.3E	88	198	19	00m35s
1758	088	-1260 Apr 03	05:04:17	30198	-40318	38	H	p-	-0.6432	1.0088	37.4S	114.0W	50	337	40	00m48s
1759	088	-1260 Sep 27	21:08:24	30188	-40312	43	A	p-	0.7766	0.9474	50.4N	10.0E	39	209	304	04m48s
1760	088	-1259 Feb 22	10:30:26	30180	-40307	10	P	-t	1.2271	0.5823	70.2N	107.9E	0	137		
1761	089	-1259 Mar 23	18:36:27	30179	-40306	48	P	t-	-1.3398	0.3669	71.6S	130.2E	0	259		
1762	089	-1259 Aug 18	04:41:25	30171	-40301	15	P	-t	-1.4779	0.1492	69.5S	157.3W	0	32		
1763	089	-1259 Sep 16	22:15:38	30169	-40300	53	P	t-	1.5024	0.1122	71.4N	83.4E	0	293		
1764	089	-1258 Feb 12	02:49:50	30161	-40295	20	T	-p	0.5430	1.0438	15.0N	98.0W	57	169	174	04m11s
1765	089	-1258 Aug 07	07:22:11	30152	-40289	25	A	-p	-0.7115	0.9717	25.1S	170.1W	44	8	144	03m15s
1766	089	-1257 Feb 01	15:23:05	30142	-40283	30	A	nn	-0.1763	0.9919	30.4S	80.8E	80	352	29	00m50s
1767	089	-1257 Jul 27	17:23:13	30133	-40277	35	T	nn	0.0662	1.0352	26.1N	45.2E	86	183	119	03m24s
1768	089	-1256 Jan 21	20:44:11	30123	-40271	40	A	p-	-0.9287	0.9302	87.0S	106.6E	21	251	722	04m45s
1769	089	-1256 Jul 16	09:02:38	30114	-40265	45	T	p-	0.7885	1.0660	75.7N	166.8E	38	177	353	03m55s
1770	089	-1256 Dec 11	01:50:04	30106	-40260	12	P	-t	1.4054	0.2739	63.7N	48.3W	0	212		
1771	089	-1255 Jun 06	18:03:43	30096	-40254	17	P	-t	-1.0300	0.9518	63.2S	72.5E	0	323		
1772	089	-1255 Jul 06	01:56:11	30095	-40253	55	Pb	t-	1.5161	0.0296	65.7N	107.5E	0	11		
1773	089	-1255 Nov 30	06:10:01	30087	-40248	22	A	-p	0.6676	0.9723	20.0N	135.5W	48	199	132	02m59s
1774	089	-1254 May 27	04:30:06	30077	-40242	27	A	-p	-0.2840	0.9882	3.2N	118.8W	74	340	43	01m19s
1775	089	-1254 Nov 19	17:39:29	30068	-40236	32	T	n-	-0.0450	1.0285	19.6S	37.0E	87	22	97	02m32s
1776	089	-1253 May 16	07:49:59	30058	-40230	37	A	p-	0.4999	0.9505	42.6N	169.7E	60	148	208	04m49s
1777	089	-1253 Nov 09	09:02:45	30049	-40224	42	T	p-	-0.7064	1.0405	52.2S	137.4E	45	42	190	02m43s
1778	089	-1252 Apr 04	18:59:23	30041	-40219	9	Pe	-t	-1.5425	0.0226	60.7S	109.6E	0	272		
1779	089	-1252 May 04	08:33:10	30039	-40218	47	P	t-	1.2528	0.5306	61.2N	62.6E	0	64		
1780	089	-1252 Sep 29	11:01:28	30031	-40213	14	P	-t	1.3727	0.3176	60.6N	127.7W	0	274		
1781	090	-1252 Oct 28	23:40:54	30030	-40212	52	P	t-	-1.4034	0.2573	61.2S	160.8W	0	111		
1782	090	-1251 Mar 25	03:43:17	30022	-40207	19	T	-p	-0.7255	1.0214	42.2S	79.6W	43	319	104	01m38s
1783	090	-1251 Sep 18	16:53:26	30012	-40201	24	A	-p	0.7145	0.9384	44.8N	80.3E	44	222	319	05m34s
1784	090	-1250 Mar 14	18:31:41	30003	-40195	29	T	nn	0.0324	1.0665	5.5S	29.8E	88	151	217	05m29s
1785	090	-1250 Sep 07	16:57:41	29993	-40189	34	A	nn	0.0126	0.9285	11.4N	51.2E	89	205	268	08m24s
1786	090	-1249 Mar 04	11:17:32	29984	-40183	39	T	p-	0.7504	1.0542	32.3N	116.2E	41	148	265	04m14s
1787	090	-1249 Aug 27	18:09:35	29974	-40177	44	A	p-	-0.7012	0.9549	26.1S	14.0E	45	27	225	04m50s
1788	090	-1248 Jan 23	12:36:42	29966	-40172	11	P	-t	-1.2527	0.5310	63.6S	96.1W	0	213		
1789	090	-1248 Feb 22	01:19:20	29965	-40171	49	Pb	t-	1.5180	0.0493	61.8N	131.1W	0	122		
1790	090	-1248 Jul 17	16:45:58	29957	-40166	16	P	-t	1.2510	0.5366	64.5N	152.8W	0	336		
1791	090	-1248 Aug 16	02:16:57	29955	-40165	54	P	t-	-1.3813	0.2970	62.2S	141.0W	0	49		
1792	090	-1247 Jan 11	15:49:47	29948	-40160	21	A	-p	-0.5821	0.9260	58.2S	81.9E	54	342	343	06m56s
1793	090	-1247 Jul 07	09:18:23	29938	-40154	26	T	-p	0.5071	1.0744	54.3N	169.8E	59	190	281	05m14s
1794	090	-1247 Dec 31	15:01:47	29929	-40148	31	A	nn	0.1084	0.9259	17.8S	80.2E	84	175	280	10m06s
1795	090	-1246 Jun 27	01:45:21	29919	-40142	36	T	n-	-0.2321	1.0522	10.2N	83.8W	77	2	178	05m13s
1796	090	-1246 Dec 20	17:51:39	29910	-40136	41	A	p-	0.7919	0.9591	29.3N	36.4E	37	179	244	04m36s
1797	090	-1245 Jun 16	13:34:06	29900	-40130	46	P	t-	-1.0381	0.9177	67.3S	105.0E	0	354		
1798	090	-1245 Nov 10	16:27:38	29892	-40125	13	P	-t	-1.3014	0.4415	70.2S	79.4W	0	136		
1799	090	-1245 Dec 10	03:56:17	29891	-40124	51	P	t-	1.4217	0.2222	67.7N	104.4W	0	190		
1800	090	-1244 May 06	03:36:52	29883	-40119	18	A	-t	0.9763	0.9374	75.7N	156.2E	12	84	1145	04m16s

Cat Num	Canon Plate	Calendar Date	TD of Greatest Eclipse	ΔT s	Luna Num	Saros Num	Ecl. Type	QLE	Gamma	Ecl. Mag.	Lat. °	Long. °	Sun Alt °	Sun Azm °	Path Width km	Central Line Dur.
1801	091	-1244 Oct 30	08:08:04	29873	-40113	23	T	-p	-0.6452	1.0377	48.6S	164.5E	50	24	165	02m48s
1802	091	-1243 Apr 25	04:56:05	29864	-40107	28	A	nn	0.2012	0.9677	20.1N	133.7W	78	163	119	03m45s
1803	091	-1243 Oct 19	21:44:13	29854	-40101	33	H	nn	0.0303	1.0006	4.6S	23.2W	88	198	2	00m04s
1804	091	-1242 Apr 14	12:13:39	29845	-40095	38	T	p-	-0.5722	1.0155	28.5S	132.8E	55	340	64	01m29s
1805	091	-1242 Oct 09	05:06:41	29835	-40089	43	A	p-	0.7632	0.9428	45.1N	114.3W	40	206	324	05m33s
1806	091	-1241 Mar 05	18:30:45	29828	-40084	10	P	-t	1.2695	0.5010	70.9N	26.8W	0	124		
1807	091	-1241 Apr 04	02:12:31	29826	-40083	48	P	t-	-1.2781	0.4845	71.7S	0.7E	0	272		
1808	091	-1241 Aug 29	11:53:01	29818	-40078	15	P	-t	-1.5203	0.0785	70.3S	80.7E	0	44		
1809	091	-1241 Sep 28	05:50:36	29817	-40077	53	P	t-	1.4788	0.1527	71.7N	45.9W	0	279		
1810	091	-1240 Feb 23	10:53:38	29809	-40072	20	T	-p	0.5822	1.0442	20.6N	137.0E	54	166	181	04m06s
1811	091	-1240 Aug 17	14:46:51	29799	-40066	25	A	-p	-0.7576	0.9709	31.2S	73.5E	41	12	159	03m08s
1812	091	-1239 Feb 11	23:16:10	29790	-40060	30	A	nn	-0.1445	0.9918	25.7S	38.3W	82	349	29	00m52s
1813	091	-1239 Aug 07	01:04:07	29780	-40054	35	T	nn	0.0136	1.0354	21.0N	71.2W	89	186	119	03m28s
1814	091	-1238 Feb 01	04:20:38	29771	-40048	40	A	p-	-0.9016	0.9321	81.7S	64.8W	25	305	596	04m52s
1815	091	-1238 Jul 27	16:48:02	29761	-40042	45	T	p-	0.7314	1.0651	69.2N	59.6E	43	189	314	04m04s
1816	091	-1238 Dec 22	09:52:16	29754	-40037	12	P	-t	1.4105	0.2645	64.6N	180.0E	0	202		
1817	091	-1237 Jun 18	01:14:59	29744	-40031	17	P	-t	-1.1075	0.8047	64.0S	46.4W	0	332		
1818	091	-1237 Jul 17	09:29:13	29742	-40030	55	P	t-	1.4588	0.1408	66.6N	17.7W	0	1		
1819	091	-1237 Dec 11	14:39:08	29735	-40025	22	A	-p	0.6673	0.9762	18.6N	93.6E	48	194	113	02m37s
1820	091	-1236 Jun 06	11:11:54	29725	-40019	27	A	-p	-0.3680	0.9838	0.4N	139.7E	68	344	61	01m54s
1821	092	-1236 Nov 30	02:28:34	29716	-40013	32	T	n-	-0.0448	1.0308	22.6S	96.0W	87	18	104	02m46s
1822	092	-1235 May 26	14:08:26	29706	-40007	37	A	p-	0.4107	0.9502	41.1N	78.1E	66	154	199	05m03s
1823	092	-1235 Nov 19	17:57:11	29697	-40001	42	T	p-	-0.7069	1.0396	56.7S	2.8E	45	41	187	02m36s
1824	092	-1234 May 15	14:57:03	29687	-39995	47	P	t-	1.1641	0.6844	61.7N	43.7W	0	56		
1825	092	-1234 Oct 10	19:19:02	29680	-39990	14	P	-t	1.3892	0.2894	60.6N	97.8E	0	265		
1826	092	-1234 Nov 09	08:23:08	29678	-39989	52	P	t-	-1.4060	0.2534	61.6S	58.4E	0	120		
1827	092	-1233 Apr 05	11:04:24	29670	-39984	19	T	-p	-0.7875	1.0261	42.6S	170.6E	38	318	140	01m59s
1828	092	-1233 Sep 30	00:39:28	29661	-39978	24	A	-p	0.7381	0.9334	42.0N	38.0W	42	221	359	06m12s
1829	092	-1232 Mar 25	02:16:17	29651	-39972	29	Tm	nn	-0.0247	1.0707	4.2S	87.3W	89	331	230	05m45s
1830	092	-1232 Sep 18	00:25:27	29642	-39966	34	A	nn	0.0445	0.9261	8.8N	61.9W	87	208	278	08m36s
1831	092	-1231 Mar 14	19:09:48	29632	-39960	39	T	p-	0.7022	1.0564	32.0N	3.7W	45	146	256	04m20s
1832	092	-1231 Sep 07	01:43:58	29623	-39954	44	A	p-	-0.6621	0.9550	26.2S	101.9W	48	30	213	04m40s
1833	092	-1230 Feb 02	20:30:49	29615	-39949	11	P	-t	-1.2827	0.4785	62.7S	134.5E	0	222		
1834	092	-1230 Mar 04	09:02:14	29613	-39948	49	P	t-	1.4798	0.1189	61.3N	102.7E	0	113		
1835	092	-1230 Jul 29	00:25:10	29606	-39943	16	P	-t	1.3033	0.4377	63.6N	81.2E	0	327		
1836	092	-1230 Aug 27	10:08:47	29604	-39942	54	P	t-	-1.3375	0.3772	61.6S	90.5E	0	58		
1837	092	-1229 Jan 22	23:29:05	29596	-39937	21	A	-p	-0.6069	0.9271	57.5S	27.9W	52	333	344	06m42s
1838	092	-1229 Jul 18	16:58:10	29587	-39931	26	T	-p	0.5689	1.0709	57.1N	60.1E	55	199	281	04m49s
1839	092	-1228 Jan 11	22:49:04	29577	-39925	31	A	nn	0.0909	0.9295	18.4S	37.2W	85	170	265	09m19s
1840	092	-1228 Jul 07	09:09:34	29568	-39919	36	T	nn	-0.1653	1.0474	14.5N	163.7E	81	6	160	04m41s
1841	093	-1228 Dec 31	02:06:33	29558	-39913	41	A	p-	0.7810	0.9641	27.5N	91.7W	38	174	208	04m02s
1842	093	-1227 Jun 26	20:27:49	29549	-39907	46	A	t-	-0.9665	0.9820	52.1S	8.8W	14	3	258	01m37s
1843	093	-1227 Nov 21	01:19:49	29541	-39902	13	P	-t	-1.2990	0.4460	69.4S	133.4E	0	149		
1844	093	-1227 Dec 20	12:37:49	29540	-39901	51	P	t-	1.4155	0.2325	66.7N	112.7E	0	179		
1845	093	-1226 May 17	09:53:35	29532	-39896	18	P	-t	1.0644	0.8521	69.9N	10.4E	0	36		
1846	093	-1226 Nov 10	16:59:00	29522	-39890	23	T	-p	-0.6465	1.0355	53.1S	31.0E	49	22	156	02m34s
1847	093	-1225 May 06	11:29:17	29513	-39884	28	A	nn	0.2843	0.9711	28.7N	124.5E	73	164	108	03m12s
1848	093	-1225 Oct 31	06:17:37	29504	-39878	33	Am	nn	0.0274	0.9963	9.0S	153.5W	88	196	13	00m24s
1849	093	-1224 Apr 24	19:19:25	29494	-39872	38	T	p-	-0.4970	1.0216	19.9S	21.1E	60	343	84	02m09s
1850	093	-1224 Oct 19	13:11:44	29485	-39866	43	A	p-	0.7557	0.9386	40.5N	120.1E	41	203	344	06m19s
1851	093	-1223 Mar 16	02:23:20	29477	-39861	10	P	-t	1.3184	0.4067	71.3N	160.0W	0	110		
1852	093	-1223 Apr 14	09:43:26	29475	-39860	48	P	t-	-1.2117	0.6122	71.5S	127.5W	0	286		
1853	093	-1223 Sep 08	19:14:20	29468	-39855	15	Pe	-t	-1.5551	0.0209	71.0S	44.3W	0	57		
1854	093	-1223 Oct 08	13:33:39	29466	-39854	53	P	t-	1.4614	0.1825	71.7N	177.3W	0	265		
1855	093	-1222 Mar 05	18:48:58	29458	-39849	20	T	-p	0.6280	1.0444	27.2N	13.5E	51	162	189	03m56s
1856	093	-1222 Aug 28	22:21:32	29449	-39843	25	A	-p	-0.7961	0.9701	37.4S	45.9W	37	17	176	03m01s
1857	093	-1221 Feb 23	06:58:23	29439	-39837	30	A	nn	-0.1046	0.9918	20.0S	155.7W	84	346	29	00m53s
1858	093	-1221 Aug 18	08:54:26	29430	-39831	35	T	nn	-0.0318	1.0351	15.6N	169.4E	88	11	118	03m27s
1859	093	-1220 Feb 12	11:47:38	29420	-39825	40	A	p-	-0.8674	0.9343	74.4S	167.9E	29	318	497	05m01s
1860	093	-1220 Aug 07	00:40:29	29411	-39819	45	T	p-	0.6800	1.0636	62.4N	55.7W	47	196	285	04m11s

Cat Num	Canon Plate	Calendar Date	TD of Greatest Eclipse	ΔT s	Luna Num	Saros Num	Ecl. Type	QLE	Gamma	Ecl. Mag.	Lat. °	Long. °	Sun Alt °	Sun Azm °	Path Width km	Central Line Dur.
1861	094	-1219 Jan 01	17:50:57	29403	-39814	12	P	-t	1.4187	0.2496	65.7N	48.8E	0	191		
1862	094	-1219 Jun 28	08:28:45	29394	-39808	17	P	-t	-1.1819	0.6646	65.0S	166.2W	0	342		
1863	094	-1219 Jul 27	17:07:31	29392	-39807	55	P	t-	1.4063	0.2420	67.6N	144.7W	0	350		
1864	094	-1219 Dec 21	23:06:55	29385	-39802	22	A	-p	0.6690	0.9806	18.1N	36.8W	48	190	93	02m10s
1865	094	-1218 Jun 17	17:53:15	29375	-39796	27	A	-p	-0.4513	0.9790	3.5S	37.8E	63	348	83	02m34s
1866	094	-1218 Dec 11	11:17:09	29366	-39790	32	T	n-	-0.0436	1.0336	24.7S	131.6E	87	14	114	03m03s
1867	094	-1217 Jun 06	20:27:30	29356	-39784	37	A	pn	0.3222	0.9495	39.1N	14.0W	71	159	196	05m24s
1868	094	-1217 Dec 01	02:50:17	29347	-39778	42	T	p-	-0.7066	1.0392	60.9S	129.8W	45	38	186	02m33s
1869	094	-1216 May 25	21:23:51	29338	-39772	47	P	t-	1.0766	0.8367	62.3N	150.9W	0	47		
1870	094	-1216 Oct 21	03:41:34	29330	-39767	14	P	-t	1.4006	0.2702	60.8N	37.9W	0	256		
1871	094	-1216 Nov 19	17:05:52	29328	-39766	52	P	t-	-1.4101	0.2466	62.2S	82.8W	0	130		
1872	094	-1215 Apr 15	18:22:31	29320	-39761	19	T	-p	-0.8528	1.0299	44.3S	61.8E	31	318	189	02m14s
1873	094	-1215 Oct 10	08:31:20	29311	-39755	24	A	-p	0.7562	0.9289	39.3N	158.6W	41	219	395	06m53s
1874	094	-1214 Apr 05	09:54:42	29302	-39749	29	T	nn	-0.0874	1.0743	3.0S	157.3E	85	331	242	06m00s
1875	094	-1214 Sep 29	08:02:25	29292	-39743	34	A	nn	0.0692	0.9239	5.7N	177.5W	86	209	287	08m49s
1876	094	-1213 Mar 26	02:51:53	29283	-39737	39	T	p-	0.6458	1.0582	32.0N	120.3W	50	145	246	04m26s
1877	094	-1213 Sep 18	09:30:30	29273	-39731	44	A	p-	-0.6319	0.9552	27.7S	139.1E	51	32	206	04m29s
1878	094	-1212 Feb 14	04:12:55	29266	-39726	11	P	-t	-1.3218	0.4101	62.0S	8.4E	0	232		
1879	094	-1212 Mar 14	16:33:47	29264	-39725	49	P	t-	1.4331	0.2040	61.0N	20.5W	0	104		
1880	094	-1212 Aug 08	08:13:20	29256	-39720	16	P	-t	1.3489	0.3516	62.8N	46.7W	0	318		
1881	095	-1212 Sep 06	18:11:01	29255	-39719	54	P	t-	-1.3014	0.4434	61.1S	40.4W	0	67		
1882	095	-1211 Feb 02	06:58:40	29247	-39714	21	A	-p	-0.6394	0.9285	56.2S	136.0W	50	326	347	06m28s
1883	095	-1211 Jul 29	00:44:16	29238	-39708	26	T	-p	0.6254	1.0668	58.4N	50.4W	51	208	280	04m25s
1884	095	-1210 Jan 22	06:30:03	29228	-39702	31	A	nn	0.0678	0.9337	18.4S	152.9W	86	165	248	08m26s
1885	095	-1210 Jul 18	16:36:28	29219	-39696	36	Tm	nn	-0.1014	1.0419	17.6N	51.1E	84	11	141	04m04s
1886	095	-1209 Jan 11	10:17:51	29209	-39690	41	A	p-	0.7663	0.9697	26.1N	141.3E	40	170	169	03m21s
1887	095	-1209 Jul 08	03:23:31	29200	-39684	46	A	p-	-0.8972	0.9796	39.9S	118.1W	26	8	165	02m03s
1888	095	-1209 Dec 02	10:12:32	29192	-39679	13	P	-t	-1.2967	0.4503	68.4S	13.4W	0	162		
1889	095	-1209 Dec 31	21:15:47	29191	-39678	51	P	t-	1.4060	0.2491	65.6N	28.8W	0	168		
1890	095	-1208 May 27	16:10:37	29183	-39673	18	P	-t	1.1519	0.7025	69.0N	97.5W	0	24		
1891	095	-1208 Nov 21	01:51:47	29174	-39667	23	T	-p	-0.6463	1.0337	57.1S	101.3W	49	19	149	02m24s
1892	095	-1207 May 16	18:03:07	29164	-39661	28	A	-p	0.3675	0.9740	37.2N	23.1E	68	165	100	02m42s
1893	095	-1207 Nov 10	14:53:08	29155	-39655	33	A	nn	0.0270	0.9923	12.9S	76.0E	89	194	27	00m49s
1894	095	-1206 May 06	02:24:27	29145	-39649	38	T	p-	-0.4199	1.0271	11.5S	89.9W	65	345	101	02m47s
1895	095	-1206 Oct 30	21:20:03	29136	-39643	43	A	p-	0.7517	0.9350	36.5N	6.3W	41	199	363	07m04s
1896	095	-1205 Mar 27	10:08:41	29128	-39638	10	P	-t	1.3732	0.3005	71.6N	68.3E	0	97		
1897	095	-1205 Apr 25	17:10:35	29127	-39637	48	P	t-	-1.1418	0.7476	71.2S	105.6E	0	299		
1898	095	-1205 Oct 19	21:23:26	29117	-39631	53	P	t-	1.4493	0.2033	71.5N	49.6E	0	250		
1899	095	-1204 Mar 16	02:35:36	29110	-39626	20	T	-p	0.6804	1.0439	34.6N	108.5W	47	159	198	03m41s
1900	095	-1204 Sep 08	06:06:41	29100	-39620	25	A	-p	-0.8269	0.9694	43.4S	168.5W	34	22	194	02m53s
1901	096	-1203 Mar 05	14:30:37	29091	-39614	30	A	nn	-0.0573	0.9916	13.5S	88.7E	87	344	30	00m55s
1902	096	-1203 Aug 28	16:54:32	29082	-39608	35	T	nn	-0.0696	1.0347	10.1N	47.2E	86	14	117	03m24s
1903	096	-1202 Feb 22	19:02:52	29072	-39602	40	A	p-	-0.8240	0.9367	66.4S	49.4E	34	325	417	05m12s
1904	096	-1202 Aug 18	08:41:01	29063	-39596	45	T	p-	0.6354	1.0613	55.8N	175.8W	50	199	261	04m15s
1905	096	-1201 Jan 13	01:42:58	29055	-39591	12	P	-t	1.4327	0.2247	66.7N	81.3W	0	180		
1906	096	-1201 Feb 11	18:52:54	29054	-39590	50	Pb	t-	-1.5351	0.0540	69.5S	170.1E	0	211		
1907	096	-1201 Jul 09	15:45:55	29046	-39585	17	P	-t	-1.2523	0.5332	66.0S	72.9E	0	352		
1908	096	-1201 Aug 08	00:53:11	29044	-39584	55	P	t-	1.3603	0.3301	68.6N	85.9E	0	339		
1909	096	-1200 Jan 02	07:29:23	29036	-39579	22	A	-p	0.6758	0.9855	18.7N	166.0W	47	185	70	01m37s
1910	096	-1200 Jun 28	00:37:20	29027	-39573	27	A	-p	-0.5311	0.9736	8.1S	65.4W	58	352	111	03m19s
1911	096	-1200 Dec 21	20:03:15	29018	-39567	32	Tm	nn	-0.0403	1.0370	25.9S	0.1E	88	10	124	03m22s
1912	096	-1199 Jun 17	02:47:15	29008	-39561	37	A	nn	0.2344	0.9482	36.1N	106.8W	76	165	196	05m51s
1913	096	-1199 Dec 11	11:43:12	28999	-39555	42	T	p-	-0.7065	1.0393	64.6S	100.1E	45	32	186	02m32s
1914	096	-1198 Jun 06	03:50:59	28990	-39549	47	An	t-	0.9885	0.9524	68.7N	114.3E	8	49	-	02m50s
1915	096	-1198 Nov 01	12:08:31	28982	-39544	14	P	-t	1.4082	0.2579	61.1N	174.8W	0	247		
1916	096	-1198 Dec 01	01:48:49	28980	-39543	52	P	t-	-1.4147	0.2388	63.0S	135.8E	0	139		
1917	096	-1197 Apr 27	01:36:01	28973	-39538	19	T	-t	-0.9229	1.0323	48.5S	44.1W	22	317	278	02m21s
1918	096	-1197 Oct 21	16:29:48	28963	-39532	24	A	-p	0.7685	0.9251	36.7N	78.3E	40	216	427	07m35s
1919	096	-1196 Apr 15	17:28:48	28954	-39526	29	T	-n	-0.1542	1.0772	2.2S	43.0E	81	331	252	06m16s
1920	096	-1196 Oct 09	15:46:43	28945	-39520	34	A	nn	0.0883	0.9223	2.1N	64.8E	85	209	295	09m03s

Cat Num	Canon Plate	Calendar Date	TD of Greatest Eclipse	ΔT s	Luna Num	Saros Num	Ecl. Type	QLE	Gamma	Ecl. Mag.	Lat. °	Long. °	Sun Alt °	Sun Azm °	Path Width km	Central Line Dur.
1921	097	-1195 Apr 05	10:28:46	28935	-39514	39	T	p-	0.5854	1.0595	32.3N	124.9E	54	145	237	04m30s
1922	097	-1195 Sep 28	17:25:34	28926	-39508	44	A	p-	-0.6079	0.9555	30.0S	18.0E	52	34	200	04m18s
1923	097	-1194 Feb 24	11:45:31	28918	-39503	11	P	-t	-1.3676	0.3297	61.4S	115.2W	0	241		
1924	097	-1194 Mar 25	23:56:37	28917	-39502	49	P	t-	1.3801	0.3006	60.8N	141.4W	0	95		
1925	097	-1194 Aug 19	16:09:22	28909	-39497	16	P	-t	1.3888	0.2764	62.1N	176.3W	0	308		
1926	097	-1194 Sep 18	02:22:03	28907	-39496	54	P	t-	-1.2717	0.4978	60.8S	173.3W	0	76		
1927	097	-1193 Feb 13	14:18:41	28900	-39491	21	A	-p	-0.6793	0.9300	54.7S	117.5E	47	321	354	06m14s
1928	097	-1193 Aug 09	08:37:50	28890	-39485	26	T	-p	0.6756	1.0623	58.2N	163.2W	47	216	276	04m02s
1929	097	-1192 Feb 02	14:02:01	28881	-39479	31	A	nn	0.0364	0.9381	18.0S	93.7E	88	160	230	07m33s
1930	097	-1192 Jul 29	00:10:28	28872	-39473	36	T	nn	-0.0440	1.0360	19.4N	62.7W	88	16	121	03m26s
1931	097	-1191 Jan 21	18:22:28	28862	-39467	41	A	p-	0.7449	0.9758	24.8N	16.3E	42	165	128	02m36s
1932	097	-1191 Jul 18	10:23:28	28853	-39461	46	A	p-	-0.8318	0.9758	32.3S	132.9E	33	12	155	02m35s
1933	097	-1191 Dec 12	19:03:24	28845	-39456	13	P	-t	-1.2966	0.4505	67.4S	159.0W	0	173		
1934	097	-1190 Jan 11	05:48:19	28844	-39455	51	P	t-	1.3915	0.2749	64.6N	168.4W	0	158		
1935	097	-1190 Jun 07	22:28:14	28836	-39450	18	P	-t	1.2388	0.5539	68.1N	155.0E	0	13		
1936	097	-1190 Dec 02	10:43:27	28827	-39444	23	T	-p	-0.6469	1.0326	60.6S	128.7E	49	13	145	02m17s
1937	097	-1189 May 28	00:37:05	28817	-39438	28	A	-p	0.4515	0.9763	45.6N	77.4W	63	168	95	02m18s
1938	097	-1189 Nov 21	23:29:47	28808	-39432	33	A	nn	0.0280	0.9891	16.3S	54.4W	88	191	38	01m11s
1939	097	-1188 May 16	09:29:05	28799	-39426	38	T	p-	-0.3414	1.0319	3.5S	159.8E	70	348	114	03m20s
1940	097	-1188 Nov 10	05:31:05	28790	-39420	43	A	p-	0.7504	0.9320	33.1N	133.3W	41	196	382	07m47s
1941	098	-1187 Apr 06	17:48:03	28782	-39415	10	P	-t	1.4330	0.1840	71.5N	62.0W	0	84		
1942	098	-1187 May 06	00:35:19	28780	-39414	48	P	t-	-1.0692	0.8888	70.6S	20.4W	0	312		
1943	098	-1187 Oct 30	05:19:26	28771	-39408	53	P	t-	1.4418	0.2161	71.0N	84.7W	0	236		
1944	098	-1186 Mar 27	10:13:06	28763	-39403	20	T	-p	0.7399	1.0429	43.0N	130.6E	42	155	211	03m23s
1945	098	-1186 Sep 19	14:02:28	28754	-39397	25	A	-p	-0.8498	0.9689	49.0S	65.9E	31	27	210	02m44s
1946	098	-1185 Mar 16	21:52:39	28745	-39391	30	A	nn	-0.0023	0.9912	6.4S	25.0W	90	347	31	00m59s
1947	098	-1185 Sep 09	01:04:17	28735	-39385	35	T	nn	-0.1000	1.0341	4.6N	77.7W	84	16	115	03m19s
1948	098	-1184 Mar 05	02:08:32	28726	-39379	40	A	p-	-0.7730	0.9391	58.0S	64.7W	39	330	355	05m26s
1949	098	-1184 Aug 28	16:50:05	28717	-39373	45	T	p-	0.5979	1.0587	49.3N	60.5E	53	201	241	04m16s
1950	098	-1183 Jan 23	09:30:09	28709	-39368	12	P	-t	1.4509	0.1921	67.8N	149.4E	0	169		
1951	098	-1183 Feb 22	02:06:01	28708	-39367	50	P	t-	-1.4909	0.1267	70.3S	47.7E	0	223		
1952	098	-1183 Jul 19	23:06:27	28700	-39362	17	P	-t	-1.3189	0.4103	67.0S	49.4W	0	2		
1953	098	-1183 Aug 18	08:44:20	28698	-39361	55	P	t-	1.3193	0.4077	69.5N	45.4W	0	327		
1954	098	-1182 Jan 12	15:48:15	28691	-39356	22	A	-p	0.6863	0.9908	20.2N	65.7E	47	180	45	01m00s
1955	098	-1182 Jul 09	07:23:58	28681	-39350	27	A	-p	-0.6074	0.9678	13.7S	169.9W	53	356	146	04m04s
1956	098	-1181 Jan 02	04:44:26	28672	-39344	32	T	nn	-0.0323	1.0408	25.8S	130.1W	88	5	137	03m45s
1957	098	-1181 Jun 28	09:11:15	28663	-39338	37	A	nn	0.1501	0.9466	32.4N	158.5E	81	171	199	06m23s
1958	098	-1181 Dec 22	20:31:34	28653	-39332	42	T	p-	-0.7031	1.0399	67.1S	25.7W	45	23	189	02m34s
1959	098	-1180 Jun 16	10:23:42	28644	-39326	47	A	t-	0.9035	0.9582	77.9N	64.7E	25	98	359	02m48s
1960	098	-1180 Nov 11	20:35:55	28636	-39321	14	P	-t	1.4146	0.2475	61.7N	48.1E	0	237		
1961	099	-1180 Dec 11	10:27:44	28635	-39320	52	P	t-	-1.4171	0.2350	63.8S	4.8W	0	149		
1962	099	-1179 May 07	08:49:20	28627	-39315	19	Ts	-t	-0.9938	1.0314	59.0S	138.7W	5	306	-	02m03s
1963	099	-1179 Nov 01	00:31:51	28618	-39309	24	A	-p	0.7772	0.9220	34.2N	46.0W	39	212	454	08m17s
1964	099	-1178 Apr 27	00:57:21	28609	-39303	29	T	-n	-0.2259	1.0793	2.0S	69.9W	77	333	262	06m31s
1965	099	-1178 Oct 20	23:38:55	28599	-39297	34	A	nn	0.1010	0.9212	1.6S	55.0W	84	209	299	09m18s
1966	099	-1177 Apr 16	17:57:28	28590	-39291	39	T	p-	0.5187	1.0599	32.7N	12.7E	59	146	228	04m33s
1967	099	-1177 Oct 10	01:30:33	28581	-39285	44	A	p-	-0.5913	0.9562	33.3S	105.5W	54	36	194	04m07s
1968	099	-1176 Mar 06	19:05:57	28573	-39280	11	P	-t	-1.4222	0.2341	61.0S	124.5E	0	250		
1969	099	-1176 Apr 05	07:09:46	28571	-39279	49	P	t-	1.3203	0.4096	60.7N	100.1E	0	87		
1970	099	-1176 Aug 30	00:15:43	28564	-39274	16	P	-t	1.4210	0.2159	61.6N	51.6E	0	299		
1971	099	-1176 Sep 28	10:43:18	28562	-39273	54	P	t-	-1.2500	0.5376	60.6S	51.2E	0	85		
1972	099	-1175 Feb 23	21:28:39	28555	-39268	21	A	-p	-0.7271	0.9315	53.3S	13.0E	43	316	367	06m01s
1973	099	-1175 Aug 19	16:37:58	28545	-39262	26	T	-p	0.7202	1.0574	56.9N	80.9E	44	223	271	03m41s
1974	099	-1174 Feb 12	21:26:57	28536	-39256	31	A	nn	-0.0013	0.9430	17.2S	18.0W	90	351	211	06m40s
1975	099	-1174 Aug 09	07:49:30	28527	-39250	36	T	nn	0.0084	1.0297	20.1N	177.7W	89	196	101	02m47s
1976	099	-1173 Feb 02	02:20:12	28517	-39244	41	A	p-	0.7170	0.9825	23.7N	106.5W	44	160	88	01m50s
1977	099	-1173 Jul 29	17:28:25	28508	-39238	46	A	p-	-0.7713	0.9712	27.2S	23.4E	39	16	160	03m10s
1978	099	-1173 Dec 24	03:51:48	28501	-39233	13	P	-t	-1.2990	0.4458	66.3S	56.5E	0	184		
1979	099	-1172 Jan 22	14:15:18	28499	-39232	51	P	t-	1.3721	0.3102	63.6N	53.7E	0	148		
1980	099	-1172 Jun 18	04:49:01	28491	-39227	18	P	-t	1.3224	0.4108	67.1N	47.3E	0	3		

Cat Num	Canon Plate	Calendar Date	TD of Greatest Eclipse	ΔT s	Luna Num	Saros Num	Ecl. Type	QLE	Gamma	Ecl. Mag.	Lat. °	Long. °	Sun Alt °	Sun Azm °	Path Width km	Central Line Dur.
1981	100	-1172 Jul 17	20:10:38	28490	-39226	56	Pb	t-	-1.5509	0.0224	64.4S	30.8W	0	24		
1982	100	-1172 Dec 12	19:33:06	28482	-39221	23	T	-p	-0.6490	1.0317	63.2S	1.3E	49	5	142	02m12s
1983	100	-1171 Jun 07	07:14:58	28473	-39215	28	A	-p	0.5331	0.9782	53.5N	177.2W	58	172	92	01m58s
1984	100	-1171 Dec 02	08:05:21	28464	-39209	33	A	nn	0.0291	0.9864	19.0S	175.9E	88	187	48	01m29s
1985	100	-1170 May 27	16:34:16	28454	-39203	38	T	n-	-0.2624	1.0360	4.0N	50.0E	75	351	125	03m46s
1986	100	-1170 Nov 21	13:42:38	28445	-39197	43	A	p-	0.7502	0.9296	30.3N	99.6E	41	191	397	08m26s
1987	100	-1169 Apr 18	01:22:14	28437	-39192	10	Pe	-t	1.4971	0.0590	71.3N	169.3E	0	70		
1988	100	-1169 May 17	07:59:08	28436	-39191	48	Ts	t-	-0.9953	1.0601	67.1S	150.8W	3	329	-	03m46s
1989	100	-1169 Nov 10	13:18:03	28427	-39185	53	P	t-	1.4363	0.2256	70.3N	140.9E	0	223		
1990	100	-1168 Apr 06	17:43:29	28419	-39180	20	T	-p	0.8047	1.0409	52.2N	9.7E	36	150	229	02m59s
1991	100	-1168 Sep 29	22:08:02	28410	-39174	25	A	-p	-0.8658	0.9689	54.3S	62.2W	30	32	222	02m34s
1992	100	-1167 Mar 27	05:04:59	28400	-39168	30	A	nn	0.0597	0.9904	1.3N	136.6W	87	162	34	01m04s
1993	100	-1167 Sep 19	09:22:26	28391	-39162	35	T	nn	-0.1239	1.0336	0.9S	155.2E	83	17	114	03m14s
1994	100	-1166 Mar 16	09:03:26	28382	-39156	40	A	p-	-0.7137	0.9415	49.2S	175.3W	44	333	307	05m40s
1995	100	-1166 Sep 09	01:07:30	28373	-39150	45	T	p-	0.5678	1.0558	43.1N	66.0W	55	202	223	04m13s
1996	100	-1165 Feb 03	17:07:22	28365	-39145	12	P	-t	1.4773	0.1451	68.8N	22.1E	0	158		
1997	100	-1165 Mar 05	09:07:32	28364	-39144	50	P	t-	-1.4370	0.2165	71.0S	72.2W	0	236		
1998	100	-1165 Jul 31	06:32:53	28356	-39139	17	P	-t	-1.3793	0.3003	68.0S	173.5W	0	13		
1999	100	-1165 Aug 29	16:43:59	28354	-39138	55	P	t-	1.2862	0.4700	70.3N	179.4W	0	315		
2000	100	-1164 Jan 23	23:59:47	28347	-39133	22	A	-p	0.7031	0.9964	22.9N	61.1W	45	176	18	00m23s
2001	101	-1164 Jul 19	14:15:09	28337	-39127	27	A	-p	-0.6788	0.9617	19.9S	83.6E	47	1	189	04m46s
2002	101	-1163 Jan 12	13:20:23	28328	-39121	32	T	nn	-0.0199	1.0449	24.5S	100.9E	89	0	150	04m08s
2003	101	-1163 Jul 08	15:39:35	28319	-39115	37	A	nn	0.0697	0.9446	27.9N	61.9E	86	176	205	07m00s
2004	101	-1162 Jan 02	05:16:03	28310	-39109	42	T	p-	-0.6968	1.0409	67.9S	148.2W	46	12	192	02m40s
2005	101	-1162 Jun 27	16:59:29	28301	-39103	47	A	p-	0.8199	0.9615	76.7N	9.0E	35	143	246	02m47s
2006	101	-1162 Nov 23	05:05:13	28293	-39098	14	P	-t	1.4187	0.2411	62.3N	89.7W	0	228		
2007	101	-1162 Dec 22	19:03:55	28291	-39097	52	P	t-	-1.4181	0.2335	64.8S	145.1W	0	159		
2008	101	-1161 May 18	16:01:23	28284	-39092	19	P	-t	-1.0662	0.8865	62.0S	111.3E	0	307		
2009	101	-1161 Jun 17	00:39:38	28282	-39091	57	Pb	t-	1.5268	0.0278	64.0N	136.6E	0	28		
2010	101	-1161 Nov 12	08:36:42	28274	-39086	24	A	-p	0.7830	0.9196	32.1N	171.4W	38	208	476	08m58s
2011	101	-1160 May 07	08:24:20	28265	-39080	29	T	-n	-0.2990	1.0806	2.4S	177.5E	73	335	271	06m45s
2012	101	-1160 Oct 31	07:36:48	28256	-39074	34	A	nn	0.1089	0.9207	5.4S	176.2W	84	207	302	09m32s
2013	101	-1159 Apr 27	01:22:00	28247	-39068	39	T	p-	0.4491	1.0598	33.1N	97.9W	63	147	218	04m35s
2014	101	-1159 Oct 20	09:42:04	28238	-39062	44	A	p-	-0.5793	0.9571	37.0S	129.6E	54	36	188	03m55s
2015	101	-1158 Mar 18	02:17:44	28230	-39057	11	P	-t	-1.4826	0.1282	60.7S	6.4E	0	259		
2016	101	-1158 Apr 16	14:16:12	28228	-39056	49	P	t-	1.2557	0.5268	60.8N	16.6W	0	78		
2017	101	-1158 Sep 10	08:30:13	28221	-39051	16	P	-t	1.4473	0.1666	61.2N	82.3W	0	290		
2018	101	-1158 Oct 09	19:11:37	28219	-39050	54	P	t-	-1.2332	0.5683	60.6S	85.9W	0	95		
2019	101	-1157 Mar 07	04:27:59	28211	-39045	21	A	-p	-0.7833	0.9328	52.4S	88.9W	38	312	395	05m49s
2020	101	-1157 Aug 31	00:46:43	28202	-39039	26	T	-p	0.7576	1.0524	54.8N	39.1W	40	227	263	03m23s
2021	102	-1156 Feb 24	04:43:07	28193	-39033	31	A	nn	-0.0471	0.9480	16.2S	127.4W	87	335	191	05m52s
2022	102	-1156 Aug 19	15:35:41	28184	-39027	36	T	nn	0.0541	1.0233	19.6N	65.5E	87	203	80	02m09s
2023	102	-1155 Feb 12	10:10:21	28175	-39021	41	A	p-	0.6817	0.9893	22.9N	133.1E	47	157	50	01m04s
2024	102	-1155 Aug 09	00:40:11	28165	-39015	46	A	p-	-0.7169	0.9663	24.0S	87.3W	44	20	172	03m44s
2025	102	-1154 Jan 03	12:34:37	28158	-39010	13	P	-t	-1.3063	0.4320	65.3S	86.0W	0	195		
2026	102	-1154 Feb 01	22:33:56	28156	-39009	51	P	t-	1.3456	0.3590	62.8N	81.8W	0	138		
2027	102	-1154 Jun 29	11:13:28	28149	-39004	18	P	-t	1.4029	0.2735	66.0N	60.8W	0	352		
2028	102	-1154 Jul 29	02:53:30	28147	-39003	56	P	t-	-1.4854	0.1352	63.5S	142.7W	0	34		
2029	102	-1154 Dec 24	04:18:39	28139	-38998	23	T	-p	-0.6543	1.0314	64.7S	123.2W	49	355	141	02m09s
2030	102	-1153 Jun 18	13:56:56	28130	-38992	28	A	-p	0.6121	0.9793	60.9N	84.5E	52	178	94	01m44s
2031	102	-1153 Dec 13	16:37:21	28121	-38986	33	A	nn	0.0277	0.9844	21.1S	47.3E	89	182	55	01m43s
2032	102	-1152 Jun 06	23:42:32	28112	-38980	38	T	n-	-0.1849	1.0392	10.8N	59.9W	79	355	134	04m04s
2033	102	-1152 Dec 01	21:54:20	28103	-38974	43	A	p-	0.7504	0.9280	28.0N	27.5W	41	187	409	08m57s
2034	102	-1151 May 27	15:21:38	28094	-38968	48	T	t-	-0.9200	1.0676	47.4S	76.5E	23	347	567	05m04s
2035	102	-1151 Nov 20	21:20:09	28084	-38962	53	P	t-	1.4332	0.2307	69.5N	6.3E	0	210		
2036	102	-1150 Apr 18	01:06:16	28077	-38957	20	T	-t	0.8751	1.0378	62.5N	113.9W	29	140	263	02m31s
2037	102	-1150 Oct 11	06:22:22	28067	-38951	25	A	-p	-0.8750	0.9693	59.2S	167.8E	29	37	227	02m24s
2038	102	-1149 Apr 07	12:07:46	28058	-38945	30	A	nn	0.1285	0.9893	9.4N	113.8E	83	162	38	01m11s
2039	102	-1149 Sep 30	17:49:36	28049	-38939	35	T	-n	-0.1406	1.0331	6.2S	25.9E	82	18	113	03m09s
2040	102	-1148 Mar 26	15:50:19	28040	-38933	40	A	p-	-0.6482	0.9436	40.3S	76.8E	49	336	271	05m55s

Cat Num	Canon Plate	Calendar Date	TD of Greatest Eclipse	ΔT s	Luna Num	Saros Num	Ecl. Type	QLE	Gamma	Ecl. Mag.	Lat. °	Long. °	Sun Alt °	Sun Azm °	Path Width km	Central Line Dur.
2041	103	-1148 Sep 19	09:31:35	28031	-38927	45	T	p-	0.5435	1.0528	37.1N	165.3E	57	202	207	04m10s
2042	103	-1147 Feb 14	00:38:45	28023	-38922	12	P	-t	1.5086	0.0890	69.7N	104.4W	0	145		
2043	103	-1147 Mar 15	16:03:24	28022	-38921	50	P	t-	-1.3779	0.3161	71.5S	168.7E	0	250		
2044	103	-1147 Aug 10	14:04:50	28014	-38916	17	P	-t	-1.4341	0.2018	69.0S	60.3E	0	24		
2045	103	-1147 Sep 09	00:50:01	28012	-38915	55	P	t-	1.2589	0.5206	71.0N	44.4E	0	302		
2046	103	-1146 Feb 03	08:04:56	28005	-38910	22	H	-p	0.7262	1.0022	26.8N	173.3E	43	171	11	00m13s
2047	103	-1146 Jul 30	21:11:33	27996	-38904	27	A	-p	-0.7450	0.9554	26.8S	25.0W	42	5	244	05m22s
2048	103	-1145 Jan 23	21:49:22	27986	-38898	32	T	nn	-0.0012	1.0493	21.9S	26.7W	90	359	164	04m33s
2049	103	-1145 Jul 19	22:15:27	27977	-38892	37	Am	nn	-0.0047	0.9423	22.9N	37.5W	90	1	213	07m36s
2050	103	-1144 Jan 13	13:52:43	27968	-38886	42	T	p-	-0.6845	1.0423	66.7S	91.4E	47	360	195	02m48s
2051	103	-1144 Jul 07	23:44:20	27959	-38880	47	A	p-	0.7428	0.9638	71.7N	70.1W	42	168	197	02m48s
2052	103	-1144 Dec 03	13:32:19	27951	-38875	14	P	-t	1.4236	0.2332	63.1N	132.9E	0	218		
2053	103	-1143 Jan 02	03:33:34	27950	-38874	52	P	t-	-1.4149	0.2394	65.8S	75.9E	0	170		
2054	103	-1143 May 28	23:14:28	27942	-38869	19	P	-t	-1.1384	0.7506	62.7S	7.6W	0	316		
2055	103	-1143 Jun 27	07:47:13	27941	-38868	57	P	t-	1.4501	0.1683	65.0N	18.5E	0	19		
2056	103	-1143 Nov 22	16:42:10	27933	-38863	24	A	-p	0.7876	0.9179	30.5N	63.0E	38	204	494	09m36s
2057	103	-1142 May 18	15:48:36	27924	-38857	29	T	-n	-0.3742	1.0809	3.7S	65.4E	68	338	280	06m56s
2058	103	-1142 Nov 11	15:38:43	27915	-38851	34	Am	nn	0.1140	0.9209	8.9S	61.8E	84	204	301	09m43s
2059	103	-1141 May 08	08:40:41	27905	-38845	39	T	n-	0.3750	1.0587	33.2N	153.4E	68	150	207	04m36s
2060	103	-1141 Oct 31	18:01:02	27896	-38839	44	A	p-	-0.5727	0.9586	41.2S	3.2E	55	36	181	03m41s
2061	104	-1140 Mar 28	09:18:50	27889	-38834	11	Pe	-t	-1.5505	0.0093	60.6S	109.0W	0	267		
2062	104	-1140 Apr 26	21:15:19	27887	-38833	49	P	t-	1.1859	0.6532	61.1N	131.6W	0	69		
2063	104	-1140 Sep 20	16:53:11	27880	-38828	16	P	-t	1.4675	0.1290	60.9N	141.7E	0	281		
2064	104	-1140 Oct 20	03:46:53	27878	-38827	54	P	t-	-1.2217	0.5894	60.8S	135.2E	0	104		
2065	104	-1139 Mar 17	11:18:25	27870	-38822	21	A	-p	-0.8464	0.9338	52.5S	171.8E	32	308	452	05m38s
2066	104	-1139 Sep 10	09:03:02	27861	-38816	26	T	-p	0.7886	1.0474	52.3N	162.9W	38	228	253	03m05s
2067	104	-1138 Mar 06	11:50:47	27852	-38810	31	Am	nn	-0.1004	0.9532	15.0S	125.2E	84	332	172	05m07s
2068	104	-1138 Aug 30	23:28:43	27843	-38804	36	H3	nn	0.0932	1.0168	18.0N	53.2W	85	206	58	01m32s
2069	104	-1137 Feb 23	17:53:24	27834	-38798	41	A	p-	0.6398	0.9965	22.5N	15.0E	50	153	16	00m20s
2070	104	-1137 Aug 20	07:58:52	27825	-38792	46	A	p-	-0.6690	0.9610	22.4S	160.5E	48	24	186	04m17s
2071	104	-1136 Jan 14	21:12:03	27817	-38787	13	P	-t	-1.3183	0.4091	64.3S	133.2E	0	205		
2072	104	-1136 Feb 13	06:45:54	27816	-38786	51	P	t-	1.3134	0.4191	62.1N	144.6E	0	129		
2073	104	-1136 Jul 09	17:44:45	27808	-38781	18	P	-t	1.4776	0.1463	65.1N	170.2W	0	343		
2074	104	-1136 Aug 08	09:44:32	27806	-38780	56	P	t-	-1.4262	0.2368	62.7S	103.8E	0	43		
2075	104	-1135 Jan 03	12:59:50	27799	-38775	23	T	-p	-0.6630	1.0314	64.9S	114.2E	48	344	142	02m08s
2076	104	-1135 Jun 28	20:43:51	27790	-38769	28	A	-p	0.6878	0.9799	67.3N	10.7W	46	188	99	01m34s
2077	104	-1135 Dec 24	01:05:28	27781	-38763	33	A	nn	0.0240	0.9828	22.4S	80.0W	89	177	61	01m54s
2078	104	-1134 Jun 18	06:54:15	27771	-38757	38	T	nn	-0.1092	1.0418	16.7N	169.7W	84	359	141	04m13s
2079	104	-1134 Dec 13	06:01:40	27762	-38751	43	A	p-	0.7477	0.9270	26.0N	153.4W	41	182	414	09m19s
2080	104	-1133 Jun 07	22:45:58	27753	-38745	48	T	p-	-0.8458	1.0708	36.5S	42.7W	32	354	433	05m48s
2081	105	-1133 Dec 02	05:21:10	27744	-38739	53	P	t-	1.4287	0.2379	68.5N	127.4W	0	198		
2082	105	-1132 Apr 28	08:24:17	27736	-38734	20	T	-t	0.9484	1.0331	72.9N	106.5E	18	114	360	01m58s
2083	105	-1132 Oct 21	14:42:46	27727	-38728	25	A	-p	-0.8802	0.9703	63.8S	36.7E	28	41	225	02m12s
2084	105	-1131 Apr 17	19:03:13	27718	-38722	30	A	nn	0.2024	0.9877	17.7N	6.1E	78	162	44	01m21s
2085	105	-1131 Oct 11	02:23:56	27709	-38716	35	T	-n	-0.1518	1.0329	11.3S	105.0W	81	18	112	03m05s
2086	105	-1130 Apr 06	22:26:34	27700	-38710	40	A	p-	-0.5743	0.9456	31.2S	28.1W	55	339	242	06m10s
2087	105	-1130 Sep 30	18:04:06	27691	-38704	45	T	p-	0.5265	1.0498	31.5N	34.3E	58	201	193	04m05s
2088	105	-1129 Feb 25	08:00:19	27683	-38699	12	Pe	-t	1.5482	0.0177	70.4N	131.0E	0	133		
2089	105	-1129 Mar 26	22:49:48	27682	-38698	50	P	t-	-1.3101	0.4314	71.8S	51.8E	0	263		
2090	105	-1129 Aug 21	21:44:25	27674	-38693	17	P	-t	-1.4816	0.1178	69.9S	68.3W	0	36		
2091	105	-1129 Sep 20	09:04:16	27673	-38692	55	P	t-	1.2390	0.5572	71.4N	94.3W	0	288		
2092	105	-1128 Feb 14	16:01:43	27665	-38687	22	H	-p	0.7569	1.0080	31.8N	49.2E	41	167	42	00m46s
2093	105	-1128 Aug 10	04:15:31	27656	-38681	27	A	-p	-0.8039	0.9490	34.1S	136.5W	36	10	314	05m49s
2094	105	-1127 Feb 03	06:11:50	27647	-38675	32	T	nn	0.0231	1.0530	18.3S	153.3W	89	171	178	04m59s
2095	105	-1127 Jul 30	04:57:21	27638	-38669	37	A	nn	-0.0740	0.9398	17.4N	139.2W	86	5	224	08m11s
2096	105	-1126 Jan 23	22:23:45	27629	-38663	42	T	p-	-0.6681	1.0440	63.6S	30.0W	48	351	198	02m58s
2097	105	-1126 Jul 19	06:35:40	27619	-38657	47	A	p-	0.6701	0.9653	65.4N	163.2W	48	181	170	02m51s
2098	105	-1126 Dec 14	21:56:41	27612	-38652	14	P	-t	1.4299	0.2226	64.0N	4.1W	0	208		
2099	105	-1125 Jan 13	11:56:16	27610	-38651	52	P	t-	-1.4071	0.2534	66.8S	61.9W	0	181		
2100	105	-1125 Jun 09	06:29:31	27603	-38646	19	P	-t	-1.2093	0.6161	63.5S	127.3W	0	326		

Cat Num	Canon Plate	Calendar Date	TD of Greatest Eclipse	ΔT s	Luna Num	Saros Num	Ecl. Type	QLE	Gamma	Ecl. Mag.	Lat. °	Long. °	Sun Alt °	Sun Azm °	Path Width km	Central Line Dur.
2101	106	-1125 Jul 08	15:01:08	27601	-38645	57	P	t-	1.3775	0.3018	65.9N	101.6W	0	9		
2102	106	-1125 Dec 04	00:47:19	27594	-38640	24	A	-p	0.7917	0.9169	29.4N	62.5W	37	199	508	10m07s
2103	106	-1124 May 28	23:13:30	27585	-38634	29	T	-p	-0.4490	1.0804	5.8S	47.1W	63	341	288	07m03s
2104	106	-1124 Nov 21	23:42:19	27575	-38628	34	A	nn	0.1179	0.9218	11.9S	60.4W	83	201	298	09m52s
2105	106	-1123 May 18	15:57:41	27566	-38622	39	T	n-	0.3003	1.0569	32.9N	45.3E	72	154	196	04m35s
2106	106	-1123 Nov 11	02:23:36	27557	-38616	44	A	p-	-0.5683	0.9606	45.4S	123.4W	55	35	172	03m26s
2107	106	-1122 May 08	04:08:41	27548	-38610	49	P	t-	1.1121	0.7859	61.5N	114.7E	0	61		
2108	106	-1122 Oct 02	01:24:08	27541	-38605	16	P	-t	1.4819	0.1024	60.8N	3.8E	0	272		
2109	106	-1122 Oct 31	12:27:26	27539	-38604	54	P	t-	-1.2142	0.6033	61.1S	5.1W	0	113		
2110	106	-1121 Mar 28	17:59:56	27532	-38599	21	A	-t	-0.9160	0.9340	54.2S	77.1E	23	303	599	05m27s
2111	106	-1121 Sep 21	17:27:06	27522	-38593	26	T	-p	0.8129	1.0426	49.7N	69.8E	35	228	240	02m50s
2112	106	-1120 Mar 16	18:50:43	27513	-38587	31	A	nn	-0.1607	0.9582	13.9S	19.8E	81	331	154	04m28s
2113	106	-1120 Sep 10	07:29:50	27504	-38581	36	H	-n	0.1245	1.0105	15.6N	174.3W	83	208	36	00m57s
2114	106	-1119 Mar 06	01:28:48	27495	-38575	41	H	p-	0.5905	1.0037	22.4N	100.6W	54	151	16	00m21s
2115	106	-1119 Aug 30	15:24:47	27486	-38569	46	A	p-	-0.6278	0.9557	22.1S	46.7E	51	27	203	04m47s
2116	106	-1118 Jan 25	05:42:04	27478	-38564	13	P	-t	-1.3367	0.3740	63.4S	5.3W	0	215		
2117	106	-1118 Feb 23	14:49:11	27477	-38563	51	P	t-	1.2737	0.4938	61.5N	13.4E	0	120		
2118	106	-1118 Jul 21	00:22:48	27469	-38558	18	Pe	-t	1.5469	0.0288	64.1N	79.0E	0	333		
2119	106	-1118 Aug 19	16:44:36	27468	-38557	56	P	t-	-1.3738	0.3261	62.0S	11.8W	0	53		
2120	106	-1117 Jan 14	21:32:42	27460	-38552	23	T	-p	-0.6783	1.0316	64.1S	6.7W	47	334	145	02m08s
2121	107	-1117 Jul 10	03:38:09	27451	-38546	28	A	-p	0.7583	0.9799	72.1N	101.1W	40	204	110	01m28s
2122	107	-1116 Jan 04	09:27:08	27442	-38540	33	A	nn	0.0153	0.9818	23.0S	154.4E	89	171	65	02m00s
2123	107	-1116 Jun 28	14:10:36	27433	-38534	38	T	nn	-0.0367	1.0436	21.8N	80.1E	88	4	146	04m16s
2124	107	-1116 Dec 23	14:05:12	27424	-38528	43	A	p-	0.7423	0.9267	24.4N	81.6E	42	178	412	09m30s
2125	107	-1115 Jun 18	06:11:48	27415	-38522	48	T	p-	-0.7726	1.0725	27.8S	159.8W	39	359	371	06m18s
2126	107	-1115 Dec 12	13:21:48	27406	-38516	53	P	t-	1.4240	0.2455	67.4N	99.6E	0	186		
2127	107	-1114 May 09	15:35:44	27398	-38511	20	P	-t	1.0262	0.9579	70.4N	75.9W	0	44		
2128	107	-1114 Jun 07	23:16:36	27397	-38510	58	Pb	t-	-1.5122	0.0354	67.9S	43.0W	0	347		
2129	107	-1114 Nov 01	23:10:13	27389	-38505	25	A	-p	-0.8802	0.9718	68.2S	94.9W	28	44	214	02m00s
2130	107	-1113 Apr 29	01:51:42	27380	-38499	30	A	-p	0.2810	0.9855	26.2N	99.7W	74	163	53	01m33s
2131	107	-1113 Oct 22	11:04:26	27371	-38493	35	Tm	-n	-0.1583	1.0330	16.2S	122.9E	81	17	113	03m04s
2132	107	-1112 Apr 17	04:57:35	27362	-38487	40	A	p-	-0.4964	0.9472	22.2S	131.1W	60	342	222	06m25s
2133	107	-1112 Oct 11	02:42:45	27353	-38481	45	T	p-	0.5147	1.0469	26.4N	98.3W	59	200	181	03m59s
2134	107	-1111 Apr 06	05:31:29	27344	-38475	50	P	t-	-1.2378	0.5555	71.8S	64.1W	0	277		
2135	107	-1111 Sep 01	05:29:52	27336	-38470	17	Pe	-t	-1.5234	0.0453	70.6S	161.1E	0	49		
2136	107	-1111 Sep 30	17:24:12	27335	-38469	55	P	t-	1.2245	0.5834	71.6N	125.2E	0	274		
2137	107	-1110 Feb 24	23:52:12	27327	-38464	22	T	-p	0.7934	1.0136	37.9N	74.1W	37	162	76	01m13s
2138	107	-1110 Aug 21	11:26:42	27318	-38458	27	A	-p	-0.8556	0.9425	41.7S	109.0E	31	16	409	06m07s
2139	107	-1109 Feb 14	14:25:24	27309	-38452	32	T	nn	0.0548	1.0583	13.5S	81.6E	87	168	192	05m23s
2140	107	-1109 Aug 10	11:49:26	27300	-38446	37	A	nn	-0.1346	0.9372	11.6N	115.9E	82	8	236	08m39s
2141	108	-1108 Feb 04	06:45:48	27291	-38440	42	T	p-	-0.6450	1.0458	59.0S	152.4W	50	344	200	03m12s
2142	108	-1108 Jul 29	13:36:42	27282	-38434	47	A	p-	0.6047	0.9663	58.8N	95.4E	53	189	153	02m56s
2143	108	-1108 Dec 25	06:15:08	27274	-38429	14	P	-t	1.4400	0.2055	65.0N	139.9W	0	198		
2144	108	-1107 Jan 23	20:10:09	27273	-38428	52	P	t-	-1.3934	0.2781	67.9S	162.1E	0	192		
2145	108	-1107 Jun 19	13:48:36	27266	-38423	19	P	-t	-1.2772	0.4869	64.4S	111.9E	0	335		
2146	108	-1107 Jul 18	22:22:57	27264	-38422	57	P	t-	1.3106	0.4253	66.9N	136.0E	0	358		
2147	108	-1107 Dec 14	08:49:04	27256	-38417	24	A	-p	0.7982	0.9166	29.2N	172.8E	37	194	522	10m27s
2148	108	-1106 Jun 09	06:37:50	27247	-38411	29	T	-p	-0.5241	1.0788	8.9S	159.9W	58	345	297	07m04s
2149	108	-1106 Dec 03	07:46:45	27238	-38405	34	A	nn	0.1215	0.9234	14.2S	177.4E	83	197	291	09m54s
2150	108	-1105 May 29	23:12:07	27229	-38399	39	T	nn	0.2242	1.0541	31.9N	62.0W	77	159	183	04m32s
2151	108	-1105 Nov 22	10:49:20	27220	-38393	44	A	p-	-0.5659	0.9632	49.4S	110.2E	55	32	161	03m09s
2152	108	-1104 May 18	10:58:05	27211	-38387	49	P	t-	1.0360	0.9221	62.0N	2.0E	0	52		
2153	108	-1104 Oct 12	10:03:00	27204	-38382	16	P	-t	1.4903	0.0870	60.8N	136.0W	0	262		
2154	108	-1104 Nov 10	21:12:57	27202	-38381	54	P	t-	-1.2105	0.6101	61.7S	146.7W	0	122		
2155	108	-1103 Apr 08	00:33:22	27195	-38376	21	As	-t	-0.9919	0.9319	59.5S	0.9W	6	286	-	05m07s
2156	108	-1103 Oct 02	01:58:07	27186	-38370	26	T	-p	0.8312	1.0379	47.2N	60.6W	33	226	224	02m36s
2157	108	-1102 Mar 28	01:44:10	27177	-38364	31	A	nn	-0.2270	0.9632	13.0S	84.0W	77	330	136	03m52s
2158	108	-1102 Sep 21	15:37:58	27168	-38358	36	H	-n	0.1494	1.0042	12.5N	62.4E	81	209	15	00m23s
2159	108	-1101 Mar 17	08:57:26	27159	-38352	41	H	p-	0.5345	1.0108	22.5N	145.9E	58	149	43	00m59s
2160	108	-1101 Sep 10	22:58:38	27150	-38346	46	A	p-	-0.5936	0.9504	23.1S	69.0W	53	30	221	05m15s

Cat Num	Canon Plate	Calendar Date	TD of Greatest Eclipse	ΔT s	Luna Num	Saros Num	Ecl. Type	QLE	Gamma	Ecl. Mag.	Lat. °	Long. °	Sun Alt °	Sun Azm °	Path Width km	Central Line Dur.
2161	109	-1100 Feb 05	14:05:47	27142	-38341	13	P	-t	-1.3608	0.3277	62.6S	142.0W	0	225		
2162	109	-1100 Mar 05	22:46:35	27140	-38340	51	P	t-	1.2287	0.5796	61.1N	116.3W	0	111		
2163	109	-1100 Aug 29	23:52:26	27131	-38334	56	P	t-	-1.3272	0.4050	61.4S	129.1W	0	62		
2164	109	-1099 Jan 25	05:59:25	27124	-38329	23	T	-p	-0.6985	1.0318	62.4S	127.4W	45	325	150	02m08s
2165	109	-1099 Jul 20	10:40:21	27115	-38323	28	A	-p	0.8232	0.9795	74.3N	173.6E	34	227	129	01m25s
2166	109	-1098 Jan 14	17:41:00	27106	-38317	33	A	nn	0.0011	0.9811	23.1S	30.9E	90	10	67	02m03s
2167	109	-1098 Jul 09	21:33:01	27097	-38311	38	T	nn	0.0316	1.0447	25.7N	30.8W	88	188	149	04m12s
2168	109	-1097 Jan 03	22:01:28	27088	-38305	43	A	p-	0.7314	0.9271	23.1N	41.4W	43	173	401	09m29s
2169	109	-1097 Jun 29	13:42:32	27079	-38299	48	T	p-	-0.7034	1.0729	21.0S	83.1E	45	3	333	06m33s
2170	109	-1097 Dec 23	21:17:02	27070	-38293	53	P	t-	1.4143	0.2611	66.3N	31.4W	0	175		
2171	109	-1096 May 19	22:45:34	27062	-38288	20	P	-t	1.1043	0.8104	69.6N	162.6E	0	32		
2172	109	-1096 Jun 18	06:37:55	27061	-38287	58	P	t-	-1.4461	0.1636	67.0S	166.0W	0	357		
2173	109	-1096 Nov 12	07:41:19	27053	-38282	25	A	-p	-0.8784	0.9740	72.5S	134.2E	28	45	196	01m47s
2174	109	-1095 May 09	08:34:13	27044	-38276	30	A	-p	0.3635	0.9828	34.9N	156.4E	68	164	66	01m46s
2175	109	-1095 Nov 01	19:50:07	27035	-38270	35	T	-n	-0.1609	1.0335	20.6S	10.1W	81	16	114	03m04s
2176	109	-1094 Apr 28	11:21:18	27026	-38264	40	A	p-	-0.4123	0.9484	13.2S	128.0E	66	344	207	06m37s
2177	109	-1094 Oct 22	11:27:40	27017	-38258	45	T	p-	0.5085	1.0444	21.8N	127.5E	59	198	171	03m54s
2178	109	-1093 Apr 17	12:06:21	27008	-38252	50	P	t-	-1.1588	0.6923	71.6S	178.2W	0	290		
2179	109	-1093 Oct 12	01:51:37	26999	-38246	55	P	t-	1.2167	0.5973	71.6N	17.2W	0	260		
2180	109	-1092 Mar 07	07:35:15	26992	-38241	22	T	-p	0.8367	1.0189	45.1N	163.0E	33	157	117	01m33s
2181	110	-1092 Aug 31	18:45:26	26983	-38235	27	A	-p	-0.9002	0.9361	49.5S	9.0W	25	22	544	06m17s
2182	110	-1091 Feb 24	22:32:15	26974	-38229	32	T	nn	0.0926	1.0627	7.9S	42.4W	85	166	206	05m45s
2183	110	-1091 Aug 20	18:50:03	26965	-38223	37	A	nn	-0.1883	0.9345	5.6N	8.3E	79	12	248	09m00s
2184	110	-1090 Feb 14	14:59:22	26956	-38217	42	T	p-	-0.6152	1.0477	53.4S	84.6E	52	341	201	03m27s
2185	110	-1090 Aug 09	20:46:57	26947	-38211	47	A	p-	0.5457	0.9669	52.0N	11.1W	57	193	142	03m02s
2186	110	-1089 Jan 05	14:28:00	26939	-38206	14	P	-t	1.4539	0.1815	66.1N	85.2E	0	187		
2187	110	-1089 Feb 04	04:15:20	26938	-38205	52	P	t-	-1.3734	0.3140	68.9S	27.8E	0	204		
2188	110	-1089 Jun 30	21:12:28	26930	-38200	19	P	-t	-1.3419	0.3633	65.3S	10.5W	0	345		
2189	110	-1089 Jul 30	05:53:04	26929	-38199	57	P	t-	1.2493	0.5386	68.0N	10.9E	0	348		
2190	110	-1089 Dec 25	16:46:10	26921	-38194	24	A	-p	0.8081	0.9168	29.9N	49.2E	36	189	536	10m34s
2191	110	-1088 Jun 19	14:05:30	26912	-38188	29	T	-p	-0.5964	1.0763	12.9S	85.9E	53	349	307	06m56s
2192	110	-1088 Dec 13	15:50:10	26903	-38182	34	A	nn	0.1263	0.9256	15.7S	55.8E	83	193	282	09m49s
2193	110	-1087 Jun 09	06:25:42	26894	-38176	39	T	nn	0.1481	1.0506	30.1N	169.3W	81	164	170	04m26s
2194	110	-1087 Dec 02	19:15:52	26885	-38170	44	A	p-	-0.5634	0.9664	52.9S	14.9W	55	28	146	02m50s
2195	110	-1086 May 29	17:44:39	26876	-38164	49	A	t-	0.9583	0.9840	71.3N	74.5W	16	76	204	00m58s
2196	110	-1086 Oct 23	18:46:56	26869	-38159	16	P	-t	1.4954	0.0776	61.1N	82.9E	0	253		
2197	110	-1086 Nov 22	05:59:18	26867	-38158	54	P	t-	-1.2071	0.6165	62.3S	71.3E	0	132		
2198	110	-1085 Apr 19	07:00:27	26860	-38153	21	P	-t	-1.0725	0.8380	60.7S	95.9W	0	284		
2199	110	-1085 Oct 13	10:35:33	26851	-38147	26	T	-p	0.8441	1.0338	44.7N	166.4E	32	223	208	02m24s
2200	110	-1084 Apr 07	08:32:24	26842	-38141	31	A	nn	-0.2985	0.9678	12.5S	173.5E	73	330	120	03m22s
2201	111	-1084 Oct 01	23:51:56	26833	-38135	36	A	-n	0.1687	0.9984	9.0N	62.6W	80	210	6	00m09s
2202	111	-1083 Mar 27	16:20:02	26824	-38129	41	T	p-	0.4725	1.0177	23.0N	34.4E	62	148	68	01m33s
2203	111	-1083 Sep 21	06:40:11	26815	-38123	46	A	p-	-0.5664	0.9454	25.2S	173.4E	55	32	239	05m40s
2204	111	-1082 Feb 15	22:19:32	26808	-38118	13	P	-t	-1.3931	0.2654	62.0S	84.0E	0	234		
2205	111	-1082 Mar 17	06:34:24	26806	-38117	51	P	t-	1.1755	0.6818	60.8N	116.6E	0	102		
2206	111	-1082 Sep 10	07:10:58	26797	-38111	56	P	t-	-1.2889	0.4693	60.9S	111.1E	0	71		
2207	111	-1081 Feb 05	14:16:06	26790	-38106	23	T	-p	-0.7266	1.0320	60.4S	113.0E	43	318	157	02m08s
2208	111	-1081 Jul 31	17:52:04	26781	-38100	28	A	-p	0.8810	0.9785	73.5N	86.9E	28	250	163	01m25s
2209	111	-1080 Jan 26	01:45:40	26772	-38094	33	A	nn	-0.0200	0.9808	22.5S	90.4W	89	345	68	02m03s
2210	111	-1080 Jul 20	05:02:51	26763	-38088	38	Tm	nn	0.0943	1.0452	28.3N	142.9W	84	193	151	04m05s
2211	111	-1079 Jan 14	05:51:35	26754	-38082	43	A	p-	0.7159	0.9280	22.0N	162.6W	44	168	384	09m17s
2212	111	-1079 Jul 09	21:15:56	26745	-38076	48	T	p-	-0.6363	1.0723	15.6S	33.6W	50	8	304	06m36s
2213	111	-1078 Jan 03	05:09:27	26736	-38070	53	P	t-	1.4020	0.2809	65.2N	161.3W	0	164		
2214	111	-1078 May 31	05:51:56	26728	-38065	20	P	-t	1.1844	0.6600	68.7N	42.7E	0	20		
2215	111	-1078 Jun 29	13:59:55	26727	-38064	58	P	t-	-1.3808	0.2894	66.0S	71.4E	0	8		
2216	111	-1078 Nov 23	16:15:17	26719	-38059	25	A	-p	-0.8749	0.9769	76.7S	6.0E	29	43	172	01m33s
2217	111	-1077 May 20	15:13:09	26710	-38053	30	A	-p	0.4478	0.9794	43.5N	54.1E	63	166	82	02m00s
2218	111	-1077 Nov 13	04:39:31	26701	-38047	35	T	-n	-0.1610	1.0345	24.5S	143.5W	81	13	118	03m08s
2219	111	-1076 May 08	17:42:37	26692	-38041	40	A	pn	-0.3266	0.9493	4.6S	28.3E	71	347	196	06m46s
2220	111	-1076 Nov 01	20:15:33	26684	-38035	45	T	p-	0.5050	1.0421	17.7N	7.3W	60	196	162	03m50s

Cat Num	Canon Plate	Calendar Date	TD of Greatest Eclipse	ΔT s	Luna Num	Saros Num	Ecl. Type	QLE	Gamma	Ecl. Mag.	Lat. °	Long. °	Sun Alt °	Sun Azm °	Path Width km	Central Line Dur.
2221	112	-1075 Apr 27	18:39:41	26675	-38029	50	P	t-	-1.0776	0.8341	71.1S	68.5E	0	303		
2222	112	-1075 Oct 22	10:23:13	26666	-38023	55	P	t-	1.2132	0.6032	71.3N	160.5W	0	245		
2223	112	-1074 Mar 18	15:11:17	26658	-38018	22	T	-p	0.8864	1.0236	53.5N	39.4E	27	150	173	01m46s
2224	112	-1074 Sep 12	02:12:47	26649	-38012	27	A	-t	-0.9368	0.9299	57.2S	131.7W	20	31	754	06m19s
2225	112	-1073 Mar 08	06:30:33	26640	-38006	32	T	-n	0.1377	1.0668	1.5S	164.9W	82	164	220	06m03s
2226	112	-1073 Sep 01	02:01:19	26631	-38000	37	A	nn	-0.2333	0.9321	0.4S	102.3W	77	14	261	09m13s
2227	112	-1072 Feb 25	23:03:35	26622	-37994	42	T	p-	-0.5781	1.0495	46.9S	38.1W	54	339	201	03m45s
2228	112	-1072 Aug 20	04:08:47	26613	-37988	47	A	p-	0.4954	0.9673	45.3N	122.2W	60	196	135	03m08s
2229	112	-1071 Jan 15	22:32:21	26606	-37983	14	P	-t	1.4737	0.1473	67.2N	48.0W	0	176		
2230	112	-1071 Feb 14	12:10:04	26605	-37982	52	P	t-	-1.3461	0.3632	69.8S	104.5W	0	216		
2231	112	-1071 Jul 11	04:41:21	26597	-37977	19	P	-t	-1.4029	0.2470	66.3S	134.5W	0	355		
2232	112	-1071 Aug 09	13:31:44	26596	-37976	57	P	t-	1.1943	0.6404	69.0N	116.8W	0	336		
2233	112	-1070 Jan 05	00:37:10	26588	-37971	24	A	-p	0.8224	0.9175	31.8N	73.1W	34	184	552	10m26s
2234	112	-1070 Jun 30	21:35:38	26579	-37965	29	T	-p	-0.6661	1.0729	17.8S	29.6W	48	353	317	06m37s
2235	112	-1070 Dec 24	23:49:37	26570	-37959	34	A	nn	0.1345	0.9286	16.1S	64.7W	82	188	270	09m35s
2236	112	-1069 Jun 20	13:40:07	26561	-37953	39	Tm	nn	0.0736	1.0463	27.5N	82.8E	86	169	155	04m16s
2237	112	-1069 Dec 14	03:42:16	26552	-37947	44	A	p-	-0.5605	0.9703	55.6S	138.6W	56	22	129	02m29s
2238	112	-1068 Jun 09	00:29:44	26543	-37941	49	A	p-	0.8802	0.9831	74.7N	142.1W	28	110	127	01m07s
2239	112	-1068 Nov 03	03:35:44	26536	-37936	16	P	-t	1.4970	0.0747	61.5N	59.5W	0	244		
2240	112	-1068 Dec 02	14:47:08	26535	-37935	54	P	t-	-1.2048	0.6210	63.1S	71.3W	0	142		
2241	113	-1067 Apr 29	13:22:50	26527	-37930	21	P	-t	-1.1563	0.6953	61.1S	158.4E	0	293		
2242	113	-1067 Oct 23	19:18:36	26518	-37924	26	T	-p	0.8522	1.0301	42.3N	31.2E	31	219	190	02m13s
2243	113	-1066 Apr 18	15:15:13	26509	-37918	31	A	-p	-0.3752	0.9722	12.6S	72.2E	68	331	106	02m56s
2244	113	-1066 Oct 13	08:12:39	26500	-37912	36	A	-n	0.1816	0.9929	5.2N	170.4E	80	209	25	00m41s
2245	113	-1065 Apr 07	23:38:13	26491	-37906	41	T	p-	0.4057	1.0242	23.5N	75.6W	66	148	89	02m06s
2246	113	-1065 Oct 02	14:28:21	26483	-37900	46	A	p-	-0.5454	0.9407	28.0S	54.3E	57	34	257	06m02s
2247	113	-1064 Feb 27	06:26:48	26475	-37895	13	P	-t	-1.4309	0.1921	61.5S	48.2W	0	243		
2248	113	-1064 Mar 27	14:17:50	26474	-37894	51	P	t-	1.1183	0.7925	60.7N	9.4W	0	93		
2249	113	-1064 Sep 20	14:38:06	26465	-37888	56	P	t-	-1.2574	0.5219	60.6S	10.7W	0	80		
2250	113	-1063 Feb 15	22:25:15	26457	-37883	23	T	-p	-0.7605	1.0320	58.2S	6.0W	40	313	165	02m07s
2251	113	-1063 Aug 11	01:12:43	26448	-37877	28	A	-p	0.9323	0.9770	70.8N	6.5W	21	268	230	01m27s
2252	113	-1062 Feb 05	09:41:12	26439	-37871	33	A	nn	-0.0479	0.9806	21.6S	150.6E	87	340	69	02m02s
2253	113	-1062 Jul 31	12:41:03	26431	-37865	38	T	nn	0.1509	1.0452	29.6N	103.2E	81	198	152	03m56s
2254	113	-1061 Jan 25	13:30:53	26422	-37859	43	A	p-	0.6917	0.9295	20.9N	79.3E	46	164	361	08m56s
2255	113	-1061 Jul 21	04:56:56	26413	-37853	48	T	p-	-0.5753	1.0709	11.6S	151.7W	55	12	281	06m28s
2256	113	-1060 Jan 14	12:54:16	26404	-37847	53	P	t-	1.3827	0.3124	64.2N	71.3E	0	154		
2257	113	-1060 Jun 10	12:58:45	26396	-37842	20	P	-t	1.2630	0.5134	67.7N	76.9W	0	9		
2258	113	-1060 Jul 09	21:26:24	26395	-37841	58	P	t-	-1.3195	0.4067	65.0S	51.9W	0	18		
2259	113	-1060 Dec 04	00:49:12	26387	-37836	25	A	-p	-0.8723	0.9803	80.7S	116.1W	29	35	145	01m17s
2260	113	-1059 May 30	21:49:35	26379	-37830	30	A	-p	0.5334	0.9755	52.1N	46.4W	58	169	104	02m15s
2261	114	-1059 Nov 23	13:30:32	26370	-37824	35	T	-n	-0.1600	1.0360	27.8S	83.3E	81	10	123	03m15s
2262	114	-1058 May 19	23:59:17	26361	-37818	40	A	nn	-0.2370	0.9496	3.8N	69.6W	76	349	190	06m52s
2263	114	-1058 Nov 13	05:07:35	26352	-37812	45	T	n-	0.5052	1.0404	14.3N	143.1W	60	193	156	03m47s
2264	114	-1057 May 09	01:09:50	26343	-37806	50	As	t-	-0.9924	0.9527	66.3S	53.3W	5	325	-	03m47s
2265	114	-1057 Nov 02	18:59:05	26334	-37800	55	P	t-	1.2138	0.6016	70.7N	55.5E	0	232		
2266	114	-1056 Mar 28	22:41:17	26327	-37795	22	T	-t	0.9420	1.0272	63.3N	89.5W	19	136	278	01m49s
2267	114	-1056 Sep 22	09:48:30	26318	-37789	27	A	-t	-0.9661	0.9240	64.5S	98.8E	14	45	1140	06m16s
2268	114	-1055 Mar 18	14:21:47	26309	-37783	32	T	-n	0.1890	1.0706	5.5N	74.0E	79	162	234	06m17s
2269	114	-1055 Sep 11	09:21:49	26300	-37777	37	A	nn	-0.2710	0.9297	6.5S	144.7E	74	16	273	09m20s
2270	114	-1054 Mar 08	06:59:28	26291	-37771	42	T	p-	-0.5344	1.0511	39.8S	160.0W	57	339	200	04m04s
2271	114	-1054 Aug 31	11:40:38	26282	-37765	47	A	p-	0.4526	0.9675	38.8N	123.2E	63	198	131	03m15s
2272	114	-1053 Jan 27	06:27:42	26275	-37760	14	P	-t	1.4997	0.1020	68.2N	179.5W	0	165		
2273	114	-1053 Feb 25	19:54:43	26273	-37759	52	P	t-	-1.3116	0.4254	70.5S	125.1E	0	229		
2274	114	-1053 Jul 22	12:17:33	26266	-37754	19	P	-t	-1.4583	0.1412	67.3S	99.2E	0	5		
2275	114	-1053 Aug 20	21:20:11	26265	-37753	57	P	t-	1.1466	0.7286	69.9N	112.5E	0	324		
2276	114	-1052 Jan 16	08:21:39	26257	-37748	24	A	-p	0.8413	0.9186	34.7N	165.9E	32	178	573	10m03s
2277	114	-1052 Jul 11	05:10:06	26248	-37742	29	T	-p	-0.7320	1.0689	23.6S	146.9W	43	358	330	06m09s
2278	114	-1051 Jan 04	07:45:10	26239	-37736	34	A	nn	0.1463	0.9321	15.4S	175.7E	82	183	256	09m13s
2279	114	-1051 Jun 30	20:56:39	26231	-37730	39	T	nn	0.0020	1.0414	24.0N	26.1W	90	177	139	03m59s
2280	114	-1051 Dec 24	12:05:18	26222	-37724	44	A	p-	-0.5537	0.9748	57.0S	100.1E	56	14	108	02m06s

Cat Num	Canon Plate	Calendar Date	TD of Greatest Eclipse	ΔT s	Luna Num	Saros Num	Ecl. Type	QLE	Gamma	Ecl. Mag.	Lat. °	Long. °	Sun Alt °	Sun Azm °	Path Width km	Central Line Dur.
2281	115	-1050 Jun 20	07:15:06	26213	-37718	49	A	p-	0.8029	0.9806	73.7N	144.7E	36	140	116	01m22s
2282	115	-1050 Nov 14	12:27:10	26205	-37713	16	P	-t	1.4971	0.0744	62.0N	157.3E	0	234		
2283	115	-1050 Dec 13	23:32:46	26204	-37712	54	P	t-	-1.2004	0.6295	64.0S	146.4E	0	152		
2284	115	-1049 May 10	19:42:41	26197	-37707	21	P	-t	-1.2419	0.5490	61.6S	53.1E	0	302		
2285	115	-1049 Nov 04	04:04:27	26188	-37701	26	T	-p	0.8579	1.0269	40.3N	105.0W	31	214	174	02m04s
2286	115	-1048 Apr 28	21:56:14	26179	-37695	31	A	-p	-0.4542	0.9760	13.4S	28.7W	63	333	95	02m33s
2287	115	-1048 Oct 23	16:38:03	26170	-37689	36	A	-n	0.1898	0.9880	1.4N	42.2E	79	208	43	01m11s
2288	115	-1047 Apr 18	06:51:18	26161	-37683	41	T	p-	0.3337	1.0303	24.0N	176.1E	70	149	108	02m36s
2289	115	-1047 Oct 12	22:23:36	26152	-37677	46	A	p-	-0.5307	0.9364	31.6S	66.6W	58	35	274	06m23s
2290	115	-1046 Mar 09	14:24:39	26145	-37672	13	P	-t	-1.4765	0.1036	61.1S	177.9W	0	252		
2291	115	-1046 Apr 07	21:53:43	26143	-37671	51	P	t-	1.0545	0.9165	60.7N	133.4W	0	84		
2292	115	-1046 Oct 01	22:15:12	26134	-37665	56	P	t-	-1.2334	0.5618	60.5S	135.0W	0	89		
2293	115	-1045 Feb 27	06:23:22	26127	-37660	23	T	-p	-0.8033	1.0315	56.4S	122.7W	36	309	177	02m05s
2294	115	-1045 Aug 22	08:44:56	26118	-37654	28	A	-p	0.9753	0.9748	66.9N	105.6W	12	282	427	01m31s
2295	115	-1044 Feb 16	17:26:53	26109	-37648	33	A	nn	-0.0834	0.9805	20.3S	33.9E	85	336	70	02m00s
2296	115	-1044 Aug 10	20:26:56	26101	-37642	38	T	-n	0.2018	1.0447	29.8N	12.6W	78	202	152	03m47s
2297	115	-1043 Feb 04	21:02:29	26092	-37636	43	A	p-	0.6615	0.9313	20.2N	36.5W	48	159	336	08m28s
2298	115	-1043 Jul 31	12:43:14	26083	-37630	48	T	p-	-0.5187	1.0687	9.0S	89.5E	59	16	260	06m11s
2299	115	-1042 Jan 24	20:32:30	26074	-37624	53	P	t-	1.3578	0.3535	63.3N	54.2W	0	144		
2300	115	-1042 Jun 21	20:05:00	26067	-37619	20	P	-t	1.3408	0.3695	66.7N	164.3E	0	359		
2301	116	-1042 Jul 21	04:56:30	26065	-37618	58	P	t-	-1.2617	0.5163	64.1S	175.7W	0	28		
2302	116	-1042 Dec 15	09:22:50	26058	-37613	25	A	-p	-0.8709	0.9842	83.7S	137.8E	29	10	115	01m00s
2303	116	-1041 Jun 11	04:25:52	26049	-37607	30	A	-p	0.6179	0.9710	60.3N	144.6W	52	173	133	02m30s
2304	116	-1041 Dec 04	22:21:18	26040	-37601	35	T	-n	-0.1597	1.0380	30.5S	49.3W	81	6	129	03m23s
2305	116	-1040 May 30	06:17:20	26031	-37595	40	A	nn	-0.1485	0.9495	11.5N	167.1W	82	353	187	06m53s
2306	116	-1040 Nov 23	14:00:18	26022	-37589	45	T	n-	0.5062	1.0390	11.4N	81.0E	60	189	151	03m45s
2307	116	-1039 May 19	07:39:42	26014	-37583	50	A	t-	-0.9059	0.9606	46.8S	174.0W	25	344	338	03m54s
2308	116	-1039 Nov 13	03:36:49	26005	-37577	55	P	t-	1.2166	0.5962	70.0N	88.4W	0	218		
2309	116	-1038 Apr 09	06:06:26	25997	-37572	22	T+	-t	1.0023	1.0034	71.6N	97.4E	0	79	-	-
2310	116	-1038 Oct 03	17:31:41	25989	-37566	27	As	-t	-0.9885	0.9182	70.6S	43.4W	7	69	-	06m07s
2311	116	-1037 Mar 29	22:05:45	25980	-37560	32	T	-n	0.2467	1.0737	12.9N	45.7W	76	161	246	06m25s
2312	116	-1037 Sep 22	16:52:58	25971	-37554	37	A	-n	-0.3001	0.9278	12.3S	29.0E	72	18	283	09m21s
2313	116	-1036 Mar 18	14:46:50	25962	-37548	42	T	p-	-0.4843	1.0523	32.4S	79.6E	61	339	197	04m22s
2314	116	-1036 Sep 10	19:22:54	25953	-37542	47	A	p-	0.4175	0.9677	32.5N	5.5E	65	199	128	03m20s
2315	116	-1035 Feb 06	14:13:01	25946	-37537	14	Pe	-t	1.5329	0.0444	69.2N	50.9E	0	153		
2316	116	-1035 Mar 08	03:29:08	25944	-37536	52	P	t-	-1.2696	0.5008	71.1S	3.2W	0	242		
2317	116	-1035 Aug 01	20:01:21	25937	-37531	19	Pe	-t	-1.5079	0.0469	68.2S	29.4W	0	17		
2318	116	-1035 Aug 31	05:18:35	25936	-37530	57	P	t-	1.1064	0.8027	70.7N	21.4W	0	311		
2319	116	-1034 Jan 26	15:55:58	25928	-37525	24	A	-p	0.8683	0.9199	39.3N	46.8E	29	173	613	09m25s
2320	116	-1034 Jul 22	12:49:59	25919	-37519	29	T	-p	-0.7932	1.0639	30.1S	93.5E	37	2	344	05m32s
2321	117	-1033 Jan 15	15:34:09	25911	-37513	34	A	nn	0.1641	0.9363	13.5S	57.5E	81	179	240	08m40s
2322	117	-1033 Jul 12	04:16:54	25902	-37507	39	T	nn	-0.0660	1.0358	19.9N	136.7W	86	358	121	03m36s
2323	117	-1032 Jan 04	20:24:42	25893	-37501	44	A	p-	-0.5432	0.9799	56.8S	19.7W	57	6	85	01m41s
2324	117	-1032 Jun 30	14:02:38	25884	-37495	49	A	p-	0.7281	0.9772	70.0N	61.5E	43	162	119	01m44s
2325	117	-1032 Nov 24	21:21:11	25877	-37490	16	P	-t	1.4957	0.0767	62.7N	13.3E	0	225		
2326	117	-1032 Dec 24	08:17:08	25875	-37489	54	P	t-	-1.1946	0.6407	65.0S	4.0E	0	162		
2327	117	-1031 May 21	01:59:28	25868	-37484	21	P	-t	-1.3295	0.3988	62.2S	51.5W	0	311		
2328	117	-1031 Jun 19	16:52:11	25866	-37483	59	Pb	t-	1.5201	0.0725	64.4N	120.2W	0	25		
2329	117	-1031 Nov 14	12:54:08	25859	-37478	26	T	-p	0.8602	1.0243	38.4N	117.5E	30	209	159	01m57s
2330	117	-1030 May 10	04:35:16	25850	-37472	31	A	-p	-0.5354	0.9794	15.2S	129.4W	58	335	85	02m14s
2331	117	-1030 Nov 04	01:07:24	25841	-37466	36	A	-n	0.1940	0.9836	2.4S	87.0W	79	206	59	01m40s
2332	117	-1029 Apr 29	14:02:41	25833	-37460	41	T	n-	0.2592	1.0357	24.2N	68.6E	75	151	124	03m04s
2333	117	-1029 Oct 24	06:24:09	25824	-37454	46	A	p-	-0.5211	0.9327	35.5S	171.6E	58	35	291	06m41s
2334	117	-1028 Mar 19	22:16:58	25817	-37449	13	Pe	-t	-1.5268	0.0054	60.9S	53.9E	0	261		
2335	117	-1028 Apr 18	05:26:57	25815	-37448	51	Tn	t-	0.9880	1.0591	62.0N	119.4E	8	90	-	03m08s
2336	117	-1028 Oct 12	05:58:51	25806	-37442	56	P	t-	-1.2144	0.5932	60.6S	99.1E	0	98		
2337	117	-1027 Mar 09	14:14:08	25799	-37437	23	T	-t	-0.8517	1.0305	55.3S	122.4E	31	306	194	02m00s
2338	117	-1027 Sep 01	16:27:11	25790	-37431	28	P	-t	1.0111	0.9614	61.3N	154.1E	0	296		
2339	117	-1026 Feb 27	01:00:56	25781	-37425	33	A	nn	-0.1278	0.9803	18.9S	79.9W	83	333	70	01m59s
2340	117	-1026 Aug 22	04:22:51	25773	-37419	38	T	-n	0.2451	1.0439	28.7N	131.2W	76	206	151	03m38s

Cat Num	Canon Plate	Calendar Date	TD of Greatest Eclipse	ΔT s	Luna Num	Saros Num	Ecl. Type	QLE	Gamma	Ecl. Mag.	Lat. °	Long. °	Sun Alt °	Sun Azm °	Path Width km	Central Line Dur.
2341	118	-1025 Feb 16	04:22:31	25764	-37413	43	A	p-	0.6221	0.9335	19.7N	148.9W	51	156	310	07m57s
2342	118	-1025 Aug 11	20:38:00	25755	-37407	48	T	p-	-0.4692	1.0660	7.7S	31.1W	62	20	242	05m49s
2343	118	-1024 Feb 05	04:01:24	25746	-37401	53	P	t-	1.3246	0.4088	62.5N	177.1W	0	135		
2344	118	-1024 Jul 02	03:14:33	25739	-37396	20	P	-t	1.4149	0.2339	65.7N	45.0E	0	349		
2345	118	-1024 Jul 31	12:33:16	25737	-37395	58	P	t-	-1.2098	0.6136	63.3S	59.1E	0	37		
2346	118	-1024 Dec 25	17:53:24	25730	-37390	25	A	-p	-0.8728	0.9887	84.0S	51.1E	29	327	83	00m43s
2347	118	-1023 Jun 21	11:01:21	25721	-37384	30	A	-p	0.7020	0.9658	68.2N	121.4E	45	182	174	02m45s
2348	118	-1023 Dec 15	07:11:01	25712	-37378	35	T	-n	-0.1607	1.0405	32.3S	178.8E	81	1	137	03m34s
2349	118	-1022 Jun 10	12:34:40	25704	-37372	40	A	nn	-0.0591	0.9489	18.7N	96.5E	87	356	188	06m50s
2350	118	-1022 Dec 04	22:52:43	25695	-37366	45	T	n-	0.5071	1.0382	9.2N	54.8W	59	185	149	03m44s
2351	118	-1021 May 30	14:10:31	25686	-37360	50	A	p-	-0.8187	0.9655	34.9S	79.7E	35	351	217	03m49s
2352	118	-1021 Nov 24	12:16:23	25677	-37354	55	P	t-	1.2213	0.5876	69.1N	127.9E	0	206		
2353	118	-1020 Apr 19	13:27:40	25670	-37349	22	P	-t	1.0665	0.8845	71.3N	28.0W	0	66		
2354	118	-1020 May 18	22:05:51	25669	-37348	60	Pb	t-	-1.5290	0.0219	69.6S	12.6W	0	328		
2355	118	-1020 Oct 14	01:21:23	25661	-37343	27	A-	-t	-1.0050	0.9429	71.7S	160.5E	0	105	-	-
2356	118	-1019 Apr 09	05:44:22	25652	-37337	32	T	-n	0.3091	1.0763	20.8N	164.2W	72	161	259	06m26s
2357	118	-1019 Oct 03	00:33:14	25644	-37331	37	A	-n	-0.3220	0.9261	18.0S	88.8W	71	19	293	09m18s
2358	118	-1018 Mar 29	22:25:34	25635	-37325	42	T	n-	-0.4276	1.0531	24.7S	39.2W	65	340	194	04m38s
2359	118	-1018 Sep 22	03:15:47	25626	-37319	47	A	p-	0.3903	0.9679	26.4N	115.2W	67	200	125	03m24s
2360	118	-1017 Mar 19	10:54:23	25617	-37313	52	P	t-	-1.2212	0.5879	71.5S	129.6W	0	255		
2361	119	-1017 Sep 11	13:25:49	25609	-37307	57	P	t-	1.0730	0.8645	71.3N	158.1W	0	298		
2362	119	-1016 Feb 06	23:22:20	25601	-37302	24	A	-p	0.9013	0.9213	45.3N	71.5W	25	167	689	08m37s
2363	119	-1016 Aug 01	20:36:28	25593	-37296	29	T	-p	-0.8486	1.0584	37.3S	28.8W	32	7	364	04m49s
2364	119	-1015 Jan 25	23:17:06	25584	-37290	34	A	nn	0.1874	0.9408	10.4S	59.6W	79	175	223	08m00s
2365	119	-1015 Jul 22	11:40:52	25575	-37284	39	T	nn	-0.1298	1.0298	15.2N	111.1E	83	2	102	03m06s
2366	119	-1014 Jan 15	04:38:44	25566	-37278	44	A	p-	-0.5275	0.9856	54.9S	138.8W	58	358	60	01m13s
2367	119	-1014 Jul 11	20:53:54	25557	-37272	49	A	p-	0.6573	0.9731	64.9N	30.9W	49	175	129	02m11s
2368	119	-1014 Dec 06	06:13:27	25550	-37267	16	P	-t	1.4958	0.0758	63.5N	130.4W	0	215		
2369	119	-1013 Jan 04	16:55:21	25549	-37266	54	P	t-	-1.1838	0.6613	66.1S	137.2W	0	173		
2370	119	-1013 Jun 01	08:18:06	25541	-37261	21	P	-t	-1.4151	0.2520	62.9S	156.7W	0	320		
2371	119	-1013 Jun 30	23:18:09	25540	-37260	59	P	t-	1.4411	0.2084	65.3N	132.0E	0	15		
2372	119	-1013 Nov 25	21:44:04	25533	-37255	26	T	-p	0.8617	1.0222	37.0N	20.1W	30	204	147	01m51s
2373	119	-1012 May 20	11:15:02	25524	-37249	31	A	-p	-0.6168	0.9821	18.0S	129.4E	52	338	79	01m57s
2374	119	-1012 Nov 14	09:38:15	25515	-37243	36	A	-n	0.1962	0.9799	5.7S	143.6E	79	203	73	02m07s
2375	119	-1011 May 09	21:12:05	25506	-37237	41	T	n-	0.1822	1.0406	24.1N	38.3W	79	154	138	03m32s
2376	119	-1011 Nov 03	14:29:28	25498	-37231	46	A	p-	-0.5157	0.9295	39.7S	49.0E	59	34	305	06m56s
2377	119	-1010 Apr 29	12:54:08	25489	-37225	51	T	t-	0.9161	1.0666	61.8N	30.9E	23	111	541	03m47s
2378	119	-1010 Oct 23	13:51:12	25480	-37219	56	P	t-	-1.2020	0.6136	60.8S	28.9W	0	107		
2379	119	-1009 Mar 20	21:54:33	25473	-37214	23	T	-t	-0.9081	1.0286	55.6S	11.6E	24	301	228	01m52s
2380	119	-1009 Sep 13	00:19:59	25464	-37208	28	P	-t	1.0391	0.9117	60.9N	25.8E	0	287		
2381	120	-1008 Mar 09	08:25:24	25455	-37202	33	A	nn	-0.1792	0.9801	17.4S	168.6E	80	331	72	01m59s
2382	120	-1008 Sep 01	12:27:39	25447	-37196	38	T	-n	0.2814	1.0428	26.7N	107.5E	74	209	149	03m30s
2383	120	-1007 Feb 26	11:33:23	25438	-37190	43	A	p-	0.5754	0.9359	19.4N	101.4E	55	153	285	07m26s
2384	120	-1007 Aug 22	04:39:03	25429	-37184	48	T	n-	-0.4251	1.0627	7.6S	153.1W	65	23	225	05m25s
2385	120	-1006 Feb 15	11:23:11	25420	-37178	53	P	t-	1.2850	0.4754	61.8N	62.1E	0	125		
2386	120	-1006 Jul 13	10:26:39	25413	-37173	20	Pe	-t	1.4859	0.1056	64.7N	74.5W	0	339		
2387	120	-1006 Aug 11	20:15:48	25412	-37172	58	P	t-	-1.1631	0.7002	62.5S	67.2W	0	46		
2388	120	-1005 Jan 06	02:19:53	25404	-37167	25	A	-p	-0.8789	0.9935	81.2S	49.1W	28	299	48	00m24s
2389	120	-1005 Jul 02	17:40:29	25396	-37161	30	A	-p	0.7822	0.9602	74.9N	35.5E	38	199	234	03m00s
2390	120	-1005 Dec 26	15:57:57	25387	-37155	35	T	-n	-0.1647	1.0434	33.4S	48.0E	80	356	147	03m46s
2391	120	-1004 Jun 20	18:55:29	25378	-37149	40	Am	nn	0.0276	0.9479	25.0N	0.1E	88	181	192	06m43s
2392	120	-1004 Dec 15	07:42:49	25369	-37143	45	T	n-	0.5063	1.0378	7.5N	170.1E	60	181	147	03m45s
2393	120	-1003 Jun 09	20:44:45	25361	-37137	50	A	p-	-0.7329	0.9694	25.3S	24.6W	43	356	162	03m38s
2394	120	-1003 Dec 04	20:53:48	25352	-37131	55	P	t-	1.2251	0.5807	68.0N	14.6W	0	194		
2395	120	-1002 Apr 30	20:45:38	25345	-37126	22	P	-t	1.1340	0.7582	70.8N	152.3W	0	53		
2396	120	-1002 May 30	05:07:56	25343	-37125	60	P	t-	-1.4485	0.1695	68.7S	131.5W	0	339		
2397	120	-1002 Oct 25	09:16:56	25336	-37120	27	A-	-t	-1.0163	0.9227	71.3S	26.2E	0	119	-	-
2398	120	-1001 Apr 20	13:18:07	25327	-37114	32	T	-n	0.3758	1.0780	28.9N	78.6E	68	161	272	06m20s
2399	120	-1001 Oct 14	08:21:12	25319	-37108	37	A	-n	-0.3380	0.9250	23.4S	151.8E	70	19	299	09m10s
2400	120	-1000 Apr 09	05:57:32	25310	-37102	42	T	n-	-0.3655	1.0533	16.8S	156.3W	68	341	189	04m51s

Cat Num	Canon Plate	Calendar Date	TD of Greatest Eclipse	ΔT s	Luna Num	Saros Num	Ecl. Type	QLE	Gamma	Ecl. Mag.	Lat. °	Long. °	Sun Alt °	Sun Azm °	Path Width km	Central Line Dur.
2401	121	-1000 Oct 02	11:18:35	25301	-37096	47	A	p-	0.3705	0.9684	20.8N	121.5E	68	199	122	03m26s
2402	121	-0999 Mar 29	18:08:50	25292	-37090	52	P	t-	-1.1646	0.6895	71.7S	106.5E	0	269		
2403	121	-0999 Sep 21	21:43:04	25284	-37084	57	P	t-	1.0470	0.9123	71.7N	62.3E	0	284		
2404	121	-0998 Feb 17	06:37:53	25276	-37079	24	A	-t	0.9430	0.9224	53.6N	170.2E	19	159	889	07m41s
2405	121	-0998 Aug 13	04:30:26	25268	-37073	29	T	-t	-0.8975	1.0524	45.0S	154.4W	26	14	394	04m04s
2406	121	-0997 Feb 06	06:51:15	25259	-37067	34	A	nn	0.2187	0.9458	6.1S	174.9W	77	171	205	07m12s
2407	121	-0997 Aug 02	19:11:20	25250	-37061	39	T	-n	-0.1874	1.0234	10.0N	3.3W	79	6	81	02m29s
2408	121	-0996 Jan 26	12:47:41	25242	-37055	44	A	p-	-0.5068	0.9917	51.6S	102.0E	59	352	34	00m43s
2409	121	-0996 Jul 22	03:48:45	25233	-37049	49	A	p-	0.5900	0.9685	58.9N	129.3W	54	184	141	02m45s
2410	121	-0996 Dec 16	15:05:32	25226	-37044	16	P	-t	1.4968	0.0734	64.4N	85.5E	0	205		
2411	121	-0995 Jan 15	01:30:15	25224	-37043	54	P	t-	-1.1698	0.6881	67.1S	82.0E	0	184		
2412	121	-0995 Jun 11	14:37:30	25217	-37038	21	Pe	-t	-1.4996	0.1068	63.8S	97.6E	0	329		
2413	121	-0995 Jul 11	05:47:29	25216	-37037	59	P	t-	1.3644	0.3397	66.3N	23.1E	0	5		
2414	121	-0995 Dec 06	06:33:58	25208	-37032	26	T	-p	0.8634	1.0207	36.2N	157.7W	30	198	139	01m47s
2415	121	-0994 May 31	17:55:43	25200	-37026	31	A	-p	-0.6985	0.9841	22.1S	27.6E	46	341	77	01m44s
2416	121	-0994 Nov 25	18:10:20	25191	-37020	36	A	-n	0.1968	0.9767	8.6S	14.2E	79	199	84	02m32s
2417	121	-0993 May 21	04:22:36	25182	-37014	41	T	nn	0.1049	1.0447	23.3N	145.4W	84	158	150	03m58s
2418	121	-0993 Nov 14	22:36:00	25174	-37008	46	A	p-	-0.5116	0.9270	43.8S	73.1W	59	32	317	07m09s
2419	121	-0992 May 09	20:20:56	25165	-37002	51	T	p-	0.8433	1.0712	61.6N	69.6W	32	123	429	04m12s
2420	121	-0992 Nov 02	21:47:52	25156	-36996	56	P	t-	-1.1926	0.6291	61.2S	158.2W	0	116		
2421	122	-0991 Mar 31	05:27:07	25149	-36991	23	T	-t	-0.9705	1.0250	58.0S	91.2W	13	293	357	01m34s
2422	122	-0991 Sep 23	08:22:26	25140	-36985	28	P	-t	1.0604	0.8739	60.7N	104.9W	0	278		
2423	122	-0990 Mar 20	15:39:10	25131	-36979	33	A	-n	-0.2387	0.9795	16.2S	59.8E	76	330	75	02m02s
2424	122	-0990 Sep 12	20:42:00	25123	-36973	38	T	-n	0.3105	1.0417	24.0N	16.7W	72	210	146	03m24s
2425	122	-0989 Mar 09	18:32:53	25114	-36967	43	A	p-	0.5194	0.9383	19.4N	4.8W	59	150	263	06m57s
2426	122	-0989 Sep 02	12:49:37	25105	-36961	48	T	n-	-0.3889	1.0592	8.7S	82.5E	67	26	210	05m00s
2427	122	-0988 Feb 26	18:36:12	25097	-36955	53	P	t-	1.2372	0.5567	61.3N	56.4W	0	116		
2428	122	-0988 Aug 22	04:05:11	25088	-36949	58	P	t-	-1.1224	0.7744	61.8S	165.0E	0	56		
2429	122	-0987 Jan 16	10:41:05	25081	-36944	25	A	-p	-0.8905	0.9986	77.4S	163.8W	27	286	11	00m05s
2430	122	-0987 Jul 13	00:22:34	25072	-36938	30	A	-t	0.8592	0.9539	78.8N	32.3W	30	234	332	03m15s
2431	122	-0986 Jan 06	00:39:41	25063	-36932	35	T	-n	-0.1735	1.0467	33.7S	81.4W	80	351	158	03m59s
2432	122	-0986 Jul 02	01:19:51	25055	-36926	40	A	nn	0.1115	0.9464	30.4N	96.0W	83	185	199	06m35s
2433	122	-0986 Dec 26	16:29:37	25046	-36920	45	T	n-	0.5028	1.0380	6.4N	35.9E	60	176	147	03m46s
2434	122	-0985 Jun 21	03:23:23	25037	-36914	50	A	p-	-0.6495	0.9724	17.4S	128.6W	49	0	130	03m25s
2435	122	-0985 Dec 16	05:28:52	25029	-36908	55	P	t-	1.2277	0.5761	67.0N	155.9W	0	182		
2436	122	-0984 May 11	04:02:43	25022	-36903	22	P	-t	1.2030	0.6278	70.1N	84.2E	0	41		
2437	122	-0984 Jun 09	12:13:01	25020	-36902	60	P	t-	-1.3689	0.3165	67.7S	109.3E	0	350		
2438	122	-0984 Nov 04	17:17:19	25013	-36897	27	A-	-t	-1.0234	0.9099	70.7S	108.8W	0	133	-	-
2439	122	-0983 Apr 30	20:47:10	25004	-36891	32	T	-p	0.4466	1.0790	37.2N	37.3W	63	161	285	06m08s
2440	122	0983 Oct 24	16:17:12	24996	-36885	37	A	-n	-0.3478	0.9245	28.4S	30.9E	69	18	303	09m01s
2441	123	-0982 Apr 20	13:23:00	24987	-36879	42	T	n-	-0.2985	1.0529	8.9S	88.2E	73	343	183	04m59s
2442	123	-0982 Oct 13	19:29:50	24978	-36873	47	A	p-	0.3566	0.9692	15.6N	4.0W	69	198	118	03m25s
2443	123	-0981 Apr 10	01:16:03	24970	-36867	52	P	t-	-1.1028	0.8002	71.6S	15.7W	0	282		
2444	123	-0981 Oct 03	06:08:45	24961	-36861	57	P	t-	1.0273	0.9486	71.8N	79.7W	0	270		
2445	123	-0980 Feb 28	13:45:50	24954	-36856	24	An	-t	0.9904	0.9221	66.3N	41.7E	7	140	-	06m33s
2446	123	-0980 Aug 23	12:30:59	24945	-36850	29	T	-t	-0.9406	1.0458	53.4S	75.9E	19	21	455	03m18s
2447	123	-0979 Feb 16	14:18:54	24936	-36844	34	A	nn	0.2559	0.9509	0.9S	70.7E	75	168	186	06m22s
2448	123	-0979 Aug 13	02:47:53	24928	-36838	39	H3	-n	-0.2387	1.0168	4.5N	119.8W	76	10	59	01m48s
2449	123	-0978 Feb 05	20:48:17	24919	-36832	44	A	p-	-0.4783	0.9982	46.9S	17.0W	61	347	7	00m09s
2450	123	-0978 Aug 02	10:50:28	24910	-36826	49	A	p-	0.5291	0.9636	52.6N	127.3E	58	190	156	03m24s
2451	123	-0978 Dec 27	23:52:42	24903	-36821	16	P	-t	1.5018	0.0629	65.4N	57.6W	0	194		
2452	123	-0977 Jan 26	09:57:00	24902	-36820	54	P	t-	-1.1492	0.7275	68.2S	57.3W	0	195		
2453	123	-0977 Jul 22	12:24:38	24893	-36814	59	P	t-	1.2935	0.4606	67.4N	88.3W	0	355		
2454	123	-0977 Dec 17	15:19:54	24886	-36809	26	T	-p	0.8680	1.0195	36.3N	65.8E	29	193	131	01m43s
2455	123	-0976 Jun 11	00:40:34	24877	-36803	31	A	-p	-0.7775	0.9854	27.6S	76.0W	39	345	81	01m34s
2456	123	-0976 Dec 06	02:40:40	24869	-36797	36	A	-n	0.1981	0.9742	10.7S	114.6W	79	195	94	02m54s
2457	123	-0975 May 31	11:32:22	24860	-36791	41	T	nn	0.0261	1.0480	21.8N	107.6E	88	163	160	04m22s
2458	123	-0975 Nov 25	06:44:22	24851	-36785	46	A	p-	-0.5096	0.9252	47.5S	165.3E	59	28	326	07m19s
2459	123	-0974 May 21	03:44:46	24843	-36779	51	T	p-	0.7680	1.0744	61.3N	171.1W	40	133	375	04m32s
2460	123	-0974 Nov 14	05:49:08	24834	-36773	56	P	t-	-1.1868	0.6387	61.8S	71.3E	0	126		

Cat Num	Canon Plate	Calendar Date	TD of Greatest Eclipse	ΔT s	Luna Num	Saros Num	Ecl. Type	QLE	Gamma	Ecl. Mag.	Lat. °	Long. °	Sun Alt °	Sun Azm °	Path Width km	Central Line Dur.
2461	124	-0973 Apr 11	12:51:12	24827	-36768	23	P	-t	-1.0390	0.9324	60.7S	173.4E	0	279		
2462	124	-0973 May 10	20:50:26	24826	-36767	61	Pb	t-	1.5114	0.0379	61.8N	150.2W	0	58		
2463	124	-0973 Oct 04	16:35:00	24818	-36762	28	P	-t	1.0741	0.8496	60.7N	122.0E	0	269		
2464	124	-0972 Mar 30	22:44:13	24810	-36756	33	A	-p	-0.3046	0.9787	15.3S	46.9W	72	329	79	02m08s
2465	124	-0972 Sep 23	05:04:16	24801	-36750	38	T	-n	0.3335	1.0406	20.7N	143.5W	70	211	143	03m19s
2466	124	-0971 Mar 20	01:24:23	24792	-36744	43	A	p-	0.4570	0.9408	19.6N	108.6W	63	149	243	06m32s
2467	124	-0971 Sep 12	21:07:23	24784	-36738	48	T	n-	-0.3588	1.0554	10.8S	43.7W	69	28	195	04m36s
2468	124	-0970 Mar 09	01:40:51	24775	-36732	53	P	t-	1.1818	0.6518	60.9N	172.7W	0	107		
2469	124	-0970 Sep 02	12:01:56	24767	-36726	58	P	t-	-1.0884	0.8356	61.3S	35.5E	0	65		
2470	124	-0969 Jan 27	18:56:26	24759	-36721	25	H	-p	-0.9078	1.0037	73.5S	77.3E	24	279	31	00m13s
2471	124	-0969 Jul 24	07:10:20	24751	-36715	30	A	-t	0.9306	0.9470	76.7N	91.5W	21	278	543	03m30s
2472	124	-0968 Jan 17	09:15:55	24742	-36709	35	T	-n	-0.1871	1.0503	33.2S	150.5E	79	345	170	04m12s
2473	124	-0968 Jul 12	07:51:00	24733	-36703	40	A	nn	0.1901	0.9447	34.5N	167.1E	79	191	208	06m28s
2474	124	-0967 Jan 06	01:10:53	24725	-36697	45	T	n-	0.4952	1.0385	5.8N	96.8W	60	172	148	03m47s
2475	124	-0967 Jul 01	10:07:18	24716	-36691	50	A	p-	-0.5693	0.9748	10.9S	127.1E	55	5	110	03m09s
2476	124	-0967 Dec 26	13:58:52	24708	-36685	55	P	t-	1.2267	0.5777	65.9N	64.5E	0	171		
2477	124	-0966 May 22	11:19:34	24701	-36680	22	P	-t	1.2729	0.4947	69.3N	38.7W	0	29		
2478	124	-0966 Jun 20	19:22:47	24699	-36679	60	P	t-	-1.2918	0.4598	66.7S	10.5W	0	1		
2479	124	-0966 Nov 16	01:19:18	24692	-36674	27	P	-t	-1.0288	0.9002	69.9S	116.4E	0	146		
2480	124	-0965 May 12	04:14:07	24683	-36668	32	T	-p	0.5194	1.0789	45.6N	152.1W	58	162	298	05m49s
2481	125	-0965 Nov 05	00:18:37	24675	-36662	37	A	-n	-0.3539	0.9246	33.0S	90.6W	69	17	303	08m48s
2482	125	-0964 Apr 30	20:43:01	24666	-36656	42	T	n-	-0.2274	1.0517	1.0S	25.7W	77	345	176	05m00s
2483	125	-0964 Oct 24	03:48:35	24657	-36650	47	A	n-	0.3481	0.9704	10.9N	131.2W	70	197	113	03m22s
2484	125	-0963 Apr 20	08:14:34	24649	-36644	52	P	t-	-1.0344	0.9224	71.3S	135.4W	0	295		
2485	125	-0963 Oct 13	14:42:55	24640	-36638	57	P	t-	1.0144	0.9726	71.7N	136.2E	0	256		
2486	125	-0962 Mar 10	20:41:34	24633	-36633	24	P	-t	1.0474	0.8764	71.4N	90.2W	0	115		
2487	125	-0962 Sep 03	20:40:19	24624	-36627	29	T	-t	-0.9762	1.0387	62.4S	61.4W	12	34	627	02m33s
2488	125	-0961 Feb 27	21:37:59	24616	-36621	34	A	nn	0.3008	0.9562	5.2N	42.0W	72	165	167	05m29s
2489	125	-0961 Aug 24	10:31:09	24607	-36615	39	H	-n	-0.2836	1.0100	1.2S	121.7E	74	13	36	01m05s
2490	125	-0960 Feb 17	04:42:54	24599	-36609	44	H	p-	-0.4440	1.0050	41.2S	136.0W	63	344	19	00m27s
2491	125	-0960 Aug 12	17:58:00	24590	-36603	49	A	p-	0.4738	0.9584	46.0N	20.5E	61	194	172	04m07s
2492	125	-0959 Jan 07	08:36:28	24583	-36598	16	P	-t	1.5101	0.0461	66.4N	159.7E	0	184		
2493	125	-0959 Feb 05	18:17:58	24582	-36597	54	P	t-	-1.1234	0.7773	69.2S	164.2E	0	207		
2494	125	-0959 Aug 01	19:07:23	24573	-36591	59	P	t-	1.2266	0.5739	68.4N	158.5E	0	344		
2495	125	-0959 Dec 28	00:03:08	24566	-36586	26	T	-p	0.8746	1.0187	37.1N	70.1W	29	187	132	01m39s
2496	125	-0958 Jun 22	07:29:45	24557	-36580	31	A	-p	-0.8536	0.9859	34.6S	178.7E	31	349	95	01m27s
2497	125	-0958 Dec 17	11:07:32	24549	-36574	36	A	-n	0.2014	0.9723	11.8S	117.6E	78	191	101	03m14s
2498	125	-0957 Jun 11	18:46:35	24540	-36568	41	T	nn	-0.0498	1.0506	19.6N	0.8W	87	345	168	04m44s
2499	125	-0957 Dec 06	14:51:13	24531	-36562	46	A	p-	-0.5070	0.9240	50.6S	45.4E	59	24	332	07m27s
2500	125	-0956 May 31	11:09:40	24523	-36556	51	T	p-	0.6932	1.0765	60.5N	86.4E	46	144	343	04m49s
2501	126	-0956 Nov 24	13:51:36	24514	-36550	56	P	t-	-1.1814	0.6476	62.5S	59.6W	0	135		
2502	126	-0955 Apr 21	20:09:28	24507	-36545	23	P	-t	-1.1115	0.7966	60.9S	53.8E	0	288		
2503	126	-0955 May 21	04:09:23	24506	-36544	61	P	t-	1.4428	0.1701	62.4N	89.6E	0	49		
2504	126	-0955 Oct 15	00:55:20	24499	-36539	28	P	-t	1.0828	0.8345	60.8N	13.1W	0	260		
2505	126	-0954 Apr 11	05:39:32	24490	-36533	33	A	-p	-0.3777	0.9774	15.0S	151.2W	68	330	86	02m18s
2506	126	-0954 Oct 04	13:35:24	24481	-36527	38	T	-n	0.3499	1.0398	17.1N	87.0E	69	211	141	03m17s
2507	126	-0953 Mar 31	08:06:38	24473	-36521	43	A	p-	0.3868	0.9430	19.9N	150.4E	67	149	226	06m12s
2508	126	-0953 Sep 24	05:32:57	24464	-36515	48	T	n-	-0.3353	1.0518	13.6S	171.8W	70	30	181	04m14s
2509	126	-0952 Mar 19	08:38:04	24456	-36509	53	P	t-	1.1192	0.7602	60.6N	73.0E	0	99		
2510	126	-0952 Sep 12	20:06:34	24447	-36503	58	P	t-	-1.0613	0.8832	60.9S	95.8W	0	74		
2511	126	-0951 Feb 07	03:04:01	24440	-36498	25	T	-p	-0.9323	1.0088	69.9S	40.5W	21	274	85	00m31s
2512	126	-0951 Aug 03	14:03:44	24431	-36492	30	An	-t	0.9965	0.9374	64.7N	155.6W	2	317	-	03m41s
2513	126	-0950 Jan 27	17:45:09	24423	-36486	35	T	-n	-0.2071	1.0541	32.2S	24.0E	78	341	183	04m25s
2514	126	-0950 Jul 23	14:29:18	24414	-36480	40	A	nn	0.2631	0.9427	37.2N	69.2E	75	196	220	06m24s
2515	126	-0949 Jan 17	09:45:26	24406	-36474	45	T	n-	0.4822	1.0393	5.6N	132.3E	61	167	150	03m49s
2516	126	-0949 Jul 12	16:59:24	24397	-36468	50	A	p-	-0.4944	0.9765	5.8S	21.6E	60	9	96	02m54s
2517	126	-0948 Jan 06	22:24:11	24389	-36462	55	P	t-	1.2226	0.5850	64.9N	73.3W	0	161		
2518	126	-0948 Jun 01	18:37:02	24381	-36457	22	P	-t	1.3428	0.3608	68.4N	161.2W	0	17		
2519	126	-0948 Jul 01	02:36:52	24380	-36456	60	P	t-	-1.2170	0.5994	65.7S	130.9W	0	11		
2520	126	-0948 Nov 26	09:23:16	24373	-36451	27	P	-t	-1.0323	0.8940	68.9S	18.3W	0	158		

Cat Num	Canon Plate	Calendar Date	TD of Greatest Eclipse	ΔT s	Luna Num	Saros Num	Ecl. Type	QLE	Gamma	Ecl. Mag.	Lat. °	Long. °	Sun Alt °	Sun Azm °	Path Width km	Central Line Dur.
2521	127	-0947 May 22	11:39:03	24364	-36445	32	T	-p	0.5940	1.0780	54.0N	94.4E	53	164	314	05m25s
2522	127	-0947 Nov 15	08:24:40	24356	-36439	37	A	-n	-0.3565	0.9253	37.0S	147.5E	69	14	301	08m32s
2523	127	-0946 May 12	03:58:30	24347	-36433	42	T	nn	-0.1532	1.0497	6.7N	138.1W	81	347	167	04m54s
2524	127	-0946 Nov 04	12:13:23	24339	-36427	47	A	n-	0.3438	0.9720	6.7N	100.3E	70	194	107	03m14s
2525	127	-0945 May 01	15:08:00	24330	-36421	52	A	t-	-0.9628	0.9812	59.3S	82.9E	15	330	252	01m30s
2526	127	-0945 Oct 24	23:21:52	24322	-36415	57	P	t-	1.0049	0.9903	71.3N	8.9W	0	242		
2527	127	-0944 Mar 21	03:30:52	24314	-36410	24	P	-t	1.1093	0.7726	71.8N	152.3E	0	101		
2528	127	-0944 Sep 14	04:56:40	24306	-36404	29	P	-t	-1.0056	0.9987	71.4S	138.9E	0	67		
2529	127	-0943 Mar 10	04:49:16	24297	-36398	34	A	-n	0.3529	0.9616	12.0N	153.4W	69	163	148	04m37s
2530	127	-0943 Sep 03	18:21:30	24289	-36392	39	H	-n	-0.3218	1.0033	7.0S	1.2E	71	15	12	00m21s
2531	127	-0942 Feb 27	12:29:02	24280	-36386	44	H	p-	-0.4016	1.0118	34.7S	105.6E	66	342	44	01m05s
2532	127	-0942 Aug 24	01:14:02	24272	-36380	49	A	p-	0.4261	0.9533	39.5N	89.6W	65	196	189	04m54s
2533	127	-0941 Jan 18	17:12:07	24265	-36375	16	Pe	-t	1.5250	0.0165	67.5N	18.5E	0	172		
2534	127	-0941 Feb 17	02:29:38	24263	-36374	54	P	t-	-1.0898	0.8423	70.1S	27.5E	0	219		
2535	127	-0941 Aug 13	02:00:51	24255	-36368	59	P	t-	1.1678	0.6727	69.4N	41.9E	0	332		
2536	127	-0940 Jan 08	08:39:54	24248	-36363	26	T	-p	0.8860	1.0181	39.2N	155.4E	27	182	134	01m36s
2537	127	-0940 Jul 02	14:24:34	24239	-36357	31	A	-p	-0.9259	0.9853	44.1S	70.9E	22	353	139	01m24s
2538	127	-0940 Dec 27	19:29:39	24231	-36351	36	A	-n	0.2080	0.9709	11.9S	8.9W	78	186	107	03m29s
2539	127	-0939 Jun 22	02:03:20	24222	-36345	41	Tm	nn	-0.1240	1.0523	16.5N	110.3W	83	350	175	05m02s
2540	127	-0939 Dec 16	22:55:10	24214	-36339	46	A	p-	-0.5023	0.9237	52.7S	72.5W	60	17	333	07m33s
2541	128	-0938 Jun 11	18:34:31	24205	-36333	51	T	p-	0.6183	1.0773	58.7N	16.9W	52	154	318	05m04s
2542	128	-0938 Dec 05	21:55:18	24196	-36327	56	P	t-	-1.1767	0.6555	63.4S	168.9E	0	145		
2543	128	-0937 May 03	03:21:34	24189	-36322	23	P	-t	-1.1882	0.6532	61.3S	64.4W	0	296		
2544	128	-0937 Jun 01	11:26:28	24188	-36321	61	P	t-	1.3729	0.3043	63.1N	30.2W	0	40		
2545	128	-0937 Oct 26	09:22:23	24181	-36316	28	P	-t	1.0869	0.8276	61.1N	149.9W	0	250		
2546	128	-0936 Apr 21	12:28:14	24172	-36310	33	A	-p	-0.4554	0.9757	15.5S	106.1E	63	331	96	02m32s
2547	128	-0936 Oct 14	22:13:31	24164	-36304	38	T	-n	0.3609	1.0391	13.4N	44.5W	69	210	139	03m17s
2548	128	-0935 Apr 10	14:40:46	24155	-36298	43	A	pn	0.3100	0.9451	20.1N	51.9E	72	149	211	05m58s
2549	128	-0935 Oct 04	14:05:36	24147	-36292	48	T	n-	-0.3180	1.0481	17.1S	58.2E	71	31	168	03m54s
2550	128	-0934 Mar 30	15:28:49	24138	-36286	53	P	t-	1.0506	0.8800	60.5N	39.7W	0	90		
2551	128	-0934 Sep 24	04:18:02	24130	-36280	58	P	t-	-1.0405	0.9188	60.7S	131.4E	0	83		
2552	128	-0933 Feb 18	11:04:47	24123	-36275	25	T	-p	-0.9632	1.0132	66.6S	154.5W	15	267	174	00m45s
2553	128	-0933 Aug 14	21:05:20	24114	-36269	30	P	-t	1.0550	0.8678	62.1N	91.8E	0	310		
2554	128	-0932 Feb 08	02:07:49	24106	-36263	35	T	-n	-0.2330	1.0579	30.6S	101.1W	76	336	196	04m38s
2555	128	-0932 Aug 02	21:15:20	24097	-36257	40	A	np	0.3301	0.9405	38.7N	30.2W	71	202	233	06m22s
2556	128	-0931 Jan 27	18:12:47	24089	-36251	45	T	n-	0.4636	1.0404	5.9N	3.2E	62	163	152	03m50s
2557	128	-0931 Jul 22	23:59:33	24080	-36245	50	A	p-	-0.4248	0.9778	2.1S	85.4W	65	13	87	02m41s
2558	128	-0930 Jan 17	06:39:57	24072	-36239	55	P	t-	1.2115	0.6049	63.9N	151.6E	0	150		
2559	128	-0930 Jun 13	01:57:00	24064	-36234	22	P	-t	1.4113	0.2293	67.4N	76.2E	0	7		
2560	128	-0930 Jul 12	09:58:18	24063	-36233	60	P	t-	-1.1471	0.7302	64.7S	107.3E	0	21		
2561	129	-0930 Dec 07	17:25:13	24056	-36228	27	P	-t	-1.0375	0.8855	67.8S	151.7W	0	170		
2562	129	-0929 Jun 02	19:04:07	24048	-36222	32	T	-p	0.6682	1.0760	62.4N	17.4W	48	167	332	04m58s
2563	129	-0929 Nov 26	16:31:30	24039	-36216	37	A	-n	-0.3590	0.9268	40.5S	26.3E	69	10	295	08m12s
2564	129	-0928 May 22	11:10:57	24030	-36210	42	Tm	nn	-0.0771	1.0470	14.0N	110.9E	86	350	157	04m39s
2565	129	-0928 Nov 14	20:42:26	24022	-36204	47	A	n-	0.3422	0.9742	3.2N	29.2W	70	191	98	03m01s
2566	129	-0927 May 11	21:53:47	24014	-36198	52	A	p-	-0.8855	0.9820	45.6S	33.9W	27	342	137	01m41s
2567	129	-0927 Nov 04	08:07:19	24005	-36192	57	P	t-	1.0005	0.9990	70.6N	155.2W	0	228		
2568	129	-0926 Apr 01	10:10:07	23998	-36187	24	P	-t	1.1791	0.6547	71.8N	37.1E	0	88		
2569	129	-0926 Sep 25	13:21:07	23989	-36181	29	P	-t	-1.0283	0.9539	71.7S	2.6W	0	81		
2570	129	-0925 Mar 21	11:53:19	23981	-36175	34	A	-p	0.4117	0.9668	19.6N	96.7E	66	161	131	03m47s
2571	129	-0925 Sep 15	02:19:48	23972	-36169	39	A	-n	-0.3526	0.9968	12.8S	121.5W	69	17	12	00m21s
2572	129	-0924 Mar 09	20:09:35	23964	-36163	44	T	p-	-0.3537	1.0187	27.7S	12.2W	69	341	68	01m45s
2573	129	-0924 Sep 03	08:36:30	23956	-36157	49	A	p-	0.3843	0.9480	33.0N	158.0E	67	198	206	05m44s
2574	129	-0923 Feb 27	10:35:25	23947	-36151	54	P	t-	-1.0511	0.9176	70.8S	108.3W	0	232		
2575	129	-0923 Aug 23	09:02:15	23939	-36145	59	P	t-	1.1154	0.7604	70.2N	77.2W	0	320		
2576	129	-0922 Jan 18	17:10:02	23932	-36140	26	T	-t	0.9024	1.0176	42.5N	22.1E	25	176	141	01m31s
2577	129	-0922 Jul 13	21:26:58	23923	-36134	31	As	-t	-0.9926	0.9822	61.1S	40.6W	5	358	-	01m27s
2578	129	-0921 Jan 08	03:45:32	23915	-36128	36	A	-n	0.2185	0.9699	10.9S	133.9W	77	182	111	03m40s
2579	129	-0921 Jul 03	09:26:14	23906	-36122	41	T	nn	-0.1939	1.0534	12.8N	138.1E	79	355	180	05m15s
2580	129	-0921 Dec 28	06:53:44	23898	-36116	46	A	p-	-0.4937	0.9238	53.3S	171.9E	60	10	331	07m38s

Cat Num	Canon Plate	Calendar Date	TD of Greatest Eclipse	ΔT s	Luna Num	Saros Num	Ecl. Type	QLE	Gamma	Ecl. Mag.	Lat. °	Long. °	Sun Alt °	Sun Azm °	Path Width km	Central Line Dur.
2581	130	-0920 Jun 22	02:03:01	23889	-36110	51	T	p-	0.5464	1.0772	56.1N	122.5W	57	164	299	05m17s
2582	130	-0920 Dec 16	05:56:25	23881	-36104	56	P	t-	-1.1695	0.6678	64.3S	37.7E	0	155		
2583	130	-0919 May 13	10:28:46	23874	-36099	23	P	-t	-1.2677	0.5051	61.8S	178.5E	0	305		
2584	130	-0919 Jun 11	18:43:06	23872	-36098	61	P	t-	1.3032	0.4376	63.9N	150.2W	0	30		
2585	130	-0919 Nov 05	17:54:59	23865	-36093	28	P	-t	1.0872	0.8275	61.5N	71.9E	0	241		
2586	130	-0918 May 02	19:10:05	23857	-36087	33	A	-p	-0.5374	0.9733	17.0S	5.0E	57	333	111	02m52s
2587	130	-0918 Oct 26	06:57:14	23848	-36081	38	T	-n	0.3675	1.0389	9.7N	177.6W	68	208	139	03m21s
2588	130	-0917 Apr 21	21:08:47	23840	-36075	43	A	nn	0.2285	0.9468	20.1N	44.8W	77	151	200	05m50s
2589	130	-0917 Oct 15	22:45:19	23831	-36069	48	T	n-	-0.3072	1.0447	21.0S	73.4W	72	31	156	03m36s
2590	130	-0916 Apr 09	22:13:31	23823	-36063	53	A	t-	0.9759	0.9512	60.3N	126.6W	12	103	841	03m26s
2591	130	-0916 Oct 04	12:36:34	23814	-36057	58	P	t-	-1.0263	0.9422	60.7S	3.3W	0	92		
2592	130	-0915 Feb 28	18:57:40	23807	-36052	25	T-	-t	-1.0012	1.0004	61.2S	110.1E	0	246	-	-
2593	130	-0915 Aug 25	04:14:54	23799	-36046	30	P	-t	1.1062	0.7790	61.5N	25.9W	0	301		
2594	130	-0914 Feb 18	10:21:12	23790	-36040	35	T	-n	-0.2669	1.0617	28.9S	135.8E	74	333	210	04m51s
2595	130	-0914 Aug 14	04:11:20	23782	-36034	40	A	-p	0.3892	0.9382	38.9N	132.3W	67	207	248	06m25s
2596	130	-0913 Feb 08	02:31:54	23773	-36028	45	T	n-	0.4384	1.0416	6.5N	123.5W	64	159	154	03m51s
2597	130	-0913 Aug 03	07:08:24	23765	-36022	50	A	p-	-0.3615	0.9786	0.2N	166.0E	69	17	81	02m29s
2598	130	-0912 Jan 28	14:48:33	23757	-36016	55	P	t-	1.1954	0.6336	63.1N	18.7E	0	141		
2599	130	-0912 Jun 23	09:20:02	23750	-36011	22	Pe	-t	1.4777	0.1017	66.4N	46.7W	0	356		
2600	130	-0912 Jul 22	17:26:11	23748	-36010	60	P	t-	-1.0815	0.8530	63.8S	15.8W	0	31		
2601	131	-0912 Dec 18	01:25:26	23741	-36005	27	P	-t	-1.0434	0.8758	66.7S	75.9E	0	182		
2602	131	-0911 Jun 13	02:28:32	23733	-35999	32	T	-p	0.7427	1.0730	70.7N	125.6W	42	173	356	04m29s
2603	131	-0911 Dec 07	00:39:47	23724	-35993	37	A	-n	-0.3608	0.9289	43.1S	94.3W	69	5	286	07m48s
2604	131	-0910 Jun 02	18:21:52	23716	-35987	42	T	nn	-0.0005	1.0435	20.9N	1.1E	90	182	145	04m16s
2605	131	-0910 Nov 26	05:13:06	23707	-35981	47	A	n-	0.3411	0.9770	0.3N	158.8W	70	188	87	02m43s
2606	131	-0909 May 23	04:37:51	23699	-35975	52	A	p-	-0.8075	0.9811	35.0S	143.2W	36	349	114	01m59s
2607	131	-0909 Nov 15	16:55:23	23691	-35969	57	T+	t-	0.9976	1.0050	69.8N	58.5E	0	215	-	-
2608	131	-0908 Apr 11	16:44:07	23683	-35964	24	P	-t	1.2526	0.5298	71.6N	76.7W	0	74		
2609	131	-0908 Oct 05	21:51:28	23675	-35958	29	P	-t	-1.0457	0.9194	71.8S	145.8W	0	95		
2610	131	-0907 Mar 31	18:51:35	23667	-35952	34	A	-p	0.4761	0.9718	27.7N	12.2W	61	160	114	03m01s
2611	131	-0907 Sep 25	10:25:04	23658	-35946	39	A	-p	-0.3764	0.9905	18.4S	114.3E	68	19	36	00m59s
2612	131	-0906 Mar 21	03:41:44	23650	-35940	44	T	n-	-0.2981	1.0254	20.2S	128.5W	73	341	90	02m26s
2613	131	-0906 Sep 14	16:08:17	23641	-35934	49	A	n-	0.3512	0.9431	26.8N	42.8E	69	199	224	06m33s
2614	131	-0905 Mar 10	18:32:35	23633	-35928	54	T-	t-	-1.0053	1.0072	71.4S	117.5E	0	245	-	-
2615	131	-0905 Sep 03	16:13:39	23625	-35922	59	P	t-	1.0707	0.8345	71.0N	160.6E	0	307		
2616	131	-0904 Jan 30	01:31:29	23617	-35917	26	T	-t	0.9253	1.0169	47.4N	110.0W	22	170	154	01m23s
2617	131	-0904 Jul 24	04:37:34	23609	-35911	31	P	-t	-1.0531	0.8902	67.6S	160.1W	0	9		
2618	131	-0903 Jan 18	11:52:49	23601	-35905	36	A	-n	0.2356	0.9694	8.7S	103.0E	76	177	113	03m46s
2619	131	-0903 Jul 13	16:53:45	23592	-35899	41	T	-n	-0.2605	1.0537	8.3N	24.8E	75	359	184	05m21s
2620	131	-0902 Jan 07	14:46:26	23584	-35893	46	A	p-	-0.4807	0.9247	52.4S	57.8E	61	3	324	07m41s
2621	132	-0902 Jul 03	09:34:32	23575	-35887	51	T	p-	0.4771	1.0761	52.4N	129.5E	61	172	281	05m26s
2622	132	-0902 Dec 27	13:54:30	23567	-35881	56	P	t-	-1.1594	0.6850	65.3S	93.1W	0	165		
2623	132	-0901 May 24	17:32:58	23560	-35876	23	P	-t	-1.3484	0.3556	62.5S	62.0E	0	314		
2624	132	-0901 Jun 23	02:01:09	23559	-35875	61	P	t-	1.2351	0.5666	64.8N	89.2E	0	21		
2625	132	-0901 Nov 17	02:32:05	23552	-35870	28	P	-t	1.0844	0.8332	62.1N	67.7W	0	231		
2626	132	-0900 May 13	01:46:44	23543	-35864	33	A	-p	-0.6227	0.9703	19.8S	95.0W	51	336	133	03m16s
2627	132	-0900 Nov 05	15:46:05	23535	-35858	38	T	-n	0.3702	1.0389	6.3N	48.0E	68	205	140	03m26s
2628	132	-0899 May 02	03:31:49	23526	-35852	43	A	nn	0.1429	0.9482	19.7N	140.0W	82	153	192	05m49s
2629	132	-0899 Oct 26	07:30:00	23518	-35846	48	T	n-	-0.3004	1.0416	25.1S	154.0E	72	30	145	03m21s
2630	132	-0898 Apr 21	04:53:45	23510	-35840	53	A	t-	0.8964	0.9591	58.2N	151.3E	26	119	330	03m04s
2631	132	-0898 Oct 15	21:00:40	23501	-35834	58	P	p-	-1.0172	0.9563	60.8S	139.2W	0	101		
2632	132	-0897 Mar 12	02:44:40	23494	-35829	25	P	-t	-1.0451	0.9213	60.9S	16.7W	0	255		
2633	132	-0897 Sep 05	11:31:46	23486	-35823	30	P	-t	1.1510	0.7020	61.0N	145.2W	0	292		
2634	132	-0896 Feb 29	18:27:53	23477	-35817	35	T	-n	-0.3068	1.0652	27.0S	14.1E	72	330	223	05m04s
2635	132	-0896 Aug 24	11:16:27	23469	-35811	40	A	-p	0.4412	0.9360	37.9N	122.9E	64	211	264	06m32s
2636	132	-0895 Feb 18	10:41:15	23461	-35805	45	T	n-	0.4055	1.0428	7.3N	112.3E	66	156	155	03m51s
2637	132	-0895 Aug 13	14:27:09	23452	-35799	50	A	p-	-0.3054	0.9792	1.2N	55.1E	72	21	78	02m19s
2638	132	-0894 Feb 07	22:46:04	23444	-35793	55	P	t-	1.1711	0.6769	62.3N	111.2W	0	131		
2639	132	-0894 Aug 03	01:03:26	23435	-35787	60	P	t-	-1.0228	0.9629	63.0S	140.8W	0	40		
2640	132	-0894 Dec 29	09:19:00	23428	-35782	27	P	-t	-1.0548	0.8575	65.6S	54.3W	0	192		

Cat Num	Canon Plate	Calendar Date	TD of Greatest Eclipse	ΔT s	Luna Num	Saros Num	Ecl. Type	QLE	Gamma	Ecl. Mag.	Lat. °	Long. °	Sun Alt °	Sun Azm °	Path Width km	Central Line Dur.
2641	133	-0893 Jun 24	09:56:27	23420	-35776	32	T	-p	0.8142	1.0689	78.3N	135.1E	35	189	391	03m59s
2642	133	-0893 Dec 18	08:46:03	23412	-35770	37	A	-n	-0.3651	0.9318	44.9S	146.3E	68	359	275	07m20s
2643	133	-0892 Jun 13	01:31:21	23403	-35764	42	T	nn	0.0765	1.0392	27.1N	107.5W	85	178	132	03m47s
2644	133	-0892 Dec 06	13:45:10	23395	-35758	47	A	n-	0.3403	0.9805	2.0S	71.3E	70	184	74	02m19s
2645	133	-0891 Jun 02	11:17:54	23386	-35752	52	A	p-	-0.7266	0.9792	25.8S	110.7E	43	354	107	02m22s
2646	133	-0891 Nov 26	01:45:53	23378	-35746	57	Tn	t-	0.9966	1.0144	67.2N	89.5W	2	200	-	00m59s
2647	133	-0890 Apr 22	23:10:40	23371	-35741	24	P	-t	1.3319	0.3944	71.2N	171.6E	0	61		
2648	133	-0890 May 22	14:07:11	23370	-35740	62	Pb	t-	-1.5229	0.0667	69.3S	92.7E	0	332		
2649	133	-0890 Oct 17	06:28:56	23363	-35735	29	P	-t	-1.0571	0.8966	71.5S	69.4E	0	109		
2650	133	-0889 Apr 12	01:45:06	23354	-35729	34	A	-p	0.5451	0.9765	36.2N	120.2W	57	158	100	02m21s
2651	133	-0889 Oct 06	18:36:18	23346	-35723	39	A	-p	-0.3944	0.9846	23.9S	11.2W	67	20	59	01m34s
2652	133	-0888 Mar 31	11:09:47	23338	-35717	44	T	n-	-0.2381	1.0318	12.5S	115.8E	76	342	110	03m06s
2653	133	-0888 Sep 24	23:47:00	23329	-35711	49	A	n-	0.3243	0.9383	20.8N	74.4W	71	199	241	07m23s
2654	133	-0887 Mar 21	02:22:47	23321	-35705	54	T	t-	-0.9534	1.0558	67.9S	64.6W	17	306	624	03m18s
2655	133	-0887 Sep 13	23:34:23	23313	-35699	59	P	t-	1.0333	0.8962	71.5N	35.4E	0	293		
2656	133	-0886 Feb 09	09:44:46	23306	-35694	26	T	-t	0.9546	1.0157	54.2N	117.8E	17	162	185	01m13s
2657	133	-0886 Aug 04	11:57:43	23297	-35688	31	P	-t	-1.1068	0.7938	68.6S	77.1E	0	20		
2658	133	-0885 Jan 29	19:51:01	23289	-35682	36	A	-n	0.2593	0.9691	5.3S	18.2W	75	173	115	03m50s
2659	133	-0885 Jul 25	00:29:59	23280	-35676	41	T	-n	-0.3204	1.0535	3.4N	91.4W	71	3	187	05m20s
2660	133	-0884 Jan 18	22:31:37	23272	-35670	46	A	p-	-0.4619	0.9261	50.0S	55.3W	62	356	314	07m44s
2661	134	-0884 Jul 13	17:10:32	23264	-35664	51	T	p-	0.4117	1.0743	47.9N	18.5E	65	179	265	05m33s
2662	134	-0883 Jan 06	21:47:16	23255	-35658	56	P	t-	-1.1446	0.7104	66.4S	137.0E	0	176		
2663	134	-0883 Jun 04	00:35:08	23248	-35653	23	P	-t	-1.4293	0.2069	63.3S	54.2W	0	323		
2664	134	-0883 Jul 03	09:21:33	23247	-35652	61	P	t-	1.1699	0.6894	65.7N	32.3W	0	11		
2665	134	-0883 Nov 27	11:10:12	23240	-35647	28	P	-t	1.0817	0.8389	62.8N	152.4E	0	222		
2666	134	-0882 May 24	08:19:43	23232	-35641	33	A	-p	-0.7099	0.9666	23.9S	165.5E	45	339	167	03m44s
2667	134	-0882 Nov 17	00:37:33	23223	-35635	38	T	-n	0.3709	1.0396	3.2N	87.0W	68	202	142	03m35s
2668	134	-0881 May 13	09:51:29	23215	-35629	43	A	nn	0.0545	0.9492	18.6N	125.6E	87	157	187	05m54s
2669	134	-0881 Nov 06	16:17:53	23207	-35623	48	T	n-	-0.2968	1.0389	29.2S	21.0E	73	28	136	03m08s
2670	134	-0880 May 01	11:30:53	23198	-35617	53	A	p-	0.8132	0.9653	56.9N	61.5E	35	128	212	02m42s
2671	134	-0880 Oct 26	05:30:05	23190	-35611	58	P	p-	-1.0132	0.9612	61.1S	83.4E	0	110		
2672	134	-0879 Mar 22	10:22:52	23183	-35606	25	P	-t	-1.0967	0.8267	60.8S	141.3W	0	264		
2673	134	-0879 Apr 20	19:29:54	23182	-35605	63	Pb	t-	1.5192	0.0414	61.0N	121.4W	0	73		
2674	134	-0879 Sep 15	18:57:47	23175	-35600	30	P	-t	1.1875	0.6397	60.7N	93.3E	0	283		
2675	134	-0878 Mar 12	02:25:29	23166	-35594	35	T	-n	-0.3543	1.0685	25.2S	105.6W	69	329	237	05m16s
2676	134	-0878 Sep 04	18:32:13	23158	-35588	40	A	-p	0.4845	0.9338	36.1N	14.5E	61	213	279	06m42s
2677	134	-0877 Mar 01	18:41:51	23150	-35582	45	T	n-	0.3660	1.0439	8.4N	9.4W	68	153	156	03m52s
2678	134	-0877 Aug 24	21:56:33	23141	-35576	50	A	n-	-0.2573	0.9795	0.9N	58.3W	75	24	75	02m12s
2679	134	-0876 Feb 19	06:35:08	23133	-35570	55	P	t-	1.1408	0.7308	61.7N	121.2E	0	122		
2680	134	-0876 Aug 13	08:47:40	23125	-35564	60	T	t-	-0.9692	1.0250	52.1S	109.5E	14	35	349	01m50s
2681	135	-0875 Jan 08	17:08:29	23118	-35559	27	P	-t	-1.0690	0.8345	64.6S	177.0E	0	203		
2682	135	-0875 Jul 04	17:26:33	23109	-35553	32	T	-t	0.8836	1.0638	83.0N	70.1E	28	239	452	03m28s
2683	135	-0875 Dec 28	16:49:08	23101	-35547	37	A	-n	-0.3725	0.9352	45.8S	28.2E	68	353	260	06m49s
2684	135	-0874 Jun 24	08:42:23	23093	-35541	42	T	nn	0.1513	1.0343	32.4N	144.5E	81	182	117	03m13s
2685	135	-0874 Dec 17	22:15:39	23084	-35535	47	A	n-	0.3373	0.9845	3.6S	58.0W	70	180	58	01m49s
2686	135	-0873 Jun 13	17:58:30	23076	-35529	52	A	p-	-0.6470	0.9766	17.9S	6.0E	50	358	109	02m51s
2687	135	-0873 Dec 07	10:35:08	23068	-35523	57	Tn	t-	0.9943	1.0174	63.5N	125.3E	4	189	-	01m13s
2688	135	-0872 May 03	05:35:02	23061	-35518	24	P	-t	1.4124	0.2562	70.6N	61.0E	0	48		
2689	135	-0872 Jun 01	20:25:59	23059	-35517	62	P	t-	-1.4382	0.2122	68.3S	15.1W	0	344		
2690	135	-0872 Oct 27	15:10:23	23052	-35512	29	P	-t	-1.0647	0.8809	71.1S	76.2W	0	123		
2691	135	-0871 Apr 22	08:33:40	23044	-35506	34	A	-p	0.6188	0.9806	45.3N	132.7E	52	157	87	01m47s
2692	135	-0871 Oct 17	02:53:28	23036	-35500	39	A	-p	-0.4066	0.9793	29.1S	137.7W	66	20	80	02m05s
2693	135	-0870 Apr 11	18:31:51	23028	-35494	44	T	n-	-0.1722	1.0377	4.6S	1.6E	80	342	128	03m42s
2694	135	-0870 Oct 06	07:33:24	23019	-35488	49	A	n-	0.3044	0.9341	15.2N	166.4E	72	199	257	08m11s
2695	135	-0869 Apr 01	10:06:04	23011	-35482	54	T	t-	-0.8955	1.0623	58.0S	157.8E	26	324	459	04m04s
2696	135	-0869 Sep 25	07:05:47	23003	-35476	59	A+	t-	1.0040	0.9441	71.8N	92.9W	0	279	-	-
2697	135	-0868 Feb 20	17:48:07	22996	-35471	26	T	-t	0.9914	1.0129	65.7N	21.5W	6	146	407	00m53s
2698	135	-0868 Aug 14	19:26:56	22987	-35465	31	P	-t	-1.1537	0.7094	69.5S	48.5W	0	32		
2699	135	-0867 Feb 09	03:39:24	22979	-35459	36	A	-n	0.2902	0.9690	0.8S	137.5W	73	170	116	03m49s
2700	135	-0867 Aug 04	08:13:12	22971	-35453	41	T	-n	-0.3746	1.0526	1.8S	150.2E	68	7	188	05m11s

Cat Num	Canon Plate	Calendar Date	TD of Greatest Eclipse	ΔT s	Luna Num	Saros Num	Ecl. Type	QLE	Gamma	Ecl. Mag.	Lat. °	Long. °	Sun Alt °	Sun Azm °	Path Width km	Central Line Dur.
2701	136	-0866 Jan 29	06:07:11	22962	-35447	46	A	p-	-0.4358	0.9280	46.0S	167.3W	64	350	301	07m46s
2702	136	-0866 Jul 25	00:52:06	22954	-35441	51	T	n-	0.3514	1.0716	42.7N	95.4W	69	185	249	05m35s
2703	136	-0865 Jan 18	05:34:46	22946	-35435	56	P	t-	-1.1254	0.7434	67.5S	8.0E	0	187		
2704	136	-0865 Jun 15	07:36:22	22939	-35430	23	Pe	-t	-1.5097	0.0606	64.1S	170.4W	0	332		
2705	136	-0865 Jul 14	16:44:39	22938	-35429	61	P	t-	1.1075	0.8054	66.7N	154.8W	0	1		
2706	136	-0865 Dec 08	19:49:05	22931	-35424	28	P	-t	1.0789	0.8450	63.7N	12.0E	0	212		
2707	136	-0864 Jun 03	14:50:41	22922	-35418	33	A	-p	-0.7974	0.9623	29.9S	66.2E	37	342	223	04m13s
2708	136	-0864 Nov 27	09:30:43	22914	-35412	38	T	-n	0.3706	1.0406	0.7N	137.7E	68	198	146	03m46s
2709	136	-0863 May 23	16:08:30	22906	-35406	43	Am	nn	-0.0362	0.9498	16.8N	31.8E	88	338	184	06m05s
2710	136	-0863 Nov 17	01:08:32	22897	-35400	48	T	n-	-0.2959	1.0367	33.1S	112.1W	73	25	129	02m58s
2711	136	-0862 May 12	18:07:22	22889	-35394	53	A	p-	0.7284	0.9706	55.9N	30.1W	43	136	152	02m21s
2712	136	-0862 Nov 06	14:01:39	22881	-35388	58	P	p-	-1.0119	0.9616	61.5S	54.6W	0	120		
2713	136	-0861 Apr 02	17:57:00	22874	-35383	25	P	-t	-1.1523	0.7235	60.7S	95.1E	0	273		
2714	136	-0861 May 02	02:38:41	22873	-35382	63	P	t-	1.4433	0.1798	61.4N	121.3E	0	64		
2715	136	-0861 Sep 27	02:31:47	22866	-35377	30	P	-t	1.2168	0.5900	60.6N	30.1W	0	274		
2716	136	-0860 Mar 22	10:16:37	22857	-35371	35	T	-n	-0.4074	1.0712	23.7S	136.2E	66	328	251	05m28s
2717	136	-0860 Sep 15	01:57:24	22849	-35365	40	A	-p	0.5206	0.9319	33.7N	97.1W	58	215	293	06m55s
2718	136	-0859 Mar 12	02:33:05	22841	-35359	45	T	n-	0.3192	1.0447	9.6N	128.5W	71	151	156	03m52s
2719	136	-0859 Sep 04	05:36:46	22833	-35353	50	A	n-	-0.2170	0.9797	0.4S	174.5W	77	27	74	02m07s
2720	136	-0858 Mar 01	14:11:36	22824	-35347	55	P	t-	1.1009	0.8017	61.2N	3.1W	0	113		
2721	137	-0858 Aug 24	16:42:49	22816	-35341	60	T	t-	-0.9238	1.0271	46.2S	8.9W	22	36	236	02m03s
2722	137	-0857 Jan 20	00:49:06	22809	-35336	27	P	-t	-1.0907	0.7993	63.6S	50.9E	0	213		
2723	137	-0857 Jul 16	01:01:26	22801	-35330	32	T	-t	0.9487	1.0572	78.3N	14.1E	18	299	615	02m55s
2724	137	-0856 Jan 09	00:47:00	22793	-35324	37	A	-n	-0.3849	0.9392	45.8S	88.5W	67	347	244	06m14s
2725	137	-0856 Jul 04	15:55:03	22784	-35318	42	T	-n	0.2239	1.0287	36.8N	37.3E	77	188	100	02m37s
2726	137	-0856 Dec 28	06:43:18	22776	-35312	47	A	n-	0.3313	0.9892	4.6S	173.6E	71	175	40	01m15s
2727	137	-0855 Jun 24	00:38:07	22768	-35306	52	A	p-	-0.5672	0.9733	11.0S	97.5W	55	2	116	03m23s
2728	137	-0855 Dec 17	19:23:47	22760	-35300	57	T	t-	0.9913	1.0208	60.5N	17.2W	6	179	641	01m30s
2729	137	-0854 May 14	11:55:47	22753	-35295	24	Pe	-t	1.4956	0.1129	69.9N	48.1W	0	36		
2730	137	-0854 Jun 13	02:44:57	22751	-35294	62	P	t-	-1.3532	0.3582	67.4S	122.5W	0	354		
2731	137	-0854 Nov 07	23:55:36	22744	-35289	29	P	-t	-1.0690	0.8718	70.4S	137.9E	0	137		
2732	137	-0853 May 03	15:20:43	22736	-35283	34	A	-p	0.6946	0.9842	54.7N	25.5E	46	155	78	01m20s
2733	137	-0853 Oct 28	11:15:54	22728	-35277	39	A	-p	-0.4137	0.9744	34.0S	95.2E	65	19	100	02m32s
2734	137	-0852 Apr 22	01:50:28	22720	-35271	44	T	nn	-0.1025	1.0432	3.3N	111.8W	84	344	145	04m14s
2735	137	-0852 Oct 16	15:26:05	22711	-35265	49	A	n-	0.2902	0.9302	10.1N	45.8E	73	197	272	08m58s
2736	137	-0851 Apr 11	17:43:49	22703	-35259	54	T	p-	-0.8327	1.0677	48.2S	31.5E	33	333	396	04m50s
2737	137	-0851 Oct 05	14:45:26	22695	-35253	59	An	t-	0.9814	0.9125	68.7N	107.9E	10	238	-	06m55s
2738	137	-0850 Mar 03	01:41:53	22688	-35248	26	P	-t	1.0352	0.9348	71.1N	164.8W	0	123		
2739	137	-0850 Aug 26	03:06:52	22680	-35242	31	P	-t	-1.1926	0.6394	70.3S	177.5W	0	44		
2740	137	-0849 Feb 20	11:17:55	22671	-35236	36	A	-p	0.3281	0.9689	4.6N	105.2E	71	167	118	03m46s
2741	138	-0849 Aug 15	16:04:24	22663	-35230	41	T	-n	-0.4226	1.0515	7.4S	29.4E	65	11	188	04m58s
2742	138	-0848 Feb 09	13:33:40	22655	-35224	46	A	p-	-0.4026	0.9301	40.9S	81.4E	66	347	286	07m48s
2743	138	-0848 Aug 04	08:40:00	22647	-35218	51	T	n-	0.2968	1.0684	37.2N	147.8E	73	189	234	05m33s
2744	138	-0847 Jan 28	13:13:23	22638	-35212	56	P	t-	-1.0983	0.7902	68.6S	119.4W	0	198		
2745	138	-0847 Jul 25	00:12:34	22630	-35206	61	P	t-	1.0497	0.9115	67.7N	81.0E	0	350		
2746	138	-0847 Dec 19	04:25:32	22623	-35201	28	P	-t	1.0788	0.8465	64.7N	128.1W	0	202		
2747	138	-0846 Jun 14	21:21:50	22615	-35195	33	A	-t	-0.8837	0.9570	38.2S	33.7W	28	345	333	04m38s
2748	138	-0846 Dec 08	18:22:12	22607	-35189	38	T	-n	0.3719	1.0423	0.9S	3.0E	68	194	152	04m01s
2749	138	-0845 Jun 03	22:26:25	22599	-35183	43	A	nn	-0.1262	0.9498	14.2N	62.5W	83	343	186	06m22s
2750	138	-0845 Nov 28	10:00:06	22590	-35177	48	T	n-	-0.2963	1.0350	36.4S	115.3E	73	22	123	02m51s
2751	138	-0844 May 23	00:42:20	22582	-35171	53	A	p-	0.6413	0.9753	54.6N	121.9W	50	145	114	02m03s
2752	138	-0844 Nov 16	22:36:23	22574	-35165	58	P	p-	-1.0139	0.9562	62.2S	166.5E	0	129		
2753	138	-0843 Apr 13	01:24:30	22567	-35160	25	P	-t	-1.2136	0.6080	60.9S	26.8W	0	281		
2754	138	-0843 May 12	09:44:12	22566	-35159	63	P	t-	1.3638	0.3263	61.9N	4.7E	0	55		
2755	138	-0843 Oct 07	10:14:11	22559	-35154	30	P	-t	1.2390	0.5526	60.6N	155.6W	0	265		
2756	138	-0842 Apr 02	17:59:38	22551	-35148	35	T	-p	-0.4674	1.0733	22.8S	19.9E	62	328	266	05m39s
2757	138	-0842 Sep 26	09:33:26	22542	-35142	40	A	-p	0.5482	0.9303	30.8N	147.7E	57	215	306	07m11s
2758	138	-0841 Mar 23	10:16:26	22534	-35136	45	T	n-	0.2662	1.0452	10.9N	114.7E	74	150	155	03m52s
2759	138	-0841 Sep 15	13:26:13	22526	-35130	50	A	n-	-0.1834	0.9799	2.7S	66.9E	79	28	73	02m02s
2760	138	-0840 Mar 11	21:39:48	22518	-35124	55	P	t-	1.0552	0.8831	60.9N	125.2W	0	104		

Cat Num	Canon Plate	Calendar Date	TD of Greatest Eclipse	ΔT s	Luna Num	Saros Num	Ecl. Type	QLE	Gamma	Ecl. Mag.	Lat. °	Long. °	Sun Alt °	Sun Azm °	Path Width km	Central Line Dur.
2761	139	-0840 Sep 04	00:45:59	22509	-35118	60	T	t-	-0.8844	1.0280	43.8S	131.2W	27	38	199	02m06s
2762	139	-0839 Jan 30	08:22:00	22503	-35113	27	P	-t	-1.1181	0.7545	62.7S	72.9W	0	222		
2763	139	-0839 Jul 26	08:40:54	22494	-35107	32	T+	-t	1.0095	1.0002	63.5N	73.2W	0	326	-	-
2764	139	-0838 Jan 19	08:39:19	22486	-35101	37	A	-n	-0.4025	0.9437	45.0S	155.9E	66	341	227	05m38s
2765	139	-0838 Jul 15	23:12:04	22478	-35095	42	T	-n	0.2919	1.0227	39.9N	70.2W	73	194	81	02m01s
2766	139	-0837 Jan 08	15:06:00	22470	-35089	47	A	n-	0.3204	0.9943	5.0S	46.4E	71	171	21	00m38s
2767	139	-0837 Jul 05	07:21:54	22461	-35083	52	A	p-	-0.4915	0.9695	5.5S	158.9E	61	6	126	03m55s
2768	139	-0837 Dec 29	04:08:34	22453	-35077	57	T	t-	0.9848	1.0249	56.8N	157.0W	9	171	530	01m50s
2769	139	-0836 Jun 23	09:06:11	22445	-35071	62	P	t-	-1.2694	0.5015	66.3S	130.1E	0	5		
2770	139	-0836 Nov 18	08:42:37	22438	-35066	29	P	-t	-1.0716	0.8661	69.5S	7.9W	0	150		
2771	139	-0835 May 13	22:06:10	22430	-35060	34	A	-p	0.7727	0.9870	64.7N	82.3W	39	152	72	00m59s
2772	139	-0835 Nov 07	19:40:57	22422	-35054	39	A	-p	-0.4178	0.9703	38.4S	31.8W	65	17	117	02m54s
2773	139	-0834 May 03	09:05:39	22414	-35048	44	T	nn	-0.0291	1.0480	11.2N	136.1E	88	345	159	04m38s
2774	139	-0834 Oct 27	23:24:23	22405	-35042	49	A	n-	0.2810	0.9270	5.4N	76.1W	74	196	285	09m41s
2775	139	-0833 Apr 23	01:16:36	22397	-35036	54	T	p-	-0.7657	1.0720	38.9S	90.1W	40	339	361	05m33s
2776	139	-0833 Oct 16	22:32:57	22389	-35030	59	A	t-	0.9646	0.9123	63.2N	27.7W	15	221	1300	07m34s
2777	139	-0832 Mar 13	09:25:58	22382	-35025	26	P	-t	1.0862	0.8403	71.5N	64.2E	0	110		
2778	139	-0832 Apr 11	18:10:34	22381	-35024	64	Pb	t-	-1.5028	0.0557	71.4S	75.5E	0	286		
2779	139	-0832 Sep 05	10:56:45	22374	-35019	31	P	-t	-1.2242	0.5826	71.0S	50.5E	0	57		
2780	139	-0831 Mar 02	18:44:13	22366	-35013	36	A	-p	0.3752	0.9688	11.1N	9.7W	68	164	121	03m41s
2781	140	-0831 Aug 26	00:04:29	22358	-35007	41	T	-n	-0.4635	1.0499	13.1S	94.0W	62	14	186	04m40s
2782	140	-0830 Feb 19	20:49:59	22349	-35001	46	A	p-	-0.3612	0.9327	34.8S	28.8W	69	344	269	07m48s
2783	140	-0830 Aug 15	16:35:06	22341	-34995	51	T	n-	0.2484	1.0646	31.4N	28.3E	75	192	218	05m25s
2784	140	-0829 Feb 08	20:45:13	22333	-34989	56	P	t-	-1.0650	0.8481	69.6S	114.3E	0	210		
2785	140	-0829 Aug 05	07:46:01	22325	-34983	61	T+	t-	0.9972	1.0064	68.7N	45.2W	0	339	-	-
2786	140	-0829 Dec 30	13:00:21	22318	-34978	28	P	-t	1.0807	0.8445	65.7N	91.8E	0	191		
2787	140	-0828 Jun 25	03:52:55	22310	-34972	33	A	-t	-0.9689	0.9501	51.8S	133.5W	14	348	763	04m51s
2788	140	-0828 Dec 19	03:13:05	22302	-34966	38	T	-n	0.3738	1.0443	1.8S	131.5W	68	189	159	04m16s
2789	140	-0827 Jun 14	04:45:39	22293	-34960	43	A	nn	-0.2148	0.9494	10.7N	157.6W	78	348	190	06m42s
2790	140	-0827 Dec 08	18:50:11	22285	-34954	48	T	n-	-0.2960	1.0338	39.0S	16.2W	73	17	119	02m47s
2791	140	-0826 Jun 03	07:20:24	22277	-34948	53	A	p-	0.5557	0.9792	52.8N	144.8E	56	153	89	01m47s
2792	140	-0826 Nov 28	07:10:28	22269	-34942	58	P	p-	-1.0166	0.9500	62.9S	27.5E	0	139		
2793	140	-0825 Apr 24	08:49:55	22262	-34937	25	P	-t	-1.2774	0.4868	61.2S	148.2W	0	290		
2794	140	-0825 May 23	16:51:36	22261	-34936	63	P	t-	1.2846	0.4735	62.5N	112.6W	0	46		
2795	140	-0825 Oct 18	18:01:53	22254	-34931	30	P	-t	1.2565	0.5234	60.7N	77.6E	0	256		
2796	140	-0824 Apr 13	01:37:47	22246	-34925	35	T	-p	-0.5315	1.0747	22.5S	95.3W	58	329	282	05m49s
2797	140	-0824 Oct 06	17:18:03	22238	-34919	40	A	-p	0.5692	0.9292	27.8N	29.8E	55	214	316	07m29s
2798	140	-0823 Apr 02	17:49:55	22229	-34913	45	T	nn	0.2058	1.0451	12.0N	0.7E	78	150	153	03m51s
2799	140	-0823 Sep 25	21:26:40	22221	-34907	50	A	n-	-0.1582	0.9802	5.7S	54.7W	81	29	71	01m58s
2800	140	-0822 Mar 23	04:56:24	22213	-34901	55	A+	t-	1.0009	0.9795	60.7N	115.6E	0	95	-	-
2801	141	-0822 Sep 15	08:59:19	22205	-34895	60	T	p-	-0.8531	1.0283	43.5S	103.1E	31	41	180	02m06s
2802	141	-0821 Feb 10	15:44:20	22198	-34890	27	P	-t	-1.1538	0.6955	62.0S	166.2E	0	231		
2803	141	-0821 Aug 06	16:27:42	22190	-34884	32	P	-t	1.0639	0.8944	62.7N	159.5E	0	317		
2804	141	-0821 Sep 05	00:54:14	22189	-34883	70	Pb	t-	-1.5106	0.0413	61.3S	171.0W	0	68		
2805	141	-0820 Jan 30	16:24:32	22182	-34878	37	A	-p	-0.4265	0.9487	43.6S	41.8E	65	335	208	05m00s
2806	141	-0820 Jul 26	06:32:15	22174	-34872	42	H3	-p	0.3561	1.0161	41.8N	177.9W	69	200	59	01m24s
2807	141	-0819 Jan 18	23:23:51	22165	-34866	47	H	n-	0.3047	1.0001	4.9S	79.4W	72	166	0	00m00s
2808	141	-0819 Jul 15	14:07:59	22157	-34860	52	A	p-	-0.4183	0.9652	1.2S	55.4E	65	11	138	04m27s
2809	141	-0818 Jan 08	12:48:46	22149	-34854	57	T	t-	0.9745	1.0296	53.0N	66.0E	12	164	464	02m13s
2810	141	0818 Jul 04	15:31:21	22141	-34848	62	P	t-	-1.1883	0.6400	65.3S	22.2E	0	15		
2811	141	-0818 Nov 29	17:30:50	22134	-34843	29	P	-t	-1.0730	0.8627	68.5S	153.4W	0	162		
2812	141	-0817 May 25	04:52:07	22126	-34837	34	A	-p	0.8508	0.9891	75.2N	165.4E	31	145	74	00m45s
2813	141	-0817 Nov 19	04:07:45	22118	-34831	39	A	-p	-0.4197	0.9667	42.2S	158.2W	65	13	132	03m14s
2814	141	-0816 May 13	16:20:00	22110	-34825	44	T	nn	0.0458	1.0521	18.8N	24.6E	07	168	173	04m56s
2815	141	-0816 Nov 07	07:26:32	22102	-34819	49	A	n-	0.2757	0.9244	1.3N	161.3E	74	193	296	10m21s
2816	141	-0815 May 03	08:45:04	22093	-34813	54	T	p-	-0.6950	1.0754	29.9S	151.1E	46	344	338	06m13s
2817	141	-0815 Oct 27	06:27:12	22085	-34807	59	A	p-	0.9531	0.9122	58.6N	157.6W	17	211	1123	08m07s
2818	141	-0814 Mar 24	17:01:29	22078	-34802	26	P	-t	1.1434	0.7344	71.8N	65.0W	0	96		
2819	141	-0814 Apr 23	01:38:07	22077	-34801	64	P	t-	-1.4421	0.1718	71.0S	51.3W	0	300		
2820	141	-0814 Sep 16	18:56:14	22070	-34796	31	P	-t	-1.2487	0.5385	71.4S	84.3W	0	71		

Cat Num	Canon Plate	Calendar Date	TD of Greatest Eclipse	ΔT s	Luna Num	Saros Num	Ecl. Type	QLE	Gamma	Ecl. Mag.	Lat. °	Long. °	Sun Alt °	Sun Azm °	Path Width km	Central Line Dur.
2821	142	-0813 Mar 14	02:01:21	22062	-34790	36	A	-p	0.4288	0.9684	18.2N	122.8W	64	162	125	03m36s
2822	142	-0813 Sep 06	08:13:35	22054	-34784	41	T	-p	-0.4973	1.0482	18.7S	140.3E	60	17	184	04m22s
2823	142	-0812 Mar 02	03:56:29	22046	-34778	46	A	nn	-0.3122	0.9353	27.9S	137.6W	72	343	253	07m46s
2824	142	-0812 Aug 26	00:37:15	22038	-34772	51	T	n-	0.2064	1.0605	25.4N	93.8W	78	195	203	05m13s
2825	142	-0811 Feb 19	04:08:00	22030	-34766	56	P	t-	-1.0238	0.9202	70.4S	10.3W	0	223		
2826	142	-0811 Aug 15	15:26:27	22022	-34760	61	T	p-	0.9514	1.0121	79.2N	127.4E	17	270	140	00m42s
2827	142	-0810 Jan 09	21:28:45	22015	-34755	28	P	-t	1.0882	0.8324	66.8N	47.1W	0	181		
2828	142	-0810 Jul 06	10:28:10	22007	-34749	33	P	-t	-1.0492	0.8815	66.0S	123.2E	0	352		
2829	142	-0810 Dec 30	11:59:37	21999	-34743	38	T	-n	0.3791	1.0467	1.6S	95.2E	68	185	168	04m33s
2830	142	-0809 Jun 25	11:08:06	21990	-34737	43	A	np	-0.3008	0.9486	6.4N	105.8E	73	352	198	07m04s
2831	142	-0809 Dec 20	03:37:52	21982	-34731	48	T	n-	-0.2942	1.0331	40.5S	146.4W	73	12	117	02m45s
2832	142	-0808 Jun 13	14:00:41	21974	-34725	53	A	p-	0.4711	0.9825	50.0N	49.9E	62	161	70	01m34s
2833	142	-0808 Dec 08	15:43:51	21966	-34719	58	P	p-	-1.0193	0.9441	63.8S	111.5W	0	149		
2834	142	-0807 May 04	16:10:16	21959	-34714	25	P	-t	-1.3456	0.3560	61.6S	91.5E	0	299		
2835	142	-0807 Jun 02	23:58:25	21958	-34713	63	P	t-	1.2038	0.6248	63.2N	130.2E	0	37		
2836	142	-0807 Oct 29	01:56:39	21951	-34708	30	P	-t	1.2678	0.5046	61.1N	51.0W	0	247		
2837	142	-0806 Apr 24	09:10:23	21943	-34702	35	T	-p	-0.6002	1.0752	23.1S	150.7E	53	331	300	05m56s
2838	142	-0806 Oct 18	01:11:04	21935	-34696	40	A	-p	0.5837	0.9286	24.7N	90.8W	54	211	322	07m47s
2839	142	-0805 Apr 14	01:17:25	21927	-34690	45	T	nn	0.1409	1.0445	13.0N	111.7W	82	151	150	03m50s
2840	142	-0805 Oct 07	05:36:11	21919	-34684	50	A	n-	-0.1398	0.9808	9.2S	178.5W	82	30	69	01m53s
2841	143	-0804 Apr 02	12:04:22	21911	-34678	55	A	p-	0.9403	0.9759	56.4N	35.5E	19	118	250	01m44s
2842	143	-0804 Sep 25	17:20:29	21902	-34672	60	T	p-	-0.8278	1.0284	44.7S	24.7W	34	43	167	02m03s
2843	143	-0803 Feb 20	22:58:47	21896	-34667	27	P	-t	-1.1957	0.6259	61.4S	47.4E	0	241		
2844	143	-0803 Aug 17	00:20:48	21888	-34661	32	P	-t	1.1127	0.7997	62.0N	30.8E	0	308		
2845	143	-0803 Sep 15	09:13:10	21886	-34660	70	P	t-	-1.4824	0.0966	61.0S	54.2E	0	77		
2846	143	-0802 Feb 10	00:02:01	21879	-34655	37	A	-p	-0.4578	0.9539	41.9S	71.0W	63	331	188	04m23s
2847	143	-0802 Aug 06	13:59:03	21871	-34649	42	H	-p	0.4140	1.0094	42.3N	72.9E	65	205	36	00m48s
2848	143	-0801 Jan 30	07:35:26	21863	-34643	47	H	n-	0.2827	1.0061	4.3S	156.3E	74	162	22	00m38s
2849	143	-0801 Jul 26	20:59:11	21855	-34637	52	A	p-	-0.3499	0.9606	1.9N	48.8W	70	15	152	04m57s
2850	143	-0800 Jan 19	21:23:02	21847	-34631	57	T	t-	0.9588	1.0346	49.4N	68.2W	16	159	417	02m37s
2851	143	-0800 Jul 14	22:02:17	21839	-34625	62	P	t-	-1.1109	0.7713	64.4S	86.7W	0	25		
2852	143	-0800 Dec 10	02:16:34	21832	-34620	29	P	-t	-1.0761	0.8565	67.4S	62.5E	0	174		
2853	143	-0799 Jun 04	11:39:33	21824	-34614	34	A	-p	0.9286	0.9897	85.1N	5.0E	21	89	100	00m38s
2854	143	-0799 Nov 29	12:34:02	21816	-34608	39	A	-p	-0.4214	0.9638	45.4S	76.5E	65	9	144	03m29s
2855	143	-0798 May 24	23:34:25	21808	-34602	44	Tm	nn	0.1215	1.0554	26.1N	86.2W	83	171	184	05m04s
2856	143	-0798 Nov 18	15:30:01	21800	-34596	49	A	n-	0.2717	0.9225	2.2S	38.6E	74	190	304	10m53s
2857	143	-0797 May 14	16:11:28	21792	-34590	54	T	p-	-0.6224	1.0778	21.6S	34.0E	51	348	320	06m47s
2858	143	-0797 Nov 07	14:27:26	21784	-34584	59	A	p-	0.9458	0.9124	54.8N	73.7E	18	203	1041	08m34s
2859	143	-0796 Apr 04	00:27:08	21777	-34579	26	P	-t	1.2078	0.6151	71.7N	168.3E	0	83		
2860	143	-0796 May 03	08:58:49	21776	-34578	64	P	t-	-1.3762	0.2978	70.5S	175.9W	0	312		
2861	144	-0796 Sep 27	03:05:45	21769	-34573	31	P	-t	-1.2660	0.5073	71.7S	138.0E	0	85		
2862	144	-0795 Mar 24	09:07:14	21761	-34567	36	A	-p	0.4910	0.9679	26.1N	126.5E	60	160	132	03m31s
2863	144	-0795 Sep 16	16:31:48	21753	-34561	41	T	-p	-0.5238	1.0465	24.4S	12.2E	58	19	181	04m03s
2864	144	-0794 Mar 13	10:52:47	21745	-34555	46	A	nn	-0.2548	0.9381	20.4S	115.4E	75	342	237	07m40s
2865	144	-0794 Sep 06	08:47:51	21737	-34549	51	T	n-	0.1718	1.0561	19.5N	141.6E	80	197	188	04m57s
2866	144	-0793 Mar 02	11:24:22	21728	-34543	56	A	t-	-0.9765	0.9399	74.5S	172.2W	12	273	1102	04m06s
2867	144	-0793 Aug 26	23:12:24	21720	-34537	61	T	p-	0.9110	1.0076	73.2N	26.9W	24	235	65	00m29s
2868	144	-0792 Jan 21	05:53:25	21714	-34532	28	P	-t	1.0995	0.8133	67.8N	174.4E	0	169		
2869	144	-0792 Jul 16	17:06:39	21706	-34526	33	P	-t	-1.1258	0.7483	67.0S	11.7E	0	2		
2870	144	-0791 Jan 09	20:41:34	21698	-34520	38	T	-n	0.3888	1.0495	0.3S	37.1W	67	180	178	04m49s
2871	144	-0791 Jul 05	17:35:32	21689	-34514	43	A	-p	-0.3830	0.9474	1.3N	7.3E	67	356	209	07m23s
2872	144	-0791 Dec 30	12:21:11	21681	-34508	48	T	n-	-0.2892	1.0328	40.7S	84.7E	73	6	116	02m46s
2873	144	-0790 Jun 24	20:47:17	21673	-34502	53	A	p-	0.3903	0.9852	46.5N	47.8W	67	169	57	01m24s
2874	144	-0790 Dec 20	00:12:19	21665	-34496	58	P	p-	-1.0187	0.9441	64.8S	110.4E	0	159		
2875	144	-0789 May 15	23:31:19	21658	-34491	25	P	-t	-1.4139	0.2244	62.2S	29.1W	0	308		
2876	144	-0789 Jun 14	07:10:12	21657	-34490	63	P	t-	1.1259	0.7713	64.1N	11.4E	0	28		
2877	144	-0789 Nov 09	09:54:51	21650	-34485	30	P	-t	1.2760	0.4910	61.6N	179.4E	0	238		
2878	144	-0788 May 04	16:38:37	21642	-34479	35	T	-p	-0.6724	1.0747	24.9S	37.6E	48	333	322	05m59s
2879	144	-0788 Oct 28	09:11:01	21634	-34473	40	A	-p	0.5931	0.9286	21.7N	146.5E	53	209	325	08m04s
2880	144	-0787 Apr 24	08:36:58	21626	-34467	45	Tm	nn	0.0704	1.0432	13.5N	138.3E	86	153	145	03m47s

Cat Num	Canon Plate	Calendar Date	TD of Greatest Eclipse	ΔT s	Luna Num	Saros Num	Ecl. Type	QLE	Gamma	Ecl. Mag.	Lat. °	Long. °	Sun Alt °	Sun Azm °	Path Width km	Central Line Dur.
2881	145	-0787 Oct 17	13:53:58	21618	-34461	50	A	n-	-0.1274	0.9817	13.1S	55.6E	83	29	65	01m47s
2882	145	-0786 Apr 13	19:02:34	21610	-34455	55	A	p-	0.8723	0.9772	54.5N	61.4W	29	125	163	01m43s
2883	145	-0786 Oct 07	01:50:51	21602	-34449	60	T	p-	-0.8099	1.0284	47.1S	155.1W	36	46	161	01m59s
2884	145	-0785 Mar 04	06:03:10	21595	-34444	27	P	-t	-1.2457	0.5420	61.0S	68.8W	0	250		
2885	145	-0785 Aug 28	08:20:47	21587	-34438	32	P	-t	1.1557	0.7171	61.5N	99.4W	0	298		
2886	145	-0785 Sep 26	17:39:36	21586	-34437	70	P	t-	-1.4605	0.1394	60.8S	82.4W	0	86		
2887	145	-0784 Feb 21	07:32:00	21579	-34432	37	A	-p	-0.4961	0.9594	40.1S	177.6E	60	327	168	03m47s
2888	145	-0784 Aug 16	21:31:24	21571	-34426	42	H	-p	0.4666	1.0024	41.8N	38.1W	62	210	9	00m12s
2889	145	-0783 Feb 09	15:39:19	21563	-34420	47	H	n-	0.2535	1.0125	3.4S	34.1E	75	158	44	01m14s
2890	145	-0783 Aug 06	03:56:12	21555	-34414	52	A	nn	-0.2870	0.9557	3.7N	154.1W	73	19	168	05m26s
2891	145	-0782 Jan 30	05:50:56	21547	-34408	57	T	t-	0.9378	1.0400	46.2N	160.3E	20	154	386	03m01s
2892	145	-0782 Jul 26	04:40:03	21539	-34402	62	P	t-	-1.0388	0.8933	63.5S	162.9E	0	34		
2893	145	-0782 Dec 21	10:59:39	21532	-34397	29	P	-t	-1.0810	0.8473	66.3S	80.4W	0	185		
2894	145	-0781 Jun 15	18:31:09	21524	-34391	34	P	-t	1.0037	0.9815	67.1N	173.0E	0	3		
2895	145	-0781 Dec 10	20:59:29	21516	-34385	39	A	-p	-0.4230	0.9615	47.7S	47.5W	65	3	154	03m41s
2896	145	-0780 Jun 04	06:49:16	21508	-34379	44	T	nn	0.1975	1.0580	32.8N	163.9E	78	174	195	05m06s
2897	145	-0780 Nov 28	23:34:55	21500	-34373	49	A	nn	0.2693	0.9212	5.0S	84.3W	74	186	310	11m17s
2898	145	-0779 May 24	23:36:32	21492	-34367	54	T	p-	-0.5483	1.0792	13.9S	81.8W	57	351	305	07m11s
2899	145	-0779 Nov 17	22:30:21	21484	-34361	59	A	p-	0.9403	0.9132	51.6N	54.6W	19	196	984	08m53s
2900	145	-0778 Apr 15	07:46:06	21477	-34356	26	P	-t	1.2767	0.4878	71.5N	43.4E	0	69		
2901	146	-0778 May 14	16:17:05	21476	-34355	64	P	t-	-1.3088	0.4264	69.7S	60.5E	0	324		
2902	146	-0778 Oct 08	11:23:56	21469	-34350	31	P	-t	-1.2773	0.4870	71.6S	2.0W	0	99		
2903	146	-0777 Apr 04	16:05:25	21461	-34344	36	A	-p	0.5584	0.9669	34.6N	17.3E	56	158	143	03m25s
2904	146	-0777 Sep 28	00:57:17	21453	-34338	41	T	-p	-0.5447	1.0449	29.9S	117.4W	57	21	177	03m46s
2905	146	-0776 Mar 23	17:40:54	21445	-34332	46	A	nn	-0.1907	0.9407	12.6S	10.0E	79	342	223	07m32s
2906	146	-0776 Sep 16	17:06:09	21437	-34326	51	T	n-	0.1441	1.0517	13.7N	14.8E	82	198	173	04m39s
2907	146	-0775 Mar 12	18:30:48	21429	-34320	56	A	t-	-0.9204	0.9479	66.8S	38.9E	23	312	493	04m01s
2908	146	-0775 Sep 06	07:07:03	21421	-34314	61	H	p-	0.8785	1.0025	66.2N	160.1W	28	223	18	00m10s
2909	146	-0774 Jan 31	14:09:54	21415	-34309	28	P	-t	1.1180	0.7809	68.8N	37.3E	0	158		
2910	146	-0774 Jul 27	23:52:12	21407	-34303	33	P	-t	-1.1957	0.6275	68.1S	102.0W	0	13		
2911	146	-0773 Jan 21	05:16:36	21398	-34297	38	T	-n	0.4043	1.0525	2.1N	167.9W	66	176	190	05m04s
2912	146	-0773 Jul 17	00:10:02	21390	-34291	43	A	-p	-0.4595	0.9459	4.4S	93.7W	63	1	224	07m36s
2913	146	-0772 Jan 10	20:59:12	21382	-34285	48	T	n-	-0.2803	1.0329	39.6S	42.9W	74	0	116	02m49s
2914	146	-0772 Jul 05	03:37:59	21374	-34279	53	A	p-	0.3119	0.9872	42.1N	147.9W	72	175	47	01m15s
2915	146	-0772 Dec 30	08:37:21	21366	-34273	58	P	p-	-1.0163	0.9477	65.8S	27.2W	0	169		
2916	146	-0771 May 26	06:50:24	21360	-34268	25	Pe	-t	-1.4838	0.0890	62.8S	149.4W	0	317		
2917	146	-0771 Jun 24	14:24:39	21358	-34267	63	P	t-	1.0495	0.9156	65.0N	108.3W	0	19		
2918	146	-0771 Nov 19	17:56:10	21352	-34262	30	P	-t	1.2813	0.4822	62.3N	48.9E	0	228		
2919	146	-0770 May 16	00:03:57	21344	-34256	35	T	-p	-0.7469	1.0732	28.1S	75.0W	41	335	352	05m54s
2920	146	-0770 Nov 08	17:17:02	21336	-34250	40	A	-p	0.5976	0.9292	18.9N	22.1E	53	205	324	08m19s
2921	147	-0769 May 05	15:52:09	21328	-34244	45	T	nn	-0.0033	1.0413	13.5N	29.4E	90	316	138	03m44s
2922	147	-0769 Oct 28	22:18:09	21320	-34238	50	A	n-	-0.1195	0.9829	17.0S	71.8W	83	27	61	01m40s
2923	147	-0768 Apr 24	01:54:31	21312	-34232	55	A	p-	0.8001	0.9776	53.7N	158.8W	37	130	131	01m44s
2924	147	-0768 Oct 17	10:27:38	21304	-34226	60	T	p-	-0.7967	1.0284	50.2S	73.0E	37	47	157	01m56s
2925	147	-0767 Mar 14	12:58:32	21297	-34221	27	P	-t	-1.3028	0.4454	60.8S	177.5E	0	258		
2926	147	-0767 Sep 07	16:28:36	21289	-34215	32	P	-t	1.1919	0.6479	61.0N	128.6E	0	289		
2927	147	-0767 Oct 07	02:13:14	21288	-34214	70	P	t-	-1.4446	0.1707	60.8S	139.3E	0	95		
2928	147	-0766 Mar 03	14:54:43	21281	-34209	37	A	-p	-0.5412	0.9648	38.3S	67.6E	57	325	149	03m14s
2929	147	-0766 Aug 28	05:11:00	21273	-34203	42	A	-p	0.5121	0.9955	40.3N	151.7W	59	213	18	00m23s
2930	147	-0765 Feb 20	23:36:23	21265	-34197	47	T	n-	0.2177	1.0190	2.2S	86.3W	77	155	66	01m49s
2931	147	-0765 Aug 17	11:00:55	21257	-34191	52	A	nn	-0.2310	0.9507	4.2N	98.8E	77	22	185	05m53s
2932	147	-0764 Feb 10	14:10:56	21249	-34185	57	T	p-	0.9100	1.0456	43.5N	31.8E	24	150	362	03m25s
2933	147	-0764 Aug 05	11:25:43	21241	-34179	62	A	t-	-0.9723	0.9282	52.4S	65.5E	13	31	1174	06m53s
2934	147	-0764 Dec 31	19:37:35	21234	-34174	29	P	-t	-1.0898	0.8308	65.3S	138.5E	0	196		
2935	147	-0763 Jun 26	01:27:44	21227	-34168	34	P	-t	1.0758	0.8521	66.1N	57.0E	0	352		
2936	147	-0763 Dec 21	05:19:37	21219	-34162	39	A	-p	-0.4284	0.9598	49.1S	169.5W	64	357	162	03m49s
2937	147	-0762 Jun 15	14:07:32	21211	-34156	44	T	-n	0.2715	1.0596	38.9N	54.3E	74	179	204	05m00s
2938	147	-0762 Dec 10	07:38:16	21203	-34150	49	A	nn	0.2654	0.9206	7.1S	153.5E	75	182	312	11m29s
2939	147	-0761 Jun 05	07:01:07	21195	-34144	54	T	p-	-0.4737	1.0797	6.8S	163.3E	62	355	292	07m25s
2940	147	-0761 Nov 29	06:36:00	21187	-34138	59	A	p-	0.9364	0.9145	49.0N	177.1E	20	190	940	09m04s

Cat Num	Canon Plate	Calendar Date	TD of Greatest Eclipse	ΔT s	Luna Num	Saros Num	Ecl. Type	QLE	Gamma	Ecl. Mag.	Lat. °	Long. °	Sun Alt °	Sun Azm °	Path Width km	Central Line Dur.
2941	148	-0760 Apr 25	14:57:16	21180	-34133	26	P	-t	1.3511	0.3506	71.0N	79.2W	0	56		
2942	148	-0760 May 24	23:31:16	21179	-34132	64	P	t-	-1.2383	0.5604	68.9S	61.5W	0	336		
2943	148	-0760 Oct 18	19:50:33	21172	-34127	31	P	-t	-1.2825	0.4777	71.3S	143.9W	0	113		
2944	148	-0759 Apr 14	22:53:03	21164	-34121	36	A	-p	0.6336	0.9654	43.8N	89.9W	50	156	160	03m20s
2945	148	-0759 Oct 08	09:31:44	21156	-34115	41	T	-p	-0.5584	1.0434	35.2S	111.1E	56	22	174	03m31s
2946	148	-0758 Apr 04	00:20:54	21148	-34109	46	A	nn	-0.1200	0.9432	4.4S	93.7W	83	342	211	07m20s
2947	148	-0758 Sep 28	01:31:40	21140	-34103	51	T	n-	0.1228	1.0474	8.1N	113.9W	83	198	159	04m20s
2948	148	-0757 Mar 24	01:32:01	21132	-34097	56	A	t-	-0.8589	0.9550	57.0S	83.4W	30	326	319	03m50s
2949	148	-0757 Sep 17	15:08:17	21124	-34091	61	A	p-	0.8522	0.9970	59.6N	70.9E	31	216	20	00m13s
2950	148	-0756 Feb 11	22:20:51	21118	-34086	28	P	-t	1.1418	0.7382	69.7N	98.9W	0	145		
2951	148	-0756 Mar 12	08:51:34	21116	-34085	66	Pb	t-	-1.5417	0.0047	71.4S	112.2W	0	250		
2952	148	-0756 Aug 07	06:42:52	21110	-34080	33	P	-t	-1.2607	0.5162	69.1S	142.4E	0	24		
2953	148	-0755 Jan 31	13:45:31	21102	-34074	38	T	-n	0.4250	1.0556	5.5N	62.5E	65	172	202	05m17s
2954	148	-0755 Jul 27	06:52:13	21094	-34068	43	A	-p	-0.5298	0.9441	10.5S	162.6E	58	5	243	07m42s
2955	148	-0754 Jan 21	05:29:27	21086	-34062	48	T	n-	-0.2655	1.0333	37.2S	169.2W	74	355	116	02m54s
2956	148	-0754 Jul 16	10:37:53	21078	-34056	53	A	n-	0.2401	0.9888	37.1N	108.4E	76	181	41	01m09s
2957	148	-0753 Jan 10	16:55:10	21070	-34050	58	P	p-	-1.0091	0.9600	66.9S	163.5W	0	180		
2958	148	-0753 Jul 05	21:44:52	21062	-34044	63	T	t-	0.9769	1.0328	77.3N	138.0E	12	16	557	01m40s
2959	148	-0753 Dec 01	01:57:14	21055	-34039	30	P	-t	1.2865	0.4736	63.1N	81.8W	0	218		
2960	148	-0752 May 26	07:27:48	21047	-34033	35	T	-p	-0.8222	1.0705	33.0S	172.5E	34	338	398	05m40s
2961	149	-0752 Nov 19	01:26:10	21039	-34027	40	A	-p	0.6005	0.9305	16.6N	103.1W	53	201	320	08m28s
2962	149	-0751 May 15	23:01:44	21031	-34021	45	T	nn	-0.0809	1.0386	12.6N	78.0W	85	337	130	03m37s
2963	149	-0751 Nov 08	06:48:08	21023	-34015	50	A	n-	-0.1155	0.9848	20.8S	159.7E	83	25	54	01m29s
2964	149	-0750 May 05	08:39:48	21015	-34009	55	A	p-	0.7229	0.9772	53.2N	105.1E	43	136	116	01m49s
2965	149	-0750 Oct 28	19:10:19	21008	-34003	60	T	p-	-0.7881	1.0288	54.0S	60.2W	38	49	156	01m55s
2966	149	-0749 Mar 25	19:45:38	21001	-33998	27	P	-t	-1.3664	0.3371	60.6S	65.8E	0	267		
2967	149	-0749 Apr 24	11:17:29	21000	-33997	65	Pb	t-	1.5140	0.0838	61.1N	10.7W	0	69		
2968	149	-0749 Sep 19	00:44:21	20993	-33992	32	P	-t	1.2213	0.5921	60.8N	5.4W	0	280		
2969	149	-0749 Oct 18	10:53:50	20992	-33991	70	P	t-	-1.4347	0.1904	61.0S	0.8W	0	104		
2970	149	-0748 Mar 13	22:09:18	20985	-33986	37	A	-p	-0.5938	0.9702	37.0S	40.7W	53	323	131	02m42s
2971	149	-0748 Sep 07	12:57:40	20977	-33980	42	A	-p	0.5509	0.9886	38.1N	92.0E	56	215	48	00m58s
2972	149	-0747 Mar 03	07:25:55	20969	-33974	47	T	n-	0.1749	1.0256	0.8S	155.3E	80	153	88	02m21s
2973	149	-0747 Aug 27	18:13:34	20961	-33968	52	A	nn	-0.1818	0.9457	3.6N	10.1W	80	25	203	06m20s
2974	149	-0746 Feb 20	22:23:27	20953	-33962	57	T	p-	0.8758	1.0511	41.4N	93.9W	29	146	344	03m47s
2975	149	-0746 Aug 16	18:21:08	20945	-33956	62	A	t-	-0.9131	0.9285	43.9S	36.9W	24	31	644	07m18s
2976	149	-0745 Jan 12	04:10:49	20939	-33951	29	P	-t	-1.1024	0.8077	64.3S	1.0W	0	206		
2977	149	-0745 Jul 07	08:29:47	20931	-33945	34	P	-t	1.1442	0.7287	65.1N	60.0W	0	343		
2978	149	-0744 Jan 01	13:36:07	20923	-33939	39	A	-p	-0.4362	0.9586	49.6S	69.8E	64	350	167	03m54s
2979	149	-0744 Jun 25	21:29:20	20915	-33933	44	T	-n	0.3434	1.0605	44.0N	54.8W	70	184	212	04m51s
2980	149	-0744 Dec 20	15:38:37	20907	-33927	49	A	nn	0.2595	0.9207	8.6S	32.2E	75	178	311	11m26s
2981	150	-0743 Jun 15	14:27:14	20899	-33921	54	T	p-	-0.4004	1.0792	0.7S	48.8E	66	359	279	07m28s
2982	150	-0743 Dec 09	14:40:59	20891	-33915	59	A	p-	0.9312	0.9166	46.6N	49.2E	21	183	881	09m04s
2983	150	-0742 May 06	22:04:25	20885	-33910	26	P	-t	1.4275	0.2104	70.4N	159.7E	0	44		
2984	150	-0742 Jun 05	06:45:52	20883	-33909	64	P	t-	-1.1686	0.6919	68.0S	177.0E	0	347		
2985	150	-0742 Oct 30	04:22:22	20877	-33904	31	P	-t	-1.2844	0.4742	70.8S	73.2E	0	127		
2986	150	-0741 Apr 26	05:35:33	20869	-33898	36	A	-p	0.7120	0.9634	53.5N	163.3E	44	153	188	03m16s
2987	150	-0741 Oct 19	18:12:23	20861	-33892	41	T	-p	-0.5676	1.0423	40.3S	21.3W	55	22	171	03m19s
2988	150	-0740 Apr 14	06:52:52	20853	-33886	46	A	nn	-0.0426	0.9455	4.1N	164.6E	88	343	201	07m04s
2989	150	-0740 Oct 08	10:04:20	20845	-33880	51	T	n-	0.1077	1.0432	2.8N	115.7E	84	198	145	04m01s
2990	150	-0739 Apr 03	08:25:25	20837	-33874	56	A	p-	-0.7897	0.9617	47.0S	162.7E	38	333	224	03m35s
2991	150	-0739 Sep 27	23:17:35	20829	-33868	61	A	p-	0.8331	0.9916	53.8N	57.9W	33	212	53	00m40s
2992	150	-0738 Feb 22	06:22:32	20823	-33863	28	P	-t	1.1738	0.6798	70.5N	126.6E	0	133		
2993	150	-0738 Mar 23	16:19:52	20821	-33862	66	P	t-	-1.4834	0.1085	71.7S	120.6E	0	263		
2994	150	-0738 Aug 18	13:43:07	20815	-33857	33	P	-t	-1.3171	0.4204	70.0S	23.8E	0	36		
2995	150	-0737 Feb 11	22:06:33	20807	-33851	38	T	-p	0.4526	1.0587	10.0N	65.6W	63	169	216	05m27s
2996	150	-0737 Aug 07	13:42:29	20799	-33845	43	A	-p	-0.5939	0.9421	17.0S	56.4E	53	9	265	07m40s
2997	150	-0736 Feb 01	13:52:44	20791	-33839	48	T	n-	-0.2453	1.0339	33.5S	65.4E	76	350	118	03m01s
2998	150	-0736 Jul 26	17:45:01	20783	-33833	53	A	nn	0.1735	0.9899	31.6N	1.7E	80	185	36	01m04s
2999	150	-0735 Jan 21	01:05:13	20775	-33827	58	As	p-	-0.9961	0.9621	70.4S	60.2E	3	193	-	02m04s
3000	150	-0735 Jul 16	05:10:58	20767	-33821	63	T	t-	0.9082	1.0373	88.5N	145.4W	24	206	305	02m06s

Cat Num	Canon Plate	Calendar Date	TD of Greatest Eclipse	ΔT s	Luna Num	Saros Num	Ecl. Type	QLE	Gamma	Ecl. Mag.	Lat. °	Long. °	Sun Alt °	Sun Azm °	Path Width km	Central Line Dur.
3001	151	-0735 Dec 11	09:58:17	20761	-33816	30	P	-t	1.2915	0.4652	64.0N	147.3E	0	209		
3002	151	-0734 Jun 06	14:51:32	20753	-33810	35	T	-t	-0.8975	1.0665	40.2S	59.8E	26	341	491	05m12s
3003	151	-0734 Nov 30	09:37:18	20745	-33804	40	A	-p	0.6024	0.9324	14.9N	131.2E	53	197	312	08m30s
3004	151	-0733 May 27	06:09:39	20737	-33798	45	T	nn	-0.1591	1.0352	11.0N	174.9E	81	341	120	03m26s
3005	151	-0733 Nov 19	15:22:07	20729	-33792	50	A	n-	-0.1140	0.9871	24.2S	30.7E	83	22	46	01m16s
3006	151	-0732 May 15	15:19:45	20721	-33786	55	A	p-	0.6419	0.9761	52.4N	10.5E	50	143	110	01m59s
3007	151	-0732 Nov 08	03:57:44	20713	-33780	60	T	p-	-0.7831	1.0293	58.2S	166.0E	38	49	158	01m54s
3008	151	-0731 Apr 05	02:25:46	20707	-33775	27	P	-t	-1.4353	0.2191	60.7S	44.2W	0	276		
3009	151	-0731 May 04	17:41:04	20706	-33774	65	P	t-	1.4323	0.2232	61.5N	116.7W	0	61		
3010	151	-0731 Sep 29	09:07:08	20699	-33769	32	P	-t	1.2448	0.5479	60.7N	140.9W	0	271		
3011	151	-0731 Oct 28	19:38:52	20698	-33768	70	P	t-	-1.4286	0.2025	61.3S	142.1W	0	113		
3012	151	-0730 Mar 25	05:17:53	20691	-33763	37	A	-p	-0.6522	0.9753	36.2S	147.6W	49	323	114	02m13s
3013	151	-0730 Sep 18	20:52:12	20683	-33757	42	A	-p	0.5823	0.9821	35.5N	27.3W	54	216	77	01m34s
3014	151	-0729 Mar 14	15:09:18	20675	-33751	47	T	n-	0.1258	1.0320	0.6N	38.6E	83	151	109	02m52s
3015	151	-0729 Sep 08	01:33:43	20667	-33745	52	A	nn	-0.1394	0.9409	2.1N	121.1W	82	27	221	06m47s
3016	151	-0728 Mar 03	06:28:10	20659	-33739	57	T	p-	0.8348	1.0566	39.9N	143.1E	33	144	331	04m07s
3017	151	-0728 Aug 27	01:26:21	20652	-33733	62	A	p-	-0.8612	0.9277	40.1S	144.2W	30	33	517	07m27s
3018	151	-0727 Jan 22	12:34:44	20645	-33728	29	P	-t	-1.1222	0.7714	63.3S	137.8W	0	216		
3019	151	-0727 Jul 17	15:40:03	20637	-33722	34	P	-t	1.2069	0.6153	64.2N	178.6W	0	333		
3020	151	-0727 Aug 16	03:50:07	20636	-33721	72	Pb	t-	-1.5437	0.0212	62.1S	153.6E	0	52		
3021	152	-0726 Jan 11	21:44:28	20629	-33716	39	A	-p	-0.4504	0.9579	49.1S	48.9W	63	343	172	03m56s
3022	152	-0726 Jul 07	04:56:39	20621	-33710	44	T	-n	0.4113	1.0607	47.9N	163.9W	65	191	219	04m38s
3023	152	-0726 Dec 31	23:33:25	20614	-33704	49	A	nn	0.2491	0.9215	9.5S	87.6W	76	173	306	11m10s
3024	152	-0725 Jun 26	21:55:33	20606	-33698	54	T	n-	-0.3288	1.0777	4.6N	65.5W	71	3	266	07m18s
3025	152	-0725 Dec 20	22:45:21	20598	-33692	59	A	p-	0.9249	0.9194	44.5N	78.1W	22	177	812	08m54s
3026	152	-0724 May 17	05:04:34	20591	-33687	26	Pe	-t	1.5084	0.0628	69.5N	40.9E	0	32		
3027	152	-0724 Jun 15	13:57:48	20590	-33686	64	P	t-	-1.0974	0.8255	67.0S	56.6E	0	358		
3028	152	-0724 Nov 09	13:00:39	20583	-33681	31	P	-t	-1.2820	0.4785	70.1S	70.7W	0	140		
3029	152	-0724 Dec 09	02:48:54	20582	-33680	69	Pb	t-	1.5521	0.0055	67.5N	130.3W	0	187		
3030	152	-0723 May 06	12:10:17	20576	-33675	36	A	-p	0.7960	0.9606	64.2N	56.1E	37	149	236	03m12s
3031	152	-0723 Oct 30	02:58:53	20568	-33669	41	T	-p	-0.5722	1.0416	45.0S	154.2W	55	21	169	03m10s
3032	152	-0722 Apr 25	13:19:51	20560	-33663	46	Am	nn	0.0391	0.9475	12.6N	64.3E	88	165	193	06m45s
3033	152	-0722 Oct 19	18:43:18	20552	-33657	51	T	nn	0.0980	1.0393	2.1S	16.1W	84	197	132	03m42s
3034	152	-0721 Apr 14	15:15:24	20544	-33651	56	A	p-	-0.7168	0.9681	37.4S	52.5E	44	338	163	03m15s
3035	152	-0721 Oct 09	07:32:07	20536	-33645	61	A	p-	0.8192	0.9863	48.5N	172.9E	35	208	84	01m09s
3036	152	-0720 Mar 04	14:18:58	20530	-33640	28	P	-t	1.2107	0.6114	71.1N	7.1W	0	119		
3037	152	-0720 Apr 02	23:44:08	20528	-33639	66	P	t-	-1.4209	0.2216	71.7S	5.7W	0	277		
3038	152	-0720 Aug 28	20:50:35	20522	-33634	33	P	-t	-1.3668	0.3368	70.7S	97.2W	0	49		
3039	152	-0719 Feb 22	06:19:28	20514	-33628	38	T	-p	0.4868	1.0617	15.4N	167.9E	61	165	231	05m32s
3040	152	-0719 Aug 17	20:43:10	20506	-33622	43	A	-p	-0.6496	0.9400	23.7S	53.1W	49	13	291	07m31s
3041	153	-0718 Feb 11	22:07:07	20498	-33616	48	T	n-	-0.2186	1.0345	28.8S	58.7W	77	347	119	03m08s
3042	153	-0718 Aug 07	01:02:14	20491	-33610	53	A	nn	0.1142	0.9906	25.8N	108.4W	83	189	33	01m01s
3043	153	-0717 Feb 01	09:05:52	20483	-33604	58	A	p-	-0.9766	0.9643	78.7S	95.2W	12	226	644	02m07s
3044	153	-0717 Jul 27	12:45:12	20475	-33598	63	T	t-	0.8458	1.0398	78.6N	96.3E	32	204	252	02m24s
3045	153	-0717 Dec 22	17:55:38	20468	-33593	30	P	-t	1.2993	0.4521	65.0N	16.9E	0	198		
3046	153	-0716 Jun 16	22:15:24	20460	-33587	35	T	-t	-0.9722	1.0601	52.4S	52.2W	13	343	874	04m21s
3047	153	-0716 Dec 10	17:48:27	20453	-33581	40	A	-p	0.6048	0.9351	13.8N	5.5E	53	193	301	08m22s
3048	153	-0715 Jun 06	13:15:20	20445	-33575	45	T	-n	-0.2380	1.0310	8.5N	68.0E	76	345	108	03m10s
3049	153	-0715 Nov 29	23:57:18	20437	-33569	50	A	n-	-0.1130	0.9900	27.0S	98.2W	83	18	35	00m59s
3050	153	-0714 May 26	21:56:48	20429	-33563	55	A	p-	0.5594	0.9743	51.1N	83.4W	56	150	110	02m14s
3051	153	-0714 Nov 19	12:48:45	20421	-33557	60	T	p-	-0.7812	1.0304	62.6S	32.3E	38	48	164	01m56s
3052	153	-0713 Apr 16	08:59:26	20415	-33552	27	Pe	-t	-1.5091	0.0917	60.9S	152.5W	0	284		
3053	153	-0713 May 16	00:00:37	20413	-33551	65	P	t-	1.3470	0.3691	62.1N	138.3E	0	52		
3054	153	-0713 Oct 10	17:36:39	20407	-33546	32	P	-t	1.2624	0.5150	60.7N	81.8E	0	262		
3055	153	-0713 Nov 09	04:28:32	20406	-33545	70	P	t-	-1.4268	0.2066	61.8S	75.4E	0	123		
3056	153	-0712 Apr 04	12:20:27	20399	-33540	37	A	-p	-0.7163	0.9801	36.4S	106.8E	44	323	98	01m47s
3057	153	-0712 Sep 29	04:54:14	20391	-33534	42	A	-p	0.6068	0.9758	32.6N	149.3W	52	216	107	02m11s
3058	153	-0711 Mar 24	22:45:09	20383	-33528	47	T	nn	0.0695	1.0382	2.0N	76.1W	86	151	128	03m21s
3059	153	-0711 Sep 18	09:03:14	20376	-33522	52	A	nn	-0.1051	0.9363	0.3S	125.5E	84	29	238	07m15s
3060	153	-0710 Mar 14	14:25:57	20368	-33516	57	T	p-	0.7877	1.0617	39.1N	22.4E	38	142	321	04m26s

Cat Num	Canon Plate	Calendar Date	TD of Greatest Eclipse	ΔT s	Luna Num	Saros Num	Ecl. Type	QLE	Gamma	Ecl. Mag.	Lat. °	Long. °	Sun Alt °	Sun Azm °	Path Width km	Central Line Dur.
3061	154	-0710 Sep 07	08:41:03	20360	-33510	62	A	p-	-0.8164	0.9266	38.6S	105.5E	35	36	462	07m29s
3062	154	-0709 Feb 02	20:51:57	20353	-33505	29	P	-t	-1.1471	0.7260	62.5S	87.4E	0	225		
3063	154	-0709 Mar 04	06:56:06	20352	-33504	67	Pb	t-	1.5025	0.0612	61.3N	92.7E	0	110		
3064	154	-0709 Jul 28	22:58:08	20346	-33499	34	P	-t	1.2640	0.5116	63.4N	61.1E	0	324		
3065	154	-0709 Aug 27	11:16:26	20344	-33498	72	P	t-	-1.4930	0.1101	61.5S	31.9E	0	61		
3066	154	-0708 Jan 23	05:45:46	20338	-33493	39	A	-p	-0.4695	0.9575	48.0S	166.2W	62	337	175	03m56s
3067	154	-0708 Jul 17	12:28:51	20330	-33487	44	T	-p	0.4756	1.0601	50.5N	86.9E	61	198	225	04m25s
3068	154	-0707 Jan 11	07:22:27	20322	-33481	49	A	nn	0.2340	0.9229	9.8S	154.2E	77	169	299	10m42s
3069	154	-0707 Jul 07	05:27:58	20314	-33475	54	T	n-	-0.2608	1.0755	8.7N	179.9E	75	8	253	07m00s
3070	154	-0707 Dec 31	06:44:38	20307	-33469	59	A	p-	0.9136	0.9230	42.2N	156.4E	24	171	718	08m34s
3071	154	-0706 Jun 26	21:13:03	20299	-33463	64	P	t-	-1.0292	0.9519	66.0S	64.1W	0	8		
3072	154	-0706 Nov 20	21:41:18	20292	-33458	31	P	-t	-1.2789	0.4842	69.2S	145.4E	0	153		
3073	154	-0706 Dec 20	11:15:50	20291	-33457	69	P	t-	1.5448	0.0165	66.4N	90.9E	0	175		
3074	154	-0705 May 17	18:41:36	20284	-33452	36	A	-t	0.8817	0.9569	75.6N	59.6W	28	134	336	03m08s
3075	154	-0705 Nov 10	11:48:56	20277	-33446	41	T	-p	-0.5744	1.0413	49.4S	73.1E	55	19	169	03m04s
3076	154	-0704 May 05	19:42:17	20269	-33440	46	A	nn	0.1249	0.9491	21.1N	34.5W	83	166	188	06m23s
3077	154	-0704 Oct 30	03:27:08	20261	-33434	51	T	nn	0.0929	1.0359	6.6S	148.8W	85	195	121	03m26s
3078	154	-0703 Apr 24	21:59:40	20253	-33428	56	A	p-	-0.6379	0.9739	28.0S	54.9W	50	342	120	02m51s
3079	154	-0703 Oct 19	15:53:39	20245	-33422	61	A	p-	0.8117	0.9813	44.1N	42.6E	35	204	113	01m40s
3080	154	-0702 Mar 15	22:07:08	20239	-33417	28	P	-t	1.2550	0.5285	71.5N	139.1W	0	106		
3081	155	-0702 Apr 14	07:02:58	20238	-33416	66	P	t-	-1.3528	0.3465	71.5S	130.6W	0	290		
3082	155	-0702 Sep 09	04:07:20	20231	-33411	33	P	-t	-1.4085	0.2675	71.3S	138.9E	0	62		
3083	155	-0701 Mar 05	14:24:23	20223	-33405	38	T	-p	0.5280	1.0643	21.7N	42.8E	58	163	246	05m33s
3084	155	-0701 Aug 29	03:53:51	20216	-33399	43	A	-p	-0.6975	0.9379	30.3S	165.6W	46	16	319	07m18s
3085	155	-0700 Feb 23	06:12:09	20208	-33393	48	T	nn	-0.1843	1.0352	23.2S	178.5E	79	345	121	03m16s
3086	155	-0700 Aug 17	08:28:33	20200	-33387	53	A	nn	0.0615	0.9910	19.8N	138.5E	86	192	32	00m59s
3087	155	-0699 Feb 11	16:57:17	20192	-33381	58	A	p-	-0.9500	0.9659	78.5S	95.9E	18	275	405	02m11s
3088	155	-0699 Aug 06	20:27:21	20184	-33375	63	T	p-	0.7890	1.0410	70.0N	21.4W	38	205	225	02m38s
3089	155	-0698 Jan 02	01:48:43	20178	-33370	30	P	-t	1.3105	0.4335	66.1N	112.8W	0	188		
3090	155	-0698 Jun 28	05:42:08	20170	-33364	35	P	-t	-1.0443	0.9358	65.4S	166.3W	0	345		
3091	155	-0698 Jul 27	12:52:31	20169	-33363	73	Pb	t-	1.4732	0.1077	68.0N	123.1W	0	347		
3092	155	-0698 Dec 22	01:59:06	20162	-33358	40	A	-p	0.6081	0.9384	13.6N	120.0W	52	188	287	08m03s
3093	155	-0697 Jun 17	20:20:39	20155	-33352	45	T	-p	-0.3164	1.0261	5.1N	39.3W	72	350	93	02m47s
3094	155	-0697 Dec 11	08:33:44	20147	-33346	50	A	n-	-0.1121	0.9935	29.0S	133.0E	83	14	23	00m39s
3095	155	-0696 Jun 06	04:32:00	20139	-33340	55	A	p-	0.4761	0.9720	49.0N	177.3W	61	158	114	02m35s
3096	155	-0696 Nov 29	21:40:32	20131	-33334	60	T	p-	-0.7793	1.0318	66.8S	99.5W	39	45	172	02m00s
3097	155	-0695 May 26	06:18:53	20124	-33328	65	P	t-	1.2598	0.5182	62.7N	33.4E	0	43		
3098	155	-0695 Oct 21	02:12:00	20117	-33323	32	P	-t	1.2752	0.4913	61.0N	56.9W	0	253		
3099	155	-0695 Nov 19	13:20:05	20116	-33322	70	P	t-	-1.4268	0.2071	62.5S	67.8W	0	132		
3100	155	-0694 Apr 15	19:19:40	20109	-33317	37	A	-p	-0.7843	0.9844	37.9S	2.1E	38	323	87	01m23s
3101	156	-0694 Oct 10	13:01:39	20101	-33311	42	A	-p	0.6263	0.9701	29.7N	86.8E	51	214	135	02m48s
3102	156	-0693 Apr 05	06:16:21	20094	-33305	47	T	nn	0.0082	1.0440	3.2N	170.6E	90	152	147	03m49s
3103	156	-0693 Sep 29	16:40:11	20086	-33299	52	A	nn	-0.0772	0.9320	3.3S	10.0E	86	29	255	07m44s
3104	156	-0692 Mar 24	22:15:10	20078	-33293	57	T	p-	0.7332	1.0664	38.8N	95.3W	43	141	313	04m42s
3105	156	-0692 Sep 17	16:06:16	20070	-33287	62	A	p-	-0.7796	0.9254	38.9S	7.6W	39	38	434	07m27s
3106	156	-0691 Feb 13	04:58:26	20064	-33282	29	P	-t	-1.1803	0.6652	61.9S	44.5W	0	235		
3107	156	-0691 Mar 14	14:48:01	20063	-33281	67	P	t-	1.4595	0.1422	61.0N	35.3W	0	101		
3108	156	-0691 Aug 08	06:26:10	20056	-33276	34	P	-t	1.3141	0.4206	62.6N	61.4W	0	314		
3109	156	-0691 Sep 06	18:54:32	20055	-33275	72	P	t-	-1.4507	0.1844	61.0S	92.7W	0	70		
3110	156	-0690 Feb 02	13:36:01	20048	-33270	39	A	-p	-0.4972	0.9573	46.4S	78.7E	60	332	178	03m55s
3111	156	-0690 Jul 28	20:09:10	20041	-33264	44	T	-p	0.5338	1.0589	51.6N	23.9W	57	206	229	04m11s
3112	156	-0689 Jan 22	15:04:06	20033	-33258	49	A	nn	0.2127	0.9248	9.6S	37.9E	78	164	290	10m05s
3113	156	-0689 Jul 18	13:03:42	20025	-33252	54	T	n-	-0.1957	1.0724	11.8N	65.0E	79	12	240	06m35s
3114	156	-0688 Jan 11	14:40:44	20017	-33246	59	A	p-	0.8989	0.9273	40.0N	32.1E	26	166	621	08m04s
3115	156	-0688 Jul 07	04:29:06	20010	-33240	64	T	t-	-0.9621	1.0234	50.3S	177.1W	15	12	299	01m53s
3116	156	-0688 Dec 01	06:24:23	20003	-33235	31	P	-t	-1.2745	0.4922	68.2S	1.5E	0	165		
3117	156	-0688 Dec 30	19:40:35	20002	-33234	69	P	t-	1.5350	0.0315	65.3N	46.8W	0	165		
3118	156	-0687 May 28	01:08:50	19995	-33229	36	A	-t	0.9697	0.9513	80.3N	112.4E	13	46	772	03m05s
3119	156	-0687 Nov 20	20:42:24	19988	-33223	41	T	-p	-0.5741	1.0414	53.1S	58.9W	55	15	170	03m01s
3120	156	-0686 May 17	02:03:10	19980	-33217	46	A	nn	0.2121	0.9503	29.4N	132.2W	78	168	187	05m59s

Cat Num	Canon Plate	Calendar Date	TD of Greatest Eclipse	ΔT s	Luna Num	Saros Num	Ecl. Type	QLE	Gamma	Ecl. Mag.	Lat. °	Long. °	Sun Alt °	Sun Azm °	Path Width km	Central Line Dur.
3121	157	-0686 Nov 10	12:13:53	19972	-33211	51	T	nn	0.0905	1.0328	10.4S	78.0E	85	193	111	03m10s
3122	157	-0685 May 06	04:43:37	19964	-33205	56	A	p-	-0.5575	0.9793	19.1S	161.2W	56	346	88	02m23s
3123	157	-0685 Oct 31	00:18:57	19957	-33199	61	A	p-	0.8080	0.9767	40.3N	88.4W	36	200	140	02m11s
3124	157	-0684 Mar 26	05:49:17	19950	-33194	28	P	-t	1.3047	0.4341	71.6N	90.2E	0	92		
3125	157	-0684 Apr 24	14:18:14	19949	-33193	66	P	t-	-1.2806	0.4804	71.0S	105.8E	0	303		
3126	157	-0684 Sep 19	11:32:18	19942	-33188	33	P	-t	-1.4430	0.2106	71.7S	12.6E	0	76		
3127	157	-0684 Oct 19	05:26:05	19941	-33187	71	Pb	t-	1.5574	0.0158	71.3N	114.3W	0	246		
3128	157	-0683 Mar 15	22:21:36	19935	-33182	38	T	-p	0.5758	1.0665	28.7N	80.7W	55	160	264	05m28s
3129	157	-0683 Sep 08	11:15:17	19927	-33176	43	A	-p	-0.7371	0.9361	36.9S	78.9E	42	20	348	07m02s
3130	157	-0682 Mar 05	14:07:49	19919	-33170	48	T	nn	-0.1429	1.0356	16.8S	57.4E	82	343	121	03m23s
3131	157	-0682 Aug 28	16:06:23	19912	-33164	53	A	nn	0.0172	0.9913	13.7N	22.0E	89	193	31	00m57s
3132	157	-0681 Feb 23	00:38:26	19904	-33158	58	A	p-	-0.9160	0.9674	72.7S	51.3W	23	305	296	02m16s
3133	157	-0681 Aug 18	04:17:49	19896	-33152	63	T	p-	0.7387	1.0416	62.1N	141.6W	42	205	207	02m50s
3134	157	-0680 Jan 13	09:35:22	19890	-33147	30	P	-t	1.3265	0.4066	67.2N	118.7E	0	177		
3135	157	-0680 Jul 08	13:12:07	19882	-33141	35	P	-t	-1.1128	0.8025	66.4S	69.6E	0	355		
3136	157	-0680 Aug 06	20:42:19	19881	-33140	73	P	t-	1.4218	0.2079	69.0N	106.6E	0	335		
3137	157	-0679 Jan 01	10:04:53	19874	-33135	40	A	-p	0.6160	0.9423	14.4N	115.7E	52	183	271	07m32s
3138	157	-0679 Jun 28	03:27:14	19866	-33129	45	T	-p	-0.3926	1.0206	0.8N	147.5W	67	354	76	02m16s
3139	157	-0679 Dec 21	17:07:55	19859	-33123	50	A	n-	-0.1090	0.9977	30.0S	5.2E	84	9	8	00m14s
3140	157	-0678 Jun 17	11:07:03	19851	-33117	55	A	p-	0.3932	0.9691	46.0N	87.9E	67	165	121	03m02s
3141	158	-0678 Dec 11	06:31:58	19843	-33111	60	T	p-	-0.7772	1.0338	70.5S	132.2E	39	38	182	02m06s
3142	158	-0677 Jun 06	12:36:38	19836	-33105	65	P	t-	1.1719	0.6687	63.5N	71.6W	0	34		
3143	158	-0677 Nov 01	10:52:30	19829	-33100	32	P	-t	1.2834	0.4762	61.4N	163.0E	0	244		
3144	158	-0677 Nov 30	22:13:15	19828	-33099	70	P	t-	-1.4286	0.2040	63.3S	148.4E	0	142		
3145	158	-0676 Apr 26	02:13:09	19821	-33094	37	A	-p	-0.8577	0.9878	41.2S	100.6W	31	324	81	01m04s
3146	158	-0676 Oct 20	21:15:40	19814	-33088	42	A	-p	0.6394	0.9648	26.8N	39.3W	50	211	161	03m27s
3147	158	-0675 Apr 15	13:41:58	19806	-33082	47	T	nn	-0.0583	1.0493	4.1N	58.7E	87	331	164	04m17s
3148	158	-0675 Oct 10	00:24:52	19798	-33076	52	A	nn	-0.0562	0.9282	6.7S	107.4W	87	29	270	08m13s
3149	158	-0674 Apr 05	05:58:48	19791	-33070	57	T	p-	0.6740	1.0706	39.0N	148.9E	47	141	306	04m58s
3150	158	-0674 Sep 28	23:41:07	19783	-33064	62	A	p-	-0.7505	0.9243	40.4S	123.3W	41	40	419	07m22s
3151	158	-0673 Feb 24	12:57:12	19776	-33059	29	P	-t	-1.2192	0.5940	61.3S	174.3W	0	244		
3152	158	-0673 Mar 25	22:33:12	19775	-33058	67	P	t-	1.4116	0.2327	60.8N	161.6W	0	92		
3153	158	-0673 Aug 19	14:02:13	19769	-33053	34	P	-t	1.3585	0.3398	62.0N	174.2E	0	305		
3154	158	-0673 Sep 18	02:41:26	19767	-33052	72	P	t-	-1.4144	0.2480	60.8S	140.7E	0	79		
3155	158	-0672 Feb 13	21:18:09	19761	-33047	39	A	-p	-0.5308	0.9573	44.5S	35.1W	58	327	182	03m54s
3156	158	-0672 Aug 08	03:56:31	19753	-33041	44	T	-p	0.5867	1.0572	51.4N	136.8W	54	212	232	03m59s
3157	158	-0671 Feb 01	22:36:20	19746	-33035	49	A	nn	0.1837	0.9271	9.0S	76.0W	79	160	278	09m24s
3158	158	-0671 Jul 28	20:46:07	19738	-33029	54	T	nn	-0.1364	1.0688	13.6N	51.1W	82	17	226	06m07s
3159	158	-0670 Jan 21	22:29:38	19730	-33023	59	A	p-	0.8774	0.9323	37.8N	89.5W	28	161	521	07m27s
3160	158	-0670 Jul 18	11:49:55	19723	-33017	64	T	p-	-0.0996	1.0208	40.2S	69.6E	26	16	162	01m50s
3161	159	-0670 Dec 12	15:05:49	19716	-33012	31	P	-t	-1.2723	0.4963	67.1S	141.4W	0	177		
3162	159	-0669 Jan 11	03:59:58	19715	-33011	69	P	t-	1.5201	0.0558	64.3N	177.2E	0	154		
3163	159	-0669 Jun 08	07:36:00	19708	-33006	36	P	-t	1.0569	0.8693	67.7N	24.4W	0	9		
3164	159	-0669 Dec 02	05:36:05	19701	-33000	41	T	-p	-0.5739	1.0421	56.2S	170.7E	55	9	172	03m01s
3165	159	-0668 May 27	08:21:30	19693	-32994	46	A	np	0.3016	0.9510	37.5N	131.7E	72	171	189	05m34s
3166	159	-0668 Nov 20	21:03:00	19685	-32988	51	T	nn	0.0904	1.0303	13.6S	55.3W	85	189	103	02m58s
3167	159	-0667 May 16	11:25:24	19678	-32982	56	A	p-	-0.4740	0.9841	10.7S	93.9E	62	349	64	01m54s
3168	159	-0667 Nov 10	08:47:36	19670	-32976	61	A	p-	0.8077	0.9726	37.3N	140.0E	36	195	166	02m42s
3169	159	-0666 Apr 06	13:24:57	19664	-32971	28	P	-t	1.3604	0.3276	71.5N	38.8W	0	79		
3170	159	-0666 May 05	21:31:04	19662	-32970	66	P	t-	-1.2052	0.6216	70.4S	16.8W	0	316		
3171	159	-0666 Sep 30	19:07:00	19656	-32965	33	P	-t	-1.4695	0.1675	71.8S	116.5W	0	90		
3172	159	-0666 Oct 30	13:27:48	19655	-32964	71	P	t-	1.5517	0.0273	70.8N	110.4E	0	232		
3173	159	-0665 Mar 27	06:11:13	19648	-32959	38	T	-p	0.6299	1.0681	36.4N	157.0E	51	157	284	05m18s
3174	159	-0665 Sep 19	18:46:45	19641	-32953	43	A	-p	-0.7690	0.9344	43.2S	39.3W	39	24	378	06m44s
3175	159	-0664 Mar 15	21:54:36	19633	-32947	48	Tm	nn	-0.0946	1.0359	10.0S	62.1W	85	342	121	03m28s
3176	159	-0664 Sep 07	23:54:24	19625	-32941	53	A	nn	-0.0191	0.9914	7.7N	97.3W	89	17	30	00m57s
3177	159	-0663 Mar 05	08:08:37	19618	-32935	58	A	p-	-0.8739	0.9688	64.8S	179.2E	29	319	231	02m22s
3178	159	-0663 Aug 28	12:17:52	19610	-32929	63	T	p-	0.6960	1.0416	54.7N	95.4E	46	205	193	02m59s
3179	159	-0662 Jan 23	17:15:52	19603	-32924	30	P	-t	1.3470	0.3720	68.2N	8.9W	0	165		
3180	159	-0662 Jul 19	20:46:12	19596	-32918	35	P	-t	-1.1775	0.6774	67.4S	56.0W	0	6		

Cat Num	Canon Plate	Calendar Date	TD of Greatest Eclipse	ΔT s	Luna Num	Saros Num	Ecl. Type	QLE	Gamma	Ecl. Mag.	Lat. °	Long. °	Sun Alt °	Sun Azm °	Path Width km	Central Line Dur.
3181	160	-0662 Aug 18	04:38:34	19595	-32917	73	P	t-	1.3757	0.2972	69.9N	26.0W	0	323		
3182	160	-0661 Jan 12	18:07:10	19588	-32912	40	A	-p	0.6274	0.9466	16.1N	7.9W	51	179	252	06m51s
3183	160	-0661 Jul 09	10:36:00	19580	-32906	45	H3	-p	-0.4659	1.0146	4.3S	103.1E	62	358	56	01m38s
3184	160	-0660 Jan 02	01:39:34	19573	-32900	50	H	n-	-0.1028	1.0023	29.7S	121.9W	84	4	8	00m14s
3185	160	-0660 Jun 27	17:42:34	19565	-32894	55	A	nn	0.3114	0.9656	42.0N	8.0W	72	172	131	03m35s
3186	160	-0660 Dec 21	15:21:15	19557	-32888	60	T	p-	-0.7731	1.0364	73.1S	9.2E	39	27	194	02m15s
3187	160	-0659 Jun 16	18:56:53	19550	-32882	65	P	t-	1.0852	0.8166	64.4N	177.4W	0	25		
3188	160	-0659 Nov 11	19:34:53	19543	-32877	32	P	-t	1.2898	0.4646	61.9N	22.3E	0	234		
3189	160	-0659 Dec 11	07:03:44	19542	-32876	70	P	t-	-1.4287	0.2038	64.2S	5.0E	0	152		
3190	160	-0658 May 07	09:06:41	19536	-32871	37	A	-t	-0.9318	0.9900	47.2S	158.0E	21	324	95	00m50s
3191	160	-0658 Nov 01	05:33:35	19528	-32865	42	A	-p	0.6482	0.9602	24.1N	166.7W	49	208	185	04m05s
3192	160	-0657 Apr 26	21:04:23	19520	-32859	47	Tm	nn	-0.1284	1.0540	4.4N	52.3W	83	333	180	04m44s
3193	160	-0657 Oct 21	08:15:36	19513	-32853	52	A	nn	-0.0405	0.9249	10.3S	133.6E	88	28	283	08m43s
3194	160	-0656 Apr 15	13:35:56	19505	-32847	57	T	p-	0.6092	1.0741	39.3N	35.4E	52	143	299	05m12s
3195	160	-0656 Oct 09	07:25:32	19497	-32841	62	A	p-	-0.7282	0.9235	43.0S	118.7E	43	42	410	07m15s
3196	160	-0655 Mar 06	20:43:54	19491	-32836	29	P	-t	-1.2674	0.5057	61.0S	59.0E	0	253		
3197	160	-0655 Apr 05	06:08:28	19490	-32835	67	P	t-	1.3562	0.3377	60.9N	74.6E	0	84		
3198	160	-0655 Aug 29	21:50:00	19483	-32830	34	P	-t	1.3946	0.2742	61.5N	47.1E	0	296		
3199	160	-0655 Sep 28	10:40:10	19482	-32829	72	P	t-	-1.3865	0.2968	60.6S	11.2E	0	88		
3200	160	-0654 Feb 24	04:48:41	19476	-32824	39	A	-p	-0.5735	0.9572	42.7S	146.4W	55	324	188	03m53s
3201	161	-0654 Aug 19	11:52:06	19468	-32818	44	T	-p	0.6332	1.0551	50.3N	107.3E	50	217	233	03m47s
3202	161	-0653 Feb 13	05:59:48	19460	-32812	49	A	nn	0.1475	0.9298	8.1S	172.4E	82	157	266	08m43s
3203	161	-0653 Aug 09	04:34:13	19453	-32806	54	T	nn	-0.0820	1.0645	14.2N	168.5W	85	21	212	05m37s
3204	161	-0652 Feb 02	06:12:01	19445	-32800	59	A	p-	0.8498	0.9378	35.9N	151.2E	32	156	429	06m42s
3205	161	-0652 Jul 28	19:14:08	19438	-32794	64	T	p-	-0.8405	1.0166	34.1S	44.4W	33	19	104	01m31s
3206	161	-0652 Dec 22	23:46:43	19431	-32789	31	P	-t	-1.2713	0.4982	66.1S	76.5E	0	188		
3207	161	-0651 Jan 21	12:14:48	19430	-32788	69	P	t-	1.5007	0.0881	63.4N	42.7E	0	145		
3208	161	-0651 Jun 18	14:02:56	19424	-32783	36	P	-t	1.1432	0.7194	66.7N	133.3W	0	358		
3209	161	-0651 Dec 12	14:29:02	19416	-32777	41	T	-p	-0.5747	1.0432	58.2S	41.9E	55	2	177	03m03s
3210	161	-0650 Jun 07	14:42:20	19408	-32771	46	A	-p	0.3893	0.9512	45.0N	36.3E	67	175	194	05m12s
3211	161	-0650 Dec 02	05:52:53	19401	-32765	51	Tm	nn	0.0911	1.0282	16.1S	171.4E	85	185	96	02m47s
3212	161	-0649 May 27	18:08:21	19393	-32759	56	A	p-	-0.3902	0.9884	2.9S	10.5W	67	353	44	01m25s
3213	161	-0649 Nov 21	17:17:39	19385	-32753	61	A	p-	0.8092	0.9690	34.9N	8.1E	36	191	189	03m11s
3214	161	-0648 Apr 16	20:56:38	19379	-32748	28	P	-t	1.4199	0.2128	71.2N	166.6W	0	66		
3215	161	-0648 May 16	04:43:15	19378	-32747	66	P	t-	-1.1279	0.7675	69.6S	138.6W	0	328		
3216	161	-0648 Oct 11	02:49:02	19371	-32742	33	P	-t	-1.4896	0.1352	71.7S	112.6E	0	104		
3217	161	-0648 Nov 09	21:32:17	19370	-32741	71	P	t-	1.5487	0.0338	70.0N	25.0W	0	219		
3218	161	-0647 Apr 06	13:53:47	19364	-32736	38	T	-p	0.6898	1.0689	44.8N	35.8E	46	155	308	05m02s
3219	161	-0647 Sep 30	02:28:46	19356	-32730	43	A	-p	-0.7927	0.9332	49.2S	160.1W	37	28	405	06m26s
3220	161	-0646 Mar 27	05:32:40	19349	-32724	48	T	nn	-0.0399	1.0356	2.6S	179.8W	88	342	120	03m29s
3221	162	-0646 Sep 19	07:51:54	19341	-32718	53	A	nn	-0.0485	0.9916	1.9N	141.0E	87	18	30	00m55s
3222	162	-0645 Mar 16	15:28:41	19333	-32712	58	A	p-	-0.8244	0.9699	56.3S	58.2E	34	327	190	02m29s
3223	162	-0645 Sep 08	20:26:44	19326	-32706	63	T	p-	0.6603	1.0412	47.9N	30.0W	48	205	183	03m06s
3224	162	-0644 Feb 04	00:46:17	19319	-32701	30	P	-t	1.3757	0.3236	69.2N	134.5W	0	154		
3225	162	-0644 Jul 30	04:25:23	19312	-32695	35	P	-t	-1.2375	0.5622	68.4S	176.7E	0	17		
3226	162	-0644 Aug 28	12:43:01	19310	-32694	73	P	t-	1.3365	0.3729	70.6N	161.1W	0	310		
3227	162	-0643 Jan 23	02:02:02	19304	-32689	40	A	-p	0.6456	0.9515	19.1N	129.9W	50	175	232	06m02s
3228	162	-0643 Jul 19	17:48:48	19297	-32683	45	H	-p	-0.5348	1.0080	9.9S	7.9W	58	2	33	00m54s
3229	162	-0642 Jan 12	10:05:24	19289	-32677	50	H	n-	-0.0912	1.0075	28.2S	112.2E	85	359	26	00m45s
3230	162	-0642 Jul 09	00:21:38	19281	-32671	55	A	nn	0.2331	0.9618	37.2N	106.1W	76	178	143	04m14s
3231	162	-0641 Jan 02	00:07:37	19274	-32665	60	T	p-	-0.7666	1.0394	73.8S	109.0W	40	11	207	02m26s
3232	162	-0641 Jun 28	01:18:35	19266	-32659	65	A+	t-	0.9992	0.9632	65.3N	76.1E	0	15	-	-
3233	162	-0641 Nov 23	04:20:11	19260	-32654	32	P	-t	1.2934	0.4580	62.6N	119.3W	0	225		
3234	162	-0641 Dec 22	15:53:17	19258	-32653	70	P	t-	-1.4287	0.2039	65.1S	138.5W	0	163		
3235	162	-0640 May 17	15:57:53	19252	-32648	37	P	-t	-1.0082	0.9746	62.3S	71.5E	0	311		
3236	162	-0640 Nov 11	13:55:10	19244	-32642	42	A	-p	0.6536	0.9563	21.7N	64.8E	49	204	206	04m43s
3237	162	-0639 May 07	04:23:33	19237	-32636	47	T	nn	-0.2020	1.0580	3.9N	162.5W	78	336	195	05m10s
3238	162	-0639 Oct 31	16:12:24	19229	-32630	52	A	nn	-0.0302	0.9222	13.9S	13.3E	88	25	294	09m12s
3239	162	-0638 Apr 26	21:09:26	19222	-32624	57	T	p-	0.5412	1.0768	39.7N	76.6W	57	145	293	05m26s
3240	162	-0638 Oct 20	15:16:38	19214	-32618	62	A	p-	-0.7109	0.9230	46.3S	0.8W	44	42	405	07m08s

Cat Num	Canon Plate	Calendar Date	TD of Greatest Eclipse	ΔT s	Luna Num	Saros Num	Ecl. Type	QLE	Gamma	Ecl. Mag.	Lat. °	Long. °	Sun Alt °	Sun Azm °	Path Width km	Central Line Dur.
3241	163	-0637 Mar 18	04:23:10	19208	-32613	29	P	-t	-1.3210	0.4076	60.7S	65.7W	0	262		
3242	163	-0637 Apr 16	13:38:29	19206	-32612	67	P	t-	1.2970	0.4499	61.0N	47.8W	0	75		
3243	163	-0637 Sep 10	05:46:18	19200	-32607	34	P	-t	1.4247	0.2194	61.1N	81.9W	0	287		
3244	163	-0637 Oct 09	18:46:43	19199	-32606	72	P	t-	-1.3641	0.3361	60.7S	120.3W	0	98		
3245	163	-0636 Mar 06	12:08:54	19193	-32601	39	A	-p	-0.6238	0.9571	41.2S	104.5E	51	322	196	03m54s
3246	163	-0636 Aug 29	19:55:51	19185	-32595	44	T	-p	0.6733	1.0528	48.3N	12.1W	47	220	234	03m36s
3247	163	-0635 Feb 23	13:13:23	19177	-32589	49	A	nn	0.1032	0.9326	7.0S	63.3E	84	154	253	08m05s
3248	163	-0635 Aug 19	12:30:05	19170	-32583	54	T	nn	-0.0345	1.0600	13.7N	72.2E	88	25	197	05m08s
3249	163	-0634 Feb 12	13:45:47	19162	-32577	59	A	p-	0.8144	0.9438	34.2N	34.9E	35	152	348	05m55s
3250	163	-0634 Aug 09	02:45:29	19155	-32571	64	T	p-	-0.7880	1.0117	30.5S	159.8W	38	23	64	01m05s
3251	163	-0633 Jan 03	08:23:21	19148	-32566	31	P	-t	-1.2746	0.4922	65.0S	64.1W	0	198		
3252	163	-0633 Feb 01	20:22:42	19147	-32565	69	P	t-	1.4748	0.1322	62.6N	89.8W	0	135		
3253	163	-0633 Jun 29	20:31:44	19141	-32560	36	P	-t	1.2270	0.5745	65.7N	117.9E	0	348		
3254	163	-0633 Dec 23	23:19:32	19133	-32554	41	T	-p	-0.5777	1.0448	59.1S	85.2W	54	353	184	03m08s
3255	163	-0632 Jun 17	21:04:47	19126	-32548	46	A	-p	0.4762	0.9509	51.9N	57.6W	61	181	205	04m53s
3256	163	-0632 Dec 12	14:40:33	19118	-32542	51	T	nn	0.0904	1.0267	17.8S	39.0E	85	181	91	02m39s
3257	163	-0631 Jun 07	00:52:54	19110	-32536	56	A	p-	-0.3066	0.9920	4.1N	114.5W	72	357	30	00m59s
3258	163	-0631 Dec 02	01:48:06	19103	-32530	61	A	p-	0.8116	0.9662	33.0N	124.0W	35	186	209	03m35s
3259	163	-0630 Apr 28	04:23:58	19097	-32525	28	Pe	-t	1.4832	0.0901	70.7N	67.0E	0	53		
3260	163	-0630 May 27	11:55:19	19095	-32524	66	P	t-	-1.0497	0.9162	68.7S	100.2E	0	340		
3261	164	-0630 Oct 22	10:37:57	19089	-32519	33	P	-t	-1.5039	0.1124	71.3S	19.8W	0	118		
3262	164	-0630 Nov 21	05:39:10	19088	-32518	71	P	t-	1.5478	0.0364	69.1N	160.3W	0	206		
3263	164	-0629 Apr 17	21:30:12	19081	-32513	38	T	-p	0.7547	1.0689	53.8N	85.0W	41	151	341	04m42s
3264	164	-0629 Oct 11	10:19:51	19074	-32507	43	A	-p	-0.8096	0.9323	54.8S	77.1E	36	32	428	06m08s
3265	164	-0628 Apr 06	13:01:12	19066	-32501	48	T	nn	0.0221	1.0349	5.0N	64.7E	89	163	118	03m26s
3266	164	-0628 Sep 29	15:59:34	19059	-32495	53	A	nn	-0.0702	0.9919	3.8S	16.7E	86	18	29	00m52s
3267	164	-0627 Mar 26	22:38:48	19051	-32489	58	A	p-	-0.7674	0.9709	47.5S	57.6W	40	333	161	02m38s
3268	164	-0627 Sep 19	04:44:18	19044	-32483	63	T	p-	0.6314	1.0405	41.5N	157.7W	51	204	174	03m11s
3269	164	-0626 Feb 14	08:09:05	19037	-32478	30	P	-t	1.4103	0.2648	70.1N	101.1E	0	141		
3270	164	-0626 Mar 16	01:05:55	19036	-32477	68	Pb	t-	-1.5403	0.0422	71.6S	9.7W	0	255		
3271	164	-0626 Aug 10	12:11:00	19030	-32472	35	P	-t	-1.2917	0.4591	69.3S	47.2E	0	28		
3272	164	-0626 Sep 08	20:55:06	19028	-32471	73	P	t-	1.3038	0.4357	71.2N	61.3E	0	297		
3273	164	-0625 Feb 03	09:52:04	19022	-32466	40	A	-p	0.6684	0.9567	23.0N	108.9E	48	170	211	05m09s
3274	164	-0625 Jul 31	01:05:12	19015	-32460	45	H	-p	-0.5995	1.0011	16.1S	120.5W	53	6	5	00m07s
3275	164	-0624 Jan 23	18:26:58	19007	-32454	50	H	nn	-0.0754	1.0131	25.5S	13.1W	86	354	45	01m18s
3276	164	-0624 Jul 19	07:04:44	18999	-32448	55	A	nn	0.1591	0.9575	31.8N	153.6E	81	182	157	04m59s
3277	164	-0623 Jan 12	08:47:53	18992	-32442	60	T	p-	-0.7552	1.0428	72.1S	133.7E	41	356	220	02m41s
3278	164	-0623 Jul 08	07:46:48	18984	-32436	65	A	t-	0.9182	0.9375	87.9N	35.2E	23	72	599	04m17s
3279	164	-0623 Dec 03	13:04:26	18978	-32431	32	P	-t	1.2971	0.4513	63.5N	99.2E	0	215		
3280	164	-0622 Jan 02	00:37:29	18977	-32430	70	P	t	-1.4255	0.2098	66.2S	79.0E	0	173		
3281	165	-0622 May 28	22:51:36	18971	-32425	37	P	-t	-1.0834	0.8397	63.0S	42.4W	0	320		
3282	165	-0622 Nov 22	22:16:40	18963	-32419	42	A	-p	0.6580	0.9530	19.8N	63.7W	49	200	224	05m19s
3283	165	-0621 May 18	11:42:11	18955	-32413	47	T	-n	-0.2765	1.0613	2.7N	87.3E	74	339	209	05m34s
3284	165	-0621 Nov 12	00:12:30	18948	-32407	52	A	nn	-0.0230	0.9202	17.3S	107.6W	89	22	302	09m40s
3285	165	-0620 May 07	04:37:21	18940	-32401	57	T	p-	0.4686	1.0787	39.7N	173.3E	62	149	286	05m40s
3286	165	-0620 Oct 30	23:15:34	18933	-32395	62	A	p-	-0.6996	0.9230	50.0S	121.8W	45	42	400	06m58s
3287	165	-0619 Mar 28	11:51:15	18927	-32390	29	P	-t	-1.3825	0.2950	60.7S	172.4E	0	270		
3288	165	-0619 Apr 26	21:00:49	18925	-32389	67	P	t-	1.2323	0.5722	61.4N	168.4W	0	66		
3289	165	-0619 Sep 20	13:53:27	18919	-32384	34	P	-t	1.4468	0.1791	60.9N	146.3E	0	278		
3290	165	-0619 Oct 20	03:02:09	18918	-32383	72	P	t-	-1.3482	0.3637	60.9S	106.0E	0	107		
3291	165	-0618 Mar 17	19:18:06	18911	-32378	39	A	-p	-0.6821	0.9567	40.5S	2.0W	47	321	211	03m57s
3292	165	-0618 Sep 10	04:09:08	18904	-32372	44	T	-p	0.7056	1.0503	45.9N	135.3W	45	222	232	03m28s
3293	165	-0617 Mar 06	20:17:43	18896	-32366	49	A	nn	0.0513	0.9357	5.7S	43.3W	87	152	240	07m30s
3294	165	-0617 Aug 30	20:32:32	18889	-32360	54	Tm	nn	0.0070	1.0551	12.3N	48.9W	90	202	102	04m39s
3295	165	-0616 Feb 23	21:12:39	18881	-32354	59	A	p-	0.7725	0.9501	33.0N	79.0W	39	149	280	05m06s
3296	165	-0616 Aug 19	10:22:40	18874	-32348	64	H	p-	-0.7411	1.0062	28.6S	83.6E	42	26	31	00m34s
3297	165	-0615 Jan 13	16:55:37	18868	-32343	31	P	-t	-1.2820	0.4784	64.1S	156.8E	0	208		
3298	165	-0615 Feb 12	04:23:39	18866	-32342	69	P	t-	1.4425	0.1884	61.9N	139.7E	0	126		
3299	165	-0615 Jul 10	03:04:02	18860	-32337	36	P	-t	1.3070	0.4372	64.7N	8.7E	0	338		
3300	165	-0615 Aug 08	17:45:17	18859	-32336	74	Pb	t-	-1.5409	0.0314	62.5S	57.0W	0	47		

Cat Num	Canon Plate	Calendar Date	TD of Greatest Eclipse	ΔT s	Luna Num	Saros Num	Ecl. Type	QLE	Gamma	Ecl. Mag.	Lat. °	Long. °	Sun Alt °	Sun Azm °	Path Width km	Central Line Dur.
3301	166	-0614 Jan 03	08:06:45	18852	-32331	41	T	-p	-0.5839	1.0467	58.9S	148.7E	54	345	192	03m13s
3302	166	-0614 Jun 29	03:31:25	18845	-32325	46	A	-p	0.5597	0.9503	57.8N	149.8W	56	189	221	04m37s
3303	166	-0614 Dec 23	23:25:35	18837	-32319	51	T	nn	0.0880	1.0256	18.8S	92.5W	85	176	87	02m32s
3304	166	-0613 Jun 18	07:41:22	18830	-32313	56	A	n-	-0.2250	0.9951	10.3N	141.4E	77	1	18	00m35s
3305	166	-0613 Dec 13	10:16:03	18822	-32307	61	A	p-	0.8129	0.9638	31.7N	104.6E	35	181	225	03m55s
3306	166	-0612 Jun 06	19:08:18	18815	-32301	66	T	t-	-0.9713	1.0379	54.8S	24.4W	13	354	558	02m52s
3307	166	-0612 Nov 01	18:32:05	18809	-32296	33	P	-t	-1.5137	0.0970	70.7S	153.0W	0	132		
3308	166	-0612 Dec 01	13:45:29	18807	-32295	71	P	t-	1.5469	0.0388	68.1N	65.1E	0	194		
3309	166	-0611 Apr 28	05:01:51	18801	-32290	38	T	-p	0.8234	1.0678	63.6N	152.8E	34	145	390	04m16s
3310	166	-0611 Oct 21	18:18:09	18794	-32284	43	A	-p	-0.8210	0.9322	60.1S	46.9W	34	34	443	05m50s
3311	166	-0610 Apr 17	20:23:07	18786	-32278	48	T	nn	0.0887	1.0336	12.9N	49.2W	85	163	114	03m18s
3312	166	-0610 Oct 11	00:15:52	18779	-32272	53	A	nn	-0.0856	0.9925	9.1S	109.5W	85	18	27	00m48s
3313	166	-0609 Apr 07	05:39:06	18771	-32266	58	A	p-	-0.7033	0.9715	38.6S	169.7W	45	337	142	02m49s
3314	166	-0609 Sep 30	13:10:06	18764	-32260	63	T	p-	0.6092	1.0398	35.6N	72.5E	52	202	167	03m16s
3315	166	-0608 Feb 25	15:21:51	18757	-32255	30	P	-t	1.4526	0.1925	70.8N	21.3W	0	128		
3316	166	-0608 Mar 26	07:53:16	18756	-32254	68	P	t-	-1.4781	0.1467	71.8S	126.7W	0	268		
3317	166	-0608 Aug 20	20:03:46	18750	-32249	35	P	-t	-1.3392	0.3698	70.2S	84.6W	0	41		
3318	166	-0608 Sep 19	05:15:43	18748	-32248	73	P	t-	1.2786	0.4838	71.6N	78.9W	0	283		
3319	166	-0607 Feb 13	17:32:25	18742	-32243	40	A	-p	0.6995	0.9620	28.1N	10.4W	45	166	191	04m14s
3320	166	-0607 Aug 10	08:28:22	18735	-32237	45	A	-p	-0.6572	0.9940	22.6S	124.5E	49	10	28	00m39s
3321	167	-0606 Feb 03	02:41:26	18727	-32231	50	T	nn	-0.0533	1.0190	21.7S	137.2W	87	351	65	01m53s
3322	167	-0606 Jul 30	13:52:42	18720	-32225	55	Am	nn	0.0900	0.9530	25.9N	51.0E	85	186	172	05m46s
3323	167	-0605 Jan 23	17:23:02	18712	-32219	60	T	p-	-0.7396	1.0467	68.4S	12.9E	42	345	232	02m59s
3324	167	-0605 Jul 19	14:19:55	18705	-32213	65	A	p-	0.8407	0.9383	79.9N	53.6E	32	191	428	04m40s
3325	167	-0605 Dec 14	21:47:46	18699	-32208	32	P	-t	1.3013	0.4437	64.4N	42.4W	0	205		
3326	167	-0604 Jan 13	09:16:42	18697	-32207	70	P	t-	-1.4187	0.2223	67.2S	62.8W	0	184		
3327	167	-0604 Jun 08	05:45:23	18691	-32202	37	P	-t	-1.1592	0.7026	63.8S	156.5W	0	329		
3328	167	-0604 Jul 07	17:02:10	18690	-32201	75	Pb	t-	1.5494	0.0066	66.4N	172.7W	0	5		
3329	167	-0604 Dec 03	06:38:55	18684	-32196	42	A	-p	0.6612	0.9504	18.5N	167.7E	48	196	240	05m51s
3330	167	-0603 May 28	19:00:44	18676	-32190	47	T	-n	-0.3517	1.0637	0.7N	23.2W	69	343	222	05m54s
3331	167	-0603 Nov 22	08:14:30	18669	-32184	52	Am	nn	-0.0176	0.9188	20.2S	131.4E	89	18	308	10m06s
3332	167	-0602 May 18	12:04:25	18661	-32178	57	T	p-	0.3953	1.0797	39.3N	63.8E	67	153	279	05m52s
3333	167	-0602 Nov 11	07:19:13	18654	-32172	62	A	p-	-0.6918	0.9235	54.1S	116.8E	46	41	395	06m47s
3334	167	-0601 Apr 08	19:12:05	18647	-32167	29	P	-t	-1.4489	0.1737	60.8S	52.3E	0	279		
3335	167	-0601 May 08	04:18:42	18646	-32166	67	P	t-	1.1645	0.7000	61.8N	72.0E	0	57		
3336	167	-0601 Oct 01	22:08:38	18640	-32161	34	P	-t	1.4634	0.1489	60.8N	12.7E	0	269		
3337	167	-0601 Oct 31	11:23:47	18639	-32160	72	P	t-	-1.3365	0.3840	61.3S	29.3W	0	116		
3338	167	-0600 Mar 28	02:18:16	18632	-32155	39	A	-p	-0.7469	0.9559	40.6S	106.2W	41	320	235	04m02s
3339	167	-0600 Sep 20	12:30:31	18625	-32149	44	T	-p	0.7318	1.0478	43.3N	98.2E	43	222	229	03m21s
3340	167	-0599 Mar 17	03:12:05	18617	-32143	49	Am	nn	-0.0090	0.9387	4.6S	147.3W	89	332	227	07m01s
3341	168	-0599 Sep 10	04:43:29	18610	-32137	54	T	nn	0.0411	1.0501	9.9N	172.5W	88	208	166	04m13s
3342	168	-0598 Mar 06	04:31:34	18602	-32131	59	A	p-	0.7228	0.9565	32.2N	169.8E	44	147	223	04m19s
3343	168	-0598 Aug 30	18:06:26	18595	-32125	64	H	p-	-0.7003	1.0005	28.2S	34.5W	45	30	2	00m02s
3344	168	-0597 Jan 25	01:21:19	18589	-32120	31	P	-t	-1.2957	0.4531	63.2S	19.7E	0	218		
3345	168	-0597 Feb 23	12:17:04	18588	-32119	69	P	t-	1.4032	0.2581	61.4N	11.3E	0	116		
3346	168	-0597 Jul 21	09:41:13	18581	-32114	36	P	-t	1.3824	0.3088	63.8N	101.5W	0	329		
3347	168	-0597 Aug 20	00:54:04	18580	-32113	74	P	t-	-1.4909	0.1203	61.9S	174.5W	0	56		
3348	168	-0596 Jan 14	16:47:19	18574	-32108	41	T	-p	-0.5957	1.0489	57.7S	23.7E	53	337	202	03m21s
3349	168	-0596 Jul 09	10:03:01	18566	-32102	46	A	-p	0.6394	0.9490	62.4N	119.8E	50	200	244	04m26s
3350	168	-0595 Jan 03	08:05:12	18559	-32096	51	T	nn	0.0815	1.0251	19.1S	137.3E	85	171	86	02m28s
3351	168	-0595 Jun 28	14:34:50	18551	-32090	56	A	nn	-0.1466	0.9974	15.5N	36.8E	82	5	9	00m18s
3352	168	-0595 Dec 23	18:39:43	18544	-32084	61	A	p-	0.8114	0.9622	30.6N	25.7W	36	176	235	04m09s
3353	168	-0594 Jun 18	02:24:37	18537	-32078	66	T	t-	-0.8947	1.0440	40.6S	141.2W	26	1	331	03m44s
3354	168	-0594 Nov 13	02:31:13	18530	-32073	33	P	-t	-1.5195	0.0880	69.9S	73.1E	0	145		
3355	168	-0594 Dec 12	21:51:21	18529	-32072	71	P	t-	1.5457	0.0411	67.1N	68.8W	0	182		
3356	168	-0593 May 09	12:28:16	18523	-32067	38	T	-t	0.8961	1.0653	73.9N	22.3E	26	129	487	03m45s
3357	168	-0593 Nov 02	02:24:04	18515	-32061	43	A	-p	-0.8268	0.9325	65.1S	171.4W	34	35	449	05m32s
3358	168	-0592 Apr 28	03:37:40	18508	-32055	48	T	-n	0.1609	1.0317	20.8N	160.9W	81	164	109	03m05s
3359	168	-0592 Oct 21	08:40:14	18500	-32049	53	A	nn	-0.0949	0.9934	14.0S	122.7E	84	17	23	00m42s
3360	168	-0591 Apr 17	12:31:12	18493	-32043	58	A	p-	-0.6331	0.9716	29.8S	81.3E	51	341	130	03m03s

Cat Num	Canon Plate	Calendar Date	TD of Greatest Eclipse	ΔT s	Luna Num	Saros Num	Ecl. Type	QLE	Gamma	Ecl. Mag.	Lat. °	Long. °	Sun Alt °	Sun Azm °	Path Width km	Central Line Dur.
3361	169	-0591 Oct 10	21:43:55	18486	-32037	63	T	p-	0.5934	1.0392	30.3N	59.2W	53	200	162	03m20s
3362	169	-0590 Mar 07	22:27:30	18479	-32032	30	P	-t	1.5004	0.1103	71.4N	142.4W	0	115		
3363	169	-0590 Apr 06	14:34:18	18478	-32031	68	P	t-	-1.4109	0.2605	71.7S	117.9E	0	282		
3364	169	-0590 Sep 01	04:02:30	18472	-32026	35	P	-t	-1.3812	0.2916	70.9S	141.4E	0	54		
3365	169	-0590 Sep 30	13:42:15	18471	-32025	73	P	t-	1.2586	0.5216	71.7N	139.2E	0	269		
3366	169	-0589 Feb 25	01:07:26	18464	-32020	40	A	-p	0.7358	0.9675	34.1N	129.0W	42	162	171	03m23s
3367	169	-0589 Aug 21	15:57:08	18457	-32014	45	A	-p	-0.7092	0.9868	29.4S	7.6E	45	15	66	01m22s
3368	169	-0588 Feb 14	10:48:38	18450	-32008	50	T	nn	-0.0244	1.0251	16.8S	99.7E	88	348	85	02m29s
3369	169	-0588 Aug 09	20:47:36	18442	-32002	55	A	nn	0.0271	0.9484	19.7N	54.0W	88	189	190	06m34s
3370	169	-0587 Feb 03	01:50:15	18435	-31996	60	T	p-	-0.7174	1.0508	63.3S	110.7W	44	339	242	03m21s
3371	169	-0587 Jul 29	21:02:02	18427	-31990	65	A	p-	0.7699	0.9382	70.6N	43.5W	39	198	362	05m06s
3372	169	-0587 Dec 25	06:25:44	18421	-31985	32	P	-t	1.3093	0.4291	65.4N	177.0E	0	194		
3373	169	-0586 Jan 23	17:47:38	18420	-31984	70	P	t-	-1.4060	0.2458	68.2S	157.0E	0	196		
3374	169	-0586 Jun 19	12:45:10	18414	-31979	37	P	-t	-1.2308	0.5723	64.7S	87.6E	0	338		
3375	169	-0586 Jul 18	23:59:19	18412	-31978	75	P	t-	1.4763	0.1354	67.4N	71.2E	0	355		
3376	169	-0586 Dec 14	14:58:15	18406	-31973	42	A	-p	0.6658	0.9484	18.0N	39.7E	48	191	253	06m18s
3377	169	-0585 Jun 09	02:19:58	18399	-31967	47	T	-p	-0.4268	1.0652	2.4S	134.2W	65	347	235	06m10s
3378	169	-0585 Dec 03	16:16:58	18391	-31961	52	A	nn	-0.0131	0.9182	22.3S	10.6E	89	14	311	10m27s
3379	169	-0584 May 28	19:28:50	18384	-31955	57	T	n-	0.3201	1.0798	38.2N	45.0W	71	158	271	06m04s
3380	169	-0584 Nov 21	15:26:35	18376	-31949	62	A	p-	-0.6866	0.9247	58.0S	4.2W	46	38	388	06m34s
3381	170	-0583 Apr 19	02:23:20	18370	-31944	29	Pe	-t	-1.5219	0.0408	61.0S	65.4W	0	288		
3382	170	-0583 May 18	11:31:36	18369	-31943	67	P	t-	1.0933	0.8334	62.4N	46.5W	0	48		
3383	170	-0583 Oct 12	06:33:42	18363	-31938	34	P	-t	1.4728	0.1316	61.0N	123.5W	0	260		
3384	170	-0583 Nov 10	19:51:52	18362	-31937	72	P	t-	-1.3298	0.3957	61.9S	166.4W	0	125		
3385	170	-0582 Apr 08	09:08:42	18355	-31932	39	A	-p	-0.8188	0.9546	42.3S	152.5E	35	319	280	04m09s
3386	170	-0582 Oct 01	20:59:24	18348	-31926	44	T	-p	0.7520	1.0455	40.6N	31.1W	41	220	225	03m16s
3387	170	-0581 Mar 28	09:58:33	18341	-31920	49	A	nn	-0.0757	0.9417	3.7S	110.8E	86	331	216	06m37s
3388	170	-0581 Sep 21	13:01:27	18333	-31914	54	T	nn	0.0690	1.0450	7.0N	62.0E	86	209	150	03m48s
3389	170	-0580 Mar 16	11:42:18	18326	-31908	59	A	p-	0.6657	0.9631	31.9N	61.3E	48	145	174	03m34s
3390	170	-0580 Sep 10	01:57:23	18318	-31902	64	A	p-	-0.6663	0.9945	29.0S	154.4W	48	32	25	00m30s
3391	170	-0579 Feb 04	09:40:57	18312	-31897	31	P	-t	-1.3147	0.4174	62.5S	115.7W	0	228		
3392	170	-0579 Mar 05	20:03:40	18311	-31896	69	P	t-	1.3576	0.3403	61.0N	115.3W	0	108		
3393	170	-0579 Jul 31	16:24:28	18305	-31891	36	P	-t	1.4518	0.1915	63.0N	147.1E	0	320		
3394	170	-0579 Aug 30	08:09:58	18303	-31890	74	P	t-	-1.4471	0.1975	61.4S	66.5E	0	65		
3395	170	-0578 Jan 25	01:22:11	18297	-31885	41	T	-p	-0.6123	1.0512	55.8S	100.9W	52	330	214	03m28s
3396	170	-0578 Jul 20	16:42:17	18290	-31879	46	A	-p	0.7131	0.9476	65.2N	30.4E	44	212	275	04m19s
3397	170	-0577 Jan 14	16:39:43	18282	-31873	51	T	nn	0.0713	1.0248	18.6S	8.5E	86	166	84	02m24s
3398	170	-0577 Jul 09	21:33:45	18275	-31867	56	A	nn	-0.0717	0.9993	19.5N	68.4W	86	10	2	00m04s
3399	170	-0576 Jan 04	02:58:21	18268	-31861	61	A	p-	0.8066	0.9611	29.9N	154.7W	36	171	238	04m17s
3400	170	-0576 Jun 28	09:45:01	18260	-31855	66	T	p-	-0.8206	1.0478	31.4S	103.9E	35	6	279	04m18s
3401	171	-0576 Nov 23	10:31:23	18254	-31850	33	P	-t	-1.5242	0.0804	68.9S	60.4W	0	158		
3402	171	-0576 Dec 23	05:51:59	18253	-31849	71	P	t-	1.5407	0.0497	66.0N	159.3E	0	171		
3403	171	-0575 May 19	19:52:45	18247	-31844	38	T	-t	0.9700	1.0606	79.3N	160.2W	13	61	865	03m06s
3404	171	-0575 Jun 18	02:27:50	18245	-31843	76	Pb	t-	-1.5192	0.0156	66.7S	143.8W	0	1		
3405	171	-0575 Nov 12	10:34:31	18239	-31838	43	A	-p	-0.8298	0.9336	69.8S	65.1E	34	34	446	05m15s
3406	171	-0574 May 09	10:47:35	18232	-31832	48	T	-n	0.2358	1.0291	28.7N	89.0E	76	166	101	02m46s
3407	171	-0574 Nov 01	17:09:53	18225	-31826	53	A	nn	-0.1006	0.9947	18.4S	6.1W	84	15	19	00m33s
3408	171	-0573 Apr 28	19:16:00	18217	-31820	58	A	p-	-0.5575	0.9713	21.1S	25.2W	56	344	123	03m19s
3409	171	-0573 Oct 22	06:24:20	18210	-31814	63	T	p-	0.5830	1.0387	25.5N	167.6E	54	198	159	03m24s
3410	171	-0572 Mar 18	05:22:04	18203	-31809	30	Pe	-t	1.5567	0.0129	71.7N	98.9E	0	101		
3411	171	-0572 Apr 16	21:05:31	18202	-31808	68	P	t-	-1.3355	0.3886	71.4S	5.1E	0	295		
3412	171	-0572 Sep 11	12:09:48	18196	-31803	35	P	-t	-1.4156	0.2284	71.4S	4.9E	0	67		
3413	171	-0572 Oct 10	22:16:56	18195	-31802	73	P	t-	1.2458	0.5456	71.6N	4.8W	0	255		
3414	171	-0571 Mar 07	08:33:18	18189	-31797	40	A	-p	0.7802	0.9728	41.2N	113.7E	38	158	154	02m36s
3415	171	-0571 Aug 31	23:33:40	18181	-31791	45	A	-p	-0.7536	0.9796	36.1S	111.8W	41	19	109	02m01s
3416	171	-0570 Feb 24	18:48:18	18174	-31785	50	T	nn	0.0115	1.0313	11.2S	22.2W	89	163	106	03m04s
3417	171	-0570 Aug 21	03:50:19	18167	-31779	55	A	nn	-0.0286	0.9437	13.3N	161.7W	88	14	208	07m18s
3418	171	-0569 Feb 14	10:10:41	18159	-31773	60	T	p-	-0.6896	1.0551	57.4S	124.1E	46	337	251	03m45s
3419	171	-0569 Aug 10	03:51:16	18152	-31767	65	A	p-	0.7045	0.9375	62.0N	146.5W	45	200	328	05m34s
3420	171	-0568 Jan 05	15:00:23	18146	-31762	32	P	-t	1.3194	0.4109	66.5N	36.8E	0	184		

Cat Num	Canon Plate	Calendar Date	TD of Greatest Eclipse	ΔT s	Luna Num	Saros Num	Ecl. Type	QLE	Gamma	Ecl. Mag.	Lat. °	Long. °	Sun Alt °	Sun Azm °	Path Width km	Central Line Dur.
3421	172	-0568 Feb 04	02:12:07	18144	-31761	70	P	t-	-1.3889	0.2776	69.2S	17.8E	0	208		
3422	172	-0568 Jun 29	19:48:30	18138	-31756	37	P	-t	-1.2998	0.4458	65.7S	29.5W	0	348		
3423	172	-0568 Jul 29	07:03:55	18137	-31755	75	P	t-	1.4081	0.2559	68.4N	47.4W	0	344		
3424	172	-0568 Dec 24	23:13:44	18131	-31750	42	A	-p	0.6726	0.9470	18.4N	87.3W	48	186	263	06m37s
3425	172	-0567 Jun 19	09:42:08	18124	-31744	47	T	-p	-0.4997	1.0659	6.2S	113.5E	60	351	248	06m17s
3426	172	-0567 Dec 14	00:18:30	18116	-31738	52	A	nn	-0.0084	0.9182	23.7S	109.7W	89	9	311	10m42s
3427	172	-0566 Jun 09	02:53:49	18109	-31732	57	T	n-	0.2456	1.0790	36.2N	154.2W	76	164	263	06m15s
3428	172	-0566 Dec 02	23:35:05	18101	-31726	62	A	p-	-0.6819	0.9265	61.5S	123.5W	47	33	377	06m20s
3429	172	-0565 May 29	18:42:22	18094	-31720	67	P	t-	1.0212	0.9676	63.1N	164.5W	0	39		
3430	172	-0565 Oct 23	15:04:39	18088	-31715	34	P	-t	1.4784	0.1212	61.2N	98.9E	0	250		
3431	172	-0565 Nov 22	04:22:19	18087	-31714	72	P	t-	-1.3240	0.4055	62.6S	55.8E	0	135		
3432	172	-0564 Apr 18	15:50:40	18080	-31709	39	A	-t	-0.8967	0.9523	46.2S	54.7E	26	318	382	04m17s
3433	172	-0564 Oct 12	05:35:35	18073	-31703	44	T	-p	0.7664	1.0435	37.9N	163.1W	40	217	221	03m13s
3434	172	-0563 Apr 07	16:37:26	18066	-31697	49	A	nn	-0.1488	0.9445	3.2S	10.9E	81	331	207	06m20s
3435	172	-0563 Oct 01	21:26:14	18058	-31691	54	T	-n	0.0905	1.0402	3.6N	65.5W	85	209	135	03m25s
3436	172	-0562 Mar 27	18:46:43	18051	-31685	59	A	p-	0.6023	0.9695	31.8N	45.0W	53	145	134	02m53s
3437	172	-0562 Sep 21	09:55:50	18044	-31679	64	A	p-	-0.6395	0.9887	31.0S	83.9E	50	34	51	01m00s
3438	172	-0561 Feb 15	17:52:30	18038	-31674	31	P	-t	-1.3409	0.3679	61.9S	111.2E	0	237		
3439	172	-0561 Mar 17	03:42:26	18036	-31673	69	P	t-	1.3048	0.4368	60.8N	120.1E	0	99		
3440	172	-0561 Aug 11	23:14:48	18030	-31668	36	Pe	-t	1.5148	0.0861	62.3N	34.2E	0	311		
3441	173	-0561 Sep 10	15:34:01	18029	-31667	74	P	t-	-1.4101	0.2622	61.0S	54.5W	0	74		
3442	173	-0560 Feb 05	09:48:53	18023	-31662	41	T	-p	-0.6358	1.0536	53.5S	135.4E	50	325	229	03m37s
3443	173	-0560 Jul 30	23:29:54	18015	-31656	46	A	-p	0.7805	0.9456	66.1N	59.8W	38	226	320	04m16s
3444	173	-0559 Jan 25	01:05:11	18008	-31650	51	T	nn	0.0540	1.0248	17.7S	118.1W	87	162	85	02m22s
3445	173	-0559 Jul 20	04:40:58	18001	-31644	56	H	nn	-0.0026	1.0006	22.3N	175.0W	90	76	2	00m04s
3446	173	-0558 Jan 14	11:10:14	17993	-31638	61	A	p-	0.7966	0.9606	29.2N	78.3E	37	166	234	04m19s
3447	173	-0558 Jul 09	17:09:51	17986	-31632	66	T	p-	-0.7494	1.0504	24.7S	10.9W	41	10	252	04m39s
3448	173	-0558 Dec 04	18:33:02	17980	-31627	33	P	-t	-1.5277	0.0747	67.9S	166.4E	0	170		
3449	173	-0557 Jan 03	13:48:54	17979	-31626	71	P	t-	1.5329	0.0626	65.0N	28.7E	0	161		
3450	173	-0557 May 31	03:14:30	17973	-31621	38	P	-t	1.0459	0.9331	68.4N	41.6E	0	16		
3451	173	-0557 Jun 29	09:56:05	17971	-31620	76	P	t-	-1.4503	0.1502	65.7S	92.4E	0	11		
3452	173	-0557 Nov 23	18:49:07	17965	-31615	43	A	-p	-0.8297	0.9354	74.0S	55.6W	34	30	434	04m57s
3453	173	-0556 May 19	17:51:44	17958	-31609	48	T	-p	0.3145	1.0258	36.6N	19.0W	71	169	92	02m22s
3454	173	-0556 Nov 12	01:45:24	17951	-31603	53	A	nn	-0.1020	0.9965	22.2S	135.7W	84	12	12	00m21s
3455	173	-0555 May 09	01:55:29	17943	-31597	58	A	p-	-0.4784	0.9704	12.7S	129.6W	61	347	120	03m38s
3456	173	-0555 Nov 01	15:09:28	17936	-31591	63	T	p-	0.5764	1.0385	21.4N	33.3E	55	195	157	03m30s
3457	173	-0554 Apr 28	03:33:20	17929	-31585	68	P	t-	-1.2569	0.5225	70.9S	106.5W	0	308		
3458	173	-0554 Sep 22	20:23:51	17922	-31580	35	P	-t	-1.4442	0.1769	71.7S	133.8W	0	81		
3459	173	-0554 Oct 22	06:56:52	17921	-31579	73	P	t-	1.2376	0.5609	71.2N	149.9W	0	241		
3460	173	-0553 Mar 18	15:53:22	17915	-31574	40	A	-p	0.8300	0.9778	49.2N	3.8W	34	152	140	01m56s
3461	174	-0553 Sep 12	07:16:43	17908	-31568	45	A	-p	-0.7916	0.9726	42.8S	126.6E	37	23	159	02m33s
3462	174	-0552 Mar 07	02:40:48	17900	-31562	50	Tm	nn	0.0541	1.0374	4.8S	142.8W	87	163	126	03m38s
3463	174	-0552 Aug 31	11:01:31	17893	-31556	55	A	nn	-0.0767	0.9391	7.0N	88.2E	86	15	227	07m59s
3464	174	-0551 Feb 24	18:21:56	17886	-31550	60	T	p-	-0.6543	1.0593	50.7S	0.9W	49	336	257	04m13s
3465	174	-0551 Aug 20	10:51:35	17878	-31544	65	A	p-	0.6477	0.9365	54.0N	106.5E	49	201	310	06m03s
3466	174	-0550 Jan 15	23:27:20	17872	-31539	32	P	-t	1.3348	0.3829	67.6N	102.0W	0	173		
3467	174	-0550 Feb 14	10:27:15	17871	-31538	70	P	t-	-1.3651	0.3219	70.1S	119.5W	0	220		
3468	174	-0550 Jul 11	02:58:25	17865	-31533	37	P	-t	-1.3642	0.3277	66.7S	148.7W	0	358		
3469	174	-0550 Aug 09	14:17:48	17864	-31532	75	P	t-	1.3463	0.3651	69.4N	168.8W	0	332		
3470	174	-0549 Jan 05	07:22:57	17858	-31527	42	A	-p	0.6837	0.9460	19.8N	147.2E	47	182	273	06m47s
3471	174	-0549 Jun 30	17:07:35	17850	-31521	47	T	-p	-0.5701	1.0657	10.9S	0.3W	55	355	261	06m16s
3472	174	-0549 Dec 25	08:15:49	17843	-31515	52	A	nn	-0.0002	0.9189	23.9S	131.3E	90	0	308	10m49s
3473	174	-0548 Jun 19	10:18:38	17836	-31509	57	T	n-	0.1712	1.0772	33.3N	96.2E	80	169	254	06m22s
3474	174	-0548 Dec 13	07:43:37	17828	-31503	62	A	p-	-0.6766	0.9291	64.2S	119.6E	47	25	362	06m05s
3475	174	-0547 Jun 09	01:51:23	17821	-31497	67	T	t-	0.9482	1.0266	76.4N	118.5E	18	69	292	01m28s
3476	174	-0547 Nov 02	23:41:55	17815	-31492	34	P	-t	1.4798	0.1182	61.7N	40.4W	0	241		
3477	174	-0547 Dec 02	12:54:50	17814	-31491	72	P	t-	-1.3193	0.4135	63.4S	82.7W	0	145		
3478	174	-0546 Apr 29	22:26:02	17808	-31486	39	A	-t	-0.9788	0.9481	55.2S	34.6W	11	311	954	04m20s
3479	174	-0546 Oct 23	14:18:31	17800	-31480	44	T	-p	0.7755	1.0418	35.3N	62.7E	39	214	216	03m11s
3480	174	-0545 Apr 18	23:09:10	17793	-31474	49	A	nn	-0.2280	0.9470	3.3S	87.2W	77	332	199	06m07s

Cat Num	Canon Plate	Calendar Date	TD of Greatest Eclipse	ΔT s	Luna Num	Saros Num	Ecl. Type	QLE	Gamma	Ecl. Mag.	Lat. °	Long. °	Sun Alt °	Sun Azm °	Path Width km	Central Line Dur.
3481	175	-0545 Oct 13	05:57:57	17786	-31468	54	T	-n	0.1060	1.0355	0.0S	165.2E	84	209	120	03m04s
3482	175	-0544 Apr 07	01:45:15	17778	-31462	59	A	p-	0.5332	0.9758	32.0N	149.2W	58	145	100	02m15s
3483	175	-0544 Oct 01	18:01:19	17771	-31456	64	A	p-	-0.6190	0.9830	33.8S	39.4W	52	36	75	01m29s
3484	175	-0543 Feb 26	01:57:02	17765	-31451	31	P	-t	-1.3735	0.3058	61.4S	20.0W	0	246		
3485	175	-0543 Mar 27	11:15:28	17764	-31450	69	P	t-	1.2463	0.5450	60.7N	3.0W	0	90		
3486	175	-0543 Sep 20	23:06:09	17757	-31444	74	P	t-	-1.3799	0.3148	60.7S	177.3W	0	83		
3487	175	-0542 Feb 15	18:09:16	17750	-31439	41	T	-p	-0.6648	1.0558	51.0S	12.2E	48	321	245	03m45s
3488	175	-0542 Aug 11	06:25:39	17743	-31433	46	A	-p	0.8418	0.9435	65.6N	152.8W	32	238	384	04m17s
3489	175	-0541 Feb 05	09:23:58	17736	-31427	51	T	nn	0.0317	1.0250	16.3S	116.8E	88	157	85	02m20s
3490	175	-0541 Jul 31	11:56:09	17729	-31421	56	H	nn	0.0611	1.0015	23.8N	76.7E	86	198	5	00m09s
3491	175	-0540 Jan 25	19:12:59	17721	-31415	61	A	p-	0.7801	0.9606	28.6N	46.1W	39	161	225	04m15s
3492	175	-0540 Jul 20	00:41:28	17714	-31409	66	T	p-	-0.6831	1.0520	19.8S	126.5W	47	14	234	04m48s
3493	175	-0540 Dec 15	02:32:19	17708	-31404	33	P	-t	-1.5331	0.0654	66.8S	34.4E	0	181		
3494	175	-0539 Jan 13	21:37:44	17707	-31403	71	P	t-	1.5187	0.0859	64.0N	99.4W	0	151		
3495	175	-0539 Jun 10	10:37:25	17701	-31398	38	P	-t	1.1203	0.7889	67.4N	81.5W	0	5		
3496	175	-0539 Jul 09	17:29:47	17699	-31397	76	P	t-	-1.3859	0.2759	64.8S	32.3W	0	21		
3497	175	-0539 Dec 04	03:03:45	17693	-31392	43	A	-p	-0.8304	0.9378	77.6S	170.2W	34	18	419	04m39s
3498	175	-0538 May 31	00:54:35	17686	-31386	48	T	-p	0.3934	1.0218	44.0N	125.4W	67	172	81	01m55s
3499	175	-0538 Nov 23	10:23:35	17679	-31380	53	A	nn	-0.1021	0.9989	25.3S	94.4E	84	8	4	00m06s
3500	175	-0537 May 20	08:29:05	17671	-31374	58	A	p-	-0.3950	0.9690	4.7S	128.1E	67	351	121	03m58s
3501	176	-0537 Nov 12	23:59:29	17664	-31368	63	T	p-	0.5735	1.0387	18.0N	102.1W	55	192	158	03m36s
3502	176	-0536 May 08	09:54:52	17657	-31362	68	P	t-	-1.1725	0.6670	70.2S	144.0E	0	321		
3503	176	-0536 Oct 03	04:46:01	17651	-31357	35	P	-t	-1.4657	0.1389	71.7S	85.3E	0	95		
3504	176	-0536 Nov 01	15:42:22	17650	-31356	73	P	t-	1.2343	0.5669	70.6N	64.1E	0	228		
3505	176	-0535 Mar 28	23:05:20	17644	-31351	40	A	-p	0.8873	0.9823	58.4N	122.9W	27	144	136	01m23s
3506	176	-0535 Sep 22	15:08:15	17636	-31345	45	A	-p	-0.8218	0.9659	49.3S	2.6E	34	28	214	03m01s
3507	176	-0534 Mar 18	10:26:34	17629	-31339	50	T	nn	0.1029	1.0433	2.1N	97.8E	84	162	145	04m08s
3508	176	-0534 Sep 11	18:20:12	17622	-31333	55	A	nn	-0.1181	0.9346	0.6N	24.0W	83	17	245	08m34s
3509	176	-0533 Mar 08	02:26:22	17614	-31327	60	T	p-	-0.6132	1.0635	43.5S	125.3W	52	337	262	04m44s
3510	176	-0533 Aug 31	18:00:44	17607	-31321	65	A	p-	0.5977	0.9353	46.5N	3.4W	53	202	299	06m34s
3511	176	-0532 Jan 27	07:46:52	17601	-31316	32	P	-t	1.3558	0.3448	68.6N	120.5E	0	161		
3512	176	-0532 Feb 25	18:33:18	17600	-31315	70	P	t-	-1.3347	0.3787	70.8S	104.8E	0	233		
3513	176	-0532 Jul 21	10:14:50	17594	-31310	37	P	-t	-1.4240	0.2175	67.7S	90.1E	0	9		
3514	176	-0532 Aug 19	21:41:26	17593	-31309	75	P	t-	1.2913	0.4624	70.2N	66.7E	0	320		
3515	176	-0531 Jan 15	15:25:39	17587	-31304	42	A	-p	0.6993	0.9455	22.3N	23.1E	45	177	281	06m47s
3516	176	-0531 Jul 11	00:38:17	17579	-31298	47	T	-p	-0.6366	1.0648	16.4S	116.0W	50	359	275	06m05s
3517	176	-0530 Jan 04	16:08:23	17572	-31292	52	A	nn	0.0118	0.9202	22.9S	13.4E	89	182	302	10m50s
3518	176	-0530 Jun 30	17:46:57	17565	-31286	57	T	nn	0.1000	1.0748	29.6N	15.0W	84	175	244	06m25s
3519	176	-0530 Dec 24	15:50:22	17557	-31280	62	A	p-	-0.6692	0.9323	65.4S	5.3E	48	15	342	05m49s
3520	176	-0529 Jun 20	08:59:47	17550	-31274	67	T	p-	0.8754	1.0253	80.1N	59.7E	29	119	180	01m30s
3521	177	-0529 Nov 14	08:23:11	17544	-31269	34	P	-t	1.4787	0.1195	62.2N	179.2E	0	231		
3522	177	-0529 Dec 13	21:27:10	17543	-31268	72	P	t-	-1.3138	0.4228	64.4S	138.4E	0	155		
3523	177	-0528 May 10	04:56:01	17537	-31263	39	P	-t	-1.0641	0.8568	61.8S	124.6W	0	305		
3524	177	-0528 Nov 02	23:05:46	17530	-31257	44	T	-p	0.7813	1.0406	33.0N	72.9W	38	210	213	03m13s
3525	177	-0527 Apr 29	05:36:29	17522	-31251	49	A	np	-0.3108	0.9492	4.1S	175.7E	72	334	195	06m01s
3526	177	-0527 Oct 23	14:35:17	17515	-31245	54	T	-n	0.1160	1.0313	3.7S	34.4E	83	207	106	02m46s
3527	177	-0526 Apr 18	08:39:23	17508	-31239	59	A	p-	0.4593	0.9818	32.2N	108.1E	62	147	72	01m41s
3528	177	-0526 Oct 13	02:12:29	17501	-31233	64	A	p-	-0.6041	0.9777	37.3S	164.0W	53	37	98	01m56s
3529	177	-0525 Mar 09	09:53:41	17495	-31228	31	P	-t	-1.4130	0.2300	61.1S	149.1W	0	255		
3530	177	-0525 Apr 07	18:42:21	17493	-31227	69	P	t-	1.1819	0.6655	60.8N	124.5W	0	81		
3531	177	-0525 Oct 02	06:46:28	17486	-31221	74	P	t-	-1.3564	0.3552	60.7S	57.8E	0	92		
3532	177	-0524 Feb 27	02:19:09	17480	-31216	41	T	-p	-0.7025	1.0578	48.9S	108.9W	45	318	265	03m52s
3533	177	-0524 Aug 21	13:32:20	17473	-31210	46	A	-p	0.8945	0.9412	64.2N	109.3E	26	248	484	04m20s
3534	177	-0523 Feb 15	17:32:20	17466	-31204	51	T	nn	0.0013	1.0252	14.6S	5.0W	90	52	[illegible]	02m19s
3535	177	-0523 Aug 10	19:20:49	17458	-31198	56	H	nn	0.1176	1.0020	24.1N	34.0W	83	202	7	00m12s
3536	177	-0522 Feb 05	03:06:44	17451	-31192	61	A	p-	0.7570	0.9609	28.2N	167.8W	41	157	212	04m08s
3537	177	-0522 Jul 31	08:20:05	17444	-31186	66	T	p-	-0.6223	1.0527	16.5S	116.6E	51	18	221	04m49s
3538	177	-0522 Dec 26	10:30:10	17438	-31181	33	P	-t	-1.5394	0.0543	65.7S	96.7W	0	192		
3539	177	-0521 Jan 25	05:20:10	17437	-31180	71	P	t-	1.4997	0.1170	63.1N	134.4E	0	141		
3540	177	-0521 Jun 21	17:58:59	17431	-31175	38	P	-t	1.1951	0.6444	66.4N	156.2E	0	355		

Cat Num	Canon Plate	Calendar Date	TD of Greatest Eclipse	ΔT s	Luna Num	Saros Num	Ecl. Type	QLE	Gamma	Ecl. Mag.	Lat. °	Long. °	Sun Alt °	Sun Azm °	Path Width km	Central Line Dur.
3541	178	-0521 Jul 21	01:06:17	17429	-31174	76	P	t-	-1.3239	0.3964	63.9S	157.4W	0	31		
3542	178	-0521 Dec 15	11:19:54	17423	-31169	43	A	-p	-0.8304	0.9409	79.7S	84.1E	34	358	397	04m20s
3543	178	-0520 Jun 10	07:54:41	17416	-31163	48	T	-p	0.4737	1.0170	51.0N	130.6E	61	177	66	01m26s
3544	178	-0520 Dec 03	19:03:27	17409	-31157	53	Hm	nn	-0.1014	1.0019	27.7S	35.3W	84	4	7	00m11s
3545	178	-0519 May 30	15:00:44	17402	-31151	58	A	pn	-0.3107	0.9671	2.8N	27.1E	72	354	125	04m19s
3546	178	-0519 Nov 23	08:51:54	17394	-31145	63	T	p-	0.5724	1.0393	15.2N	122.0E	55	188	160	03m44s
3547	178	-0518 May 19	16:14:51	17387	-31139	68	P	t-	-1.0865	0.8144	69.3S	35.5E	0	333		
3548	178	-0518 Oct 14	13:12:47	17381	-31134	35	P	-t	-1.4830	0.1090	71.5S	56.6W	0	109		
3549	178	-0518 Nov 13	00:30:08	17380	-31133	73	P	t-	1.2334	0.5684	69.8N	81.8W	0	215		
3550	178	-0517 Apr 09	06:13:34	17374	-31128	40	A	-t	0.9484	0.9858	68.4N	108.2E	18	125	161	00m59s
3551	178	-0517 Oct 03	23:05:52	17367	-31122	45	A	-p	-0.8458	0.9597	55.4S	123.3W	32	33	273	03m23s
3552	178	-0516 Mar 28	18:05:08	17360	-31116	50	T	nn	0.1583	1.0489	9.5N	20.1W	81	162	164	04m34s
3553	178	-0516 Sep 22	01:47:59	17352	-31110	55	A	nn	-0.1513	0.9305	5.5S	138.5W	81	18	263	09m04s
3554	178	-0515 Mar 18	10:22:33	17345	-31104	60	T	p-	-0.5654	1.0673	36.1S	111.8E	55	338	265	05m15s
3555	178	-0515 Sep 11	01:20:22	17338	-31098	65	A	p-	0.5559	0.9343	39.4N	116.1W	56	202	293	07m03s
3556	178	-0514 Feb 06	15:56:48	17332	-31093	32	P	-t	1.3837	0.2942	69.6N	15.2W	0	149		
3557	178	-0514 Mar 08	02:30:09	17331	-31092	70	P	t-	-1.2977	0.4481	71.3S	29.1W	0	246		
3558	178	-0514 Aug 01	17:40:32	17325	-31087	37	P	-t	-1.4768	0.1200	68.6S	34.0W	0	20		
3559	178	-0514 Aug 31	05:16:01	17323	-31086	75	P	t-	1.2443	0.5456	71.0N	61.2W	0	307		
3560	178	-0513 Jan 26	23:19:08	17317	-31081	42	A	-p	0.7217	0.9452	25.9N	99.1W	44	173	291	06m38s
3561	179	-0513 Jul 22	08:14:07	17310	-31075	47	T	-p	-0.6990	1.0630	22.4S	126.2E	45	4	289	05m44s
3562	179	-0512 Jan 15	23:54:16	17303	-31069	52	A	nn	0.0290	0.9222	20.7S	103.2W	88	177	294	10m41s
3563	179	-0512 Jul 11	01:18:21	17296	-31063	57	Tm	nn	0.0320	1.0714	25.2N	127.7W	88	179	232	06m21s
3564	179	-0511 Jan 03	23:52:51	17289	-31057	62	A	p-	-0.6579	0.9362	64.9S	107.1W	49	4	317	05m31s
3565	179	-0511 Jun 30	16:09:58	17281	-31051	67	T	p-	0.8051	1.0223	77.0N	9.2W	36	160	129	01m26s
3566	179	-0511 Nov 24	17:08:19	17275	-31046	34	P	-t	1.4749	0.1257	63.0N	37.6E	0	222		
3567	179	-0511 Dec 24	05:58:52	17274	-31045	72	P	t-	-1.3076	0.4335	65.4S	0.6W	0	165		
3568	179	-0510 May 21	11:21:36	17268	-31040	39	P	-t	-1.1519	0.7048	62.5S	128.8E	0	314		
3569	179	-0510 Nov 14	07:57:17	17261	-31034	44	T	-p	0.7837	1.0397	31.0N	150.2E	38	205	211	03m16s
3570	179	-0509 May 10	12:00:03	17254	-31028	49	A	-p	-0.3967	0.9510	5.8S	79.4E	67	337	194	05m57s
3571	179	-0509 Nov 03	23:17:35	17247	-31022	54	T	-n	0.1217	1.0275	7.3S	97.6W	83	205	94	02m29s
3572	179	-0508 Apr 28	15:29:32	17239	-31016	59	A	p-	0.3809	0.9874	32.1N	6.9E	67	150	48	01m11s
3573	179	-0508 Oct 23	10:29:43	17232	-31010	64	A	p-	-0.5949	0.9727	41.2S	70.1E	53	37	120	02m20s
3574	179	-0507 Mar 19	17:44:28	17226	-31005	31	P	-t	-1.4579	0.1431	60.9S	83.3E	0	264		
3575	179	-0507 Apr 18	02:06:04	17225	-31004	69	P	t-	1.1138	0.7942	61.0N	114.7E	0	72		
3576	179	-0507 Oct 12	14:32:48	17218	-30998	74	P	t-	-1.3381	0.3867	60.8S	68.5W	0	101		
3577	179	-0506 Mar 09	10:22:57	17212	-30993	41	T	-p	-0.7454	1.0594	47.3S	130.9E	42	316	288	03m59s
3578	179	-0506 Sep 01	20:49:02	17205	-30987	46	A	-p	0.9396	0.9387	62.7N	7.5E	20	255	664	04m25s
3579	179	-0505 Feb 27	01:32:20	17197	-30981	51	T	nn	-0.0357	1.0254	12.7S	126.3W	88	334	86	02m18s
3580	179	-0505 Aug 22	02:54:54	17190	-30975	56	Hm	nn	0.1674	1.0023	23.3N	147.1W	80	206	8	00m13s
3581	180	-0504 Feb 16	10:50:13	17183	-30969	61	A	p-	0.7261	0.9616	28.0N	73.6E	43	153	197	03m59s
3582	180	-0504 Aug 10	16:07:21	17176	-30963	66	T	p-	-0.5679	1.0528	14.8S	2.1W	55	21	210	04m43s
3583	180	-0503 Jan 05	18:21:11	17170	-30958	33	P	-t	-1.5515	0.0332	64.7S	134.4E	0	202		
3584	180	-0503 Feb 04	12:51:39	17169	-30957	71	P	t-	1.4717	0.1630	62.3N	11.2E	0	131		
3585	180	-0503 Jul 02	01:24:31	17163	-30952	38	P	-t	1.2662	0.5077	65.4N	33.3E	0	345		
3586	180	-0503 Jul 31	08:50:27	17162	-30951	76	P	t-	-1.2681	0.5044	63.0S	76.0E	0	40		
3587	180	-0503 Dec 25	19:33:10	17156	-30946	43	A	-p	-0.8340	0.9446	79.6S	16.0W	33	333	374	03m59s
3588	180	-0502 Jun 21	14:54:51	17148	-30940	48	H	-p	0.5529	1.0116	57.3N	28.9E	56	184	48	00m56s
3589	180	-0502 Dec 15	03:42:43	17141	-30934	53	H	nn	-0.1022	1.0054	29.2S	164.5W	84	360	19	00m32s
3590	180	-0501 Jun 10	21:30:01	17134	-30928	58	A	nn	-0.2247	0.9647	9.6N	72.5W	77	358	131	04m40s
3591	180	-0501 Dec 04	17:44:58	17127	-30922	63	T	p-	0.5717	1.0404	13.1N	14.0W	55	184	165	03m53s
3592	180	-0500 May 29	22:32:09	17118	-30916	68	A-	t-	-0.9975	0.9670	68.4S	71.7W	0	344	-	-
3593	180	-0500 Oct 24	21:46:30	17111	-30911	35	P	-t	-1.4943	0.0902	71.0S	160.0E	0	123		
3594	180	-0500 Nov 23	09:20:49	17110	-30910	73	P	t-	1.2353	0.5650	68.8N	132.1E	0	202		
3595	180	-0499 Apr 19	13:16:39	17102	-30905	40	P	-t	1.0144	0.9617	71.2N	65.8W	0	61		
3596	180	-0499 Oct 14	07:09:56	17094	-30899	45	A	-p	-0.8637	0.9540	61.2S	109.1E	30	38	333	03m40s
3597	180	-0498 Apr 09	01:38:45	17085	-30893	50	T	-n	0.2186	1.0540	17.1N	136.8W	77	162	183	04m54s
3598	180	-0498 Oct 03	09:23:57	17076	-30887	55	A	nn	-0.1775	0.9266	11.4S	105.1E	80	18	280	09m29s
3599	180	-0497 Mar 29	18:11:18	17067	-30881	60	T	p-	-0.5113	1.0708	28.4S	9.6W	59	339	267	05m47s
3600	180	-0497 Sep 22	08:50:00	17059	-30875	65	A	p-	0.5216	0.9332	32.9N	128.4E	58	201	290	07m32s

Cat Num	Canon Plate	Calendar Date	TD of Greatest Eclipse	ΔT s	Luna Num	Saros Num	Ecl. Type	QLE	Gamma	Ecl. Mag.	Lat. °	Long. °	Sun Alt °	Sun Azm °	Path Width km	Central Line Dur.
3601	181	-0496 Feb 17	23:58:10	17052	-30870	32	P	-t	1.4180	0.2319	70.4N	149.4W	0	136		
3602	181	-0496 Mar 18	10:18:15	17050	-30869	70	P	t-	-1.2540	0.5302	71.6S	161.1W	0	260		
3603	181	-0496 Aug 12	01:14:42	17043	-30864	37	Pe	-t	-1.5237	0.0335	69.5S	160.8W	0	32		
3604	181	-0496 Sep 10	13:00:50	17041	-30863	75	P	t-	1.2046	0.6161	71.5N	167.8E	0	294		
3605	181	-0495 Feb 06	07:03:27	17034	-30858	42	A	-p	0.7509	0.9451	30.7N	140.3E	41	168	305	06m22s
3606	181	-0495 Aug 01	15:57:30	17026	-30852	47	T	-p	-0.7555	1.0607	29.0S	5.6E	41	8	304	05m17s
3607	181	-0494 Jan 26	07:33:51	17017	-30846	52	A	nn	0.0513	0.9245	17.4S	141.3E	87	173	285	10m24s
3608	181	-0494 Jul 22	08:53:54	17008	-30840	57	T	nn	-0.0322	1.0674	20.2N	117.6E	88	4	220	06m11s
3609	181	-0493 Jan 15	07:51:11	17000	-30834	62	A	p-	-0.6424	0.9407	62.5S	140.1E	50	355	288	05m12s
3610	181	-0493 Jul 11	23:22:27	16991	-30828	67	T	p-	0.7375	1.0183	71.0N	100.9W	42	179	93	01m16s
3611	181	-0493 Dec 06	01:53:20	16984	-30823	34	P	-t	1.4722	0.1296	63.8N	104.3W	0	212		
3612	181	-0492 Jan 04	14:25:59	16982	-30822	72	P	t-	-1.2968	0.4524	66.5S	139.0W	0	176		
3613	181	-0492 May 31	17:45:10	16975	-30817	39	P	-t	-1.2404	0.5518	63.3S	22.4E	0	323		
3614	181	-0492 Nov 24	16:50:18	16966	-30811	44	T	-p	0.7851	1.0395	29.4N	12.7E	38	200	211	03m21s
3615	181	-0491 May 20	18:22:59	16958	-30805	49	A	-p	-0.4836	0.9522	8.6S	17.2W	61	340	197	05m58s
3616	181	-0491 Nov 14	08:01:38	16949	-30799	54	T	-n	0.1254	1.0241	10.5S	130.1E	83	202	83	02m15s
3617	181	-0490 May 09	22:18:35	16941	-30793	59	A	p-	0.3004	0.9924	31.6N	94.0W	72	153	28	00m43s
3618	181	-0490 Nov 03	18:51:11	16932	-30787	64	A	p-	-0.5900	0.9682	45.4S	56.3W	54	36	140	02m42s
3619	181	-0489 Mar 31	01:26:56	16925	-30782	31	Pe	-t	-1.5100	0.0418	60.9S	42.3W	0	273		
3620	181	-0489 Apr 29	09:24:17	16924	-30781	69	P	t-	1.0402	0.9342	61.4N	5.0W	0	64		
3621	182	-0489 Oct 23	22:26:39	16915	-30775	74	P	t-	-1.3260	0.4072	61.1S	163.1E	0	110		
3622	182	-0488 Mar 19	18:16:39	16908	-30770	41	T	-p	-0.7965	1.0603	46.6S	13.2E	37	315	321	04m03s
3623	182	-0488 Sep 12	04:17:06	16899	-30764	46	A	-p	0.9761	0.9359	61.6N	97.0W	12	262	1132	04m30s
3624	182	-0487 Mar 09	09:21:28	16891	-30758	51	T	nn	-0.0810	1.0253	10.9S	115.8E	85	332	86	02m17s
3625	182	-0487 Sep 01	10:39:47	16882	-30752	56	H	nn	0.2091	1.0023	21.5N	96.4E	78	208	8	00m13s
3626	182	-0486 Feb 26	18:24:09	16874	-30746	61	A	p-	0.6880	0.9624	28.0N	42.2W	46	150	182	03m48s
3627	182	-0486 Aug 22	00:01:27	16865	-30740	66	T	p-	-0.5188	1.0522	14.2S	122.5W	59	25	200	04m33s
3628	182	-0485 Jan 17	02:08:17	16858	-30735	33	Pe	-t	-1.5666	0.0066	63.7S	6.6E	0	212		
3629	182	-0485 Feb 15	20:15:44	16857	-30734	71	P	t-	1.4379	0.2191	61.7N	110.1W	0	122		
3630	182	-0485 Jul 13	08:51:40	16849	-30729	38	P	-t	1.3356	0.3754	64.5N	89.8W	0	335		
3631	182	-0485 Aug 11	16:39:41	16848	-30728	76	P	t-	-1.2165	0.6036	62.3S	51.9W	0	50		
3632	182	-0484 Jan 06	03:43:31	16841	-30723	43	A	-p	-0.8405	0.9488	77.3S	121.4W	32	313	350	03m38s
3633	182	-0484 Jul 01	21:55:42	16832	-30717	48	H	-p	0.6304	1.0055	62.5N	69.9W	51	194	25	00m25s
3634	182	-0484 Dec 25	12:20:30	16824	-30711	53	H	nn	-0.1049	1.0095	29.9S	66.6E	84	355	33	00m55s
3635	182	-0483 Jun 21	04:00:44	16815	-30705	58	A	nn	-0.1407	0.9618	15.6N	171.8W	82	2	140	04m58s
3636	182	-0483 Dec 15	02:36:40	16807	-30699	63	T	p-	0.5698	1.0419	11.6N	149.8W	55	179	170	04m03s
3637	182	-0482 Jun 10	04:51:57	16798	-30693	68	A	t-	-0.9098	0.9417	43.3S	177.8E	24	357	524	06m26s
3638	182	-0482 Nov 05	06:23:22	16791	-30688	35	P	-t	-1.5027	0.0767	70.4S	16.1E	0	137		
3639	182	-0482 Dec 04	18:11:07	16790	-30687	73	P	t-	1.2373	0.5611	67.8N	13.4W	0	190		
3640	182	-0481 Apr 30	20:16:46	16783	-30682	40	P	-t	1.0836	0.8385	70.6N	174.6E	0	48		
3641	183	-0481 Oct 25	15:19:12	16774	-30676	45	A	-p	-0.8766	0.9489	66.7S	19.9W	28	42	391	03m54s
3642	183	-0480 Apr 19	09:07:31	16766	-30670	50	T	-n	0.2839	1.0585	25.0N	107.5E	73	162	201	05m07s
3643	183	-0480 Oct 13	17:07:20	16758	-30664	55	A	nn	-0.1973	0.9234	17.0S	13.0W	79	18	295	09m49s
3644	183	-0479 Apr 09	01:53:28	16749	-30658	60	T	p-	-0.4515	1.0737	20.6S	129.6W	63	341	267	06m16s
3645	183	-0479 Oct 02	16:30:01	16741	-30652	65	A	p-	0.4951	0.9324	26.8N	10.2E	60	200	289	07m57s
3646	183	-0478 Feb 28	07:49:20	16734	-30647	32	P	-t	1.4597	0.1563	71.1N	78.2E	0	123		
3647	183	-0478 Mar 29	17:57:45	16732	-30646	70	P	t-	-1.2039	0.6240	71.7S	68.7E	0	273		
3648	183	-0478 Sep 21	20:55:38	16724	-30640	75	P	t-	1.1720	0.6736	71.8N	33.7E	0	280		
3649	183	-0477 Feb 17	14:37:28	16717	-30635	42	A	-p	0.7877	0.9451	36.7N	21.4E	38	163	326	06m00s
3650	183	-0477 Aug 12	23:48:23	16708	-30629	47	T	-p	-0.8059	1.0578	35.8S	117.8W	36	13	321	04m46s
3651	183	-0476 Feb 06	15:03:05	16700	-30623	52	A	nn	0.0820	0.9274	12.9S	27.5E	85	169	274	10m00s
3652	183	-0476 Aug 01	16:35:33	16692	-30617	57	T	nn	-0.0907	1.0627	14.7N	0.6E	85	8	207	05m53s
3653	183	-0475 Jan 25	15:43:04	16683	-30611	62	A	p-	-0.6212	0.9459	58.5S	26.1E	51	348	255	04m52s
3654	183	-0475 Jul 22	06:39:02	16675	-30605	67	T	p-	0.6740	1.0136	64.3N	156.2E	47	188	64	01m00s
3655	183	-0475 Dec 16	10:38:23	16668	-30600	34	P	-t	1.4701	0.1323	64.8N	113.4E	0	202		
3656	183	-0474 Jan 14	22:49:31	16666	-30599	72	P	t-	-1.2825	0.4778	67.5S	82.9E	0	187		
3657	183	-0474 Jun 12	00:08:10	16659	-30594	39	P	-t	-1.3282	0.4005	64.1S	84.2W	0	332		
3658	183	-0474 Jul 11	14:09:39	16658	-30593	77	Pb	t-	1.4887	0.1193	66.8N	142.4W	0	1		
3659	183	-0474 Dec 06	01:44:51	16651	-30588	44	T	-p	0.7855	1.0397	28.4N	125.2W	38	195	213	03m27s
3660	183	-0473 Jun 01	00:43:55	16643	-30582	49	A	-p	-0.5724	0.9530	12.6S	113.8W	55	344	207	05m59s

Cat Num	Canon Plate	Calendar Date	TD of Greatest Eclipse	ΔT s	Luna Num	Saros Num	Ecl. Type	QLE	Gamma	Ecl. Mag.	Lat. °	Long. °	Sun Alt °	Sun Azm °	Path Width km	Central Line Dur.
3661	184	-0473 Nov 25	16:48:43	16634	-30576	54	T	-n	0.1262	1.0213	13.2S	2.7W	83	198	73	02m03s
3662	184	-0472 May 20	05:06:50	16626	-30570	59	A	nn	0.2182	0.9970	30.5N	165.4E	77	157	11	00m17s
3663	184	-0472 Nov 14	03:15:19	16618	-30564	64	A	p-	-0.5880	0.9643	49.5S	177.4E	54	34	158	03m01s
3664	184	-0471 May 09	16:41:49	16609	-30558	69	T	t-	0.9651	1.0386	67.3N	92.6W	15	84	509	02m06s
3665	184	-0471 Nov 03	06:24:52	16601	-30552	74	P	t-	-1.3180	0.4208	61.5S	33.5E	0	120		
3666	184	-0470 Mar 31	02:04:46	16594	-30547	41	T	-p	-0.8523	1.0604	47.1S	102.9W	31	314	372	04m03s
3667	184	-0470 Sep 23	11:54:07	16586	-30541	46	A+	-p	1.0061	0.9514	60.6N	163.4E	0	275	-	-
3668	184	-0469 Mar 20	17:02:24	16577	-30535	51	T	-n	-0.1325	1.0250	9.2S	0.2W	82	331	86	02m15s
3669	184	-0469 Sep 12	18:34:28	16569	-30529	56	H	-n	0.2437	1.0023	19.1N	23.0W	76	210	8	00m13s
3670	184	-0468 Mar 09	01:46:01	16561	-30523	61	A	p-	0.6406	0.9633	28.1N	154.3W	50	148	169	03m39s
3671	184	-0468 Sep 01	08:05:35	16553	-30517	66	T	p-	-0.4775	1.0513	14.9S	114.7E	61	27	191	04m21s
3672	184	-0467 Feb 26	03:28:46	16544	-30511	71	P	t-	1.3949	0.2908	61.2N	131.5E	0	113		
3673	184	-0467 Jul 23	16:24:32	16537	-30506	38	P	-t	1.3997	0.2544	63.6N	146.1E	0	326		
3674	184	-0467 Aug 22	00:36:58	16536	-30505	76	P	t-	-1.1716	0.6889	61.7S	178.4E	0	59		
3675	184	-0466 Jan 16	11:47:55	16529	-30500	43	A	-p	-0.8528	0.9533	74.1S	127.4E	31	301	328	03m16s
3676	184	-0466 Jul 13	04:59:52	16521	-30494	48	A	-p	0.7042	0.9989	66.2N	165.8W	45	207	5	00m05s
3677	184	-0465 Jan 05	20:54:17	16513	-30488	53	H	nn	-0.1117	1.0141	29.8S	61.1W	83	349	49	01m20s
3678	184	-0465 Jul 02	10:31:43	16504	-30482	58	Am	nn	-0.0573	0.9584	20.6N	89.7E	87	7	152	05m16s
3679	184	-0465 Dec 26	11:26:22	16496	-30476	63	T	p-	0.5660	1.0440	10.7N	75.1E	55	175	178	04m14s
3680	184	-0464 Jun 20	11:13:11	16488	-30470	68	A	p-	-0.8224	0.9439	32.0S	76.9E	34	2	365	06m51s
3681	185	-0464 Nov 15	15:03:43	16481	-30465	35	P	-t	-1.5080	0.0687	69.5S	128.0W	0	150		
3682	185	-0464 Dec 15	03:00:05	16480	-30464	73	P	t-	1.2387	0.5586	66.7N	158.0W	0	179		
3683	185	-0463 May 11	03:14:59	16473	-30459	40	P	-t	1.1553	0.7094	69.9N	56.0E	0	36		
3684	185	-0463 Jun 09	14:14:46	16471	-30458	78	Pb	t-	-1.5420	0.0174	67.4S	39.2E	0	354		
3685	185	-0463 Nov 04	23:33:41	16465	-30453	45	A	-p	-0.8847	0.9445	71.8S	149.3W	27	46	442	04m06s
3686	185	-0462 Apr 30	16:32:38	16456	-30447	50	T	-n	0.3529	1.0624	32.9N	6.8W	69	163	218	05m13s
3687	185	-0462 Oct 25	00:57:28	16448	-30441	55	A	nn	-0.2114	0.9206	22.1S	132.3W	78	17	307	10m05s
3688	185	-0461 Apr 20	09:30:02	16440	-30435	60	T	n-	-0.3870	1.0760	12.8S	112.0E	67	343	266	06m42s
3689	185	-0461 Oct 14	00:19:05	16432	-30429	65	A	p-	0.4756	0.9318	21.3N	110.1W	61	198	288	08m20s
3690	185	-0460 Mar 10	15:30:30	16425	-30424	32	Pe	-t	1.5085	0.0679	71.5N	52.1W	0	110		
3691	185	-0460 Apr 09	01:28:57	16423	-30423	70	P	t-	-1.1479	0.7291	71.5S	59.3W	0	287		
3692	185	-0460 Oct 02	05:00:36	16415	-30417	75	P	t-	1.1470	0.7180	71.9N	103.1W	0	265		
3693	185	-0459 Feb 27	22:02:12	16408	-30412	42	A	-p	0.8311	0.9449	43.8N	96.4W	33	158	362	05m35s
3694	185	-0459 Aug 23	07:46:54	16400	-30406	47	T	-p	-0.8502	1.0546	42.8S	115.9E	31	19	341	04m14s
3695	185	-0458 Feb 16	22:25:01	16392	-30400	52	A	nn	0.1187	0.9304	7.5S	85.0W	83	167	262	09m29s
3696	185	-0458 Aug 13	00:23:29	16384	-30394	57	T	-n	-0.1434	1.0577	9.0N	118.5W	82	11	192	05m29s
3697	185	-0457 Feb 05	23:28:11	16376	-30388	62	A	p-	-0.5934	0.9515	53.3S	88.8W	53	343	221	04m29s
3698	185	-0457 Aug 02	14:00:01	16368	-30382	67	H	p-	0.6149	1.0083	57.4N	47.8E	52	194	37	00m39s
3699	185	-0457 Dec 27	19:20:31	16361	-30377	34	P	-t	1.4710	0.1292	65.8N	28.5W	0	191		
3700	185	-0456 Jan 26	07:06:25	16360	-30376	72	P	t-	-1.2621	0.5146	68.6S	54.1W	0	198		
3701	186	-0456 Jun 22	06:32:48	16353	-30371	39	P	-t	-1.4136	0.2541	65.1S	168.5E	0	342		
3702	186	-0456 Jul 21	20:57:12	16351	-30370	77	P	t-	1.4221	0.2366	67.8N	103.5E	0	350		
3703	186	-0456 Dec 16	10:36:32	16345	-30365	44	T	-p	0.7883	1.0403	28.2N	97.7E	38	190	219	03m34s
3704	186	-0455 Jun 11	07:08:11	16336	-30359	49	A	-p	-0.6586	0.9531	17.7S	148.1E	49	348	226	05m58s
3705	186	-0455 Dec 06	01:34:59	16328	-30353	54	T	-n	0.1270	1.0191	15.2S	135.1W	83	194	66	01m53s
3706	186	-0454 May 31	11:55:38	16320	-30347	59	H	nn	0.1354	1.0010	28.6N	64.5E	82	162	3	00m06s
3707	186	-0454 Nov 25	11:40:41	16312	-30341	64	A	p-	-0.5878	0.9611	53.4S	52.1E	54	30	174	03m18s
3708	186	-0453 May 20	23:57:02	16304	-30335	69	T	t-	0.8874	1.0460	69.4N	175.5W	27	110	333	02m40s
3709	186	-0453 Nov 14	14:27:24	16296	-30329	74	P	t-	-1.3134	0.4287	62.1S	97.3W	0	129		
3710	186	-0452 Apr 10	09:43:44	16289	-30324	41	T	-t	-0.9156	1.0592	49.7S	145.0E	23	313	476	03m55s
3711	186	-0452 Oct 03	19:42:34	16281	-30318	46	P	-t	1.0275	0.9146	60.7N	36.4E	0	266		
3712	186	-0451 Mar 31	00:33:55	16273	-30312	51	T	-n	-0.1914	1.0243	7.9S	113.8W	79	330	84	02m12s
3713	186	-0451 Sep 23	02:38:36	16265	-30306	56	H	-n	0.2713	1.0024	16.1N	145.2W	74	210	9	00m13s
3714	186	-0450 Mar 20	08:59:06	16257	-30300	61	A	p-	0.5865	0.9641	28.6N	96.5E	54	146	157	03m31s
3715	186	-0450 Sep 12	16:17:47	16249	-30294	66	T	p-	-0.4426	1.0501	16.6S	10.0W	64	30	184	04m09s
3716	186	-0449 Mar 09	10:32:43	16241	-30288	71	P	t-	1.3445	0.3753	60.9N	15.5E	0	104		
3717	186	-0449 Aug 04	00:01:15	16234	-30283	38	P	-t	1.4601	0.1417	62.8N	21.2E	0	317		
3718	186	-0449 Sep 02	08:41:00	16233	-30282	76	P	t-	-1.1326	0.7625	61.2S	47.2E	0	68		
3719	186	-0448 Jan 27	19:47:31	16226	-30277	43	A	-p	-0.8698	0.9583	70.6S	12.6E	29	294	308	02m53s
3720	186	-0448 Jul 23	12:07:49	16218	-30271	48	A	-p	0.7738	0.9918	68.0N	100.3E	39	222	45	00m35s

Cat Num	Canon Plate	Calendar Date	TD of Greatest Eclipse	ΔT s	Luna Num	Saros Num	Ecl. Type	QLE	Gamma	Ecl. Mag.	Lat. °	Long. °	Sun Alt °	Sun Azm °	Path Width km	Central Line Dur.
3721	187	-0447 Jan 16	05:23:22	16210	-30265	53	T	nn	-0.1232	1.0191	29.0S	172.2E	83	345	66	01m46s
3722	187	-0447 Jul 12	17:07:51	16202	-30259	58	A	nn	0.0214	0.9546	24.4N	9.2W	89	190	166	05m31s
3723	187	-0446 Jan 05	20:12:29	16194	-30253	63	T	p-	0.5588	1.0465	10.2N	59.1W	56	170	186	04m24s
3724	187	-0446 Jul 01	17:38:34	16186	-30247	68	A	p-	-0.7375	0.9451	23.7S	23.4W	42	7	299	07m03s
3725	187	-0446 Nov 26	23:44:34	16179	-30242	35	P	-t	-1.5126	0.0618	68.5S	88.4E	0	162		
3726	187	-0446 Dec 26	11:45:57	16178	-30241	73	P	t-	1.2379	0.5601	65.7N	58.7E	0	168		
3727	187	-0445 May 22	10:13:27	16171	-30236	40	P	-t	1.2277	0.5777	69.0N	62.0W	0	24		
3728	187	-0445 Jun 20	20:59:47	16170	-30235	78	P	t-	-1.4589	0.1643	66.4S	74.1W	0	5		
3729	187	-0445 Nov 16	07:50:05	16163	-30230	45	A	-p	-0.8905	0.9409	76.8S	82.0E	27	49	487	04m15s
3730	187	-0444 May 10	23:55:15	16155	-30224	50	T	-p	0.4247	1.0654	40.9N	120.0W	65	165	236	05m11s
3731	187	-0444 Nov 04	08:52:49	16147	-30218	55	A	nn	-0.2212	0.9185	26.7S	107.7E	77	14	317	10m16s
3732	187	-0443 Apr 30	17:02:20	16139	-30212	60	T	n-	-0.3188	1.0774	5.1S	5.0W	71	346	264	07m01s
3733	187	-0443 Oct 24	08:15:20	16131	-30206	65	A	p-	0.4612	0.9318	16.4N	128.0E	62	196	286	08m38s
3734	187	-0442 Apr 20	08:53:18	16123	-30200	70	P	t-	-1.0866	0.8437	71.2S	174.6E	0	300		
3735	187	-0442 Oct 13	13:14:46	16115	-30194	75	P	t-	1.1285	0.7510	71.6N	117.9E	0	251		
3736	187	-0441 Mar 11	05:14:24	16109	-30189	42	A	-t	0.8841	0.9444	52.5N	146.4E	27	151	436	05m08s
3737	187	-0441 Sep 03	15:54:13	16101	-30183	47	T	-p	-0.8876	1.0509	49.9S	13.8W	27	25	366	03m41s
3738	187	-0440 Feb 28	05:36:11	16093	-30177	52	A	nn	0.1647	0.9337	1.2S	164.6E	81	165	250	08m54s
3739	187	-0440 Aug 23	08:18:57	16085	-30171	57	T	-n	-0.1895	1.0523	3.1N	120.2E	79	14	176	04m59s
3740	187	-0439 Feb 16	07:05:14	16077	-30165	62	A	p-	-0.5579	0.9575	47.1S	156.3E	56	341	186	04m04s
3741	188	-0439 Aug 12	21:27:24	16069	-30159	67	H	p-	0.5617	1.0026	50.4N	64.3W	56	197	11	00m13s
3742	188	-0438 Jan 07	04:00:16	16062	-30154	34	P	-t	1.4745	0.1211	66.8N	170.3W	0	180		
3743	188	-0438 Feb 05	15:18:18	16061	-30153	72	P	t-	-1.2368	0.5607	69.6S	169.6E	0	211		
3744	188	-0438 Jul 03	12:58:27	16055	-30148	39	Pe	-t	-1.4974	0.1112	66.1S	60.6E	0	352		
3745	188	-0438 Aug 02	03:48:23	16053	-30147	77	P	t-	1.3586	0.3475	68.8N	12.0W	0	339		
3746	188	-0438 Dec 27	19:27:14	16047	-30142	44	T	-p	0.7922	1.0413	28.7N	39.2W	37	185	227	03m40s
3747	188	-0437 Jun 22	13:34:11	16039	-30136	49	A	-p	-0.7436	0.9527	24.2S	48.9E	42	352	258	05m53s
3748	188	-0437 Dec 17	10:19:56	16031	-30130	54	H3	-n	0.1286	1.0173	16.2S	93.1E	83	190	60	01m45s
3749	188	-0436 Jun 10	18:47:03	16023	-30124	59	H	nn	0.0536	1.0043	25.8N	37.5W	87	167	15	00m27s
3750	188	-0436 Dec 05	20:05:47	16015	-30118	64	A	p-	-0.5881	0.9584	56.6S	71.5W	54	25	187	03m32s
3751	188	-0435 May 31	07:14:08	16007	-30112	69	T	p-	0.8101	1.0514	69.3N	94.5E	36	130	291	03m07s
3752	188	-0435 Nov 24	22:30:25	15999	-30106	74	P	t-	-1.3093	0.4356	62.9S	131.5E	0	139		
3753	188	-0434 Apr 21	17:18:50	15993	-30101	41	T	-t	-0.9821	1.0556	56.5S	40.8E	10	306	1017	03m29s
3754	188	-0434 May 21	00:03:26	15991	-30100	79	Pb	t-	1.5128	0.0279	62.7N	112.6E	0	45		
3755	188	-0434 Oct 15	03:38:57	15985	-30095	46	P	-t	1.0434	0.8875	60.8N	92.6W	0	257		
3756	188	-0433 Apr 11	07:56:43	15977	-30089	51	T	-n	-0.2567	1.0230	7.2S	134.9E	75	331	81	02m07s
3757	188	-0433 Oct 04	10:52:17	15969	-30083	56	H	-n	0.2920	1.0026	12.7N	89.9E	73	210	9	00m14s
3758	188	-0432 Mar 30	16:01:21	15961	-30077	61	A	p-	0.5243	0.9646	29.0N	9.4W	58	146	147	03m27s
3759	188	-0432 Sep 23	00:39:06	15953	-30071	66	T	n-	-0.4149	1.0489	19.2S	137.0W	65	31	177	03m57s
3760	188	-0431 Mar 19	17:26:30	15946	-30065	71	P	t-	1.2856	0.4745	60.7N	97.9W	0	95		
3761	189	-0431 Aug 14	07:45:52	15939	-30060	38	Pe	-t	1.5136	0.0432	62.1N	105.3W	0	307		
3762	189	-0431 Sep 12	16:53:58	15938	-30059	76	P	t-	-1.1012	0.8209	60.9S	86.0W	0	77		
3763	189	-0430 Feb 07	03:39:58	15931	-30054	43	A	-p	-0.8937	0.9633	67.4S	102.0W	26	290	296	02m30s
3764	189	-0430 Aug 03	19:20:15	15923	-30048	48	A	-p	0.8388	0.9843	68.0N	5.9E	33	238	102	01m04s
3765	189	-0429 Jan 27	13:46:44	15916	-30042	53	T	nn	-0.1404	1.0245	27.6S	46.8E	82	340	84	02m12s
3766	189	-0429 Jul 23	23:47:53	15908	-30036	58	A	nn	0.0966	0.9505	27.1N	108.5W	84	196	182	05m48s
3767	189	-0428 Jan 17	04:52:35	15900	-30030	63	T	p-	0.5466	1.0494	10.2N	168.4E	57	166	195	04m35s
3768	189	-0428 Jul 12	00:09:36	15892	-30024	68	A	p-	-0.6564	0.9456	17.3S	124.1W	49	11	264	07m07s
3769	189	-0428 Dec 07	08:26:02	15886	-30019	35	P	-t	-1.5165	0.0560	67.5S	54.6W	0	174		
3770	189	-0427 Jan 05	20:27:49	15884	-30018	73	P	t-	1.2343	0.5669	64.7N	83.0W	0	157		
3771	189	-0427 Jun 01	17:12:55	15878	-30013	40	P	-t	1.2999	0.4453	68.1N	179.7W	0	13		
3772	189	-0427 Jul 01	03:49:51	15877	-30012	78	P	t-	-1.3784	0.3073	65.4S	171.8E	0	15		
3773	189	-0427 Nov 26	16:07:50	15870	-30007	45	A	-p	-0.8945	0.9379	81.6S	43.6W	26	48	524	04m22s
3774	189	-0426 May 22	07:16:55	15862	-30001	50	T	-p	0.4979	1.0676	48.7N	128.2E	60	167	255	05m04s
3775	189	-0426 Nov 15	16:52:47	15855	-29995	55	A	nn	-0.2269	0.9170	30.6S	12.7W	77	11	324	10m23s
3776	189	-0425 May 12	00:29:44	15847	-29989	60	T	n-	-0.2467	1.0782	2.4N	120.3W	76	349	260	07m12s
3777	189	-0425 Nov 04	16:19:10	15839	-29983	65	A	p-	0.4525	0.9322	12.2N	4.4E	63	193	283	08m51s
3778	189	-0424 Apr 30	16:11:23	15831	-29977	70	P	t-	-1.0208	0.9663	70.6S	50.6E	0	313		
3779	189	-0424 Oct 23	21:36:01	15824	-29971	75	P	t-	1.1150	0.7752	71.2N	22.5W	0	237		
3780	189	-0423 Mar 21	12:18:06	15817	-29966	42	A	-t	0.9433	0.9430	62.8N	24.3E	19	138	640	04m39s

Cat Num	Canon Plate	Calendar Date	TD of Greatest Eclipse	ΔT s	Luna Num	Saros Num	Ecl. Type	QLE	Gamma	Ecl. Mag.	Lat. °	Long. °	Sun Alt °	Sun Azm °	Path Width km	Central Line Dur.
3781	190	-0423 Sep 14	00:09:47	15809	-29960	47	T	-p	-0.9186	1.0472	56.7S	147.3W	23	33	399	03m12s
3782	190	-0422 Mar 10	12:40:07	15802	-29954	52	A	nn	0.2165	0.9371	5.7N	55.5E	77	163	239	08m14s
3783	190	-0422 Sep 03	16:20:38	15794	-29948	57	T	-n	-0.2298	1.0468	3.0S	3.0W	77	16	160	04m26s
3784	190	-0421 Feb 27	14:35:29	15786	-29942	62	A	p-	-0.5161	0.9637	40.3S	41.6E	59	340	152	03m36s
3785	190	-0421 Aug 24	05:01:15	15778	-29936	67	A	p-	0.5151	0.9966	43.6N	179.2W	59	199	14	00m18s
3786	190	-0420 Jan 18	12:33:13	15772	-29931	34	P	-t	1.4838	0.1019	67.8N	49.2E	0	169		
3787	190	-0420 Feb 16	23:21:21	15771	-29930	72	P	t-	-1.2039	0.6214	70.4S	35.0E	0	223		
3788	190	-0420 Aug 12	10:47:49	15763	-29924	77	P	t-	1.3022	0.4450	69.7N	130.1W	0	327		
3789	190	-0419 Jan 07	04:12:19	15757	-29919	44	T	-p	0.8004	1.0426	30.2N	174.7W	37	180	238	03m46s
3790	190	-0419 Jul 02	20:06:15	15749	-29913	49	A	-p	-0.8239	0.9517	31.9S	52.8W	34	356	312	05m43s
3791	190	-0419 Dec 27	19:00:22	15741	-29907	54	H	-n	0.1333	1.0160	16.2S	37.5W	82	185	55	01m39s
3792	190	-0418 Jun 22	01:42:32	15733	-29901	59	H	nn	-0.0257	1.0071	22.3N	141.1W	89	351	25	00m46s
3793	190	-0418 Dec 17	04:28:25	15726	-29895	64	A	p-	-0.5871	0.9564	58.8S	167.3E	54	17	197	03m43s
3794	190	-0417 Jun 11	14:30:53	15718	-29889	69	T	p-	0.7318	1.0555	67.4N	1.2E	43	148	269	03m32s
3795	190	-0417 Dec 06	06:34:48	15710	-29883	74	P	t-	-1.3066	0.4401	63.8S	0.1W	0	148		
3796	190	-0416 May 02	00:47:10	15704	-29878	41	P	-t	-1.0535	0.9180	61.6S	64.2W	0	299		
3797	190	-0416 May 31	07:28:25	15703	-29877	79	P	t-	1.4411	0.1676	63.4N	9.2W	0	36		
3798	190	-0416 Oct 25	11:43:57	15696	-29872	46	P	-t	1.0527	0.8718	61.2N	136.3E	0	247		
3799	190	-0415 Apr 21	15:12:09	15689	-29866	51	T	-p	-0.3272	1.0211	7.1S	25.5E	71	333	76	02m00s
3800	190	-0415 Oct 14	19:14:55	15681	-29860	56	H	-n	0.3057	1.0031	9.2N	37.5W	72	209	11	00m17s
3801	191	-0414 Apr 10	22:55:21	15673	-29854	61	A	p-	0.4558	0.9650	29.5N	112.7W	63	147	140	03m25s
3802	191	-0414 Oct 04	09:07:54	15666	-29848	66	T	n-	-0.3933	1.0475	22.4S	94.3E	67	32	170	03m47s
3803	191	-0413 Mar 31	00:12:50	15658	-29842	71	P	t-	1.2205	0.5850	60.6N	150.7E	0	87		
3804	191	-0413 Sep 24	01:14:10	15650	-29836	76	P	t-	-1.0760	0.8671	60.8S	139.0E	0	86		
3805	191	-0412 Feb 18	11:25:21	15644	-29831	43	A	-p	-0.9242	0.9683	64.7S	145.4E	22	285	299	02m07s
3806	191	-0412 Aug 14	02:39:14	15636	-29825	48	A	-t	0.8974	0.9765	66.7N	91.5W	26	251	190	01m34s
3807	191	-0411 Feb 06	22:04:04	15629	-29819	53	T	nn	-0.1635	1.0300	25.8S	77.3W	80	336	103	02m38s
3808	191	-0411 Aug 03	06:34:26	15621	-29813	58	A	nn	0.1660	0.9464	28.5N	150.9E	80	201	200	06m05s
3809	191	-0410 Jan 27	13:26:55	15614	-29807	63	T	p-	0.5294	1.0525	10.7N	37.4E	58	162	203	04m44s
3810	191	-0410 Jul 23	06:47:50	15606	-29801	68	A	p-	-0.5804	0.9456	12.7S	134.2E	54	15	243	07m03s
3811	191	-0410 Dec 18	17:04:14	15600	-29796	35	P	-t	-1.5225	0.0462	66.4S	163.8E	0	185		
3812	191	-0409 Jan 17	05:02:42	15598	-29795	73	P	t-	1.2256	0.5831	63.7N	137.4E	0	147		
3813	191	-0409 Jun 13	00:14:15	15592	-29790	40	P	-t	1.3713	0.3136	67.1N	62.7E	0	3		
3814	191	-0409 Jul 12	10:46:02	15591	-29789	78	P	t-	-1.3014	0.4445	64.4S	56.6E	0	25		
3815	191	-0409 Dec 08	00:24:23	15585	-29784	45	A	-p	-0.8989	0.9357	86.2S	156.3W	26	35	557	04m27s
3816	191	-0408 Jun 01	14:38:54	15577	-29778	50	T	-p	0.5714	1.0689	56.3N	18.0E	55	172	274	04m51s
3817	191	-0408 Nov 26	00:53:19	15569	-29772	55	A	nn	-0.2323	0.9163	33.8S	132.5W	76	7	327	10m22s
3818	191	-0407 May 22	07:55:23	15562	-29766	60	T	n-	-0.1732	1.0779	9.4N	125.5E	80	352	256	07m13s
3819	191	-0407 Nov 15	00:27:37	15554	-29760	65	A	p-	0.4466	0.9332	8.7N	120.2W	63	190	278	08m55s
3820	191	-0406 May 11	23:23:29	15547	-29754	70	T	t-	-0.9507	1.0249	54.1S	88.5W	18	340	278	01m56s
3821	192	-0406 Nov 04	06:04:15	15539	-29748	75	P	t-	1.1064	0.7909	70.5N	164.2W	0	224		
3822	192	-0405 Apr 01	19:10:51	15533	-29743	42	A+	-t	1.0107	0.9459	71.9N	138.8W	0	82	-	-
3823	192	-0405 Sep 25	08:34:09	15525	-29737	47	T	-p	-0.9423	1.0435	62.9S	74.7E	19	43	440	02m45s
3824	192	-0404 Mar 20	19:33:16	15518	-29731	52	A	nn	0.2774	0.9405	13.4N	51.3W	74	162	229	07m32s
3825	192	-0404 Sep 14	00:30:59	15510	-29725	57	T	-n	-0.2623	1.0412	8.9S	128.3W	75	17	142	03m52s
3826	192	-0403 Mar 09	21:58:29	15503	-29719	62	A	p-	-0.4674	0.9702	33.0S	72.2W	62	340	121	03m04s
3827	192	-0403 Sep 03	12:41:28	15495	-29713	67	A	p-	0.4746	0.9905	37.0N	63.7E	61	200	38	00m54s
3828	192	-0402 Jan 28	21:01:43	15489	-29708	34	P	-t	1.4974	0.0746	68.8N	90.8W	0	157		
3829	192	-0402 Feb 27	07:18:58	15488	-29707	72	P	t-	-1.1656	0.6926	71.1S	98.9W	0	236		
3830	192	-0402 Aug 23	17:53:19	15480	-29701	77	P	t-	1.2510	0.5327	70.5N	109.7E	0	315		
3831	192	-0401 Jan 18	12:52:53	15474	-29696	44	T	-p	0.8128	1.0442	32.8N	50.6E	35	175	253	03m49s
3832	192	-0401 Jul 14	02:42:55	15466	-29690	49	A	-p	-0.9009	0.9499	41.7S	157.1W	25	1	427	05m26s
3833	192	-0400 Jan 08	03:36:41	15459	-29684	54	H	-n	0.1409	1.0152	15.1S	167.2W	82	180	53	01m35s
3834	192	-0400 Jul 02	08:43:41	15451	-29678	59	H	nn	-0.1015	1.0092	18.0N	113.3E	84	356	32	01m01s
3835	192	-0400 Dec 27	12:46:23	15444	-29672	64	A	p-	-0.5829	0.9549	59.4S	48.5E	54	9	203	03m54s
3836	192	-0399 Jun 21	21:52:27	15437	-29666	69	T	p-	0.6567	1.0585	64.1N	97.5W	49	163	256	03m53s
3837	192	-0399 Dec 16	14:36:52	15429	-29660	74	P	t-	-1.3022	0.4472	64.7S	131.5W	0	159		
3838	192	-0398 May 13	08:12:32	15423	-29655	41	P	-t	-1.1272	0.7758	62.1S	174.3E	0	308		
3839	192	-0398 Jun 11	14:54:03	15422	-29654	79	P	t-	1.3697	0.3067	64.3N	131.3W	0	27		
3840	192	-0398 Nov 05	19:54:43	15415	-29649	46	P	-t	1.0582	0.8627	61.7N	3.6E	0	238		

Cat Num	Canon Plate	Calendar Date	TD of Greatest Eclipse	ΔT s	Luna Num	Saros Num	Ecl. Type	QLE	Gamma	Ecl. Mag.	Lat. °	Long. °	Sun Alt °	Sun Azm °	Path Width km	Central Line Dur.
3841	193	-0397 May 02	22:21:16	15408	-29643	51	T	-p	-0.4020	1.0186	7.9S	82.5W	66	335	69	01m49s
3842	193	-0397 Oct 26	03:44:28	15400	-29637	56	H	-n	0.3148	1.0040	5.7N	166.7W	72	207	14	00m23s
3843	193	-0396 Apr 21	05:40:33	15393	-29631	61	A	p-	0.3807	0.9649	29.7N	146.8E	67	149	136	03m29s
3844	193	-0396 Oct 14	17:44:49	15385	-29625	66	T	n-	-0.3782	1.0463	26.1S	36.3W	68	32	165	03m39s
3845	193	-0395 Apr 10	06:51:24	15378	-29619	71	P	t-	1.1483	0.7080	60.8N	41.2E	0	78		
3846	193	-0395 Oct 04	09:41:25	15371	-29613	76	P	t-	-1.0569	0.9014	60.8S	2.4E	0	95		
3847	193	-0394 Feb 28	19:03:05	15364	-29608	43	A	-p	-0.9619	0.9727	63.0S	37.8E	15	278	362	01m46s
3848	193	-0394 Aug 25	10:04:55	15357	-29602	48	A	-t	0.9496	0.9682	64.9N	169.0E	18	263	369	02m04s
3849	193	-0393 Feb 18	06:12:56	15350	-29596	53	T	-n	-0.1944	1.0358	23.8S	160.4E	79	333	123	03m03s
3850	193	-0393 Aug 14	13:27:45	15342	-29590	58	A	nn	0.2293	0.9419	28.8N	48.6E	77	205	220	06m25s
3851	193	-0392 Feb 07	21:53:26	15335	-29584	63	T	p-	0.5057	1.0560	11.4N	91.3W	60	158	212	04m53s
3852	193	-0392 Aug 02	13:34:16	15327	-29578	68	A	p-	-0.5104	0.9452	9.7S	31.0E	59	19	232	06m56s
3853	193	-0392 Dec 29	01:38:40	15321	-29573	35	P	-t	-1.5311	0.0315	65.3S	23.6E	0	196		
3854	193	-0391 Jan 27	13:30:34	15320	-29572	73	P	t-	1.2119	0.6086	62.9N	0.2W	0	138		
3855	193	-0391 Jun 23	07:19:55	15314	-29567	40	P	-t	1.4400	0.1861	66.1N	55.5W	0	352		
3856	193	-0391 Jul 22	17:50:17	15313	-29566	78	P	t-	-1.2296	0.5730	63.5S	60.3W	0	34		
3857	193	-0391 Dec 18	08:39:37	15306	-29561	45	A	-p	-0.9039	0.9339	87.5S	170.1E	25	303	587	04m31s
3858	193	-0390 Jun 12	22:01:00	15299	-29555	50	T	-p	0.6452	1.0693	63.4N	89.3W	50	178	297	04m35s
3859	193	-0390 Dec 07	08:55:37	15292	-29549	55	A	nn	-0.2361	0.9162	36.2S	107.9E	76	3	328	10m16s
3860	193	-0389 Jun 02	15:19:08	15284	-29543	60	T	nn	-0.0980	1.0769	15.9N	12.4E	84	355	250	07m04s
3861	194	-0389 Nov 26	08:39:30	15277	-29537	65	A	p-	0.4429	0.9349	5.9N	114.6E	64	186	270	08m51s
3862	194	-0388 May 22	06:32:18	15269	-29531	70	T	p-	-0.8785	1.0251	41.8S	153.6E	28	349	179	02m12s
3863	194	-0388 Nov 14	14:37:03	15262	-29525	75	P	t-	1.1007	0.8017	69.6N	53.7E	0	211		
3864	194	-0387 Apr 12	01:57:00	15256	-29520	42	P	-t	1.0823	0.8230	71.6N	104.5E	0	69		
3865	194	-0387 Oct 05	17:04:43	15249	-29514	47	T	-p	-0.9612	1.0400	68.5S	68.6W	15	57	500	02m22s
3866	194	-0386 Apr 01	02:20:48	15241	-29508	52	A	-p	0.3431	0.9437	21.4N	156.8W	70	161	220	06m49s
3867	194	-0386 Sep 25	08:48:00	15234	-29502	57	T	-n	-0.2886	1.0357	14.7S	104.7E	73	19	125	03m19s
3868	194	-0385 Mar 21	05:13:33	15227	-29496	62	A	p-	-0.4111	0.9766	25.3S	175.5E	66	340	91	02m29s
3869	194	-0385 Sep 14	20:29:28	15219	-29490	67	A	p-	0.4414	0.9845	30.6N	55.8W	64	200	61	01m32s
3870	194	-0384 Feb 09	05:21:47	15213	-29485	34	Pe	-t	1.5182	0.0332	69.7N	130.8E	0	145		
3871	194	-0384 Mar 09	15:07:57	15212	-29484	72	P	t-	-1.1195	0.7792	71.6S	129.0E	0	250		
3872	194	-0384 Sep 03	01:08:00	15204	-29478	77	P	t-	1.2073	0.6067	71.2N	13.4W	0	302		
3873	194	-0383 Jan 28	21:25:04	15198	-29473	44	T	-p	0.8320	1.0456	36.6N	82.4W	33	170	274	03m49s
3874	194	-0383 Jul 24	09:29:05	15191	-29467	49	A	-t	-0.9706	0.9468	55.1S	93.1E	13	8	844	05m02s
3875	194	-0382 Jan 18	12:05:58	15184	-29461	54	H	-n	0.1538	1.0147	12.9S	64.7E	81	176	51	01m33s
3876	194	-0382 Jul 13	15:50:17	15177	-29455	59	Hm	nn	-0.1735	1.0107	13.0N	5.6E	80	1	38	01m12s
3877	194	-0381 Jan 07	20:59:04	15169	-29449	64	A	p-	-0.5753	0.9541	58.5S	69.0W	55	0	205	04m03s
3878	194	-0381 Jul 03	05:16:57	15162	-29443	69	T	p-	0.5837	1.0606	59.5N	159.1E	54	173	246	04m13s
3879	194	-0381 Dec 27	22:35:31	15155	-29437	74	P	t-	-1.2954	0.4584	65.8S	97.6E	0	169		
3880	194	-0380 May 23	15:33:57	15148	-29432	41	P	-t	-1.2034	0.6288	62.8S	53.6E	0	317		
3881	195	-0380 Jun 21	22:20:47	15147	-29431	79	P	t-	1.2994	0.4435	65.2N	106.0E	0	18		
3882	195	-0380 Nov 16	04:11:34	15141	-29426	46	P	-t	1.0592	0.8616	62.3N	130.7W	0	229		
3883	195	-0379 May 13	05:25:04	15134	-29420	51	H3	-p	-0.4803	1.0155	9.8S	170.7E	61	338	60	01m34s
3884	195	-0379 Nov 05	12:19:59	15127	-29414	56	H	-n	0.3195	1.0052	2.5N	62.6E	71	204	19	00m31s
3885	195	-0378 May 02	12:19:43	15119	-29408	61	A	nn	0.3012	0.9645	29.5N	48.2E	72	152	134	03m37s
3886	195	-0378 Oct 26	02:27:56	15112	-29402	66	T	n-	-0.3681	1.0453	30.0S	168.1W	68	31	161	03m32s
3887	195	-0377 Apr 21	13:22:31	15105	-29396	71	P	t-	1.0698	0.8424	61.0N	66.4W	0	69		
3888	195	-0377 Oct 15	18:15:21	15098	-29390	76	P	t-	-1.0438	0.9244	60.9S	135.9W	0	104		
3889	195	-0376 Mar 11	02:34:06	15092	-29385	43	P	-t	-1.0057	0.9726	60.8S	52.2W	0	258		
3890	195	-0376 Sep 04	17:37:31	15084	-29379	48	An	-t	0.9951	0.9584	62.2N	77.3E	4	282	-	02m34s
3891	195	-0375 Feb 28	14:15:18	15077	-29373	53	T	-n	-0.2312	1.0415	21.7S	39.6E	77	331	142	03m28s
3892	195	-0375 Aug 24	20:29:30	15070	-29367	58	A	nn	0.2851	0.9376	28.0N	56.1W	73	208	241	06m47s
3893	195	-0374 Feb 18	06:12:52	15063	-29361	63	T	p-	0.4761	1.0594	12.4N	142.0E	61	155	220	05m02s
3894	195	-0374 Aug 13	20:28:56	15055	-29355	68	A	p-	-0.4165	0.9445	8.0S	73.9W	63	22	220	06m47s
3895	195	-0373 Jan 09	10:07:20	15049	-29350	35	Pe	-t	-1.5438	0.0092	64.3S	114.7W	0	206		
3896	195	-0373 Feb 07	21:50:08	15048	-29349	73	P	t-	1.1920	0.6457	62.2N	135.4W	0	128		
3897	195	-0373 Jul 04	14:30:41	15042	-29344	40	Pe	-t	1.5055	0.0643	65.2N	174.6W	0	343		
3898	195	-0373 Aug 03	01:03:18	15041	-29343	78	P	t-	-1.1635	0.6914	62.7S	179.1W	0	43		
3899	195	-0373 Dec 29	16:48:43	15035	-29338	45	A	-p	-0.9133	0.9327	83.4S	81.2E	24	268	629	04m33s
3900	195	-0372 Jun 23	05:26:41	15028	-29332	50	T	-p	0.7165	1.0686	69.5N	167.3E	44	190	323	04m17s

Cat Num	Canon Plate	Calendar Date	TD of Greatest Eclipse	ΔT s	Luna Num	Saros Num	Ecl. Type	QLE	Gamma	Ecl. Mag.	Lat. °	Long. °	Sun Alt °	Sun Azm °	Path Width km	Central Line Dur.
3901	196	-0372 Dec 17	16:55:17	15020	-29326	55	A	nn	-0.2424	0.9169	37.7S	10.5W	76	357	326	10m02s
3902	196	-0371 Jun 12	22:43:10	15013	-29320	60	Tm	nn	-0.0234	1.0749	21.7N	99.8W	89	359	243	06m46s
3903	196	-0371 Dec 06	16:52:06	15006	-29314	65	A	p-	0.4391	0.9372	3.7N	10.6W	64	182	260	08m35s
3904	196	-0370 Jun 02	13:37:55	14999	-29308	70	T	p-	-0.8041	1.0238	32.1S	40.6E	36	355	136	02m17s
3905	196	-0370 Nov 25	23:13:25	14992	-29302	75	P	t-	1.0974	0.8084	68.6N	88.6W	0	198		
3906	196	-0369 Apr 23	08:32:44	14986	-29297	42	P	-t	1.1614	0.6870	71.1N	9.2W	0	56		
3907	196	-0369 Oct 17	01:43:35	14979	-29291	47	T	-t	-0.9735	1.0368	72.5S	142.1E	12	75	570	02m04s
3908	196	-0368 Apr 11	08:59:55	14971	-29285	52	A	-p	0.4160	0.9467	30.0N	99.6E	65	160	215	06m06s
3909	196	-0368 Oct 05	17:12:24	14964	-29279	57	T	-n	-0.3083	1.0304	20.3S	23.7W	72	19	108	02m48s
3910	196	-0367 Mar 31	12:23:03	14957	-29273	62	A	p-	-0.3492	0.9830	17.4S	64.4E	69	341	64	01m52s
3911	196	-0367 Sep 25	04:24:51	14950	-29267	67	A	n-	0.4150	0.9786	24.6N	177.3W	65	200	83	02m14s
3912	196	-0366 Mar 20	22:51:24	14943	-29261	72	P	t-	-1.0682	0.8767	71.8S	2.0W	0	263		
3913	196	-0366 Sep 14	08:29:47	14936	-29255	77	P	t-	1.1697	0.6697	71.7N	138.7W	0	288		
3914	196	-0365 Feb 09	05:51:31	14930	-29250	44	T	-p	0.8559	1.0471	41.5N	145.4E	31	165	302	03m45s
3915	196	-0365 Aug 04	16:22:57	14922	-29244	49	P	-t	-1.0341	0.9075	69.1S	28.1W	0	24		
3916	196	-0364 Jan 29	20:27:34	14915	-29238	54	H	-n	0.1722	1.0144	9.5S	61.9W	80	172	50	01m31s
3917	196	-0364 Jul 23	23:05:36	14908	-29232	59	H	nn	-0.2391	1.0118	7.6N	105.0W	76	5	42	01m19s
3918	196	-0363 Jan 18	05:04:49	14901	-29226	64	A	p-	-0.5631	0.9538	55.9S	174.0E	55	353	205	04m11s
3919	196	-0363 Jul 13	12:47:19	14894	-29220	69	T	p-	0.5151	1.0618	54.2N	51.0E	59	181	237	04m30s
3920	196	-0362 Jan 07	06:28:31	14887	-29214	74	P	t-	-1.2844	0.4766	66.9S	32.3W	0	180		
3921	197	-0362 Jun 03	22:54:45	14881	-29209	41	P	-t	-1.2794	0.4826	63.6S	67.1W	0	326		
3922	197	-0362 Jul 03	05:50:34	14880	-29208	79	P	t-	1.2318	0.5745	66.1N	17.8W	0	8		
3923	197	-0362 Nov 27	12:30:33	14874	-29203	46	P	-t	1.0592	0.8622	63.1N	94.2E	0	219		
3924	197	-0361 May 24	12:23:59	14867	-29197	51	H	-p	-0.5616	1.0115	12.8S	64.9E	56	341	47	01m12s
3925	197	-0361 Nov 16	20:59:44	14860	-29191	56	H	-n	0.3213	1.0071	0.4S	69.1W	71	200	26	00m42s
3926	197	-0360 May 12	18:53:04	14852	-29185	61	A	nn	0.2177	0.9636	28.6N	48.7W	77	156	135	03m52s
3927	197	-0360 Nov 05	11:15:17	14845	-29179	66	T	n-	-0.3614	1.0446	33.9S	59.5E	69	29	159	03m28s
3928	197	-0359 May 01	19:49:10	14838	-29173	71	An	t-	0.9871	0.9357	64.7N	156.3W	8	76	-	04m13s
3929	197	-0359 Oct 26	02:55:14	14831	-29167	76	P	t-	-1.0362	0.9369	61.3S	84.3E	0	113		
3930	197	-0358 Mar 22	09:57:33	14825	-29162	43	P	-t	-1.0565	0.8844	60.7S	172.8W	0	267		
3931	197	-0358 Sep 16	01:17:33	14818	-29156	48	P	-t	1.0334	0.9131	60.7N	39.7W	0	280		
3932	197	-0357 Mar 11	22:09:30	14811	-29150	53	T	-n	-0.2756	1.0471	19.7S	79.3W	74	330	162	03m53s
3933	197	-0357 Sep 05	03:40:10	14804	-29144	58	A	-p	0.3334	0.9333	26.4N	163.5W	70	211	262	07m13s
3934	197	-0356 Feb 29	14:23:05	14797	-29138	63	T	n-	0.4386	1.0628	13.6N	17.9E	64	152	227	05m11s
3935	197	-0356 Aug 24	03:34:30	14790	-29132	68	A	p-	-0.3907	0.9437	7.7S	178.6E	67	25	224	06m39s
3936	197	-0355 Feb 18	06:01:40	14783	-29126	73	P	t-	1.1661	0.6939	61.6N	91.7E	0	119		
3937	197	-0355 Aug 13	08:24:46	14776	-29120	78	P	t-	-1.1031	0.7997	62.0S	60.3E	0	52		
3938	197	-0354 Jan 09	00:53:20	14770	-29115	45	A	-t	-0.9257	0.9319	78.9S	35.4W	22	262	686	04m34s
3939	197	-0354 Jul 04	12:55:08	14763	-29109	50	T	-p	0.7859	1.0671	74.1N	71.5E	38	208	357	03m58s
3940	197	-0354 Dec 29	00:52:27	14756	-29103	55	A	nn	-0.2505	0.9182	38.3S	127.9W	75	352	321	09m42s
3941	198	-0353 Jun 24	06:07:24	14749	-29097	60	T	nn	0.0505	1.0721	26.7N	148.8E	87	184	235	06m22s
3942	198	-0353 Dec 18	01:04:52	14742	-29091	65	A	p-	0.4347	0.9402	2.3N	135.7W	64	178	246	08m07s
3943	198	-0352 Jun 12	20:43:07	14735	-29085	70	T	p-	-0.7303	1.0213	24.1S	70.5W	43	359	106	02m11s
3944	198	-0352 Dec 06	07:49:53	14728	-29079	75	P	t-	1.0934	0.8165	67.5N	129.7E	0	187		
3945	198	-0351 May 03	15:04:44	14722	-29074	42	P	-t	1.2425	0.5475	70.5N	121.4W	0	43		
3946	198	-0351 Oct 27	10:27:30	14715	-29068	47	T	-t	-0.9819	1.0341	74.9S	13.0W	10	97	664	01m49s
3947	198	-0350 Apr 22	15:34:12	14708	-29062	52	A	-p	0.4930	0.9494	38.8N	2.8W	60	160	213	05m25s
3948	198	-0350 Oct 17	01:42:35	14701	-29056	57	T	-n	-0.3226	1.0254	25.6S	153.1W	71	18	91	02m19s
3949	198	-0349 Apr 11	19:26:44	14694	-29050	62	A	p-	-0.2812	0.9891	9.3S	45.3W	74	343	40	01m13s
3950	198	-0349 Oct 06	12:27:26	14687	-29044	67	A	n-	0.3955	0.9730	19.1N	59.5E	67	199	105	02m56s
3951	198	-0348 Mar 31	06:27:06	14680	-29038	72	P	t-	-1.0096	0.9887	71.8S	131.1W	0	277		
3952	198	-0348 Sep 24	16:01:22	14673	-29032	77	P	t-	1.1402	0.7185	71.9N	93.1E	0	274		
3953	198	-0347 Feb 19	14:08:39	14667	-29027	44	T	-p	0.8873	1.0481	47.7N	14.1E	27	159	346	03m36s
3954	198	-0347 Aug 14	23:26:34	14660	-29021	49	P	-t	-1.0902	0.8105	70.0S	147.4W	0	36		
3955	198	-0346 Feb 09	04:40:18	14653	-29015	54	H	-n	0.1973	1.0142	5.1S	173.2E	79	169	50	01m30s
3956	198	-0346 Aug 04	06:28:34	14646	-29009	59	H	-n	-0.2990	1.0123	1.9N	142.0E	73	8	44	01m22s
3957	198	-0345 Jan 29	13:01:02	14639	-29003	64	A	p-	-0.5436	0.9540	51.9S	57.5E	57	347	201	04m18s
3958	198	-0345 Jul 24	20:22:43	14632	-28997	69	T	p-	0.4502	1.0621	48.2N	60.8W	63	187	229	04m44s
3959	198	-0344 Jan 18	14:15:11	14625	-28991	74	P	t-	-1.2682	0.5035	68.0S	161.1W	0	191		
3960	198	-0344 Jun 14	06:14:22	14619	-28986	41	P	-t	-1.3557	0.3365	64.5S	172.1E	0	336		

Cat Num	Canon Plate	Calendar Date	TD of Greatest Eclipse	ΔT s	Luna Num	Saros Num	Ecl. Type	QLE	Gamma	Ecl. Mag.	Lat. °	Long. °	Sun Alt °	Sun Azm °	Path Width km	Central Line Dur.
3961	199	-0344 Jul 13	13:23:45	14618	-28985	79	P	t-	1.1671	0.6993	67.1N	142.9W	0	357		
3962	199	-0344 Dec 07	20:51:06	14613	-28980	46	P	-t	1.0588	0.8640	64.0N	41.4W	0	209		
3963	199	-0343 Jun 03	19:20:33	14606	-28974	51	H	-p	-0.6437	1.0069	17.1S	40.9W	50	345	31	00m44s
3964	199	-0343 Nov 27	05:42:53	14599	-28968	56	H	-n	0.3210	1.0094	2.7S	158.6E	71	197	34	00m57s
3965	199	-0342 May 24	01:21:41	14592	-28962	61	Am	nn	0.1308	0.9622	26.9N	144.4W	82	160	138	04m13s
3966	199	-0342 Nov 16	20:07:00	14585	-28956	66	T	n-	-0.3583	1.0443	37.5S	73.3W	69	26	158	03m26s
3967	199	-0341 May 13	02:12:00	14578	-28950	71	A	t-	0.9010	0.9417	67.4N	141.8E	25	108	497	04m15s
3968	199	-0341 Nov 06	11:39:10	14571	-28944	76	P	t-	-1.0323	0.9426	61.8S	56.7W	0	123		
3969	199	-0340 Apr 01	17:15:28	14565	-28939	43	P	-t	-1.1126	0.7853	60.8S	67.9E	0	276		
3970	199	-0340 Sep 26	09:05:02	14558	-28933	48	P	-t	1.0650	0.8560	60.6N	166.3W	0	271		
3971	199	-0339 Mar 22	05:58:20	14551	-28927	53	T	-n	-0.3252	1.0523	17.9S	163.0E	71	329	182	04m17s
3972	199	-0339 Sep 15	10:58:24	14544	-28921	58	A	-p	0.3753	0.9293	24.2N	86.7E	68	212	283	07m41s
3973	199	-0338 Mar 11	22:26:24	14538	-28915	63	T	n-	0.3954	1.0660	15.0N	104.2W	67	150	233	05m19s
3974	199	-0338 Sep 04	10:49:39	14531	-28909	68	A	p-	-0.3420	0.9427	8.5S	68.8E	70	28	223	06m33s
3975	199	-0337 Mar 01	14:02:23	14524	-28903	73	P	t-	1.1321	0.7573	61.2N	38.4W	0	110		
3976	199	-0337 Aug 24	15:56:49	14517	-28897	78	P	t-	-1.0503	0.8944	61.5S	62.7W	0	61		
3977	199	-0336 Jan 20	08:48:59	14511	-28892	45	A	-t	-0.9449	0.9312	74.4S	151.1W	19	257	804	04m33s
3978	199	-0336 Jul 14	20:29:37	14504	-28886	50	T	-p	0.8505	1.0646	75.8N	17.3W	31	235	405	03m38s
3979	199	-0335 Jan 08	08:42:56	14497	-28880	55	A	nn	-0.2643	0.9200	38.1S	116.4E	74	346	314	09m16s
3980	199	-0335 Jul 04	13:35:01	14491	-28874	60	T	nn	0.1213	1.0685	30.6N	37.4E	83	189	225	05m53s
3981	200	-0335 Dec 28	09:15:22	14484	-28868	65	A	p-	0.4276	0.9438	1.5N	99.9E	65	174	229	07m28s
3982	200	-0334 Jun 24	03:46:52	14477	-28862	70	T	p-	-0.6560	1.0179	17.3S	179.8E	49	4	81	01m55s
3983	200	-0334 Dec 17	16:27:19	14470	-28856	75	P	t-	1.0896	0.8246	66.4N	11.6W	0	175		
3984	200	-0333 May 14	21:29:55	14464	-28851	42	P	-t	1.3284	0.3998	69.6N	128.7E	0	31		
3985	200	-0333 Jun 13	11:20:58	14463	-28850	80	Pb	t-	-1.4808	0.1318	67.0S	70.1E	0	358		
3986	200	-0333 Nov 07	19:16:57	14457	-28845	47	T	-t	-0.9860	1.0321	75.8S	169.1W	8	120	736	01m39s
3987	200	-0332 May 02	22:03:21	14451	-28839	52	A	-p	0.5746	0.9516	48.0N	103.7W	55	160	216	04m47s
3988	200	-0332 Oct 27	10:19:14	14444	-28833	57	T	-n	-0.3312	1.0209	30.4S	76.5E	70	17	75	01m53s
3989	200	-0331 Apr 22	02:26:56	14437	-28827	62	A	nn	-0.2095	0.9949	1.3S	153.9W	78	344	18	00m35s
3990	200	-0331 Oct 16	20:35:36	14430	-28821	67	A	n-	0.3814	0.9677	14.0N	65.0W	67	197	125	03m40s
3991	200	-0330 Apr 11	13:58:37	14423	-28815	72	T	t-	-0.9468	1.0360	60.1S	65.0E	18	324	381	02m28s
3992	200	-0330 Oct 05	23:39:58	14416	-28809	77	P	t-	1.1169	0.7567	71.8N	36.8W	0	260		
3993	200	-0329 Mar 02	22:17:43	14411	-28804	44	T	-t	0.9252	1.0486	55.3N	117.9W	22	151	427	03m22s
3994	200	-0329 Aug 26	06:39:50	14404	-28798	49	P	-t	-1.1389	0.7263	70.7S	90.3E	0	49		
3995	200	-0328 Feb 20	12:44:07	14397	-28792	54	H	-n	0.2290	1.0141	0.1N	50.1E	77	166	50	01m29s
3996	200	-0328 Aug 14	14:00:58	14390	-28786	59	H	-n	-0.3519	1.0126	4.1S	26.2E	69	12	46	01m21s
3997	200	-0327 Feb 08	20:48:05	14383	-28780	64	A	p-	-0.5174	0.9544	46.8S	58.6W	59	343	195	04m27s
3998	200	-0327 Aug 04	04:06:08	14377	-28774	69	T	p-	0.3918	1.0618	41.9N	176.2W	67	191	221	04m54s
3999	200	-0326 Jan 28	21:53:54	14370	-28768	74	P	t-	-1.2458	0.5408	69.0S	71.5E	0	203		
4000	200	-0326 Jun 25	13:34:41	14364	-28763	41	P	-t	-1.4306	0.1943	65.5S	51.0E	0	346		
4001	201	-0326 Jul 24	21:01.18	14363	-28762	79	P	t-	1.1063	0.8155	68.1N	90.5E	0	346		
4002	201	-0326 Dec 19	05:10:48	14357	-28757	46	P	-t	1.0595	0.8640	65.0N	177.2W	0	199		
4003	201	-0325 Jun 15	02:15:43	14351	-28751	51	H	-p	-0.7254	1.0014	22.7S	146.9W	43	349	7	00m09s
4004	201	-0325 Dec 08	14:25:47	14344	-28745	56	H	-n	0.3213	1.0123	4.3S	26.5E	71	192	45	01m16s
4005	201	-0324 Jun 03	07:48:31	14337	-28739	61	A	nn	0.0434	0.9603	24.3N	120.0E	87	165	144	04m41s
4006	201	-0324 Nov 27	05:00:21	14330	-28733	66	T	n-	-0.3570	1.0445	40.7S	154.4E	69	22	159	03m28s
4007	201	-0323 May 23	08:32:52	14323	-28727	71	A	p-	0.8126	0.9453	66.8N	66.3E	35	128	345	04m15s
4008	201	-0323 Nov 16	20:25:59	14317	-28721	76	P	t-	-1.0315	0.9427	62.4S	161.5E	0	132		
4009	201	-0322 Apr 13	00:27:44	14311	-28716	43	P	-t	-1.1739	0.6756	61.0S	50.0W	0	285		
4010	201	-0322 May 12	11:41:16	14310	-28715	81	Pb	t-	1.5329	0.0338	62.1N	62.2W	0	52		
4011	201	-0322 Oct 07	16:59:55	14304	-28710	48	P	-t	1.0895	0.8118	60.7N	65.3E	0	262		
4012	201	-0321 Apr 02	13:38:29	14298	-28704	53	T	-n	-0.3826	1.0571	16.8S	47.4E	67	330	202	04m41s
4013	201	-0321 Sep 26	18:26:32	14291	-28698	58	A	-p	0.4086	0.9256	21.6N	26.2W	66	212	303	08m14s
4014	201	-0320 Mar 22	06:21:05	14284	-28692	63	T	n-	0.3452	1.0689	16.4N	136.3E	70	149	238	05m28s
4015	201	-0320 Sep 14	18:15:18	14277	-28686	68	A	p-	-0.3015	0.9418	10.3S	43.6W	72	29	224	06m29s
4016	201	-0319 Mar 11	21:54:53	14271	-28680	73	P	t-	1.0922	0.8314	60.9N	166.4W	0	101		
4017	201	-0319 Sep 03	23:38:48	14264	-28674	78	P	t-	-1.0048	0.9758	61.0S	172.0E	0	70		
4018	201	-0318 Jan 30	16:37:39	14258	-28669	45	A	-t	-0.9688	0.9305	69.9S	96.1E	14	252	1092	04m31s
4019	201	-0318 Jul 26	04:08:04	14252	-28663	50	T	-t	0.9118	1.0610	74.2N	105.6W	24	263	494	03m17s
4020	201	-0317 Jan 19	16:28:44	14245	-28657	55	A	-n	-0.2819	0.9225	37.1S	1.6E	73	341	305	08m47s

Cat Num	Canon Plate	Calendar Date	TD of Greatest Eclipse	ΔT s	Luna Num	Saros Num	Ecl. Type	QLE	Gamma	Ecl. Mag.	Lat. °	Long. °	Sun Alt °	Sun Azm °	Path Width km	Central Line Dur.
4021	202	-0317 Jul 15	21:05:41	14238	-28651	60	T	-n	0.1890	1.0642	33.4N	74.0W	79	194	214	05m21s
4022	202	-0316 Jan 08	17:21:51	14231	-28645	65	A	p-	0.4165	0.9481	1.2N	23.4W	65	169	209	06m41s
4023	202	-0316 Jul 04	10:53:21	14225	-28639	70	H3	p-	-0.5847	1.0137	12.0S	70.4E	54	8	58	01m30s
4024	202	-0316 Dec 28	01:02:05	14218	-28633	75	P	t-	1.0827	0.8386	65.4N	151.7W	0	165		
4025	202	-0315 May 25	03:53:17	14212	-28628	42	P	-t	1.4145	0.2521	68.7N	19.9E	0	20		
4026	202	-0315 Jun 23	17:55:20	14211	-28627	80	P	t-	-1.4037	0.2677	66.0S	40.2W	0	9		
4027	202	-0315 Nov 18	04:08:18	14206	-28622	47	T	-t	-0.9887	1.0306	75.6S	35.5E	7	141	807	01m32s
4028	202	-0314 May 14	04:30:35	14199	-28616	52	A	-p	0.6581	0.9534	57.5N	156.4E	49	161	227	04m13s
4029	202	-0314 Nov 07	18:59:06	14192	-28610	57	H3	-n	-0.3363	1.0168	34.7S	53.9W	70	15	61	01m30s
4030	202	-0313 May 03	09:22:29	14186	-28604	62	H	nn	-0.1330	1.0002	6.6N	98.9E	82	347	1	00m02s
4031	202	-0313 Oct 28	04:49:09	14179	-28598	67	A	n-	0.3727	0.9630	9.5N	169.3E	68	195	144	04m22s
4032	202	-0312 Apr 21	21:24:48	14172	-28592	72	T	t-	-0.8788	1.0438	48.6S	61.6W	28	336	306	03m20s
4033	202	-0312 Oct 16	07:25:31	14166	-28586	77	P	t-	1.0994	0.7848	71.5N	168.4W	0	246		
4034	202	-0311 Mar 13	06:17:39	14160	-28581	44	T	-t	0.9704	1.0476	64.9N	103.7E	13	135	680	02m58s
4035	202	-0311 Sep 05	14:04:12	14153	-28575	49	P	-t	-1.1791	0.6569	71.3S	35.3W	0	62		
4036	202	-0310 Mar 02	20:37:22	14147	-28569	54	H	-n	0.2687	1.0139	6.2N	70.9W	74	164	49	01m26s
4037	202	-0310 Aug 25	21:42:10	14140	-28563	59	H	-n	-0.3981	1.0124	10.2S	92.1W	66	14	47	01m18s
4038	202	-0309 Feb 20	04:24:40	14133	-28557	64	A	p-	-0.4834	0.9551	40.7S	173.7W	61	341	186	04m34s
4039	202	-0309 Aug 15	11:56:49	14127	-28551	69	T	n-	0.3394	1.0609	35.5N	65.4E	70	194	212	05m00s
4040	202	-0308 Feb 09	05:23:28	14120	-28545	74	P	t-	-1.2161	0.5904	69.9S	54.2W	0	215		
4041	203	-0308 Jul 05	20:57:08	14115	-28540	41	Pe	-t	-1.5027	0.0586	66.5S	71.1W	0	356		
4042	203	-0308 Aug 04	04:44:56	14113	-28539	79	P	t-	1.0511	0.9201	69.1N	38.1W	0	335		
4043	203	-0308 Dec 29	13:29:23	14108	-28534	46	P	-t	1.0614	0.8622	66.1N	46.9E	0	188		
4044	203	-0307 Jun 25	09:10:15	14101	-28528	51	A	-p	-0.8063	0.9953	29.9S	106.4E	36	353	28	00m30s
4045	203	-0307 Dec 18	23:08:59	14095	-28522	56	H	-n	0.3218	1.0156	5.0S	105.6W	71	188	57	01m38s
4046	203	-0306 Jun 14	14:14:06	14088	-28516	61	A	nn	-0.0442	0.9579	20.8N	24.2E	88	348	153	05m14s
4047	203	-0306 Dec 08	13:54:19	14081	-28510	66	T	n-	-0.3559	1.0452	43.0S	22.7E	69	16	161	03m31s
4048	203	-0305 Jun 03	14:53:00	14075	-28504	71	A	p-	0.7228	0.9481	64.8N	13.5W	43	145	276	04m17s
4049	203	-0305 Nov 28	05:14:09	14068	-28498	76	P	t-	-1.0324	0.9400	63.2S	19.2E	0	142		
4050	203	-0304 Apr 23	07:37:10	14063	-28493	43	P	-t	-1.2383	0.5588	61.3S	167.2W	0	293		
4051	203	-0304 May 22	18:26:38	14062	-28492	81	P	t-	1.4496	0.1804	62.8N	173.7W	0	43		
4052	203	-0304 Oct 18	00:59:44	14056	-28487	48	P	-t	1.1092	0.7764	60.9N	64.3W	0	253		
4053	203	-0303 Apr 12	21:15:09	14049	-28481	53	T	-p	-0.4435	1.0613	16.2S	67.4W	64	331	222	05m04s
4054	203	-0303 Oct 07	02:02:17	14043	-28475	58	A	-p	0.4353	0.9223	18.8N	141.4W	64	211	321	08m49s
4055	203	-0302 Apr 02	14:08:34	14036	-28469	63	T	n-	0.2890	1.0713	17.8N	19.0E	73	149	241	05m37s
4056	203	-0302 Sep 26	01:50:57	14030	-28463	68	A	n-	-0.2686	0.9411	12.9S	158.6W	74	30	225	06m26s
4057	203	-0301 Mar 23	05:37:13	14023	-28457	73	P	t-	1.0448	0.9196	60.8N	68.3E	0	92		
4058	203	-0301 Sep 15	07:31:43	14016	-28451	78	A	t-	-0.9674	0.9823	55.7S	69.1E	14	58	248	01m16s
4059	203	-0300 Feb 11	00:14:24	14011	-28446	45	A-	-t	-1.0021	0.9559	61.8S	1.0E	0	235	-	-
4060	203	-0300 Aug 05	11:55:06	14004	-28440	50	T	-t	0.9664	1.0562	69.9N	159.6E	14	285	748	02m52s
4061	204	-0299 Jan 30	00:06:10	13998	-28434	55	A	-n	-0.3069	0.9254	35.7S	111.3W	72	336	295	08m15s
4062	204	-0299 Jul 26	04:40:36	13991	-28428	60	T	-n	0.2527	1.0593	35.0N	174.0E	75	199	201	04m50s
4063	204	-0298 Jan 19	01:23:21	13985	-28422	65	A	p-	0.4003	0.9531	1.4N	145.3W	66	165	186	05m48s
4064	204	-0298 Jul 15	18:01:16	13978	-28416	70	H	p-	-0.5152	1.0089	7.7S	38.6W	59	12	36	00m59s
4065	204	-0297 Jan 08	09:33:14	13972	-28410	75	P	t-	1.0724	0.8588	64.3N	69.6E	0	154		
4066	204	-0297 Jun 05	10:12:52	13966	-28405	42	Pe	-t	1.5027	0.1011	67.7N	87.4W	0	9		
4067	204	-0297 Jul 05	00:30:09	13965	-28404	80	P	t-	-1.3270	0.4017	65.0S	150.2W	0	19		
4068	204	-0297 Nov 29	13:02:25	13960	-28399	47	T	-t	-0.9892	1.0298	75.1S	117.3W	7	159	815	01m28s
4069	204	-0296 May 24	10:56:12	13953	-28393	52	A	-p	0.7430	0.9545	67.3N	57.6E	42	162	250	03m45s
4070	204	-0296 Nov 18	03:41:28	13946	-28387	57	H	-n	-0.3389	1.0133	38.3S	176.0E	70	11	49	01m11s
4071	204	-0295 May 13	16:17:41	13940	-28381	62	H	nn	-0.0550	1.0051	14.2N	7.6W	87	349	18	00m34s
4072	204	-0295 Nov 07	13:06:53	13933	-28375	67	A	n-	0.3682	0.9587	5.7N	42.8E	68	192	161	05m03s
4073	204	-0294 May 03	04:47:27	13927	-28369	72	T	p-	-0.8072	1.0505	38.4S	178.9E	36	343	282	04m11s
4074	204	-0294 Oct 27	15:17:24	13920	-28363	77	P	t-	1.0873	0.8041	71.0N	58.9E	0	232		
4075	204	-0293 Mar 24	14:10:04	13915	-28358	44	P	-t	1.0217	0.9751	71.8N	60.9W	0	92		
4076	204	-0293 Apr 22	21:31:56	13914	-28357	82	Pb	t-	-1.5048	0.0445	70.9S	28.3W	0	304		
4077	204	-0293 Sep 16	21:38:41	13908	-28352	49	P	-t	-1.2116	0.6008	71.7S	163.8W	0	76		
4078	204	-0292 Mar 13	04:21:15	13902	-28346	54	H	-p	0.3154	1.0134	13.0N	170.0E	72	162	48	01m22s
4079	204	-0292 Sep 05	05:33:58	13895	-28340	59	H	-n	-0.4363	1.0123	16.2S	146.8E	64	17	47	01m14s
4080	204	-0291 Mar 02	11:51:44	13889	-28334	64	A	p-	-0.4425	0.9559	34.0S	72.5E	64	340	178	04m42s

Cat Num	Canon Plate	Calendar Date	TD of Greatest Eclipse	ΔT s	Luna Num	Saros Num	Ecl. Type	QLE	Gamma	Ecl. Mag.	Lat. °	Long. °	Sun Alt °	Sun Azm °	Path Width km	Central Line Dur.
4081	205	-0291 Aug 25	19:55:06	13882	-28328	69	T	n-	0.2931	1.0595	29.1N	55.7W	73	196	205	05m02s
4082	205	-0290 Feb 19	12:44:05	13876	-28322	74	P	t-	-1.1792	0.6525	70.8S	178.3W	0	228		
4083	205	-0290 Aug 15	12:34:54	13869	-28316	79	T+	t-	1.0017	1.0127	70.0N	168.9W	0	323	-	-
4084	205	-0289 Jan 09	21:42:33	13864	-28311	46	P	-t	1.0685	0.8515	67.2N	88.2W	0	177		
4085	205	-0289 Jul 06	16:06:35	13857	-28305	51	A	-t	-0.8846	0.9881	39.0S	1.9W	27	358	90	01m11s
4086	205	-0289 Dec 30	07:48:35	13851	-28299	56	T	-n	0.3258	1.0196	4.7S	123.3E	71	183	71	02m04s
4087	205	-0288 Jun 24	20:41:21	13844	-28293	61	A	nn	-0.1300	0.9551	16.4N	72.7W	83	354	165	05m53s
4088	205	-0288 Dec 18	22:45:47	13838	-28287	66	T	n-	-0.3529	1.0463	44.2S	107.6W	69	11	165	03m38s
4089	205	-0287 Jun 13	21:15:08	13831	-28281	71	A	p-	0.6341	0.9501	61.4N	97.6W	50	158	237	04m22s
4090	205	-0287 Dec 08	14:02:31	13825	-28275	76	P	t-	-1.0341	0.9359	64.1S	123.4W	0	152		
4091	205	-0286 May 04	14:41:45	13820	-28270	43	P	-t	-1.3071	0.4326	61.8S	76.7E	0	302		
4092	205	-0286 Jun 03	01:10:51	13818	-28269	81	P	t-	1.3641	0.3323	63.6N	74.9E	0	34		
4093	205	-0286 Oct 29	09:06:06	13813	-28264	48	P	-t	1.1226	0.7523	61.3N	164.4E	0	244		
4094	205	-0285 Apr 24	04:45:23	13807	-28258	53	T	-p	-0.5100	1.0649	16.5S	179.3E	59	333	244	05m25s
4095	205	-0285 Oct 18	09:46:36	13800	-28252	58	A	-p	0.4547	0.9195	15.9N	100.9E	63	209	337	09m27s
4096	205	-0284 Apr 12	21:48:54	13794	-28246	63	T	n-	0.2270	1.0731	18.8N	96.2W	77	150	243	05m47s
4097	205	-0284 Oct 06	09:37:04	13787	-28240	68	A	n-	-0.2438	0.9406	16.0S	83.8E	76	30	226	06m25s
4098	205	-0283 Apr 02	13:11:49	13781	-28234	73	T	t-	0.9917	1.0166	61.1N	42.5W	6	94	519	00m59s
4099	205	-0283 Sep 25	15:33:09	13774	-28228	78	A	t-	-0.9361	0.9842	54.6S	49.6W	20	57	157	01m08s
4100	205	-0282 Feb 21	07:43:16	13769	-28223	45	P	-t	-1.0409	0.8905	61.3S	121.0W	0	244		
4101	206	-0282 Aug 16	19:47:49	13763	-28217	50	P	-t	1.0166	0.9877	61.9N	62.2E	0	305		
4102	206	-0281 Feb 10	07:35:25	13756	-28211	55	A	-p	-0.3386	0.9286	33.9S	137.4E	70	333	283	07m43s
4103	206	-0281 Aug 06	12:20:45	13750	-28205	60	T	-n	0.3114	1.0539	35.4N	60.7E	72	204	187	04m18s
4104	206	-0280 Jan 30	09:18:46	13743	-28199	65	A	p-	0.3783	0.9584	2.0N	94.4E	68	161	162	04m55s
4105	206	-0280 Jul 26	01:13:50	13737	-28193	70	H	p-	-0.4506	1.0037	4.9S	148.3W	63	16	14	00m24s
4106	206	-0279 Jan 18	17:58:54	13730	-28187	75	P	t-	1.0570	0.8888	63.4N	67.4W	0	144		
4107	206	-0279 Jul 15	07:09:25	13724	-28181	80	P	t-	-1.2541	0.5282	64.1S	99.1E	0	28		
4108	206	-0279 Dec 09	21:55:28	13719	-28176	47	Ts	-t	-0.9907	1.0294	73.7S	91.4E	7	176	-	01m25s
4109	206	-0278 Jun 04	17:21:55	13712	-28170	52	A	-p	0.8282	0.9548	77.8N	39.4W	34	164	297	03m22s
4110	206	-0278 Nov 29	12:24:41	13706	-28164	57	H	-n	-0.3401	1.0103	41.1S	46.6E	70	7	38	00m55s
4111	206	-0277 May 24	23:12:00	13699	-28158	62	H	nn	0.0249	1.0093	21.4N	113.1W	88	173	32	01m00s
4112	206	-0277 Nov 18	21:26:20	13693	-28152	67	A	n-	0.3658	0.9551	2.6N	83.9W	69	189	176	05m40s
4113	206	-0276 May 13	12:07:52	13687	-28146	72	T	p-	-0.7328	1.0559	29.2S	62.4E	43	348	270	04m57s
4114	206	-0276 Nov 06	23:14:33	13680	-28140	77	P	t-	1.0793	0.8166	70.2N	74.6W	0	219		
4115	206	-0275 Apr 03	21:54:21	13675	-28135	44	P	-t	1.0792	0.8654	71.7N	167.9E	0	78		
4116	206	-0275 May 03	05:03:29	13674	-28134	82	P	t-	-1.4392	0.1713	70.3S	155.2W	0	317		
4117	206	-0275 Sep 27	05:23:15	13669	-28129	49	P	-t	-1.2366	0.5576	71.8S	64.9E	0	90		
4118	206	-0274 Mar 24	11:55:07	13662	-28123	54	H	-p	0.3695	1.0126	20.3N	53.2E	68	161	47	01m15s
4119	206	-0274 Sep 16	13:35:23	13656	-28117	59	H	-n	-0.4671	1.0120	22.2S	23.3E	62	19	46	01m09s
4120	206	-0273 Mar 13	19:06:33	13649	-28111	64	A	p-	-0.3923	0.9569	26.7S	39.1W	67	340	170	04m50s
4121	207	-0273 Sep 06	04:02:20	13643	-28105	69	T	n-	0.2543	1.0578	22.8N	179.4W	75	198	197	05m00s
4122	207	-0272 Mar 01	19:55:28	13637	-28099	74	P	t-	-1.1349	0.7275	71.4S	59.4E	0	241		
4123	207	-0272 Aug 25	20:31:42	13630	-28093	79	T	t-	0.9583	1.0435	75.2N	3.3E	16	258	525	02m25s
4124	207	-0271 Jan 20	05:51:47	13625	-28088	46	P	-t	1.0794	0.8342	68.2N	137.3E	0	166		
4125	207	-0271 Jul 16	23:05:36	13619	-28082	51	A	-t	-0.9592	0.9798	51.7S	113.2W	16	4	261	01m49s
4126	207	-0270 Jan 09	16:25:37	13612	-28076	56	T	-n	0.3324	1.0240	3.4S	7.3W	71	179	86	02m30s
4127	207	-0270 Jul 06	03:09:49	13606	-28070	61	A	nn	-0.2141	0.9519	11.2N	170.6W	78	358	180	06m34s
4128	207	-0270 Dec 30	07:35:31	13600	-28064	66	T	n-	-0.3486	1.0479	44.2S	122.7E	69	4	170	03m47s
4129	207	-0269 Jun 25	03:40:35	13593	-28058	71	A	p-	0.5476	0.9515	56.8N	173.8E	57	169	213	04m31s
4130	207	-0269 Dec 19	22:47:56	13587	-28052	76	P	t-	-1.0343	0.9349	65.1S	94.3E	0	162		
4131	207	-0268 May 14	21:46:32	13582	-28047	43	P	-t	-1.3763	0.3045	62.4S	39.6W	0	311		
4132	207	-0268 Jun 13	07:59:41	13581	-28046	81	P	t-	1.2811	0.4807	64.5N	37.9W	0	25		
4133	207	-0268 Nov 08	17:15:46	13575	-28041	48	P	-t	1.1322	0.7349	61.9N	32.2E	0	234		
4134	207	-0267 May 04	12:13:28	13569	-28035	53	T	-p	-0.5788	1.0677	17.8S	66.3E	55	335	267	05m43s
4135	207	-0267 Oct 28	17:36:16	13563	-28029	58	A	-p	0.4695	0.9174	13.1N	18.4W	62	206	350	10m06s
4136	207	-0266 Apr 24	05:23:32	13556	-28023	63	T	nn	0.1606	1.0742	19.5N	150.4E	81	152	244	05m57s
4137	207	-0266 Oct 17	17:31:31	13550	-28017	68	A	n-	-0.2251	0.9405	19.5S	35.7W	77	29	225	06m24s
4138	207	-0265 Apr 13	20:36:22	13544	-28011	73	T	p-	0.9310	1.0201	59.7N	132.9W	21	113	187	01m16s
4139	207	-0265 Oct 06	23:44:55	13537	-28005	78	A	t-	-0.9125	0.9855	55.6S	173.7W	24	59	123	01m01s
4140	207	-0264 Mar 03	14:59:58	13532	-28000	45	P	-t	-1.0891	0.8090	60.9S	120.1E	0	253		

Cat Num	Canon Plate	Calendar Date	TD of Greatest Eclipse	ΔT s	Luna Num	Saros Num	Ecl. Type	QLE	Gamma	Ecl. Mag.	Lat. °	Long. °	Sun Alt °	Sun Azm °	Path Width km	Central Line Dur.
4141	208	-0264 Aug 27	03:49:10	13526	-27994	50	P	-t	1.0597	0.9038	61.4N	68.0W	0	295		
4142	208	-0263 Feb 20	14:55:12	13520	-27988	55	A	-p	-0.3782	0.9321	32.1S	28.2E	68	330	272	07m12s
4143	208	-0263 Aug 16	20:07:28	13513	-27982	60	T	-n	0.3639	1.0480	34.8N	54.5W	68	208	171	03m48s
4144	208	-0262 Feb 09	17:06:51	13507	-27976	65	A	p-	0.3495	0.9643	3.0N	23.9W	70	157	137	04m02s
4145	208	-0262 Aug 06	08:30:11	13501	-27970	70	A	n-	-0.3900	0.9979	3.2S	101.5E	67	20	8	00m13s
4146	208	-0261 Jan 30	02:19:05	13494	-27964	75	P	t-	1.0365	0.9286	62.6N	157.3E	0	135		
4147	208	-0261 Jul 26	13:52:50	13488	-27958	80	P	t-	-1.1846	0.6477	63.2S	12.3W	0	38		
4148	208	-0261 Dec 21	06:47:13	13483	-27953	47	Ts	-t	-0.9933	1.0291	71.0S	56.5W	5	190	-	01m24s
4149	208	-0260 Jun 14	23:50:04	13477	-27947	52	A	-p	0.9118	0.9541	89.1N	75.8W	24	227	416	03m05s
4150	208	-0260 Dec 09	21:08:14	13470	-27941	57	H	-n	-0.3409	1.0079	43.0S	82.2W	70	1	29	00m42s
4151	208	-0259 Jun 04	06:07:27	13464	-27935	62	Hm	nn	0.1047	1.0130	28.1N	142.0E	84	176	45	01m21s
4152	208	-0259 Nov 29	05:47:16	13458	-27929	67	A	n-	0.3653	0.9521	0.2N	149.2E	69	185	188	06m12s
4153	208	-0258 May 24	19:27:21	13452	-27923	72	T	p-	-0.6568	1.0605	20.9S	52.5W	49	353	263	05m37s
4154	208	-0258 Nov 18	07:14:57	13445	-27917	77	P	t-	1.0744	0.8243	69.3N	151.8E	0	206		
4155	208	-0257 Apr 15	05:31:25	13440	-27912	44	P	-t	1.1422	0.7449	71.4N	38.8E	0	65		
4156	208	-0257 May 14	12:31:26	13439	-27911	82	P	t-	-1.3708	0.3041	69.5S	79.2E	0	329		
4157	208	-0257 Oct 08	13:17:29	13434	-27906	49	P	-t	-1.2544	0.5270	71.7S	68.8W	0	104		
4158	208	-0256 Apr 03	19:20:53	13428	-27900	54	H	-p	0.4293	1.0114	28.2N	61.8W	64	160	43	01m05s
4159	208	-0256 Sep 26	21:45:25	13421	-27894	59	H	-p	-0.4915	1.0119	27.9S	102.0W	60	20	47	01m06s
4160	208	-0255 Mar 24	02:12:47	13415	-27888	64	A	nn	-0.3357	0.9576	19.1S	149.0W	70	341	163	04m58s
4161	209	-0255 Sep 16	12:17:51	13409	-27882	69	T	n-	0.2223	1.0559	16.7N	54.5E	77	198	189	04m55s
4162	209	-0254 Mar 13	02:56:47	13403	-27876	74	P	t-	-1.0821	0.8173	71.8S	60.8W	0	255		
4163	209	-0254 Sep 06	04:36:03	13396	-27870	79	T	p-	0.9216	1.0404	68.8N	146.4W	22	231	353	02m27s
4164	209	-0253 Jan 31	13:53:47	13391	-27865	46	P	-t	1.0968	0.8054	69.2N	4.0E	0	154		
4165	209	-0253 Jul 28	06:09:43	13385	-27859	51	P	-t	-1.0284	0.9293	68.5S	125.9E	0	17		
4166	209	-0252 Jan 21	00:55:45	13379	-27853	56	T	-n	0.3453	1.0287	0.9S	136.3W	70	175	104	02m57s
4167	209	-0252 Jul 16	09:43:49	13373	-27847	61	A	np	-0.2930	0.9483	5.4N	89.4E	73	2	199	07m13s
4168	209	-0251 Jan 09	16:20:26	13366	-27841	66	T	n-	-0.3405	1.0499	42.9S	5.9W	70	358	176	03m59s
4169	209	-0251 Jul 05	10:09:32	13360	-27835	71	A	p-	0.4633	0.9524	51.3N	81.3E	62	177	197	04m44s
4170	209	-0251 Dec 30	07:30:44	13354	-27829	76	P	t-	-1.0329	0.9369	66.2S	47.6W	0	173		
4171	209	-0250 May 26	04:49:31	13349	-27824	43	P	-t	-1.4474	0.1720	63.2S	155.6W	0	320		
4172	209	-0250 Jun 24	14:50:42	13348	-27823	81	P	t-	1.1987	0.6289	65.4N	151.6W	0	15		
4173	209	-0250 Nov 20	01:28:50	13343	-27818	48	P	-t	1.1388	0.7230	62.6N	101.1W	0	225		
4174	209	-0249 May 15	19:37:08	13336	-27812	53	T	-p	-0.6518	1.0695	20.4S	45.8W	49	338	295	05m56s
4175	209	-0249 Nov 09	01:32:18	13330	-27806	58	A	-p	0.4792	0.9158	10.6N	139.4W	61	203	360	10m44s
4176	209	-0248 May 04	12:53:34	13324	-27800	63	T	nn	0.0902	1.0746	19.6N	38.4E	85	155	243	06m07s
4177	209	-0248 Oct 28	01:33:12	13318	-27794	68	A	n-	-0.2115	0.9408	23.1S	156.8W	78	28	224	06m21s
4178	209	-0247 Apr 24	03:55:23	13312	-27788	73	T	p-	0.8664	1.0211	59.4N	126.5E	30	121	143	01m23s
4179	209	-0247 Oct 17	08:04:31	13305	-27782	78	A	p-	-0.8945	0.9867	57.8S	59.4E	26	61	103	00m55s
4180	209	-0246 Mar 14	22:07:25	13300	-27777	45	P	-t	-1.1439	0.7160	60.7S	3.6E	0	262		
4181	210	-0246 Sep 07	11:57:32	13294	-27771	50	P	-t	1.0974	0.8308	61.0N	160.1E	0	286		
4182	210	-0246 Oct 06	21:20:58	13293	-27770	88	Pb	t-	-1.5237	0.0238	60.9S	176.1E	0	98		
4183	210	-0245 Mar 03	22:06:59	13288	-27765	55	A	-p	-0.4246	0.9358	30.3S	79.3W	65	328	261	06m44s
4184	210	-0245 Aug 28	04:01:10	13282	-27759	60	T	-p	0.4102	1.0420	33.2N	172.1W	66	211	153	03m19s
4185	210	-0244 Feb 21	00:47:37	13276	-27753	65	A	p-	0.3135	0.9704	4.2N	140.2W	72	154	111	03m12s
4186	210	-0244 Aug 16	15:53:47	13269	-27747	70	A	n-	-0.3361	0.9920	2.8S	10.4W	70	23	30	00m51s
4187	210	-0243 Feb 09	10:32:49	13263	-27741	75	P	t-	1.0098	0.9803	61.9N	23.8E	0	126		
4188	210	-0243 Aug 05	20:41:50	13257	-27735	80	P	t-	-1.1196	0.7583	62.5S	124.7W	0	47		
4189	210	-0243 Dec 31	15:35:17	13252	-27730	47	T-	-t	-0.9993	1.0102	65.0S	162.5E	0	199	-	-
4190	210	-0242 Jun 26	06:22:06	13246	-27724	52	An	-t	0.9929	0.9507	71.0N	57.1W	5	345	-	02m50s
4191	210	-0242 Dec 21	05:48:07	13240	-27718	57	H	-n	-0.3441	1.0060	44.0S	150.5E	70	355	22	00m32s
4192	210	-0241 Jun 15	13:05:26	13234	-27712	62	H2	nn	0.1836	1.0160	34.0N	37.5E	79	181	56	01m34s
4193	210	-0241 Dec 10	14:06:33	13227	-27706	67	A	n-	0.3639	0.9498	1.5S	22.9E	69	181	198	06m35s
4194	210	-0240 Jun 04	02:47:32	13221	-27700	72	T	p-	-0.5807	1.0641	13.6S	166.5W	54	357	258	06m08s
4195	210	-0240 Nov 28	15:16:22	13215	-27694	77	P	t-	1.0702	0.8310	68.2N	18.7E	0	194		
4196	210	-0239 Apr 25	13:02:25	13210	-27689	44	P	-t	1.2099	0.6153	70.8N	88.3W	0	52		
4197	210	-0239 May 24	19:57:18	13209	-27688	82	P	t-	-1.3006	0.4408	68.6S	45.2W	0	340		
4198	210	-0239 Oct 18	21:21:02	13204	-27683	49	P	-t	-1.2656	0.5077	71.3S	155.4E	0	118		
4199	210	-0238 Apr 15	02:36:22	13198	-27677	54	H	-p	0.4970	1.0096	36.5N	174.4W	60	159	38	00m53s
4200	210	-0238 Oct 08	06:05:19	13192	-27671	59	H	-p	-0.5085	1.0118	33.4S	130.6E	59	21	47	01m03s

Cat Num	Canon Plate	Calendar Date	TD of Greatest Eclipse	ΔT s	Luna Num	Saros Num	Ecl. Type	QLE	Gamma	Ecl. Mag.	Lat. °	Long. °	Sun Alt °	Sun Azm °	Path Width km	Central Line Dur.
4201	211	-0237 Apr 04	09:08:11	13186	-27665	64	A	nn	-0.2706	0.9582	11.2S	103.7E	74	342	157	05m05s
4202	211	-0237 Sep 27	20:42:09	13180	-27659	69	T	n-	0.1973	1.0539	10.8N	73.8W	79	198	181	04m49s
4203	211	-0236 Mar 23	09:49:30	13173	-27653	74	P	t-	-1.0219	0.9204	72.0S	179.0W	0	268		
4204	211	-0236 Sep 16	12:48:23	13167	-27647	79	T	p-	0.8919	1.0366	62.0N	77.2E	26	220	272	02m24s
4205	211	-0235 Feb 10	21:50:55	13162	-27642	46	P	-t	1.1190	0.7678	70.1N	128.7W	0	141		
4206	211	-0235 Aug 07	13:17:17	13156	-27636	51	P	-t	-1.0935	0.8117	69.5S	6.2E	0	29		
4207	211	-0234 Jan 31	09:21:39	13150	-27630	56	T	-n	0.3623	1.0337	2.5N	95.3E	69	171	122	03m24s
4208	211	-0234 Jul 27	16:21:57	13144	-27624	61	A	-p	-0.3678	0.9445	0.9S	12.4W	68	6	220	07m46s
4209	211	-0233 Jan 21	00:59:37	13138	-27618	66	T	n-	-0.3277	1.0522	40.2S	133.9W	71	353	183	04m13s
4210	211	-0233 Jul 16	16:45:14	13132	-27612	71	A	p-	0.3839	0.9528	45.2N	15.1W	67	183	187	04m58s
4211	211	-0232 Jan 10	16:07:48	13126	-27606	76	P	t-	-1.0278	0.9461	67.3S	171.4E	0	184		
4212	211	-0232 Jun 05	11:55:23	13121	-27601	43	Pe	-t	-1.5169	0.0413	64.0S	87.4E	0	329		
4213	211	-0232 Jul 04	21:48:29	13120	-27600	81	P	t-	1.1203	0.7704	66.4N	92.7E	0	5		
4214	211	-0232 Nov 30	09:40:44	13115	-27595	48	P	-t	1.1455	0.7110	63.4N	125.8E	0	215		
4215	211	-0231 May 26	03:01:25	13109	-27589	53	T	-p	-0.7246	1.0704	24.2S	158.4W	43	341	329	06m01s
4216	211	-0231 Nov 19	09:31:10	13102	-27583	58	A	-p	0.4864	0.9150	8.5N	98.9E	61	199	366	11m19s
4217	211	-0230 May 15	20:18:47	13096	-27577	63	Tm	nn	0.0162	1.0742	18.8N	72.4W	89	161	241	06m16s
4218	211	-0230 Nov 08	09:41:29	13090	-27571	68	A	n-	-0.2028	0.9417	26.6S	80.8E	78	25	220	06m17s
4219	211	-0229 May 05	11:06:48	13084	-27565	73	T	p-	0.7965	1.0210	59.2N	26.6E	37	129	118	01m25s
4220	211	-0229 Oct 28	16:31:34	13078	-27559	78	A	p-	-0.8822	0.9882	61.0S	70.0W	28	62	88	00m47s
4221	212	-0228 Mar 25	05:03:37	13073	-27554	45	P	-t	-1.2070	0.6086	60.6S	110.1W	0	270		
4222	212	-0228 Sep 17	20:15:24	13067	-27548	50	P	-t	1.1272	0.7730	60.8N	26.0E	0	277		
4223	212	-0228 Oct 17	05:57:25	13066	-27547	88	P	t-	-1.5079	0.0539	61.1S	37.4E	0	107		
4224	212	-0227 Mar 14	05:09:43	13061	-27542	55	A	-p	-0.4788	0.9395	28.9S	175.3E	61	327	252	06m18s
4225	212	-0227 Sep 07	12:01:32	13055	-27536	60	T	-p	0.4503	1.0359	31.1N	68.0E	63	213	134	02m52s
4226	212	-0226 Mar 03	08:21:06	13049	-27530	65	A	n-	0.2706	0.9769	5.6N	105.5E	74	152	85	02m24s
4227	212	-0226 Aug 27	23:23:22	13043	-27524	70	A	n-	-0.2879	0.9858	3.4S	123.6W	73	26	52	01m28s
4228	212	-0225 Feb 20	18:38:42	13037	-27518	75	T	t-	0.9760	1.0159	54.7N	88.8W	12	132	255	01m06s
4229	212	-0225 Aug 17	03:38:08	13031	-27512	80	P	t-	-1.0606	0.8575	61.8S	121.2E	0	56		
4230	212	-0224 Jan 12	00:19:29	13026	-27507	47	P	-t	-1.0087	0.9932	64.0S	20.6E	0	209		
4231	212	-0224 Jul 06	12:59:15	13020	-27501	52	P	-t	1.0700	0.8486	64.8N	164.2W	0	339		
4232	212	-0224 Dec 31	14:24:47	13014	-27495	57	H	-n	-0.3495	1.0046	43.9S	24.1E	69	349	17	00m24s
4233	212	-0223 Jun 25	20:07:48	13008	-27489	62	T	nn	0.2598	1.0184	38.9N	66.8W	75	186	65	01m42s
4234	212	-0223 Dec 20	22:24:11	13002	-27483	67	A	n-	0.3619	0.9481	2.5S	102.9W	69	176	205	06m50s
4235	212	-0222 Jun 15	10:08:27	12996	-27477	72	T	p-	-0.5047	1.0668	7.1S	80.3E	60	1	253	06m28s
4236	212	-0222 Dec 09	23:18:20	12990	-27471	77	P	t-	1.0664	0.8371	67.1N	114.0W	0	182		
4237	212	-0221 May 06	20:28:38	12985	-27466	44	P	-t	1.2811	0.4790	70.1N	146.3E	0	39		
4238	212	-0221 Jun 05	03:22:34	12984	-27465	82	P	t-	-1.2298	0.5784	67.7S	168.9W	0	351		
4239	212	-0221 Oct 30	05:31:21	12979	-27460	49	P	-t	-1.2721	0.4966	70.7S	18.3E	0	131		
4240	212	-0220 Apr 25	09:45:57	12973	-27454	54	H	-p	0.5684	1.0072	45.2N	74.6E	55	159	30	00m37s
4241	213	-0220 Oct 10	14:32:51	12967	-27448	59	H	-p	-0.5201	1.0122	38.6S	2.0E	58	21	49	01m03s
4242	213	-0219 Apr 14	15:55:53	12961	-27442	64	A	nn	-0.1999	0.9585	3.1S	1.7W	78	343	154	05m11s
4243	213	-0219 Oct 08	05:13:14	12955	-27436	69	T	n-	0.1780	1.0519	5.3N	156.4E	80	198	175	04m43s
4244	213	-0218 Apr 03	16:34:07	12949	-27430	74	A	t-	-0.9546	0.9326	63.1S	26.3E	17	318	858	05m40s
4245	213	-0218 Sep 27	21:08:11	12943	-27424	79	T	p-	0.8691	1.0324	55.8N	56.1W	29	213	220	02m16s
4246	213	-0217 Feb 22	05:38:32	12938	-27419	46	P	-t	1.1494	0.7153	70.8N	100.4E	0	128		
4247	213	-0217 Aug 18	20:32:33	12932	-27413	51	P	-t	-1.1510	0.7089	70.3S	116.1W	0	41		
4248	213	-0216 Feb 11	17:39:22	12926	-27407	56	T	-n	0.3864	1.0389	6.9N	31.4W	67	168	141	03m48s
4249	213	-0216 Aug 06	23:07:26	12920	-27401	61	A	-p	-0.4360	0.9404	7.6S	116.6W	64	10	244	08m12s
4250	213	-0215 Jan 31	09:31:56	12914	-27395	66	T	n-	-0.3093	1.0548	36.3S	98.9E	72	349	190	04m30s
4251	213	-0215 Jul 26	23:28:05	12908	-27389	71	A	p-	0.3097	0.9528	38.7N	115.1W	72	187	182	05m14s
4252	213	-0214 Jan 21	00:38:49	12902	-27383	76	P	t-	-1.0182	0.9634	68.3S	31.4E	0	195		
4253	213	-0214 Jul 16	04:51:12	12896	-27377	81	P	t-	1.0448	0.9072	67.4N	24.7W	0	355		
4254	213	-0214 Dec 11	17:53:27	12891	-27372	48	P	-t	1.1509	0.7015	64.4N	7.9W	0	205		
4255	213	-0213 Jun 06	10:24:42	12885	-27366	53	T	-p	-0.7985	1.0702	29.5S	88.6E	37	345	377	05m55s
4256	213	-0213 Nov 30	17:32:23	12879	-27360	58	A	-p	0.4916	0.9148	7.0N	23.3W	61	195	370	11m47s
4257	213	-0212 May 26	03:42:22	12873	-27354	63	T	nn	-0.0588	1.0729	17.3N	177.2E	87	342	237	06m23s
4258	213	-0212 Nov 18	17:55:03	12867	-27348	68	A	n-	-0.1981	0.9431	29.7S	42.4W	78	22	214	06m10s
4259	213	-0211 May 15	18:13:53	12861	-27342	73	T	p-	0.7236	1.0201	58.9N	72.3W	43	138	99	01m25s
4260	213	-0211 Nov 08	01:04:24	12856	-27336	78	A	p-	-0.8739	0.9898	64.8S	159.3E	29	64	74	00m40s

Cat Num	Canon Plate	Calendar Date	TD of Greatest Eclipse	ΔT s	Luna Num	Saros Num	Ecl. Type	QLE	Gamma	Ecl. Mag.	Lat. °	Long. °	Sun Alt °	Sun Azm °	Path Width km	Central Line Dur.
4261	214	-0210 Apr 05	11:52:15	12851	-27331	45	P	-t	-1.2754	0.4918	60.7S	138.2E	0	279		
4262	214	-0210 May 05	02:07:49	12850	-27330	83	Pb	t-	1.5500	0.0109	61.9N	80.4E	0	57		
4263	214	-0210 Sep 29	04:40:21	12845	-27325	50	P	-t	1.1518	0.7256	60.8N	109.9W	0	268		
4264	214	-0210 Oct 28	14:39:29	12844	-27324	88	P	t-	-1.4962	0.0761	61.5S	102.9W	0	116		
4265	214	-0209 Mar 25	12:04:25	12839	-27319	55	A	-p	-0.5401	0.9431	28.1S	71.8E	57	326	245	05m55s
4266	214	-0209 Sep 18	20:09:46	12833	-27313	60	T	-p	0.4835	1.0298	28.5N	54.5W	61	213	114	02m25s
4267	214	-0208 Mar 13	15:48:10	12827	-27307	65	A	n-	0.2211	0.9832	7.0N	6.9W	77	151	61	01m41s
4268	214	-0208 Sep 07	07:00:00	12821	-27301	70	A	nn	-0.2462	0.9797	4.9S	121.3E	76	28	74	02m05s
4269	214	-0207 Mar 03	02:37:59	12815	-27295	75	T	t-	0.9359	1.0242	50.5N	154.2E	20	134	230	01m42s
4270	214	-0207 Aug 27	10:42:23	12809	-27289	80	A-	t-	-1.0082	0.9449	61.3S	5.3E	0	65	-	-
4271	214	-0206 Jan 22	08:56:33	12804	-27284	47	P	-t	-1.0240	0.9651	63.2S	119.0W	0	219		
4272	214	-0206 Jul 17	19:43:04	12798	-27278	52	P	-t	1.1423	0.7229	63.9N	84.3E	0	329		
4273	214	-0205 Jan 11	22:55:00	12792	-27272	57	H	-n	-0.3598	1.0036	43.0S	100.8W	69	343	13	00m19s
4274	214	-0205 Jul 07	03:16:01	12786	-27266	62	T	-n	0.3323	1.0200	42.7N	171.5W	70	192	73	01m46s
4275	214	-0204 Jan 01	06:35:33	12781	-27260	67	A	n-	0.3552	0.9471	3.0S	133.0E	69	172	208	06m55s
4276	214	-0204 Jun 25	17:33:02	12775	-27254	72	T	p-	-0.4309	1.0686	1.7S	33.1W	64	5	247	06m36s
4277	214	-0204 Dec 20	07:18:35	12769	-27248	77	P	t-	1.0610	0.8464	66.0N	114.4E	0	171		
4278	214	-0203 May 17	03:49:42	12764	-27243	44	P	-t	1.3559	0.3360	69.3N	22.8E	0	27		
4279	214	-0203 Jun 15	10:46:45	12763	-27242	82	P	t-	-1.1581	0.7173	66.7S	68.1E	0	2		
4280	214	-0203 Nov 09	13:48:43	12758	-27237	49	P	-t	-1.2738	0.4937	69.9S	119.9W	0	145		
4281	215	-0202 May 06	16:47:22	12752	-27231	54	H	-p	0.6460	1.0041	54.4N	34.3W	49	159	19	00m20s
4282	215	-0202 Oct 29	23:07:46	12746	-27225	59	H	-p	-0.5261	1.0129	43.3S	127.5W	58	20	52	01m05s
4283	215	-0201 Apr 25	22:34:26	12740	-27219	64	Am	nn	-0.1221	0.9586	5.1N	104.6W	83	345	152	05m16s
4284	215	-0201 Oct 19	13:52:09	12735	-27213	69	T	n-	0.1652	1.0501	0.4N	24.7E	80	196	169	04m37s
4285	215	-0200 Apr 13	23:12:36	12729	-27207	74	A	t-	-0.8818	0.9382	50.9S	91.4W	28	333	483	06m00s
4286	215	-0200 Oct 08	05:34:03	12723	-27201	79	T	p-	0.8517	1.0281	50.4N	171.0E	31	208	181	02m06s
4287	215	-0199 Mar 04	13:21:05	12718	-27196	46	P	-t	1.1848	0.6531	71.4N	29.7W	0	115		
4288	215	-0199 Aug 29	03:53:28	12712	-27190	51	P	-t	-1.2029	0.6171	71.0S	119.6E	0	54		
4289	215	-0198 Feb 22	01:50:40	12706	-27184	56	T	-n	0.4166	1.0441	12.2N	156.9W	65	165	161	04m10s
4290	215	-0198 Aug 18	05:59:42	12700	-27178	61	A	-p	-0.4984	0.9363	14.4S	137.0E	60	13	272	08m28s
4291	215	-0197 Feb 11	17:57:05	12694	-27172	66	T	n-	-0.2849	1.0574	31.5S	27.6W	73	346	197	04m48s
4292	215	-0197 Aug 07	06:20:16	12689	-27166	71	A	nn	0.2424	0.9525	32.0N	141.4E	76	191	179	05m29s
4293	215	-0196 Feb 01	09:01:17	12683	-27160	76	P	p-	-1.0024	0.9924	69.3S	107.1W	0	207		
4294	215	-0196 Jul 26	12:03:26	12677	-27154	81	A	t-	0.9758	0.9909	79.3N	163.0W	12	326	157	00m31s
4295	215	-0196 Dec 22	02:02:20	12672	-27149	48	P	-t	1.1585	0.6883	65.4N	140.9W	0	195		
4296	215	-0195 Jun 16	17:49:40	12666	-27143	53	T	-p	-0.8712	1.0687	36.7S	25.4W	29	349	457	05m36s
4297	215	-0195 Dec 11	01:33:19	12660	-27137	58	A	-p	0.4971	0.9153	6.3N	145.5W	60	191	370	12m04s
4298	215	-0194 Jun 06	11:04:05	12655	-27131	63	T	-n	-0.1348	1.0707	14.9N	66.9E	82	346	232	06m26s
4299	215	-0194 Nov 30	02:10:48	12649	-27125	68	A	n-	-0.1943	0.9452	32.2S	165.6W	79	18	206	05m58s
4300	215	-0193 May 27	01:16:17	12643	-27119	73	T	p-	0.6476	1.0182	57.8N	170.2W	49	148	81	01m20s
4301	216	-0193 Nov 19	09:41:51	12637	-27113	78	A	p-	-0.8689	0.9919	68.9S	27.9E	29	65	58	00m31s
4302	216	-0192 Apr 15	18:32:04	12632	-27108	45	P	-t	-1.3502	0.3636	61.0S	28.6E	0	288		
4303	216	-0192 May 15	08:44:22	12631	-27107	83	P	t-	1.4737	0.1451	62.5N	28.7W	0	48		
4304	216	-0192 Oct 09	13:12:43	12626	-27102	50	P	-t	1.1705	0.6895	60.9N	112.5E	0	259		
4305	216	-0192 Nov 07	23:26:28	12626	-27101	88	P	t-	-1.4885	0.0906	62.1S	115.5E	0	126		
4306	216	-0191 Apr 04	18:52:02	12621	-27096	55	A	-p	-0.6076	0.9464	28.2S	30.0W	52	327	242	05m35s
4307	216	-0191 Sep 29	04:25:30	12615	-27090	60	T	-p	0.5099	1.0238	25.7N	179.6W	59	213	93	01m59s
4308	216	-0190 Mar 24	23:07:08	12609	-27084	65	A	nn	0.1639	0.9897	8.3N	117.0W	81	150	37	01m01s
4309	216	-0190 Sep 18	14:44:22	12603	-27078	70	A	nn	-0.2118	0.9736	7.3S	4.3E	78	29	96	02m41s
4310	216	-0189 Mar 14	10:29:38	12597	-27072	75	T	t-	0.8889	1.0320	48.3N	37.7E	27	134	231	02m15s
4311	216	-0189 Sep 07	17:55:21	12592	-27066	80	A	t-	-0.9629	0.9257	54.1S	87.3W	15	52	1028	06m21s
4312	216	-0188 Feb 02	17:27:14	12587	-27061	47	P	-t	-1.0444	0.9271	62.4S	103.1E	0	228		
4313	216	-0188 Jul 28	02:35:11	12581	-27055	52	P	-t	1.2085	0.6078	63.1N	29.1W	0	320		
4314	216	-0187 Jan 22	07:19:53	12575	-27049	57	H	-p	-0.3741	1.0028	41.4S	135.2E	68	338	11	00m15s
4315	216	-0187 Jul 17	10:29:18	12569	-27043	62	T	-n	0.4014	1.0212	45.2N	83.4E	66	199	79	01m47s
4316	216	-0186 Jan 11	14:42:38	12564	-27037	67	A	n-	0.3456	0.9465	2.8S	10.0E	70	167	209	06m51s
4317	216	-0186 Jul 07	01:01:07	12558	-27031	72	T	n-	-0.3594	1.0695	2.6N	146.7W	69	9	242	06m35s
4318	216	-0186 Dec 31	15:14:40	12552	-27025	77	P	t-	1.0521	0.8615	65.0N	15.7W	0	161		
4319	216	-0185 May 28	11:08:37	12547	-27020	44	P	-t	1.4317	0.1917	68.4N	99.6W	0	16		
4320	216	-0185 Jun 26	18:13:18	12546	-27019	82	P	t-	-1.0885	0.8517	65.7S	55.0W	0	12		

Cat Num	Canon Plate	Calendar Date	TD of Greatest Eclipse	ΔT s	Luna Num	Saros Num	Ecl. Type	QLE	Gamma	Ecl. Mag.	Lat. °	Long. °	Sun Alt °	Sun Azm °	Path Width km	Central Line Dur.
4321	217	-0185 Nov 20	22:09:44	12542	-27014	49	P	-t	-1.2734	0.4944	69.0S	101.6E	0	157		
4322	217	-0184 May 16	23:45:33	12536	-27008	54	H	-p	0.7249	1.0003	63.9N	141.9W	43	159	2	00m01s
4323	217	-0184 Nov 09	07:46:40	12530	-27002	59	H	-p	-0.5296	1.0140	47.6S	103.1E	58	17	57	01m08s
4324	217	-0183 May 06	05:08:22	12524	-26996	64	A	nn	-0.0411	0.9581	13.1N	154.1E	88	347	152	05m20s
4325	217	-0183 Oct 29	22:36:26	12519	-26990	69	T	n-	0.1566	1.0486	4.1S	107.9W	81	194	163	04m32s
4326	217	-0182 Apr 25	05:43:43	12513	-26984	74	A	p-	-0.8023	0.9429	39.9S	161.0E	36	341	350	06m13s
4327	217	-0182 Oct 19	14:07:02	12507	-26978	79	T	p-	0.8407	1.0239	45.8N	37.3E	32	204	149	01m53s
4328	217	-0181 Mar 15	20:54:50	12502	-26973	46	P	-t	1.2282	0.5757	71.7N	158.0W	0	101		
4329	217	-0181 Apr 14	08:48:52	12501	-26972	84	Pb	t-	-1.5134	0.0692	71.4S	165.8E	0	295		
4330	217	-0181 Sep 09	11:23:18	12496	-26967	51	P	-t	-1.2466	0.5407	71.6S	7.4W	0	67		
4331	217	-0180 Mar 04	09:53:23	12491	-26961	56	T	-p	0.4546	1.0492	18.3N	79.3E	63	162	182	04m27s
4332	217	-0180 Aug 28	13:02:00	12485	-26955	61	A	-p	-0.5523	0.9323	21.3S	27.7E	56	16	301	08m37s
4333	217	-0179 Feb 22	02:14:20	12479	-26949	66	T	n-	-0.2540	1.0602	25.8S	153.1W	75	344	204	05m08s
4334	217	-0179 Aug 17	13:20:51	12474	-26943	71	A	nn	0.1813	0.9519	25.2N	34.9E	79	194	179	05m43s
4335	217	-0178 Feb 11	17:16:19	12468	-26937	76	T	p-	-0.9813	1.0078	76.5S	87.2E	10	247	153	00m26s
4336	217	-0178 Aug 06	19:23:28	12462	-26931	81	A	t-	0.9123	0.9949	79.2N	1.3W	24	240	44	00m19s
4337	217	-0177 Jan 02	10:07:17	12457	-26926	48	P	-t	1.1687	0.6709	66.5N	86.6E	0	184		
4338	217	-0177 Jun 28	01:16:50	12452	-26920	53	T	-t	-0.9423	1.0656	46.9S	141.0W	19	353	650	04m59s
4339	217	-0177 Dec 22	09:33:34	12446	-26914	58	A	-p	0.5030	0.9165	6.4N	92.6E	60	186	367	12m08s
4340	217	-0176 Jun 16	18:26:21	12440	-26908	63	T	-n	-0.2101	1.0678	11.5N	44.0W	78	351	226	06m23s
4341	218	-0176 Dec 10	10:28:09	12434	-26902	68	A	n-	-0.1911	0.9479	33.9S	71.3E	79	13	195	05m43s
4342	218	-0175 Jun 06	08:17:11	12429	-26896	73	T	p-	0.5711	1.0156	55.8N	91.8E	55	157	65	01m13s
4343	218	-0175 Nov 29	18:22:07	12423	-26890	78	A	p-	-0.8654	0.9944	73.3S	102.5W	30	64	39	00m21s
4344	218	-0174 Apr 27	01:04:42	12418	-26885	45	P	-t	-1.4299	0.2270	61.4S	79.3W	0	297		
4345	218	-0174 May 26	15:16:29	12417	-26884	83	P	t-	1.3940	0.2847	63.2N	136.8W	0	39		
4346	218	-0174 Oct 20	21:51:48	12413	-26879	50	P	-t	1.1840	0.6635	61.2N	26.9W	0	250		
4347	218	-0174 Nov 19	08:17:16	12412	-26878	88	P	t-	-1.4838	0.0993	62.8S	27.2W	0	135		
4348	218	-0173 Apr 16	01:33:51	12407	-26873	55	A	-p	-0.6800	0.9494	29.3S	130.5W	47	328	246	05m18s
4349	218	-0173 Oct 10	12:47:45	12401	-26867	60	T	-p	0.5303	1.0183	22.7N	53.2E	58	211	73	01m35s
4350	218	-0172 Apr 04	06:21:06	12395	-26861	65	A	nn	0.1016	0.9959	9.5N	134.4E	84	151	14	00m24s
4351	218	-0172 Sep 28	22:36:11	12390	-26855	70	A	nn	-0.1846	0.9679	10.3S	114.6W	79	30	117	03m17s
4352	218	-0171 Mar 24	18:14:50	12384	-26849	75	T	p-	0.8357	1.0394	47.2N	77.2W	33	134	235	02m45s
4353	218	-0171 Sep 18	01:16:16	12378	-26843	80	A	p-	-0.9241	0.9233	51.8S	164.7E	22	52	739	06m40s
4354	218	-0170 Feb 13	01:49:34	12374	-26838	47	P	-t	-1.0719	0.8758	61.8S	32.3W	0	238		
4355	218	-0170 Mar 14	10:28:38	12373	-26837	85	Pb	t-	1.5120	0.0356	61.0N	5.4W	0	98		
4356	218	-0170 Aug 08	09:36:26	12368	-26832	52	P	-t	1.2681	0.5041	62.4N	144.5W	0	311		
4357	218	-0169 Feb 02	15:34:27	12362	-26826	57	H	-p	-0.3964	1.0023	39.4S	13.2E	66	333	9	00m12s
4358	218	-0169 Jul 28	17:51:19	12357	-26820	62	T	-p	0.4644	1.0217	46.3N	23.5W	62	205	83	01m45s
4359	218	-0168 Jan 22	22:41:06	12351	-26814	67	A	n-	0.3291	0.9465	2.2S	110.7W	71	163	208	06m40s
4360	218	-0168 Jul 17	08:34:21	12345	-26808	72	T	n-	-0.2917	1.0695	5.7N	99.1E	73	13	236	06m25s
4361	219	-0167 Jan 10	23:05:42	12340	-26802	77	P	t-	1.0389	0.8841	64.0N	144.1W	0	151		
4362	219	-0167 Jun 07	18:25:15	12335	-26797	44	Pe	-t	1.5089	0.0459	67.4N	139.1E	0	5		
4363	219	-0167 Jul 07	01:41:42	12334	-26796	82	P	t-	-1.0204	0.9821	64.8S	178.1W	0	22		
4364	219	-0167 Dec 01	06:34:42	12329	-26791	49	P	-t	-1.2707	0.4992	67.9S	37.2W	0	169		
4365	219	-0166 May 28	06:37:14	12323	-26785	54	A	-t	0.8082	0.9955	74.2N	112.2E	36	159	27	00m19s
4366	219	-0166 Nov 20	16:31:00	12318	-26779	59	T	-p	-0.5293	1.0157	51.1S	26.1W	58	12	64	01m15s
4367	219	-0165 May 17	11:36:29	12312	-26773	64	A	nn	0.0443	0.9573	21.0N	54.9E	87	171	156	05m21s
4368	219	-0165 Nov 10	07:25:27	12307	-26767	69	T	n-	0.1519	1.0475	7.9S	118.6E	81	191	160	04m28s
4369	219	-0164 May 05	12:12:03	12301	-26761	74	A	p-	-0.7198	0.9469	29.9S	57.2E	44	346	279	06m20s
4370	219	-0164 Oct 29	22:45:19	12295	-26755	79	T	p-	0.8343	1.0199	41.9N	97.1W	33	199	123	01m39s
4371	219	-0163 Mar 26	04:23:52	12291	-26750	46	P	-t	1.2760	0.4891	71.7N	74.9E	0	88		
4372	219	-0163 Apr 24	15:44:48	12290	-26749	84	P	t-	-1.4394	0.1987	70.9S	47.5E	0	308		
4373	219	-0163 Sep 19	18:59:07	12285	-26744	51	P	-t	-1.2845	0.4752	71.9S	136.3W	0	81		
4374	219	-0162 Mar 15	17:50:15	12279	-26738	56	T	-p	0.4980	1.0540	25.1N	43.3W	60	160	205	04m40s
4375	219	-0162 Sep 08	20:12:38	12274	-26732	61	A	-p	-0.5988	0.9284	28.2S	83.9W	53	19	332	08m39s
4376	219	-0161 Mar 05	10:23:04	12268	-26726	66	T	n-	-0.2163	1.0627	19.6S	82.8E	77	342	210	05m27s
4377	219	-0161 Aug 28	20:32:49	12262	-26720	71	A	nn	0.1290	0.9513	18.5N	74.9W	82	196	180	05m54s
4378	219	-0160 Feb 23	01:22:05	12257	-26714	76	T	p-	-0.9536	1.0099	74.2S	77.5W	17	288	117	00m36s
4379	219	-0160 Aug 17	02:53:04	12251	-26708	81	A	t-	0.8557	0.9973	70.2N	137.1W	31	220	18	00m11s
4380	219	-0159 Jan 12	18:05:07	12246	-26703	48	P	-t	1.1839	0.6453	67.6N	44.7W	0	173		

Cat Num	Canon Plate	Calendar Date	TD of Greatest Eclipse	ΔT s	Luna Num	Saros Num	Ecl. Type	QLE	Gamma	Ecl. Mag.	Lat. °	Long. °	Sun Alt °	Sun Azm °	Path Width km	Central Line Dur.
4381	220	-0159 Jul 08	08:48:26	12241	-26697	53	T-	-t	-1.0096	1.0051	66.8S	100.1E	0	359	-	-
4382	220	-0159 Aug 06	16:17:02	12240	-26696	91	Pb	t-	1.5260	0.0126	69.4N	137.1E	0	331		
4383	220	-0158 Jan 01	17:29:28	12235	-26691	58	A	-p	0.5128	0.9184	7.4N	28.4W	59	182	361	11m54s
4384	220	-0158 Jun 28	01:49:01	12230	-26685	63	T	-n	-0.2844	1.0639	7.3N	155.5W	74	355	218	06m13s
4385	220	-0158 Dec 21	18:44:20	12224	-26679	68	A	n-	-0.1862	0.9514	34.5S	51.2W	79	8	181	05m22s
4386	220	-0157 Jun 17	15:16:42	12218	-26673	73	H	p-	0.4943	1.0121	52.7N	7.2W	60	166	48	01m00s
4387	220	-0157 Dec 11	03:02:37	12213	-26667	78	A	p-	-0.8616	0.9976	77.4S	131.6E	30	58	17	00m09s
4388	220	-0156 May 07	07:31:45	12208	-26662	45	Pe	-t	-1.5131	0.0844	62.0S	174.2E	0	305		
4389	220	-0156 Jun 05	21:46:46	12207	-26661	83	P	t-	1.3130	0.4259	64.0N	115.3E	0	30		
4390	220	-0156 Oct 31	06:37:13	12202	-26656	50	P	-t	1.1922	0.6476	61.6N	168.0W	0	241		
4391	220	-0156 Nov 29	17:10:43	12202	-26655	88	P	t-	-1.4815	0.1033	63.6S	170.8W	0	145		
4392	220	-0155 Apr 26	08:09:37	12197	-26650	55	A	-p	-0.7577	0.9520	32.0S	130.6E	41	330	261	05m02s
4393	220	-0155 Oct 20	21:16:23	12191	-26644	60	H	-p	0.5448	1.0130	19.8N	75.9W	57	209	53	01m10s
4394	220	-0154 Apr 15	13:29:12	12186	-26638	65	H	nn	0.0334	1.0019	10.2N	27.4E	88	153	7	00m11s
4395	220	-0154 Oct 10	06:35:16	12180	-26632	70	A	nn	-0.1637	0.9624	13.7S	124.7E	80	29	138	03m52s
4396	220	-0153 Apr 05	01:53:17	12175	-26626	75	T	p-	0.7761	1.0463	46.8N	170.1E	39	135	240	03m13s
4397	220	-0153 Sep 29	08:46:34	12169	-26620	80	A	p-	-0.8929	0.9205	52.1S	51.8E	26	53	651	06m53s
4398	220	-0152 Feb 24	10:04:50	12164	-26615	47	P	-t	-1.1053	0.8131	61.3S	165.9W	0	247		
4399	220	-0152 Mar 24	18:24:00	12163	-26614	85	P	t-	1.4645	0.1261	60.9N	133.9W	0	89		
4400	220	-0152 Aug 18	16:46:22	12159	-26609	52	P	-t	1.3214	0.4113	61.8N	98.1E	0	302		
4401	221	-0151 Feb 12	23:42:12	12153	-26603	57	H	-p	-0.4239	1.0019	37.1S	107.6W	65	330	7	00m10s
4402	221	-0151 Aug 08	01:20:46	12148	-26597	62	T	-p	0.5222	1.0218	46.3N	132.5W	58	210	87	01m43s
4403	221	-0150 Feb 02	06:31:32	12142	-26591	67	A	n-	0.3067	0.9469	1.1S	130.6E	72	159	204	06m25s
4404	221	-0150 Jul 28	16:13:20	12136	-26585	72	T	n-	-0.2286	1.0689	7.7N	16.2W	77	17	230	06m11s
4405	221	-0149 Jan 22	06:49:55	12131	-26579	77	A+	t-	1.0200	0.9166	63.1N	89.6E	0	141	-	-
4406	221	-0149 Jul 18	09:15:06	12125	-26573	82	T	t-	-0.9567	1.0553	49.1S	71.0E	16	20	635	04m06s
4407	221	-0149 Dec 12	14:58:47	12121	-26568	49	P	-t	-1.2695	0.5013	66.9S	175.2W	0	180		
4408	221	-0148 Jun 07	13:28:56	12115	-26562	54	A	-t	0.8904	0.9897	85.5N	4.6E	27	157	81	00m39s
4409	221	-0148 Dec 01	01:16:43	12110	-26556	59	T	-p	-0.5288	1.0179	53.7S	154.3W	58	7	72	01m23s
4410	221	-0147 May 27	18:01:25	12104	-26550	64	A	nn	0.1319	0.9560	28.5N	42.7W	82	174	162	05m21s
4411	221	-0147 Nov 20	16:17:33	12099	-26544	69	T	n-	0.1496	1.0468	11.0S	15.4W	81	188	157	04m26s
4412	221	-0146 May 16	18:36:34	12093	-26538	74	A	p-	-0.6331	0.9504	20.6S	44.1W	51	350	234	06m22s
4413	221	-0146 Nov 10	07:27:50	12087	-26532	79	T	p-	0.8320	1.0163	38.9N	127.7E	33	194	101	01m25s
4414	221	-0145 Apr 06	11:45:06	12083	-26527	46	P	-t	1.3309	0.3886	71.6N	50.2W	0	74		
4415	221	-0145 May 05	22:36:34	12082	-26526	84	P	t-	-1.3604	0.3385	70.2S	69.3W	0	321		
4416	221	-0145 Oct 01	02:44:19	12077	-26521	51	P	-t	-1.3139	0.4249	71.9S	92.3E	0	95		
4417	221	-0144 Mar 26	01:39:38	12072	-26515	56	T	-p	0.5482	1.0584	32.4N	164.5W	57	159	229	04m47s
4418	221	-0144 Sep 19	03:32:37	12066	-26509	61	A	-p	-0.6376	0.9247	34.8S	162.2E	50	22	364	08m36s
4419	221	-0143 Mar 15	18:24:02	12061	-26503	66	T	n-	-0.1720	1.0650	12.8S	39.9W	80	342	216	05m45s
4420	221	-0143 Sep 08	03:54:50	12055	-26497	71	A	nn	0.0843	0.9505	11.9N	172.5E	85	197	182	06m03s
4421	222	-0142 Mar 05	09:17:45	12050	-26491	76	T	p-	-0.9180	1.0118	67.5S	138.9E	23	311	103	00m46s
4422	222	-0142 Aug 28	10:32:25	12044	-26485	81	A	p-	0.8060	0.9988	61.7N	99.0E	36	213	7	00m05s
4423	222	-0141 Jan 24	01:56:29	12039	-26480	48	P	-t	1.2040	0.6116	68.7N	174.8W	0	161		
4424	222	-0141 Jul 19	16:24:37	12034	-26474	53	P	-t	-1.0736	0.8805	67.8S	25.8W	0	9		
4425	222	-0141 Aug 18	00:07:27	12033	-26473	91	P	t-	1.4734	0.1129	70.2N	6.2E	0	319		
4426	222	-0140 Jan 13	01:20:54	12028	-26468	58	A	-p	0.5265	0.9208	9.6N	148.4W	58	178	352	11m25s
4427	222	-0140 Jul 08	09:14:45	12023	-26462	63	T	-p	-0.3557	1.0594	2.3N	91.6E	69	360	209	05m53s
4428	222	-0139 Jan 01	02:59:33	12017	-26456	68	A	n-	-0.1796	0.9554	34.0S	173.4W	79	2	166	04m57s
4429	222	-0139 Jun 27	22:15:44	12012	-26450	73	H	p-	0.4178	1.0081	48.5N	107.6W	65	173	31	00m42s
4430	222	-0139 Dec 21	11:43:23	12006	-26444	78	H	p-	-0.8574	1.0012	80.8S	15.2E	31	43	8	00m05s
4431	222	-0138 Jun 17	04:16:00	12001	-26438	83	P	t-	1.2315	0.5673	64.9N	7.4E	0	21		
4432	222	-0138 Nov 11	15:26:07	11996	-26433	50	P	-t	1.1981	0.6362	62.2N	50.0E	0	231		
4433	222	-0138 Dec 11	02:03:35	11995	-26432	88	P	t-	-1.4787	0.1081	64.5S	45.5E	0	155		
4434	222	-0137 May 07	14:42:41	11991	-26427	55	A	-p	-0.8381	0.9538	36.5S	32.4E	33	331	301	04m46s
4435	222	-0137 Nov 01	05:49:52	11985	-26421	60	H	-p	0.5549	1.0083	17.2N	153.6E	56	206	34	00m47s
4436	222	-0136 Apr 25	20:34:51	11980	-26415	65	H	nn	-0.0381	1.0074	10.3N	78.8W	88	333	26	00m44s
4437	222	-0136 Oct 20	14:39:14	11974	-26409	70	A	nn	-0.1476	0.9574	17.2S	2.9E	81	28	157	04m26s
4438	222	-0135 Apr 15	09:27:08	11969	-26403	75	T	p-	0.7119	1.0526	46.8N	58.9E	44	138	244	03m39s
4439	222	-0135 Oct 09	16:24:33	11963	-26397	80	A	p-	-0.8680	0.9177	53.7S	63.8W	29	54	614	07m02s
4440	222	-0134 Mar 06	18:09:42	11959	-26392	47	P	-t	-1.1471	0.7342	61.0S	63.2E	0	256		

Cat Num	Canon Plate	Calendar Date	TD of Greatest Eclipse	ΔT s	Luna Num	Saros Num	Ecl. Type	QLE	Gamma	Ecl. Mag.	Lat. °	Long. °	Sun Alt °	Sun Azm °	Path Width km	Central Line Dur.
4441	223	-0134 Apr 05	02:10:19	11958	-26391	85	P	t-	1.4096	0.2311	61.0N	99.9E	0	81		
4442	223	-0134 Aug 30	00:07:33	11953	-26386	52	P	-t	1.3664	0.3330	61.3N	21.9W	0	293		
4443	223	-0134 Sep 28	15:55:39	11952	-26385	90	Pb	t-	-1.5411	0.0433	60.7S	102.3W	0	92		
4444	223	-0133 Feb 24	07:38:33	11948	-26380	57	H	-p	-0.4601	1.0015	34.9S	134.0E	62	327	6	00m08s
4445	223	-0133 Aug 19	09:00:20	11942	-26374	62	T	-p	0.5723	1.0215	45.3N	115.1E	55	215	89	01m40s
4446	223	-0132 Feb 13	14:11:16	11937	-26368	67	A	nn	0.2761	0.9476	0.1N	14.8E	74	156	200	06m09s
4447	223	-0132 Aug 07	23:59:55	11931	-26362	72	T	n-	-0.1714	1.0676	8.4N	133.2W	80	21	224	05m53s
4448	223	-0131 Feb 01	14:27:14	11926	-26356	77	An	t-	0.9951	0.9121	59.7N	29.0W	4	136	-	07m45s
4449	223	-0131 Jul 28	16:51:40	11921	-26350	82	T	p-	-0.8959	1.0539	40.6S	44.3W	26	23	398	04m14s
4450	223	-0131 Dec 22	23:24:08	11916	-26345	49	P	-t	-1.2683	0.5036	65.8S	47.1E	0	191		
4451	223	-0130 Jun 18	20:17:43	11911	-26339	54	A	-t	0.9739	0.9819	78.6N	91.6E	12	348	303	01m02s
4452	223	-0130 Dec 12	10:03:40	11905	-26333	59	T	-p	-0.5279	1.0206	55.3S	78.4E	58	359	83	01m35s
4453	223	-0129 Jun 08	00:24:36	11900	-26327	64	A	nn	0.2205	0.9542	35.5N	138.7W	77	178	172	05m19s
4454	223	-0129 Dec 02	01:12:00	11894	-26321	69	T	n-	0.1490	1.0465	13.3S	149.6W	82	184	156	04m25s
4455	223	-0128 May 27	01:01:11	11889	-26315	74	A	p-	-0.5459	0.9534	12.3S	144.3W	57	354	203	06m16s
4456	223	-0128 Nov 20	16:12:27	11883	-26309	79	T	p-	0.8319	1.0131	36.5N	7.8W	33	190	81	01m10s
4457	223	-0127 Apr 16	19:03:40	11879	-26304	46	P	-t	1.3888	0.2815	71.2N	174.4W	0	61		
4458	223	-0127 May 16	05:28:25	11878	-26303	84	P	t-	-1.2800	0.4823	69.4S	174.5E	0	333		
4459	223	-0127 Oct 11	10:35:28	11873	-26298	51	P	-t	-1.3377	0.3849	71.7S	40.5W	0	109		
4460	223	-0126 Apr 06	09:22:42	11868	-26292	56	T	-p	0.6041	1.0623	40.3N	75.6E	53	157	255	04m48s
4461	224	-0126 Sep 30	11:01:41	11862	-26286	61	A	-p	-0.6688	0.9214	41.3S	46.2E	48	24	395	08m30s
4462	224	-0125 Mar 27	02:17:01	11857	-26280	66	T	nn	-0.1213	1.0669	5.7S	161.0W	83	342	220	05m59s
4463	224	-0125 Sep 19	11:27:28	11852	-26274	71	A	nn	0.0475	0.9500	5.6N	57.1E	87	198	184	06m08s
4464	224	-0124 Mar 15	17:04:11	11846	-26268	76	T	p-	-0.8757	1.0131	59.5S	8.0E	28	323	93	00m56s
4465	224	-0124 Sep 07	18:22:16	11841	-26262	81	A	p-	0.7642	0.9999	54.1N	24.4W	40	209	1	00m00s
4466	224	-0123 Feb 03	09:38:33	11836	-26257	48	P	-t	1.2307	0.5665	69.7N	56.7E	0	149		
4467	224	-0123 Jul 30	00:06:03	11831	-26251	53	P	-t	-1.1331	0.7648	68.8S	153.5W	0	21		
4468	224	-0123 Aug 28	08:05:47	11830	-26250	91	P	t-	1.4272	0.2012	70.9N	127.3W	0	306		
4469	224	-0122 Jan 23	09:05:39	11825	-26245	58	A	-p	0.5456	0.9237	12.7N	93.0E	57	173	343	10m42s
4470	224	-0122 Jul 19	16:43:53	11820	-26239	63	T	-p	-0.4232	1.0541	3.2S	22.9W	65	4	197	05m24s
4471	224	-0121 Jan 12	11:09:11	11815	-26233	68	A	n-	-0.1679	0.9601	32.1S	65.5E	80	357	147	04m27s
4472	224	-0121 Jul 09	05:17:01	11809	-26227	73	H	n-	0.3444	1.0034	43.5N	149.8E	70	180	12	00m19s
4473	224	-0120 Jan 01	20:21:20	11804	-26221	78	H	p-	-0.8509	1.0056	81.9S	86.0W	31	13	37	00m21s
4474	224	-0120 Jun 27	10:45:54	11798	-26215	83	P	t-	1.1507	0.7067	65.8N	100.9W	0	11		
4475	224	-0120 Nov 22	00:18:07	11794	-26210	50	P	-t	1.2013	0.6299	62.9N	93.0W	0	222		
4476	224	-0120 Dec 21	10:55:43	11793	-26209	88	P	t-	-1.4757	0.1131	65.5S	98.4W	0	166		
4477	224	-0119 May 17	21:13:08	11788	-26204	55	A	-t	-0.9208	0.9546	44.2S	64.3W	23	333	419	04m28s
4478	224	-0119 Nov 11	14:27:46	11783	-26198	60	H	-p	0.5607	1.0041	14.8N	21.8E	56	202	17	00m24s
4479	224	-0118 May 07	03:35:48	11778	-26192	65	Hm	nn	-0.1145	1.0125	9.7N	176.2E	83	337	43	01m15s
4480	224	-0118 Oct 31	22:49:30	11772	-26186	70	A	nn	-0.1375	0.9529	20.8S	120.2W	82	26	174	05m00s
4481	225	-0117 Apr 26	16:56:27	11767	-26180	75	T	p-	0.6433	1.0583	47.0N	50.5W	50	141	248	04m04s
4482	225	-0117 Oct 21	00:09:43	11761	-26174	80	A	p-	-0.8491	0.9153	56.4S	178.5E	32	55	599	07m08s
4483	225	-0116 Mar 17	02:07:30	11757	-26169	47	P	-t	-1.1943	0.6448	60.8S	65.8W	0	265		
4484	225	-0116 Apr 15	09:51:58	11756	-26168	85	P	t-	1.3512	0.3437	61.2N	25.2W	0	72		
4485	225	-0116 Sep 09	07:38:46	11752	-26163	52	P	-t	1.4039	0.2680	61.0N	144.3W	0	284		
4486	225	-0116 Oct 08	23:39:19	11751	-26162	90	P	t-	-1.5142	0.0890	60.8S	132.3E	0	101		
4487	225	-0115 Mar 06	15:26:42	11746	-26157	57	H	-p	-0.5024	1.0009	32.9S	17.3E	60	326	4	00m05s
4488	225	-0115 Aug 29	16:47:48	11741	-26151	62	T	-p	0.6167	1.0210	43.6N	0.3W	52	218	90	01m36s
4489	225	-0114 Feb 23	21:42:08	11735	-26145	67	A	nn	0.2386	0.9485	1.6N	98.7W	76	154	194	05m53s
4490	225	-0114 Aug 19	07:54:00	11730	-26139	72	T	n-	-0.1203	1.0658	8.0N	108.0E	83	24	216	05m36s
4491	225	-0113 Feb 12	21:54:51	11725	-26133	77	A	t-	0.9620	0.9179	51.6N	135.4W	15	140	1135	07m43s
4492	225	-0113 Aug 09	00:35:30	11719	-26127	82	T	p-	-0.8414	1.0510	36.0S	162.0W	32	26	308	04m05s
4493	225	-0112 Jan 03	07:45:39	11715	-26122	49	P	-t	-1.2712	0.4985	64.8S	89.3W	0	202		
4494	225	-0112 Jun 29	03:08:17	11709	-26116	54	P	-t	1.0548	0.8844	65.4N	16.0W	0	344		
4495	225	-0112 Dec 22	18:48:32	11704	-26110	59	T	-p	-0.5295	1.0238	55.7S	47.7W	58	352	96	01m47s
4496	225	-0111 Jun 18	06:48:07	11699	-26104	64	A	np	0.3085	0.9519	41.7N	126.6E	72	183	185	05m17s
4497	225	-0111 Dec 12	10:05:34	11693	-26098	69	T	n-	0.1478	1.0467	14.8S	76.7E	82	179	157	04m26s
4498	225	-0110 Jun 07	07:24:32	11688	-26092	74	A	p-	-0.4568	0.9557	4.7S	116.8E	63	358	181	06m05s
4499	225	-0110 Dec 02	00:58:31	11683	-26086	79	T	p-	0.8335	1.0104	34.8N	143.6W	33	184	65	00m58s
4500	225	-0109 Apr 28	02:16:52	11678	-26081	46	P	-t	1.4516	0.1641	70.6N	63.3E	0	48		

Cat Num	Canon Plate	Calendar Date	TD of Greatest Eclipse	ΔT s	Luna Num	Saros Num	Ecl. Type	QLE	Gamma	Ecl. Mag.	Lat. °	Long. °	Sun Alt °	Sun Azm °	Path Width km	Central Line Dur.
4501	226	-0109 May 27	12:19:21	11677	-26080	84	P	t-	-1.1974	0.6312	68.4S	59.1E	0	344		
4502	226	-0109 Oct 22	18:33:27	11673	-26075	51	P	-t	-1.3552	0.3557	71.2S	174.7W	0	123		
4503	226	-0108 Apr 16	16:59:52	11668	-26069	56	T	-p	0.6655	1.0654	48.8N	43.1W	48	155	286	04m43s
4504	226	-0108 Oct 10	18:39:51	11662	-26063	61	A	-p	-0.6925	0.9186	47.3S	71.7W	46	26	424	08m21s
4505	226	-0107 Apr 06	10:01:59	11657	-26057	66	Tm	nn	-0.0640	1.0684	1.7N	79.8E	86	343	223	06m10s
4506	226	-0107 Sep 29	19:10:01	11652	-26051	71	A	nn	0.0183	0.9495	0.4S	60.6W	89	198	185	06m12s
4507	226	-0106 Mar 27	00:41:08	11646	-26045	76	T	p-	-0.8261	1.0142	51.0S	116.0W	34	331	86	01m07s
4508	226	-0106 Sep 19	02:21:57	11641	-26039	81	H	p-	0.7298	1.0006	47.1N	149.0W	43	207	3	00m03s
4509	226	-0105 Feb 14	17:11:14	11636	-26034	48	P	-t	1.2643	0.5099	70.5N	70.0W	0	137		
4510	226	-0105 Aug 10	07:54:22	11631	-26028	53	P	-t	-1.1869	0.6604	69.7S	76.5E	0	33		
4511	226	-0105 Sep 08	16:13:09	11630	-26027	91	P	t-	1.3881	0.2759	71.4N	96.5E	0	293		
4512	226	-0104 Feb 03	16:44:32	11626	-26022	58	A	-p	0.5696	0.9270	16.8N	24.5W	55	170	332	09m50s
4513	226	-0104 Jul 30	00:16:56	11620	-26016	63	T	-p	-0.4867	1.0484	9.2S	138.9W	61	7	184	04m48s
4514	226	-0103 Jan 22	19:15:23	11615	-26010	68	A	nn	-0.1527	0.9652	29.1S	55.3W	81	353	127	03m53s
4515	226	-0103 Jul 19	12:20:49	11610	-26004	73	A	nn	0.2740	0.9982	37.8N	45.0E	74	185	6	00m11s
4516	226	-0102 Jan 12	04:55:26	11604	-25998	78	T	p-	-0.8403	1.0104	79.5S	170.3E	32	346	67	00m39s
4517	226	-0102 Jul 08	17:18:06	11599	-25992	83	P	t-	1.0718	0.8418	66.8N	149.7E	0	1		
4518	226	-0102 Dec 03	09:10:52	11595	-25987	50	P	-t	1.2041	0.6244	63.8N	123.6E	0	212		
4519	226	-0101 Jan 01	19:44:17	11594	-25986	88	P	t-	-1.4699	0.1233	66.5S	118.1E	0	176		
4520	226	-0101 May 29	03:44:34	11589	-25981	55	A-	-t	-1.0032	0.9654	63.3S	149.6W	0	323	-	-
4521	227	-0101 Nov 22	23:06:25	11584	-25975	60	H	-p	0.5654	1.0005	12.9N	110.1W	55	198	2	00m03s
4522	227	-0100 May 17	10:37:31	11579	-25969	65	H2	nn	-0.1912	1.0170	8.4N	70.9E	79	340	59	01m44s
4523	227	-0100 Nov 11	07:02:52	11573	-25963	70	A	nn	-0.1308	0.9490	24.1S	116.3E	82	23	189	05m32s
4524	227	-0099 May 07	00:22:02	11568	-25957	75	T	p-	0.5708	1.0633	46.9N	158.3W	55	146	251	04m28s
4525	227	-0099 Oct 31	08:01:17	11563	-25951	80	A	p-	-0.8354	0.9132	59.8S	59.2E	33	56	594	07m11s
4526	227	-0098 Mar 28	09:55:44	11558	-25946	47	P	-t	-1.2491	0.5408	60.8S	167.6E	0	273		
4527	227	-0098 Apr 26	17:26:44	11558	-25945	85	P	t-	1.2874	0.4672	61.6N	148.5W	0	63		
4528	227	-0098 Sep 20	15:21:31	11553	-25940	52	P	-t	1.4331	0.2172	60.9N	90.5E	0	275		
4529	227	-0098 Oct 20	07:32:27	11552	-25939	90	P	t-	-1.4940	0.1233	61.1S	4.5E	0	110		
4530	227	-0097 Mar 17	23:02:59	11548	-25934	57	H	-p	-0.5539	1.0001	31.4S	96.6W	56	325	0	00m00s
4531	227	-0097 Sep 10	00:46:41	11543	-25928	62	T	-p	0.6525	1.0204	41.3N	119.8W	49	219	90	01m34s
4532	227	-0096 Mar 06	05:02:17	11537	-25922	67	A	nn	0.1928	0.9495	3.1N	150.7E	79	152	188	05m38s
4533	227	-0096 Aug 29	15:55:39	11532	-25916	72	T	nn	-0.0752	1.0636	6.7N	12.8W	86	27	209	05m18s
4534	227	-0095 Feb 23	05:14:37	11527	-25910	77	A	p-	0.9222	0.9229	47.7N	115.3E	22	139	732	07m19s
4535	227	-0095 Aug 19	08:24:57	11521	-25904	82	T	p-	-0.7919	1.0472	33.5S	78.9E	37	29	252	03m47s
4536	227	-0094 Jan 13	16:04:15	11517	-25899	49	P	-t	-1.2771	0.4883	63.8S	135.5E	0	212		
4537	227	-0094 Jul 10	09:59:11	11512	-25893	54	P	-t	1.1341	0.7419	64.4N	129.5W	0	335		
4538	227	-0093 Jan 03	03:31:36	11507	-25887	59	T	-p	-0.5331	1.0275	55.0S	173.5W	58	344	110	02m02s
4539	227	-0093 Jun 29	13:13:16	11501	-25881	64	A	-p	0.3945	0.9493	46.9N	33.2E	67	189	203	05m17s
4540	227	-0093 Dec 23	18:57:31	11496	-25875	69	T	n-	0.1453	1.0474	15.5S	56.5W	82	175	159	04m28s
4541	228	-0092 Jun 17	13:51:37	11491	-25869	74	A	p-	-0.3699	0.9576	1.8N	18.0E	68	2	166	05m48s
4542	228	-0092 Dec 12	09:44:07	11485	-25863	79	H	p-	0.8354	1.0081	33.7N	80.7E	33	179	51	00m46s
4543	228	-0091 May 08	09:27:57	11481	-25858	46	Pe	-t	1.5167	0.0414	69.8N	57.9W	0	36		
4544	228	-0091 Jun 06	19:11:26	11480	-25857	84	P	t-	-1.1144	0.7821	67.4S	56.1W	0	355		
4545	228	-0091 Nov 02	02:35:52	11476	-25852	51	P	-t	-1.3683	0.3341	70.5S	50.5E	0	137		
4546	228	-0090 Apr 28	00:32:53	11471	-25846	56	T	-p	0.7309	1.0677	57.7N	161.4W	43	153	324	04m32s
4547	228	-0090 Oct 22	02:25:08	11465	-25840	61	A	-p	-0.7103	0.9164	53.1S	169.6E	44	26	450	08m11s
4548	228	-0089 Apr 17	17:40:32	11460	-25834	66	T	nn	-0.0015	1.0692	9.3N	37.7W	90	314	225	06m15s
4549	228	-0089 Oct 11	03:02:29	11455	-25828	71	A	nn	-0.0036	0.9494	6.0S	179.4E	90	14	186	06m12s
4550	228	-0088 Apr 06	08:09:48	11450	-25822	76	T	p-	-0.7707	1.0148	42.5S	124.2E	39	336	79	01m16s
4551	228	-0088 Sep 29	10:30:23	11444	-25816	81	H	p-	0.7020	1.0013	40.8N	84.7E	45	204	6	00m07s
4552	228	-0087 Feb 25	00:33:44	11440	-25811	48	P	-t	1.3053	0.4406	71.2N	165.3E	0	123		
4553	228	-0087 Aug 20	15:49:58	11435	-25805	53	P	-t	-1.2346	0.5682	70.5S	55.9W	0	45		
4554	228	-0087 Sep 19	00:29:08	11434	-25804	91	P	t-	1.3559	0.3372	71.7N	42.3W	0	279		
4555	228	-0086 Feb 14	00:13:10	11429	-25799	58	A	-p	0.6022	0.9307	22.1N	140.0W	53	166	323	08m51s
4556	228	-0086 Aug 10	07:56:02	11424	-25793	63	T	-p	-0.5444	1.0422	15.6S	103.1E	57	11	168	04m06s
4557	228	-0085 Feb 03	03:13:49	11419	-25787	68	A	nn	-0.1302	0.9709	25.0S	175.0W	82	349	105	03m15s
4558	228	-0085 Jul 30	19:29:21	11414	-25781	73	A	nn	0.2084	0.9926	31.7N	62.1W	78	189	27	00m46s
4559	228	-0084 Jan 23	13:23:46	11409	-25775	78	T	p-	-0.8246	1.0158	74.8S	53.0E	34	334	97	01m01s
4560	228	-0084 Jul 18	23:54:35	11403	-25769	83	An	t-	0.9965	0.9375	69.9N	38.0E	2	349	-	03m39s

Cat Num	Canon Plate	Calendar Date	TD of Greatest Eclipse	ΔT s	Luna Num	Saros Num	Ecl. Type	QLE	Gamma	Ecl. Mag.	Lat. °	Long. °	Sun Alt °	Sun Azm °	Path Width km	Central Line Dur.
4561	229	-0084 Dec 13	18:04:07	11399	-25764	50	P	-t	1.2064	0.6199	64.8N	20.2W	0	202		
4562	229	-0083 Jan 12	04:29:23	11398	-25763	88	P	t-	-1.4617	0.1379	67.6S	24.9W	0	188		
4563	229	-0083 Jun 08	10:14:58	11394	-25758	55	P	-t	-1.0867	0.8208	64.2S	102.4E	0	332		
4564	229	-0083 Dec 03	07:47:15	11388	-25752	60	A	-p	0.5675	0.9974	11.6N	117.4E	55	194	11	00m16s
4565	229	-0082 May 28	17:37:45	11383	-25746	65	T	nn	-0.2698	1.0209	6.0N	34.3W	74	344	74	02m11s
4566	229	-0082 Nov 22	15:19:03	11378	-25740	70	A	nn	-0.1272	0.9458	27.0S	7.5W	83	19	201	06m01s
4567	229	-0081 May 18	07:45:45	11373	-25734	75	T	p-	0.4964	1.0674	46.2N	94.6E	60	152	253	04m52s
4568	229	-0081 Nov 11	15:58:12	11368	-25728	80	A	p-	-0.8266	0.9117	63.7S	60.8W	34	55	595	07m12s
4569	229	-0080 Apr 07	17:37:49	11363	-25723	47	P	-t	-1.3085	0.4277	60.9S	42.6E	0	282		
4570	229	-0080 May 07	00:58:19	11362	-25722	85	P	t-	1.2210	0.5958	62.1N	88.8E	0	54		
4571	229	-0080 Sep 30	23:12:44	11358	-25717	52	P	-t	1.4562	0.1769	60.9N	36.8W	0	266		
4572	229	-0080 Oct 30	15:31:59	11357	-25716	90	P	t-	-1.4782	0.1498	61.5S	124.9W	0	119		
4573	229	-0079 Mar 28	06:31:33	11353	-25711	57	A	-p	-0.6108	0.9989	30.7S	151.3E	52	325	5	00m06s
4574	229	-0079 Sep 20	08:53:49	11348	-25705	62	T	-p	0.6823	1.0197	38.9N	117.8E	47	219	90	01m32s
4575	229	-0078 Mar 17	12:11:29	11342	-25699	67	A	nn	0.1385	0.9506	4.5N	43.0E	82	151	183	05m27s
4576	229	-0078 Sep 10	00:05:45	11337	-25693	72	T	nn	-0.0372	1.0611	4.6N	135.8W	88	29	201	05m02s
4577	229	-0077 Mar 06	12:24:40	11332	-25687	77	A	p-	0.8741	0.9279	45.2N	8.4E	29	138	539	06m50s
4578	229	-0077 Aug 30	16:21:46	11327	-25681	82	T	p-	-0.7491	1.0429	32.7S	42.0W	41	32	211	03m25s
4579	229	-0076 Jan 25	00:16:04	11323	-25676	49	P	-t	-1.2891	0.4668	63.0S	2.4E	0	221		
4580	229	-0076 Feb 23	14:20:30	11322	-25675	87	Pb	t-	1.5513	0.0121	61.3N	53.0W	0	113		
4581	230	-0076 Jul 20	16:54:35	11317	-25670	54	P	-t	1.2084	0.6094	63.5N	116.3E	0	325		
4582	230	-0076 Aug 19	05:01:57	11316	-25669	92	Pb	t-	-1.5314	0.0299	61.8S	90.1E	0	59		
4583	230	-0075 Jan 13	12:09:37	11312	-25664	59	T	-p	-0.5414	1.0315	53.4S	61.3E	57	337	127	02m19s
4584	230	-0075 Jul 09	19:40:40	11307	-25658	64	A	-p	0.4782	0.9461	51.0N	59.2W	61	197	226	05m18s
4585	230	-0074 Jan 03	03:45:55	11302	-25652	69	T	n-	0.1398	1.0485	15.5S	171.3E	82	170	163	04m30s
4586	230	-0074 Jun 28	20:21:24	11297	-25646	74	A	pn	-0.2842	0.9589	7.3N	80.8W	74	6	156	05m29s
4587	230	-0074 Dec 23	18:26:24	11291	-25640	79	H	p-	0.8350	1.0066	32.9N	54.2W	33	174	41	00m37s
4588	230	-0073 Jun 18	02:06:24	11286	-25634	84	P	t-	-1.0323	0.9320	66.4S	171.4W	0	5		
4589	230	-0073 Nov 13	10:43:12	11282	-25629	51	P	-t	-1.3771	0.3199	69.6S	84.9W	0	150		
4590	230	-0072 May 08	08:01:55	11277	-25623	56	T	-p	0.7997	1.0690	67.2N	79.8E	37	150	376	04m16s
4591	230	-0072 Nov 01	10:17:08	11272	-25617	61	A	-p	-0.7225	0.9147	58.4S	50.6E	43	26	470	08m01s
4592	230	-0071 Apr 28	01:12:51	11266	-25611	66	T	nn	0.0658	1.0694	16.8N	153.3W	86	166	226	06m13s
4593	230	-0071 Oct 21	11:03:45	11261	-25605	71	A	nn	-0.0186	0.9495	11.1S	57.5E	89	15	185	06m09s
4594	230	-0070 Apr 17	15:29:01	11256	-25599	76	T	p-	-0.7084	1.0149	34.0S	8.1E	45	341	72	01m22s
4595	230	-0070 Oct 10	18:48:28	11251	-25593	81	H	p-	0.6815	1.0020	35.2N	43.5W	47	201	9	00m11s
4596	230	-0069 Mar 08	07:46:43	11247	-25588	48	P	-t	1.3531	0.3595	71.7N	42.5E	0	110		
4597	230	-0069 Apr 06	23:01:31	11246	-25587	86	Pb	t-	-1.5229	0.0607	71.6S	44.3W	0	287		
4598	230	-0069 Aug 31	23:52:47	11241	-25582	53	P	-t	-1.2764	0.4878	71.1S	169.3E	0	58		
4599	230	-0069 Sep 30	08:52:47	11241	-25581	91	P	t-	1.3300	0.3866	71.8N	176.8E	0	265		
4600	230	-0068 Feb 25	07:35:07	11236	-25576	58	A	-p	0.6405	0.9345	28.2N	105.7E	50	162	315	07m51s
4601	231	-0068 Aug 20	15:41:05	11231	-25570	63	T	-p	-0.5964	1.0357	22.1S	16.9W	53	15	149	03m22s
4602	231	-0067 Feb 13	11:06:56	11226	-25564	68	A	nn	-0.1026	0.9769	20.0S	65.9E	84	346	83	02m35s
4603	231	-0067 Aug 10	02:42:07	11221	-25558	73	A	nn	0.1474	0.9867	25.3N	171.2W	81	192	47	01m26s
4604	231	-0066 Feb 02	21:46:36	11216	-25552	78	T	p-	-0.8039	1.0217	69.1S	71.0W	36	331	125	01m26s
4605	231	-0066 Jul 30	06:36:44	11211	-25546	83	A	t-	0.9262	0.9386	82.5N	158.4W	22	256	617	04m17s
4606	231	-0066 Dec 25	02:53:42	11206	-25541	50	P	-t	1.2115	0.6103	65.8N	163.5W	0	191		
4607	231	-0065 Jan 23	13:07:21	11205	-25540	88	P	t-	-1.4482	0.1625	68.6S	166.7W	0	199		
4608	231	-0065 Jun 19	16:50:19	11201	-25535	55	P	-t	-1.1662	0.6822	65.1S	7.1W	0	342		
4609	231	-0065 Dec 14	16:26:07	11196	-25529	60	A	-p	0.5702	0.9949	11.0N	14.5W	55	189	22	00m33s
4610	231	-0064 Jun 08	00:40:35	11191	-25523	65	T	-n	-0.3471	1.0241	2.8N	140.7W	70	348	87	02m34s
4611	231	-0064 Dec 02	23:34:49	11186	-25517	70	Am	nn	-0.1241	0.9432	29.1S	130.7W	83	15	212	06m28s
4612	231	-0063 May 28	15:08:22	11181	-25511	75	T	p-	0.4207	1.0707	44.8N	12.2W	65	158	253	05m15s
4613	231	-0063 Nov 21	23:58:15	11175	-25505	80	A	p-	-0.8201	0.9108	67.8S	179.9E	35	53	595	07m12s
4614	231	-0062 Apr 19	01:10:25	11171	-25500	47	P	-t	-1.3750	0.3011	61.2S	80.2W	0	291		
4615	231	-0062 May 18	08:24:31	11170	-25499	85	P	t-	1.1504	0.7326	62.7N	32.7W	0	45		
4616	231	-0062 Oct 12	07:15:02	11166	-25494	52	P	-t	1.4713	0.1506	61.0N	166.8W	0	257		
4617	231	-0062 Nov 10	23:38:44	11165	-25493	90	P	t-	-1.4676	0.1675	62.2S	103.7E	0	128		
4618	231	-0061 Apr 08	13:49:15	11161	-25488	57	A	-p	-0.6756	0.9971	31.0S	41.9E	47	326	14	00m16s
4619	231	-0061 Oct 01	17:10:28	11156	-25482	62	T	-p	0.7048	1.0192	36.3N	8.0W	45	217	91	01m31s
4620	231	-0060 Mar 27	19:10:48	11151	-25476	67	Am	nn	0.0768	0.9515	5.6N	61.9W	86	151	178	05m21s

Cat Num	Canon Plate	Calendar Date	TD of Greatest Eclipse	ΔT s	Luna Num	Saros Num	Ecl. Type	QLE	Gamma	Ecl. Mag.	Lat. °	Long. °	Sun Alt °	Sun Azm °	Path Width km	Central Line Dur.
4621	232	-0060 Sep 20	08:24:07	11146	-25470	72	T	nn	-0.0062	1.0584	1.8N	98.9E	90	30	192	04m48s
4622	232	-0059 Mar 16	19:25:48	11140	-25464	77	A	p-	0.8184	0.9329	43.7N	96.0W	35	138	422	06m19s
4623	232	-0059 Sep 10	00:25:09	11135	-25458	82	T	p-	-0.7123	1.0380	33.3S	164.5W	44	34	178	02m59s
4624	232	-0058 Feb 04	08:23:05	11131	-25453	49	P	-t	-1.3058	0.4370	62.3S	129.4W	0	231		
4625	232	-0058 Mar 05	21:50:17	11130	-25452	87	P	t-	1.5057	0.0889	61.0N	175.0W	0	104		
4626	232	-0058 Jul 31	23:53:34	11126	-25447	54	P	-t	1.2787	0.4855	62.7N	1.4E	0	316		
4627	232	-0058 Aug 30	12:36:27	11125	-25446	92	P	t-	-1.4902	0.1066	61.4S	33.2W	0	68		
4628	232	-0057 Jan 24	20:42:31	11121	-25441	59	T	-p	-0.5546	1.0359	51.1S	63.7W	56	331	145	02m36s
4629	232	-0057 Jul 21	02:13:29	11116	-25435	64	A	-p	0.5572	0.9427	53.7N	151.6W	56	205	254	05m23s
4630	232	-0056 Jan 14	12:30:30	11111	-25429	69	T	n-	0.1310	1.0499	14.7S	39.9E	83	166	167	04m33s
4631	232	-0056 Jul 09	02:56:21	11105	-25423	74	A	nn	-0.2021	0.9599	11.7N	179.9E	78	11	149	05m09s
4632	232	-0055 Jan 03	03:05:23	11100	-25417	79	H	p-	0.8326	1.0054	32.5N	171.8E	33	169	34	00m31s
4633	232	-0055 Jun 28	09:05:34	11095	-25411	84	A	t-	-0.9523	0.9991	48.5S	79.5E	17	9	10	00m05s
4634	232	-0055 Nov 23	18:51:48	11091	-25406	51	P	-t	-1.3840	0.3090	68.6S	140.0E	0	162		
4635	232	-0054 May 19	15:28:23	11086	-25400	56	T	-p	0.8709	1.0691	77.4N	44.6W	29	140	463	03m55s
4636	232	-0054 Nov 12	18:13:57	11081	-25394	61	A	-p	-0.7309	0.9137	63.2S	67.7W	43	23	483	07m49s
4637	232	-0053 May 09	08:41:04	11076	-25388	66	T	-n	0.1361	1.0686	24.2N	92.5E	82	168	226	06m03s
4638	232	-0053 Nov 01	19:11:14	11071	-25382	71	A	nn	-0.0294	0.9503	15.7S	65.4W	88	13	182	06m02s
4639	232	-0052 Apr 27	22:41:58	11066	-25376	76	T	p-	-0.6416	1.0143	25.7S	105.4W	50	345	63	01m25s
4640	232	-0052 Oct 21	03:14:09	11060	-25370	81	H	p-	0.6666	1.0028	30.3N	173.4W	48	198	13	00m16s
4641	233	-0051 Mar 18	14:48:25	11056	-25365	48	P	-t	1.4092	0.2640	71.9N	77.8W	0	96		
4642	233	-0051 Apr 17	05:51:00	11055	-25364	86	P	t-	-1.4574	0.1746	71.2S	161.1W	0	300		
4643	233	-0051 Sep 11	08:03:38	11051	-25359	53	P	-t	-1.3117	0.4204	71.6S	32.0E	0	72		
4644	233	-0051 Oct 10	17:24:36	11050	-25358	91	P	t-	1.3105	0.4235	71.6N	33.9E	0	251		
4645	233	-0050 Mar 07	14:46:58	11046	-25353	58	A	-p	0.6874	0.9383	35.2N	6.8W	46	159	312	06m51s
4646	233	-0050 Aug 31	23:33:54	11041	-25347	63	T	-p	-0.6410	1.0290	28.7S	139.2W	50	18	127	02m38s
4647	233	-0049 Feb 24	18:50:54	11036	-25341	68	A	nn	-0.0666	0.9831	14.1S	51.6W	86	344	60	01m52s
4648	233	-0049 Aug 21	10:02:16	11031	-25335	73	Am	nn	0.0935	0.9807	18.8N	77.2E	85	194	69	02m10s
4649	233	-0048 Feb 14	06:02:38	11026	-25329	78	T	p-	-0.7773	1.0279	62.7S	163.1E	39	330	150	01m54s
4650	233	-0048 Aug 09	13:24:18	11021	-25323	83	A	p-	0.8605	0.9367	73.2N	60.8E	30	220	466	04m55s
4651	233	-0047 Jan 04	11:41:07	11016	-25318	50	P	-t	1.2182	0.5976	66.9N	53.3E	0	180		
4652	233	-0047 Feb 02	21:40:10	11016	-25317	88	P	t-	-1.4308	0.1946	69.6S	52.2E	0	211		
4653	233	-0047 Jun 29	23:28:06	11011	-25312	55	P	-t	-1.2440	0.5461	66.1S	117.6W	0	352		
4654	233	-0047 Dec 25	01:02:53	11006	-25306	60	A	-p	0.5744	0.9930	11.3N	145.9W	55	185	30	00m47s
4655	233	-0046 Jun 19	07:44:32	11001	-25300	65	T	-p	-0.4244	1.0265	1.3S	112.2E	65	352	99	02m52s
4656	233	-0046 Dec 14	07:50:16	10996	-25294	70	A	nn	-0.1213	0.9413	30.4S	106.5E	83	10	219	06m50s
4657	233	-0045 Jun 08	22:31:40	10991	-25288	75	T	n-	0.3449	1.0731	42.5N	119.5W	70	164	253	05m37s
4658	233	-0045 Dec 03	07:59:17	10986	-25282	80	A	p-	-0.8143	0.9105	71.7S	63.3E	35	48	592	07m10s
4659	233	-0044 Apr 29	08:38:41	10982	-25277	47	P	-t	-1.4445	0.1690	61.7S	158.1E	0	300		
4660	233	-0044 May 28	15:49:54	10981	-25276	85	P	t-	1.0793	0.8701	63.4N	154.2W	0	36		
4661	234	-0044 Oct 22	15:24:22	10977	-25271	52	P	-t	1.4815	0.1325	61.4N	61.3E	0	247		
4662	234	-0044 Nov 21	07:49:10	10976	-25270	90	P	t-	-1.4592	0.1812	62.9S	28.8W	0	138		
4663	234	-0043 Apr 18	20:59:08	10972	-25265	57	A	-p	-0.7460	0.9947	32.7S	65.6W	42	328	27	00m29s
4664	234	-0043 Oct 12	01:34:43	10967	-25259	62	T	-p	0.7217	1.0189	33.8N	136.4W	44	215	91	01m32s
4665	234	-0042 Apr 08	02:01:04	10962	-25253	67	A	nn	0.0083	0.9522	6.4N	164.3W	89	154	175	05m19s
4666	234	-0042 Oct 01	16:50:13	10957	-25247	72	T	nn	0.0186	1.0558	1.4S	28.5W	89	209	184	04m35s
4667	234	-0041 Mar 28	02:18:20	10952	-25241	77	A	p-	0.7551	0.9377	42.9N	162.3E	41	138	343	05m50s
4668	234	-0041 Sep 21	08:36:41	10947	-25235	82	T	p-	-0.6828	1.0331	35.0S	71.0E	47	36	150	02m34s
4669	234	-0040 Feb 15	16:22:41	10942	-25230	49	P	-t	-1.3294	0.3942	61.7S	101.0E	0	240		
4670	234	-0040 Mar 16	05:12:45	10942	-25229	87	P	t-	1.4529	0.1791	60.8N	64.9E	0	96		
4671	234	-0040 Aug 11	06:58:01	10937	-25224	54	P	-t	1.3435	0.3726	62.1N	114.6W	0	307		
4672	234	-0040 Sep 09	20:17:29	10936	-25223	92	P	t-	-1.4549	0.1720	61.0S	158.0W	0	77		
4673	234	-0039 Feb 04	05:08:41	10932	-25218	59	T	-p	-0.5739	1.0404	48.5S	172.1E	55	327	164	02m54s
4674	234	-0039 Jul 31	08:52:03	10927	-25212	64	A	-p	0.6312	0.9390	55.1N	115.1E	51	214	290	05m31s
4675	234	-0038 Jan 24	21:07:27	10922	-25206	69	T	n-	0.1161	1.0517	13.5S	89.5W	83	161	172	04m37s
4676	234	-0038 Jul 20	09:37:47	10917	-25200	74	A	nn	-0.1246	0.9603	14.8N	79.6E	83	15	145	04m52s
4677	234	-0037 Jan 14	11:38:03	10912	-25194	79	H	p-	0.8256	1.0048	32.3N	39.6E	34	164	30	00m27s
4678	234	-0037 Jul 09	16:10:20	10907	-25188	84	H	t-	-0.8758	1.0041	37.3S	29.6W	29	13	30	00m24s
4679	234	-0037 Dec 05	03:00:55	10903	-25183	51	P	-t	-1.3899	0.2996	67.5S	5.5E	0	174		
4680	234	-0036 May 29	22:53:45	10898	-25177	56	T	-t	0.9432	1.0674	84.8N	122.3E	19	60	683	03m28s

Cat Num	Canon Plate	Calendar Date	TD of Greatest Eclipse	ΔT s	Luna Num	Saros Num	Ecl. Type	QLE	Gamma	Ecl. Mag.	Lat. °	Long. °	Sun Alt °	Sun Azm °	Path Width km	Central Line Dur.
4681	235	-0036 Nov 23	02:15:09	10893	-25171	61	A	-p	-0.7359	0.9134	67.2S	176.0E	42	17	490	07m38s
4682	235	-0035 May 19	16:03:49	10888	-25165	66	T	-n	0.2106	1.0672	31.4N	19.5W	78	171	225	05m46s
4683	235	-0035 Nov 12	03:25:54	10883	-25159	71	A	nn	-0.0348	0.9515	19.4S	170.4E	88	10	178	05m51s
4684	235	-0034 May 09	05:47:57	10878	-25153	76	H	p-	-0.5697	1.0131	17.7S	143.6E	55	348	55	01m23s
4685	235	-0034 Nov 01	11:46:46	10873	-25147	81	H	p-	0.6569	1.0040	26.1N	55.2E	49	195	18	00m23s
4686	235	-0033 Mar 29	21:41:30	10869	-25142	48	P	-t	1.4714	0.1577	71.8N	164.1E	0	82		
4687	235	-0033 Apr 28	12:34:02	10868	-25141	86	P	t-	-1.3870	0.2970	70.6S	84.1E	0	313		
4688	235	-0033 Sep 22	16:22:35	10864	-25136	53	P	-t	-1.3406	0.3658	71.8S	107.5W	0	85		
4689	235	-0033 Oct 22	02:03:25	10863	-25135	91	P	t-	1.2967	0.4498	71.1N	110.4W	0	237		
4690	235	-0032 Mar 17	21:52:49	10859	-25130	58	A	-p	0.7394	0.9421	43.1N	118.6W	42	155	314	05m54s
4691	235	-0032 Sep 11	07:32:25	10854	-25124	63	T	-p	-0.6801	1.0224	35.3S	97.0E	47	21	103	01m57s
4692	235	-0031 Mar 07	02:29:23	10849	-25118	68	A	nn	-0.0252	0.9895	7.7S	168.3W	89	343	37	01m09s
4693	235	-0031 Aug 31	17:28:31	10844	-25112	73	A	nn	0.0458	0.9745	12.3N	36.3W	87	196	91	02m55s
4694	235	-0030 Feb 24	14:10:39	10839	-25106	78	T	p-	-0.7435	1.0343	55.7S	37.6E	42	332	172	02m27s
4695	235	-0030 Aug 20	20:20:09	10834	-25100	83	A	p-	0.8018	0.9341	64.0N	53.3W	36	212	412	05m38s
4696	235	-0029 Jan 15	20:21:59	10829	-25095	50	P	-t	1.2300	0.5754	68.0N	88.8W	0	169		
4697	235	-0029 Feb 14	06:04:21	10829	-25094	88	P	t-	-1.4069	0.2392	70.4S	87.3W	0	224		
4698	235	-0029 Jul 11	06:13:03	10824	-25089	55	P	-t	-1.3164	0.4190	67.1S	129.7E	0	2		
4699	235	-0029 Aug 09	20:27:28	10824	-25088	93	Pb	t-	1.5170	0.0782	69.7N	64.8E	0	328		
4700	235	-0028 Jan 05	09:33:30	10819	-25083	60	A	-p	0.5827	0.9915	12.6N	84.1E	54	180	37	00m58s
4701	236	-0028 Jun 29	14:53:57	10814	-25077	65	T	-p	-0.4978	1.0283	6.2S	3.0E	60	357	110	03m03s
4702	236	-0028 Dec 24	16:01:49	10809	-25071	70	A	nn	-0.1161	0.9400	30.5S	15.1W	83	5	224	07m08s
4703	236	-0027 Jun 19	05:55:09	10804	-25065	75	T	n-	0.2690	1.0746	39.2N	132.4E	74	170	251	05m57s
4704	236	-0027 Dec 13	16:00:18	10799	-25059	80	A	p-	-0.8087	0.9109	74.9S	48.1W	36	38	583	07m08s
4705	236	-0026 May 10	15:59:57	10795	-25054	47	Pe	-t	-1.5186	0.0287	62.3S	37.9E	0	309		
4706	236	-0026 Jun 08	23:13:01	10795	-25053	85	T+	t-	1.0068	1.0095	64.3N	84.7E	0	27	-	-
4707	236	-0026 Nov 02	23:41:43	10790	-25048	52	P	-t	1.4859	0.1244	61.9N	72.7W	0	238		
4708	236	-0026 Dec 02	16:02:20	10790	-25047	90	P	t-	-1.4526	0.1916	63.8S	162.3W	0	148		
4709	236	-0025 Apr 30	04:00:34	10785	-25042	57	A	-t	-0.8220	0.9915	36.3S	170.7W	34	329	51	00m47s
4710	236	-0025 Oct 23	10:07:36	10780	-25036	62	T	-p	0.7318	1.0189	31.2N	92.5E	43	211	93	01m35s
4711	236	-0024 Apr 18	08:42:47	10775	-25030	67	A	nn	-0.0667	0.9527	6.6N	95.5E	86	333	173	05m22s
4712	236	-0024 Oct 12	01:23:20	10770	-25024	72	T	nn	0.0373	1.0532	4.8S	157.7W	88	209	176	04m25s
4713	236	-0023 Apr 07	09:03:45	10765	-25018	77	A	p-	0.6856	0.9425	42.5N	62.9E	46	140	285	05m23s
4714	236	-0023 Oct 01	16:55:10	10760	-25012	82	T	p-	-0.6597	1.0281	37.6S	55.1W	49	38	124	02m09s
4715	236	-0022 Feb 26	00:15:02	10756	-25007	49	P	-t	-1.3595	0.3393	61.3S	26.7W	0	249		
4716	236	-0022 Mar 27	12:27:48	10756	-25006	87	P	t-	1.3932	0.2829	60.7N	53.3W	0	87		
4717	236	-0022 Aug 22	14:08:43	10751	-25001	54	P	-t	1.4020	0.2719	61.5N	128.0E	0	298		
4718	236	-0022 Sep 21	04:06:06	10751	-25000	92	P	t-	-1.4264	0.2245	60.9S	75.3E	0	86		
4719	236	-0021 Feb 15	13:28:32	10746	-24995	59	T	-p	-0.5991	1.0450	45.8S	48.5E	53	324	186	03m13s
4720	236	-0021 Aug 11	15:37:32	10741	-24989	64	A	-p	0.6990	0.9351	55.3N	19.8E	45	221	334	05m42s
4721	237	-0020 Feb 05	05:38:33	10736	-24983	69	Tm	nn	0.0964	1.0536	11.8S	142.4E	85	158	178	04m40s
4722	237	-0020 Jul 30	16:27:14	10731	-24977	74	A	nn	-0.0529	0.9604	16.6N	22.4W	87	20	144	04m38s
4723	237	-0019 Jan 24	20:04:24	10727	-24971	79	H	p-	0.8142	1.0047	32.2N	90.7W	35	159	27	00m26s
4724	237	-0019 Jul 19	23:20:57	10722	-24965	84	H	p-	-0.8030	1.0076	30.2S	139.5W	36	17	44	00m45s
4725	237	-0019 Dec 15	11:08:27	10717	-24960	51	P	-t	-1.3965	0.2890	66.5S	128.0W	0	185		
4726	237	-0018 Jun 10	06:19:35	10712	-24954	56	P	-t	1.0154	0.9954	67.1N	49.8W	0	2		
4727	237	-0018 Jul 09	13:12:23	10712	-24953	94	Pb	t-	-1.4930	0.0727	64.5S	0.4W	0	25		
4728	237	-0018 Dec 04	10:16:40	10708	-24948	61	A	-p	-0.7407	0.9139	70.2S	63.3E	42	7	492	07m25s
4729	237	-0017 May 30	23:25:29	10703	-24942	66	T	-n	0.2855	1.0649	38.2N	130.2W	73	175	222	05m23s
4730	237	-0017 Nov 23	11:44:10	10698	-24936	71	A	nn	-0.0384	0.9533	22.5S	45.7E	88	7	171	05m35s
4731	237	-0016 May 19	12:49:09	10693	-24930	76	H	p-	-0.4948	1.0112	10.1S	34.6E	60	352	44	01m15s
4732	237	-0016 Nov 11	20:24:21	10688	-24924	81	H	p-	0.6508	1.0054	22.7N	77.2W	49	191	25	00m33s
4733	237	-0015 Apr 09	04:25:09	10684	-24919	48	Pe	-t	1.5403	0.0397	71.6N	48.5E	0	69		
4734	237	-0015 May 08	19:09:41	10683	-24918	86	P	t-	-1.3108	0.4294	69.9S	28.4W	0	326		
4735	237	-0015 Oct 03	00:49:29	10679	-24913	53	P	-t	-1.3627	0.3242	71.7S	110.8E	0	100		
4736	237	-0015 Nov 01	10:48:19	10678	-24912	91	P	t-	1.2881	0.4661	70.4N	104.3E	0	224		
4737	237	-0014 Mar 29	04:48:09	10674	-24907	58	A	-p	0.8002	0.9456	52.0N	130.3E	37	150	331	05m02s
4738	237	-0014 Sep 22	15:39:28	10669	-24901	63	T	-p	-0.7112	1.0158	41.6S	29.0W	44	24	77	01m19s
4739	237	-0013 Mar 18	09:59:34	10664	-24895	68	A	nn	0.0238	0.9959	0.8S	76.6E	89	162	14	00m27s
4740	237	-0013 Sep 12	01:01:55	10659	-24889	73	A	nn	0.0050	0.9685	5.9N	151.7W	90	195	113	03m41s

Cat Num	Canon Plate	Calendar Date	TD of Greatest Eclipse	ΔT s	Luna Num	Saros Num	Ecl. Type	QLE	Gamma	Ecl. Mag.	Lat. °	Long. °	Sun Alt °	Sun Azm °	Path Width km	Central Line Dur.
4741	238	-0012 Mar 06	22:11:44	10654	-24883	78	T	p-	-0.7036	1.0407	48.4S	86.9W	45	334	191	03m03s
4742	238	-0012 Aug 31	03:23:34	10649	-24877	83	A	p-	0.7495	0.9311	55.6N	165.2W	41	209	388	06m24s
4743	238	-0011 Jan 26	04:57:33	10645	-24872	50	P	-t	1.2460	0.5453	69.0N	129.8E	0	157		
4744	238	-0011 Feb 24	14:21:12	10644	-24871	88	P	t-	-1.3773	0.2949	71.0S	134.5E	0	237		
4745	238	-0011 Jul 21	13:02:52	10640	-24866	55	P	-t	-1.3853	0.2978	68.1S	15.3E	0	13		
4746	238	-0011 Aug 20	03:25:56	10639	-24865	93	P	t-	1.4557	0.1833	70.6N	53.2W	0	315		
4747	238	-0010 Jan 15	17:59:33	10635	-24860	60	A	-p	0.5941	0.9903	14.9N	44.8W	53	176	42	01m05s
4748	238	-0010 Jul 10	22:07:51	10630	-24854	65	T	-p	-0.5681	1.0294	11.8S	108.0W	55	1	121	03m06s
4749	238	-0009 Jan 05	00:08:42	10625	-24848	70	A	nn	-0.1079	0.9393	29.4S	135.6W	84	0	227	07m21s
4750	238	-0009 Jun 30	13:22:34	10620	-24842	75	T	n-	0.1963	1.0753	35.1N	22.4E	78	176	249	06m14s
4751	238	-0009 Dec 24	23:59:49	10615	-24836	80	A	p-	-0.8021	0.9119	76.6S	152.8W	36	21	568	07m05s
4752	238	-0008 Jun 19	06:36:42	10610	-24830	85	T	t-	0.9352	1.0602	82.0N	10.9E	20	64	573	03m08s
4753	238	-0008 Nov 13	08:03:40	10606	-24825	52	P	-t	1.4874	0.1211	62.5N	152.0E	0	229		
4754	238	-0008 Dec 13	00:16:16	10605	-24824	90	P	t-	-1.4460	0.2020	64.7S	63.8E	0	158		
4755	238	-0007 May 10	10:56:39	10601	-24819	57	A	-t	-0.9012	0.9872	42.4S	86.1E	25	330	102	01m09s
4756	238	-0007 Nov 02	18:45:46	10596	-24813	62	T	-p	0.7385	1.0194	29.0N	40.1W	42	207	96	01m40s
4757	238	-0006 Apr 29	15:17:08	10592	-24807	67	A	nn	-0.1473	0.9528	6.0N	2.7W	82	335	174	05m32s
4758	238	-0006 Oct 23	10:03:04	10587	-24801	72	T	nn	0.0506	1.0509	8.3S	71.4E	87	207	169	04m17s
4759	238	-0005 Apr 18	15:43:14	10582	-24795	77	A	p-	0.6104	0.9470	42.3N	34.4W	52	143	242	05m00s
4760	238	-0005 Oct 13	01:19:37	10577	-24789	82	T	p-	-0.6420	1.0233	40.9S	177.6E	50	38	102	01m46s
4761	239	-0004 Mar 08	07:59:58	10573	-24784	49	P	-t	-1.3963	0.2713	61.0S	152.5W	0	258		
4762	239	-0004 Apr 06	19:37:20	10572	-24783	87	P	t-	1.3277	0.3983	60.9N	170.1W	0	78		
4763	239	-0004 Sep 01	21:27:04	10568	-24778	54	P	-t	1.4533	0.1846	61.1N	8.9E	0	289		
4764	239	-0004 Oct 01	12:01:46	10567	-24777	92	P	t-	-1.4041	0.2653	60.9S	53.1W	0	95		
4765	239	-0003 Feb 25	21:39:39	10563	-24772	59	T	-p	-0.6319	1.0495	43.4S	73.5W	51	322	209	03m31s
4766	239	-0003 Aug 21	22:31:23	10558	-24766	64	A	-p	0.7597	0.9311	54.7N	78.7W	40	227	390	05m57s
4767	239	-0002 Feb 15	14:00:36	10553	-24760	69	T	nn	0.0693	1.0557	9.8S	16.5E	86	155	184	04m45s
4768	239	-0002 Aug 10	23:26:07	10548	-24754	74	A	nn	0.0118	0.9602	17.3N	126.5W	89	200	144	04m28s
4769	239	-0001 Feb 05	04:21:20	10543	-24748	79	H	p-	0.7956	1.0049	32.2N	141.9E	37	154	27	00m26s
4770	239	-0001 Jul 31	06:40:18	10538	-24742	84	H	p-	-0.7362	1.0101	25.6S	108.8E	42	20	50	00m59s
4771	239	-0001 Dec 26	19:13:58	10534	-24737	51	P	-t	-1.4044	0.2759	65.4S	99.5E	0	196		
4772	239	0000 Jun 20	13:45:32	10529	-24731	56	P	-t	1.0876	0.8552	66.1N	172.7W	0	351		
4773	239	0000 Jul 19	20:44:11	10529	-24730	94	P	t-	-1.4254	0.2024	63.6S	123.8W	0	34		
4774	239	0000 Dec 14	18:19:36	10525	-24725	61	A	-p	-0.7446	0.9150	71.8S	46.2W	42	354	488	07m12s
4775	239	0001 Jun 10	06:44:16	10520	-24719	66	T	-p	0.3625	1.0617	44.4N	121.2E	69	179	218	04m56s
4776	239	0001 Dec 03	20:05:38	10515	-24713	71	A	nn	-0.0397	0.9558	24.7S	79.2W	88	3	161	05m14s
4777	239	0002 May 30	19:45:43	10510	-24707	76	H	p-	-0.4168	1.0087	3.1S	72.5W	65	356	33	01m00s
4778	239	0002 Nov 23	05:06:15	10505	-24701	81	H	p-	0.6476	1.0074	20.0N	149.5E	49	187	33	00m45s
4779	239	0003 May 20	01:41:30	10500	-24695	86	P	t-	-1.2319	0.5663	69.0S	139.3W	0	337		
4780	239	0003 Oct 14	09:21:59	10496	-24690	53	P	-t	-1.3804	0.2914	71.4S	32.1W	0	114		
4781	240	0003 Nov 12	19:36:29	10495	-24689	91	P	t-	1.2822	0.4772	69.6N	41.2W	0	211		
4782	240	0004 Apr 08	11:39:14	10491	-24684	58	A	-p	0.8647	0.9486	61.7N	16.8E	30	142	375	04m15s
4783	240	0004 Oct 02	23:52:25	10486	-24678	63	H	-p	-0.7368	1.0095	47.8S	156.3W	42	27	48	00m46s
4784	240	0005 Mar 28	17:23:34	10481	-24672	68	H	nn	0.0789	1.0022	6.6N	37.2W	85	162	8	00m14s
4785	240	0005 Sep 22	08:42:23	10476	-24666	73	A	nn	-0.0291	0.9626	0.4S	91.0E	88	18	135	04m25s
4786	240	0006 Mar 18	06:05:10	10472	-24660	78	T	p-	-0.6566	1.0470	40.8S	150.3E	49	336	206	03m42s
4787	240	0006 Sep 11	10:36:21	10467	-24654	83	A	p-	0.7049	0.9281	48.0N	82.0E	45	206	378	07m13s
4788	240	0007 Feb 06	13:24:00	10463	-24649	50	P	-t	1.2693	0.5016	69.9N	9.8W	0	145		
4789	240	0007 Mar 07	22:28:54	10462	-24648	88	P	t-	-1.3406	0.3647	71.5S	1.9W	0	251		
4790	240	0007 Aug 01	20:02:56	10458	-24643	55	P	-t	-1.4464	0.1902	69.1S	102.2W	0	24		
4791	240	0007 Aug 31	10:35:56	10457	-24642	93	P	t-	1.4027	0.2739	71.2N	174.7W	0	302		
4792	240	0008 Jan 27	02:17:23	10453	-24637	60	A	-p	0.6115	0.9895	18.2N	172.1W	52	172	47	01m09s
4793	240	0008 Jul 21	05:28:07	10448	-24631	65	T	-p	-0.6339	1.0298	18.0S	138.8E	51	5	130	03m02s
4794	240	0009 Jan 15	08:08:55	10443	-24625	70	A	nn	-0.0949	0.9392	27.2S	105.2E	84	356	227	07m30s
4795	240	0009 Jul 10	20:53:09	10438	-24619	75	T	n-	0.1261	1.0750	30.2N	89.4W	83	181	245	06m25s
4796	240	0010 Jan 04	07:54:31	10433	-24613	80	A	p-	-0.7914	0.9138	75.7S	106.1E	37	1	542	07m03s
4797	240	0010 Jun 30	14:01:10	10429	-24607	85	T	p-	0.8645	1.0599	83.1N	11.6W	30	154	397	03m22s
4798	240	0010 Nov 24	16:30:45	10425	-24602	52	P	-t	1.4856	0.1234	63.3N	15.3E	0	219		
4799	240	0010 Dec 24	08:29:33	10424	-24601	90	P	t-	-1.4384	0.2138	65.8S	70.4W	0	168		
4800	240	0011 May 21	17:47:07	10420	-24596	57	A	-t	-0.9838	0.9804	55.4S	10.8W	9	327	411	01m36s

Cat Num	Canon Plate	Calendar Date	TD of Greatest Eclipse	ΔT s	Luna Num	Saros Num	Ecl. Type	QLE	Gamma	Ecl. Mag.	Lat. °	Long. °	Sun Alt °	Sun Azm °	Path Width km	Central Line Dur.
4801	241	0011 Nov 14	03:29:00	10415	-24590	62	T	-p	0.7415	1.0203	27.0N	174.2W	42	203	101	01m48s
4802	241	0012 May 09	21:45:58	10410	-24584	67	A	nn	-0.2319	0.9525	4.5N	99.6W	77	338	178	05m48s
4803	241	0012 Nov 02	18:48:38	10405	-24578	72	Tm	-n	0.0590	1.0488	11.6S	60.7W	87	205	162	04m11s
4804	241	0013 Apr 28	22:16:18	10400	-24572	77	A	p-	0.5293	0.9512	41.8N	129.6W	58	146	208	04m41s
4805	241	0013 Oct 23	09:50:43	10395	-24566	82	T	p-	-0.6305	1.0186	44.7S	48.9E	51	38	81	01m23s
4806	241	0014 Mar 19	15:38:24	10391	-24561	49	P	-t	-1.4389	0.1919	60.9S	83.4E	0	267		
4807	241	0014 Apr 18	02:41:51	10391	-24560	87	P	t-	1.2571	0.5240	61.2N	74.2E	0	69		
4808	241	0014 Sep 13	04:52:58	10386	-24555	54	P	-t	1.4976	0.1105	60.8N	112.1W	0	280		
4809	241	0014 Oct 12	20:04:10	10386	-24554	92	P	t-	-1.3880	0.2948	61.0S	176.9E	0	104		
4810	241	0015 Mar 09	05:44:12	10382	-24549	59	T	-p	-0.6707	1.0537	41.3S	165.7E	48	321	236	03m49s
4811	241	0015 Sep 02	05:34:31	10377	-24543	64	A	-p	0.8127	0.9272	53.8N	179.0E	35	230	460	06m15s
4812	241	0016 Feb 26	22:16:04	10372	-24537	69	T	nn	0.0369	1.0576	7.7S	107.9W	88	152	190	04m50s
4813	241	0016 Aug 21	06:33:33	10367	-24531	74	A	nn	0.0703	0.9598	16.9N	127.1E	86	205	146	04m22s
4814	241	0017 Feb 15	12:30:45	10362	-24525	79	H	p-	0.7717	1.0053	32.4N	16.8E	39	150	28	00m28s
4815	241	0017 Aug 10	14:07:44	10357	-24519	84	H	p-	-0.6750	1.0118	22.9S	4.6W	47	24	54	01m08s
4816	241	0018 Jan 06	03:13:06	10353	-24514	51	P	-t	-1.4168	0.2553	64.4S	30.9W	0	206		
4817	241	0018 Feb 04	20:20:34	10353	-24513	89	Pb	t-	1.5494	0.0205	62.2N	132.7W	0	128		
4818	241	0018 Jul 01	21:14:39	10349	-24508	56	P	-t	1.1572	0.7197	65.1N	64.1E	0	342		
4819	241	0018 Jul 31	04:23:15	10348	-24507	94	P	t-	-1.3632	0.3220	62.8S	111.3E	0	44		
4820	241	0018 Dec 26	02:19:23	10344	-24502	61	A	-p	-0.7515	0.9167	71.7S	153.7W	41	339	483	06m56s
4821	242	0019 Jun 21	14:04:18	10339	-24496	66	T	-p	0.4376	1.0577	49.7N	13.9E	64	186	212	04m26s
4822	242	0019 Dec 15	04:26:24	10334	-24490	71	A	nn	-0.0425	0.9589	26.0S	156.3E	87	358	150	04m47s
4823	242	0020 Jun 10	02:40:35	10329	-24484	76	H	n-	-0.3383	1.0055	3.2N	178.4W	70	360	20	00m38s
4824	242	0020 Dec 03	13:49:49	10324	-24478	81	H	p-	0.6453	1.0098	18.0N	15.9E	50	183	44	01m01s
4825	242	0021 May 30	08:07:52	10320	-24472	86	P	t-	-1.1487	0.7102	68.0S	111.8E	0	348		
4826	242	0021 Oct 24	18:01:51	10316	-24467	53	P	-t	-1.3922	0.2698	70.9S	176.5W	0	127		
4827	242	0021 Nov 23	04:28:42	10315	-24466	91	P	t-	1.2797	0.4820	68.6N	172.9E	0	198		
4828	242	0022 Apr 19	18:22:41	10311	-24461	58	A	-t	0.9361	0.9506	72.1N	108.8W	20	120	523	03m35s
4829	242	0022 Oct 14	08:12:53	10306	-24455	63	H	-p	-0.7556	1.0037	53.6S	75.1E	41	29	19	00m17s
4830	242	0023 Apr 09	00:41:09	10301	-24449	68	Hm	nn	0.1404	1.0082	14.2N	149.4W	82	163	29	00m51s
4831	242	0023 Oct 03	16:31:02	10296	-24443	73	A	nn	-0.0558	0.9570	6.3S	28.2W	87	18	157	05m07s
4832	242	0024 Mar 28	13:52:19	10291	-24437	78	T	p-	-0.6039	1.0531	33.0S	29.0E	53	339	219	04m24s
4833	242	0024 Sep 21	17:57:07	10287	-24431	83	A	p-	0.6669	0.9250	40.9N	32.2W	48	204	375	08m05s
4834	242	0025 Feb 16	21:44:01	10283	-24426	50	P	-t	1.2974	0.4485	70.7N	148.5W	0	132		
4835	242	0025 Mar 18	06:29:39	10282	-24425	88	P	t-	-1.2985	0.4452	71.7S	136.8W	0	264		
4836	242	0025 Aug 12	03:10:44	10278	-24420	55	P	-t	-1.5017	0.0928	69.9S	137.9E	0	36		
4837	242	0025 Sep 10	17:55:34	10277	-24419	93	P	t-	1.3567	0.3525	71.7N	60.9E	0	289		
4838	242	0026 Feb 06	10:26:56	10273	-24414	60	A	-p	0.6348	0.9888	22.5N	62.4E	50	168	51	01m12s
4839	242	0026 Aug 01	12:55:45	10268	-24408	65	T	-p	-0.6941	1.0296	24.7S	22.9E	46	9	139	02m53s
4840	242	0027 Jan 26	16:02:14	10263	-24402	70	A	nn	-0.0771	0.9395	23.8S	13.0W	85	352	225	07m34s
4841	243	0027 Jul 22	04:29:07	10258	-24396	75	T	nn	0.0603	1.0741	24.8N	156.5E	86	185	241	06m31s
4842	243	0028 Jan 15	15:44:31	10254	-24390	80	A	p-	-0.7769	0.9162	72.5S	0.9E	39	347	509	07m01s
4843	243	0028 Jul 10	21:28:53	10249	-24384	85	T	p-	0.7971	1.0582	75.9N	95.4W	37	184	320	03m30s
4844	243	0028 Dec 05	00:59:19	10245	-24379	52	P	-t	1.4833	0.1263	64.1N	122.1W	0	209		
4845	243	0029 Jan 03	16:40:00	10244	-24378	90	P	t-	-1.4276	0.2309	66.9S	155.7E	0	179		
4846	243	0029 Jun 01	00:33:40	10240	-24373	57	P	-t	-1.0684	0.8606	63.7S	111.2W	0	327		
4847	243	0029 Nov 24	12:15:21	10235	-24367	62	T	-p	0.7424	1.0217	25.5N	50.8E	42	198	109	01m59s
4848	243	0030 May 21	04:10:57	10230	-24361	67	A	np	-0.3188	0.9517	2.0N	164.1E	71	342	185	06m09s
4849	243	0030 Nov 14	03:37:26	10226	-24355	72	T	-n	0.0644	1.0473	14.6S	166.6E	86	202	158	04m08s
4850	243	0031 May 10	04:46:51	10221	-24349	77	A	p-	0.4456	0.9549	41.0N	136.1E	63	151	183	04m26s
4851	243	0031 Nov 03	18:26:45	10216	-24343	82	T	p-	-0.6241	1.0143	48.8S	80.4W	51	37	62	01m04s
4852	243	0032 Mar 29	23:10:01	10212	-24338	49	Pe	-t	-1.4877	0.1002	61.0S	39.0W	0	276		
4853	243	0032 Apr 28	09:42:10	10211	-24337	87	P	t-	1.1818	0.6596	61.6N	40.4W	0	61		
4854	243	0032 Sep 23	12:26:34	10207	-24332	54	Pe	-t	1.5349	0.0491	60.7N	125.2E	0	271		
4855	243	0032 Oct 23	04:12:28	10206	-24331	92	P	t-	-1.3773	0.3144	61.4S	45.3E	0	113		
4856	243	0033 Mar 19	13:40:16	10202	-24326	59	T	-p	-0.7168	1.0576	40.1S	46.9E	44	321	267	04m06s
4857	243	0033 Sep 12	12:47:30	10198	-24320	64	A	-p	0.8576	0.9233	52.8N	72.9E	31	232	549	06m36s
4858	243	0034 Mar 09	06:20:44	10193	-24314	69	T	nn	-0.0045	1.0594	5.6S	130.6E	90	335	195	04m56s
4859	243	0034 Sep 01	13:52:18	10188	-24308	74	A	nn	0.1203	0.9593	15.5N	17.6E	83	208	149	04m20s
4860	243	0035 Feb 26	20:29:54	10183	-24302	79	H	p-	0.7398	1.0059	32.8N	105.0W	42	147	30	00m31s

Cat Num	Canon Plate	Calendar Date	TD of Greatest Eclipse	ΔT s	Luna Num	Saros Num	Ecl. Type	QLE	Gamma	Ecl. Mag.	Lat. °	Long. °	Sun Alt °	Sun Azm °	Path Width km	Central Line Dur.
4861	244	0035 Aug 21	21:43:50	10178	-24296	84	H2	p-	-0.6201	1.0130	21.7S	119.8W	52	27	56	01m13s
4862	244	0036 Jan 17	11:06:51	10174	-24291	51	P	-t	-1.4329	0.2281	63.4S	159.7W	0	216		
4863	244	0036 Feb 16	04:03:39	10174	-24290	89	P	t-	1.5240	0.0641	61.7N	101.8E	0	119		
4864	244	0036 Jul 12	04:46:32	10170	-24285	56	P	-t	1.2245	0.5888	64.2N	59.5W	0	332		
4865	244	0036 Aug 10	12:08:25	10169	-24284	94	P	t-	-1.3057	0.4325	62.2S	15.0W	0	53		
4866	244	0037 Jan 05	10:17:02	10165	-24279	61	A	-p	-0.7602	0.9191	70.1S	97.6E	40	327	474	06m40s
4867	244	0037 Jul 01	21:23:24	10160	-24273	66	T	-p	0.5128	1.0529	54.1N	91.2W	59	193	204	03m54s
4868	244	0037 Dec 25	12:47:35	10155	-24267	71	A	nn	-0.0458	0.9626	26.4S	31.9E	87	353	136	04m15s
4869	244	0038 Jun 21	09:34:16	10151	-24261	76	H	nn	-0.2595	1.0016	8.7N	76.8E	75	4	6	00m11s
4870	244	0038 Dec 14	22:33:40	10146	-24255	81	H	p-	0.6427	1.0128	16.7N	117.8W	50	178	57	01m19s
4871	244	0039 Jun 10	14:34:18	10141	-24249	86	P	t-	-1.0656	0.8534	67.0S	3.3E	0	359		
4872	244	0039 Nov 05	02:46:19	10137	-24244	53	P	-t	-1.4007	0.2547	70.2S	38.5E	0	141		
4873	244	0039 Dec 04	13:22:13	10136	-24243	91	P	t-	1.2781	0.4850	67.5N	27.3E	0	187		
4874	244	0040 Apr 30	01:03:17	10132	-24238	58	A+	-t	1.0100	0.9522	70.5N	70.2E	0	43	-	-
4875	244	0040 Oct 24	16:37:57	10127	-24232	63	A	-p	-0.7699	0.9982	59.2S	53.7W	39	30	10	00m08s
4876	244	0041 Apr 19	07:54:46	10123	-24226	68	H	nn	0.2062	1.0139	22.0N	99.5E	78	163	49	01m24s
4877	244	0041 Oct 14	00:26:01	10118	-24220	73	A	nn	-0.0762	0.9519	11.8S	148.6W	86	17	177	05m47s
4878	244	0042 Apr 08	21:32:29	10113	-24214	78	T	p-	-0.5449	1.0588	25.1S	90.4W	57	341	229	05m05s
4879	244	0042 Oct 03	01:27:27	10108	-24208	83	A	p-	0.6369	0.9222	34.6N	148.4W	50	202	377	08m56s
4880	244	0043 Feb 28	05:54:27	10104	-24203	50	P	-t	1.3329	0.3816	71.3N	74.7E	0	119		
4881	245	0043 Mar 29	14:22:23	10104	-24202	88	P	t-	-1.2502	0.5381	71.7S	90.2E	0	278		
4882	245	0043 Aug 23	10:28:50	10100	-24197	55	Pe	-t	-1.5493	0.0091	70.6S	14.7E	0	49		
4883	245	0043 Sep 22	01:25:13	10099	-24196	93	P	t-	1.3181	0.4183	71.9N	66.4W	0	275		
4884	245	0044 Feb 17	18:26:46	10095	-24191	60	A	-p	0.6653	0.9881	27.9N	61.2W	48	164	56	01m13s
4885	245	0044 Aug 11	20:31:57	10090	-24185	65	T	-p	-0.7477	1.0290	31.5S	95.8W	41	14	147	02m39s
4886	245	0045 Feb 05	23:45:29	10085	-24179	70	A	nn	-0.0514	0.9402	19.2S	129.3W	87	349	222	07m34s
4887	245	0045 Aug 01	12:10:30	10080	-24173	75	Tm	nn	-0.0011	1.0724	18.9N	40.2E	90	169	235	06m30s
4888	245	0046 Jan 25	23:27:16	10076	-24167	80	A	p-	-0.7561	0.9194	67.6S	109.2W	41	339	469	06m59s
4889	245	0046 Jul 22	05:00:32	10071	-24161	85	T	p-	0.7334	1.0552	68.1N	158.1E	43	193	270	03m34s
4890	245	0046 Dec 16	09:28:43	10067	-24156	52	P	-t	1.4810	0.1290	65.1N	100.0E	0	199		
4891	245	0047 Jan 15	00:46:09	10066	-24155	90	P	t-	-1.4127	0.2553	68.0S	22.4E	0	191		
4892	245	0047 Jun 12	07:18:22	10062	-24150	57	P	-t	-1.1530	0.7087	64.6S	137.2E	0	336		
4893	245	0047 Jul 11	18:20:49	10061	-24149	95	Pb	t-	1.5350	0.0174	67.2N	123.1E	0	356		
4894	245	0047 Dec 05	21:04:34	10057	-24144	62	T	-p	0.7414	1.0236	24.4N	84.8W	42	193	119	02m13s
4895	245	0048 May 31	10:32:07	10053	-24138	67	A	-p	-0.4085	0.9506	1.6S	68.4E	66	346	198	06m33s
4896	245	0048 Nov 24	12:30:09	10048	-24132	72	T	-n	0.0667	1.0461	17.1S	33.2E	86	198	154	04m07s
4897	245	0049 May 20	11:14:25	10043	-24126	77	A	p-	0.3588	0.9583	39.4N	42.6E	69	156	162	04m15s
4898	245	0049 Nov 14	03:06:57	10038	-24120	82	H	p-	-0.6212	1.0104	52.8S	150.3E	51	34	46	00m46s
4899	245	0050 May 09	16:39:47	10034	-24114	87	P	t-	1.1028	0.8033	62.1N	154.6W	0	52		
4900	245	0050 Nov 03	12:25:52	10029	-24108	92	P	t-	-1.3712	0.3258	61.9S	87.6W	0	123		
4901	246	0051 Mar 30	21:31:07	10025	-24103	59	T	-p	-0.7681	1.0609	39.8S	70.7W	40	321	305	04m21s
4902	246	0051 Sep 23	20:08:25	10020	-24097	64	A	-p	0.8962	0.9197	52.2N	36.3W	26	233	667	06m58s
4903	246	0052 Mar 19	14:19:19	10015	-24091	69	T	nn	-0.0509	1.0609	3.7S	10.6E	87	331	200	05m02s
4904	246	0052 Sep 11	21:20:36	10011	-24085	74	Am	nn	0.1634	0.9588	13.4N	94.6W	81	209	151	04m20s
4905	246	0053 Mar 09	04:19:20	10006	-24079	79	H	p-	0.7006	1.0065	33.4N	136.3E	45	145	31	00m33s
4906	246	0053 Sep 01	05:29:38	10001	-24073	84	H2	p-	-0.5723	1.0138	21.8S	122.6E	55	29	57	01m15s
4907	246	0054 Jan 27	18:51:58	9997	-24068	51	P	-t	-1.4557	0.1898	62.6S	74.0E	0	225		
4908	246	0054 Feb 26	11:35:35	9996	-24067	89	P	t-	1.4903	0.1220	61.2N	20.8W	0	110		
4909	246	0054 Jul 23	12:24:26	9993	-24062	56	P	-t	1.2871	0.4677	63.4N	175.7E	0	323		
4910	246	0054 Aug 21	20:02:32	9992	-24061	94	P	t-	-1.2550	0.5296	61.6S	143.2W	0	62		
4911	246	0055 Jan 16	18:07:44	9988	-24056	61	A	-p	-0.7752	0.9219	67.7S	12.1W	39	317	467	06m22s
4912	246	0055 Jul 13	04:47:03	9983	-24050	66	T	-p	0.5837	1.0475	57.0N	164.2E	54	202	195	03m23s
4913	246	0056 Jan 05	21:05:29	9978	-24044	71	Am	nn	-0.0529	0.9669	26.1S	91.7W	87	349	119	03m39s
4914	246	0056 Jul 01	16:27:58	9974	-24038	76	A	nn	-0.1816	0.9972	13.2N	27.3W	80	8	10	00m20s
4915	246	0056 Dec 25	07:16:33	9969	-24032	81	T	p-	0.6386	1.0163	15.9N	108.9E	50	173	73	01m40s
4916	246	0057 Jun 20	20:59:17	9964	-24026	86	As	t-	-0.9809	0.9434	55.8S	101.3W	10	7	-	05m25s
4917	246	0057 Nov 15	11:35:19	9960	-24021	53	P	-t	-1.4055	0.2463	69.3S	107.1W	0	153		
4918	246	0057 Dec 14	22:15:33	9959	-24020	91	P	t-	1.2766	0.4881	66.5N	117.6W	0	175		
4919	246	0058 May 11	07:39:16	9955	-24015	58	P	-t	1.0882	0.8176	69.7N	42.0W	0	31		
4920	246	0058 Nov 05	01:09:07	9951	-24009	63	A	-p	-0.7787	0.9934	64.3S	177.8E	39	29	37	00m29s

Cat Num	Canon Plate	Calendar Date	TD of Greatest Eclipse	ΔT s	Luna Num	Saros Num	Ecl. Type	QLE	Gamma	Ecl. Mag.	Lat. °	Long. °	Sun Alt °	Sun Azm °	Path Width km	Central Line Dur.
4921	247	0059 Apr 30	15:04:34	9946	-24003	68	T	-n	0.2762	1.0191	29.8N	10.2W	74	165	68	01m50s
4922	247	0059 Oct 25	08:26:50	9941	-23997	73	A	nn	-0.0913	0.9472	16.8S	90.0E	85	15	195	06m23s
4923	247	0060 Apr 19	05:08:12	9936	-23991	78	T	p-	-0.4815	1.0640	17.3S	151.7E	61	344	238	05m45s
4924	247	0060 Oct 13	09:05:51	9932	-23985	83	A	p-	0.6134	0.9196	28.9N	93.7E	52	199	381	09m47s
4925	247	0061 Mar 10	13:56:24	9928	-23980	50	P	-t	1.3751	0.3017	71.7N	60.3W	0	105		
4926	247	0061 Apr 08	22:07:32	9927	-23979	88	P	t-	-1.1958	0.6431	71.5S	40.7W	0	291		
4927	247	0061 Oct 02	09:05:28	9922	-23973	93	P	t-	1.2870	0.4713	71.9N	163.5E	0	260		
4928	247	0062 Feb 28	02:17:27	9918	-23968	60	A	-p	0.7028	0.9874	34.1N	176.9E	45	160	62	01m14s
4929	247	0062 Aug 23	04:17:02	9914	-23962	65	T	-p	-0.7950	1.0280	38.5S	142.5E	37	18	155	02m24s
4930	247	0063 Feb 17	07:19:50	9909	-23956	70	A	nn	-0.0188	0.9412	13.8S	115.9E	89	347	218	07m30s
4931	247	0063 Aug 12	19:59:34	9904	-23950	75	T	nn	-0.0563	1.0702	12.8N	78.6W	87	12	229	06m22s
4932	247	0064 Feb 06	07:03:55	9899	-23944	80	A	p-	-0.7300	0.9230	61.7S	137.8E	43	336	426	06m57s
4933	247	0064 Aug 01	12:36:13	9895	-23938	85	T	p-	0.6735	1.0515	60.5N	45.5E	47	198	231	03m33s
4934	247	0064 Dec 26	17:56:59	9891	-23933	52	P	-t	1.4805	0.1283	66.1N	38.1W	0	188		
4935	247	0065 Jan 25	08:47:30	9890	-23932	90	P	t-	-1.3930	0.2879	69.0S	110.4W	0	202		
4936	247	0065 Jun 22	14:02:12	9886	-23927	57	P	-t	-1.2369	0.5592	65.5S	25.5E	0	346		
4937	247	0065 Jul 22	01:29:10	9885	-23926	95	P	t-	1.4729	0.1325	68.2N	4.2E	0	346		
4938	247	0065 Dec 16	05:52:32	9881	-23921	62	T	-p	0.7420	1.0261	24.1N	139.9E	42	188	132	02m28s
4939	247	0066 Jun 11	16:53:07	9877	-23915	67	A	-p	-0.4979	0.9488	6.3S	27.9W	60	350	216	06m58s
4940	247	0066 Dec 05	21:23:09	9872	-23909	72	T	-n	0.0684	1.0454	18.8S	100.0W	86	193	152	04m07s
4941	248	0067 May 31	17:42:22	9867	-23903	77	A	pn	0.2714	0.9612	37.0N	51.4W	74	162	146	04m08s
4942	248	0067 Nov 25	11:48:04	9862	-23897	82	H	p-	-0.6200	1.0071	56.5S	22.2E	51	30	31	00m31s
4943	248	0068 May 19	23:36:21	9858	-23891	87	P	t-	1.0218	0.9519	62.8N	91.4E	0	43		
4944	248	0068 Nov 13	20:42:55	9853	-23885	92	P	t-	-1.3687	0.3308	62.5S	138.4E	0	132		
4945	248	0069 Apr 10	05:13:17	9849	-23880	59	T	-p	-0.8268	1.0633	41.0S	174.1E	34	322	361	04m31s
4946	248	0069 Oct 04	03:39:59	9844	-23874	64	A	-p	0.9260	0.9164	51.9N	149.6W	22	232	820	07m21s
4947	248	0070 Mar 30	22:07:59	9840	-23868	69	T	-n	-0.1051	1.0619	2.1S	106.9W	84	331	204	05m08s
4948	248	0070 Sep 23	05:00:07	9835	-23862	74	A	nn	0.1978	0.9584	10.8N	150.0E	79	210	154	04m22s
4949	248	0071 Mar 20	11:58:48	9830	-23856	79	H	p-	0.6541	1.0069	34.1N	20.8E	49	144	31	00m35s
4950	248	0071 Sep 12	13:25:34	9826	-23850	84	H2	p-	-0.5324	1.0142	23.1S	2.5E	58	31	57	01m15s
4951	248	0072 Feb 08	02:30:05	9822	-23845	51	P	-t	-1.4835	0.1427	62.0S	50.2W	0	235		
4952	248	0072 Mar 08	18:58:25	9821	-23844	89	P	t-	1.4497	0.1916	60.9N	140.9W	0	101		
4953	248	0072 Aug 02	20:06:10	9817	-23839	56	P	-t	1.3464	0.3533	62.6N	50.2E	0	313		
4954	248	0072 Sep 01	04:03:21	9816	-23838	94	P	t-	-1.2096	0.6162	61.2S	87.0E	0	71		
4955	248	0073 Jan 27	01:54:29	9812	-23833	61	A	-p	-0.7938	0.9252	64.7S	123.8W	37	311	461	06m02s
4956	248	0073 Jul 23	12:12:48	9808	-23827	66	T	-p	0.6522	1.0413	58.6N	60.1E	49	212	182	02m52s
4957	248	0074 Jan 16	05:19:43	9803	-23821	71	A	nn	-0.0638	0.9718	25.0S	145.5E	86	344	101	03m01s
4958	248	0074 Jul 12	23:23:58	9798	-23815	76	Am	nn	-0.1064	0.9922	16.7N	131.3W	84	13	27	00m52s
4959	248	0075 Jan 05	15:57:02	9793	-23809	81	T	p-	0.6317	1.0204	15.6N	23.7W	51	169	89	02m01s
4960	248	0075 Jul 02	03:26:47	9789	-23803	86	A	p-	-0.8984	0.9446	40.1S	158.7E	26	10	466	06m10s
4961	249	0075 Nov 26	20:25:25	9785	-23798	53	P	-t	-1.4094	0.2394	68.3S	107.8E	0	165		
4962	249	0075 Dec 26	07:06:47	9784	-23797	91	P	t-	1.2734	0.4942	65.4N	98.4E	0	165		
4963	249	0076 May 21	14:15:48	9780	-23792	58	P	-t	1.1664	0.6818	68.8N	153.8W	0	20		
4964	249	0076 Nov 15	09:42:44	9775	-23786	63	A	-p	-0.7848	0.9891	68.9S	51.4E	38	25	62	00m46s
4965	249	0077 May 10	22:11:37	9771	-23780	68	T	-n	0.3493	1.0238	37.5N	118.6W	69	167	86	02m10s
4966	249	0077 Nov 04	16:32:39	9766	-23774	73	A	nn	-0.1015	0.9431	21.1S	32.2W	84	13	211	06m55s
4967	249	0078 Apr 30	12:39:31	9761	-23768	78	T	p-	-0.4136	1.0685	9.7S	35.2E	66	347	244	06m19s
4968	249	0078 Oct 24	16:51:28	9757	-23762	83	A	p-	0.5957	0.9175	24.0N	25.8W	53	196	386	10m35s
4969	249	0079 Mar 21	21:49:07	9753	-23757	50	P	-t	1.4245	0.2084	71.8N	166.7E	0	91		
4970	249	0079 Apr 20	05:46:38	9752	-23756	88	P	t-	-1.1365	0.7577	71.0S	169.8W	0	305		
4971	249	0079 Oct 13	16:55:33	9747	-23750	93	P	t-	1.2629	0.5123	71.6N	31.2E	0	246		
4972	249	0080 Mar 10	09:57:24	9743	-23745	60	A	-p	0.7482	0.9864	41.4N	56.7E	41	156	72	01m15s
4973	249	0080 Sep 02	12:11:02	9739	-23739	65	T	-p	-0.8355	1.0267	45.5S	17.7E	33	24	164	02m08s
4974	249	0081 Feb 27	14:43:32	9734	-23733	70	A	nn	0.0220	0.9424	7.6S	3.1E	89	163	213	07m23s
4975	249	0081 Aug 23	03:55:59	9729	-23727	75	T	nn	-0.1052	1.0675	6.6N	160.4E	84	15	221	06m08s
4976	249	0082 Feb 16	14:30:25	9725	-23721	80	A	p-	-0.6954	0.9272	55.0S	24.7E	46	335	380	06m55s
4977	249	0082 Aug 12	20:18:16	9720	-23715	85	T	p-	0.6196	1.0470	53.0N	70.7W	51	200	199	03m27s
4978	249	0083 Jan 07	02:23:03	9716	-23710	52	P	-t	1.4820	0.1238	67.2N	176.0W	0	177		
4979	249	0083 Feb 05	16:42:51	9715	-23709	90	P	t-	-1.3681	0.3299	69.9S	117.8E	0	215		
4980	249	0083 Jul 03	20:46:30	9711	-23704	57	P	-t	-1.3191	0.4139	66.5S	86.7W	0	356		

Cat Num	Canon Plate	Calendar Date	TD of Greatest Eclipse	ΔT s	Luna Num	Saros Num	Ecl. Type	QLE	Gamma	Ecl. Mag.	Lat. °	Long. °	Sun Alt °	Sun Azm °	Path Width km	Central Line Dur.
4981	250	0083 Aug 02	08:41:17	9710	-23703	95	P	t-	1.4145	0.2399	69.2N	116.2W	0	334		
4982	250	0083 Dec 27	14:39:57	9707	-23698	62	T	-p	0.7435	1.0290	24.5N	4.7E	42	184	147	02m44s
4983	250	0084 Jun 21	23:14:05	9702	-23692	67	A	-p	-0.5867	0.9466	12.1S	124.9W	54	354	242	07m19s
4984	250	0084 Dec 16	06:16:36	9697	-23686	72	T	-n	0.0699	1.0451	19.7S	126.9E	86	188	151	04m10s
4985	250	0085 Jun 11	00:09:59	9693	-23680	77	A	nn	0.1828	0.9636	33.6N	145.9W	79	168	134	04m04s
4986	250	0085 Dec 05	20:30:56	9688	-23674	82	H	p-	-0.6207	1.0042	59.5S	104.5W	51	24	19	00m19s
4987	250	0086 May 31	06:33:47	9683	-23668	87	H	t-	0.9401	1.0022	75.5N	24.1E	19	78	23	00m08s
4988	250	0086 Nov 25	05:01:20	9678	-23662	92	P	t-	-1.3680	0.3327	63.3S	3.8E	0	142		
4989	250	0087 Apr 21	12:52:07	9675	-23657	59	T	-p	-0.8888	1.0647	44.1S	60.2E	27	323	453	04m34s
4990	250	0087 Oct 15	11:19:28	9670	-23651	64	A	-p	0.9493	0.9136	52.2N	94.1E	18	229	1029	07m44s
4991	250	0088 Apr 10	05:50:43	9665	-23645	69	T	-n	-0.1642	1.0625	1.0S	137.2E	81	332	207	05m15s
4992	250	0088 Oct 03	12:48:46	9661	-23639	74	A	nn	0.2256	0.9582	7.8N	32.0E	77	209	155	04m25s
4993	250	0089 Mar 30	19:29:18	9656	-23633	79	H	p-	0.6008	1.0071	35.0N	91.8W	53	144	30	00m36s
4994	250	0089 Sep 22	21:31:16	9651	-23627	84	H2	p-	-0.4994	1.0146	25.2S	119.9W	60	33	57	01m15s
4995	250	0090 Feb 18	09:56:40	9647	-23622	51	P	-t	-1.5202	0.0803	61.4S	171.5W	0	244		
4996	250	0090 Mar 20	02:09:19	9646	-23621	89	P	t-	1.3998	0.2774	60.8N	101.9E	0	92		
4997	250	0090 Aug 14	03:56:10	9643	-23616	56	P	-t	1.3992	0.2521	62.0N	77.1W	0	304		
4998	250	0090 Sep 12	12:14:01	9642	-23615	94	P	t-	-1.1718	0.6879	60.9S	45.1W	0	80		
4999	250	0091 Feb 07	09:32:43	9638	-23610	61	A	-p	-0.8201	0.9288	61.9S	125.2E	35	306	463	05m42s
5000	250	0091 Aug 03	19:44:06	9633	-23604	66	T	-p	0.7154	1.0347	58.9N	45.5W	44	221	167	02m21s
5001	251	0092 Jan 27	13:28:08	9629	-23598	71	A	nn	-0.0807	0.9772	23.4S	24.0E	85	340	82	02m22s
5002	251	0092 Jul 23	06:23:12	9624	-23592	76	A	nn	-0.0345	0.9868	19.0N	124.4E	88	18	46	01m26s
5003	251	0093 Jan 16	00:32:32	9619	-23586	81	T	p-	0.6202	1.0250	15.7N	155.0W	52	164	107	02m25s
5004	251	0093 Jul 12	09:56:16	9615	-23580	86	A	p-	-0.8175	0.9437	31.2S	58.8E	35	14	358	06m41s
5005	251	0093 Dec 07	05:17:33	9611	-23575	53	P	-t	-1.4120	0.2352	67.2S	37.3W	0	177		
5006	251	0094 Jan 05	15:55:02	9610	-23574	91	P	t-	1.2679	0.5046	64.4N	44.3W	0	154		
5007	251	0094 Jun 01	20:51:32	9606	-23569	58	P	-t	1.2458	0.5428	67.8N	95.1E	0	9		
5008	251	0094 Jul 01	10:27:36	9605	-23568	96	Pb	t-	-1.5566	0.0070	65.1S	42.5E	0	19		
5009	251	0094 Nov 26	18:18:40	9601	-23563	63	A	-p	-0.7886	0.9855	72.8S	71.4W	38	17	84	01m01s
5010	251	0095 May 22	05:17:54	9597	-23557	68	T	-p	0.4242	1.0277	45.0N	134.3E	65	170	104	02m22s
5011	251	0095 Nov 16	00:43:03	9592	-23551	73	A	nn	-0.1075	0.9396	24.8S	154.9W	84	9	225	07m23s
5012	251	0096 May 10	20:07:02	9587	-23545	78	T	n-	-0.3421	1.0723	2.3S	79.8W	70	350	250	06m47s
5013	251	0096 Nov 04	00:44:00	9583	-23539	83	A	p-	0.5836	0.9159	19.8N	146.7W	54	193	392	11m18s
5014	251	0097 Apr 01	05:33:53	9579	-23534	50	Pe	-t	1.4800	0.1036	71.7N	35.7E	0	78		
5015	251	0097 Apr 30	13:19:49	9578	-23533	88	P	t-	-1.0723	0.8818	70.4S	63.1E	0	317		
5016	251	0097 Oct 24	00:54:11	9573	-23527	93	P	t-	1.2450	0.5427	71.0N	103.0W	0	233		
5017	251	0098 Mar 21	17:27:30	9569	-23522	60	A	-p	0.8008	0.9850	49.6N	62.4W	36	152	88	01m16s
5018	251	0098 Sep 13	20:14:32	9565	-23516	65	T	-p	-0.8688	1.0253	52.3S	110.3W	29	30	173	01m53s
5019	251	0099 Mar 10	21:58:35	9560	-23510	70	A	nn	0.0692	0.9437	0.8S	108.0W	86	163	208	07m12s
5020	251	0099 Sep 03	11:59:15	9556	-23504	75	T	-n	-0.1483	1.0645	0.3N	37.3E	81	16	213	05m50s
5021	252	0100 Feb 27	21:50:10	9551	-23498	80	A	p-	-0.6548	0.9316	47.8S	88.1W	49	336	336	06m52s
5022	252	0100 Aug 23	04:06:15	9546	-23492	85	T	p-	0.5713	1.0421	45.8N	170.6E	55	201	171	03m17s
5023	252	0101 Jan 17	10:44:09	9542	-23487	52	P	-t	1.4885	0.1101	68.2N	46.7E	0	165		
5024	252	0101 Feb 16	00:30:33	9542	-23486	90	P	t-	-1.3358	0.3853	70.7S	12.8W	0	228		
5025	252	0101 Jul 14	03:33:01	9538	-23481	57	P	-t	-1.3982	0.2753	67.5S	160.1E	0	6		
5026	252	0101 Aug 12	15:58:57	9537	-23480	95	P	t-	1.3610	0.3372	70.0N	121.4E	0	322		
5027	252	0102 Jan 06	23:23:12	9533	-23475	62	T	-p	0.7490	1.0323	25.9N	129.5W	41	179	165	03m00s
5028	252	0102 Jul 03	05:38:53	9528	-23469	67	A	-p	-0.6720	0.9439	18.9S	136.3E	48	358	280	07m33s
5029	252	0102 Dec 27	15:06:16	9524	-23463	72	T	-n	0.0743	1.0454	19.4S	5.3W	86	183	152	04m15s
5030	252	0103 Jun 22	06:42:09	9519	-23457	77	A	nn	0.0971	0.9654	29.4N	117.6E	84	173	125	04m02s
5031	252	0103 Dec 17	05:12:14	9514	-23451	82	H	p-	-0.6209	1.0019	61.5S	131.0E	51	16	9	00m09s
5032	252	0104 Jun 10	13:31:44	9510	-23445	87	H2	t-	0.8576	1.0087	77.6N	34.9W	31	125	59	00m33s
5033	252	0104 Dec 05	13:20:54	9505	-23439	92	P	t-	-1.3687	0.3318	64.2S	131.3W	0	152		
5034	252	0105 May 01	20:24:44	9501	-23434	59	T	-t	-0.9559	1.0643	50.7S	49.6W	17	321	716	04m21s
5035	252	0105 Oct 25	19:07:56	9497	-23428	64	A	-p	0.9657	0.9114	53.0N	25.8W	14	226	1299	08m04s
5036	252	0106 Apr 21	13:25:03	9492	-23422	69	T	-n	-0.2300	1.0623	0.7S	23.5E	77	334	209	05m22s
5037	252	0106 Oct 14	20:48:09	9487	-23416	74	A	nn	0.2454	0.9583	4.7N	88.9W	76	208	156	04m29s
5038	252	0107 Apr 11	02:51:12	9483	-23410	79	H	p-	0.5410	1.0069	35.8N	158.6E	57	145	28	00m35s
5039	252	0107 Oct 04	05:45:26	9478	-23404	84	H2	p-	-0.4727	1.0149	28.1S	115.6E	62	33	57	01m15s
5040	252	0108 Feb 29	17:15:40	9474	-23399	51	Pe	-t	-1.5625	0.0082	61.1S	69.3E	0	253		

Cat Num	Canon Plate	Calendar Date	TD of Greatest Eclipse	ΔT s	Luna Num	Saros Num	Ecl. Type	QLE	Gamma	Ecl. Mag.	Lat. °	Long. °	Sun Alt °	Sun Azm °	Path Width km	Central Line Dur.
5041	253	0108 Mar 30	09:12:09	9473	-23398	89	P	t-	1.3438	0.3738	60.8N	13.2W	0	83		
5042	253	0108 Aug 24	11:52:11	9469	-23393	56	P	-t	1.4473	0.1608	61.5N	154.2E	0	295		
5043	253	0108 Sep 22	20:32:21	9469	-23392	94	P	t-	-1.1403	0.7476	60.8S	179.1W	0	89		
5044	253	0109 Feb 17	17:04:01	9465	-23387	61	A	-p	-0.8523	0.9325	59.5S	15.2E	31	303	475	05m21s
5045	253	0109 Aug 14	03:20:13	9460	-23381	66	T	-p	0.7739	1.0276	58.4N	153.2W	39	228	147	01m51s
5046	253	0110 Feb 06	21:31:05	9456	-23375	71	A	nn	-0.1029	0.9829	21.4S	96.4W	84	336	61	01m43s
5047	253	0110 Aug 03	13:27:32	9451	-23369	76	A	nn	0.0323	0.9812	20.1N	19.1E	88	200	67	02m00s
5048	253	0111 Jan 27	09:03:03	9446	-23363	81	T	p-	0.6041	1.0299	16.3N	75.2E	53	160	126	02m47s
5049	253	0111 Jul 23	16:31:55	9442	-23357	86	A	p-	-0.7415	0.9421	25.2S	42.0W	42	18	314	07m03s
5050	253	0111 Dec 18	14:08:22	9438	-23352	53	P	-t	-1.4157	0.2285	66.1S	178.5E	0	188		
5051	253	0112 Jan 17	00:38:49	9437	-23351	91	P	t-	1.2589	0.5217	63.5N	174.4E	0	144		
5052	253	0112 Jun 12	03:29:03	9433	-23346	58	P	-t	1.3243	0.4048	66.8N	15.8W	0	358		
5053	253	0112 Jul 11	17:00:26	9432	-23345	96	P	t-	-1.4751	0.1475	64.2S	66.3W	0	28		
5054	253	0112 Dec 07	02:54:33	9428	-23340	63	A	-p	-0.7919	0.9825	75.5S	171.7E	37	3	103	01m13s
5055	253	0113 Jun 01	12:24:30	9424	-23334	68	T	-p	0.5000	1.0310	52.1N	28.7E	60	175	121	02m29s
5056	253	0113 Nov 26	08:54:28	9419	-23328	73	A	nn	-0.1122	0.9369	27.6S	82.7E	83	5	236	07m44s
5057	253	0114 May 22	03:32:43	9414	-23322	78	T	n-	-0.2684	1.0753	4.6N	166.2E	74	353	253	07m06s
5058	253	0114 Nov 15	08:41:03	9410	-23316	83	A	p-	0.5746	0.9149	16.3N	91.5E	55	189	395	11m52s
5059	253	0115 May 11	20:48:54	9405	-23310	88	T-	t-	-1.0051	1.0117	69.6S	62.5W	0	329	-	-
5060	253	0115 Nov 04	08:59:23	9401	-23304	93	P	t-	1.2315	0.5657	70.3N	121.8E	0	219		
5061	254	0116 Apr 01	00:47:48	9397	-23299	60	A	-t	0.8606	0.9830	58.9N	177.9E	30	144	118	01m19s
5062	254	0116 Sep 24	04:27:28	9392	-23293	65	T	-p	-0.8950	1.0239	58.6S	118.2E	26	36	182	01m39s
5063	254	0117 Mar 21	05:01:14	9387	-23287	70	A	nn	0.1263	0.9449	6.6N	143.6E	83	162	205	06m59s
5064	254	0117 Sep 13	20:11:32	9383	-23281	75	T	-n	-0.1840	1.0612	5.9S	88.0W	79	18	204	05m29s
5065	254	0118 Mar 10	04:59:55	9378	-23275	80	A	p-	-0.6055	0.9363	40.1S	160.9E	53	337	295	06m47s
5066	254	0118 Sep 03	12:01:49	9374	-23269	85	T	p-	0.5298	1.0368	38.9N	49.5E	58	201	145	03m02s
5067	254	0119 Jan 28	19:00:14	9370	-23264	52	P	-t	1.4998	0.0877	69.2N	89.9W	0	153		
5068	254	0119 Feb 27	08:11:25	9369	-23263	90	P	t-	-1.2970	0.4527	71.4S	142.2W	0	241		
5069	254	0119 Jul 25	10:23:34	9365	-23258	57	P	-t	-1.4729	0.1459	68.5S	45.4E	0	17		
5070	254	0119 Aug 23	23:23:07	9364	-23257	95	P	t-	1.3133	0.4232	70.8N	3.2W	0	310		
5071	254	0120 Jan 18	08:03:34	9360	-23252	62	T	-p	0.7574	1.0359	28.2N	96.9E	41	174	185	03m16s
5072	254	0120 Jul 13	12:05:37	9356	-23246	67	A	-p	-0.7554	0.9407	26.9S	36.0E	41	2	336	07m36s
5073	254	0121 Jan 06	23:54:07	9351	-23240	72	T	-n	0.0802	1.0460	18.2S	137.0W	86	179	154	04m20s
5074	254	0121 Jul 02	13:17:36	9346	-23234	77	A	nn	0.0136	0.9668	24.3N	19.4E	89	178	120	04m02s
5075	254	0121 Dec 27	13:50:54	9342	-23228	82	H	p-	-0.6196	1.0002	62.0S	8.3E	51	6	1	00m01s
5076	254	0122 Jun 21	20:33:47	9337	-23222	87	T	p-	0.7773	1.0136	74.0N	109.2W	39	158	75	00m55s
5077	254	0122 Dec 16	21:38:59	9333	-23216	92	P	t-	-1.3693	0.3311	65.2S	93.6E	0	162		
5078	254	0123 May 13	03:55:44	9329	-23211	59	P	-t	-1.0248	0.9772	62.4S	149.9W	0	311		
5079	254	0123 Jun 11	10:40:13	9328	-23210	97	Pb	t-	1.4766	0.1028	64.6N	96.4W	0	24		
5080	254	0123 Nov 06	03:01:20	9324	-23205	64	An	-p	0.9783	0.9098	54.4N	147.6W	11	221	-	08m20s
5081	255	0124 May 01	20:55:09	9319	-23199	69	T	-n	-0.2990	1.0615	1.1S	89.2W	73	337	211	05m26s
5082	255	0124 Oct 25	04:55:04	9315	-23193	74	A	-n	0.2599	0.9588	1.6N	148.3E	75	206	154	04m31s
5083	255	0125 Apr 21	10:04:01	9310	-23187	79	H	p-	0.4745	1.0063	36.3N	51.8E	61	147	24	00m32s
5084	255	0125 Oct 14	14:08:53	9306	-23181	84	H2	p-	-0.4532	1.0153	31.5S	10.9W	63	33	58	01m16s
5085	255	0126 Apr 10	16:04:42	9301	-23175	89	P	t-	1.2800	0.4837	61.0N	125.8W	0	75		
5086	255	0126 Sep 04	19:56:53	9297	-23170	56	P	-t	1.4882	0.0838	61.2N	23.5E	0	286		
5087	255	0126 Oct 04	04:59:04	9296	-23169	94	P	t-	-1.1156	0.7938	60.8S	44.8E	0	98		
5088	255	0127 Mar 01	00:25:56	9292	-23164	61	A	-p	-0.8926	0.9361	58.1S	92.1W	26	299	518	05m00s
5089	255	0127 Aug 25	11:03:53	9288	-23158	66	T	-p	0.8252	1.0203	57.2N	95.5E	34	233	121	01m22s
5090	255	0128 Feb 18	05:26:41	9283	-23152	71	A	nn	-0.1321	0.9890	19.2S	144.8E	82	333	39	01m04s
5091	255	0128 Aug 13	20:36:30	9279	-23146	76	A	nn	0.0943	0.9752	20.3N	87.4W	84	204	89	02m35s
5092	255	0129 Feb 06	17:27:01	9274	-23140	81	T	p-	0.5820	1.0354	17.1N	52.8W	54	156	144	03m10s
5093	255	0129 Aug 02	23:12:37	9269	-23134	86	A	p-	-0.6694	0.9398	21.2S	143.6W	48	21	295	07m20s
5094	255	0129 Dec 28	22:56:40	9266	-23129	53	P	-t	-1.4215	0.2179	65.1S	35.5E	0	199		
5095	255	0130 Jan 27	09:15:57	9265	-23128	91	P	t-	1.2446	0.5488	62.7N	35.1E	0	135		
5096	255	0130 Jun 23	10:09:31	9261	-23123	58	P	-t	1.4009	0.2693	65.8N	127.1W	0	348		
5097	255	0130 Jul 22	23:40:13	9260	-23122	96	P	t-	-1.3977	0.2808	63.3S	176.5W	0	38		
5098	255	0130 Dec 18	11:29:55	9256	-23117	63	A	-p	-0.7953	0.9800	76.2S	59.4E	37	345	118	01m22s
5099	255	0131 Jun 12	19:32:12	9252	-23111	68	T	-p	0.5756	1.0335	58.6N	74.8W	55	181	139	02m31s
5100	255	0131 Dec 07	17:06:49	9247	-23105	73	A	nn	-0.1155	0.9347	29.6S	39.4W	83	1	245	07m59s

Cat Num	Canon Plate	Calendar Date	TD of Greatest Eclipse	ΔT s	Luna Num	Saros Num	Ecl. Type	QLE	Gamma	Ecl. Mag.	Lat. °	Long. °	Sun Alt °	Sun Azm °	Path Width km	Central Line Dur.
5101	256	0132 Jun 01	10:57:16	9242	-23099	78	T	n-	-0.1932	1.0775	10.9N	53.3E	79	357	255	07m14s
5102	256	0132 Nov 25	16:42:02	9238	-23093	83	A	p-	0.5691	0.9144	13.6N	31.1W	55	185	396	12m16s
5103	256	0133 May 22	04:13:35	9233	-23087	88	T	t-	-0.9345	1.0601	48.9S	163.6E	20	350	562	04m32s
5104	256	0133 Nov 14	17:11:10	9229	-23081	93	P	t-	1.2227	0.5809	69.4N	14.4W	0	206		
5105	256	0134 Apr 12	07:59:44	9225	-23076	60	A	-t	0.9262	0.9801	69.0N	50.7E	22	128	190	01m24s
5106	256	0134 Oct 05	12:48:19	9220	-23070	65	T	-p	-0.9153	1.0226	64.5S	16.5W	23	45	193	01m29s
5107	256	0135 Apr 01	11:56:36	9215	-23064	70	A	nn	0.1888	0.9460	14.3N	36.9E	79	162	202	06m43s
5108	256	0135 Sep 25	04:31:21	9211	-23058	75	T	-n	-0.2133	1.0578	11.9S	144.9E	78	18	195	05m08s
5109	256	0136 Mar 20	12:02:35	9206	-23052	80	A	p-	-0.5499	0.9411	32.2S	51.3E	56	339	259	06m38s
5110	256	0136 Sep 13	20:03:55	9202	-23046	85	T	p-	0.4944	1.0314	32.4N	73.4W	60	201	121	02m43s
5111	256	0137 Feb 08	03:09:58	9198	-23041	52	P	-t	1.5167	0.0546	70.1N	134.5E	0	141		
5112	256	0137 Mar 09	15:44:51	9197	-23040	90	P	t-	-1.2511	0.5335	71.8S	89.9E	0	254		
5113	256	0137 Aug 04	17:19:28	9193	-23035	57	Pe	-t	-1.5418	0.0280	69.5S	71.2W	0	29		
5114	256	0137 Sep 03	06:54:33	9192	-23034	95	P	t-	1.2722	0.4962	71.4N	130.2W	0	297		
5115	256	0138 Jan 28	16:36:21	9189	-23029	62	T	-p	0.7723	1.0396	31.7N	35.2W	39	169	209	03m28s
5116	256	0138 Jul 24	18:39:58	9184	-23023	67	A	-p	-0.8320	0.9370	35.7S	67.5W	33	7	422	07m29s
5117	256	0139 Jan 18	08:35:48	9179	-23017	72	T	-n	0.0906	1.0469	15.8S	92.4E	85	174	157	04m27s
5118	256	0139 Jul 13	19:59:21	9175	-23011	77	A	nn	-0.0655	0.9677	18.7N	81.3W	86	2	116	04m02s
5119	256	0140 Jan 07	22:24:43	9170	-23005	82	A	p-	-0.6151	0.9989	60.9S	113.5W	52	357	5	00m05s
5120	256	0140 Jul 02	03:39:41	9165	-22999	87	T	p-	0.6994	1.0175	68.0N	159.3E	45	175	84	01m15s
5121	257	0140 Dec 27	05:53:56	9161	-22993	92	P	t-	-1.3677	0.3339	66.2S	41.1W	0	173		
5122	257	0141 May 23	11:22:17	9157	-22988	59	P	-t	-1.0973	0.8366	63.2S	88.5E	0	321		
5123	257	0141 Jun 21	18:02:55	9156	-22987	97	P	t-	1.4015	0.2470	65.6N	142.2E	0	14		
5124	257	0141 Nov 16	11:01:39	9152	-22982	64	An	-p	0.9854	0.9089	55.9N	87.2E	9	216	-	08m31s
5125	257	0142 May 13	04:19:27	9148	-22976	69	T	-p	-0.3722	1.0598	2.5S	159.4E	68	340	211	05m28s
5126	257	0142 Nov 05	13:09:35	9143	-22970	74	A	-n	0.2688	0.9598	1.3S	23.5E	74	203	151	04m31s
5127	257	0143 May 02	17:10:28	9138	-22964	79	H	p-	0.4035	1.0051	36.4N	52.7W	66	151	19	00m27s
5128	257	0143 Oct 25	22:40:07	9134	-22958	84	H2	p-	-0.4397	1.0158	35.1S	139.1W	64	32	60	01m18s
5129	257	0144 Apr 20	22:49:07	9129	-22952	89	P	t-	1.2098	0.6047	61.3N	123.6E	0	66		
5130	257	0144 Sep 15	04:08:10	9125	-22947	56	Pe	-t	1.5239	0.0176	61.0N	108.8W	0	277		
5131	257	0144 Oct 14	13:32:59	9125	-22946	94	P	t-	-1.0968	0.8287	61.1S	93.1W	0	107		
5132	257	0145 Mar 11	07:41:10	9121	-22941	61	A	-p	-0.9387	0.9395	57.9S	164.2E	20	295	643	04m38s
5133	257	0145 Sep 04	18:53:49	9116	-22935	66	T	-p	0.8709	1.0128	56.1N	18.7W	29	236	88	00m52s
5134	257	0146 Feb 28	13:15:23	9112	-22929	71	A	nn	-0.1681	0.9953	17.0S	27.6E	80	331	17	00m27s
5135	257	0146 Aug 25	03:52:42	9107	-22923	76	A	nn	0.1497	0.9692	19.4N	164.1E	81	207	112	03m11s
5136	257	0147 Feb 18	01:45:03	9102	-22917	81	T	p-	0.5541	1.0409	18.2N	179.0W	56	153	162	03m32s
5137	257	0147 Aug 14	06:00:10	9098	-22911	86	A	p-	-0.6027	0.9372	18.8S	113.5E	53	24	286	07m31s
5138	257	0148 Jan 09	07:40:46	9094	-22906	53	P	-t	-1.4308	0.2007	64.1S	106.1W	0	209		
5139	257	0148 Feb 07	17:46:55	9093	-22905	91	P	t-	1.2255	0.5854	62.0N	102.4W	0	125		
5140	257	0148 Jul 03	16:55:07	9089	-22900	58	P	-t	1.4741	0.1394	64.9N	120.8E	0	339		
5141	258	0148 Aug 02	06:27:42	9089	-22899	96	P	t-	-1.3252	0.4058	62.5S	71.7E	0	47		
5142	258	0148 Dec 28	20:00:51	9085	-22894	63	A	-p	-0.8017	0.9780	75.1S	52.5W	36	327	132	01m30s
5143	258	0149 Jun 23	02:42:52	9080	-22888	68	T	-p	0.6495	1.0352	64.1N	175.6W	49	191	156	02m29s
5144	258	0149 Dec 18	01:16:52	9075	-22882	73	A	nn	-0.1204	0.9332	30.7S	160.7W	83	356	251	08m07s
5145	258	0150 Jun 12	18:23:03	9071	-22876	78	T	nn	-0.1187	1.0787	16.5N	59.3W	83	1	256	07m13s
5146	258	0150 Dec 07	00:43:01	9066	-22870	83	A	p-	0.5630	0.9147	11.6N	153.6W	56	181	393	12m23s
5147	258	0151 Jun 02	11:36:59	9062	-22864	88	T	p-	-0.8630	1.0613	37.8S	45.5E	30	356	400	05m06s
5148	258	0151 Nov 26	01:27:03	9057	-22858	93	P	t-	1.2161	0.5923	68.3N	151.0W	0	194		
5149	258	0152 Apr 22	15:01:57	9053	-22853	60	A+	-t	0.9989	0.9832	71.0N	136.2W	0	51	-	-
5150	258	0152 Oct 15	21:18:07	9049	-22847	65	T	-p	-0.9291	1.0217	69.7S	154.7W	21	54	203	01m20s
5151	258	0153 Apr 11	18:41:20	9044	-22841	70	A	np	0.2599	0.9470	22.5N	67.2W	75	162	202	06m24s
5152	258	0153 Oct 05	12:59:47	9039	-22835	75	T	-n	-0.2355	1.0545	17.6S	15.9E	76	18	185	04m47s
5153	258	0154 Mar 31	18:56:08	9035	-22829	80	A	p-	-0.4860	0.9460	24.0S	56.0W	61	341	226	06m25s
5154	258	0154 Sep 25	04:14:23	9030	-22823	85	T	n-	0.4664	1.0258	26.2N	161.4E	62	200	99	02m20s
5155	258	0155 Feb 19	11:13:58	9026	-22818	52	Pe	-t	1.5389	0.0115	70.8N	0.2W	0	128		
5156	258	0155 Mar 20	23:12:18	9026	-22817	90	P	t-	-1.1993	0.6259	71.9S	36.8W	0	268		
5157	258	0155 Sep 14	14:32:13	9021	-22811	95	P	t-	1.2368	0.5584	71.7N	100.8E	0	283		
5158	258	0156 Feb 09	01:04:40	9017	-22806	62	T	-p	0.7914	1.0435	36.0N	166.6W	37	165	237	03m38s
5159	258	0156 Aug 04	01:19:27	9013	-22800	67	A	-p	-0.9040	0.9328	46.1S	174.5W	25	14	588	07m11s
5160	258	0157 Jan 28	17:11:51	9008	-22794	72	T	-n	0.1059	1.0482	12.4S	37.2W	84	171	161	04m34s

Cat Num	Canon Plate	Calendar Date	TD of Greatest Eclipse	ΔT s	Luna Num	Saros Num	Ecl. Type	QLE	Gamma	Ecl. Mag.	Lat. °	Long. °	Sun Alt °	Sun Azm °	Path Width km	Central Line Dur.
5161	259	0157 Jul 24	02:47:35	9003	-22788	77	Am	nn	-0.1401	0.9682	12.6N	175.5E	82	6	115	04m01s
5162	259	0158 Jan 18	06:53:14	8999	-22782	82	A	p-	-0.6069	0.9981	58.2S	124.6E	52	350	8	00m09s
5163	259	0158 Jul 13	10:52:07	8994	-22776	87	T	p-	0.6255	1.0206	61.3N	57.9E	51	184	90	01m33s
5164	259	0159 Jan 07	14:03:20	8989	-22770	92	P	t-	-1.3623	0.3432	67.3S	174.9W	0	184		
5165	259	0159 Jun 03	18:49:41	8986	-22765	59	P	-t	-1.1694	0.6964	64.0S	33.6W	0	330		
5166	259	0159 Jul 03	01:30:16	8985	-22764	97	P	t-	1.3295	0.3856	66.5N	19.3E	0	4		
5167	259	0159 Nov 27	19:04:23	8981	-22759	64	An	-p	0.9908	0.9087	58.0N	39.4W	7	209	-	08m34s
5168	259	0160 May 23	11:40:03	8976	-22753	69	T	-p	-0.4481	1.0573	5.0S	48.6E	63	343	210	05m24s
5169	259	0160 Nov 15	21:29:22	8972	-22747	74	A	-n	0.2741	0.9614	3.9S	102.5W	74	199	145	04m27s
5170	259	0161 May 13	00:10:12	8967	-22741	79	H	n-	0.3280	1.0034	35.8N	155.3W	71	155	12	00m18s
5171	259	0161 Nov 05	07:17:14	8963	-22735	84	T	p-	-0.4303	1.0168	38.8S	91.9E	64	29	63	01m22s
5172	259	0162 May 02	05:26:00	8958	-22729	89	P	t-	1.1337	0.7359	61.8N	14.9E	0	57		
5173	259	0162 Oct 25	22:13:52	8953	-22723	94	P	t-	-1.0839	0.8525	61.5S	127.2E	0	116		
5174	259	0163 Mar 22	14:48:14	8950	-22718	61	As	-t	-0.9922	0.9409	60.3S	75.0E	6	281	-	04m11s
5175	259	0163 Sep 16	02:50:40	8945	-22712	66	H	-t	0.9101	1.0053	55.3N	135.8W	24	238	43	00m21s
5176	259	0164 Mar 10	20:56:48	8940	-22706	71	H	nn	-0.2112	1.0016	14.9S	88.0W	78	330	6	00m09s
5177	259	0164 Sep 04	11:15:41	8936	-22700	76	A	nn	0.1987	0.9632	17.8N	53.6E	78	209	136	03m48s
5178	259	0165 Feb 28	09:54:09	8931	-22694	81	T	p-	0.5184	1.0467	19.4N	57.5E	59	151	179	03m54s
5179	259	0165 Aug 24	12:55:58	8927	-22688	86	A	p-	-0.5429	0.9343	17.9S	8.7E	57	27	286	07m42s
5180	259	0166 Jan 19	16:19:33	8923	-22683	53	P	-t	-1.4443	0.1756	63.3S	114.0E	0	219		
5181	260	0166 Feb 18	02:09:50	8922	-22682	91	P	t-	1.2001	0.6340	61.5N	122.2E	0	116		
5182	260	0166 Jul 14	23:46:42	8918	-22677	58	Pe	-t	1.5428	0.0170	64.0N	7.4E	0	329		
5183	260	0166 Aug 13	13:24:10	8917	-22676	96	P	t-	-1.2589	0.5199	61.9S	42.2W	0	56		
5184	260	0167 Jan 09	04:27:22	8914	-22671	63	A	-p	-0.8110	0.9765	72.5S	168.1W	35	315	144	01m37s
5185	260	0167 Jul 04	09:57:52	8909	-22665	68	T	-p	0.7207	1.0361	68.2N	87.0E	44	205	176	02m24s
5186	260	0167 Dec 29	09:25:01	8904	-22659	73	A	nn	-0.1262	0.9324	30.8S	78.7E	83	351	255	08m08s
5187	260	0168 Jun 23	01:48:53	8900	-22653	78	T	nn	-0.0441	1.0792	21.3N	171.0W	88	6	256	07m03s
5188	260	0168 Dec 17	08:45:18	8895	-22647	83	A	p-	0.5579	0.9156	10.3N	83.7E	56	176	387	12m14s
5189	260	0169 Jun 12	18:58:55	8891	-22641	88	T	p-	-0.7904	1.0610	29.2S	69.8W	38	1	328	05m25s
5190	260	0169 Dec 06	09:45:08	8886	-22635	93	P	t-	1.2106	0.6021	67.2N	72.6E	0	183		
5191	260	0170 May 03	21:58:07	8882	-22630	60	P	-t	1.0752	0.8467	70.2N	106.1E	0	38		
5192	260	0170 Oct 27	05:54:29	8878	-22624	65	T	-p	-0.9382	1.0212	74.4S	63.7E	20	65	213	01m15s
5193	260	0171 Apr 23	01:20:33	8873	-22618	70	A	-p	0.3349	0.9476	30.9N	169.7W	70	163	204	06m04s
5194	260	0171 Oct 16	21:33:43	8868	-22612	75	T	-n	-0.2529	1.0514	22.9S	114.1W	75	17	176	04m27s
5195	260	0172 Apr 11	01:44:29	8864	-22606	80	A	p-	-0.4172	0.9506	15.8S	162.0W	65	343	198	06m07s
5196	260	0172 Oct 05	12:31:27	8859	-22600	85	T	n-	0.4449	1.0204	20.6N	34.7E	63	199	77	01m56s
5197	260	0173 Mar 31	06:31:47	8855	-22594	90	P	t-	-1.1400	0.7329	71.9S	161.5W	0	282		
5198	260	0173 Sep 24	22:18:14	8850	-22588	95	P	t-	1.2090	0.6067	71.9N	30.6W	0	269		
5199	260	0174 Feb 19	09:23:59	8846	-22583	62	T	-p	0.8182	1.0472	41.4N	63.4E	35	160	272	03m43s
5200	260	0174 Aug 15	08:08:21	8841	-22577	67	A	-t	-0.9681	0.9277	58.7S	70.7E	14	25	1115	06m42s
5201	261	0175 Feb 09	01:39:26	8837	-22571	72	T	-n	0.1278	1.0495	8.0S	165.3W	83	167	165	04m41s
5202	261	0175 Aug 04	09:44:59	8832	-22565	77	A	nn	-0.2079	0.9684	6.2N	69.4E	78	10	116	03m59s
5203	261	0176 Jan 29	15:13:49	8828	-22559	82	A	p-	-0.5930	0.9978	54.1S	2.3E	53	344	10	00m11s
5204	261	0176 Jul 23	18:09:39	8823	-22553	87	T	p-	0.5550	1.0228	54.2N	48.8W	56	190	94	01m48s
5205	261	0177 Jan 17	22:06:44	8819	-22547	92	P	t-	-1.3529	0.3592	68.4S	52.2E	0	195		
5206	261	0177 Jun 14	02:15:28	8815	-22542	59	P	-t	-1.2424	0.5543	64.9S	155.6W	0	339		
5207	261	0177 Jul 13	09:00:47	8814	-22541	97	P	t-	1.2600	0.5196	67.5N	104.9W	0	354		
5208	261	0177 Dec 08	03:09:25	8810	-22536	64	An	-p	0.9944	0.9093	60.5N	167.5W	4	202	-	08m28s
5209	261	0178 Jun 03	18:57:36	8805	-22530	69	T	-p	-0.5255	1.0540	8.6S	61.9W	58	347	209	05m13s
5210	261	0178 Nov 27	05:53:56	8801	-22524	74	A	-n	0.2758	0.9635	5.9S	130.5E	74	195	137	04m18s
5211	261	0179 May 24	07:05:09	8796	-22518	79	H	nn	0.2493	1.0011	34.4N	103.3E	75	160	4	00m06s
5212	261	0179 Nov 16	15:59:19	8792	-22512	84	T	p-	-0.4248	1.0180	42.3S	37.6W	65	26	68	01m27s
5213	261	0180 May 12	11:57:30	8787	-22506	89	P	t-	1.0536	0.8738	62.4N	92.7W	0	48		
5214	261	0180 Nov 05	07:00:07	8782	-22500	94	P	t-	-1.0755	0.8678	62.0S	14.0W	0	120		
5215	261	0181 Apr 01	21:47:59	8779	-22495	61	P	-t	-1.0519	0.8780	60.8S	27.5W	0	279		
5216	261	0181 Sep 26	10:54:43	8774	-22489	66	A	-t	0.9426	0.9978	55.1N	104.4E	19	238	23	00m09s
5217	261	0182 Mar 22	04:32:06	8769	-22483	71	H	-n	-0.2602	1.0078	13.1S	158.0E	75	330	28	00m44s
5218	261	0182 Sep 15	18:45:44	8765	-22477	76	A	-n	0.2409	0.9574	15.6N	59.1W	76	210	159	04m27s
5219	261	0183 Mar 11	17:57:10	8760	-22471	81	T	p-	0.4771	1.0523	20.9N	64.2W	61	149	195	04m14s
5220	261	0183 Sep 04	20:00:32	8756	-22465	86	A	p-	-0.4903	0.9315	18.2S	98.2W	61	29	289	07m51s

Cat Num	Canon Plate	Calendar Date	TD of Greatest Eclipse	ΔT s	Luna Num	Saros Num	Ecl. Type	QLE	Gamma	Ecl. Mag.	Lat. °	Long. °	Sun Alt °	Sun Azm °	Path Width km	Central Line Dur.
5221	262	0184 Jan 31	00:51:23	8752	-22460	53	P	-t	-1.4631	0.1402	62.5S	23.9W	0	228		
5222	262	0184 Feb 29	10:24:45	8751	-22459	91	P	t-	1.1684	0.6947	61.1N	11.0W	0	107		
5223	262	0184 Aug 23	20:29:42	8746	-22453	96	P	t-	-1.1989	0.6231	61.3S	158.2W	0	65		
5224	262	0185 Jan 19	12:46:54	8743	-22448	63	A	-p	-0.8256	0.9752	69.3S	73.9E	34	306	157	01m42s
5225	262	0185 Jul 14	17:18:39	8738	-22442	68	T	-p	0.7879	1.0362	70.2N	7.9W	38	222	198	02m17s
5226	262	0186 Jan 08	17:26:06	8733	-22436	73	A	nn	-0.1376	0.9321	30.3S	40.2W	82	346	256	08m01s
5227	262	0186 Jul 04	09:18:54	8729	-22430	78	Tm	nn	0.0275	1.0787	25.0N	77.0E	88	189	254	06m47s
5228	262	0186 Dec 28	16:44:33	8724	-22424	83	A	p-	0.5496	0.9173	9.6N	38.1W	57	172	375	11m49s
5229	262	0187 Jun 24	02:20:57	8720	-22418	88	T	p-	-0.7183	1.0595	22.2S	176.2E	44	6	281	05m31s
5230	262	0187 Dec 17	18:03:55	8715	-22412	93	P	t-	1.2047	0.6124	66.1N	63.5W	0	172		
5231	262	0188 May 14	04:47:09	8711	-22407	60	P	-t	1.1564	0.7019	69.3N	9.2W	0	26		
5232	262	0188 Jun 12	15:45:40	8710	-22406	98	Pb	t-	-1.5291	0.0272	66.7S	23.7W	0	2		
5233	262	0188 Nov 06	14:37:31	8707	-22401	65	T	-p	-0.9423	1.0212	78.4S	80.9W	19	78	222	01m13s
5234	262	0189 May 03	07:50:35	8702	-22395	70	A	-p	0.4170	0.9478	39.5N	90.5E	65	164	211	05m43s
5235	262	0189 Oct 27	06:15:28	8697	-22389	75	T	-n	-0.2636	1.0485	27.7S	114.7E	75	16	167	04m10s
5236	262	0190 Apr 22	08:26:19	8693	-22383	80	A	p-	-0.3424	0.9551	7.7S	94.0E	70	345	174	05m44s
5237	262	0190 Oct 16	20:54:58	8688	-22377	85	H3	n-	0.4293	1.0152	15.6N	93.5W	64	197	57	01m30s
5238	262	0191 Apr 11	13:46:44	8684	-22371	90	P	t-	-1.0759	0.8499	71.5S	75.1E	0	296		
5239	262	0191 Oct 06	06:10:58	8679	-22365	95	P	t-	1.1871	0.6441	71.7N	163.6W	0	255		
5240	262	0192 Mar 01	17:37:08	8675	-22360	62	T	-p	0.8505	1.0507	47.8N	66.2W	31	155	318	03m44s
5241	263	0192 Aug 25	15:04:06	8671	-22354	67	P	-t	-1.0264	0.9112	71.1S	64.5W	0	53		
5242	263	0193 Feb 19	10:00:11	8666	-22348	72	T	-n	0.1552	1.0509	2.8S	67.9E	81	165	171	04m47s
5243	263	0193 Aug 14	16:50:56	8661	-22342	77	A	nn	-0.2692	0.9683	0.4S	39.4W	74	12	118	03m56s
5244	263	0194 Feb 08	23:26:02	8657	-22336	82	A	p-	-0.5730	0.9977	49.0S	119.9W	55	341	10	00m12s
5245	263	0194 Aug 04	01:36:08	8652	-22330	87	T	p-	0.4912	1.0245	47.0N	159.9W	60	194	96	02m02s
5246	263	0195 Jan 29	06:02:37	8648	-22324	92	P	t-	-1.3383	0.3840	69.4S	79.3W	0	207		
5247	263	0195 Jun 25	09:43:24	8644	-22319	59	P	-t	-1.3136	0.4157	65.9S	81.5E	0	349		
5248	263	0195 Jul 24	16:36:40	8643	-22318	97	P	t-	1.1945	0.6457	68.5N	129.1E	0	343		
5249	263	0195 Dec 19	11:13:25	8639	-22313	64	A+	-p	0.9991	0.9519	65.4N	64.8E	0	195	-	-
5250	263	0196 Jun 14	02:14:22	8634	-22307	69	T	-p	-0.6028	1.0498	13.3S	172.8W	53	351	207	04m54s
5251	263	0196 Dec 07	14:19:38	8630	-22301	74	A	-n	0.2776	0.9662	7.2S	3.3E	74	191	127	04m02s
5252	263	0197 Jun 03	13:56:11	8625	-22295	79	Am	nn	0.1683	0.9981	31.9N	2.7E	80	165	7	00m11s
5253	263	0197 Nov 27	00:44:28	8621	-22289	84	T	n-	-0.4213	1.0198	45.2S	167.0W	65	21	75	01m35s
5254	263	0198 May 23	18:24:58	8616	-22283	89	A	t-	0.9702	0.9440	71.7N	172.1W	13	65	892	03m34s
5255	263	0198 Nov 16	15:49:37	8612	-22277	94	P	t-	-1.0699	0.8779	62.7S	156.2W	0	135		
5256	263	0199 Apr 13	04:41:40	8608	-22272	61	P	-t	-1.1170	0.7667	61.1S	140.3W	0	288		
5257	263	0199 Oct 07	19:06:10	8603	-22266	66	A	-t	0.9683	0.9905	55.8N	17.8W	14	238	134	00m40s
5258	263	0200 Apr 01	11:59:38	8598	-22260	71	H	-n	-0.3166	1.0139	11.8S	45.9E	71	331	50	01m17s
5259	263	0200 Sep 26	02:23:46	8594	-22254	76	A	-n	0.2758	0.9517	13.0N	174.1W	74	210	183	05m08s
5260	263	0201 Mar 22	01:51:41	8589	-22248	81	T	p-	0.4284	1.0578	22.3N	176.7E	65	148	209	04m35s
5261	264	0201 Sep 15	03:14:40	8585	-22242	86	A	p-	-0.4453	0.9286	19.6S	152.6E	63	31	295	08m01s
5262	264	0202 Feb 10	09:15:07	8581	-22237	53	P	-t	-1.4884	0.0928	61.9S	159.6W	0	238		
5263	264	0202 Mar 11	18:30:53	8580	-22236	91	P	t-	1.1299	0.7688	60.9N	142.0W	0	98		
5264	264	0202 Sep 04	03:46:22	8575	-22230	96	P	t-	-1.1469	0.7123	61.0S	83.2E	0	74		
5265	264	0203 Jan 30	20:59:58	8572	-22225	63	A	-p	-0.8449	0.9742	66.0S	45.4W	32	301	172	01m47s
5266	264	0203 Jul 26	00:44:37	8567	-22219	68	T	-p	0.8516	1.0355	70.3N	102.6W	31	240	229	02m09s
5267	264	0204 Jan 20	01:22:31	8562	-22213	73	A	nn	-0.1522	0.9324	29.0S	158.2W	81	341	256	07m51s
5268	264	0204 Jul 14	16:51:47	8558	-22207	78	T	nn	0.0968	1.0774	27.6N	35.2W	84	195	252	06m27s
5269	264	0205 Jan 08	00:40:31	8553	-22201	83	A	p-	0.5386	0.9196	9.4N	159.0W	57	167	359	11m09s
5270	264	0205 Jul 04	09:44:16	8549	-22195	88	T	p-	-0.6478	1.0570	16.7S	62.7E	50	10	246	05m23s
5271	264	0205 Dec 28	02:21:27	8544	-22189	93	P	t-	1.1971	0.6261	65.1N	161.2E	0	161		
5272	264	0206 May 25	11:32:45	8540	-22184	60	P	-t	1.2389	0.5555	68.4N	123.1W	0	15		
5273	264	0206 Jun 23	22:44:50	8539	-22183	98	P	t-	-1.4594	0.1560	65.7S	139.8W	0	12		
5274	264	0206 Nov 17	23:23:19	8536	-22178	65	T	-p	-0.9446	1.0218	82.0S	128.4E	19	96	233	01m12s
5275	264	0207 May 14	14:17:53	8531	-22172	70	A	-p	0.5008	0.9476	48.1N	7.7W	60	167	223	05m22s
5276	264	0207 Nov 07	15:01:09	8526	-22166	75	T	-n	-0.2711	1.0460	31.9S	16.8W	74	13	159	03m55s
5277	264	0208 May 02	15:03:13	8522	-22160	80	A	pn	-0.2628	0.9593	0.4N	8.4W	75	348	153	05m17s
5278	264	0208 Oct 27	05:23:55	8517	-22154	85	H	n-	0.4189	1.0103	11.2N	137.1E	65	194	39	01m03s
5279	264	0209 Apr 21	20:55:53	8513	-22148	90	P	t-	-1.0056	0.9794	71.0S	46.4W	0	309		
5280	264	0209 Oct 16	14:10:27	8508	-22142	95	P	t-	1.1714	0.6705	71.3N	61.8E	0	241		

Cat Num	Canon Plate	Calendar Date	TD of Greatest Eclipse	ΔT s	Luna Num	Saros Num	Ecl. Type	QLE	Gamma	Ecl. Mag.	Lat. °	Long. °	Sun Alt °	Sun Azm °	Path Width km	Central Line Dur.
5281	265	0210 Mar 13	01:40:48	8504	-22137	62	T	-p	0.8909	1.0536	55.5N	164.2E	27	147	390	03m38s
5282	265	0210 Sep 05	22:11:04	8500	-22131	67	P	-t	-1.0756	0.8274	71.6S	174.4E	0	67		
5283	265	0211 Mar 02	18:11:38	8495	-22125	72	T	-n	0.1899	1.0521	3.1N	57.1W	79	163	176	04m51s
5284	265	0211 Aug 26	00:06:18	8490	-22119	77	A	-n	-0.3234	0.9679	7.0S	150.9W	71	15	122	03m51s
5285	265	0212 Feb 20	07:28:52	8486	-22113	82	A	p-	-0.5462	0.9978	43.2S	118.6E	57	339	9	00m11s
5286	265	0212 Aug 14	09:10:09	8481	-22107	87	T	p-	0.4330	1.0255	39.9N	85.7E	64	196	96	02m12s
5287	265	0213 Feb 08	13:48:25	8477	-22101	92	P	t-	-1.3162	0.4215	70.3S	151.0E	0	220		
5288	265	0213 Jul 05	17:12:35	8473	-22096	59	P	-t	-1.3835	0.2801	66.9S	42.0W	0	359		
5289	265	0213 Aug 04	00:18:25	8472	-22095	97	P	t-	1.1337	0.7625	69.5N	1.1E	0	331		
5290	265	0213 Dec 29	19:16:25	8468	-22090	64	A+	-p	1.0047	0.9436	66.5N	67.0W	0	185	-	-
5291	265	0214 Jun 25	09:30:41	8463	-22084	69	T	-p	-0.6796	1.0448	19.2S	75.6E	47	355	203	04m23s
5292	265	0214 Dec 18	22:46:04	8459	-22078	74	A	-n	0.2794	0.9696	7.6S	124.0W	74	186	114	03m41s
5293	265	0215 Jun 14	20:45:22	8454	-22072	79	A	nn	0.0867	0.9946	28.5N	98.1W	85	171	19	00m34s
5294	265	0215 Dec 08	09:31:55	8450	-22066	84	T	n-	-0.4194	1.0220	47.3S	64.0E	65	16	83	01m46s
5295	265	0216 Jun 03	00:48:22	8445	-22060	89	A	p-	0.8836	0.9464	76.5N	137.3E	28	110	426	03m48s
5296	265	0216 Nov 27	00:42:30	8440	-22054	94	P	t-	-1.0674	0.8823	63.6S	60.6E	0	145		
5297	265	0217 Apr 23	11:30:26	8437	-22049	61	P	-t	-1.1862	0.6472	61.5S	108.0E	0	297		
5298	265	0217 Oct 18	03:23:22	8432	-22043	66	A	-t	0.9884	0.9833	57.9N	140.4W	8	238	425	01m10s
5299	265	0218 Apr 12	19:22:42	8427	-22037	71	T	-n	-0.3774	1.0195	11.2S	65.2W	68	332	71	01m49s
5300	265	0218 Oct 07	10:08:56	8423	-22031	76	A	-n	0.3038	0.9465	10.2N	68.7E	72	209	205	05m51s
5301	266	0219 Apr 02	09:40:55	8418	-22025	81	T	n-	0.3748	1.0629	23.7N	59.3E	68	148	221	04m56s
5302	266	0219 Sep 26	10:36:54	8414	-22019	86	A	p-	-0.4070	0.9259	21.7S	41.4E	66	32	301	08m13s
5303	266	0220 Feb 21	17:30:46	8410	-22014	53	Pe	-t	-1.5199	0.0338	61.4S	66.9E	0	247		
5304	266	0220 Mar 22	02:29:25	8409	-22013	91	P	t-	1.0855	0.8543	60.8N	88.9E	0	89		
5305	266	0220 Sep 14	11:13:10	8404	-22007	96	P	t-	-1.1022	0.7890	60.7S	37.9W	0	83		
5306	266	0221 Feb 10	05:02:06	8400	-22002	63	A	-p	-0.8724	0.9731	63.2S	162.7W	29	297	196	01m52s
5307	266	0221 Aug 05	08:18:57	8396	-21996	68	T	-p	0.9089	1.0339	68.8N	159.0E	24	256	276	01m58s
5308	266	0222 Jan 30	09:09:29	8391	-21990	73	A	nn	-0.1743	0.9331	27.3S	86.0E	80	337	254	07m37s
5309	266	0222 Jul 26	00:30:26	8387	-21984	78	T	-n	0.1612	1.0754	29.0N	148.5W	81	200	248	06m06s
5310	266	0223 Jan 19	08:29:58	8382	-21978	83	A	p-	0.5218	0.9226	9.6N	81.8E	58	163	339	10m21s
5311	266	0223 Jul 15	17:10:13	8377	-21972	88	T	p-	-0.5800	1.0536	12.5S	50.7W	54	14	216	05m05s
5312	266	0224 Jan 08	10:36:12	8373	-21966	93	P	t-	1.1865	0.6450	64.1N	27.0E	0	151		
5313	266	0224 Jun 04	18:12:32	8369	-21961	60	P	-t	1.3248	0.4037	67.4N	125.1E	0	4		
5314	266	0224 Jul 04	05:42:31	8368	-21960	98	P	t-	-1.3891	0.2851	64.8S	105.0E	0	22		
5315	266	0224 Nov 28	08:13:29	8364	-21955	65	T	-p	-0.9439	1.0230	85.1S	35.1W	19	126	244	01m15s
5316	266	0225 May 24	20:39:31	8360	-21949	70	A	-p	0.5890	0.9468	56.8N	103.0W	54	170	243	05m02s
5317	266	0225 Nov 17	23:51:08	8355	-21943	75	T	-n	-0.2749	1.0440	35.4S	148.6W	74	9	153	03m43s
5318	266	0226 May 13	21:37:07	8351	-21937	80	A	nn	-0.1799	0.9631	8.1N	109.4W	80	351	136	04m47s
5319	266	0226 Nov 07	13:58:09	8346	-21931	85	H	n-	0.4133	1.0058	7.5N	6.7E	66	191	22	00m37s
5320	266	0227 May 03	04:02:14	8341	-21925	90	H	t-	-0.9320	1.0011	51.9S	172.2E	21	340	10	00m05s
5321	267	0227 Oct 27	22:15:17	8337	-21919	95	P	t-	1.1606	0.6881	70.7N	73.6W	0	227		
5322	267	0228 Mar 23	09:38:46	8333	-21914	62	T	-t	0.9363	1.0557	64.0N	30.3E	20	135	529	03m27s
5323	267	0228 Sep 16	05:26:27	8328	-21908	67	P	-t	-1.1176	0.7560	71.9S	50.8E	0	81		
5324	267	0229 Mar 13	02:14:35	8324	-21902	72	T	-n	0.2312	1.0532	9.6N	179.6E	77	162	180	04m52s
5325	267	0229 Sep 05	07:32:08	8319	-21896	77	A	-n	-0.3696	0.9676	13.6S	94.8E	68	17	125	03m46s
5326	267	0230 Mar 02	15:22:39	8314	-21890	82	A	p-	-0.5130	0.9980	36.8S	1.8W	59	339	8	00m11s
5327	267	0230 Aug 25	16:53:17	8310	-21884	87	T	p-	0.3818	1.0261	32.9N	31.6W	67	198	96	02m20s
5328	267	0231 Feb 19	21:25:10	8305	-21878	92	P	t-	-1.2876	0.4702	71.0S	23.0E	0	233		
5329	267	0231 Jul 17	00:46:30	8301	-21873	59	P	-t	-1.4492	0.1531	67.9S	167.3W	0	10		
5330	267	0231 Aug 15	08:07:30	8301	-21872	97	P	t-	1.0789	0.8673	70.3N	129.4W	0	319		
5331	267	0232 Jan 10	03:14:59	8297	-21867	64	A+	-t	1.0141	0.9292	67.6N	161.8E	0	174	-	-
5332	267	0232 Jul 05	16:47:56	8292	-21861	69	T	-p	-0.7547	1.0389	26.1S	37.0W	41	360	199	03m44s
5333	267	0232 Dec 29	07:10:36	8287	-21855	74	A	-n	0.2833	0.9736	7.0S	109.1E	74	182	99	03m12s
5334	267	0233 Jun 25	03:34:11	8283	-21849	79	A	nn	0.0061	0.9905	24.1N	160.5E	90	177	34	01m04s
5335	267	0233 Dec 18	18:17:54	8278	-21843	84	T	n-	-0.4161	1.0249	48.2S	64.0W	65	9	93	02m00s
5336	267	0234 Jun 14	07:11:48	8274	-21837	89	A	p-	0.7974	0.9469	74.4N	76.6E	37	146	326	04m05s
5337	267	0234 Dec 08	09:36:15	8269	-21831	94	P	t-	-1.0662	0.8843	64.5S	83.2W	0	155		
5338	267	0235 May 04	18:14:55	8265	-21826	61	P	-t	-1.2592	0.5197	62.0S	2.8W	0	305		
5339	267	0235 Jun 03	07:49:57	8264	-21825	99	Pb	t-	1.5417	0.0318	64.0N	51.6W	0	30		
5340	267	0235 Oct 29	11:46:24	8261	-21820	66	A+	-t	1.0029	0.9781	61.6N	98.0E	0	241	-	-

Cat Num	Canon Plate	Calendar Date	TD of Greatest Eclipse	ΔT s	Luna Num	Saros Num	Ecl. Type	QLE	Gamma	Ecl. Mag.	Lat. °	Long. °	Sun Alt °	Sun Azm °	Path Width km	Central Line Dur.
5341	268	0236 Apr 23	02:40:05	8256	-21814	71	T	-p	-0.4436	1.0248	11.5S	175.0W	64	334	93	02m20s
5342	268	0236 Oct 17	18:01:50	8251	-21808	76	A	-n	0.3248	0.9417	7.4N	50.6W	71	207	226	06m37s
5343	268	0237 Apr 12	17:21:56	8247	-21802	81	T	n-	0.3138	1.0677	24.8N	55.4W	72	150	232	05m18s
5344	268	0237 Oct 06	18:09:22	8242	-21796	86	A	p-	-0.3770	0.9234	24.5S	72.3W	68	32	309	08m25s
5345	268	0238 Apr 02	10:19:52	8237	-21790	91	P	t-	1.0347	0.9523	60.9N	38.1W	0	80		
5346	268	0238 Sep 25	18:49:44	8233	-21784	96	P	t-	-1.0645	0.8534	60.7S	161.3W	0	92		
5347	268	0239 Feb 21	12:56:11	8229	-21779	63	A	-t	-0.9056	0.9719	61.1S	81.9E	25	294	235	01m57s
5348	268	0239 Aug 16	16:00:15	8224	-21773	68	T	-t	0.9611	1.0313	66.4N	58.8E	15	270	392	01m45s
5349	268	0240 Feb 10	16:48:43	8220	-21767	73	A	nn	-0.2021	0.9342	25.2S	28.2W	78	334	250	07m21s
5350	268	0240 Aug 05	08:13:15	8215	-21761	78	T	-n	0.2221	1.0728	29.4N	97.2E	77	204	242	05m45s
5351	268	0241 Jan 29	16:13:53	8211	-21755	83	A	p-	0.5002	0.9261	10.2N	35.8W	60	159	317	09m27s
5352	268	0241 Jul 26	00:40:00	8206	-21749	88	T	p-	-0.5160	1.0495	9.7S	164.5W	59	18	190	04m40s
5353	268	0242 Jan 18	18:46:02	8201	-21743	93	P	t-	1.1710	0.6727	63.2N	105.6W	0	141		
5354	268	0242 Jun 16	00:52:20	8197	-21738	60	P	-t	1.4092	0.2555	66.4N	13.7E	0	354		
5355	268	0242 Jul 15	12:43:49	8197	-21737	98	P	t-	-1.3222	0.4067	63.9S	10.9W	0	32		
5356	268	0242 Dec 09	17:03:55	8193	-21732	65	T	-p	-0.9438	1.0246	85.7S	134.1E	19	183	261	01m19s
5357	268	0243 Jun 05	03:00:14	8188	-21726	70	A	-p	0.6774	0.9456	65.4N	164.8E	47	176	274	04m45s
5358	268	0243 Nov 29	08:42:30	8184	-21720	75	T	-n	-0.2774	1.0424	37.9S	80.1E	74	5	148	03m33s
5359	268	0244 May 24	04:09:35	8179	-21714	80	A	nn	-0.0948	0.9665	15.4N	150.6E	85	354	121	04m16s
5360	268	0244 Nov 17	22:34:57	8174	-21708	85	H	n-	0.4108	1.0019	4.6N	124.2W	66	188	7	00m12s
5361	269	0245 May 13	11:05:04	8170	-21702	90	H2	t-	-0.8543	1.0086	39.9S	56.8E	31	348	57	00m48s
5362	269	0245 Nov 07	06:25:01	8165	-21696	95	P	t-	1.1545	0.6978	69.9N	150.3E	0	214		
5363	269	0246 Apr 03	17:28:38	8161	-21691	62	Tn	-t	0.9885	1.0553	72.5N	129.1W	7	96	-	02m59s
5364	269	0246 Sep 27	12:51:42	8157	-21685	67	P	-t	-1.1517	0.6984	72.0S	75.5W	0	95		
5365	269	0247 Mar 24	10:08:36	8152	-21679	72	T	-n	0.2796	1.0538	16.6N	58.4E	74	161	185	04m50s
5366	269	0247 Sep 16	15:08:27	8147	-21673	77	A	-n	-0.4079	0.9671	20.0S	22.1W	66	19	129	03m40s
5367	269	0248 Mar 12	23:05:01	8143	-21667	82	A	p-	-0.4710	0.9982	29.9S	120.1W	62	340	7	00m10s
5368	269	0248 Sep 05	00:45:17	8138	-21661	87	T	n-	0.3372	1.0263	26.1N	151.6W	70	199	95	02m25s
5369	269	0249 Mar 02	04:51:05	8133	-21655	92	P	t-	-1.2506	0.5333	71.5S	102.7W	0	246		
5370	269	0249 Jul 27	08:24:29	8130	-21650	59	Pe	-t	-1.5113	0.0341	68.8S	66.0E	0	21		
5371	269	0249 Aug 25	16:04:09	8129	-21649	97	P	t-	1.0303	0.9597	71.0N	97.6E	0	306		
5372	269	0250 Jan 20	11:08:46	8125	-21644	64	P	-t	1.0277	0.9080	68.7N	31.2E	0	162		
5373	269	0250 Jul 17	00:07:55	8120	-21638	69	T	-p	-0.8264	1.0322	34.2S	151.5W	34	4	194	02m57s
5374	269	0251 Jan 09	15:33:30	8116	-21632	74	A	-n	0.2892	0.9781	5.5S	17.4W	73	178	82	02m38s
5375	269	0251 Jul 06	10:22:43	8111	-21626	79	A	nn	-0.0738	0.9859	19.0N	58.3E	86	360	50	01m40s
5376	269	0251 Dec 30	03:03:44	8106	-21620	84	T	n-	-0.4123	1.0281	48.0S	168.3E	65	3	105	02m16s
5377	269	0252 Jun 24	13:35:01	8102	-21614	89	A	p-	0.7112	0.9466	68.9N	0.1W	44	168	281	04m27s
5378	269	0252 Dec 18	18:29:19	8097	-21608	94	P	t-	-1.0645	0.8874	65.5S	132.8E	0	165		
5379	269	0253 May 15	00:56:59	8093	-21603	61	P	-t	-1.3347	0.3868	62.7S	113.2W	0	314		
5380	269	0253 Jun 13	14:16:07	8093	-21602	99	P	t-	1.4545	0.1817	64.9N	158.5W	0	21		
5381	270	0253 Nov 08	20:13:33	8089	-21597	66	P	-t	1.0136	0.9570	62.2N	38.4W	0	231		
5382	270	0254 May 04	09:55:34	8084	-21591	71	T	-p	-0.5125	1.0294	12.7S	75.5E	59	337	114	02m49s
5383	270	0254 Oct 29	01:58:53	8079	-21585	76	A	-p	0.3418	0.9374	4.7N	171.1W	70	205	246	07m23s
5384	270	0255 Apr 24	00:59:20	8075	-21579	81	T	n-	0.2494	1.0718	25.5N	169.0W	75	152	241	05m39s
5385	270	0255 Oct 18	01:49:03	8070	-21573	86	A	p-	-0.3529	0.9214	27.7S	172.5E	69	31	315	08m37s
5386	270	0256 Apr 12	18:02:10	8065	-21567	91	T	t-	0.9776	1.0522	62.9N	139.0W	11	93	857	02m50s
5387	270	0256 Oct 06	02:36:21	8061	-21561	96	P	t-	-1.0342	0.9052	60.8S	72.7E	0	101		
5388	270	0257 Mar 03	20:38:34	8057	-21556	63	A	-t	-0.9476	0.9702	60.2S	27.9W	18	289	335	02m03s
5389	270	0257 Aug 26	23:51:08	8052	-21550	68	P	-t	1.0060	0.9969	61.3N	34.3W	0	292		
5390	270	0258 Feb 21	00:16:25	8048	-21544	73	A	nn	-0.2392	0.9355	23.2S	139.8W	76	331	247	07m06s
5391	270	0258 Aug 16	16:04:16	8043	-21538	78	T	-n	0.2761	1.0696	28.7N	19.4W	74	207	235	05m25s
5392	270	0259 Feb 09	23:49:56	8038	-21532	83	A	p-	0.4715	0.9302	11.1N	151.2W	62	156	292	08m33s
5393	270	0259 Aug 06	08:13:39	8034	-21526	88	T	p-	-0.4558	1.0447	8.0S	81.0E	63	21	166	04m09s
5394	270	0260 Jan 30	02:51:04	8029	-21520	93	P	t-	1.1508	0.7091	62.4N	123.3E	0	132		
5395	270	0260 Jun 26	07:30:13	8025	-21515	60	Pe	-t	1.4942	0.1074	65.4N	96.7W	0	344		
5396	270	0260 Jul 25	19:47:10	8025	-21514	98	P	t-	-1.2572	0.5236	63.1S	126.9W	0	41		
5397	270	0260 Dec 20	01:54:30	8021	-21509	65	T	-p	-0.9438	1.0267	83.2S	39.8W	19	223	283	01m25s
5398	270	0261 Jun 15	09:19:24	8016	-21503	70	A	-p	0.7667	0.9436	73.7N	79.7E	40	188	326	04m31s
5399	270	0261 Dec 09	17:35:29	8011	-21497	75	T	-n	-0.2785	1.0413	39.5S	51.1W	74	359	144	03m26s
5400	270	0262 Jun 04	10:42:17	8007	-21491	80	A	nn	-0.0093	0.9694	22.1N	51.4E	90	358	110	03m45s

Cat Num	Canon Plate	Calendar Date	TD of Greatest Eclipse	ΔT s	Luna Num	Saros Num	Ecl. Type	QLE	Gamma	Ecl. Mag.	Lat. °	Long. °	Sun Alt °	Sun Azm °	Path Width km	Central Line Dur.
5401	271	0262 Nov 29	07:13:23	8002	-21485	85	A	n-	0.4102	0.9984	2.5N	104.6E	66	184	6	00m11s
5402	271	0263 May 24	18:08:00	7997	-21479	90	T	p-	-0.7755	1.0151	30.1S	55.0W	39	354	82	01m30s
5403	271	0263 Nov 18	14:38:13	7993	-21473	95	P	t-	1.1518	0.7018	69.0N	14.0E	0	202		
5404	271	0264 Apr 14	01:12:18	7989	-21468	62	P	-t	1.0461	0.9330	71.3N	76.4E	0	60		
5405	271	0264 May 13	08:17:18	7988	-21467	100	Pb	t-	-1.4730	0.1096	69.3S	115.9E	0	333		
5406	271	0264 Oct 07	20:25:53	7984	-21462	67	P	-t	-1.1786	0.6530	71.7S	156.1E	0	109		
5407	271	0265 Apr 03	17:54:54	7980	-21456	72	T	-n	0.3341	1.0540	24.1N	61.0W	70	161	189	04m44s
5408	271	0265 Sep 26	22:54:42	7975	-21450	77	A	-p	-0.4387	0.9669	26.2S	141.2W	64	20	132	03m32s
5409	271	0266 Mar 24	06:38:31	7970	-21444	82	A	p-	-0.4226	0.9982	22.7S	123.4E	65	341	7	00m11s
5410	271	0266 Sep 16	08:47:23	7966	-21438	87	T	n-	0.3004	1.0264	19.7N	85.7E	72	199	94	02m29s
5411	271	0267 Mar 13	12:07:39	7961	-21432	92	P	t-	-1.2067	0.6083	71.9S	133.4E	0	260		
5412	271	0267 Sep 06	00:08:11	7956	-21426	97	Tn	t-	0.9878	1.0472	73.1N	62.9W	8	269	-	02m32s
5413	271	0268 Jan 31	18:56:05	7953	-21421	64	P	-t	1.0468	0.8775	69.7N	98.5W	0	150		
5414	271	0268 Jul 27	07:31:46	7948	-21415	69	T	-t	-0.8938	1.0247	43.5S	91.3E	26	10	188	02m07s
5415	271	0269 Jan 19	23:50:15	7943	-21409	74	A	-n	0.3008	0.9831	2.9S	142.8W	73	174	63	01m59s
5416	271	0269 Jul 16	17:14:24	7939	-21403	79	A	nn	-0.1498	0.9808	13.2N	45.5W	81	4	69	02m21s
5417	271	0270 Jan 09	11:45:03	7934	-21397	84	T	n-	-0.4047	1.0320	46.3S	41.2E	66	356	118	02m35s
5418	271	0270 Jul 05	20:00:46	7929	-21391	89	A	p-	0.6271	0.9456	62.2N	87.5W	51	179	258	04m54s
5419	271	0270 Dec 30	03:19:31	7925	-21385	94	P	t-	-1.0610	0.8943	66.6S	10.9W	0	176		
5420	271	0271 May 26	07:37:54	7921	-21380	61	P	-t	-1.4113	0.2507	63.5S	136.5E	0	323		
5421	272	0271 Jun 24	20:44:32	7920	-21379	99	P	t-	1.3681	0.3308	65.9N	93.6E	0	11		
5422	272	0271 Nov 20	04:44:21	7916	-21374	66	P	-t	1.0204	0.9430	62.9N	175.9W	0	222		
5423	272	0272 May 14	17:06:05	7912	-21368	71	T	-p	-0.5860	1.0334	15.2S	33.0W	54	340	137	03m13s
5424	272	0272 Nov 08	10:02:12	7907	-21362	76	A	-p	0.3528	0.9337	2.2N	66.8E	69	201	262	08m10s
5425	272	0273 May 04	08:30:52	7902	-21356	81	T	n-	0.1800	1.0753	25.5N	79.3E	79	155	248	06m02s
5426	272	0273 Oct 28	09:36:39	7898	-21350	86	A	n-	-0.3353	0.9197	31.0S	55.6E	70	29	321	08m49s
5427	272	0274 Apr 24	01:38:24	7893	-21344	91	T	p-	0.9159	1.0564	63.5N	124.8E	23	109	463	03m14s
5428	272	0274 Oct 17	10:32:32	7888	-21338	96	A-	t-	-1.0113	0.9444	61.1S	55.7W	0	110	-	-
5429	272	0275 Mar 15	04:12:26	7884	-21333	63	As	-t	-0.9957	0.9661	61.0S	120.4W	3	271	-	02m10s
5430	272	0275 Sep 07	07:49:10	7880	-21327	68	P	-t	1.0456	0.9222	61.0N	163.3W	0	284		
5431	272	0276 Mar 03	07:36:11	7875	-21321	73	A	-n	-0.2822	0.9370	21.2S	110.5E	73	330	243	06m53s
5432	272	0276 Aug 27	00:01:24	7870	-21315	78	T	-n	0.3249	1.0660	27.3N	137.9W	71	210	227	05m06s
5433	272	0277 Feb 20	07:17:51	7866	-21309	83	A	p-	0.4355	0.9346	12.3N	95.7E	64	153	267	07m40s
5434	272	0277 Aug 16	15:53:34	7861	-21303	88	T	n-	-0.4014	1.0395	7.6S	34.9W	66	24	143	03m37s
5435	272	0278 Feb 09	10:49:47	7856	-21297	93	P	t-	1.1243	0.7569	61.8N	6.1W	0	122		
5436	272	0278 Aug 06	02:55:27	7852	-21291	98	P	t-	-1.1970	0.6308	62.3S	116.1E	0	50		
5437	272	0278 Dec 31	10:41:52	7848	-21286	65	T	-p	-0.9471	1.0290	79.4S	170.8E	18	240	316	01m32s
5438	272	0279 Jun 26	15:41:17	7843	-21280	70	A	-p	0.8534	0.9410	80.3N	15.7E	31	222	423	04m20s
5439	272	0279 Dec 21	02:26:27	7839	-21274	75	T	-n	-0.2811	1.0406	40.1S	178.6E	73	354	142	03m21s
5440	272	0280 Jun 14	17:15:49	7834	-21268	80	A	nn	0.0764	0.9718	28.0N	47.0W	85	183	102	03m18s
5441	273	0280 Dec 09	15:51:37	7829	-21262	85	A	n-	0.4102	0.9955	1.1N	26.3W	66	179	17	00m30s
5442	273	0281 Jun 04	01:11:00	7825	-21256	90	T	p-	-0.6954	1.0205	21.7S	165.1W	46	358	97	02m08s
5443	273	0281 Nov 28	22:52:13	7820	-21250	95	P	t-	1.1503	0.7038	67.9N	121.8W	0	190		
5444	273	0282 Apr 25	08:50:06	7816	-21245	62	P	-t	1.1087	0.8128	70.7N	52.1W	0	47		
5445	273	0282 May 24	15:41:33	7815	-21244	100	P	t-	-1.4000	0.2495	68.4S	7.7W	0	344		
5446	273	0282 Oct 19	04:09:45	7811	-21239	67	P	-t	-1.1981	0.6202	71.3S	25.5E	0	123		
5447	273	0283 Apr 15	01:32:54	7807	-21233	72	T	-p	0.3951	1.0537	31.8N	178.3W	67	161	193	04m32s
5448	273	0283 Oct 08	06:51:00	7802	-21227	77	A	-p	-0.4620	0.9669	32.1S	97.6E	62	20	134	03m24s
5449	273	0284 Apr 03	14:01:36	7797	-21221	82	A	n-	-0.3662	0.9980	15.2S	9.3E	68	342	7	00m13s
5450	273	0284 Sep 26	16:58:52	7793	-21215	87	T	n-	0.2710	1.0263	13.6N	39.4W	74	198	92	02m31s
5451	273	0285 Mar 23	19:12:28	7788	-21209	92	P	t-	-1.1538	0.6989	71.9S	12.4E	0	274		
5452	273	0285 Sep 16	08:21:02	7783	-21203	97	T	t-	0.9528	1.0475	67.4N	135.4E	17	233	531	02m51s
5453	273	0286 Feb 11	02:37:17	7780	-21198	64	P	-t	1.0709	0.8382	70.5N	132.8E	0	137		
5454	273	0286 Aug 07	14:59:28	7775	-21192	69	T	-t	-0.9570	1.0161	55.1S	30.7W	16	10	194	01m15s
5455	273	0287 Jan 31	08:03:05	7770	-21186	74	A	-n	0.3163	0.9886	0.7N	92.6E	72	170	42	01m18s
5456	273	0287 Jul 28	00:08:38	7765	-21180	79	A	nn	-0.2228	0.9754	6.9N	150.7W	77	7	90	03m04s
5457	273	0288 Jan 20	20:22:45	7761	-21174	84	T	n-	-0.3936	1.0362	43.4S	85.8W	67	351	133	02m57s
5458	273	0288 Jul 16	02:29:07	7756	-21168	89	A	p-	0.5451	0.9441	54.9N	179.3E	57	186	247	05m26s
5459	273	0289 Jan 09	12:06:27	7752	-21162	94	P	t-	-1.0551	0.9057	67.7S	154.3W	0	187		
5460	273	0289 Jun 05	14:20:04	7748	-21157	61	Pe	-t	-1.4873	0.1149	64.3S	25.6E	0	333		

Cat Num	Canon Plate	Calendar Date	TD of Greatest Eclipse	ΔT s	Luna Num	Saros Num	Ecl. Type	QLE	Gamma	Ecl. Mag.	Lat. °	Long. °	Sun Alt °	Sun Azm °	Path Width km	Central Line Dur.
5461	274	0289 Jul 05	03:18:00	7747	-21156	99	P	t-	1.2845	0.4754	66.9N	16.0W	0	1		
5462	274	0289 Nov 30	13:15:06	7743	-21151	66	P	-t	1.0264	0.9307	63.8N	46.3E	0	212		
5463	274	0290 May 26	00:17:58	7738	-21145	71	T	-p	-0.6592	1.0365	18.8S	142.4W	49	344	162	03m32s
5464	274	0290 Nov 19	18:07:55	7734	-21139	76	A	-p	0.3609	0.9307	0.3N	55.8W	69	198	276	08m56s
5465	274	0291 May 15	16:00:04	7729	-21133	81	T	nn	0.1081	1.0781	24.7N	31.8W	84	159	254	06m24s
5466	274	0291 Nov 08	17:29:47	7724	-21127	86	A	n-	-0.3220	0.9186	34.2S	62.2W	71	26	325	09m00s
5467	274	0292 May 04	09:08:37	7720	-21121	91	T	p-	0.8497	1.0586	63.9N	24.2E	31	121	364	03m29s
5468	274	0292 Oct 27	18:36:49	7715	-21115	96	As	t-	-0.9939	0.9382	63.6S	177.0W	5	111	-	03m50s
5469	274	0293 Mar 25	11:33:58	7711	-21110	63	P	-t	-1.0531	0.8842	60.7S	126.8E	0	274		
5470	274	0293 Sep 17	15:57:50	7706	-21104	68	P	-t	1.0773	0.8624	60.9N	65.1E	0	275		
5471	274	0294 Mar 14	14:44:43	7702	-21098	73	A	-p	-0.3343	0.9385	19.5S	3.5E	70	329	240	06m43s
5472	274	0294 Sep 07	08:06:29	7697	-21092	78	T	-n	0.3671	1.0621	25.2N	101.0E	68	211	218	04m48s
5473	274	0295 Mar 03	14:37:14	7692	-21086	83	A	p-	0.3920	0.9393	13.5N	14.9W	67	151	242	06m51s
5474	274	0295 Aug 27	23:39:35	7688	-21080	88	T	n-	-0.3527	1.0338	8.2S	152.1W	69	27	121	03m03s
5475	274	0296 Feb 20	18:40:52	7683	-21074	93	P	t-	1.0908	0.8177	61.3N	133.4W	0	113		
5476	274	0296 Aug 16	10:08:36	7678	-21068	98	P	t-	-1.1412	0.7288	61.7S	1.9W	0	59		
5477	274	0297 Jan 10	19:26:43	7674	-21063	65	T	-p	-0.9528	1.0318	75.5S	30.2E	17	248	364	01m40s
5478	274	0297 Jul 06	22:05:39	7670	-21057	70	A	-t	0.9379	0.9373	78.8N	15.1W	20	288	688	04m12s
5479	274	0297 Dec 31	11:15:04	7665	-21051	75	T	-n	-0.2855	1.0404	39.7S	48.9E	73	348	141	03m19s
5480	274	0298 Jun 25	23:53:31	7660	-21045	80	Am	nn	0.1595	0.9736	32.9N	145.4W	81	188	96	02m55s
5481	275	0298 Dec 21	00:29:21	7656	-21039	85	A	n-	0.4101	0.9932	0.5N	157.1W	66	175	26	00m46s
5482	275	0299 Jun 15	08:15:24	7651	-21033	90	T	p-	-0.6154	1.0251	14.6S	85.6E	52	2	108	02m40s
5483	275	0299 Dec 10	07:07:16	7646	-21027	95	P	t-	1.1501	0.7037	66.8N	102.7E	0	179		
5484	275	0300 May 05	16:23:42	7642	-21022	62	P	-t	1.1751	0.6846	69.9N	178.9W	0	35		
5485	275	0300 Jun 03	23:05:01	7642	-21021	100	P	t-	-1.3256	0.3929	67.4S	130.6W	0	355		
5486	275	0300 Oct 29	12:00:54	7638	-21016	67	P	-t	-1.2118	0.5973	70.6S	106.4W	0	136		
5487	275	0301 Apr 25	09:04:15	7633	-21010	72	T	-p	0.4610	1.0526	39.8N	66.3E	62	162	196	04m16s
5488	275	0301 Oct 18	14:56:19	7628	-21004	77	A	-p	-0.4787	0.9672	37.5S	25.1W	61	20	134	03m15s
5489	275	0302 Apr 14	21:16:59	7624	-20998	82	A	n-	-0.3045	0.9974	7.6S	102.8W	72	344	9	00m17s
5490	275	0302 Oct 08	01:17:57	7619	-20992	87	T	n-	0.2474	1.0263	7.9N	166.3W	76	197	92	02m34s
5491	275	0303 Apr 04	02:09:00	7614	-20986	92	P	t-	-1.0945	0.8007	71.8S	106.5W	0	287		
5492	275	0303 Sep 27	16:41:24	7610	-20980	97	T	p-	0.9240	1.0463	60.6N	5.2W	22	219	405	03m00s
5493	275	0304 Feb 22	10:09:17	7606	-20975	64	P	-t	1.1031	0.7851	71.2N	5.8E	0	124		
5494	275	0304 Aug 17	22:33:03	7601	-20969	69	P	-t	-1.0145	0.9705	70.7S	173.1W	0	45		
5495	275	0305 Feb 10	16:07:59	7596	-20963	74	A	-n	0.3393	0.9944	5.2N	30.6W	70	167	21	00m37s
5496	275	0305 Aug 07	07:08:56	7592	-20957	79	A	-p	-0.2898	0.9697	0.2N	102.0E	73	11	114	03m48s
5497	275	0306 Jan 31	04:53:16	7587	-20951	84	T	n-	-0.3761	1.0408	39.3S	147.7E	68	347	147	03m23s
5498	275	0306 Jul 27	09:03:59	7582	-20945	89	A	p-	0.4683	0.9422	47.5N	81.8E	62	191	242	06m03s
5499	275	0307 Jan 20	20:48:20	7578	-20939	94	P	t-	-1.0455	0.9243	68.7S	63.0E	0	199		
5500	275	0307 Jul 16	09:55:38	7573	-20933	99	P	t-	1.2031	0.6164	67.9N	127.0W	0	350		
5501	276	0307 Dec 11	21:47:09	7569	-20928	66	P	-t	1.0308	0.9217	64.8N	92.1W	0	202		
5502	276	0308 Jun 05	07:28:21	7564	-20922	71	T	-p	-0.7341	1.0389	23.7S	108.1E	43	347	191	03m42s
5503	276	0308 Nov 30	02:16:27	7560	-20916	76	A	-p	0.3664	0.9284	1.1S	179.1W	69	194	288	09m37s
5504	276	0309 May 25	23:26:08	7555	-20910	81	T	nn	0.0334	1.0799	23.1N	142.3W	88	164	258	06m45s
5505	276	0309 Nov 19	01:28:42	7550	-20904	86	A	n-	-0.3133	0.9181	37.1S	179.1E	72	22	327	09m08s
5506	276	0310 May 15	16:34:57	7545	-20898	91	T	p-	0.7804	1.0596	63.9N	76.4W	38	133	313	03m41s
5507	276	0310 Nov 08	02:47:32	7541	-20892	96	As	t-	-0.9810	0.9405	67.3S	62.6E	10	109	-	03m43s
5508	276	0311 Apr 05	18:47:59	7537	-20887	63	P	-t	-1.1156	0.7737	60.9S	8.9E	0	283		
5509	276	0311 Sep 29	00:14:07	7532	-20881	68	P	-t	1.1035	0.8130	60.9N	68.3W	0	266		
5510	276	0312 Mar 24	21:43:57	7527	-20875	73	A	-p	-0.3936	0.9401	18.4S	101.3W	67	329	239	06m36s
5511	276	0312 Sep 17	16:18:56	7523	-20869	78	T	-n	0.4032	1.0581	22.8N	22.5W	66	212	207	04m33s
5512	276	0313 Mar 13	21:48:58	7518	-20863	83	A	p-	0.3416	0.9442	14.9N	123.3W	70	150	217	06m08s
5513	276	0313 Sep 07	07:32:57	7513	-20857	88	T	n-	-0.3108	1.0280	9.8S	88.7E	72	29	99	02m30s
5514	276	0314 Mar 03	02:25:09	7509	-20851	93	P	t-	1.0507	0.8910	60.9N	101.1E	0	104		
5515	276	0314 Aug 27	17:29:13	7504	-20845	98	P	t-	-1.0922	0.8136	61.3S	121.7W	0	68		
5516	276	0315 Jan 22	04:06:13	7500	-20840	65	T	-p	-0.9636	1.0344	71.5S	104.1W	15	252	449	01m48s
5517	276	0315 Jul 18	04:34:43	7495	-20834	70	P	-t	1.0183	0.9296	63.5N	74.6W	0	325		
5518	276	0316 Jan 11	19:59:24	7491	-20828	75	T	-n	-0.2935	1.0406	38.5S	80.1W	73	342	142	03m18s
5519	276	0316 Jul 06	06:35:51	7486	-20822	80	A	nn	0.2400	0.9750	36.7N	115.9E	76	193	92	02m37s
5520	276	0316 Dec 31	09:02:13	7481	-20816	85	A	n-	0.4067	0.9915	0.4N	73.3E	66	170	33	00m58s

Cat Num	Canon Plate	Calendar Date	TD of Greatest Eclipse	ΔT s	Luna Num	Saros Num	Ecl. Type	QLE	Gamma	Ecl. Mag.	Lat. °	Long. °	Sun Alt °	Sun Azm °	Path Width km	Central Line Dur.
5521	277	0317 Jun 25	15:23:13	7477	-20810	90	T	p-	-0.5370	1.0288	8.7S	23.8W	57	6	116	03m04s
5522	277	0317 Dec 20	15:19:59	7472	-20804	95	P	t-	1.1482	0.7066	65.8N	31.7W	0	168		
5523	277	0318 May 16	23:53:36	7468	-20799	62	P	-t	1.2444	0.5501	69.0N	55.7E	0	23		
5524	277	0318 Jun 15	06:29:09	7467	-20798	100	P	t-	-1.2512	0.5368	66.4S	106.9E	0	6		
5525	277	0318 Nov 09	19:58:19	7463	-20793	67	P	-t	-1.2209	0.5822	69.7S	120.8E	0	149		
5526	277	0319 May 06	16:29:08	7458	-20787	72	T	-p	0.5318	1.0508	48.0N	46.8W	58	163	199	03m56s
5527	277	0319 Oct 29	23:10:06	7454	-20781	77	A	-p	-0.4892	0.9679	42.3S	149.0W	60	18	132	03m04s
5528	277	0320 Apr 25	04:21:58	7449	-20775	82	A	nn	-0.2349	0.9965	0.0N	147.9E	76	346	13	00m24s
5529	277	0320 Oct 18	09:46:09	7444	-20769	87	T	n-	0.2308	1.0263	2.9N	64.8E	77	196	92	02m36s
5530	277	0321 Apr 14	08:55:50	7440	-20763	92	P	t-	-1.0274	0.9159	71.4S	137.3E	0	301		
5531	277	0321 Oct 08	01:09:39	7435	-20757	97	T	p-	0.9018	1.0445	54.6N	141.7W	25	211	343	03m05s
5532	277	0322 Mar 04	17:34:33	7431	-20752	64	P	-t	1.1412	0.7212	71.7N	120.0W	0	110		
5533	277	0322 Aug 29	06:13:02	7426	-20746	69	P	-t	-1.0658	0.8736	71.3S	57.9E	0	58		
5534	277	0323 Feb 22	00:07:59	7422	-20740	74	H	-n	0.3671	1.0004	10.5N	152.9W	68	164	2	00m03s
5535	277	0323 Aug 18	14:13:22	7417	-20734	79	A	-p	-0.3525	0.9638	6.7S	6.8W	69	14	139	04m29s
5536	277	0324 Feb 11	13:18:54	7412	-20728	84	T	n-	-0.3543	1.0457	34.3S	21.2E	69	344	163	03m50s
5537	277	0324 Aug 06	15:44:50	7407	-20722	89	A	p-	0.3967	0.9399	39.9N	18.8W	66	194	243	06m42s
5538	277	0325 Jan 31	05:23:25	7403	-20716	94	P	t-	-1.0308	0.9526	69.7S	78.6W	0	211		
5539	277	0325 Jul 26	16:41:44	7398	-20710	99	P	t-	1.1276	0.7472	68.9N	119.3E	0	339		
5540	277	0325 Dec 22	06:16:14	7394	-20705	66	P	-t	1.0365	0.9106	65.9N	129.9E	0	191		
5541	278	0326 Jun 16	14:42:01	7389	-20699	71	T	-p	-0.8071	1.0403	30.0S	3.0W	36	352	228	03m43s
5542	278	0326 Dec 11	10:23:27	7385	-20693	76	A	-p	0.3724	0.9267	1.7S	58.0E	68	189	296	10m11s
5543	278	0327 Jun 06	06:52:21	7380	-20687	81	Tm	nn	-0.0413	1.0810	20.5N	106.9E	88	347	261	07m03s
5544	278	0327 Nov 30	09:29:55	7375	-20681	86	A	n-	-0.3063	0.9183	39.3S	60.5E	72	18	326	09m12s
5545	278	0328 May 25	23:56:30	7370	-20675	91	T	p-	0.7077	1.0596	63.0N	175.9W	45	145	277	03m50s
5546	278	0328 Nov 18	11:04:09	7366	-20669	96	A	t-	-0.9722	0.9426	70.6S	66.0W	13	113	951	03m33s
5547	278	0329 Apr 16	01:51:19	7362	-20664	63	P	-t	-1.1855	0.6500	61.2S	106.4W	0	291		
5548	278	0329 May 15	13:16:33	7361	-20663	101	Pb	t-	1.5194	0.0460	62.9N	122.2W	0	45		
5549	278	0329 Oct 09	08:39:31	7357	-20658	68	P	-t	1.1227	0.7769	61.0N	155.9E	0	256		
5550	278	0330 Apr 05	04:33:27	7352	-20652	73	A	-p	-0.4603	0.9413	18.1S	156.3E	62	330	241	06m34s
5551	278	0330 Sep 29	00:40:17	7348	-20646	78	T	-n	0.4313	1.0540	20.0N	148.8W	64	211	196	04m19s
5552	278	0331 Mar 25	04:52:12	7343	-20640	83	A	p-	0.2836	0.9493	16.1N	130.8E	73	150	193	05m28s
5553	278	0331 Sep 18	15:33:09	7338	-20634	88	T	n-	-0.2754	1.0221	12.1S	32.1W	74	30	78	01m58s
5554	278	0332 Mar 13	10:01:57	7333	-20628	93	A+	t-	1.0036	0.9779	60.7N	22.5W	0	95	-	-
5555	278	0332 Sep 07	00:56:45	7329	-20622	98	P	t-	-1.0491	0.8869	60.9S	117.0E	0	77		
5556	278	0333 Feb 01	12:39:53	7325	-20617	65	T	-t	-0.9795	1.0368	67.6S	127.7E	11	252	655	01m54s
5557	278	0333 Jul 28	11:09:49	7320	-20611	70	P	-t	1.0937	0.8007	62.8N	176.6E	0	316		
5558	278	0334 Jan 22	04:39:12	7315	-20605	75	T	-n	-0.3052	1.0410	36.5S	151.7E	72	338	144	03m19s
5559	278	0334 Jul 17	13:23:32	7311	-20599	80	A	nn	0.3168	0.9759	39.4N	16.7E	71	199	91	02m23s
5560	278	0335 Jan 11	17:31:30	7306	-20593	85	A	n-	0.4009	0.9901	1.1N	55.3W	66	166	38	01m05s
5561	279	0335 Jul 06	22:34:59	7301	-20587	90	T	p-	-0.4608	1.0319	4.0S	133.3W	63	11	121	03m20s
5562	279	0335 Dec 31	23:29:52	7296	-20581	95	P	t-	1.1448	0.7123	64.7N	165.0W	0	157		
5563	279	0336 May 27	07:20:31	7293	-20576	62	P	-t	1.3160	0.4107	68.1N	68.3W	0	12		
5564	279	0336 Jun 25	13:54:25	7292	-20575	100	P	t-	-1.1774	0.6799	65.4S	15.6W	0	16		
5565	279	0336 Nov 20	04:00:39	7288	-20570	67	P	-t	-1.2263	0.5732	68.7S	12.6W	0	161		
5566	279	0337 May 16	23:50:04	7283	-20564	72	T	-p	0.6053	1.0481	56.2N	157.8W	52	166	201	03m31s
5567	279	0337 Nov 09	07:29:17	7278	-20558	77	A	-p	-0.4961	0.9691	46.6S	87.0E	60	15	128	02m52s
5568	279	0338 May 06	11:22:03	7274	-20552	82	Am	nn	-0.1621	0.9949	7.4N	40.2E	81	349	18	00m35s
5569	279	0338 Oct 29	18:20:54	7269	-20546	87	T	n-	0.2190	1.0266	1.6S	65.5W	77	193	92	02m39s
5570	279	0339 Apr 25	15:35:16	7264	-20540	92	A	t-	-0.9545	0.9409	57.0S	0.8E	17	334	745	05m24s
5571	279	0339 Oct 19	09:44:42	7259	-20534	97	T	p-	0.8852	1.0425	49.5N	82.6E	27	204	305	03m07s
5572	279	0340 Mar 15	00:51:25	7255	-20529	64	P	-t	1.1866	0.6442	71.9N	116.0E	0	96		
5573	279	0340 Sep 08	14:00:38	7251	-20523	69	P	-t	-1.1098	0.7912	71.7S	73.6W	0	72		
5574	279	0341 Mar 04	07:58:43	7246	-20517	74	H	-n	0.4033	1.0065	16.6N	86.7E	66	162	25	00m40s
5575	279	0341 Aug 28	21:26:32	7241	-20511	79	A	-p	-0.4068	0.9579	13.5S	118.1W	66	16	167	05m06s
5576	279	0342 Feb 21	21:36:39	7237	-20505	84	T	n-	-0.3256	1.0507	28.6S	104.3W	71	342	177	04m20s
5577	279	0342 Aug 17	22:33:14	7232	-20499	89	A	p-	0.3313	0.9374	32.4N	122.3W	70	196	246	07m21s
5578	279	0343 Feb 11	13:51:49	7227	-20493	94	P	t-	-1.0110	0.9909	70.5S	140.8E	0	224		
5579	279	0343 Aug 06	23:35:10	7222	-20487	99	P	t-	1.0571	0.8695	69.8N	3.1E	0	328		
5580	279	0344 Jan 02	14:42:39	7218	-20482	66	P	-t	1.0440	0.8967	67.0N	8.0W	0	180		

Cat Num	Canon Plate	Calendar Date	TD of Greatest Eclipse	ΔT s	Luna Num	Saros Num	Ecl. Type	QLE	Gamma	Ecl. Mag.	Lat. °	Long. °	Sun Alt °	Sun Azm °	Path Width km	Central Line Dur.
5581	280	0344 Jun 26	21:56:50	7214	-20476	71	T	-p	-0.8798	1.0405	38.3S	115.5W	28	356	286	03m32s
5582	280	0344 Dec 21	18:30:19	7209	-20470	76	A	-p	0.3783	0.9258	1.5S	64.8W	68	185	302	10m34s
5583	280	0345 Jun 16	14:18:48	7204	-20464	81	T	nn	-0.1162	1.0811	17.0N	4.5W	83	352	263	07m17s
5584	280	0345 Dec 10	17:32:54	7199	-20458	86	A	n-	-0.3003	0.9191	40.6S	57.9W	72	12	322	09m13s
5585	280	0346 Jun 06	07:17:19	7195	-20452	91	T	p-	0.6346	1.0586	60.9N	83.5E	50	157	250	03m58s
5586	280	0346 Nov 29	19:25:12	7190	-20446	96	A	t-	-0.9664	0.9449	74.1S	160.6E	14	121	825	03m23s
5587	280	0347 Apr 27	08:47:40	7186	-20441	63	P	-t	-1.2601	0.5183	61.6S	139.9E	0	300		
5588	280	0347 May 26	20:15:24	7185	-20440	101	P	t-	1.4489	0.1753	63.6N	123.1E	0	35		
5589	280	0347 Oct 20	17:12:08	7181	-20435	68	P	-t	1.1367	0.7504	61.4N	18.3E	0	247		
5590	280	0348 Apr 15	11:15:44	7176	-20429	73	A	-p	-0.5326	0.9424	18.8S	55.6E	58	332	247	06m35s
5591	280	0348 Oct 09	09:08:38	7172	-20423	78	T	-n	0.4540	1.0501	17.2N	82.8E	63	210	185	04m07s
5592	280	0349 Apr 04	11:48:32	7167	-20417	83	A	nn	0.2189	0.9542	17.0N	27.0E	77	150	171	04m54s
5593	280	0349 Sep 28	23:41:03	7162	-20411	88	H3	n-	-0.2467	1.0163	15.0S	154.9W	76	30	57	01m27s
5594	280	0350 Mar 24	17:33:04	7158	-20405	93	A	t-	0.9504	0.9890	57.3N	110.8W	18	116	124	00m45s
5595	280	0350 Sep 18	08:30:51	7153	-20399	98	A-	t-	-1.0119	0.9493	60.8S	6.0W	0	86	-	-
5596	280	0351 Feb 12	21:06:45	7149	-20394	65	T-	-t	-1.0018	1.0102	61.7S	13.6E	0	241	-	-
5597	280	0351 Aug 08	17:52:21	7144	-20388	70	P	-t	1.1632	0.6821	62.1N	66.2E	0	307		
5598	280	0352 Feb 02	13:10:52	7139	-20382	75	T	-n	-0.3236	1.0417	34.2S	24.9E	71	334	147	03m21s
5599	280	0352 Jul 27	20:18:57	7135	-20376	80	A	-p	0.3881	0.9763	40.7N	84.2W	67	204	91	02m15s
5600	280	0353 Jan 22	01:53:03	7130	-20370	85	A	n-	0.3893	0.9894	2.1N	178.0E	67	162	40	01m09s
5601	281	0353 Jul 17	05:53:15	7125	-20364	90	T	p-	-0.3890	1.0341	0.6S	116.0E	67	15	124	03m27s
5602	281	0354 Jan 11	07:33:10	7120	-20358	95	P	t-	1.1363	0.7266	63.8N	63.8E	0	147		
5603	281	0354 Jun 07	14:46:56	7116	-20353	62	P	-t	1.3882	0.2704	67.1N	168.3E	0	1		
5604	281	0354 Jul 06	21:23:38	7116	-20352	100	P	t-	-1.1062	0.8178	64.5S	138.6W	0	25		
5605	281	0354 Dec 01	12:07:06	7112	-20347	67	P	-t	-1.2292	0.5686	67.7S	146.5W	0	173		
5606	281	0355 May 28	07:05:38	7107	-20341	72	T	-p	0.6827	1.0446	64.6N	94.5E	47	170	204	03m04s
5607	281	0355 Nov 20	15:55:14	7102	-20335	77	A	-p	-0.4984	0.9708	49.9S	37.4W	60	10	121	02m38s
5608	281	0356 May 16	18:14:27	7097	-20329	82	A	nn	-0.0835	0.9929	14.7N	65.0W	85	352	25	00m49s
5609	281	0356 Nov 09	03:02:18	7093	-20323	87	T	n-	0.2122	1.0272	5.3S	162.9E	78	190	94	02m44s
5610	281	0357 May 05	22:07:36	7088	-20317	92	A	p-	-0.8758	0.9440	43.6S	109.9W	29	345	427	05m58s
5611	281	0357 Oct 29	18:26:42	7083	-20311	97	T	p-	0.8744	1.0406	45.3N	53.7W	29	199	280	03m07s
5612	281	0358 Mar 26	08:02:37	7079	-20306	64	P	-t	1.2370	0.5574	71.9N	6.6W	0	83		
5613	281	0358 Sep 19	21:53:58	7074	-20300	69	P	-t	-1.1483	0.7198	71.9S	153.2E	0	86		
5614	281	0359 Mar 15	15:44:41	7070	-20294	74	H	-p	0.4444	1.0126	23.2N	32.8W	63	161	48	01m13s
5615	281	0359 Sep 09	04:45:34	7065	-20288	79	A	-p	-0.4555	0.9520	20.4S	128.9E	63	18	196	05m38s
5616	281	0360 Mar 04	05:47:15	7060	-20282	84	T	n-	-0.2905	1.0557	22.4S	131.2E	73	342	192	04m50s
5617	281	0360 Aug 28	05:30:14	7055	-20276	89	A	nn	0.2727	0.9348	25.1N	131.5E	74	197	252	07m59s
5618	281	0361 Feb 21	22:12:14	7051	-20270	94	T	t-	-0.9850	1.0395	74.3S	26.7W	9	264	853	02m07s
5619	281	0361 Aug 17	06:38:48	7046	-20264	99	An	t-	0.9936	0.9481	73.8N	128.0W	5	304	-	03m12s
5620	281	0362 Jan 12	23:02:10	7042	-20259	66	P	-t	1.0565	0.8743	68.0N	144.7W	0	169		
5621	282	0362 Jul 08	05:17:57	7037	-20253	71	T	-t	-0.9480	1.0393	49.2S	128.7E	18	1	421	03m07s
5622	282	0363 Jan 02	02:32:58	7032	-20247	76	A	-p	0.3870	0.9254	0.3S	173.3E	67	180	305	10m44s
5623	282	0363 Jun 27	21:46:29	7028	-20241	81	T	-n	-0.1899	1.0804	12.7N	116.8W	79	357	264	07m24s
5624	282	0363 Dec 22	01:35:17	7023	-20235	86	A	n-	-0.2937	0.9207	40.8S	175.9W	73	6	315	09m08s
5625	282	0364 Jun 16	14:36:12	7018	-20229	91	T	p-	0.5608	1.0566	57.4N	18.6W	56	167	226	04m02s
5626	282	0364 Dec 10	03:47:33	7013	-20223	96	A	t-	-0.9609	0.9479	77.7S	24.3E	15	131	721	03m11s
5627	282	0365 May 07	15:35:42	7009	-20218	63	P	-t	-1.3402	0.3772	62.2S	28.2E	0	309		
5628	282	0365 Jun 06	03:10:31	7009	-20217	101	P	t-	1.3761	0.3084	64.4N	9.0E	0	26		
5629	282	0365 Oct 31	01:52:16	7005	-20212	68	P	-t	1.1450	0.7347	61.9N	121.4W	0	238		
5630	282	0366 Apr 26	17:50:18	7000	-20206	73	A	-p	-0.6110	0.9430	20.6S	43.3W	52	334	260	06m38s
5631	282	0366 Oct 20	17:43:30	6995	-20200	78	T	-p	0.4710	1.0464	14.4N	47.5W	62	207	173	03m56s
5632	282	0367 Apr 15	18:38:26	6990	-20194	83	A	nn	0.1482	0.9590	17.5N	74.9W	81	152	150	04m24s
5633	282	0367 Oct 10	07:55:49	6986	-20188	88	H	n-	-0.2245	1.0105	18.3S	80.7E	77	30	37	00m57s
5634	282	0368 Apr 04	00:56:17	6981	-20182	93	A	t-	0.8896	0.9977	55.8N	146.6E	27	123	17	00m10s
5635	282	0368 Sep 28	16:13:03	6976	-20176	98	As	p-	-0.9818	0.9459	60.2S	110.3W	10	77	-	03m47s
5636	282	0369 Feb 23	05:26:33	6972	-20171	65	P	-t	-1.0301	0.9583	61.2S	120.9W	0	250		
5637	282	0369 Aug 19	00:43:14	6967	-20165	70	P	-t	1.2258	0.5758	61.5N	46.2W	0	298		
5638	282	0370 Feb 12	21:36:02	6962	-20159	75	T	-n	-0.3470	1.0425	31.6S	100.7W	70	331	151	03m23s
5639	282	0370 Aug 08	03:22:48	6958	-20153	80	A	-p	0.4533	0.9765	41.0N	172.6E	63	209	94	02m09s
5640	282	0371 Feb 02	10:08:42	6953	-20147	85	A	n-	0.3735	0.9889	3.7N	52.8E	68	158	42	01m10s

Cat Num	Canon Plate	Calendar Date	TD of Greatest Eclipse	ΔT s	Luna Num	Saros Num	Ecl. Type	QLE	Gamma	Ecl. Mag.	Lat. °	Long. °	Sun Alt °	Sun Azm °	Path Width km	Central Line Dur.
5641	283	0371 Jul 28	13:16:47	6948	-20141	90	T	p-	-0.3207	1.0357	1.6N	4.5E	71	19	126	03m30s
5642	283	0372 Jan 22	15:31:22	6943	-20135	95	P	t-	1.1241	0.7473	62.9N	65.8W	0	138		
5643	283	0372 Jun 17	22:13:25	6939	-20130	62	Pe	-t	1.4603	0.1305	66.1N	45.3E	0	351		
5644	283	0372 Jul 17	04:56:56	6939	-20129	100	P	t-	-1.0379	0.9497	63.6S	97.7E	0	35		
5645	283	0372 Dec 11	20:14:00	6935	-20124	67	P	-t	-1.2320	0.5641	66.6S	80.2E	0	184		
5646	283	0373 Jun 07	14:20:16	6930	-20118	72	T	-p	0.7601	1.0401	72.7N	8.0W	40	179	208	02m36s
5647	283	0373 Dec 01	00:23:43	6925	-20112	77	A	-p	-0.4996	0.9732	52.3S	161.1W	60	4	111	02m22s
5648	283	0374 May 28	01:04:03	6920	-20106	82	A	nn	-0.0035	0.9903	21.4N	168.7W	90	347	34	01m06s
5649	283	0374 Nov 20	11:46:54	6916	-20100	87	T	n-	0.2076	1.0283	8.3S	30.8E	78	187	98	02m51s
5650	283	0375 May 17	04:35:36	6911	-20094	92	A	p-	-0.7934	0.9459	32.9S	146.0E	37	351	327	06m26s
5651	283	0375 Nov 10	03:13:23	6906	-20088	97	T	p-	0.8677	1.0388	42.0N	169.5E	29	194	262	03m07s
5652	283	0376 Apr 05	15:05:04	6902	-20083	64	P	-t	1.2949	0.4567	71.6N	126.9W	0	69		
5653	283	0376 May 05	05:09:34	6901	-20082	102	Pb	t-	-1.5322	0.0487	69.9S	166.1E	0	325		
5654	283	0376 Sep 30	05:55:50	6897	-20077	69	P	-t	-1.1791	0.6633	71.9S	17.8E	0	100		
5655	283	0377 Mar 25	23:22:23	6893	-20071	74	T	-p	0.4932	1.0184	30.4N	150.6W	60	159	72	01m41s
5656	283	0377 Sep 19	12:13:48	6888	-20065	79	A	-p	-0.4959	0.9464	27.1S	13.8E	60	20	225	06m05s
5657	283	0378 Mar 15	13:50:01	6883	-20059	84	T	n-	-0.2483	1.0606	15.6S	8.1E	76	342	205	05m21s
5658	283	0378 Sep 08	12:37:19	6878	-20053	89	A	nn	0.2222	0.9322	18.0N	22.5E	77	198	259	08m35s
5659	283	0379 Mar 05	06:24:24	6873	-20047	94	T	p-	-0.9524	1.0438	69.7S	170.8E	17	301	491	02m35s
5660	283	0379 Aug 28	13:51:24	6869	-20041	99	A	t-	0.9364	0.9517	71.7N	55.2E	20	239	511	03m28s
5661	284	0380 Jan 24	07:16:41	6865	-20036	66	P	-t	1.0722	0.8464	69.1N	79.3E	0	157		
5662	284	0380 Jul 18	12:43:15	6860	-20030	71	P	-t	-1.0132	0.9869	68.2S	4.3E	0	13		
5663	284	0381 Jan 12	10:30:55	6855	-20024	76	A	-p	0.3991	0.9256	1.9N	52.3E	66	176	305	10m40s
5664	284	0381 Jul 08	05:17:09	6850	-20018	81	T	-n	-0.2612	1.0788	7.6N	129.4E	75	1	264	07m22s
5665	284	0382 Jan 01	09:36:33	6846	-20012	86	A	n-	-0.2861	0.9228	39.9S	66.2E	73	1	305	08m59s
5666	284	0382 Jun 27	21:55:37	6841	-20006	91	T	p-	0.4878	1.0538	52.9N	123.1W	61	175	204	04m03s
5667	284	0382 Dec 21	12:10:55	6836	-20000	96	A	t-	-0.9555	0.9513	81.4S	117.6W	17	146	628	02m58s
5668	284	0383 May 18	22:19:09	6832	-19995	63	P	-t	-1.4227	0.2323	63.0S	82.6W	0	318		
5669	284	0383 Jun 17	10:04:04	6831	-19994	101	P	t-	1.3025	0.4418	65.3N	105.0W	0	17		
5670	284	0383 Nov 11	10:37:14	6827	-19989	68	P	-t	1.1502	0.7249	62.5N	97.7E	0	228		
5671	284	0384 May 07	00:18:55	6822	-19983	73	A	-p	-0.6941	0.9432	24.0S	140.9W	46	336	284	06m42s
5672	284	0384 Oct 31	02:24:17	6818	-19977	78	T	-p	0.4831	1.0431	11.9N	179.4W	61	204	163	03m47s
5673	284	0385 Apr 26	01:24:13	6813	-19971	83	A	nn	0.0732	0.9636	17.5N	175.6W	86	155	132	03m59s
5674	284	0385 Oct 20	16:16:17	6808	-19965	88	H	n-	-0.2077	1.0052	21.7S	44.9W	78	28	18	00m28s
5675	284	0386 Apr 15	08:15:41	6803	-19959	93	H	p-	0.8246	1.0055	55.3N	42.7E	34	128	33	00m23s
5676	284	0386 Oct 10	00:02:21	6798	-19953	98	A	p-	-0.9583	0.9422	60.9S	135.8E	16	75	751	04m06s
5677	284	0387 Mar 06	13:38:22	6795	-19948	65	P	-t	-1.0654	0.8923	61.0S	106.7E	0	259		
5678	284	0387 Apr 04	21:52:16	6794	-19947	103	Pb	t-	1.5046	0.0533	61.1N	140.0E	0	78		
5679	284	0387 Aug 30	07:43:13	6790	-19942	70	P	-t	1.2812	0.4820	61.1N	160.7W	0	289		
5680	284	0388 Feb 24	05:52:00	6785	-19936	75	T	-n	-0.3781	1.0432	29.0S	135.6E	68	329	155	03m27s
5681	285	0388 Aug 18	10:36:49	6780	-19930	80	A	-p	0.5112	0.9763	40.3N	66.3E	59	213	98	02m08s
5682	285	0389 Feb 12	18:13:26	6775	-19924	85	A	n-	0.3491	0.9887	5.3N	69.6W	70	155	42	01m10s
5683	285	0389 Aug 07	20:49:26	6771	-19918	90	T	n-	-0.2590	1.0366	2.6N	109.1W	75	22	127	03m28s
5684	285	0390 Feb 01	23:20:58	6766	-19912	95	P	t-	1.1050	0.7797	62.2N	166.9E	0	128		
5685	285	0390 Jul 28	12:35:00	6761	-19906	100	T	t-	-0.9732	1.0595	52.8S	12.4W	13	32	873	04m06s
5686	285	0390 Dec 23	04:21:36	6757	-19901	67	P	-t	-1.2349	0.5594	65.5S	52.8W	0	195		
5687	285	0391 Jun 18	21:32:22	6752	-19895	72	T	-t	0.8390	1.0346	80.3N	94.3W	33	203	216	02m06s
5688	285	0391 Dec 12	08:55:08	6747	-19889	77	A	-p	-0.4991	0.9761	53.5S	75.4E	60	357	99	02m04s
5689	285	0392 Jun 07	07:48:24	6743	-19883	82	A	nn	0.0801	0.9872	27.7N	89.8E	85	180	45	01m26s
5690	285	0392 Nov 30	20:35:47	6738	-19877	87	T	n-	0.2060	1.0298	10.4S	102.1W	78	183	103	03m00s
5691	285	0393 May 27	11:00:05	6733	-19871	92	A	p-	-0.7081	0.9471	23.6S	44.9E	45	356	276	06m50s
5692	285	0393 Nov 20	12:03:39	6728	-19865	97	T	p-	0.8640	1.0373	39.4N	32.1E	30	188	250	03m05s
5693	285	0394 Apr 16	22:03:59	6724	-19860	64	P	-t	1.3563	0.3488	71.1N	114.1E	0	56		
5694	285	0394 May 16	11:41:58	6724	-19859	102	P	t-	-1.4503	0.1889	69.1S	55.1E	0	337		
5695	285	0394 Oct 11	14:04:04	6720	-19854	69	P	-t	-1.2040	0.6179	71.6S	119.1W	0	114		
5696	285	0395 Apr 06	06:55:08	6715	-19848	74	T	-p	0.5470	1.0240	38.1N	92.8E	57	158	97	02m04s
5697	285	0395 Sep 30	19:48:39	6710	-19842	79	A	-p	-0.5302	0.9411	33.6S	102.8W	58	21	255	06m27s
5698	285	0396 Mar 25	21:46:08	6705	-19836	84	T	n-	-0.2001	1.0653	8.6S	113.6W	78	342	218	05m50s
5699	285	0396 Sep 18	19:53:40	6700	-19830	89	A	nn	0.1792	0.9297	11.2N	88.9W	80	198	267	09m07s
5700	285	0397 Mar 15	14:27:52	6696	-19824	94	T	p-	-0.9131	1.0473	62.1S	30.3E	24	319	388	03m02s

Cat Num	Canon Plate	Calendar Date	TD of Greatest Eclipse	ΔT s	Luna Num	Saros Num	Ecl. Type	QLE	Gamma	Ecl. Mag.	Lat. °	Long. °	Sun Alt °	Sun Azm °	Path Width km	Central Line Dur.
5701	286	0397 Sep 07	21:15:23	6691	-19818	99	A	t-	0.8876	0.9531	63.2N	76.1W	27	221	372	03m43s
5702	286	0398 Feb 03	15:22:22	6687	-19813	66	P	-t	1.0942	0.8077	70.0N	55.2W	0	145		
5703	286	0398 Jul 29	20:15:04	6682	-19807	71	P	-t	-1.0735	0.8727	69.2S	121.2W	0	25		
5704	286	0399 Jan 23	18:21:42	6677	-19801	76	A	-p	0.4165	0.9263	5.2N	67.1W	65	172	304	10m25s
5705	286	0399 Jul 19	12:51:41	6672	-19795	81	T	-n	-0.3290	1.0764	2.0N	13.9E	71	5	262	07m11s
5706	286	0400 Jan 12	17:32:37	6668	-19789	86	A	n-	-0.2735	0.9257	37.6S	50.8W	74	355	291	08m44s
5707	286	0400 Jul 08	05:16:04	6663	-19783	91	T	p-	0.4163	1.0502	47.5N	129.9E	65	182	183	04m00s
5708	286	0400 Dec 31	20:32:08	6658	-19777	96	A	t-	-0.9475	0.9556	85.0S	86.8E	18	175	523	02m44s
5709	286	0401 May 29	04:57:52	6654	-19772	63	Pe	-t	-1.5078	0.0836	63.8S	167.6E	0	327		
5710	286	0401 Jun 27	16:57:24	6653	-19771	101	P	t-	1.2292	0.5739	66.3N	140.7E	0	7		
5711	286	0401 Nov 21	19:26:03	6649	-19766	68	P	-t	1.1528	0.7201	63.3N	44.5W	0	219		
5712	286	0402 May 18	06:43:34	6644	-19760	73	A	-p	-0.7800	0.9428	29.1S	122.1E	39	339	331	06m44s
5713	286	0402 Nov 11	11:10:32	6640	-19754	78	T	-p	0.4904	1.0401	9.6N	47.2E	61	201	153	03m39s
5714	286	0403 May 07	08:04:37	6635	-19748	83	A	nn	-0.0070	0.9679	16.6N	85.2E	90	330	116	03m36s
5715	286	0403 Nov 01	00:43:05	6630	-19742	88	H	n-	-0.1968	1.0001	25.1S	171.8W	79	26	1	00m01s
5716	286	0404 Apr 25	15:29:35	6625	-19736	93	T	p-	0.7541	1.0127	55.1N	59.9W	41	134	66	00m55s
5717	286	0404 Oct 20	07:58:52	6620	-19730	98	A	p-	-0.9410	0.9382	63.1S	15.7E	19	76	678	04m23s
5718	286	0405 Mar 16	21:42:37	6617	-19725	65	P	-t	-1.1073	0.8133	60.9S	23.8W	0	268		
5719	286	0405 Apr 15	05:32:35	6616	-19724	103	P	t-	1.4444	0.1668	61.3N	15.4E	0	69		
5720	286	0405 Sep 09	14:53:32	6612	-19719	70	P	-t	1.3285	0.4022	60.9N	82.3E	0	280		
5721	287	0406 Mar 06	14:00:59	6607	-19713	75	T	-p	-0.4147	1.0438	26.6S	13.3E	65	328	159	03m31s
5722	287	0406 Aug 29	17:59:08	6602	-19707	80	A	-p	0.5632	0.9759	39.0N	42.9W	56	216	103	02m08s
5723	287	0407 Feb 24	02:11:01	6597	-19701	85	A	n-	0.3195	0.9887	7.3N	170.0E	71	153	42	01m08s
5724	287	0407 Aug 19	04:29:24	6593	-19695	90	T	n-	-0.2024	1.0371	2.5N	135.6E	78	25	127	03m24s
5725	287	0408 Feb 13	07:01:38	6588	-19689	95	P	t-	1.0792	0.8238	61.6N	42.1E	0	119		
5726	287	0408 Aug 07	20:19:36	6583	-19683	100	T	t-	-0.9138	1.0609	44.3S	126.4W	24	31	487	04m26s
5727	287	0409 Jan 02	12:26:14	6579	-19678	67	P	-t	-1.2407	0.5499	64.5S	175.3E	0	205		
5728	287	0409 Jun 29	04:46:23	6574	-19672	72	T	-t	0.9153	1.0279	82.0N	135.1W	23	273	239	01m35s
5729	287	0409 Dec 22	17:24:51	6569	-19666	77	A	-p	-0.5009	0.9797	53.6S	47.3W	60	350	84	01m44s
5730	287	0410 Jun 18	14:33:19	6565	-19660	82	A	nn	0.1623	0.9835	33.1N	10.7W	80	185	59	01m47s
5731	287	0410 Dec 12	05:25:03	6560	-19654	87	T	n-	0.2041	1.0318	11.7S	125.1E	78	178	110	03m11s
5732	287	0411 Jun 07	17:21:39	6555	-19648	92	A	p-	-0.6202	0.9476	15.5S	54.1W	52	360	246	07m08s
5733	287	0411 Dec 01	20:56:26	6550	-19642	97	T	p-	0.8625	1.0363	37.6N	105.7W	30	183	242	03m04s
5734	287	0412 Apr 27	04:56:34	6546	-19637	64	P	-t	1.4234	0.2296	70.4N	2.9W	0	43		
5735	287	0412 May 26	18:11:23	6545	-19636	102	P	t-	-1.3644	0.3368	68.1S	54.6W	0	348		
5736	287	0412 Oct 21	22:19:21	6541	-19631	69	P	-t	-1.2225	0.5845	71.0S	102.6E	0	128		
5737	287	0413 Apr 16	14:20:56	6537	-19625	74	T	-p	0.6075	1.0291	46.3N	22.2W	52	158	124	02m20s
5738	287	0413 Oct 11	03:32:32	6532	-19619	79	A	-p	-0.5566	0.9362	39.7S	139.0E	56	22	284	06m44s
5739	287	0414 Apr 06	05:35:10	6527	-19613	84	T	n-	-0.1457	1.0696	1.4S	126.3E	82	343	229	06m16s
5740	287	0414 Sep 30	03:19:02	6522	-19607	89	A	nn	0.1434	0.9274	4.9N	157.5E	82	198	275	09m35s
5741	288	0415 Mar 26	22:23:20	6517	-19601	94	T	p-	-0.8676	1.0503	53.7S	100.5W	29	329	334	03m31s
5742	288	0415 Sep 19	04:49:16	6513	-19595	99	A	p-	0.8460	0.9539	55.3N	160.3E	32	212	314	03m58s
5743	288	0416 Feb 14	23:19:01	6509	-19590	66	P	-t	1.1229	0.7575	70.8N	171.9E	0	132		
5744	288	0416 Aug 09	03:53:39	6504	-19584	71	P	-t	-1.1288	0.7677	70.0S	111.1E	0	37		
5745	288	0416 Sep 07	13:26:34	6503	-19583	109	Pb	t-	1.5077	0.0637	71.6N	112.1E	0	288		
5746	288	0417 Feb 03	02:05:28	6499	-19578	76	A	-p	0.4393	0.9274	9.4N	174.8E	64	169	302	09m59s
5747	288	0417 Jul 29	20:31:08	6494	-19572	81	T	-n	-0.3928	1.0734	4.1S	103.3W	67	9	259	06m50s
5748	288	0418 Jan 23	01:24:23	6489	-19566	86	A	n-	-0.2570	0.9291	34.2S	167.6W	75	351	275	08m26s
5749	288	0418 Jul 19	12:39:47	6485	-19560	91	T	n-	0.3481	1.0459	41.5N	20.4E	69	187	163	03m52s
5750	288	0419 Jan 12	04:51:32	6480	-19554	96	A	t-	-0.9373	0.9605	85.8S	108.7W	20	245	423	02m29s
5751	288	0419 Jul 08	23:50:44	6475	-19548	101	P	t-	1.1565	0.7035	67.3N	26.0E	0	357		
5752	288	0419 Dec 03	04:17:38	6471	-19543	68	P	-t	1.1536	0.7187	64.2N	172.4E	0	209		
5753	288	0420 May 28	13:05:35	6466	-19537	73	A	-p	-0.8673	0.9415	36.7S	25.5E	30	342	430	06m38s
5754	288	0420 Nov 21	19:59:16	6461	-19531	78	T	-p	0.4956	1.0377	7.9N	86.8W	60	196	145	03m33s
5755	288	0421 May 17	14:44:02	6457	-19525	83	Am	nn	-0.0888	0.9716	14.9N	14.0W	85	341	102	03m17s
5756	288	0421 Nov 11	09:13:49	6452	-19519	88	A	n-	-0.1901	0.9956	28.3S	60.7E	79	23	16	00m25s
5757	288	0422 May 06	22:41:20	6447	-19513	93	T	p-	0.6806	1.0193	54.8N	161.7W	47	141	89	01m25s
5758	288	0422 Oct 31	16:00:07	6442	-19507	98	A	p-	-0.9282	0.9344	66.1S	107.0W	21	78	659	04m37s
5759	288	0423 Mar 28	05:39:26	6438	-19502	65	P	-t	-1.1556	0.7213	60.9S	152.5W	0	276		
5760	288	0423 Apr 26	13:07:50	6437	-19501	103	P	t-	1.3796	0.2901	61.8N	108.0W	0	60		

Cat Num	Canon Plate	Calendar Date	TD of Greatest Eclipse	ΔT s	Luna Num	Saros Num	Ecl. Type	QLE	Gamma	Ecl. Mag.	Lat. °	Long. °	Sun Alt °	Sun Azm °	Path Width km	Central Line Dur.
5761	289	0423 Sep 20	22:13:51	6433	-19496	70	P	-t	1.3680	0.3358	60.8N	37.2W	0	271		
5762	289	0424 Mar 16	21:58:57	6429	-19490	75	T	-p	-0.4601	1.0441	24.8S	106.4W	62	328	164	03m34s
5763	289	0424 Sep 09	01:33:22	6424	-19484	80	A	-p	0.6062	0.9755	37.1N	156.0W	52	217	108	02m10s
5764	289	0425 Mar 06	09:57:04	6419	-19478	85	A	n-	0.2811	0.9887	9.2N	52.7E	74	151	41	01m07s
5765	289	0425 Aug 29	12:19:01	6414	-19472	90	T	n-	-0.1534	1.0371	1.3N	17.8E	81	27	126	03m19s
5766	289	0426 Feb 23	14:32:13	6409	-19466	95	P	t-	1.0456	0.8811	61.1N	80.0W	0	110		
5767	289	0426 Aug 19	04:11:14	6405	-19460	100	T	p-	-0.8601	1.0605	40.3S	115.3E	30	33	382	04m27s
5768	289	0427 Jan 13	20:28:41	6401	-19455	67	P	-t	-1.2488	0.5366	63.6S	44.4E	0	215		
5769	289	0427 Jul 10	11:59:29	6396	-19449	72	T	-t	0.9913	1.0180	69.4N	167.3E	6	323	576	00m55s
5770	289	0428 Jan 03	01:54:51	6391	-19443	77	A	-p	-0.5033	0.9838	52.5S	170.4W	60	342	66	01m22s
5771	289	0428 Jun 28	21:16:15	6386	-19437	82	A	np	0.2453	0.9792	37.6N	109.7W	76	190	76	02m10s
5772	289	0428 Dec 22	14:14:19	6381	-19431	87	T	n-	0.2019	1.0343	12.1S	7.5W	78	173	118	03m22s
5773	289	0429 Jun 17	23:43:27	6377	-19425	92	A	p-	-0.5325	0.9475	8.6S	152.2W	58	4	228	07m19s
5774	289	0429 Dec 12	05:50:23	6372	-19419	97	T	p-	0.8620	1.0356	36.4N	116.2E	30	178	237	03m02s
5775	289	0430 May 08	11:47:21	6368	-19414	64	Pe	-t	1.4922	0.1063	69.6N	118.8W	0	31		
5776	289	0430 Jun 07	00:41:34	6367	-19413	102	P	t-	-1.2780	0.4865	67.1S	164.0W	0	359		
5777	289	0430 Nov 02	06:38:40	6363	-19408	69	P	-t	-1.2369	0.5587	70.3S	36.2W	0	141		
5778	289	0431 Apr 27	21:43:54	6358	-19402	74	T	-p	0.6712	1.0337	54.7N	136.4W	48	157	153	02m31s
5779	289	0431 Oct 22	11:22:11	6353	-19396	79	A	-p	-0.5773	0.9318	45.4S	20.1E	54	21	311	06m58s
5780	289	0432 Apr 16	13:17:57	6348	-19390	84	T	nn	-0.0858	1.0734	5.9N	7.9E	85	344	239	06m37s
5781	290	0432 Oct 10	10:53:48	6344	-19384	89	A	nn	0.1154	0.9254	1.0S	41.8E	83	197	283	10m00s
5782	290	0433 Apr 06	06:11:16	6339	-19378	94	T	p-	-0.8163	1.0527	45.3S	133.9E	35	336	300	03m59s
5783	290	0433 Sep 29	12:33:08	6334	-19372	99	A	p-	0.8118	0.9546	48.3N	37.5E	35	207	281	04m11s
5784	290	0434 Feb 25	07:05:34	6330	-19367	66	P	-t	1.1588	0.6947	71.5N	41.0E	0	119		
5785	290	0434 Aug 20	11:40:53	6325	-19361	71	P	-t	-1.1776	0.6752	70.8S	19.3W	0	49		
5786	290	0434 Sep 18	21:26:54	6325	-19360	109	P	t-	1.4688	0.1352	71.9N	22.7W	0	275		
5787	290	0435 Feb 14	09:39:34	6320	-19355	76	A	-p	0.4697	0.9289	14.5N	58.7E	62	166	300	09m26s
5788	290	0435 Aug 10	04:16:06	6316	-19349	81	T	-p	-0.4516	1.0697	10.6S	137.5E	63	12	254	06m22s
5789	290	0436 Feb 03	09:08:51	6311	-19343	86	A	nn	-0.2341	0.9331	29.8S	76.4E	76	347	257	08m03s
5790	290	0436 Jul 29	20:07:35	6306	-19337	91	T	n-	0.2842	1.0409	35.1N	91.5W	73	190	143	03m37s
5791	290	0437 Jan 22	13:04:52	6301	-19331	96	A	t-	-0.9216	0.9662	81.8S	80.8E	22	290	321	02m11s
5792	290	0437 Jul 19	06:47:24	6296	-19325	101	P	t-	1.0873	0.8256	68.3N	90.1W	0	346		
5793	290	0437 Dec 13	13:10:27	6293	-19320	68	P	-t	1.1533	0.7193	65.1N	28.6E	0	199		
5794	290	0438 Jun 08	19:25:47	6288	-19314	73	A	-t	-0.9558	0.9388	48.9S	70.3W	17	345	783	06m17s
5795	290	0438 Dec 03	04:50:38	6283	-19308	78	T	-p	0.4982	1.0357	6.7N	138.5E	60	192	138	03m27s
5796	290	0439 May 28	21:21:24	6278	-19302	83	A	nn	-0.1728	0.9750	12.2N	113.0W	80	345	90	02m59s
5797	290	0439 Nov 22	17:48:16	6273	-19296	88	A	nn	-0.1870	0.9916	31.1S	67.3W	79	19	30	00m48s
5798	290	0440 May 17	05:49:32	6268	-19290	93	T	p-	0.6030	1.0253	53.8N	97.6E	53	149	107	01m54s
5799	290	0440 Nov 11	00:07:01	6264	-19284	98	A	p-	-0.9205	0.9309	69.7S	127.6E	23	81	665	04m49s
5800	290	0441 Apr 07	13:29:51	6260	-19279	65	P	-t	-1.2093	0.6182	61.1S	80.4E	0	285		
5801	291	0441 May 06	20:39:42	6259	-19278	103	P	t-	1.3114	0.4208	62.3N	129.2E	0	51		
5802	291	0441 Oct 01	05:43:30	6255	-19273	70	P	-t	1.4003	0.2818	60.9N	159.0W	0	263		
5803	291	0442 Mar 28	05:50:14	6250	-19267	75	T	-p	-0.5107	1.0440	23.5S	135.4E	59	328	169	03m37s
5804	291	0442 Sep 20	09:16:50	6245	-19261	80	A	-p	0.6424	0.9751	35.0N	87.6E	50	216	114	02m14s
5805	291	0443 Mar 17	17:33:56	6241	-19255	85	A	nn	0.2358	0.9887	11.0N	62.0W	76	150	41	01m06s
5806	291	0443 Sep 09	20:16:39	6236	-19249	90	T	n-	-0.1104	1.0368	0.5S	102.1W	84	29	124	03m14s
5807	291	0444 Mar 05	21:53:09	6231	-19243	95	A+	t-	1.0045	0.9513	60.8N	160.3E	0	101	-	-
5808	291	0444 Aug 29	12:10:31	6226	-19237	100	T	p-	-0.8127	1.0590	38.6S	5.5W	35	36	326	04m19s
5809	291	0445 Jan 24	04:24:13	6222	-19232	67	P	-t	-1.2630	0.5126	62.7S	84.5W	0	225		
5810	291	0445 Jul 20	19:17:00	6217	-19226	72	P	-t	1.0629	0.8841	63.3N	56.1E	0	322		
5811	291	0446 Jan 13	10:20:25	6213	-19220	77	A	-p	-0.5105	0.9884	50.6S	66.9E	59	336	47	00m58s
5812	291	0446 Jul 10	04:00:55	6208	-19214	82	A	-p	0.3260	0.9745	41.1N	151.9E	71	196	97	02m32s
5813	291	0447 Jan 02	23:00:54	6203	-19208	87	T	n-	0.1969	1.0373	11.8S	139.5W	79	169	128	03m35s
5814	291	0447 Jun 29	06:05:57	6198	-19202	92	A	p-	-0.4449	0.9469	2.7S	110.4E	64	8	218	07m23s
5815	291	0447 Dec 23	14:42:21	6193	-19196	97	T	p-	0.8603	1.0355	35.6N	21.3W	30	172	235	03m02s
5816	291	0448 Jun 17	07:12:15	6189	-19190	102	P	t-	-1.1908	0.6385	66.1S	87.0E	0	9		
5817	291	0448 Nov 12	15:03:08	6185	-19185	69	P	-t	-1.2465	0.5416	69.4S	175.6W	0	154		
5818	291	0449 May 08	05:02:39	6180	-19179	74	T	-p	0.7391	1.0374	63.7N	110.5E	42	157	187	02m35s
5819	291	0449 Nov 01	19:17:57	6175	-19173	79	A	-p	-0.5925	0.9280	50.6S	99.0W	53	19	335	07m09s
5820	291	0450 Apr 27	20:55:28	6170	-19167	84	T	nn	-0.0211	1.0765	13.2N	109.0W	89	346	248	06m50s

Cat Num	Canon Plate	Calendar Date	TD of Greatest Eclipse	ΔT s	Luna Num	Saros Num	Ecl. Type	QLE	Gamma	Ecl. Mag.	Lat. °	Long. °	Sun Alt °	Sun Azm °	Path Width km	Central Line Dur.
5821	292	0450 Oct 21	18:37:15	6165	-19161	89	A	nn	0.0940	0.9238	6.2S	75.7W	85	195	289	10m20s
5822	292	0451 Apr 17	13:50:39	6161	-19155	94	T	p-	-0.7582	1.0545	36.8S	12.3E	40	341	274	04m27s
5823	292	0451 Oct 10	20:27:02	6156	-19149	99	A	p-	0.7850	0.9551	42.2N	86.3W	38	203	262	04m24s
5824	292	0452 Mar 07	14:42:20	6152	-19144	66	P	-t	1.2019	0.6192	71.8N	87.9W	0	105		
5825	292	0452 Aug 30	19:36:11	6147	-19138	71	P	-t	-1.2202	0.5944	71.3S	152.3W	0	62		
5826	292	0452 Sep 29	05:36:43	6146	-19137	109	P	t-	1.4369	0.1938	71.8N	160.1W	0	260		
5827	292	0453 Feb 24	17:04:40	6142	-19132	76	A	-p	0.5070	0.9305	20.6N	55.6W	59	163	299	08m48s
5828	292	0453 Aug 20	12:07:55	6137	-19126	81	T	-p	-0.5045	1.0656	17.2S	16.3E	60	15	247	05m50s
5829	292	0454 Feb 13	16:47:54	6133	-19120	86	A	nn	-0.2063	0.9375	24.5S	39.0W	78	345	237	07m36s
5830	292	0454 Aug 10	03:39:07	6128	-19114	91	T	n-	0.2243	1.0355	28.5N	154.6E	77	193	122	03m17s
5831	292	0455 Feb 02	21:14:29	6123	-19108	96	A	t-	-0.9020	0.9723	76.1S	59.9W	25	307	233	01m52s
5832	292	0455 Jul 30	13:47:01	6118	-19102	101	P	t-	1.0212	0.9406	69.3N	152.6E	0	334		
5833	292	0455 Dec 24	22:01:42	6114	-19097	68	P	-t	1.1551	0.7163	66.2N	115.2W	0	188		
5834	292	0456 Jan 23	08:46:30	6113	-19096	106	Pb	t-	-1.5332	0.0112	69.0S	126.5W	0	203		
5835	292	0456 Jun 19	01:47:00	6109	-19091	73	P	-t	-1.0431	0.8887	65.6S	166.1W	0	346		
5836	292	0456 Dec 13	13:41:28	6105	-19085	78	T	-p	0.5010	1.0342	6.3N	4.0E	60	188	133	03m23s
5837	292	0457 Jun 08	04:01:10	6100	-19079	83	A	nn	-0.2559	0.9779	8.5N	146.9E	75	350	81	02m43s
5838	292	0457 Dec 03	02:22:23	6095	-19073	88	A	nn	-0.1844	0.9882	33.0S	165.4E	79	14	42	01m09s
5839	292	0458 May 28	12:58:30	6090	-19067	93	T	p-	0.5250	1.0305	52.0N	3.7W	58	157	121	02m21s
5840	292	0458 Nov 22	08:16:28	6086	-19061	98	A	p-	-0.9155	0.9280	73.8S	1.2E	23	84	679	05m00s
5841	293	0459 Apr 18	21:12:31	6082	-19056	65	P	-t	-1.2694	0.5020	61.5S	44.8W	0	294		
5842	293	0459 May 18	04:07:08	6081	-19055	103	P	t-	1.2392	0.5603	63.0N	7.4E	0	42		
5843	293	0459 Oct 12	13:23:10	6077	-19050	70	P	-t	1.4248	0.2409	61.1N	76.7E	0	253		
5844	293	0460 Apr 07	13:31:13	6072	-19044	75	T	-p	-0.5691	1.0434	23.2S	19.7E	55	330	173	03m38s
5845	293	0460 Sep 30	17:11:26	6067	-19038	80	A	-p	0.6702	0.9749	32.8N	32.4W	48	215	118	02m18s
5846	293	0461 Mar 28	00:59:22	6062	-19032	85	A	nn	0.1820	0.9885	12.6N	173.6W	79	150	41	01m08s
5847	293	0461 Sep 20	04:24:35	6058	-19026	90	T	nn	-0.0755	1.0364	3.1S	135.2E	86	29	123	03m09s
5848	293	0462 Mar 17	05:04:02	6053	-19020	95	A	p-	0.9555	0.9323	56.3N	74.2E	17	119	850	05m24s
5849	293	0462 Sep 09	20:16:30	6048	-19014	100	T	p-	-0.7709	1.0569	38.5S	128.1W	39	38	288	04m07s
5850	293	0463 Feb 04	12:15:23	6044	-19009	67	P	-t	-1.2813	0.4816	62.1S	147.9E	0	234		
5851	293	0463 Aug 01	02:36:42	6039	-19003	72	P	-t	1.1318	0.7553	62.5N	63.8W	0	313		
5852	293	0463 Aug 30	12:11:41	6039	-19002	110	Pb	t-	-1.4959	0.0774	61.4S	51.2W	0	71		
5853	293	0464 Jan 24	18:42:06	6035	-18997	77	A	-p	-0.5215	0.9935	48.0S	55.9W	58	331	27	00m32s
5854	293	0464 Jul 20	10:47:29	6030	-18991	82	A	-p	0.4042	0.9693	43.4N	53.7E	66	202	120	02m57s
5855	293	0465 Jan 13	07:44:43	6025	-18985	87	T	n-	0.1892	1.0407	10.9S	89.1E	79	165	139	03m49s
5856	293	0465 Jul 09	12:31:50	6020	-18979	92	A	p-	-0.3603	0.9459	1.9N	12.9E	69	12	213	07m21s
5857	293	0466 Jan 02	23:31:56	6015	-18973	97	T	p-	0.8569	1.0357	35.2N	158.1W	31	167	233	03m02s
5858	293	0466 Jun 28	13:47:21	6011	-18967	102	P	t-	-1.1058	0.7870	65.1S	22.7W	0	19		
5859	293	0466 Nov 23	23:29:39	6007	-18962	69	P	-t	-1.2537	0.5290	68.3S	45.2E	0	166		
5860	293	0467 May 19	12:19:15	6002	-18956	74	T	-p	0.8094	1.0403	73.1N	2.2W	36	156	231	02m33s
5861	294	0467 Nov 13	03:18:02	5997	-18950	79	A	-p	-0.6035	0.9248	55.1S	142.5E	53	15	356	07m17s
5862	294	0468 May 08	04:28:58	5992	-18944	84	Tm	nn	0.0474	1.0789	20.2N	135.7E	87	170	255	06m56s
5863	294	0468 Nov 01	02:27:24	5988	-18938	89	A	nn	0.0780	0.9227	10.8S	165.5E	86	193	293	10m34s
5864	294	0469 Apr 27	21:24:18	5983	-18932	94	T	p-	-0.6958	1.0555	28.7S	106.7W	46	345	254	04m52s
5865	294	0469 Oct 21	04:30:00	5978	-18926	99	A	p-	0.7646	0.9559	36.9N	148.4E	40	199	247	04m33s
5866	294	0470 Mar 18	22:08:44	5974	-18921	66	P	-t	1.2523	0.5311	72.0N	145.7E	0	91		
5867	294	0470 Apr 17	10:26:56	5973	-18920	104	Pb	t-	-1.4981	0.0867	71.0S	103.9E	0	305		
5868	294	0470 Sep 11	03:39:39	5969	-18915	71	P	-t	-1.2566	0.5255	71.7S	72.3E	0	76		
5869	294	0470 Oct 10	13:54:40	5968	-18914	109	P	t-	1.4111	0.2412	71.5N	60.6E	0	246		
5870	294	0471 Mar 08	00:19:40	5964	-18909	76	A	-p	0.5522	0.9323	27.4N	167.8W	56	160	301	08m07s
5871	294	0471 Aug 31	20:07:11	5960	-18903	81	T	-p	-0.5507	1.0611	23.8S	107.1W	56	18	239	05m15s
5872	294	0472 Feb 25	00:16:40	5955	-18897	86	A	nn	-0.1696	0.9424	18.4S	152.8W	80	343	216	07m05s
5873	294	0472 Aug 20	11:17:31	5950	-18891	91	T	nn	0.1710	1.0296	21.8N	38.5E	80	195	102	02m50s
5874	294	0473 Feb 13	05:16:16	5945	-18885	96	A	p-	-0.8754	0.9790	69.5S	168.0E	29	317	156	01m29s
5875	294	0473 Aug 09	20:51:51	5941	-18879	101	A	t-	0.9601	0.9654	78.2N	19.0W	16	272	463	02m13s
5876	294	0474 Jan 04	06:50:38	5937	-18874	68	P	-t	1.1589	0.7095	67.3N	101.1E	0	177		
5877	294	0474 Feb 02	17:16:01	5936	-18873	106	P	t-	-1.5145	0.0438	69.9S	93.1E	0	215		
5878	294	0474 Jun 30	08:10:05	5932	-18868	73	P	-t	-1.1285	0.7425	66.6S	87.0E	0	356		
5879	294	0474 Dec 24	22:31:52	5927	-18862	78	T	-p	0.5039	1.0332	6.7N	130.4W	60	183	129	03m20s
5880	294	0475 Jun 19	10:40:23	5922	-18856	83	A	np	-0.3401	0.9802	3.9N	46.3E	70	354	75	02m29s

Cat Num	Canon Plate	Calendar Date	TD of Greatest Eclipse	ΔT s	Luna Num	Saros Num	Ecl. Type	QLE	Gamma	Ecl. Mag.	Lat. °	Long. °	Sun Alt °	Sun Azm °	Path Width km	Central Line Dur.
5881	295	0475 Dec 14	10:57:44	5918	-18850	88	A	nn	-0.1835	0.9854	34.1S	38.2E	79	9	53	01m27s
5882	295	0476 Jun 07	20:06:49	5913	-18844	93	T	p-	0.4457	1.0350	49.1N	105.6W	63	165	132	02m48s
5883	295	0476 Dec 02	16:27:33	5908	-18838	98	A	p-	-0.9125	0.9256	78.0S	125.5W	24	87	694	05m09s
5884	295	0477 Apr 29	04:50:23	5904	-18833	65	P	-t	-1.3332	0.3782	62.0S	169.0W	0	303		
5885	295	0477 May 28	11:33:37	5903	-18832	103	P	t-	1.1660	0.7023	63.8N	114.4W	0	33		
5886	295	0477 Oct 22	21:11:13	5899	-18827	70	P	-t	1.4426	0.2113	61.5N	49.8W	0	244		
5887	295	0478 Apr 18	21:06:05	5895	-18821	75	T	-p	-0.6320	1.0421	23.9S	94.6W	51	332	178	03m36s
5888	295	0478 Oct 12	01:13:46	5890	-18815	80	A	-p	0.6924	0.9748	30.7N	154.9W	46	213	121	02m21s
5889	295	0479 Apr 08	08:16:42	5885	-18809	85	Am	nn	0.1223	0.9881	13.8N	77.2E	83	152	42	01m11s
5890	295	0479 Oct 01	12:40:26	5880	-18803	90	T	nn	-0.0466	1.0358	6.0S	10.6E	87	29	121	03m06s
5891	295	0480 Mar 27	12:04:20	5876	-18797	95	A	p-	0.8980	0.9358	54.3N	23.7W	26	124	530	05m16s
5892	295	0480 Sep 20	04:31:03	5871	-18791	100	T	p-	-0.7364	1.0544	39.8S	107.1E	42	39	260	03m52s
5893	295	0481 Feb 14	19:58:22	5867	-18786	67	P	-t	-1.3068	0.4379	61.5S	22.6E	0	244		
5894	295	0481 Aug 11	10:02:04	5862	-18780	72	P	-t	1.1949	0.6383	61.9N	175.0E	0	304		
5895	295	0481 Sep 09	20:08:13	5861	-18779	110	P	t-	-1.4589	0.1488	61.1S	179.9W	0	80		
5896	295	0482 Feb 04	02:57:02	5857	-18774	77	A	-p	-0.5389	0.9989	45.2S	178.0W	57	327	4	00m05s
5897	295	0482 Jul 31	17:38:50	5853	-18768	82	A	-p	0.4775	0.9639	44.5N	45.5W	61	208	148	03m23s
5898	295	0483 Jan 24	16:22:34	5848	-18762	87	T	n-	0.1763	1.0446	9.4S	40.8W	80	161	151	04m03s
5899	295	0483 Jul 20	19:01:07	5843	-18756	92	A	pn	-0.2782	0.9444	5.3N	84.9W	74	16	213	07m18s
5900	295	0484 Jan 14	08:16:49	5838	-18750	97	T	p-	0.8499	1.0366	35.0N	66.5E	32	161	232	03m04s
5901	296	0484 Jul 08	20:26:40	5834	-18744	102	P	t-	-1.0229	0.9321	64.1S	133.0W	0	29		
5902	296	0484 Dec 04	07:57:06	5830	-18739	69	P	-t	-1.2595	0.5187	67.2S	93.7W	0	178		
5903	296	0485 May 29	19:34:46	5825	-18733	74	T	-p	0.8814	1.0420	83.4N	118.0W	28	151	301	02m26s
5904	296	0485 Nov 23	11:22:19	5820	-18727	79	A	-p	-0.6109	0.9222	58.6S	24.9E	52	10	372	07m23s
5905	296	0486 May 19	11:58:26	5815	-18721	84	T	nn	0.1193	1.0806	27.0N	22.1E	83	173	262	06m54s
5906	296	0486 Nov 12	10:24:14	5811	-18715	89	A	nn	0.0672	0.9221	14.6S	45.6E	86	189	295	10m43s
5907	296	0487 May 09	04:51:29	5806	-18709	94	T	p-	-0.6283	1.0559	20.9S	136.9E	51	349	236	05m11s
5908	296	0487 Nov 01	12:41:08	5801	-18703	99	A	p-	0.7505	0.9568	32.6N	21.6E	41	195	236	04m39s
5909	296	0488 Mar 29	05:24:28	5797	-18698	66	P	-t	1.3102	0.4298	71.9N	21.9E	0	77		
5910	296	0488 Apr 27	17:36:45	5796	-18697	104	P	t-	-1.4368	0.1983	70.3S	17.3W	0	318		
5911	296	0488 Sep 21	11:52:21	5793	-18692	71	P	-t	-1.2858	0.4701	71.8S	65.7W	0	90		
5912	296	0488 Oct 20	22:21:32	5792	-18691	109	P	t-	1.3921	0.2761	71.0N	80.5W	0	233		
5913	296	0489 Mar 18	07:26:06	5788	-18686	76	A	-p	0.6038	0.9340	35.0N	81.6E	53	158	306	07m25s
5914	296	0489 Sep 11	04:13:02	5783	-18680	81	T	-p	-0.5912	1.0564	30.4S	127.8E	54	20	229	04m40s
5915	296	0490 Mar 07	07:39:25	5778	-18674	86	A	nn	-0.1274	0.9474	11.8S	94.5E	83	342	195	06m30s
5916	296	0490 Aug 31	19:01:26	5774	-18668	91	T	nn	0.1232	1.0235	15.1N	79.5W	83	197	81	02m19s
5917	296	0491 Feb 24	13:11:57	5769	-18662	96	A	p-	-0.8429	0.9860	62.3S	40.3E	32	324	92	01m02s
5918	296	0491 Aug 21	04:01:41	5764	-18656	101	A	p-	0.9036	0.9621	71.1N	170.7W	25	230	323	02m44s
5919	296	0492 Jan 15	15:35:24	5760	-18651	68	P	-t	1.1665	0.6957	68.3N	42.1W	0	165		
5920	296	0492 Feb 14	01:38:44	5759	-18650	106	P	t-	-1.4900	0.0876	70.7S	46.2W	0	228		
5921	297	0492 Jul 10	14:37:35	5755	-18645	73	P	-t	-1.2099	0.6032	67.6S	21.4W	0	6		
5922	297	0493 Jan 04	07:17:22	5751	-18639	78	T	-p	0.5104	1.0326	8.1N	96.3E	59	179	128	03m16s
5923	297	0493 Jun 29	17:25:33	5746	-18633	83	A	-p	-0.4200	0.9819	1.5S	56.5W	65	358	71	02m16s
5924	297	0493 Dec 24	19:30:00	5741	-18627	88	Am	nn	-0.1813	0.9831	34.1S	88.2W	79	4	61	01m43s
5925	297	0494 Jun 19	03:17:12	5737	-18621	93	T	p-	0.3672	1.0388	45.2N	150.7E	68	172	140	03m14s
5926	297	0494 Dec 14	00:37:53	5732	-18615	98	A	p-	-0.9095	0.9239	82.5S	110.7E	24	87	703	05m17s
5927	297	0495 May 10	12:22:50	5728	-18610	65	P	-t	-1.4013	0.2458	62.6S	68.0E	0	312		
5928	297	0495 Jun 08	18:58:41	5727	-18609	103	P	t-	1.0914	0.8472	64.7N	124.0E	0	24		
5929	297	0495 Nov 03	05:07:15	5723	-18604	70	P	-t	1.4546	0.1912	62.0N	178.4W	0	235		
5930	297	0496 Apr 29	04:31:53	5718	-18598	75	T	-p	-0.7019	1.0400	26.0S	153.3E	45	334	185	03m29s
5931	297	0496 Oct 22	09:26:26	5714	-18592	80	A	-p	0.7070	0.9752	28.5N	79.5E	45	209	122	02m23s
5932	297	0497 Apr 18	15:24:14	5709	-18586	85	A	nn	0.0554	0.9873	14.3N	29.3W	87	154	45	01m18s
5933	297	0497 Oct 11	21:04:41	5704	-18580	90	T	nn	-0.0243	1.0354	9.2S	116.2W	89	27	119	03m04s
5934	297	0498 Apr 07	18:55:52	5700	-18574	95	A	p-	0.8337	0.9389	53.5N	121.2W	33	129	401	05m08s
5935	297	0498 Oct 01	12:53:06	5695	-18568	100	T	p-	-0.7082	1.0515	42.0S	19.5W	45	40	238	03m37s
5936	297	0499 Feb 26	03:34:26	5691	-18563	67	P	-t	-1.3382	0.3834	61.2S	100.9W	0	253		
5937	297	0499 Mar 27	19:05:53	5690	-18562	105	Pb	t-	1.5545	0.0158	60.9N	177.1W	0	83		
5938	297	0499 Aug 22	17:31:58	5686	-18557	72	P	-t	1.2532	0.5314	61.4N	52.9E	0	295		
5939	297	0499 Sep 21	04:11:45	5685	-18556	110	P	t-	-1.4281	0.2079	61.0S	49.8E	0	89		
5940	297	0500 Feb 15	11:06:27	5681	-18551	77	H	-p	-0.5616	1.0046	42.2S	60.6E	56	325	19	00m22s

Cat Num	Canon Plate	Calendar Date	TD of Greatest Eclipse	ΔT s	Luna Num	Saros Num	Ecl. Type	QLE	Gamma	Ecl. Mag.	Lat. °	Long. °	Sun Alt °	Sun Azm °	Path Width km	Central Line Dur.
5941	298	0500 Aug 11	00:35:02	5677	-18545	82	A	-p	0.5458	0.9582	44.7N	146.1W	57	213	180	03m50s
5942	298	0501 Feb 04	00:54:51	5672	-18539	87	T	n-	0.1585	1.0487	7.4S	169.5W	81	157	164	04m18s
5943	298	0501 Jul 31	01:37:16	5667	-18533	92	A	nn	-0.2016	0.9427	7.4N	176.1E	78	20	216	07m14s
5944	298	0502 Jan 24	16:57:17	5662	-18527	97	T	p-	0.8396	1.0378	35.1N	67.5W	33	157	230	03m06s
5945	298	0502 Jul 20	03:11:36	5657	-18521	102	A	t-	-0.9434	0.9595	46.9S	132.5E	19	24	442	03m52s
5946	298	0502 Dec 15	16:23:51	5653	-18516	69	P	-t	-1.2652	0.5088	66.2S	128.2E	0	189		
5947	298	0503 Jun 10	02:51:04	5649	-18510	74	T	-t	0.9535	1.0420	83.7N	27.9W	17	353	483	02m12s
5948	298	0503 Dec 04	19:27:10	5644	-18504	79	A	-p	-0.6172	0.9204	61.1S	90.9W	52	2	384	07m27s
5949	298	0504 May 29	19:26:16	5639	-18498	84	T	-n	0.1927	1.0813	33.3N	90.2W	79	177	267	06m44s
5950	298	0504 Nov 22	18:25:08	5634	-18492	89	A	nn	0.0592	0.9222	17.6S	75.0W	87	186	295	10m41s
5951	298	0505 May 19	12:15:05	5630	-18486	94	T	p-	-0.5585	1.0553	13.7S	22.1E	56	353	220	05m23s
5952	298	0505 Nov 11	20:57:42	5625	-18480	99	A	p-	0.7399	0.9582	29.0N	106.2W	42	190	225	04m39s
5953	298	0506 Apr 09	12:30:50	5621	-18475	66	P	-t	1.3745	0.3174	71.5N	99.3W	0	64		
5954	298	0506 May 09	00:40:11	5620	-18474	104	P	t-	-1.3709	0.3179	69.5S	136.5W	0	330		
5955	298	0506 Oct 02	20:13:20	5616	-18469	71	P	-t	-1.3089	0.4266	71.7S	154.2E	0	104		
5956	298	0506 Nov 01	06:55:05	5615	-18468	109	P	t-	1.3780	0.3019	70.3N	137.2E	0	219		
5957	298	0507 Mar 29	14:20:55	5611	-18463	76	A	-p	0.6647	0.9356	43.3N	26.8W	48	156	318	06m42s
5958	298	0507 Sep 22	12:27:05	5606	-18457	81	T	-p	-0.6247	1.0516	36.8S	0.7E	51	23	217	04m06s
5959	298	0508 Mar 17	14:52:12	5602	-18451	86	A	nn	-0.0762	0.9525	4.7S	16.3W	86	342	174	05m52s
5960	298	0508 Sep 11	02:53:05	5597	-18445	91	H3	nn	0.0826	1.0173	8.6N	160.4E	85	198	59	01m45s
5961	299	0509 Mar 06	20:59:05	5592	-18439	96	A	p-	-0.8024	0.9933	54.6S	83.9W	36	329	39	00m32s
5962	299	0509 Aug 31	11:18:53	5587	-18433	101	A	p-	0.8538	0.9579	62.5N	65.1E	31	217	294	03m22s
5963	299	0510 Jan 26	00:15:39	5583	-18428	68	P	-t	1.1778	0.6746	69.3N	175.2E	0	153		
5964	299	0510 Feb 24	09:54:50	5583	-18427	106	P	t-	-1.4600	0.1419	71.3S	175.6E	0	241		
5965	299	0510 Jul 21	21:08:35	5579	-18422	73	P	-t	-1.2882	0.4692	68.6S	131.3W	0	17		
5966	299	0511 Jan 15	16:00:05	5574	-18416	78	T	-p	0.5189	1.0323	10.3N	36.4W	59	175	128	03m14s
5967	299	0511 Jul 11	00:13:47	5569	-18410	83	A	-p	-0.4979	0.9831	7.6S	160.8W	60	2	69	02m05s
5968	299	0512 Jan 05	03:59:03	5564	-18404	88	A	nn	-0.1769	0.9815	33.0S	146.1E	80	359	67	01m55s
5969	299	0512 Jun 29	10:30:12	5560	-18398	93	T	n-	0.2901	1.0418	40.4N	45.0E	73	178	146	03m37s
5970	299	0512 Dec 24	08:46:57	5555	-18392	98	A	p-	-0.9062	0.9228	86.7S	2.2E	25	72	702	05m24s
5971	299	0513 May 20	19:52:49	5551	-18387	65	Pe	-t	-1.4713	0.1095	63.3S	54.5W	0	321		
5972	299	0513 Jun 19	02:25:05	5550	-18386	103	P	t-	1.0176	0.9908	65.6N	1.6E	0	14		
5973	299	0513 Nov 13	13:08:18	5546	-18381	70	P	-t	1.4630	0.1770	62.7N	51.6E	0	225		
5974	299	0514 May 10	11:53:33	5541	-18375	75	T	-p	-0.7744	1.0371	29.6S	42.0E	39	337	194	03m16s
5975	299	0514 Nov 02	17:45:19	5537	-18369	80	A	-p	0.7173	0.9760	26.7N	48.1W	44	205	120	02m23s
5976	299	0515 Apr 29	22:23:38	5532	-18363	85	A	nn	-0.0175	0.9861	14.1N	133.6W	89	334	49	01m28s
5977	299	0515 Oct 23	05:36:13	5527	-18357	90	T	nn	-0.0076	1.0351	12.5S	115.2E	89	22	118	03m04s
5978	299	0516 Apr 18	01:38:56	5522	-18351	95	A	p-	0.7628	0.9415	53.1N	143.2E	40	134	329	05m01s
5979	299	0516 Oct 11	21:22:23	5518	-18345	100	T	p-	-0.6864	1.0487	44.9S	147.7W	46	41	219	03m23s
5980	299	0517 Mar 08	11:01:54	5514	-18340	67	P	-t	-1.3770	0.3155	60.9S	137.8E	0	261		
5981	300	0517 Apr 07	01:59:05	5513	-18339	105	P	t-	1.4893	0.1253	61.0N	70.1E	0	75		
5982	300	0517 Sep 02	01:09:31	5509	-18334	72	P	-t	1.3044	0.4388	61.1N	71.0W	0	286		
5983	300	0517 Oct 01	12:22:50	5508	-18333	110	P	t-	-1.4042	0.2535	61.0S	82.5W	0	98		
5984	300	0518 Feb 25	19:08:25	5504	-18328	77	H	-p	-0.5912	1.0104	39.6S	59.6W	54	323	44	00m50s
5985	300	0518 Aug 22	07:37:14	5499	-18322	82	A	-p	0.6084	0.9524	44.1N	111.0E	52	217	216	04m21s
5986	300	0519 Feb 15	09:19:56	5495	-18316	87	T	n-	0.1343	1.0530	5.2S	63.6E	82	154	177	04m33s
5987	300	0519 Aug 11	08:20:20	5490	-18310	92	A	nn	-0.1304	0.9406	8.4N	75.5E	83	24	222	07m14s
5988	300	0520 Feb 05	01:29:14	5485	-18304	97	T	p-	0.8227	1.0395	35.2N	161.1E	34	152	228	03m10s
5989	300	0520 Jul 30	10:04:27	5480	-18298	102	A	t-	-0.8693	0.9626	38.3S	29.5E	29	25	268	03m45s
5990	300	0520 Dec 26	00:48:33	5476	-18293	69	P	-t	-1.2721	0.4969	65.1S	9.0W	0	199		
5991	300	0521 Jun 20	10:08:44	5472	-18287	74	P	-t	1.0248	0.9670	65.8N	143.0W	0	348		
5992	300	0521 Dec 15	03:32:14	5467	-18281	79	A	-p	-0.6230	0.9193	62.2S	154.8E	51	352	393	07m28s
5993	300	0522 Jun 10	02:52:34	5462	-18275	84	T	-n	0.2675	1.0812	38.9N	159.1E	74	181	272	06m28s
5994	300	0522 Dec 04	02:29:54	5457	-18269	89	A	nn	0.0543	0.9229	19.7S	163.9E	87	181	292	10m31s
5995	300	0523 May 30	19:33:08	5453	-18263	94	T	p-	-0.4847	1.0541	6.9S	90.5W	61	357	204	05m25s
5996	300	0523 Nov 23	05:20:29	5448	-18257	99	A	p-	0.7339	0.9601	26.3N	124.6E	43	186	213	04m33s
5997	300	0524 Apr 19	19:28:06	5444	-18252	66	P	-t	1.4451	0.1940	70.9N	142.2E	0	51		
5998	300	0524 May 19	07:37:25	5443	-18251	104	P	t-	-1.3006	0.4452	68.7S	106.5E	0	341		
5999	300	0524 Oct 13	04:41:57	5439	-18246	71	P	-t	-1.3264	0.3939	71.3S	12.4E	0	118		
6000	300	0524 Nov 11	15:34:10	5438	-18245	109	P	t-	1.3680	0.3201	69.4N	5.9W	0	207		

Cat Num	Canon Plate	Calendar Date	TD of Greatest Eclipse	ΔT s	Luna Num	Saros Num	Ecl. Type	QLE	Gamma	Ecl. Mag.	Lat. °	Long. °	Sun Alt °	Sun Azm °	Path Width km	Central Line Dur.
6001	301	0525 Apr 08	21:08:45	5434	-18240	76	A	-p	0.7309	0.9369	52.4N	134.3W	43	153	342	06m01s
6002	301	0525 Oct 02	20:48:13	5430	-18234	81	T	-p	-0.6521	1.0469	43.1S	127.7W	49	24	205	03m35s
6003	301	0526 Mar 28	21:59:31	5425	-18228	86	A	nn	-0.0200	0.9578	2.7N	125.8W	89	342	154	05m11s
6004	301	0526 Sep 22	10:50:19	5420	-18222	91	H	nn	0.0476	1.0111	2.3N	38.9E	87	198	38	01m08s
6005	301	0527 Mar 18	04:40:28	5415	-18216	96	H	p-	-0.7564	1.0006	46.8S	154.4E	41	333	3	00m03s
6006	301	0527 Sep 11	18:42:58	5411	-18210	101	A	p-	0.8106	0.9532	54.6N	54.3W	36	211	290	04m05s
6007	301	0528 Feb 06	08:48:21	5407	-18205	68	P	-t	1.1955	0.6417	70.2N	33.8E	0	141		
6008	301	0528 Mar 06	18:02:29	5406	-18204	106	P	t-	-1.4232	0.2100	71.7S	39.1E	0	255		
6009	301	0528 Aug 01	03:47:42	5402	-18199	73	P	-t	-1.3595	0.3476	69.5S	116.3E	0	29		
6010	301	0528 Aug 30	20:01:34	5401	-18198	111	Pb	t-	1.5569	0.0164	71.4N	17.5E	0	297		
6011	301	0529 Jan 26	00:35:32	5397	-18193	78	T	-p	0.5327	1.0323	13.6N	167.6W	58	171	129	03m10s
6012	301	0529 Jul 21	07:09:42	5393	-18187	83	A	-p	-0.5701	0.9838	14.2S	92.2E	55	6	70	01m56s
6013	301	0530 Jan 15	12:21:32	5388	-18181	88	A	nn	-0.1682	0.9802	30.6S	21.6E	80	354	71	02m05s
6014	301	0530 Jul 10	17:48:09	5383	-18175	93	T	n-	0.2166	1.0440	35.0N	63.4W	77	183	151	03m57s
6015	301	0531 Jan 04	16:51:18	5378	-18169	98	A	p-	-0.8996	0.9225	86.9S	25.1W	25	337	682	05m32s
6016	301	0531 Jun 30	09:51:59	5374	-18163	103	T	t-	0.9440	1.0666	85.2N	106.3W	19	19	680	03m23s
6017	301	0531 Nov 24	21:15:03	5370	-18158	70	P	-t	1.4671	0.1697	63.5N	80.1W	0	216		
6018	301	0532 May 20	19:08:48	5365	-18152	75	T	-t	-0.8510	1.0330	35.4S	68.0W	31	340	210	02m53s
6019	301	0532 Nov 13	02:11:04	5360	-18146	80	A	-p	0.7224	0.9774	25.0N	177.6W	44	201	115	02m19s
6020	301	0533 May 10	05:15:53	5356	-18140	85	A	nn	-0.0950	0.9844	13.0N	123.8E	85	339	56	01m43s
6021	302	0533 Nov 02	14:15:16	5351	-18134	90	T	nn	0.0028	1.0350	15.6S	15.0W	90	233	118	03m06s
6022	302	0534 Apr 29	08:14:13	5346	-18128	95	A	p-	0.6856	0.9438	52.6N	50.0E	46	140	282	04m58s
6023	302	0534 Oct 23	05:58:25	5341	-18122	100	T	p-	-0.6704	1.0459	48.4S	82.8E	48	40	204	03m09s
6024	302	0535 Mar 19	18:22:57	5338	-18117	67	P	-t	-1.4213	0.2371	60.9S	18.1E	0	270		
6025	302	0535 Apr 18	08:45:54	5337	-18116	105	P	t-	1.4182	0.2460	61.4N	41.1W	0	66		
6026	302	0535 Sep 13	08:53:31	5333	-18111	72	P	-t	1.3493	0.3585	60.9N	163.6E	0	277		
6027	302	0535 Oct 12	20:41:01	5332	-18110	110	P	t-	-1.3865	0.2873	61.3S	143.5E	0	107		
6028	302	0536 Mar 08	03:03:15	5328	-18105	77	T	-p	-0.6276	1.0162	37.4S	178.3W	51	323	70	01m17s
6029	302	0536 Sep 01	14:46:51	5323	-18099	82	A	-p	0.6640	0.9466	43.0N	5.4E	48	220	258	04m54s
6030	302	0537 Feb 25	17:38:41	5319	-18093	87	T	n-	0.1043	1.0574	2.9S	61.7W	84	152	190	04m49s
6031	302	0537 Aug 21	15:10:51	5314	-18087	92	A	nn	-0.0653	0.9384	8.4N	27.0W	86	27	229	07m17s
6032	302	0538 Feb 15	09:55:14	5309	-18081	97	T	p-	0.8012	1.0412	35.7N	31.8E	37	148	226	03m14s
6033	302	0538 Aug 10	17:05:14	5305	-18075	102	A	p-	-0.8005	0.9646	33.5S	76.3W	37	28	208	03m35s
6034	302	0539 Jan 06	09:08:45	5301	-18070	69	P	-t	-1.2818	0.4803	64.1S	144.5W	0	209		
6035	302	0539 Jul 01	17:28:39	5296	-18064	74	P	-t	1.0948	0.8345	64.9N	96.2E	0	338		
6036	302	0539 Jul 31	01:59:02	5295	-18063	112	Pb	t-	-1.4924	0.0889	62.6S	123.1E	0	47		
6037	302	0539 Dec 26	11:34:40	5291	-18058	79	A	-p	-0.6305	0.9188	62.0S	41.4E	51	343	398	07m26s
6038	302	0540 Jun 20	10:19:58	5286	-18052	84	T	-n	0.3414	1.0801	43.7N	49.4E	70	187	275	06m07s
6039	302	0540 Dec 14	10:34:02	5282	-18046	89	A	nn	0.0482	0.9243	20.9S	43.2E	87	176	286	10m10s
6040	302	0541 Jun 10	02:50:29	5277	-18040	94	T	p-	-0.4108	1.0519	1.0S	157.9E	66	1	188	05m17s
6041	303	0541 Dec 03	13:45:53	5272	-18034	99	A	p-	0.7288	0.9625	24.3N	5.0W	43	181	199	04m19s
6042	303	0542 May 01	02:16:58	5268	-18029	66	Pe	-t	1.5211	0.0615	70.2N	26.4E	0	38		
6043	303	0542 May 30	14:29:35	5268	-18028	104	P	t-	-1.2268	0.5781	67.7S	8.7W	0	352		
6044	303	0542 Oct 24	13:17:39	5264	-18023	71	P	-t	-1.3386	0.3711	70.7S	130.8W	0	131		
6045	303	0542 Nov 23	00:17:37	5263	-18022	109	P	t-	1.3612	0.3325	68.4N	149.3W	0	195		
6046	303	0543 Apr 20	03:47:29	5259	-18017	76	A	-p	0.8044	0.9378	62.4N	118.1E	36	148	389	05m23s
6047	303	0543 Oct 14	05:17:10	5254	-18011	81	T	-p	-0.6726	1.0424	48.9S	102.6E	47	25	191	03m07s
6048	303	0544 Apr 08	04:57:45	5250	-18005	86	A	nn	0.0445	0.9629	10.4N	126.9E	87	164	135	04m29s
6049	303	0544 Oct 02	18:56:04	5245	-17999	91	H	nn	0.0206	1.0050	3.5S	84.5W	89	198	17	00m31s
6050	303	0545 Mar 28	12:14:40	5240	-17993	96	H	p-	-0.7036	1.0079	38.7S	35.1E	45	337	38	00m42s
6051	303	0545 Sep 22	02:14:00	5235	-17987	101	A	p-	0.7737	0.9485	47.4N	173.0W	39	207	296	04m53s
6052	303	0546 Feb 16	17:15:08	5232	-17982	68	P	-t	1.2181	0.5989	70.9N	106.7W	0	128		
6053	303	0546 Mar 18	02:03:46	5231	-17981	106	P	t-	-1.3808	0.2891	71.9S	96.0W	0	269		
6054	303	0546 Aug 12	10:33:29	5227	-17976	73	P	-t	-1.4253	0.2357	70.4S	1.6E	0	41		
6055	303	0546 Sep 11	03:08:22	5226	-17975	111	P	t-	1.5085	0.0994	71.8N	103.8W	0	283		
6056	303	0547 Feb 06	09:04:44	5222	-17970	78	T	-p	0.5516	1.0325	17.7N	62.5E	56	167	131	03m07s
6057	303	0547 Aug 01	14:11:29	5217	-17964	83	A	-p	-0.6384	0.9840	21.3S	17.0W	50	10	73	01m48s
6058	303	0548 Jan 26	20:38:16	5213	-17958	88	A	nn	-0.1553	0.9794	27.2S	102.2W	81	350	74	02m12s
6059	303	0548 Jul 21	01:11:32	5208	-17952	93	T	n-	0.1468	1.0455	29.0N	174.2W	81	187	153	04m12s
6060	303	0549 Jan 15	00:50:28	5203	-17946	98	A	p-	-0.8891	0.9228	82.1S	127.7W	27	318	646	05m40s

Cat Num	Canon Plate	Calendar Date	TD of Greatest Eclipse	ΔT s	Luna Num	Saros Num	Ecl. Type	QLE	Gamma	Ecl. Mag.	Lat. °	Long. °	Sun Alt °	Sun Azm °	Path Width km	Central Line Dur.
6061	304	0549 Jul 10	17:23:12	5199	-17940	103	T	t-	0.8739	1.0688	83.1N	38.7W	29	201	468	03m48s
6062	304	0549 Dec 05	05:24:33	5195	-17935	70	P	-t	1.4693	0.1655	64.5N	147.2E	0	206		
6063	304	0550 Jan 04	00:04:07	5194	-17934	108	Pb	t-	-1.5674	0.0067	67.3S	19.5E	0	183		
6064	304	0550 Jun 01	02:21:04	5190	-17929	75	T	-t	-0.9293	1.0276	44.5S	177.2W	21	342	254	02m18s
6065	304	0550 Nov 24	10:41:00	5185	-17923	80	A	-p	0.7249	0.9793	23.7N	51.8E	43	197	106	02m11s
6066	304	0551 May 21	12:03:02	5181	-17917	85	A	nn	-0.1757	0.9822	10.9N	22.4E	80	343	64	02m03s
6067	304	0551 Nov 13	22:58:50	5176	-17911	90	T	nn	0.0101	1.0354	18.4S	146.1W	90	205	119	03m10s
6068	304	0552 May 09	14:43:48	5171	-17905	95	A	p-	0.6037	0.9455	51.6N	41.6W	53	146	250	05m00s
6069	304	0552 Nov 02	14:40:25	5167	-17899	100	T	p-	-0.6597	1.0433	52.2S	47.4W	48	38	191	02m57s
6070	304	0553 Mar 30	01:36:53	5163	-17894	67	P	-t	-1.4721	0.1465	61.0S	99.8W	0	279		
6071	304	0553 Apr 28	15:28:06	5162	-17893	105	P	t-	1.3421	0.3764	61.8N	151.2W	0	57		
6072	304	0553 Sep 23	16:44:18	5158	-17888	72	P	-t	1.3880	0.2903	60.8N	36.5E	0	268		
6073	304	0553 Oct 23	05:04:28	5157	-17887	110	P	t-	-1.3737	0.3117	61.6S	8.1E	0	117		
6074	304	0554 Mar 19	10:50:48	5153	-17882	77	T	-p	-0.6708	1.0217	35.9S	64.5E	48	323	98	01m44s
6075	304	0554 Sep 12	22:04:09	5149	-17876	82	A	-p	0.7127	0.9409	41.8N	103.1W	44	220	304	05m30s
6076	304	0555 Mar 09	01:47:58	5144	-17870	87	T	nn	0.0662	1.0618	0.7S	175.4E	86	151	203	05m05s
6077	304	0555 Sep 01	22:10:46	5139	-17864	92	A	nn	-0.0076	0.9360	7.4N	131.9W	90	32	238	07m25s
6078	304	0556 Feb 26	18:11:26	5135	-17858	97	T	p-	0.7721	1.0433	36.2N	94.4W	39	145	222	03m19s
6079	304	0556 Aug 21	00:15:45	5130	-17852	102	A	p-	-0.7387	0.9659	31.0S	175.3E	42	30	177	03m24s
6080	304	0557 Jan 16	17:22:42	5126	-17847	69	P	-t	-1.2959	0.4558	63.2S	81.8E	0	219		
6081	305	0557 Feb 15	07:19:28	5125	-17846	107	Pb	t-	1.5320	0.0300	61.6N	28.6E	0	116		
6082	305	0557 Jul 12	00:53:11	5121	-17841	74	P	-t	1.1616	0.7075	64.0N	25.4W	0	329		
6083	305	0557 Aug 10	09:28:40	5121	-17840	112	P	t-	-1.4288	0.2066	62.0S	0.9E	0	56		
6084	305	0558 Jan 05	19:34:45	5117	-17835	79	A	-p	-0.6398	0.9189	60.5S	72.3W	50	334	400	07m23s
6085	305	0558 Jul 01	17:47:13	5112	-17829	84	T	-p	0.4153	1.0783	47.5N	59.0W	65	194	278	05m45s
6086	305	0558 Dec 25	18:39:21	5107	-17823	89	A	nn	0.0427	0.9263	21.2S	77.7W	88	171	278	09m41s
6087	305	0559 Jun 21	10:05:31	5103	-17817	94	T	n-	-0.3354	1.0490	4.1N	47.6E	70	5	173	04m59s
6088	305	0559 Dec 14	22:13:23	5098	-17811	99	A	p-	0.7250	0.9655	23.0N	135.1W	43	176	181	03m58s
6089	305	0560 Jun 09	21:18:56	5093	-17805	104	P	t-	-1.1512	0.7134	66.7S	122.7W	0	3		
6090	305	0560 Nov 03	21:59:47	5089	-17800	71	P	-t	-1.3461	0.3570	69.9S	85.0E	0	144		
6091	305	0560 Dec 03	09:04:10	5089	-17799	109	P	t-	1.3564	0.3411	67.3N	67.0E	0	183		
6092	305	0561 Apr 30	10:21:38	5085	-17794	76	A	-p	0.8811	0.9379	73.1N	3.9E	28	135	491	04m48s
6093	305	0561 Oct 24	13:50:59	5080	-17788	81	T	-p	-0.6888	1.0381	54.5S	27.3W	46	24	176	02m43s
6094	305	0562 Apr 19	11:52:31	5075	-17782	86	Am	nn	0.1125	0.9678	18.2N	20.6E	83	165	117	03m47s
6095	305	0562 Oct 14	03:07:26	5071	-17776	91	A	nn	-0.0010	0.9992	9.0S	150.9E	90	8	3	00m05s
6096	305	0563 Apr 08	19:42:28	5066	-17770	96	T	p-	-0.6445	1.0150	30.6S	82.0W	50	341	67	01m25s
6097	305	0563 Oct 03	09:52:51	5061	-17764	101	A	p-	0.7438	0.9438	41.0N	67.4E	42	203	307	05m44s
6098	305	0564 Feb 28	01:33:17	5058	-17759	68	P	-t	1.2481	0.5419	71.5N	114.5E	0	114		
6099	305	0564 Mar 28	09:57:15	5057	-17758	106	P	t-	-1.3317	0.3821	71.8S	130.8E	0	283		
6100	305	0564 Aug 22	17:29:17	5053	-17753	73	P	-t	-1.4832	0.1376	71.0S	116.1W	0	54		
6101	306	0564 Sep 21	10:24:56	5052	-17752	111	P	t-	1.4676	0.1693	72.0N	132.2E	0	269		
6102	306	0565 Feb 16	17:24:37	5048	-17747	78	T	-p	0.5777	1.0327	22.8N	65.5W	55	164	134	03m02s
6103	306	0565 Aug 11	21:23:39	5044	-17741	83	A	-p	-0.6988	0.9839	28.6S	129.5W	45	14	79	01m42s
6104	306	0566 Feb 06	04:46:17	5039	-17735	88	A	nn	-0.1361	0.9790	22.8S	135.5E	82	347	75	02m17s
6105	306	0566 Aug 01	08:41:01	5034	-17729	93	T	nn	0.0815	1.0464	22.6N	72.6E	85	190	155	04m22s
6106	306	0567 Jan 26	08:42:29	5030	-17723	98	A	p-	-0.8734	0.9238	76.4S	112.3E	29	318	596	05m50s
6107	306	0567 Jul 22	00:57:59	5025	-17717	103	T	p-	0.8068	1.0692	73.4N	153.4W	36	202	385	04m07s
6108	306	0567 Dec 16	13:35:23	5021	-17712	70	P	-t	1.4708	0.1621	65.5N	13.9E	0	195		
6109	306	0568 Jan 15	07:57:59	5020	-17711	108	P	t-	-1.5538	0.0280	68.4S	110.9W	0	195		
6110	306	0568 Jun 11	09:30:04	5016	-17706	75	P	-t	-1.0092	0.9867	65.0S	79.7E	0	340		
6111	306	0568 Dec 04	19:15:00	5012	-17700	80	A	-p	0.7245	0.9818	22.9N	79.9W	43	192	93	01m57s
6112	306	0569 May 31	18:45:18	5007	-17694	85	A	-p	-0.2596	0.9794	7.7N	78.2W	75	347	76	02m29s
6113	306	0569 Nov 24	07:46:44	5002	-17688	90	T	nn	0.0140	1.0361	20.6S	82.0E	89	199	121	03m17s
6114	306	0570 May 20	21:08:33	4998	-17682	95	A	p-	0.5179	0.9470	49.8N	132.0W	59	154	227	05m06s
6115	306	0570 Nov 13	23:27:26	4993	-17676	100	T	p-	-0.6534	1.0409	56.0S	177.8W	49	35	180	02m46s
6116	306	0571 Apr 10	08:43:42	4989	-17671	67	Pe	-t	-1.5289	0.0441	61.3S	144.0E	0	288		
6117	306	0571 May 09	22:04:44	4989	-17670	105	P	t-	1.2604	0.5175	62.5N	99.9E	0	48		
6118	306	0571 Oct 05	00:42:13	4985	-17665	72	P	-t	1.4200	0.2348	60.9N	92.3W	0	259		
6119	306	0571 Nov 03	13:34:02	4984	-17664	110	P	t-	-1.3665	0.3256	62.2S	128.9W	0	126		
6120	306	0572 Mar 29	18:32:10	4980	-17659	77	T	-p	-0.7198	1.0270	35.3S	51.2W	44	324	129	02m10s

Cat Num	Canon Plate	Calendar Date	TD of Greatest Eclipse	ΔT s	Luna Num	Saros Num	Ecl. Type	QLE	Gamma	Ecl. Mag.	Lat. °	Long. °	Sun Alt °	Sun Azm °	Path Width km	Central Line Dur.
6121	307	0572 Sep 23	05:28:53	4975	-17653	82	A	-p	0.7545	0.9354	40.6N	145.5E	41	220	355	06m08s
6122	307	0573 Mar 19	09:51:15	4971	-17647	87	Tm	nn	0.0228	1.0659	1.5N	54.2E	89	151	215	05m22s
6123	307	0573 Sep 12	05:19:53	4966	-17641	92	A	nn	0.0425	0.9338	5.7N	120.7E	88	209	247	07m36s
6124	307	0574 Mar 09	02:20:43	4961	-17635	97	T	p-	0.7376	1.0452	37.1N	141.8E	42	143	218	03m23s
6125	307	0574 Sep 01	07:35:04	4957	-17629	102	A	p-	-0.6831	0.9668	30.1S	64.7E	47	32	160	03m15s
6126	307	0575 Jan 28	01:30:01	4953	-17624	69	P	-t	-1.3147	0.4234	62.4S	49.9W	0	229		
6127	307	0575 Feb 26	15:18:32	4952	-17623	107	P	t-	1.5067	0.0756	61.3N	100.7W	0	107		
6128	307	0575 Jul 23	08:23:00	4948	-17618	74	P	-t	1.2247	0.5871	63.2N	148.1W	0	319		
6129	307	0575 Aug 21	17:06:51	4947	-17617	112	P	t-	-1.3710	0.3137	61.5S	123.3W	0	65		
6130	307	0576 Jan 17	03:27:40	4944	-17612	79	A	-p	-0.6547	0.9196	58.4S	174.5E	49	327	402	07m17s
6131	307	0576 Jul 12	01:18:44	4939	-17606	84	T	-p	0.4856	1.0755	50.0N	167.5W	61	201	280	05m21s
6132	307	0577 Jan 05	02:41:04	4934	-17600	89	A	nn	0.0334	0.9290	20.7S	162.3E	88	166	266	09m03s
6133	307	0577 Jul 01	17:21:52	4930	-17594	94	T	n-	-0.2618	1.0453	8.3N	62.4W	75	10	156	04m33s
6134	307	0577 Dec 25	06:39:51	4925	-17588	99	A	p-	0.7194	0.9691	22.2N	95.2E	44	172	160	03m30s
6135	307	0578 Jun 21	04:06:31	4920	-17582	104	P	t-	-1.0745	0.8494	65.7S	124.2E	0	13		
6136	307	0578 Nov 15	06:46:52	4917	-17577	71	P	-t	-1.3501	0.3496	69.0S	59.8W	0	157		
6137	307	0578 Dec 14	17:51:31	4916	-17576	109	P	t-	1.3521	0.3488	66.2N	76.2W	0	172		
6138	307	0579 May 11	16:48:05	4912	-17571	76	A	-t	0.9638	0.9365	80.3N	163.3W	15	67	926	04m15s
6139	307	0579 Nov 04	22:31:45	4907	-17565	81	T	-p	-0.6991	1.0344	59.4S	157.1W	45	22	162	02m24s
6140	307	0580 Apr 29	18:41:09	4903	-17559	86	A	nn	0.1866	0.9724	25.9N	83.6W	79	167	101	03m08s
6141	308	0580 Oct 24	11:25:33	4898	-17553	91	A	nn	-0.0163	0.9938	13.9S	25.2E	89	13	22	00m40s
6142	308	0581 Apr 19	03:05:14	4893	-17547	96	T	p-	-0.5800	1.0218	22.6S	162.8E	54	344	91	02m09s
6143	308	0581 Oct 13	17:39:23	4889	-17541	101	A	p-	0.7205	0.9393	35.4N	53.5W	44	200	321	06m38s
6144	308	0582 Mar 10	09:44:43	4885	-17536	68	P	-t	1.2837	0.4737	71.8N	23.1W	0	101		
6145	308	0582 Apr 08	17:44:36	4884	-17535	106	P	t-	-1.2773	0.4861	71.4S	0.6W	0	296		
6146	308	0582 Sep 03	00:33:28	4880	-17530	73	Pe	-t	-1.5345	0.0511	71.5S	123.5E	0	67		
6147	308	0582 Oct 02	17:50:20	4879	-17529	111	P	t-	1.4333	0.2276	71.8N	6.0E	0	255		
6148	308	0583 Feb 28	01:37:28	4876	-17524	78	T	-p	0.6092	1.0329	28.7N	167.8E	52	161	139	02m55s
6149	308	0583 Aug 23	04:44:22	4871	-17518	83	A	-p	-0.7529	0.9834	36.1S	115.2E	41	19	89	01m38s
6150	308	0584 Feb 17	12:45:26	4866	-17512	88	A	nn	-0.1105	0.9788	17.5S	14.7E	84	345	76	02m21s
6151	308	0584 Aug 11	16:18:35	4862	-17506	93	T	nn	0.0224	1.0467	16.1N	43.4W	89	192	155	04m27s
6152	308	0585 Feb 05	16:27:40	4857	-17500	98	A	p-	-0.8529	0.9253	70.2S	8.9W	31	321	541	06m01s
6153	308	0585 Aug 01	08:37:54	4852	-17494	103	T	p-	0.7441	1.0687	64.6N	89.6E	42	203	336	04m22s
6154	308	0585 Dec 26	21:45:28	4849	-17489	70	P	-t	1.4733	0.1569	66.5N	119.7W	0	185		
6155	308	0586 Jan 25	15:47:13	4848	-17488	108	P	t-	-1.5361	0.0559	69.4S	119.2E	0	207		
6156	308	0586 Jun 22	16:38:32	4844	-17483	75	P	-t	-1.0883	0.8380	66.0S	38.3W	0	350		
6157	308	0586 Jul 22	01:07:19	4843	-17482	113	Pb	t-	1.4880	0.0869	68.6N	15.5W	0	342		
6158	308	0586 Dec 16	03:49:22	4839	-17477	80	A	-p	0.7247	0.9848	22.7N	148.3E	43	187	78	01m38s
6159	308	0587 Jun 12	01:24:29	4835	-17471	85	A	-p	-0.3451	0.9761	3.4N	178.6W	70	351	91	03m01s
6160	308	0587 Dec 05	16:36:28	4830	-17465	90	T	nn	0.0167	1.0373	22.1S	50.1W	89	194	125	03m26s
6161	309	0588 May 31	03:31:10	4825	-17459	95	A	p-	0.4303	0.9478	47.1N	137.4E	64	161	212	05m18s
6162	309	0588 Nov 24	08:16:26	4821	-17453	100	T	p-	-0.6490	1.0390	59.3S	53.1E	49	30	172	02m38s
6163	309	0589 May 20	04:40:20	4816	-17447	105	P	t-	1.1766	0.6632	63.2N	8.9W	0	39		
6164	309	0589 Oct 15	08:47:11	4812	-17442	72	P	-t	1.4453	0.1917	61.2N	137.0E	0	250		
6165	309	0589 Nov 13	22:07:32	4812	-17441	110	P	t-	-1.3634	0.3320	62.9S	92.9E	0	136		
6166	309	0590 Apr 10	02:06:40	4808	-17436	77	T	-p	-0.7752	1.0318	36.0S	165.2W	39	326	166	02m33s
6167	309	0590 Oct 04	13:02:00	4803	-17430	82	A	-p	0.7888	0.9303	39.5N	31.1E	38	218	411	06m50s
6168	309	0591 Mar 30	17:45:50	4799	-17424	87	T	nn	-0.0281	1.0697	3.2N	64.8W	88	331	227	05m41s
6169	309	0591 Sep 23	12:39:49	4794	-17418	92	Am	nn	0.0843	0.9317	3.4N	10.3E	85	209	256	07m51s
6170	309	0592 Mar 19	10:19:26	4789	-17412	97	T	p-	0.6948	1.0470	38.0N	21.5E	46	142	213	03m29s
6171	309	0592 Sep 11	15:06:02	4785	-17406	102	A	p-	-0.6360	0.9673	30.6S	48.8W	50	34	149	03m07s
6172	309	0593 Feb 07	09:29:17	4781	-17401	69	P	-t	-1.3396	0.3804	61.8S	179.5W	0	238		
6173	309	0593 Mar 08	23:07:54	4780	-17400	107	P	t-	1.4742	0.1341	61.0N	132.5E	0	98		
6174	309	0593 Aug 02	15:58:20	4776	-17395	74	P	-t	1.2039	0.4743	62.5N	88.2E	0	310		
6175	309	0593 Sep 01	00:53:15	4776	-17394	112	P	t-	-1.3190	0.4100	61.1S	110.6E	0	74		
6176	309	0594 Jan 27	11:15:35	4772	-17389	79	A	-p	-0.6735	0.9207	55.7S	60.9E	47	322	403	07m10s
6177	309	0594 Jul 23	08:52:46	4767	-17383	84	T	-p	0.5537	1.0720	51.3N	83.9E	56	208	280	04m58s
6178	309	0595 Jan 16	10:39:56	4762	-17377	89	A	nn	0.0214	0.9324	19.5S	42.9E	89	161	253	08m20s
6179	309	0595 Jul 13	00:38:24	4758	-17371	94	T	nn	-0.1891	1.0409	11.4N	171.9W	79	14	139	04m02s
6180	309	0596 Jan 05	15:05:31	4753	-17365	99	A	p-	0.7123	0.9733	22.0N	34.2W	44	167	135	02m56s

Cat Num	Canon Plate	Calendar Date	TD of Greatest Eclipse	ΔT s	Luna Num	Saros Num	Ecl. Type	QLE	Gamma	Ecl. Mag.	Lat. °	Long. °	Sun Alt °	Sun Azm °	Path Width km	Central Line Dur.
6181	310	0596 Jul 01	10:54:33	4749	-17359	104	A-	t-	-0.9988	0.9827	64.7S	11.5E	0	23	-	-
6182	310	0596 Nov 25	15:36:49	4745	-17354	71	P	-t	-1.3524	0.3455	68.0S	155.2E	0	169		
6183	310	0596 Dec 25	02:38:01	4744	-17353	109	P	t-	1.3467	0.3584	65.2N	141.2E	0	161		
6184	310	0597 May 21	23:13:25	4740	-17348	76	P	-t	1.0472	0.8815	68.5N	46.0E	0	15		
6185	310	0597 Nov 15	07:16:12	4736	-17342	81	T	-p	-0.7064	1.0310	63.7S	74.5E	45	18	148	02m07s
6186	310	0598 May 11	01:27:21	4731	-17336	86	A	nn	0.2635	0.9766	33.6N	173.4E	75	169	86	02m31s
6187	310	0598 Nov 04	19:48:16	4726	-17330	91	A	nn	-0.0271	0.9888	18.0S	101.3W	88	11	40	01m13s
6188	310	0599 Apr 30	10:23:56	4722	-17324	96	T	p-	-0.5110	1.0281	14.8S	49.1E	59	348	110	02m52s
6189	310	0599 Oct 25	01:32:36	4717	-17318	101	A	p-	0.7033	0.9352	30.6N	175.5W	45	196	337	07m32s
6190	310	0600 Mar 20	17:47:01	4713	-17313	68	P	-t	1.3269	0.3907	71.8N	158.5W	0	87		
6191	310	0600 Apr 19	01:25:29	4713	-17312	106	P	t-	-1.2171	0.6021	70.9S	130.0W	0	309		
6192	310	0600 Oct 13	01:25:21	4708	-17306	111	P	t-	1.4062	0.2734	71.5N	122.5W	0	241		
6193	310	0601 Mar 10	09:40:34	4704	-17301	78	T	-p	0.6483	1.0327	35.3N	43.1E	49	158	144	02m47s
6194	310	0601 Sep 02	12:14:51	4700	-17295	83	A	-p	-0.7997	0.9828	43.5S	3.3W	37	23	101	01m34s
6195	310	0602 Feb 27	20:34:40	4695	-17289	88	A	nn	-0.0774	0.9787	11.6S	104.3W	85	343	76	02m23s
6196	310	0602 Aug 23	00:04:00	4690	-17283	93	T	nn	-0.0306	1.0465	9.5N	161.7W	88	16	155	04m26s
6197	310	0603 Feb 17	00:02:03	4686	-17277	98	A	p-	-0.8236	0.9274	63.4S	128.4W	34	325	481	06m13s
6198	310	0603 Aug 12	16:23:34	4681	-17271	103	T	p-	0.6860	1.0671	56.4N	29.3W	46	203	301	04m33s
6199	310	0604 Jan 07	05:53:30	4677	-17266	70	P	-t	1.4780	0.1475	67.6N	106.7E	0	173		
6200	310	0604 Feb 05	23:29:00	4677	-17265	108	P	t-	-1.5118	0.0948	70.3S	9.5W	0	219		
6201	311	0604 Jul 02	23:46:58	4673	-17260	75	P	-t	-1.1665	0.6918	67.0S	156.6W	0	360		
6202	311	0604 Aug 01	08:37:51	4672	-17259	113	P	t-	1.4301	0.1981	69.5N	140.7W	0	330		
6203	311	0604 Dec 26	12:23:39	4668	-17254	80	A	-p	0.7257	0.9884	23.3N	16.6E	43	182	59	01m14s
6204	311	0605 Jun 22	08:02:39	4664	-17248	85	A	-p	-0.4305	0.9722	1.8S	80.6E	64	356	110	03m35s
6205	311	0605 Dec 16	01:28:09	4659	-17242	90	Tm	nn	0.0181	1.0389	22.7S	177.6E	89	188	131	03m37s
6206	311	0606 Jun 11	09:50:53	4655	-17236	95	A	pn	0.3401	0.9483	43.1N	46.5E	70	168	202	05m35s
6207	311	0606 Dec 05	17:08:42	4650	-17230	100	T	p-	-0.6476	1.0374	62.1S	75.0W	49	23	165	02m31s
6208	311	0607 May 31	11:14:14	4645	-17224	105	P	t-	1.0906	0.8140	64.0N	117.5W	0	30		
6209	311	0607 Oct 26	16:58:21	4642	-17219	72	P	-t	1.4652	0.1585	61.7N	4.7E	0	241		
6210	311	0607 Nov 25	06:44:16	4641	-17218	110	P	t-	-1.3633	0.3327	63.7S	46.4W	0	145		
6211	311	0608 Apr 20	09:36:37	4637	-17213	77	T	-p	-0.8352	1.0358	38.3S	82.0E	33	327	214	02m51s
6212	311	0608 Oct 14	20:42:44	4632	-17207	82	A	-p	0.8163	0.9257	38.5N	86.0W	35	215	468	07m33s
6213	311	0609 Apr 10	01:35:40	4628	-17201	87	T	nn	-0.0835	1.0730	4.5N	177.6E	85	332	238	06m00s
6214	311	0609 Oct 03	20:07:39	4623	-17195	92	A	nn	0.1201	0.9298	0.9N	102.1W	83	209	265	08m09s
6215	311	0610 Mar 30	18:11:46	4619	-17189	97	T	p-	0.6469	1.0485	39.2N	96.4W	49	142	208	03m34s
6216	311	0610 Sep 22	22:46:30	4614	-17183	102	A	p-	-0.5956	0.9678	32.2S	164.6W	53	35	142	03m00s
6217	311	0611 Feb 18	17:18:20	4610	-17178	69	P	-t	-1.3719	0.3245	61.3S	53.8E	0	247		
6218	311	0611 Mar 20	06:46:33	4610	-17177	107	P	t-	1.4339	0.2065	61.0N	8.4E	0	89		
6219	311	0611 Aug 13	23:41:16	4606	-17172	74	P	-t	1.3374	0.3723	61.9N	37.3W	0	301		
6220	311	0611 Sep 12	08:49:34	4605	-17171	112	P	t-	-1.2744	0.4929	60.9S	17.9W	0	83		
6221	312	0612 Feb 07	18:54:14	4601	-17166	79	A	-p	-0.6997	0.9221	53.0S	51.4W	45	318	407	07m02s
6222	312	0612 Aug 02	16:33:18	4597	-17160	84	T	-p	0.6163	1.0679	51.5N	26.5W	52	215	280	04m35s
6223	312	0613 Jan 26	18:31:59	4592	-17154	89	A	nn	0.0030	0.9363	17.9S	74.9W	90	128	237	07m35s
6224	312	0613 Jul 23	07:59:27	4587	-17148	94	T	nn	-0.1207	1.0358	13.4N	78.0E	83	18	121	03m28s
6225	312	0614 Jan 15	23:27:23	4583	-17142	99	A	p-	0.7011	0.9782	22.2N	162.5W	45	162	108	02m19s
6226	312	0614 Jul 12	17:42:54	4578	-17136	104	A	p-	-0.9237	0.9722	43.7S	84.8W	22	19	259	02m41s
6227	312	0614 Dec 07	00:29:40	4575	-17131	71	P	-t	-1.3529	0.3445	67.0S	10.2E	0	180		
6228	312	0615 Jan 05	11:22:56	4574	-17130	109	P	t-	1.3397	0.3712	64.2N	0.6W	0	151		
6229	312	0615 Jun 02	05:35:00	4570	-17125	76	P	-t	1.1337	0.7339	67.4N	61.3W	0	4		
6230	312	0615 Nov 26	16:04:23	4565	-17119	81	T	-p	-0.7105	1.0282	67.0S	51.7W	44	10	137	01m54s
6231	312	0616 May 21	08:10:42	4561	-17113	86	A	-p	0.3438	0.9803	40.9N	72.2E	70	173	74	01m59s
6232	312	0616 Nov 15	04:16:21	4556	-17107	91	A	nn	-0.0331	0.9843	21.4S	131.5E	88	8	56	01m43s
6233	312	0617 May 10	17:39:29	4552	-17101	96	T	p-	-0.4384	1.0340	7.4S	63.2W	64	351	127	03m32s
6234	312	0617 Nov 04	09:31:18	4547	-17095	101	A	p-	0.6910	0.9314	26.6N	61.4E	46	192	353	08m25s
6235	312	0618 Apr 01	01:42:57	4543	-17090	68	P	-t	1.3754	0.2969	71.6N	67.9E	0	73		
6236	312	0618 Apr 30	09:01:45	4543	-17089	106	P	t-	-1.1526	0.7270	70.2S	102.3E	0	322		
6237	312	0618 Oct 24	09:08:27	4538	-17083	111	P	t-	1.3853	0.3086	70.8N	107.5E	0	228		
6238	312	0619 Mar 21	17:34:36	4534	-17078	78	T	-p	0.6944	1.0323	42.8N	80.0W	46	155	150	02m36s
6239	312	0619 Sep 13	19:55:18	4530	-17072	83	A	-p	-0.8391	0.9819	50.7S	125.2W	33	29	117	01m31s
6240	312	0620 Mar 10	04:14:30	4525	-17066	88	A	nn	-0.0373	0.9787	5.1S	138.5E	88	343	76	02m25s

Cat Num	Canon Plate	Calendar Date	TD of Greatest Eclipse	ΔT s	Luna Num	Saros Num	Ecl. Type	QLE	Gamma	Ecl. Mag.	Lat. °	Long. °	Sun Alt °	Sun Azm °	Path Width km	Central Line Dur.
6241	313	0620 Sep 02	07:57:50	4521	-17060	93	T	nn	-0.0772	1.0459	2.9N	77.6E	86	17	153	04m21s
6242	313	0621 Feb 27	07:28:15	4516	-17054	98	A	p-	-0.7879	0.9297	56.1S	114.2E	38	329	426	06m27s
6243	313	0621 Aug 23	00:16:24	4512	-17048	103	T	p-	0.6340	1.0648	48.6N	150.4W	50	203	274	04m40s
6244	313	0622 Jan 17	13:58:14	4508	-17043	70	P	-t	1.4858	0.1326	68.7N	26.7W	0	162		
6245	313	0622 Feb 16	07:04:22	4507	-17042	108	P	t-	-1.4819	0.1434	71.0S	137.1W	0	232		
6246	313	0622 Jul 14	06:56:26	4503	-17037	75	P	-t	-1.2426	0.5507	68.0S	84.3E	0	10		
6247	313	0622 Aug 12	16:12:43	4502	-17036	113	P	t-	1.3763	0.3006	70.3N	92.4E	0	318		
6248	313	0623 Jan 06	20:55:51	4499	-17031	80	A	-p	0.7292	0.9926	24.7N	114.7W	43	177	38	00m47s
6249	313	0623 Jul 03	14:41:23	4494	-17025	85	A	-p	-0.5142	0.9678	8.0S	21.1W	59	360	135	04m11s
6250	313	0623 Dec 27	10:17:25	4490	-17019	90	T	nn	0.0217	1.0412	22.2S	45.8E	89	182	138	03m51s
6251	313	0624 Jun 21	16:12:34	4485	-17013	95	A	pn	0.2517	0.9481	38.3N	46.2W	75	174	197	05m56s
6252	313	0624 Dec 16	02:00:22	4480	-17007	100	T	p-	-0.6465	1.0364	63.6S	159.0E	49	14	161	02m28s
6253	313	0625 Jun 10	17:49:19	4476	-17001	105	A+	t-	1.0042	0.9661	64.9N	133.3E	0	21	-	-
6254	313	0625 Nov 06	01:14:00	4472	-16996	72	P	-t	1.4805	0.1333	62.3N	128.8W	0	231		
6255	313	0625 Dec 05	15:21:37	4471	-16995	110	P	t-	-1.3647	0.3308	64.6S	173.9E	0	155		
6256	313	0626 May 01	17:01:46	4468	-16990	77	T	-p	-0.8997	1.0389	42.9S	29.1W	26	329	294	03m01s
6257	313	0626 Oct 26	04:31:02	4463	-16984	82	A	-p	0.8372	0.9216	37.8N	154.2E	33	212	525	08m17s
6258	313	0627 Apr 21	09:16:52	4459	-16978	87	T	-n	-0.1462	1.0758	5.0N	62.3E	82	335	248	06m19s
6259	313	0627 Oct 15	03:46:30	4454	-16972	92	A	nn	0.1470	0.9282	1.9S	142.4E	82	207	272	08m30s
6260	313	0628 Apr 10	01:54:42	4449	-16966	97	T	p-	0.5919	1.0496	40.2N	148.8E	53	143	202	03m40s
6261	314	0628 Oct 03	06:37:26	4445	-16960	102	A	p-	-0.5632	0.9682	34.6S	77.1E	56	35	136	02m54s
6262	314	0629 Mar 01	00:58:09	4441	-16955	69	P	-t	-1.4107	0.2572	61.0S	70.6W	0	256		
6263	314	0629 Mar 30	14:16:12	4440	-16954	107	P	t-	1.3872	0.2905	61.1N	113.4W	0	80		
6264	314	0629 Aug 24	07:31:59	4437	-16949	74	P	-t	1.3849	0.2818	61.5N	164.6W	0	292		
6265	314	0629 Sep 22	16:55:16	4436	-16948	112	P	t-	-1.2369	0.5622	60.8S	148.7W	0	92		
6266	314	0630 Feb 18	02:26:01	4432	-16943	79	A	-p	-0.7313	0.9239	50.5S	162.9W	43	316	413	06m53s
6267	314	0630 Aug 14	00:17:53	4428	-16937	84	T	-p	0.6751	1.0631	50.9N	138.7W	47	220	278	04m13s
6268	314	0631 Feb 07	02:19:34	4423	-16931	89	A	nn	-0.0196	0.9407	15.8S	168.2E	89	337	220	06m49s
6269	314	0631 Aug 03	15:23:41	4418	-16925	94	Tm	nn	-0.0557	1.0303	14.3N	32.8W	87	22	102	02m52s
6270	314	0632 Jan 27	07:45:01	4414	-16919	99	A	p-	0.6856	0.9836	22.7N	70.5E	47	158	78	01m40s
6271	314	0632 Jul 23	00:35:01	4409	-16913	104	A	p-	-0.8522	0.9696	35.9S	170.9E	31	22	205	03m07s
6272	314	0632 Dec 17	09:22:54	4406	-16908	71	P	-t	-1.3539	0.3426	65.9S	134.4W	0	191		
6273	314	0633 Jan 15	20:04:43	4405	-16907	109	P	t-	1.3295	0.3897	63.3N	141.2W	0	141		
6274	314	0633 Jun 12	11:57:22	4401	-16902	76	P	-t	1.2191	0.5877	66.4N	168.2W	0	354		
6275	314	0633 Dec 07	00:53:04	4397	-16896	81	T	-p	-0.7139	1.0258	68.9S	175.0W	44	359	126	01m43s
6276	314	0634 Jun 01	14:54:37	4392	-16890	86	A	-p	0.4246	0.9836	47.8N	27.7W	65	178	64	01m33s
6277	314	0634 Nov 26	12:46:01	4388	-16884	91	A	nn	-0.0369	0.9804	23.9S	4.4E	88	4	70	02m09s
6278	314	0635 May 22	00:52:37	4383	-16878	96	T	p-	-0.3629	1.0391	0.4S	174.3W	69	355	141	04m06s
6279	314	0635 Nov 15	17:34:25	4379	-16872	101	A	p-	0.6828	0.9283	23.4N	62.5W	47	188	367	09m12s
6280	314	0636 Apr 11	09:31:19	4375	-16867	68	P	-t	1.4301	0.1909	71.2N	63.6W	0	60		
6281	315	0636 May 10	16:34:17	4374	-16866	106	P	t-	-1.0847	0.8593	69.3S	23.9W	0	334		
6282	315	0636 Nov 03	16:57:50	4370	-16860	111	P	t-	1.3690	0.3360	70.0N	23.5W	0	215		
6283	315	0637 Apr 01	01:19:52	4366	-16855	78	T	-p	0.7473	1.0312	50.9N	158.3E	41	152	158	02m21s
6284	315	0637 Sep 24	03:46:29	4361	-16849	83	A	-p	-0.8707	0.9813	57.5S	109.4E	29	35	135	01m28s
6285	315	0638 Mar 21	11:43:16	4357	-16843	88	A	nn	0.0112	0.9786	1.9N	23.8E	89	162	76	02m27s
6286	315	0638 Sep 13	16:00:38	4352	-16837	93	Tm	nn	-0.1169	1.0450	3.6S	45.4W	83	18	151	04m14s
6287	315	0639 Mar 10	14:43:34	4348	-16831	98	A	p-	-0.7434	0.9323	48.3S	0.3W	42	333	375	06m42s
6288	315	0639 Sep 03	08:17:01	4343	-16825	103	T	p-	0.5887	1.0620	41.3N	86.5E	54	203	250	04m42s
6289	315	0640 Jan 28	21:57:14	4339	-16820	70	P	-t	1.4986	0.1089	69.6N	159.2W	0	150		
6290	315	0640 Feb 27	14:30:29	4339	-16819	108	P	t-	-1.4439	0.2060	71.6S	97.0E	0	246		
6291	315	0640 Jul 24	14:09:19	4335	-16814	75	P	-t	-1.3147	0.4185	69.0S	36.2W	0	22		
6292	315	0640 Aug 22	23:54:28	4334	-16813	113	P	t-	1.3287	0.3906	71.0N	36.8W	0	305		
6293	315	0641 Jan 17	05:25:38	4330	-16808	80	A	-p	0.7350	0.9972	27.0N	114.5E	42	173	15	00m17s
6294	315	0641 Jul 13	21:20:57	4326	-16802	85	A	p	0.5062	0.9630	15.0S	123.9W	53	4	167	04m45s
6295	315	0642 Jan 06	19:05:52	4321	-16796	90	T	nn	0.0263	1.0437	20.8S	85.9W	89	177	146	04m06s
6296	315	0642 Jul 02	22:35:07	4317	-16790	95	A	nn	0.1639	0.9477	32.6N	140.5W	80	180	195	06m21s
6297	315	0642 Dec 27	10:51:27	4312	-16784	100	T	p-	-0.6448	1.0358	63.7S	34.0E	50	4	158	02m27s
6298	315	0643 Jun 22	00:25:51	4308	-16778	105	A	t-	0.9176	0.9648	85.5N	94.4E	23	81	326	02m17s
6299	315	0643 Nov 17	09:34:13	4304	-16773	72	P	-t	1.4917	0.1155	63.0N	96.3E	0	222		
6300	315	0643 Dec 16	23:59:22	4303	-16772	110	P	t-	-1.3669	0.3272	65.6S	33.8E	0	166		

Cat Num	Canon Plate	Calendar Date	TD of Greatest Eclipse	ΔT s	Luna Num	Saros Num	Ecl. Type	QLE	Gamma	Ecl. Mag.	Lat. °	Long. °	Sun Alt °	Sun Azm °	Path Width km	Central Line Dur.
6301	316	0644 May 12	00:25:07	4300	-16767	77	T	-t	-0.9666	1.0400	51.6S	137.6W	14	329	531	02m56s
6302	316	0644 Nov 05	12:24:02	4295	-16761	82	A	-p	0.8539	0.9181	37.4N	32.6E	31	207	582	08m59s
6303	316	0645 May 01	16:55:24	4291	-16755	87	T	-n	-0.2115	1.0779	4.7N	52.4W	78	338	257	06m38s
6304	316	0645 Oct 25	11:32:44	4286	-16749	92	A	nn	0.1682	0.9271	4.5S	25.2E	80	205	277	08m51s
6305	316	0646 Apr 21	09:31:04	4282	-16743	97	T	p-	0.5316	1.0502	41.0N	36.5E	58	146	195	03m45s
6306	316	0646 Oct 14	14:37:46	4277	-16737	102	A	p-	-0.5375	0.9688	37.6S	43.2W	57	35	132	02m48s
6307	316	0647 Mar 12	08:27:23	4273	-16732	69	P	-t	-1.4573	0.1764	60.8S	167.7E	0	265		
6308	316	0647 Apr 10	21:36:12	4273	-16731	107	P	t-	1.3334	0.3871	61.3N	127.1E	0	71		
6309	316	0647 Sep 04	15:31:41	4269	-16726	74	P	-t	1.4260	0.2039	61.2N	65.9E	0	283		
6310	316	0647 Oct 04	01:10:30	4268	-16725	112	P	t-	-1.2065	0.6186	60.9S	78.1E	0	101		
6311	316	0648 Feb 29	09:46:24	4265	-16720	79	A	-p	-0.7722	0.9257	48.8S	88.3E	39	315	430	06m44s
6312	316	0648 Aug 24	08:10:54	4260	-16714	84	T	-p	0.7270	1.0579	49.8N	105.7E	43	223	274	03m52s
6313	316	0649 Feb 17	09:59:12	4256	-16708	89	A	nn	-0.0499	0.9455	13.6S	53.2E	87	334	201	06m05s
6314	316	0649 Aug 13	22:52:54	4251	-16702	94	T	nn	0.0045	1.0243	14.2N	144.8W	90	199	83	02m16s
6315	316	0650 Feb 06	15:56:44	4247	-16696	99	A	p-	0.6641	0.9896	23.5N	54.7W	48	154	48	01m01s
6316	316	0650 Aug 03	07:30:19	4242	-16690	104	A	p-	-0.7836	0.9660	30.9S	66.0E	38	25	193	03m34s
6317	316	0650 Dec 28	18:14:40	4238	-16685	71	P	-t	-1.3564	0.3379	64.9S	81.8E	0	202		
6318	316	0651 Jan 27	04:40:40	4238	-16684	109	P	t-	1.3144	0.4175	62.5N	79.9E	0	132		
6319	316	0651 Jun 23	18:19:23	4234	-16679	76	P	-t	1.3048	0.4409	65.4N	85.3E	0	344		
6320	316	0651 Jul 23	09:28:50	4233	-16678	114	Pb	t-	-1.5588	0.0086	63.1S	11.9E	0	41		
6321	317	0651 Dec 18	09:42:29	4229	-16673	81	T	-p	-0.7167	1.0241	69.2S	63.2E	44	346	118	01m36s
6322	317	0652 Jun 11	21:39:10	4225	-16667	86	A	-p	0.5057	0.9862	54.0N	125.7W	59	184	56	01m13s
6323	317	0652 Dec 06	21:16:51	4220	-16661	91	A	nn	-0.0391	0.9771	25.5S	122.6W	88	359	82	02m31s
6324	317	0653 Jun 01	08:05:39	4216	-16655	96	T	n-	-0.2861	1.0436	5.9N	75.5E	73	358	152	04m32s
6325	317	0653 Nov 26	01:41:15	4212	-16649	101	A	p-	0.6779	0.9257	21.1N	172.8E	47	184	380	09m51s
6326	317	0654 Apr 22	17:13:12	4208	-16644	68	Pe	-t	1.4901	0.0743	70.6N	167.0E	0	47		
6327	317	0654 May 22	00:02:55	4207	-16643	106	P	t-	-1.0131	0.9990	68.4S	148.6W	0	345		
6328	317	0654 Nov 15	00:54:10	4203	-16637	111	P	t-	1.3578	0.3547	69.1N	155.6W	0	202		
6329	317	0655 Apr 12	08:56:58	4199	-16632	78	T	-p	0.8065	1.0296	59.8N	36.8E	36	147	169	02m05s
6330	317	0655 Oct 05	11:47:14	4194	-16626	83	A	-p	-0.8953	0.9807	63.9S	19.5W	26	42	154	01m24s
6331	317	0656 Mar 31	19:02:49	4190	-16620	88	A	nn	0.0666	0.9783	9.2N	88.7W	86	163	77	02m29s
6332	317	0656 Sep 24	00:12:25	4185	-16614	93	T	-n	-0.1496	1.0441	9.7S	170.5W	81	18	149	04m05s
6333	317	0657 Mar 20	21:51:04	4181	-16608	98	A	p-	-0.6929	0.9350	40.4S	112.4W	46	336	332	06m56s
6334	317	0657 Sep 13	16:24:06	4177	-16602	103	T	p-	0.5490	1.0588	34.4N	38.3W	56	202	230	04m42s
6335	317	0658 Feb 08	05:51:03	4173	-16597	70	P	-t	1.5160	0.0767	70.4N	69.0E	0	137		
6336	317	0658 Mar 09	21:49:57	4172	-16596	108	P	t-	-1.3999	0.2794	71.9S	27.5W	0	260		
6337	317	0658 Aug 04	21:25:50	4168	-16591	75	P	-t	-1.3825	0.2955	69.9S	158.2W	0	33		
6338	317	0658 Sep 03	07:42:11	4168	-16590	113	P	t-	1.2866	0.4693	71.5N	167.9W	0	292		
6339	317	0659 Jan 28	13:49:17	4164	-16585	80	H	-p	0.7468	1.0021	30.3N	15.0W	41	168	11	00m12s
6340	317	0659 Jul 25	04:04:28	4159	-16579	85	A	-p	-0.6740	0.9577	22.7S	131.5E	47	8	208	05m13s
6341	318	0660 Jan 18	03:49:03	4155	-16573	90	T	nn	0.0354	1.0468	18.2S	143.3E	88	173	156	04m22s
6342	318	0660 Jul 13	05:02:57	4151	-16567	95	A	nn	0.0801	0.9468	26.4N	122.7E	85	184	197	06m46s
6343	318	0661 Jan 06	19:37:59	4146	-16561	100	T	p-	-0.6398	1.0357	62.1S	90.6W	50	354	157	02m29s
6344	318	0661 Jul 02	07:07:22	4142	-16555	105	A	t-	0.8339	0.9692	79.9N	88.8E	33	178	203	02m11s
6345	318	0661 Nov 27	17:56:51	4138	-16550	72	P	-t	1.5001	0.1023	63.9N	39.4W	0	212		
6346	318	0661 Dec 27	08:34:50	4137	-16549	110	P	t-	-1.3681	0.3254	66.7S	106.2W	0	177		
6347	318	0662 May 23	07:44:42	4133	-16544	77	P	-t	-1.0371	0.9440	63.5S	118.9E	0	324		
6348	318	0662 Jun 21	15:58:36	4133	-16543	115	Pb	t-	1.5377	0.0030	66.0N	148.9E	0	11		
6349	318	0662 Nov 16	20:23:11	4129	-16538	82	A	-p	0.8650	0.9153	37.2N	91.0W	30	202	630	09m37s
6350	318	0663 May 13	00:27:42	4125	-16532	87	T	-n	-0.2818	1.0792	3.4N	165.7W	74	341	266	06m56s
6351	318	0663 Nov 05	19:27:43	4120	-16526	92	A	nn	0.1826	0.9265	7.1S	94.3W	80	202	281	09m12s
6352	318	0664 May 01	16:59:55	4116	-16520	97	T	p-	0.4658	1.0501	41.3N	73.2W	62	150	187	03m50s
6353	318	0664 Oct 24	22:47:46	4111	-16514	102	A	p-	-0.5193	0.9695	40.9S	165.6W	59	33	127	02m42s
6354	318	0665 Mar 22	15:47:20	4108	-16509	69	Pe	-t	-1.5106	0.0840	60.8S	48.4E	0	274		
6355	318	0665 Apr 21	04:48:01	4107	-16508	107	P	t-	1.2736	0.4945	61.7N	9.7E	0	62		
6356	318	0665 Sep 14	23:38:28	4103	-16503	74	P	-t	1.4618	0.1363	61.0N	65.2W	0	274		
6357	318	0665 Oct 14	09:33:22	4102	-16502	112	P	t-	-1.1819	0.6641	61.2S	57.0W	0	110		
6358	318	0666 Mar 11	16:59:40	4099	-16497	79	A	-p	-0.8187	0.9275	47.9S	18.8W	35	314	461	06m34s
6359	318	0666 Sep 04	16:09:34	4094	-16491	84	T	-p	0.7740	1.0524	48.5N	12.3W	39	225	269	03m31s
6360	318	0667 Feb 28	17:31:36	4090	-16485	89	Am	nn	-0.0868	0.9506	11.3S	60.1W	85	332	182	05m23s

Cat Num	Canon Plate	Calendar Date	TD of Greatest Eclipse	ΔT s	Luna Num	Saros Num	Ecl. Type	QLE	Gamma	Ecl. Mag.	Lat. °	Long. °	Sun Alt °	Sun Azm °	Path Width km	Central Line Dur.
6361	319	0667 Aug 25	06:27:37	4085	-16479	94	T	nn	0.0594	1.0180	13.3N	101.8E	87	207	62	01m40s
6362	319	0668 Feb 18	00:02:44	4081	-16473	99	A	p-	0.6371	0.9959	24.5N	178.1W	50	151	18	00m23s
6363	319	0668 Aug 13	14:31:17	4076	-16467	104	A	p-	-0.7202	0.9619	28.1S	40.3W	44	28	194	04m00s
6364	319	0669 Jan 08	03:03:18	4073	-16462	71	P	-t	-1.3617	0.3275	63.9S	60.8W	0	212		
6365	319	0669 Feb 06	13:11:22	4072	-16461	109	P	t-	1.2949	0.4541	61.9N	57.5W	0	122		
6366	319	0669 Jul 04	00:46:07	4068	-16456	76	P	-t	1.3865	0.3008	64.5N	21.9W	0	335		
6367	319	0669 Aug 02	16:06:06	4068	-16455	114	P	t-	-1.4861	0.1336	62.4S	97.3W	0	50		
6368	319	0669 Dec 28	18:29:34	4064	-16450	81	T	-p	-0.7212	1.0227	68.0S	58.9W	44	334	112	01m30s
6369	319	0670 Jun 23	04:26:06	4060	-16444	86	A	-p	0.5856	0.9883	59.2N	138.4E	54	193	51	00m58s
6370	319	0670 Dec 18	05:46:46	4055	-16438	91	A	nn	-0.0415	0.9744	26.1S	110.8E	87	355	92	02m49s
6371	319	0671 Jun 12	15:19:39	4051	-16432	96	T	n-	-0.2090	1.0473	11.4N	34.3W	78	2	161	04m48s
6372	319	0671 Dec 07	09:48:18	4046	-16426	101	A	p-	0.6733	0.9238	19.4N	48.2E	48	179	389	10m18s
6373	319	0672 Jun 01	07:30:46	4042	-16420	106	T	t-	-0.9404	1.0680	48.1S	85.4E	19	358	663	05m06s
6374	319	0672 Nov 25	08:54:26	4037	-16414	111	P	t-	1.3490	0.3694	68.0N	72.0E	0	190		
6375	319	0673 Apr 22	16:26:17	4034	-16409	78	T	-t	0.8714	1.0270	69.6N	88.0W	29	138	188	01m44s
6376	319	0673 Oct 15	19:56:59	4029	-16403	83	A	-p	-0.9133	0.9805	69.6S	151.9W	24	51	172	01m20s
6377	319	0674 Apr 12	02:12:24	4025	-16397	88	A	nn	0.1295	0.9778	16.8N	161.4E	82	164	80	02m31s
6378	319	0674 Oct 05	08:33:20	4020	-16391	93	T	-n	-0.1750	1.0430	15.5S	62.4E	80	18	146	03m56s
6379	319	0675 Apr 01	04:46:39	4016	-16385	98	A	p-	-0.6326	0.9377	32.1S	138.6E	51	340	296	07m10s
6380	319	0675 Sep 25	00:40:11	4012	-16379	103	T	p-	0.5170	1.0553	28.0N	165.2W	59	200	212	04m37s
6381	320	0676 Feb 19	13:37:43	4008	-16374	70	Pe	-t	1.5396	0.0335	71.1N	61.6W	0	124		
6382	320	0676 Mar 20	05:00:44	4007	-16373	108	P	t-	-1.3481	0.3667	72.0S	150.1W	0	273		
6383	320	0676 Aug 15	04:47:42	4004	-16368	75	P	-t	-1.4450	0.1836	70.7S	77.9E	0	46		
6384	320	0676 Sep 13	15:37:12	4003	-16367	113	P	t-	1.2509	0.5352	71.8N	58.7E	0	278		
6385	320	0677 Feb 07	22:08:16	3999	-16362	80	H	-p	0.7632	1.0073	34.4N	143.7W	40	164	39	00m40s
6386	320	0677 Aug 04	10:52:22	3995	-16356	85	A	-p	-0.7474	0.9521	31.0S	24.7E	41	12	262	05m33s
6387	320	0678 Jan 28	12:28:34	3990	-16350	90	T	nn	0.0480	1.0501	14.8S	13.0E	87	169	166	04m40s
6388	320	0678 Jul 24	11:34:07	3986	-16344	95	Am	nn	-0.0011	0.9455	19.6N	24.1E	90	172	201	07m10s
6389	320	0679 Jan 18	04:21:40	3982	-16338	100	T	p-	-0.6329	1.0360	59.1S	143.5E	50	347	157	02m33s
6390	320	0679 Jul 13	13:53:46	3977	-16332	105	A	p-	0.7531	0.9724	70.6N	2.3W	41	191	151	02m07s
6391	320	0679 Dec 09	02:20:02	3973	-16327	72	P	-t	1.5072	0.0911	64.9N	175.5W	0	202		
6392	320	0680 Jan 07	17:06:18	3973	-16326	110	P	t-	-1.3669	0.3276	67.7S	114.3E	0	188		
6393	320	0680 Jun 02	15:04:59	3969	-16321	77	P	-t	-1.1075	0.8109	64.4S	1.5W	0	333		
6394	320	0680 Jul 01	23:10:18	3968	-16320	115	P	t-	1.4605	0.1456	67.0N	29.8E	0	0		
6395	320	0680 Nov 27	04:24:53	3965	-16315	82	A	-p	0.8734	0.9133	37.3N	144.4E	29	197	673	10m08s
6396	320	0681 May 23	07:58:17	3960	-16309	87	T	-n	-0.3538	1.0797	1.2N	81.2E	69	345	274	07m10s
6397	320	0681 Nov 16	03:27:16	3956	-16303	92	A	nn	0.1936	0.9264	9.2S	145.3E	79	198	282	09m29s
6398	320	0682 May 13	00:23:49	3951	-16297	97	T	n-	0.3964	1.0494	40.8N	178.7E	66	155	178	03m54s
6399	320	0682 Nov 05	07:04:37	3947	-16291	102	A	p-	-0.5056	0.9706	44.3S	71.0E	59	31	122	02m35s
6400	320	0683 May 02	11:50:52	3943	-16285	107	P	t-	1.2071	0.6135	62.2N	105.7W	0	54		
6401	321	0683 Sep 26	07:54:43	3939	-16280	74	P	-t	1.4907	0.0822	61.0N	161.3E	0	265		
6402	321	0683 Oct 25	18:04:33	3938	-16279	112	P	t-	-1.1635	0.6981	61.7S	165.7E	0	119		
6403	321	0684 Mar 22	00:02:14	3935	-16274	79	A	-p	-0.8738	0.9290	48.6S	122.4W	29	313	531	06m23s
6404	321	0684 Sep 15	00:16:17	3930	-16268	84	T	-p	0.8140	1.0468	47.2N	133.6W	35	225	263	03m11s
6405	321	0685 Mar 11	00:55:43	3926	-16262	89	A	nn	-0.1313	0.9559	9.2S	171.3W	82	331	162	04m44s
6406	321	0685 Sep 04	14:09:25	3922	-16256	94	H	nn	0.1074	1.0116	11.6N	13.7W	84	209	40	01m04s
6407	321	0686 Feb 28	08:00:49	3917	-16250	99	H	p-	0.6026	1.0026	25.7N	61.0E	53	149	11	00m14s
6408	321	0686 Aug 24	21:37:46	3913	-16244	104	A	p-	-0.6619	0.9573	26.8S	147.8W	48	30	202	04m27s
6409	321	0687 Jan 19	11:48:03	3909	-16239	71	P	-t	-1.3707	0.3102	63.1S	157.9E	0	222		
6410	321	0687 Feb 17	21:34:47	3908	-16238	109	P	t-	1.2691	0.5026	61.4N	167.1E	0	113		
6411	321	0687 Jul 15	07:16:25	3905	-16233	76	P	-t	1.4655	0.1655	63.7N	129.7W	0	325		
6412	321	0687 Aug 13	22:50:13	3904	-16232	114	P	t-	-1.4180	0.2500	61.7S	152.0E	0	59		
6413	321	0688 Jan 09	03:13:09	3900	-16227	81	T	-p	-0.7284	1.0218	65.6S	177.5E	43	325	108	01m27s
6414	321	0688 Jul 03	11:17:17	3896	-16221	86	A	-p	0.6632	0.9897	63.1N	44.4E	40	204	49	00m48s
6415	321	0688 Dec 28	14:15:19	3892	-16215	91	A	nn	-0.0447	0.9723	25.8S	15.4W	87	350	99	03m02s
6416	321	0689 Jun 22	22:34:45	3887	-16209	96	T	nn	-0.1319	1.0503	16.1N	143.6W	83	7	168	04m56s
6417	321	0689 Dec 17	17:56:06	3883	-16203	101	A	p-	0.6698	0.9225	18.4N	76.4W	48	175	394	10m31s
6418	321	0690 Jun 12	14:57:16	3878	-16197	106	T	t-	-0.8662	1.0718	36.8S	31.7W	30	3	469	05m52s
6419	321	0690 Dec 06	16:57:53	3874	-16191	111	P	t-	1.3426	0.3801	66.9N	60.6W	0	179		
6420	321	0691 May 03	23:48:12	3870	-16186	78	T	-t	0.9415	1.0229	78.9N	123.0E	19	102	238	01m20s

Cat Num	Canon Plate	Calendar Date	TD of Greatest Eclipse	ΔT	Luna Num	Saros Num	Ecl. Type	QLE	Gamma	Ecl. Mag.	Lat.	Long.	Sun Alt	Sun Azm	Path Width	Central Line Dur.
				s							°	°	°	°	km	
6421	322	0691 Oct 27	04:15:17	3866	-16180	83	A	-p	-0.9252	0.9807	74.7S	72.2E	22	61	184	01m16s
6422	322	0692 Apr 22	09:14:36	3862	-16174	88	A	nn	0.1975	0.9769	24.5N	53.6E	78	165	84	02m35s
6423	322	0692 Oct 15	17:00:55	3857	-16168	93	T	-n	-0.1952	1.0421	21.0S	65.9W	79	17	143	03m48s
6424	322	0693 Apr 11	11:36:01	3853	-16162	98	A	p-	-0.5674	0.9403	23.9S	31.8E	55	343	266	07m21s
6425	322	0693 Oct 05	09:03:11	3849	-16156	103	T	p-	0.4910	1.0517	22.2N	66.3E	60	198	196	04m30s
6426	322	0694 Mar 31	12:04:07	3844	-16150	108	P	t-	-1.2894	0.4669	71.8S	89.3E	0	287		
6427	322	0694 Aug 26	12:15:04	3841	-16145	75	Pe	-t	-1.5020	0.0832	71.3S	47.9W	0	59		
6428	322	0694 Sep 24	23:39:10	3840	-16144	113	P	t-	1.2216	0.5889	71.8N	76.6W	0	264		
6429	322	0695 Feb 19	06:19:39	3836	-16139	80	T	-p	0.7865	1.0126	39.6N	88.8E	38	160	70	01m05s
6430	322	0695 Aug 15	17:47:00	3832	-16133	85	A	-p	-0.8142	0.9463	39.7S	84.9W	35	18	338	05m46s
6431	322	0696 Feb 08	20:59:58	3828	-16127	90	T	nn	0.0674	1.0537	10.3S	115.8W	86	166	178	04m58s
6432	322	0696 Aug 03	18:13:52	3823	-16121	95	A	nn	-0.0754	0.9439	12.6N	77.4W	86	11	208	07m30s
6433	322	0697 Jan 28	12:58:36	3819	-16115	100	T	p-	-0.6210	1.0367	54.9S	16.9E	51	342	158	02m41s
6434	322	0697 Jul 23	20:46:01	3815	-16109	105	A	p-	0.6761	0.9749	61.7N	103.5W	47	195	122	02m04s
6435	322	0697 Dec 19	10:42:19	3811	-16104	72	P	-t	1.5144	0.0798	65.9N	48.1E	0	191		
6436	322	0698 Jan 18	01:32:29	3810	-16103	110	P	t-	-1.3624	0.3359	68.7S	24.4W	0	199		
6437	322	0698 Jun 13	22:24:22	3807	-16098	77	P	-t	-1.1789	0.6750	65.3S	122.0W	0	343		
6438	322	0698 Jul 13	06:25:23	3806	-16097	115	P	t-	1.3851	0.2855	68.0N	90.6W	0	350		
6439	322	0698 Dec 08	12:28:45	3802	-16092	82	A	-p	0.8799	0.9120	37.8N	19.0E	28	192	707	10m28s
6440	322	0699 Jun 03	15:24:55	3798	-16086	87	T	-p	-0.4291	1.0792	2.2S	31.3W	65	349	282	07m17s
6441	323	0699 Nov 27	11:32:43	3794	-16080	92	A	nn	0.2003	0.9270	10.7S	23.5E	79	194	280	09m40s
6442	323	0700 May 23	07:42:50	3789	-16074	97	T	n-	0.3233	1.0480	39.4N	72.0E	71	161	168	03m56s
6443	323	0700 Nov 15	15:27:33	3785	-16068	102	A	p-	-0.4962	0.9721	47.4S	53.0W	60	27	115	02m26s
6444	323	0701 May 12	18:48:06	3781	-16062	107	P	t-	1.1366	0.7392	62.8N	140.2E	0	45		
6445	323	0701 Oct 06	16:18:38	3777	-16057	74	P	-t	1.5138	0.0391	61.2N	25.9E	0	256		
6446	323	0701 Nov 05	02:42:23	3776	-16056	112	P	t-	-1.1502	0.7226	62.3S	26.6E	0	128		
6447	323	0702 Apr 02	06:56:41	3773	-16051	79	A	-t	-0.9353	0.9299	51.7S	138.2E	20	310	725	06m09s
6448	323	0702 Sep 26	08:29:49	3768	-16045	84	T	-p	0.8481	1.0410	46.2N	102.4E	32	224	253	02m51s
6449	323	0703 Mar 22	08:13:08	3764	-16039	89	A	nn	-0.1819	0.9612	7.5S	79.2E	79	331	143	04m08s
6450	323	0703 Sep 15	21:57:57	3760	-16033	94	H	-n	0.1492	1.0052	9.4N	131.1W	81	209	18	00m29s
6451	323	0704 Mar 10	15:52:29	3755	-16027	99	H	p-	0.5619	1.0093	27.1N	57.8W	56	147	38	00m49s
6452	323	0704 Sep 04	04:52:06	3751	-16021	104	A	p-	-0.6104	0.9526	26.8S	102.8E	52	32	214	04m53s
6453	323	0705 Jan 29	20:27:40	3747	-16016	71	P	-t	-1.3844	0.2840	62.4S	18.1E	0	231		
6454	323	0705 Feb 28	05:52:06	3747	-16015	109	P	t-	1.2380	0.5617	61.1N	33.3E	0	104		
6455	323	0705 Jul 25	13:52:37	3743	-16010	76	Pe	-t	1.5397	0.0384	62.9N	121.2E	0	316		
6456	323	0705 Aug 24	05:41:35	3742	-16009	114	P	t-	-1.3552	0.3570	61.3S	39.7E	0	68		
6457	323	0706 Jan 19	11:51:48	3739	-16004	81	T	-p	-0.7395	1.0211	62.5S	52.6E	42	319	107	01m24s
6458	323	0706 Jul 14	18:13:55	3735	-15998	86	A	-p	0.7372	0.9904	65.4N	48.4W	42	217	50	00m42s
6459	323	0707 Jan 08	22:38:11	3730	-15992	91	A	nn	-0.0520	0.9708	24.8S	140.3W	87	345	105	03m09s
6460	323	0707 Jul 04	05:53:44	3726	-15986	96	T	nn	-0.0573	1.0525	19.8N	106.8E	87	12	174	04m57s
6461	324	0707 Dec 29	02:00:34	3722	-15980	101	A	p-	0.6636	0.9219	17.9N	159.8E	48	170	392	10m30s
6462	324	0708 Jun 22	22:25:31	3717	-15974	106	T	p-	-0.7933	1.0737	28.8S	147.5W	37	8	393	06m18s
6463	324	0708 Dec 17	01:00:57	3713	-15968	111	P	t-	1.3352	0.3923	65.8N	167.5E	0	168		
6464	324	0709 May 14	07:04:35	3709	-15963	78	P	-t	1.0154	0.9736	69.1N	68.7W	0	22		
6465	324	0709 Nov 06	12:41:08	3705	-15957	83	A	-p	-0.9319	0.9814	79.2S	67.4W	21	73	186	01m10s
6466	324	0710 May 03	16:06:47	3701	-15951	88	A	-p	0.2731	0.9755	32.3N	51.2W	74	167	91	02m38s
6467	324	0710 Oct 27	01:37:19	3696	-15945	93	T	-n	-0.2087	1.0414	25.7S	164.3E	78	14	141	03m41s
6468	324	0711 Apr 22	18:15:28	3692	-15939	98	A	p-	-0.4937	0.9427	15.6S	72.2W	60	346	242	07m29s
6469	324	0711 Oct 16	17:34:15	3688	-15933	103	T	n-	0.4719	1.0482	17.1N	64.0W	62	196	181	04m21s
6470	324	0712 Apr 10	18:59:59	3684	-15927	108	P	t-	-1.2234	0.5805	71.4S	29.1W	0	301		
6471	324	0712 Oct 05	07:48:54	3679	-15921	113	P	t-	1.1989	0.6297	71.6N	146.2E	0	250		
6472	324	0713 Mar 01	14:25:35	3676	-15916	80	T	-p	0.8151	1.0179	45.7N	38.0W	35	155	105	01m27s
6473	324	0713 Aug 26	00:46:30	3671	-15910	85	A	-p	-0.8762	0.9401	49.0S	162.1E	28	24	458	05m52s
6474	324	0714 Feb 19	05:26:24	3667	-15904	90	T	nn	0.0914	1.0574	5.2S	116.1E	85	164	190	05m15s
6475	324	0714 Aug 15	00:59:30	3663	-15898	95	A	nn	-0.1449	0.9420	5.4N	179.0E	82	14	217	07m46s
6476	324	0715 Feb 08	21:28:51	3659	-15892	100	T	p-	-0.6040	1.0377	49.8S	110.1W	53	339	159	02m50s
6477	324	0715 Aug 04	03:46:01	3654	-15886	105	A	p-	0.6044	0.9767	53.3N	150.8E	53	198	104	02m03s
6478	324	0715 Dec 30	19:02:01	3651	-15881	72	P	-t	1.5228	0.0660	67.0N	88.0W	0	180		
6479	324	0716 Jan 29	09:52:15	3650	-15880	110	P	t-	-1.3537	0.3514	69.7S	162.2W	0	212		
6480	324	0716 Jun 24	05:46:51	3646	-15875	77	P	-t	-1.2484	0.5420	66.3S	116.4E	0	353		

Cat Num	Canon Plate	Calendar Date	TD of Greatest Eclipse	ΔT s	Luna Num	Saros Num	Ecl. Type	QLE	Gamma	Ecl. Mag.	Lat. °	Long. °	Sun Alt °	Sun Azm °	Path Width km	Central Line Dur.
6481	325	0716 Jul 23	13:47:09	3646	-15874	115	P	t-	1.3140	0.4180	68.9N	146.8E	0	339		
6482	325	0716 Dec 18	20:30:55	3642	-15869	82	A	-p	0.8874	0.9112	39.0N	106.1W	27	186	740	10m35s
6483	325	0717 Jun 13	22:52:28	3638	-15863	87	T	-p	-0.5035	1.0779	6.5S	144.6W	60	353	291	07m15s
6484	325	0717 Dec 07	19:39:54	3634	-15857	92	A	nn	0.2059	0.9282	11.6S	98.6W	78	190	275	09m43s
6485	325	0718 Jun 03	14:57:54	3629	-15851	97	T	n-	0.2478	1.0458	36.9N	34.1W	75	166	157	03m56s
6486	325	0718 Nov 26	23:55:08	3625	-15845	102	A	p-	-0.4900	0.9742	50.1S	177.1W	60	22	106	02m15s
6487	325	0719 May 24	01:39:12	3621	-15839	107	P	t-	1.0620	0.8714	63.6N	27.5E	0	35		
6488	325	0719 Oct 18	00:50:25	3617	-15834	74	Pe	-t	1.5311	0.0071	61.5N	111.5W	0	247		
6489	325	0719 Nov 16	11:25:02	3616	-15833	112	P	t-	-1.1408	0.7402	63.0S	113.8W	0	138		
6490	325	0720 Apr 12	13:42:10	3613	-15828	79	A-	-t	-1.0037	0.9531	61.2S	60.5E	0	291	-	-
6491	325	0720 Oct 06	16:51:39	3609	-15822	84	T	-p	0.8749	1.0355	45.5N	24.8W	29	222	241	02m32s
6492	325	0721 Apr 01	15:23:02	3604	-15816	89	A	nn	-0.2396	0.9666	6.3S	28.5W	76	331	124	03m34s
6493	325	0721 Sep 26	05:53:18	3600	-15810	94	A	-n	0.1845	0.9988	6.8N	109.5E	79	209	4	00m07s
6494	325	0722 Mar 21	23:36:39	3596	-15804	99	T	p-	0.5141	1.0162	28.5N	174.2W	59	146	64	01m23s
6495	325	0722 Sep 15	12:13:45	3592	-15798	104	A	p-	-0.5655	0.9477	28.0S	8.2W	55	33	228	05m19s
6496	325	0723 Feb 10	04:59:52	3588	-15793	71	P	-t	-1.4042	0.2457	61.8S	119.6W	0	241		
6497	325	0723 Mar 11	14:00:20	3587	-15792	109	P	t-	1.1992	0.6358	60.9N	98.1W	0	95		
6498	325	0723 Sep 04	12:42:29	3583	-15786	114	P	t-	-1.2996	0.4512	60.9S	74.9W	0	77		
6499	325	0724 Jan 30	20:24:34	3580	-15781	81	T	-p	-0.7552	1.0207	59.2S	72.8W	41	315	107	01m23s
6500	325	0724 Jul 25	01:16:50	3575	-15775	86	A	-p	0.8066	0.9906	66.1N	141.6W	36	231	56	00m39s
6501	326	0725 Jan 19	06:56:11	3571	-15769	91	A	nn	-0.0624	0.9696	23.1S	95.8E	86	341	109	03m14s
6502	326	0725 Jul 14	13:16:42	3567	-15763	96	T	nn	0.0146	1.0540	22.3N	3.2W	89	194	179	04m53s
6503	326	0726 Jan 08	10:02:27	3563	-15757	101	A	p-	0.6558	0.9219	18.0N	36.8E	49	165	387	10m16s
6504	326	0726 Jul 04	05:53:59	3558	-15751	106	T	p-	-0.7206	1.0745	22.6S	97.7E	44	12	347	06m31s
6505	326	0726 Dec 28	09:04:22	3554	-15745	111	P	t-	1.3278	0.4046	64.8N	35.9E	0	158		
6506	326	0727 May 25	14:15:57	3551	-15740	78	P	-t	1.0925	0.8295	68.1N	171.2E	0	11		
6507	326	0727 Jun 23	22:34:35	3550	-15739	116	Pb	t-	-1.4763	0.1082	65.4S	161.5W	0	16		
6508	326	0727 Nov 17	21:12:00	3546	-15734	83	A	-p	-0.9352	0.9827	83.3S	146.8E	20	89	177	01m03s
6509	326	0728 May 13	22:54:08	3542	-15728	88	A	-p	0.3517	0.9737	39.9N	153.9W	69	170	101	02m43s
6510	326	0728 Nov 06	10:18:56	3538	-15722	93	T	-n	-0.2182	1.0409	29.8S	33.8E	77	12	140	03m36s
6511	326	0729 May 03	00:49:47	3534	-15716	98	A	p-	-0.4160	0.9448	7.7S	174.3W	65	349	223	07m31s
6512	326	0729 Oct 27	02:10:21	3529	-15710	103	T	n-	0.4574	1.0448	12.6N	164.7E	63	193	167	04m11s
6513	326	0730 Apr 22	01:50:35	3525	-15704	108	P	t-	-1.1522	0.7044	70.8S	145.8W	0	314		
6514	326	0730 Oct 16	16:04:50	3521	-15698	113	P	t-	1.1824	0.6590	71.2N	7.8E	0	236		
6515	326	0731 Mar 12	22:22:19	3517	-15693	80	T	-p	0.8516	1.0229	52.7N	164.2W	31	150	148	01m43s
6516	326	0731 Sep 06	07:54:58	3513	-15687	85	A	-t	-0.9299	0.9338	58.5S	43.3E	21	34	673	05m51s
6517	326	0732 Mar 01	13:43:50	3509	-15681	90	T	-n	0.1226	1.0611	0.7N	10.2W	83	163	202	05m32s
6518	326	0732 Aug 25	07:54:41	3505	-15675	95	A	nn	-0.2066	0.9401	1.8S	72.7E	78	16	226	07m55s
6519	326	0733 Feb 19	05:50:33	3500	-15669	100	T	p-	-0.5806	1.0389	44.0S	123.5E	54	338	160	03m03s
6520	326	0733 Aug 14	10:54:42	3496	-15663	105	A	p-	0.5391	0.9780	45.2N	42.0E	57	199	93	02m03s
6521	327	0734 Jan 10	03:17:02	3493	-15658	72	P	-t	1.5341	0.0469	68.1N	136.5E	0	169		
6522	327	0734 Feb 08	18:03:13	3492	-15657	110	P	t-	-1.3386	0.3781	70.5S	61.7E	0	224		
6523	327	0734 Jul 05	13:10:19	3489	-15652	77	P	-t	-1.3172	0.4098	67.3S	5.9W	0	3		
6524	327	0734 Aug 03	21:14:27	3488	-15651	115	P	t-	1.2467	0.5437	69.9N	22.2E	0	327		
6525	327	0734 Dec 30	04:32:13	3484	-15646	82	A	-p	0.8952	0.9112	40.9N	128.8E	26	181	768	10m28s
6526	327	0735 Jun 25	06:19:09	3480	-15640	87	T	-p	-0.5781	1.0756	11.9S	101.6E	55	357	300	07m02s
6527	327	0735 Dec 19	03:48:28	3476	-15634	92	A	nn	0.2107	0.9301	11.6S	139.0E	78	185	268	09m35s
6528	327	0736 Jun 13	22:11:01	3472	-15628	97	T	nn	0.1715	1.0428	33.5N	140.4W	80	172	145	03m52s
6529	327	0736 Dec 07	08:26:24	3467	-15622	102	A	p-	-0.4866	0.9767	51.9S	58.8E	61	15	95	02m01s
6530	327	0737 Jun 03	08:25:50	3463	-15616	107	A	t-	0.9844	0.9737	72.3N	71.2W	9	39	595	01m30s
6531	327	0737 Nov 26	20:12:02	3459	-15610	112	P	t-	-1.1347	0.7516	63.9S	104.4E	0	148		
6532	327	0738 Apr 23	20:21:34	3456	-15605	79	P	-t	-1.0767	0.8301	61.7S	48.8W	0	300		
6533	327	0738 Oct 18	01:19:49	3451	-15599	84	T	-p	0.8964	1.0302	45.0N	154.4W	26	218	226	02m13s
6534	327	0739 Apr 12	22:26:36	3447	-15593	89	A	nn	-0.3033	0.9717	5.9S	134.5W	72	333	106	03m03s
6535	327	0739 Oct 07	13:55:43	3443	-15587	94	A	-n	0.2131	0.9926	4.1N	11.9W	78	208	26	00m43s
6536	327	0740 Apr 01	07:15:31	3439	-15581	99	T	p-	0.4609	1.0229	29.8N	71.4E	62	147	87	01m55s
6537	327	0740 Sep 25	19:42:24	3435	-15575	104	A	p-	-0.5269	0.9431	29.9S	120.9W	58	34	243	05m45s
6538	327	0741 Feb 20	13:25:25	3431	-15570	71	P	-t	-1.4296	0.1966	61.4S	104.5E	0	250		
6539	327	0741 Mar 21	22:02:37	3430	-15569	109	P	t-	1.1553	0.7203	60.9N	132.0E	0	86		
6540	327	0741 Sep 14	19:52:17	3426	-15563	114	P	t-	-1.2506	0.5335	60.8S	168.3E	0	86		

Cat Num	Canon Plate	Calendar Date	TD of Greatest Eclipse	ΔT s	Luna Num	Saros Num	Ecl. Type	QLE	Gamma	Ecl. Mag.	Lat. °	Long. °	Sun Alt °	Sun Azm °	Path Width km	Central Line Dur.
6541	328	0742 Feb 10	04:48:57	3423	-15558	81	T	-p	-0.7774	1.0203	56.2S	162.8E	39	312	109	01m22s
6542	328	0742 Aug 05	08:27:12	3419	-15552	86	A	-p	0.8707	0.9901	65.6N	122.9E	29	243	71	00m40s
6543	328	0743 Jan 30	15:05:55	3414	-15546	91	A	nn	-0.0792	0.9690	21.0S	26.3W	85	337	112	03m15s
6544	328	0743 Jul 25	20:46:00	3410	-15540	96	Tm	nn	0.0819	1.0547	23.7N	114.5W	85	200	181	04m46s
6545	328	0744 Jan 19	17:56:37	3406	-15534	101	A	p-	0.6418	0.9227	18.4N	84.1W	50	161	375	09m52s
6546	328	0744 Jul 14	13:26:41	3402	-15528	106	T	p-	-0.6512	1.0741	18.0S	17.5W	49	16	314	06m30s
6547	328	0745 Jan 07	17:04:29	3398	-15522	111	P	t-	1.3168	0.4226	63.8N	94.4W	0	148		
6548	328	0745 Jun 04	21:22:42	3394	-15517	78	P	-t	1.1724	0.6809	67.1N	52.9E	0	0		
6549	328	0745 Jul 04	05:53:18	3394	-15516	116	P	t-	-1.4085	0.2381	64.5S	78.2E	0	26		
6550	328	0745 Nov 28	05:47:42	3390	-15511	83	A	-p	-0.9354	0.9847	86.7S	19.2W	20	125	157	00m54s
6551	328	0746 May 25	05:34:14	3386	-15505	88	A	-p	0.4356	0.9713	47.5N	106.5E	64	174	114	02m48s
6552	328	0746 Nov 17	19:06:17	3382	-15499	93	T	-n	-0.2232	1.0409	33.1S	97.3W	77	8	140	03m33s
6553	328	0747 May 14	07:16:41	3378	-15493	98	A	pn	-0.3320	0.9466	0.1N	86.1E	71	352	208	07m27s
6554	328	0747 Nov 07	10:53:06	3373	-15487	103	T	n-	0.4487	1.0416	8.9N	32.0E	63	190	155	04m00s
6555	328	0748 May 02	08:36:33	3369	-15481	108	P	t-	-1.0762	0.8378	70.0S	99.3E	0	326		
6556	328	0748 Oct 27	00:26:16	3365	-15475	113	P	t-	1.1709	0.6789	70.5N	131.5W	0	223		
6557	328	0749 Mar 23	06:13:53	3362	-15470	80	T	-p	0.8932	1.0275	60.6N	68.0E	26	141	208	01m53s
6558	328	0749 Sep 16	15:10:13	3357	-15464	85	As	-t	-0.9773	0.9269	68.3S	89.2W	11	56	-	05m42s
6559	328	0750 Mar 12	21:54:22	3353	-15458	90	T	-n	0.1600	1.0647	7.0N	135.1W	81	162	214	05m46s
6560	328	0750 Sep 05	14:57:59	3349	-15452	95	A	nn	-0.2620	0.9381	8.9S	35.8W	75	17	238	08m00s
6561	329	0751 Mar 02	14:04:34	3345	-15446	100	T	p-	-0.5512	1.0401	37.8S	2.1W	56	338	160	03m16s
6562	329	0751 Aug 25	18:12:50	3341	-15440	105	A	p-	0.4805	0.9789	37.4N	69.8W	61	200	85	02m04s
6563	329	0752 Jan 21	11:25:25	3337	-15435	72	Pe	-t	1.5500	0.0200	69.1N	2.1E	0	157		
6564	329	0752 Feb 20	02:05:24	3337	-15434	110	P	t-	-1.3174	0.4159	71.2S	72.7W	0	238		
6565	329	0752 Jul 15	20:39:39	3333	-15429	77	P	-t	-1.3816	0.2859	68.3S	130.1W	0	14		
6566	329	0752 Aug 14	04:50:37	3333	-15428	115	P	t-	1.1859	0.6576	70.7N	105.2W	0	315		
6567	329	0753 Jan 09	12:29:12	3329	-15423	82	A	-p	0.9060	0.9116	43.6N	4.4E	25	175	805	10m06s
6568	329	0753 Jul 05	13:47:40	3325	-15417	87	T	-p	-0.6509	1.0725	18.1S	13.4W	49	1	310	06m38s
6569	329	0753 Dec 29	11:55:43	3321	-15411	92	A	nn	0.2170	0.9326	10.6S	16.8E	78	181	258	09m16s
6570	329	0754 Jun 25	05:23:15	3317	-15405	97	Tm	nn	0.0955	1.0391	29.1N	112.7E	84	177	132	03m43s
6571	329	0754 Dec 18	16:57:41	3313	-15399	102	A	p-	-0.4824	0.9799	52.5S	64.5W	61	8	82	01m45s
6572	329	0755 Jun 14	15:09:50	3308	-15393	107	A	p-	0.9053	0.9746	83.1N	112.7W	25	98	217	01m38s
6573	329	0755 Dec 08	05:00:55	3304	-15387	112	P	t-	-1.1301	0.7604	64.8S	38.2W	0	158		
6574	329	0756 May 04	02:54:59	3301	-15382	79	P	-t	-1.1544	0.6982	62.3S	156.8W	0	309		
6575	329	0756 Oct 28	09:53:47	3297	-15376	84	T	-p	0.9127	1.0254	44.9N	73.6E	24	214	208	01m55s
6576	329	0757 Apr 23	05:24:52	3293	-15370	89	A	-p	-0.3722	0.9766	6.3S	120.8E	68	335	89	02m34s
6577	329	0757 Oct 17	22:05:15	3289	-15364	94	A	-n	0.2349	0.9868	1.3N	135.1W	76	206	48	01m20s
6578	329	0758 Apr 12	14:46:19	3284	-15358	99	T	p-	0.4002	1.0295	30.8N	40.4W	66	149	108	02m27s
6579	329	0758 Oct 07	03:19:35	3280	-15352	104	A	p-	-0.4959	0.9385	32.6S	124.4E	60	34	259	06m12s
6580	329	0759 Mar 03	21:42:22	3277	-15347	71	P	-t	-1.4620	0.1340	61.1S	29.1W	0	259		
6581	330	0759 Apr 02	05:56:17	3276	-15346	109	P	t-	1.1044	0.8190	61.0N	4.2E	0	77		
6582	330	0759 Sep 26	03:11:57	3272	-15340	114	P	t-	-1.2093	0.6027	60.8S	49.1E	0	95		
6583	330	0760 Feb 21	13:05:15	3269	-15335	81	T	-p	-0.8057	1.0199	53.6S	39.6E	36	310	113	01m22s
6584	330	0760 Aug 15	15:46:22	3265	-15329	86	A	-p	0.9284	0.9888	64.5N	24.9E	21	255	106	00m43s
6585	330	0761 Feb 09	23:08:52	3260	-15323	91	A	nn	-0.1007	0.9686	18.5S	146.9W	84	334	113	03m15s
6586	330	0761 Aug 05	04:19:51	3256	-15317	96	T	nn	0.1459	1.0548	24.1N	133.1E	81	204	183	04m38s
6587	330	0762 Jan 30	01:45:52	3252	-15311	101	A	p-	0.6242	0.9239	19.2N	156.4E	51	157	360	09m22s
6588	330	0762 Jul 25	21:02:22	3248	-15305	106	T	p-	-0.5842	1.0729	14.8S	132.8W	54	19	289	06m20s
6589	330	0763 Jan 19	01:00:28	3244	-15299	111	P	t-	1.3022	0.4470	63.0N	136.6E	0	138		
6590	330	0763 Jun 16	04:27:29	3241	-15294	78	P	-t	1.2526	0.5326	66.1N	64.5W	0	350		
6591	330	0763 Jul 15	13:14:24	3240	-15293	116	P	t-	-1.3432	0.3625	63.7S	42.4W	0	35		
6592	330	0763 Dec 09	14:25:24	3237	-15288	83	A	-p	-0.9348	0.9872	86.6S	127.7E	20	208	131	00m45s
6593	330	0764 Jun 04	12:12:18	3232	-15282	88	A	-p	0.5199	0.9685	54.5N	9.5E	58	180	133	02m55s
6594	330	0764 Nov 28	03:55:17	3228	-15276	93	T	-n	-0.2270	1.0412	35.4S	131.9E	77	3	142	03m32s
6595	330	0765 May 24	13:41:46	3224	-15270	98	A	nn	-0.2465	0.9480	7.3N	12.3W	76	356	197	07m17s
6596	330	0765 Nov 17	19:39:02	3220	-15264	103	T	n-	0.4430	1.0389	6.0N	101.2W	64	186	145	03m49s
6597	330	0766 May 13	15:18:08	3216	-15258	108	As	t-	-0.9955	0.9562	66.1S	17.0W	3	341	-	03m35s
6598	330	0766 Nov 07	08:52:52	3212	-15252	113	P	t-	1.1644	0.6897	69.6N	88.5E	0	210		
6599	330	0767 Apr 03	13:57:12	3209	-15247	80	T	-t	0.9422	1.0311	69.2N	67.0W	19	125	319	01m55s
6600	330	0767 Sep 27	22:34:38	3205	-15241	85	A-	-t	-1.0165	0.9270	72.0S	112.6E	0	100	-	-

Cat Num	Canon Plate	Calendar Date	TD of Greatest Eclipse	ΔT s	Luna Num	Saros Num	Ecl. Type	QLE	Gamma	Ecl. Mag.	Lat. °	Long. °	Sun Alt °	Sun Azm °	Path Width km	Central Line Dur.
6601	331	0768 Mar 23	05:55:58	3200	-15235	90	T	-n	0.2050	1.0679	13.9N	102.1E	78	162	226	05m56s
6602	331	0768 Sep 15	22:12:37	3196	-15229	95	A	nn	-0.3084	0.9362	15.8S	147.1W	72	18	249	08m00s
6603	331	0769 Mar 12	22:09:04	3192	-15223	100	T	p-	-0.5145	1.0413	31.1S	126.0W	59	339	160	03m32s
6604	331	0769 Sep 05	01:40:10	3188	-15217	105	A	p-	0.4288	0.9795	30.0N	175.9E	64	200	80	02m06s
6605	331	0770 Mar 02	09:57:41	3184	-15211	110	P	t-	-1.2892	0.4661	71.7S	154.8E	0	251		
6606	331	0770 Jul 27	04:12:53	3181	-15206	77	P	-t	-1.4428	0.1682	69.2S	104.3E	0	25		
6607	331	0770 Aug 25	12:34:31	3180	-15205	115	P	t-	1.1309	0.7603	71.3N	125.0E	0	302		
6608	331	0771 Jan 20	20:20:40	3177	-15200	82	A	-p	0.9210	0.9125	47.5N	119.4W	22	169	863	09m31s
6609	331	0771 Jul 16	21:18:25	3173	-15194	87	T	-p	-0.7214	1.0684	25.2S	129.8W	44	6	322	06m04s
6610	331	0772 Jan 09	20:01:18	3169	-15188	92	A	nn	0.2248	0.9358	8.7S	105.1W	77	176	245	08m46s
6611	331	0772 Jul 05	12:35:35	3165	-15182	97	T	nn	0.0204	1.0347	23.9N	4.9E	89	181	117	03m27s
6612	331	0772 Dec 29	01:29:11	3160	-15176	102	A	p-	-0.4778	0.9836	51.9S	172.2E	61	1	66	01m26s
6613	331	0773 Jun 24	21:52:22	3156	-15170	107	A	p-	0.8257	0.9734	79.1N	148.0W	34	165	171	01m53s
6614	331	0773 Dec 18	13:51:01	3152	-15164	112	P	t-	-1.1264	0.7678	65.9S	178.5E	0	168		
6615	331	0774 May 15	09:22:59	3149	-15159	79	P	-t	-1.2361	0.5588	63.0S	96.4E	0	318		
6616	331	0774 Nov 08	18:33:10	3145	-15153	84	T	-p	0.9242	1.0210	45.0N	60.4W	22	209	186	01m38s
6617	331	0775 May 04	12:19:13	3141	-15147	89	A	-p	-0.4448	0.9812	7.7S	16.8E	64	338	74	02m06s
6618	331	0775 Oct 29	06:19:41	3137	-15141	94	A	-n	0.2517	0.9814	1.3S	100.4E	75	204	68	01m57s
6619	331	0776 Apr 22	22:13:29	3133	-15135	99	T	p-	0.3357	1.0356	31.4N	150.8W	70	151	127	02m58s
6620	331	0776 Oct 17	11:03:26	3129	-15129	104	A	p-	-0.4712	0.9343	35.6S	8.5E	62	33	275	06m37s
6621	332	0777 Mar 14	05:52:01	3125	-15124	71	Pe	-t	-1.5004	0.0595	61.0S	160.9W	0	268		
6622	332	0777 Apr 12	13:44:14	3125	-15123	109	P	t-	1.0485	0.9279	61.3N	122.2W	0	69		
6623	332	0777 Oct 06	10:39:54	3121	-15117	114	P	t-	-1.1742	0.6611	60.9S	72.1W	0	104		
6624	332	0778 Mar 03	21:12:24	3117	-15112	81	T	-p	-0.8411	1.0192	51.9S	81.3W	32	310	119	01m20s
6625	332	0778 Aug 26	23:14:42	3113	-15106	86	A	-p	0.9795	0.9864	63.1N	71.8W	11	269	250	00m50s
6626	332	0779 Feb 21	07:00:09	3109	-15100	91	A	nn	-0.1312	0.9685	16.1S	95.2E	82	332	114	03m14s
6627	332	0779 Aug 16	12:02:26	3105	-15094	96	T	-n	0.2033	1.0544	23.5N	18.3E	78	207	183	04m29s
6628	332	0780 Feb 10	09:25:34	3101	-15088	101	A	p-	0.5986	0.9257	20.2N	39.7E	53	154	341	08m49s
6629	332	0780 Aug 05	04:43:27	3097	-15082	106	T	p-	-0.5220	1.0708	13.0S	110.9E	58	23	267	06m03s
6630	332	0781 Jan 29	08:50:22	3093	-15076	111	P	t-	1.2820	0.4807	62.2N	9.5E	0	128		
6631	332	0781 Jun 26	11:30:45	3090	-15071	78	P	-t	1.3330	0.3849	65.1N	178.9E	0	340		
6632	332	0781 Jul 25	20:37:53	3089	-15070	116	P	t-	-1.2802	0.4816	62.9S	163.3W	0	44		
6633	332	0781 Dec 19	23:04:25	3086	-15065	83	A	-p	-0.9337	0.9903	83.2S	37.2W	21	242	97	00m34s
6634	332	0782 Jun 15	18:45:24	3082	-15059	88	A	-p	0.6075	0.9649	61.0N	83.3W	52	188	160	03m02s
6635	332	0782 Dec 09	12:47:43	3078	-15053	93	T	-n	-0.2283	1.0421	36.7S	0.7E	77	358	145	03m34s
6636	332	0783 Jun 04	20:03:18	3074	-15047	98	A	nn	-0.1578	0.9489	13.9N	109.0W	81	360	190	07m01s
6637	332	0783 Nov 29	04:28:09	3070	-15041	103	T	n-	0.4403	1.0365	3.9N	124.9E	64	182	137	03m39s
6638	332	0784 May 23	21:58:34	3066	-15035	108	A	t-	-0.9126	0.9653	44.6S	132.5W	24	354	309	03m30s
6639	332	0784 Nov 17	17:23:51	3062	-15029	113	P	t-	1.1619	0.6934	68.7N	52.0W	0	198		
6640	332	0785 Apr 13	21:35:47	3059	-15024	80	Tn	-t	0.9956	1.0317	72.8N	113.4E	3	63	-	01m40s
6641	333	0785 Oct 08	06:06:06	3055	-15018	85	P	-t	-1.0495	0.8699	71.7S	15.0W	0	114		
6642	333	0786 Apr 03	13:51:28	3051	-15012	90	T	-n	0.2552	1.0709	21.0N	19.3W	75	162	238	06m02s
6643	333	0786 Sep 27	05:36:17	3047	-15006	95	A	-n	-0.3475	0.9344	22.4S	99.6E	70	19	260	07m57s
6644	333	0787 Mar 24	06:05:01	3043	-15000	100	T	p-	-0.4715	1.0422	24.2S	111.9E	62	341	159	03m46s
6645	333	0787 Sep 16	09:18:23	3039	-14994	105	A	p-	0.3853	0.9799	23.1N	58.8E	67	199	77	02m07s
6646	333	0788 Mar 12	17:41:00	3035	-14988	110	P	t-	-1.2547	0.5273	71.9S	24.3E	0	265		
6647	333	0788 Aug 06	11:52:52	3031	-14983	77	Pe	-t	-1.4989	0.0607	70.0S	23.7W	0	37		
6648	333	0788 Sep 04	20:26:59	3031	-14982	115	P	t-	1.0824	0.8508	71.8N	7.6W	0	288		
6649	333	0789 Jan 31	04:05:08	3027	-14977	82	A	-p	0.9412	0.9136	52.7N	117.2E	19	162	984	08m47s
6650	333	0789 Jul 27	04:53:37	3023	-14971	87	T	-p	-0.7876	1.0636	32.9S	111.7E	38	10	338	05m22s
6651	333	0790 Jan 20	04:01:48	3019	-14965	92	A	nn	0.2375	0.9396	5.8S	133.9E	76	172	230	08m07s
6652	333	0790 Jul 16	19:49:25	3015	-14959	97	T	nn	-0.0525	1.0296	18.0N	104.1W	87	6	100	03m03s
6653	333	0791 Jan 09	09:57:36	3011	-14953	102	A	p-	-0.4700	0.9880	49.8S	48.9E	62	354	48	01m03s
6654	333	0791 Jul 06	04:35:40	3007	-14947	107	A	p-	0.7473	0.9710	71.2N	127.5E	41	184	158	02m14s
6655	333	0791 Dec 29	22:38:40	3004	-14941	112	P	t-	-1.1206	0.7792	67.0S	35.4E	0	179		
6656	333	0792 May 25	15:48:34	3000	-14936	79	P	-t	-1.3193	0.4161	63.8S	10.0W	0	327		
6657	333	0792 Jun 24	06:42:26	3000	-14935	117	Pb	t-	1.5320	0.0523	66.4N	80.5W	0	7		
6658	333	0792 Nov 19	03:16:47	2996	-14930	84	T	-t	0.9316	1.0172	45.2N	163.8E	21	203	162	01m23s
6659	333	0793 May 14	19:09:27	2992	-14924	89	A	-p	-0.5216	0.9852	10.3S	86.4W	58	342	61	01m41s
6660	333	0793 Nov 08	14:39:47	2988	-14918	94	A	-n	0.2631	0.9765	3.7S	25.5W	75	200	87	02m34s

Cat Num	Canon Plate	Calendar Date	TD of Greatest Eclipse	ΔT s	Luna Num	Saros Num	Ecl. Type	QLE	Gamma	Ecl. Mag.	Lat. °	Long. °	Sun Alt °	Sun Azm °	Path Width km	Central Line Dur.
6661	334	0794 May 04	05:34:59	2984	-14912	99	T	n-	0.2658	1.0413	31.2N	100.6E	74	155	143	03m30s
6662	334	0794 Oct 28	18:54:46	2980	-14906	104	A	p-	-0.4528	0.9305	38.8S	108.9W	63	31	290	07m02s
6663	334	0795 Apr 23	21:24:56	2976	-14900	109	Tn	t-	0.9863	1.0587	65.2N	130.7E	9	76	-	02m58s
6664	334	0795 Oct 17	18:17:55	2973	-14894	114	P	t-	-1.1468	0.7065	61.3S	164.1E	0	113		
6665	334	0796 Mar 14	05:11:10	2969	-14889	81	T	-t	-0.8830	1.0181	51.4S	160.4E	28	309	129	01m16s
6666	334	0796 Sep 06	06:51:48	2965	-14883	86	P	-t	1.0242	0.9435	61.0N	172.4W	0	280		
6667	334	0797 Mar 03	14:43:35	2961	-14877	91	A	nn	-0.1675	0.9685	13.7S	20.8W	80	331	115	03m14s
6668	334	0797 Aug 26	19:51:37	2957	-14871	96	T	-n	0.2556	1.0535	22.2N	98.4W	75	209	182	04m21s
6669	334	0798 Feb 20	16:57:03	2954	-14865	101	A	p-	0.5668	0.9279	21.3N	74.5W	55	151	321	08m15s
6670	334	0798 Aug 16	12:29:40	2950	-14859	106	T	p-	-0.4642	1.0680	12.4S	6.5W	62	26	248	05m42s
6671	334	0799 Feb 09	16:34:07	2946	-14853	111	P	t-	1.2562	0.5241	61.6N	116.0W	0	119		
6672	334	0799 Jul 07	18:35:08	2942	-14848	78	P	-t	1.4112	0.2426	64.2N	62.4E	0	331		
6673	334	0799 Aug 06	04:06:08	2942	-14847	116	P	t-	-1.2217	0.5910	62.2S	74.8E	0	54		
6674	334	0799 Dec 31	07:41:20	2938	-14842	83	A	-p	-0.9351	0.9939	79.1S	179.4W	20	254	62	00m21s
6675	334	0800 Jun 26	01:19:56	2935	-14836	88	A	-p	0.6928	0.9608	66.4N	171.9W	46	201	198	03m11s
6676	334	0800 Dec 19	21:39:25	2931	-14830	93	T	-n	-0.2307	1.0434	37.0S	130.0W	76	352	149	03m38s
6677	334	0801 Jun 15	02:24:30	2927	-14824	98	A	nn	-0.0686	0.9495	19.8N	155.3E	86	4	186	06m41s
6678	334	0801 Dec 09	13:18:07	2923	-14818	103	T	n-	0.4386	1.0346	2.7N	9.0W	64	178	130	03m29s
6679	334	0802 Jun 04	04:38:09	2919	-14812	108	A	t-	-0.8275	0.9711	33.2S	121.9E	34	360	185	03m12s
6680	334	0802 Nov 29	01:56:35	2915	-14806	113	P	t-	1.1616	0.6933	67.6N	167.8E	0	186		
6681	335	0803 Apr 25	05:07:36	2912	-14801	80	P	-t	1.0553	0.9074	70.5N	21.3W	0	43		
6682	335	0803 May 24	13:35:52	2911	-14800	118	Pb	t-	-1.5325	0.0122	68.1S	0.7W	0	348		
6683	335	0803 Oct 19	13:46:27	2908	-14795	85	P	-t	-1.0747	0.8264	71.2S	144.4W	0	128		
6684	335	0804 Apr 13	21:39:23	2904	-14789	90	T	-n	0.3118	1.0732	28.3N	138.5W	72	163	250	06m03s
6685	335	0804 Oct 07	13:09:48	2900	-14783	95	A	-n	-0.3789	0.9329	28.6S	15.7W	68	19	270	07m52s
6686	335	0805 Apr 03	13:51:56	2896	-14777	100	T	n-	-0.4214	1.0429	17.1S	8.1W	65	342	157	04m00s
6687	335	0805 Sep 26	17:06:36	2892	-14771	105	A	p-	0.3492	0.9802	16.6N	60.7W	69	198	75	02m09s
6688	335	0806 Mar 24	01:12:11	2888	-14765	110	P	t-	-1.2111	0.6048	71.9S	103.2W	0	279		
6689	335	0806 Sep 16	04:28:29	2884	-14759	115	P	t-	1.0409	0.9284	72.0N	142.7W	0	274		
6690	335	0807 Feb 11	11:42:08	2881	-14754	82	A	-t	0.9674	0.9147	59.8N	7.8W	14	152	1318	07m53s
6691	335	0807 Aug 07	12:33:13	2877	-14748	87	T	-p	-0.8498	1.0579	41.4S	9.4W	31	16	361	04m35s
6692	335	0808 Jan 31	11:57:31	2873	-14742	92	A	nn	0.2548	0.9439	1.9S	13.8E	75	169	214	07m22s
6693	335	0808 Jul 27	03:06:05	2869	-14736	97	T	nn	-0.1222	1.0241	11.7N	145.4E	83	9	83	02m33s
6694	335	0809 Jan 19	18:23:53	2866	-14730	102	A	p-	-0.4598	0.9929	46.6S	75.0W	62	349	28	00m38s
6695	335	0809 Jul 16	11:18:57	2862	-14724	107	A	p-	0.6696	0.9679	62.9N	31.3E	48	191	156	02m41s
6696	335	0810 Jan 09	07:25:19	2858	-14718	112	P	t-	-1.1139	0.7924	68.0S	108.0W	0	191		
6697	335	0810 Jun 05	22:12:29	2855	-14713	79	P	-t	-1.4034	0.2714	64.7S	116.3W	0	336		
6698	335	0810 Jul 05	13:05:52	2854	-14712	117	P	t-	1.4484	0.1957	67.4N	172.4E	0	357		
6699	335	0810 Nov 30	12:02:24	2851	-14707	84	T	-t	0.9373	1.0139	45.7N	27.2E	20	197	138	01m08s
6700	335	0811 May 26	01:59:01	2847	-14701	89	A	-p	-0.5998	0.9887	14.1S	170.1E	53	346	49	01m17s
6701	336	0811 Nov 19	23:02:52	2843	-14695	94	A	-n	0.2710	0.9722	5.6S	152.0W	74	196	103	03m10s
6702	336	0812 May 14	12:55:16	2839	-14689	99	T	n-	0.1939	1.0464	30.3N	7.6W	79	160	157	04m00s
6703	336	0812 Nov 08	02:49:59	2835	-14683	104	A	p-	-0.4381	0.9272	41.9S	133.5E	64	27	303	07m26s
6704	336	0813 May 04	05:01:53	2831	-14677	109	T	t-	0.9209	1.0659	68.6N	43.5E	23	101	556	03m34s
6705	336	0813 Oct 28	02:03:14	2828	-14671	114	P	t-	-1.1249	0.7426	61.8S	38.4E	0	122		
6706	336	0814 Mar 25	12:59:04	2824	-14666	81	T	-t	-0.9332	1.0161	52.9S	46.7E	21	307	152	01m07s
6707	336	0814 Sep 17	14:39:24	2821	-14660	86	P	-t	1.0612	0.8771	60.9N	61.4E	0	272		
6708	336	0815 Mar 14	22:15:01	2817	-14654	91	A	-n	-0.2128	0.9684	11.7S	133.7W	78	330	116	03m15s
6709	336	0815 Sep 07	03:50:18	2813	-14648	96	T	-n	0.3003	1.0522	20.3N	142.1E	72	210	180	04m14s
6710	336	0816 Mar 03	00:17:49	2809	-14642	101	A	p-	0.5262	0.9304	22.6N	174.4E	58	149	300	07m45s
6711	336	0816 Aug 26	20:23:34	2805	-14636	106	T	n-	-0.4132	1.0647	12.8S	125.7W	66	28	230	05m20s
6712	336	0817 Feb 20	00:10:20	2801	-14630	111	P	t-	1.2237	0.5794	61.2N	120.6E	0	110		
6713	336	0817 Jul 18	01:39:41	2798	-14625	78	Pe	-t	1.4880	0.1042	63.4N	53.8W	0	322		
6714	336	0817 Aug 16	11:38:18	2797	-14624	116	P	t-	-1.1670	0.6921	61.7S	47.8W	0	63		
6715	336	0818 Jan 10	16:17:09	2794	-14619	83	A	-p	-0.9382	0.9979	75.1S	43.4E	20	261	22	00m07s
6716	336	0818 Jul 07	07:53:36	2790	-14613	88	A	-p	0.7782	0.9561	70.2N	105.9E	39	219	256	03m22s
6717	336	0818 Dec 31	06:30:18	2787	-14607	93	T	-n	-0.2339	1.0452	36.3S	99.3E	76	347	155	03m44s
6718	336	0819 Jun 26	08:46:33	2783	-14601	98	Am	nn	0.0201	0.9495	24.7N	60.3E	89	188	185	06m22s
6719	336	0819 Dec 20	22:08:51	2779	-14595	103	T	n-	0.4377	1.0332	2.2N	143.0W	64	174	124	03m19s
6720	336	0820 Jun 14	11:19:09	2775	-14589	108	A	p-	-0.7426	0.9759	24.5S	17.9E	42	4	128	02m49s

Cat Num	Canon Plate	Calendar Date	TD of Greatest Eclipse	ΔT s	Luna Num	Saros Num	Ecl. Type	QLE	Gamma	Ecl. Mag.	Lat. °	Long. °	Sun Alt °	Sun Azm °	Path Width km	Central Line Dur.
6721	337	0820 Dec 09	10:30:28	2771	-14583	113	P	t-	1.1629	0.6904	66.5N	27.9E	0	175		
6722	337	0821 May 05	12:36:38	2768	-14578	80	P	-t	1.1177	0.7903	69.7N	146.6W	0	31		
6723	337	0821 Jun 03	20:45:44	2768	-14577	118	P	t-	-1.4545	0.1561	67.1S	119.9W	0	359		
6724	337	0821 Oct 29	21:33:18	2764	-14572	85	P	-t	-1.0942	0.7929	70.4S	85.1E	0	141		
6725	337	0822 Apr 25	05:20:58	2761	-14566	90	T	-n	0.3737	1.0750	35.8N	104.2E	68	164	262	05m58s
6726	337	0822 Oct 18	20:52:23	2757	-14560	95	A	-p	-0.4031	0.9318	34.4S	132.5W	66	18	278	07m45s
6727	337	0823 Apr 14	21:31:15	2753	-14554	100	T	n-	-0.3655	1.0431	9.9S	126.1W	68	345	154	04m11s
6728	337	0823 Oct 08	01:04:03	2749	-14548	105	A	n-	0.3201	0.9807	10.7N	177.7E	71	197	72	02m09s
6729	337	0824 Apr 03	08:35:03	2745	-14542	110	P	t-	-1.1616	0.6928	71.6S	131.6E	0	293		
6730	337	0824 Sep 26	12:39:01	2742	-14536	115	P	t-	1.0062	0.9929	71.9N	79.8E	0	260		
6731	337	0825 Feb 21	19:10:32	2738	-14531	82	A+	-t	1.0002	0.9516	71.5N	154.8W	0	119	-	-
6732	337	0825 Aug 17	20:18:04	2735	-14525	87	T	-t	-0.9071	1.0515	50.7S	134.1W	24	23	406	03m46s
6733	337	0826 Feb 10	19:46:24	2731	-14519	92	A	nn	0.2780	0.9487	2.8N	105.1W	74	166	196	06m32s
6734	337	0826 Aug 07	10:27:15	2727	-14513	97	T	-n	-0.1869	1.0180	5.0N	33.1E	79	12	63	01m56s
6735	337	0827 Jan 31	02:43:18	2723	-14507	102	A	p-	-0.4435	0.9983	42.3S	161.5E	63	345	7	00m09s
6736	337	0827 Jul 27	18:06:50	2719	-14501	107	A	p-	0.5966	0.9642	54.7N	69.8W	53	195	161	03m15s
6737	337	0828 Jan 20	16:07:03	2716	-14495	112	P	t-	-1.1035	0.8129	69.1S	109.3E	0	203		
6738	337	0828 Jun 16	04:36:44	2712	-14490	79	Pe	-t	-1.4870	0.1273	65.7S	137.0E	0	346		
6739	337	0828 Jul 15	19:32:47	2712	-14489	117	P	t-	1.3671	0.3348	68.4N	64.0E	0	346		
6740	337	0828 Dec 10	20:49:00	2709	-14484	84	T	-t	0.9416	1.0113	46.6N	110.0W	19	191	117	00m56s
6741	338	0829 Jun 05	08:47:46	2705	-14478	89	A	-p	-0.6791	0.9916	19.2S	66.2E	47	349	40	00m57s
6742	338	0829 Nov 30	07:29:09	2701	-14472	94	A	-n	0.2757	0.9685	6.9S	80.8E	74	192	118	03m43s
6743	338	0830 May 25	20:11:21	2697	-14466	99	T	nn	0.1178	1.0508	28.4N	114.9W	83	164	170	04m31s
6744	338	0830 Nov 19	10:51:10	2694	-14460	104	A	p-	-0.4289	0.9245	44.7S	15.1E	64	23	314	07m47s
6745	338	0831 May 15	12:34:04	2690	-14454	109	T	p-	0.8514	1.0705	69.6N	50.4W	31	120	439	04m00s
6746	338	0831 Nov 08	09:55:47	2686	-14448	114	P	t-	-1.1085	0.7696	62.4S	89.2W	0	131		
6747	338	0832 Apr 04	20:39:15	2683	-14443	81	T	-t	-0.9888	1.0120	58.4S	56.1W	8	298	305	00m47s
6748	338	0832 Sep 27	22:36:40	2679	-14437	86	P	-t	1.0908	0.8238	60.9N	67.2W	0	263		
6749	338	0833 Mar 25	05:37:48	2675	-14431	91	A	-n	-0.2645	0.9683	10.1S	115.4E	75	331	117	03m18s
6750	338	0833 Sep 17	11:56:03	2672	-14425	96	T	-n	0.3394	1.0507	18.0N	20.3E	70	211	178	04m08s
6751	338	0834 Mar 14	07:30:30	2668	-14419	101	A	p-	0.4792	0.9331	23.9N	65.8E	61	148	280	07m17s
6752	338	0834 Sep 07	04:24:14	2664	-14413	106	T	n-	-0.3680	1.0609	14.2S	113.6E	68	30	214	04m57s
6753	338	0835 Mar 03	07:38:42	2661	-14407	111	P	t-	1.1839	0.6476	60.9N	0.7W	0	101		
6754	338	0835 Aug 27	19:17:30	2657	-14401	116	P	t-	-1.1186	0.7805	61.3S	172.0W	0	71		
6755	338	0836 Jan 22	00:48:35	2654	-14396	83	H	-p	-0.9459	1.0021	71.2S	89.1W	18	265	23	00m07s
6756	338	0836 Jul 17	14:30:27	2650	-14390	88	A	-t	0.8599	0.9508	71.6N	29.0E	30	242	354	03m34s
6757	338	0837 Jan 10	15:17:40	2646	-14384	93	T	-n	-0.2405	1.0473	34.7S	30.7W	76	342	162	03m52s
6758	338	0837 Jul 06	15:11:47	2642	-14378	98	A	nn	0.1066	0.9491	28.5N	34.6W	84	194	188	06m05s
6759	338	0837 Dec 31	06:56:29	2639	-14372	103	T	n-	0.4348	1.0323	2.5N	83.7E	64	169	121	03m12s
6760	338	0838 Jun 25	18:01:51	2635	-14366	108	A	p-	-0.6578	0.9798	17.4S	85.4W	49	8	95	02m24s
6761	339	0838 Dec 20	19:02:58	2631	-14360	113	P	t-	1.1638	0.6885	65.5N	111.2W	0	164		
6762	339	0839 May 16	20:01:39	2628	-14355	80	P	-t	1.1841	0.6646	68.7N	89.6E	0	19		
6763	339	0839 Jun 15	03:56:24	2628	-14354	118	P	t-	-1.3758	0.3023	66.1S	121.2E	0	9		
6764	339	0839 Nov 10	05:25:44	2624	-14349	85	P	-t	-1.1088	0.7679	69.5S	46.1W	0	154		
6765	339	0840 May 05	12:57:06	2621	-14343	90	T	-p	0.4402	1.0759	43.3N	11.0W	64	166	274	05m46s
6766	339	0840 Oct 29	04:43:55	2617	-14337	95	A	-p	-0.4208	0.9311	39.4S	109.3E	65	16	284	07m35s
6767	339	0841 Apr 25	05:01:45	2613	-14331	100	T	n-	-0.3029	1.0429	2.8S	118.4E	72	347	150	04m17s
6768	339	0841 Oct 18	09:11:07	2610	-14325	105	A	n-	0.2980	0.9812	5.5N	53.9E	73	195	70	02m07s
6769	339	0842 Apr 14	15:47:20	2606	-14319	110	P	t-	-1.1039	0.7950	71.1S	9.3E	0	306		
6770	339	0842 Oct 07	20:58:30	2602	-14313	115	T	t-	0.9787	1.0229	65.2N	84.3W	11	223	403	01m30s
6771	339	0843 Mar 05	02:29:36	2599	-14308	82	P	-t	1.0404	0.8856	71.9N	80.9E	0	105		
6772	339	0843 Aug 29	04:09:48	2595	-14302	87	T	-t	-0.9582	1.0442	60.8S	94.1E	16	35	526	02m56s
6773	339	0844 Feb 22	03:29:45	2592	-14296	92	A	-n	0.3065	0.9538	8.3N	137.0E	72	164	177	05m41s
6774	339	0844 Aug 17	17:52:11	2588	-14290	97	H	-n	-0.2475	1.0117	1.8S	80.5W	76	14	42	01m15s
6775	339	0845 Feb 10	10:58:55	2584	-14284	102	H	p-	-0.4232	1.0041	37.2S	37.6E	65	343	16	00m22s
6776	339	0845 Aug 07	00:57:55	2581	-14278	107	A	p-	0.5268	0.9602	46.6N	173.5W	58	197	170	03m54s
6777	339	0846 Jan 31	00:44:07	2577	-14272	112	P	t-	-1.0889	0.8414	70.0S	32.9W	0	215		
6778	339	0846 Jul 27	02:04:20	2573	-14266	117	P	t-	1.2888	0.4684	69.3N	46.2W	0	335		
6779	339	0846 Dec 22	05:34:51	2570	-14261	84	T	-t	0.9462	1.0090	48.0N	112.8E	18	185	98	00m45s
6780	339	0847 Jun 16	15:39:22	2567	-14255	89	A	-p	-0.7570	0.9937	25.5S	39.1W	41	354	34	00m42s

Cat Num	Canon Plate	Calendar Date	TD of Greatest Eclipse	ΔT s	Luna Num	Saros Num	Ecl. Type	QLE	Gamma	Ecl. Mag.	Lat. °	Long. °	Sun Alt °	Sun Azm °	Path Width km	Central Line Dur.
6781	340	0847 Dec 11	15:54:14	2563	-14249	94	A	-n	0.2806	0.9654	7.4S	46.0W	74	188	130	04m13s
6782	340	0848 Jun 05	03:29:17	2559	-14243	99	T	nn	0.0424	1.0545	25.6N	137.0E	87	170	180	04m59s
6783	340	0848 Nov 29	18:54:07	2556	-14237	104	A	p-	-0.4215	0.9224	46.7S	102.8W	65	18	323	08m06s
6784	340	0849 May 25	20:03:24	2552	-14231	109	T	p-	0.7794	1.0738	69.0N	145.3W	38	139	383	04m22s
6785	340	0849 Nov 18	17:53:57	2548	-14225	114	P	t-	-1.0962	0.7898	63.2S	141.5E	0	141		
6786	340	0850 Apr 16	04:09:53	2545	-14220	81	P	-t	-1.0515	0.9045	61.4S	164.2W	0	295		
6787	340	0850 May 15	12:49:29	2545	-14219	119	Pb	t-	1.5295	0.0066	63.2N	138.7W	0	41		
6788	340	0850 Oct 09	06:43:55	2542	-14214	86	P	-t	1.1133	0.7832	61.2N	161.7E	0	254		
6789	340	0851 Apr 05	12:48:54	2538	-14208	91	A	-p	-0.3251	0.9680	9.3S	7.6E	71	332	121	03m25s
6790	340	0851 Sep 28	20:12:09	2534	-14202	96	T	-n	0.3705	1.0492	15.4N	104.4W	68	210	174	04m03s
6791	340	0852 Mar 24	14:33:08	2531	-14196	101	A	p-	0.4238	0.9359	25.1N	39.7W	65	148	260	06m53s
6792	340	0852 Sep 17	12:32:17	2527	-14190	106	T	n-	-0.3291	1.0569	16.3S	9.0W	71	30	197	04m35s
6793	340	0853 Mar 13	14:59:08	2523	-14184	111	P	t-	1.1368	0.7290	60.8N	120.0W	0	92		
6794	340	0853 Sep 07	03:02:42	2520	-14178	116	P	t-	-1.0755	0.8578	61.0S	62.3E	0	80		
6795	340	0854 Feb 01	09:15:03	2517	-14173	83	T	-p	-0.9582	1.0065	67.8S	142.0E	16	267	80	00m22s
6796	340	0854 Jul 28	21:09:52	2513	-14167	88	A	-t	0.9386	0.9444	70.2N	43.9W	20	269	603	03m47s
6797	340	0855 Jan 22	00:01:28	2509	-14161	93	T	-n	-0.2502	1.0499	32.5S	160.3W	75	338	171	04m01s
6798	340	0855 Jul 17	21:41:01	2506	-14155	98	A	nn	0.1899	0.9484	31.2N	130.0W	79	199	193	05m52s
6799	340	0856 Jan 11	15:41:03	2502	-14149	103	T	n-	0.4296	1.0318	3.4N	48.7W	65	165	118	03m05s
6800	340	0856 Jul 06	00:49:20	2499	-14143	108	A	p-	-0.5758	0.9831	12.0S	171.0E	55	12	73	02m00s
6801	341	0856 Dec 31	03:34:01	2495	-14137	113	P	t-	1.1642	0.6875	64.5N	110.5E	0	154		
6802	341	0857 May 27	03:24:35	2492	-14132	80	P	-t	1.2525	0.5338	67.8N	33.1W	0	8		
6803	341	0857 Jun 25	11:08:23	2491	-14131	118	P	t-	-1.2972	0.4494	65.1S	2.5E	0	19		
6804	341	0857 Nov 20	13:23:00	2488	-14126	85	P	-t	-1.1192	0.7502	68.4S	177.9W	0	166		
6805	341	0858 May 16	20:29:15	2485	-14120	90	T	-p	0.5102	1.0760	50.7N	124.0W	59	170	287	05m30s
6806	341	0858 Nov 09	12:41:59	2481	-14114	95	A	-p	-0.4334	0.9309	43.7S	9.3W	64	12	287	07m24s
6807	341	0859 May 06	12:26:38	2478	-14108	100	T	nn	-0.2362	1.0420	4.2N	4.7E	76	350	144	04m17s
6808	341	0859 Oct 29	17:26:19	2474	-14102	105	A	n-	0.2818	0.9821	1.0N	71.5W	74	192	66	02m03s
6809	341	0860 Apr 24	22:52:40	2470	-14096	110	P	t-	-1.0414	0.9057	70.5S	110.7W	0	318		
6810	341	0860 Oct 18	05:25:01	2467	-14090	115	T	t-	0.9565	1.0241	58.5N	134.0E	16	209	286	01m42s
6811	341	0861 Mar 15	09:40:04	2464	-14085	82	P	-t	1.0873	0.8079	72.1N	41.5W	0	91		
6812	341	0861 Sep 08	12:08:25	2460	-14079	87	T-	-t	-1.0032	1.0053	71.9S	70.3W	0	77	-	-
6813	341	0862 Mar 04	11:03:38	2457	-14073	92	A	-n	0.3434	0.9591	14.5N	21.2E	70	162	158	04m49s
6814	341	0862 Aug 29	01:24:08	2453	-14067	97	H	-n	-0.3013	1.0051	8.7S	163.9E	72	16	19	00m33s
6815	341	0863 Feb 21	19:06:16	2449	-14061	102	H	p-	-0.3954	1.0103	31.3S	85.3W	67	341	39	00m57s
6816	341	0863 Aug 18	07:55:50	2446	-14055	107	A	p-	0.4631	0.9558	38.8N	80.3E	62	198	181	04m38s
6817	341	0864 Feb 11	09:14:03	2442	-14049	112	P	t-	-1.0684	0.8814	70.8S	173.9W	0	228		
6818	341	0864 Aug 06	08:42:56	2439	-14043	117	P	t-	1.2155	0.5928	70.2N	158.6W	0	323		
6819	341	0865 Jan 01	14:19:03	2436	-14038	84	T	-t	0.9518	1.0073	50.1N	24.4W	17	178	84	00m36s
6820	341	0865 Jun 26	22:31:49	2432	-14032	89	A	-p	-0.8348	0.9949	33.5S	145.7W	33	358	32	00m31s
6821	342	0865 Dec 22	00:20:03	2429	-14026	94	A	-n	0.2840	0.9629	7.1S	173.0W	74	183	140	04m37s
6822	342	0866 Jun 16	10:46:25	2425	-14020	99	T	nn	-0.0341	1.0574	21.8N	28.4E	88	353	189	05m24s
6823	342	0866 Dec 11	02:59:11	2421	-14014	104	A	p-	-0.4163	0.9210	47.9S	139.5E	65	12	329	08m22s
6824	342	0867 Jun 06	03:30:59	2418	-14008	109	T	p-	0.7058	1.0760	66.6N	117.5E	45	155	349	04m43s
6825	342	0867 Nov 30	01:57:13	2414	-14002	114	P	t-	-1.0880	0.8035	64.1S	10.7E	0	151		
6826	342	0868 Apr 26	11:34:04	2411	-13997	81	P	-t	-1.1188	0.7799	61.9S	75.3E	0	303		
6827	342	0868 May 25	20:11:13	2411	-13996	119	P	t-	1.4636	0.1327	64.0N	100.9E	0	32		
6828	342	0868 Oct 19	14:59:08	2408	-13991	86	P	-t	1.1302	0.7528	61.6N	28.6E	0	244		
6829	342	0869 Apr 15	19:52:37	2404	-13985	91	A	-p	-0.3909	0.9674	9.3S	98.4W	67	334	126	03m35s
6830	342	0869 Oct 09	04:35:25	2401	-13979	96	T	-n	0.3960	1.0475	12.8N	128.8E	67	208	171	04m00s
6831	342	0870 Apr 04	21:26:39	2397	-13973	101	A	p-	0.3608	0.9388	26.0N	142.3W	69	149	242	06m35s
6832	342	0870 Sep 28	20:47:50	2394	-13967	106	T	n-	-0.2968	1.0527	19.0S	133.4W	73	30	182	04m15s
6833	342	0871 Mar 24	22:12:34	2390	-13961	111	P	t-	1.0832	0.8227	60.8N	122.4E	0	83		
6834	342	0871 Sep 18	10:54:51	2387	-13955	116	P	t-	-1.0389	0.9224	60.9S	64.9W	0	89		
6835	342	0872 Feb 12	17:35:06	2384	-13950	83	T	-p	-0.9765	1.0105	64.9S	18.6E	12	266	175	00m36s
6836	342	0872 Aug 08	03:55:30	2380	-13944	88	A+	-t	1.0114	0.9437	61.9N	108.0W	0	303	-	-
6837	342	0873 Feb 01	08:39:14	2377	-13938	93	T	-n	-0.2652	1.0527	29.9S	71.2E	74	334	181	04m12s
6838	342	0873 Jul 28	04:15:40	2373	-13932	98	A	nn	0.2692	0.9473	32.8N	133.6E	74	204	201	05m44s
6839	342	0874 Jan 22	00:20:14	2370	-13926	103	T	n-	0.4203	1.0318	4.8N	179.8W	65	161	117	03m01s
6840	342	0874 Jul 17	07:42:26	2366	-13920	108	A	p-	-0.4970	0.9856	8.0S	66.7E	60	16	58	01m39s

Cat Num	Canon Plate	Calendar Date	TD of Greatest Eclipse	ΔT s	Luna Num	Saros Num	Ecl. Type	QLE	Gamma	Ecl. Mag.	Lat. °	Long. °	Sun Alt °	Sun Azm °	Path Width km	Central Line Dur.
6841	343	0875 Jan 11	11:59:11	2363	-13914	113	P	t-	1.1603	0.6941	63.6N	25.9W	0	144		
6842	343	0875 Jun 07	10:46:55	2360	-13909	80	P	-t	1.3221	0.4002	66.8N	155.1W	0	358		
6843	343	0875 Jul 06	18:24:50	2359	-13908	118	P	t-	-1.2210	0.5929	64.2S	117.0W	0	29		
6844	343	0875 Dec 01	21:23:41	2356	-13903	85	P	-t	-1.1270	0.7370	67.3S	50.1E	0	177		
6845	343	0876 May 27	03:57:56	2353	-13897	90	T	-p	0.5830	1.0753	57.9N	125.9E	54	175	301	05m10s
6846	343	0876 Nov 19	20:46:20	2349	-13891	95	A	-p	-0.4415	0.9313	47.1S	128.2W	64	8	287	07m11s
6847	343	0877 May 16	19:44:56	2346	-13885	100	T	nn	-0.1645	1.0406	10.9N	106.7W	81	353	138	04m09s
6848	343	0877 Nov 09	01:49:24	2342	-13879	105	A	n-	0.2715	0.9832	2.7S	161.5E	74	189	62	01m57s
6849	343	0878 May 06	05:48:12	2339	-13873	110	A	t-	-0.9715	0.9711	57.7S	120.3E	13	341	455	02m28s
6850	343	0878 Oct 29	13:59:59	2335	-13867	115	T	t-	0.9411	1.0246	53.5N	3.7W	19	201	250	01m50s
6851	343	0879 Mar 26	16:42:00	2332	-13862	82	P	-t	1.1411	0.7181	72.0N	161.7W	0	77		
6852	343	0879 Sep 19	20:14:10	2329	-13856	87	P	-t	-1.0419	0.9299	72.0S	153.5E	0	91		
6853	343	0880 Mar 14	18:31:57	2325	-13850	92	A	-p	0.3857	0.9645	21.2N	93.5W	67	161	138	03m59s
6854	343	0880 Sep 08	09:02:03	2322	-13844	97	A	-p	-0.3493	0.9985	15.6S	46.8E	69	18	6	00m09s
6855	343	0881 Mar 04	03:08:22	2318	-13838	102	H2	p-	-0.3624	1.0167	25.1S	152.4E	69	341	61	01m35s
6856	343	0881 Aug 28	14:59:00	2315	-13832	107	A	p-	0.4042	0.9512	31.1N	27.7W	66	199	195	05m26s
6857	343	0882 Feb 21	17:38:16	2311	-13826	112	P	t-	-1.0431	0.9309	71.4S	46.0E	0	241		
6858	343	0882 Aug 17	15:29:23	2308	-13820	117	P	t-	1.1482	0.7066	71.0N	86.3E	0	310		
6859	343	0883 Jan 12	22:58:13	2305	-13815	84	T	-t	0.9609	1.0057	53.3N	160.9W	15	171	73	00m27s
6860	343	0883 Jul 08	05:30:38	2302	-13809	89	A	-p	-0.9080	0.9951	43.3S	104.5E	24	3	41	00m27s
6861	344	0884 Jan 02	08:42:03	2298	-13803	94	A	-n	0.2896	0.9610	5.8S	60.9E	73	179	148	04m55s
6862	344	0884 Jun 26	18:06:43	2295	-13797	99	Tm	nn	-0.1087	1.0596	17.2N	81.6W	84	358	197	05m43s
6863	344	0884 Dec 21	11:02:39	2291	-13791	104	A	p-	-0.4102	0.9202	47.9S	22.5E	66	5	332	08m35s
6864	344	0885 Jun 16	10:58:05	2288	-13785	109	T	p-	0.6320	1.0772	62.6N	16.4E	51	168	323	05m02s
6865	344	0885 Dec 10	10:02:13	2284	-13779	114	P	t-	-1.0804	0.8164	65.2S	120.9W	0	161		
6866	344	0886 May 07	18:49:14	2282	-13774	81	P	-t	-1.1926	0.6435	62.6S	43.1W	0	312		
6867	344	0886 Jun 06	03:28:26	2281	-13773	119	P	t-	1.3944	0.2647	64.8N	18.7W	0	23		
6868	344	0886 Oct 30	23:23:09	2278	-13768	86	P	-t	1.1408	0.7338	62.1N	106.8W	0	235		
6869	344	0887 Apr 27	02:46:27	2275	-13762	91	A	-p	-0.4641	0.9664	10.4S	157.9E	62	336	135	03m50s
6870	344	0887 Oct 20	13:06:40	2271	-13756	96	T	-n	0.4151	1.0461	10.2N	0.2W	65	206	167	03m59s
6871	344	0888 Apr 15	04:11:50	2268	-13750	101	A	pn	0.2910	0.9414	26.5N	117.7E	73	151	225	06m22s
6872	344	0888 Oct 09	05:11:15	2264	-13744	106	T	n-	-0.2715	1.0484	22.1S	100.4E	74	30	167	03m55s
6873	344	0889 Apr 04	05:17:45	2261	-13738	111	P	t-	1.0218	0.9306	61.0N	6.9E	0	75		
6874	344	0889 Sep 28	18:54:00	2258	-13732	116	P	t-	-1.0086	0.9748	61.0S	166.1E	0	98		
6875	344	0890 Feb 23	01:48:59	2255	-13727	83	T-	-t	-1.0005	1.0005	61.2S	89.1W	0	253	-	-
6876	344	0890 Aug 19	10:47:01	2251	-13721	88	P	-t	1.0789	0.8268	61.4N	139.8E	0	295		
6877	344	0891 Feb 12	17:10:42	2248	-13715	93	T	-n	-0.2857	1.0557	27.0S	56.2W	73	331	191	04m23s
6878	344	0891 Aug 08	10:57:55	2244	-13709	98	A	np	0.3427	0.9458	33.4N	35.3E	70	208	211	05m42s
6879	344	0892 Feb 02	08:54:33	2241	-13703	103	T	n-	0.4071	1.0320	6.6N	50.3E	66	157	117	02m57s
6880	344	0892 Jul 27	14:41:15	2238	-13697	108	A	p-	-0.4218	0.9877	5.2S	38.6W	65	20	48	01m22s
6881	345	0893 Jan 21	20:20:19	2234	-13691	113	P	t-	1.1538	0.7054	62.8N	161.0W	0	134		
6882	345	0893 Jun 17	18:09:48	2231	-13686	80	P	-t	1.3917	0.2658	65.8N	83.2E	0	348		
6883	345	0893 Jul 17	01:45:18	2231	-13685	118	P	t-	-1.1469	0.7327	63.4S	122.8E	0	38		
6884	345	0893 Dec 12	05:25:39	2228	-13680	85	P	-t	-1.1334	0.7264	66.2S	81.6W	0	188		
6885	345	0894 Jun 07	11:24:26	2225	-13674	90	T	-p	0.6575	1.0736	64.6N	19.5E	49	183	318	04m47s
6886	345	0894 Dec 01	04:54:33	2221	-13668	95	A	-p	-0.4470	0.9324	49.4S	113.1E	63	2	283	06m56s
6887	345	0895 May 28	03:00:09	2218	-13662	100	Tm	nn	-0.0909	1.0383	17.0N	143.3E	85	357	129	03m54s
6888	345	0895 Nov 20	10:16:39	2214	-13656	105	A	n-	0.2637	0.9849	5.6S	33.8E	75	185	55	01m45s
6889	345	0896 May 16	12:39:28	2211	-13650	110	A	p-	-0.8986	0.9727	43.7S	6.4E	26	351	224	02m43s
6890	345	0896 Nov 08	22:40:33	2208	-13644	115	T	t-	0.9297	1.0251	49.5N	140.8W	21	194	234	01m57s
6891	345	0897 Apr 05	23:34:59	2205	-13639	82	P	-t	1.2019	0.6157	71.6N	80.6E	0	64		
6892	345	0897 Sep 30	04:27:23	2202	-13633	87	P	-t	-1.0742	0.8672	71.9S	15.5E	0	105		
6893	345	0898 Mar 26	01:51:40	2198	-13627	92	A	-p	0.4361	0.9700	28.5N	153.7E	64	160	119	03m12s
6894	345	0898 Sep 19	16:48:21	2195	-13621	97	A	-p	-0.3894	0.9919	22.2S	72.4W	67	19	31	00m50s
6895	345	0899 Mar 15	11:01:39	2191	-13615	102	T	n-	-0.3212	1.0232	18.3S	31.7E	71	341	83	02m14s
6896	345	0899 Sep 08	22:11:30	2188	-13609	107	A	p-	0.3536	0.9465	23.8N	138.2W	69	199	210	06m15s
6897	345	0900 Mar 04	01:54:43	2185	-13603	112	P	t-	-1.0112	0.9933	71.8S	92.6W	0	255		
6898	345	0900 Aug 27	22:23:35	2181	-13597	117	P	t-	1.0865	0.8104	71.6N	31.2W	0	297		
6899	345	0901 Jan 23	07:33:16	2179	-13592	84	T	-t	0.9731	1.0042	57.8N	62.0E	13	163	67	00m19s
6900	345	0901 Jul 18	12:33:26	2175	-13586	89	A	-t	-0.9786	0.9934	58.0S	10.5W	11	12	119	00m32s

Cat Num	Canon Plate	Calendar Date	TD of Greatest Eclipse	ΔT s	Luna Num	Saros Num	Ecl. Type	QLE	Gamma	Ecl. Mag.	Lat. °	Long. °	Sun Alt °	Sun Azm °	Path Width km	Central Line Dur.
6901	346	0902 Jan 12	17:00:22	2172	-13580	94	A	-n	0.2978	0.9597	3.6S	64.5W	73	175	153	05m06s
6902	346	0902 Jul 08	01:28:49	2169	-13574	99	T	nn	-0.1825	1.0609	11.7N	167.1E	80	2	203	05m55s
6903	346	0903 Jan 01	19:04:58	2165	-13568	104	A	p-	-0.4032	0.9202	46.6S	94.4W	66	358	331	08m45s
6904	346	0903 Jun 27	18:26:02	2162	-13562	109	T	p-	0.5585	1.0773	57.4N	88.8W	56	177	302	05m18s
6905	346	0903 Dec 21	18:08:11	2159	-13556	114	P	t-	-1.0730	0.8293	66.2S	106.9E	0	172		
6906	346	0904 May 18	02:00:21	2156	-13551	81	P	-t	-1.2687	0.5031	63.3S	160.7W	0	321		
6907	346	0904 Jun 16	10:45:03	2155	-13550	119	P	t-	1.3251	0.3964	65.8N	138.4W	0	13		
6908	346	0904 Nov 10	07:53:36	2152	-13545	86	P	-t	1.1471	0.7227	62.8N	116.0E	0	226		
6909	346	0905 May 07	09:33:22	2149	-13539	91	A	-p	-0.5422	0.9649	12.8S	55.8E	57	339	149	04m07s
6910	346	0905 Oct 30	21:44:30	2146	-13533	96	T	-n	0.4293	1.0449	7.8N	131.0W	65	203	164	03m59s
6911	346	0906 Apr 26	10:50:32	2142	-13527	101	A	nn	0.2158	0.9440	26.4N	19.6E	77	154	211	06m15s
6912	346	0906 Oct 20	13:41:16	2139	-13521	106	T	n-	-0.2518	1.0444	25.3S	27.1W	75	28	153	03m37s
6913	346	0907 Apr 15	12:17:35	2136	-13515	111	A	t-	0.9549	0.9550	63.9N	70.9W	17	99	557	03m04s
6914	346	0907 Oct 10	03:00:37	2133	-13509	116	A	p-	-0.9851	0.9869	62.7S	54.1E	9	91	291	00m47s
6915	346	0908 Mar 05	09:56:14	2130	-13504	83	P	-t	-1.0310	0.9465	61.0S	139.9E	0	262		
6916	346	0908 Aug 29	17:45:24	2126	-13498	88	P	-t	1.1402	0.7214	61.0N	25.9E	0	286		
6917	346	0909 Feb 23	01:35:04	2123	-13492	93	T	-n	-0.3127	1.0587	24.2S	177.8E	72	330	203	04m36s
6918	346	0909 Aug 18	17:48:28	2120	-13486	98	A	-p	0.4100	0.9442	33.1N	65.4W	66	211	224	05m44s
6919	346	0910 Feb 12	17:19:42	2117	-13480	103	T	n-	0.3867	1.0325	8.6N	77.1W	67	154	118	02m56s
6920	346	0910 Aug 07	21:48:49	2113	-13474	108	A	p-	-0.3526	0.9891	3.8S	145.8W	69	23	41	01m10s
6921	347	0911 Feb 02	04:32:54	2110	-13468	113	P	t-	1.1410	0.7279	62.1N	66.3E	0	125		
6922	347	0911 Jun 29	01:34:54	2107	-13463	80	P	-t	1.4599	0.1340	64.9N	38.6W	0	338		
6923	347	0911 Jul 28	09:12:23	2107	-13462	118	P	t-	-1.0774	0.8640	62.6S	1.2E	0	47		
6924	347	0911 Dec 23	13:26:37	2104	-13457	85	P	-t	-1.1406	0.7146	65.2S	147.4E	0	199		
6925	347	0912 Jun 17	18:50:25	2101	-13451	90	T	-p	0.7322	1.0708	70.4N	81.2W	43	196	340	04m21s
6926	347	0912 Dec 11	13:06:23	2097	-13445	95	A	-p	-0.4504	0.9341	50.4S	5.7W	63	355	276	06m38s
6927	347	0913 Jun 07	10:09:57	2094	-13439	100	T	nn	-0.0134	1.0354	22.6N	35.5E	89	2	119	03m33s
6928	347	0913 Nov 30	18:50:01	2091	-13433	105	A	n-	0.2602	0.9870	7.6S	95.2W	75	181	48	01m30s
6929	347	0914 May 27	19:24:02	2088	-13427	110	A	p-	-0.8205	0.9728	33.2S	100.7W	35	357	171	03m00s
6930	347	0914 Nov 20	07:26:39	2084	-13421	115	T	p-	0.9225	1.0258	46.5N	81.8E	22	188	229	02m04s
6931	347	0915 Apr 17	06:21:01	2082	-13416	82	P	-t	1.2684	0.5029	71.0N	34.9W	0	51		
6932	347	0915 Oct 11	12:48:25	2078	-13410	87	P	-t	-1.0999	0.8174	71.5S	124.3W	0	119		
6933	347	0916 Apr 05	09:06:41	2075	-13404	92	A	-p	0.4911	0.9753	36.1N	42.2E	60	160	101	02m28s
6934	347	0916 Sep 30	00:40:10	2072	-13398	97	A	-p	-0.4242	0.9855	28.7S	167.5E	65	20	56	01m27s
6935	347	0917 Mar 25	18:49:52	2069	-13392	102	T	n-	-0.2749	1.0296	11.4S	87.9W	74	342	104	02m53s
6936	347	0917 Sep 19	05:30:32	2065	-13386	107	A	n-	0.3089	0.9419	16.9N	109.6E	72	199	225	07m06s
6937	347	0918 Mar 15	10:03:27	2062	-13380	112	T	t-	-0.9730	1.0489	68.4S	94.4E	13	303	737	02m53s
6938	347	0918 Sep 08	05:27:33	2059	-13374	117	P	t-	1.0322	0.9012	72.0N	151.7W	0	283		
6939	347	0919 Feb 03	16:00:53	2056	-13369	84	T	-t	0.9909	1.0020	65.2N	78.9W	6	150	62	00m09s
6940	347	0919 Jul 29	19:44:05	2053	-13363	89	P	-t	-1.0437	0.9112	69.6S	136.8W	0	29		
6941	348	0920 Jan 24	01:11:14	2050	-13357	94	A	-n	0.3115	0.9587	0.3S	171.6E	72	171	158	05m12s
6942	348	0920 Jul 18	08:56:43	2047	-13351	99	T	-n	-0.2519	1.0615	5.8N	53.7E	75	6	209	05m59s
6943	348	0921 Jan 12	03:02:26	2043	-13345	104	A	p-	-0.3925	0.9207	44.0S	149.1E	67	353	327	08m51s
6944	348	0921 Jul 08	01:55:03	2040	-13339	109	T	p-	0.4862	1.0765	51.4N	162.7E	61	184	284	05m32s
6945	348	0922 Jan 01	02:12:56	2037	-13333	114	P	t-	-1.0642	0.8447	67.3S	25.5W	0	183		
6946	348	0922 May 29	09:05:23	2034	-13328	81	P	-t	-1.3483	0.3570	64.2S	83.0E	0	331		
6947	348	0922 Jun 27	18:00:27	2034	-13327	119	P	t-	1.2555	0.5277	66.7N	101.8E	0	3		
6948	348	0922 Nov 21	16:29:24	2031	-13322	86	P	-t	1.1496	0.7185	63.6N	22.7W	0	216		
6949	348	0923 May 18	16:13:31	2028	-13316	91	A	-p	-0.6247	0.9629	16.6S	44.9W	51	343	170	04m28s
6950	348	0923 Nov 11	06:29:02	2025	-13310	96	T	-n	0.4377	1.0440	5.8N	96.5E	64	199	162	04m01s
6951	348	0924 May 06	17:22:41	2021	-13304	101	A	nn	0.1351	0.9462	25.4N	76.7W	82	158	200	06m12s
6952	348	0924 Oct 30	22:17:32	2018	-13298	106	T	n-	-0.2378	1.0405	28.5S	155.9W	76	25	139	03m20s
6953	348	0925 Apr 25	19:11:24	2015	-13292	111	A	t-	0.8822	0.9628	64.2N	156.5W	28	115	285	02m40s
6954	348	0925 Oct 20	11:14:00	2012	-13286	116	A	p-	-0.9677	0.9827	64.9S	67.0W	14	89	250	01m03s
6955	348	0926 Mar 16	17:55:33	2009	-13281	83	P	-t	-1.0686	0.8785	60.9S	10.8E	0	271		
6956	348	0926 Sep 10	00:52:08	2006	-13275	88	P	-t	1.1941	0.6293	60.8N	89.9W	0	277		
6957	348	0927 Mar 06	09:52:27	2003	-13269	93	T	-n	-0.3456	1.0617	21.6S	53.4E	70	329	214	04m50s
6958	348	0927 Aug 30	00:47:33	2000	-13263	98	A	-p	0.4706	0.9425	32.2N	168.8W	62	213	238	05m51s
6959	348	0928 Feb 24	01:38:26	1996	-13257	103	T	n-	0.3615	1.0331	10.9N	157.3E	69	152	119	02m55s
6960	348	0928 Aug 18	05:04:32	1993	-13251	108	A	p-	-0.2893	0.9902	3.5S	105.1E	73	26	36	01m01s

Cat Num	Canon Plate	Calendar Date	TD of Greatest Eclipse	ΔT s	Luna Num	Saros Num	Ecl. Type	QLE	Gamma	Ecl. Mag.	Lat. °	Long. °	Sun Alt °	Sun Azm °	Path Width km	Central Line Dur.
6961	349	0929 Feb 12	12:38:40	1990	-13245	113	P	t-	1.1233	0.7590	61.5N	64.5W	0	116		
6962	349	0929 Jul 09	09:02:03	1988	-13240	80	Pe	-t	1.5267	0.0049	64.0N	160.7W	0	329		
6963	349	0929 Aug 07	16:45:08	1987	-13239	118	P	t-	-1.0118	0.9880	62.0S	121.6W	0	56		
6964	349	0930 Jan 02	21:26:03	1984	-13234	85	P	-t	-1.1490	0.7010	64.2S	17.3E	0	209		
6965	349	0930 Jun 29	02:17:19	1981	-13228	90	T	-p	0.8061	1.0671	74.4N	173.2W	36	218	372	03m55s
6966	349	0930 Dec 22	21:17:36	1978	-13222	95	A	-p	-0.4550	0.9364	50.3S	124.2W	63	348	266	06m17s
6967	349	0931 Jun 18	17:19:48	1975	-13216	100	T	nn	0.0633	1.0318	27.3N	71.4W	86	186	108	03m06s
6968	349	0931 Dec 12	03:24:53	1972	-13210	105	A	n-	0.2568	0.9896	8.7S	135.7E	75	177	38	01m11s
6969	349	0932 Jun 07	02:05:46	1969	-13204	110	A	p-	-0.7406	0.9719	24.7S	154.8E	42	2	150	03m19s
6970	349	0932 Nov 30	16:15:34	1965	-13198	115	T	p-	0.9174	1.0267	44.3N	55.9W	23	182	230	02m11s
6971	349	0933 Apr 27	13:00:25	1963	-13193	82	P	-t	1.3402	0.3803	70.3N	148.1W	0	38		
6972	349	0933 May 27	04:10:41	1962	-13192	120	Pb	t-	-1.5258	0.0630	67.7S	132.3E	0	353		
6973	349	0933 Oct 21	21:16:05	1960	-13187	87	P	-t	-1.1197	0.7793	70.8S	94.8E	0	132		
6974	349	0934 Apr 16	16:13:20	1957	-13181	92	A	-p	0.5538	0.9804	44.2N	67.2W	56	160	83	01m50s
6975	349	0934 Oct 11	08:40:25	1953	-13175	97	A	-p	-0.4510	0.9794	34.6S	45.8E	63	19	82	02m01s
6976	349	0935 Apr 06	02:30:30	1950	-13169	102	T	n-	-0.2215	1.0358	4.3S	154.4E	77	344	123	03m32s
6977	349	0935 Sep 30	12:58:02	1947	-13163	107	A	nn	0.2716	0.9375	10.5N	4.5W	74	198	241	07m57s
6978	349	0936 Mar 25	18:04:49	1944	-13157	112	T	t-	-0.9284	1.0555	59.4S	48.5W	21	324	496	03m37s
6979	349	0936 Sep 18	12:41:00	1941	-13151	117	An	t-	0.9851	0.9215	70.1N	58.4E	9	244	-	05m59s
6980	349	0937 Feb 14	00:21:18	1938	-13146	84	P	-t	1.0142	0.9686	71.1N	132.6E	0	127		
6981	350	0937 Aug 09	03:00:57	1935	-13140	89	P	-t	-1.1044	0.8013	70.4S	101.0E	0	41		
6982	350	0938 Feb 03	09:16:04	1932	-13134	94	A	-p	0.3294	0.9583	3.8N	48.9E	71	168	160	05m11s
6983	350	0938 Jul 29	16:29:29	1929	-13128	99	T	-n	-0.3177	1.0614	0.5S	61.5W	71	10	212	05m55s
6984	350	0939 Jan 23	10:54:43	1926	-13122	104	A	p-	-0.3778	0.9219	40.3S	32.8E	68	348	319	08m56s
6985	350	0939 Jul 19	09:27:56	1923	-13116	109	T	p-	0.4172	1.0748	44.9N	51.1E	65	189	267	05m42s
6986	350	0940 Jan 12	10:16:11	1920	-13110	114	P	t-	-1.0538	0.8631	68.4S	158.0W	0	194		
6987	350	0940 Jun 08	16:07:41	1917	-13105	81	P	-t	-1.4291	0.2096	65.1S	32.9W	0	340		
6988	350	0940 Jul 08	01:16:23	1917	-13104	119	P	t-	1.1868	0.6564	67.7N	18.5W	0	352		
6989	350	0940 Dec 02	01:09:04	1914	-13099	86	P	-t	1.1498	0.7185	64.5N	162.7W	0	206		
6990	350	0941 May 28	22:49:33	1911	-13093	91	A	-p	-0.7096	0.9603	21.9S	145.2W	45	347	203	04m49s
6991	350	0941 Nov 21	15:17:26	1908	-13087	96	T	-n	0.4437	1.0436	4.2N	36.9W	64	195	162	04m05s
6992	350	0942 May 17	23:50:31	1905	-13081	101	A	nn	0.0504	0.9481	23.4N	172.0W	87	163	191	06m15s
6993	350	0942 Nov 11	06:58:47	1902	-13075	106	T	n-	-0.2281	1.0370	31.4S	74.6E	77	22	127	03m05s
6994	350	0943 May 07	02:02:38	1899	-13069	111	A	p-	0.8059	0.9693	64.0N	114.2E	36	129	185	02m17s
6995	350	0943 Oct 31	19:32:08	1896	-13063	116	A	p-	-0.9548	0.9783	67.8S	166.7E	17	91	265	01m19s
6996	350	0944 Mar 27	01:48:39	1893	-13058	83	P	-t	-1.1120	0.7986	61.0S	116.6W	0	279		
6997	350	0944 Apr 25	10:56:13	1893	-13057	121	Pb	t-	1.5044	0.0666	62.0N	97.1W	0	57		
6998	350	0944 Sep 20	08:07:18	1890	-13052	88	P	-t	1.2406	0.5505	60.8N	152.2E	0	268		
6999	350	0945 Mar 16	18:00:58	1887	-13046	93	T	-n	-0.3861	1.0644	19.4S	69.0W	67	329	227	05m05s
7000	350	0945 Sep 09	07:56:57	1884	-13040	98	A	-p	0.5232	0.9407	30.8N	84.6E	58	214	253	06m03s
7001	351	0946 Mar 06	09:46:58	1881	-13034	103	T	n-	0.3283	1.0338	13.1N	34.4E	71	150	120	02m56s
7002	351	0946 Aug 29	12:31:00	1878	-13028	108	A	n-	-0.2337	0.9909	4.3S	6.6W	76	28	33	00m55s
7003	351	0947 Feb 23	20:33:12	1875	-13022	113	P	t-	1.0972	0.8049	61.2N	167.6E	0	107		
7004	351	0947 Aug 19	00:26:53	1872	-13016	118	T	t-	-0.9527	1.0357	51.3S	139.1E	17	44	393	02m29s
7005	351	0948 Jan 14	05:22:02	1869	-13011	85	P	-t	-1.1604	0.6824	63.3S	111.6W	0	219		
7006	351	0948 Jul 09	09:44:37	1866	-13005	90	T	-t	0.8793	1.0621	75.4N	104.3E	28	248	431	03m27s
7007	351	0949 Jan 02	05:29:26	1863	-12999	95	A	-p	-0.4598	0.9394	49.1S	116.8E	62	341	253	05m54s
7008	351	0949 Jun 29	00:26:52	1860	-12993	100	T	nn	0.1417	1.0274	31.3N	176.8W	82	191	94	02m36s
7009	351	0949 Dec 22	12:01:30	1857	-12987	105	A	n-	0.2543	0.9929	8.9S	6.2E	75	172	26	00m48s
7010	351	0950 Jun 18	08:43:41	1854	-12981	110	A	p-	-0.6580	0.9703	17.5S	52.5E	49	6	142	03m40s
7011	351	0950 Dec 12	01:07:14	1851	-12975	115	T	p-	0.9143	1.0281	42.9N	166.1E	24	176	237	02m19s
7012	351	0951 May 08	19:35:30	1848	-12970	82	P	-t	1.4152	0.2514	69.4N	100.4E	0	26		
7013	351	0951 Jun 07	10:33:36	1848	-12969	120	P	t-	-1.4414	0.2073	66.7S	25.1E	0	3		
7014	351	0951 Nov 02	05:48:45	1845	-12964	87	P	-t	-1.1351	0.7496	70.0S	46.8W	0	145		
7015	351	0952 Apr 26	23:17:22	1842	-12958	92	A	-p	0.6195	0.9851	52.5N	175.4W	51	160	67	01m17s
7016	351	0952 Oct 21	16:46:04	1839	-12952	97	A	-p	-0.4727	0.9736	40.2S	76.4W	62	18	107	02m32s
7017	351	0953 Apr 16	10:05:48	1836	-12946	102	T	n-	-0.1628	1.0418	2.8N	38.1E	81	345	142	04m07s
7018	351	0953 Oct 10	20:32:48	1833	-12940	107	A	nn	0.2407	0.9333	4.6N	120.1W	76	196	256	08m46s
7019	351	0954 Apr 06	01:59:15	1830	-12934	112	T	p-	-0.8778	1.0611	50.3S	178.9W	28	334	417	04m20s
7020	351	0954 Sep 29	20:04:21	1827	-12928	117	A	t-	0.9454	0.9220	61.4N	80.1W	18	218	912	06m47s

Cat Num	Canon Plate	Calendar Date	TD of Greatest Eclipse	ΔT s	Luna Num	Saros Num	Ecl. Type	QLE	Gamma	Ecl. Mag.	Lat. °	Long. °	Sun Alt °	Sun Azm °	Path Width km	Central Line Dur.
7021	352	0955 Feb 25	08:32:13	1825	-12923	84	P	-t	1.0447	0.9127	71.7N	4.3W	0	114		
7022	352	0955 Aug 20	10:27:56	1822	-12917	89	P	-t	-1.1576	0.7047	71.1S	24.3W	0	54		
7023	352	0956 Feb 14	17:11:45	1819	-12911	94	A	-p	0.3540	0.9580	8.8N	71.8W	69	165	163	05m06s
7024	352	0956 Aug 09	00:08:26	1816	-12905	99	T	-n	-0.3787	1.0607	7.2S	178.7W	68	13	215	05m43s
7025	352	0957 Feb 02	18:39:45	1813	-12899	104	A	n-	-0.3573	0.9236	35.6S	82.9W	69	345	308	08m57s
7026	352	0957 Jul 29	17:04:35	1810	-12893	109	T	n-	0.3518	1.0723	38.0N	63.0W	69	192	251	05m46s
7027	352	0958 Jan 22	18:13:32	1807	-12887	114	P	t-	-1.0380	0.8911	69.5S	70.3E	0	206		
7028	352	0958 Jun 19	23:06:54	1805	-12882	81	Pe	-t	-1.5110	0.0612	66.1S	148.3W	0	350		
7029	352	0958 Jul 19	08:34:18	1804	-12881	119	P	t-	1.1204	0.7797	68.7N	139.8W	0	341		
7030	352	0958 Dec 13	09:50:52	1802	-12876	86	P	-t	1.1488	0.7206	65.5N	56.4E	0	195		
7031	352	0959 Jun 09	05:22:03	1799	-12870	91	A	-p	-0.7965	0.9569	29.1S	114.7E	37	350	259	05m06s
7032	352	0959 Dec 03	00:08:45	1796	-12864	96	T	-n	0.4473	1.0435	3.3N	171.0W	63	191	162	04m10s
7033	352	0960 May 28	06:15:21	1793	-12858	101	Am	nn	-0.0369	0.9497	20.3N	93.1E	88	346	185	06m21s
7034	352	0960 Nov 21	15:44:42	1790	-12852	106	T	n-	-0.2228	1.0338	33.8S	55.6W	77	18	117	02m52s
7035	352	0961 May 17	08:48:53	1787	-12846	111	A	p-	0.7245	0.9753	62.8N	25.1E	43	142	128	01m54s
7036	352	0961 Nov 11	03:56:16	1784	-12840	116	A	p-	-0.9475	0.9739	71.3S	35.8E	18	96	298	01m35s
7037	352	0962 Apr 07	09:34:49	1781	-12835	83	P	-t	-1.1615	0.7065	61.3S	117.6E	0	288		
7038	352	0962 May 06	18:15:23	1781	-12834	121	P	t-	1.4334	0.1965	62.6N	143.6E	0	48		
7039	352	0962 Oct 01	15:31:30	1778	-12829	88	P	-t	1.2793	0.4853	60.9N	32.1E	0	259		
7040	352	0963 Mar 28	02:02:32	1775	-12823	93	T	-p	-0.4324	1.0668	17.8S	170.3E	64	330	240	05m20s
7041	353	0963 Sep 20	15:16:29	1773	-12817	98	A	-p	0.5678	0.9391	29.1N	25.3W	55	214	269	06m18s
7042	353	0964 Mar 16	17:48:19	1770	-12811	103	T	n-	0.2895	1.0344	15.3N	86.3W	73	150	120	02m57s
7043	353	0964 Sep 08	20:05:22	1767	-12805	108	A	n-	-0.1838	0.9914	5.8S	120.3W	79	29	31	00m51s
7044	353	0965 Mar 06	04:19:59	1764	-12799	113	P	t-	1.0655	0.8607	60.9N	41.8E	0	98		
7045	353	0965 Aug 29	08:15:51	1761	-12793	118	T	t-	-0.8990	1.0377	47.1S	24.8E	26	43	283	02m41s
7046	353	0966 Jan 24	13:11:50	1758	-12788	85	P	-t	-1.1766	0.6559	62.5S	121.3E	0	228		
7047	353	0966 Jul 20	17:15:22	1755	-12782	90	T	-t	0.9492	1.0556	72.4N	23.0E	18	279	597	02m55s
7048	353	0967 Jan 13	13:37:19	1753	-12776	95	A	-p	-0.4687	0.9430	47.0S	1.9W	62	336	238	05m28s
7049	353	0967 Jul 10	07:36:08	1750	-12770	100	T	-n	0.2173	1.0225	34.1N	78.0E	77	196	79	02m04s
7050	353	0968 Jan 02	20:35:37	1747	-12764	105	A	n-	0.2490	0.9967	8.4S	122.7W	76	168	12	00m21s
7051	353	0968 Jun 28	15:22:10	1744	-12758	110	A	p-	-0.5764	0.9680	11.7S	49.0W	55	10	140	04m01s
7052	353	0968 Dec 22	09:58:17	1741	-12752	115	T	p-	0.9105	1.0300	41.8N	28.3E	24	170	246	02m28s
7053	353	0969 May 19	02:05:07	1739	-12747	82	Pe	-t	1.4945	0.1145	68.4N	9.1W	0	15		
7054	353	0969 Jun 17	16:54:26	1738	-12746	120	P	t-	-1.3547	0.3558	65.7S	81.0W	0	13		
7055	353	0969 Nov 12	14:26:54	1736	-12741	87	P	-t	-1.1457	0.7291	69.1S	170.8E	0	158		
7056	353	0970 May 08	06:15:56	1733	-12735	92	A	-p	0.6907	0.9892	61.3N	78.3E	46	161	52	00m51s
7057	353	0970 Nov 02	00:58:17	1730	-12729	97	A	-p	-0.4883	0.9684	45.1S	161.0E	61	16	130	02m59s
7058	353	0971 Apr 27	17:35:25	1727	-12723	102	T	nn	-0.0983	1.0473	9.9N	76.5W	84	348	158	04m38s
7059	353	0971 Oct 22	04:16:15	1724	-12717	107	A	nn	0.2171	0.9295	0.6S	122.4E	77	194	270	09m32s
7060	353	0972 Apr 16	09:46:46	1721	-12711	112	T	p-	-0.8213	1.0660	41.4S	56.2E	34	341	376	05m04s
7061	354	0972 Oct 10	03:36:50	1718	-12705	117	A	p-	0.9127	0.9216	54.0N	155.2E	24	208	724	07m31s
7062	354	0973 Mar 07	16:34:55	1716	-12700	84	P	-t	1.0813	0.8457	72.0N	139.5W	0	100		
7063	354	0973 Aug 30	18:03:13	1713	-12694	89	P	-t	-1.2046	0.6191	71.6S	152.2W	0	67		
7064	354	0974 Feb 25	00:57:51	1710	-12688	94	A	-p	0.3857	0.9580	14.6N	169.4E	67	163	165	04m58s
7065	354	0974 Aug 20	07:54:38	1707	-12682	99	T	-n	-0.4339	1.0594	13.9S	61.9E	64	15	216	05m25s
7066	354	0975 Feb 14	02:17:42	1705	-12676	104	A	n-	-0.3314	0.9258	30.2S	162.2E	70	343	296	08m56s
7067	354	0975 Aug 10	00:46:07	1702	-12670	109	T	n-	0.2907	1.0692	31.0N	179.2W	73	195	236	05m45s
7068	354	0976 Feb 03	02:06:44	1699	-12664	114	A-	t-	-1.0182	0.9262	70.4S	61.0W	0	219	-	-
7069	354	0976 Jul 29	15:55:20	1696	-12658	119	P	t-	1.0571	0.8957	69.6N	97.6E	0	330		
7070	354	0976 Dec 23	18:33:17	1694	-12653	86	P	-t	1.1481	0.7224	66.6N	85.0W	0	185		
7071	354	0977 Jun 19	11:52:12	1691	-12647	91	A	-t	-0.8844	0.9526	38.8S	14.2E	27	355	373	05m16s
7072	354	0977 Dec 13	09:01:41	1688	-12641	96	T	-n	0.4499	1.0439	3.0N	54.5E	63	186	164	04m16s
7073	354	0978 Jun 08	12:39:24	1685	-12635	101	A	nn	-0.1248	0.9507	16.3N	2.2W	83	351	182	06m30s
7074	354	0978 Dec 03	00:31:51	1682	-12629	106	T	n-	-0.2191	1.0311	35.4S	174.5E	77	13	108	02m41s
7075	354	0979 May 28	15:35:35	1679	-12623	111	A	p-	0.6424	0.9806	60.5N	65.9W	50	154	90	01m34s
7076	354	0979 Nov 22	12:23:31	1677	-12617	116	A	p-	-0.9439	0.9699	75.0S	98.4W	19	102	335	01m49s
7077	354	0980 Apr 17	17:15:17	1674	-12612	83	P	-t	-1.2162	0.6033	61.7S	6.8W	0	297		
7078	354	0980 May 17	01:32:03	1674	-12611	121	P	t-	1.3592	0.3338	63.3N	24.8E	0	39		
7079	354	0980 Oct 11	23:03:22	1671	-12606	88	P	-t	1.3115	0.4317	61.2N	90.0W	0	250		
7080	354	0981 Apr 07	09:55:46	1669	-12600	93	T	-p	-0.4857	1.0687	17.1S	51.7E	61	332	253	05m34s

Cat Num	Canon Plate	Calendar Date	TD of Greatest Eclipse	ΔT s	Luna Num	Saros Num	Ecl. Type	QLE	Gamma	Ecl. Mag.	Lat. °	Long. °	Sun Alt °	Sun Azm °	Path Width km	Central Line Dur.
7081	355	0981 Sep 30	22:46:30	1666	-12594	98	A	-p	0.6044	0.9376	27.4N	138.5W	53	212	284	06m35s
7082	355	0982 Mar 28	01:38:12	1663	-12588	103	T	n-	0.2418	1.0347	17.2N	156.3E	76	150	120	02m59s
7083	355	0982 Sep 20	03:51:25	1660	-12582	108	A	n-	-0.1426	0.9916	8.0S	122.9E	82	29	30	00m49s
7084	355	0983 Mar 17	11:55:18	1657	-12576	113	P	t-	1.0251	0.9317	60.9N	81.1W	0	89		
7085	355	0983 Sep 09	16:13:17	1655	-12570	118	T	p-	-0.8518	1.0386	45.7S	93.8W	31	44	242	02m44s
7086	355	0984 Feb 04	20:55:19	1652	-12565	85	P	-t	-1.1979	0.6210	61.8S	3.9W	0	238		
7087	355	0984 Jul 31	00:49:12	1649	-12559	90	P	-t	1.0161	0.9871	62.4N	58.3W	0	309		
7088	355	0984 Aug 29	08:35:40	1649	-12558	128	Pb	t-	-1.5263	0.0085	61.3S	18.5W	0	74		
7089	355	0985 Jan 23	21:42:16	1647	-12553	95	A	-p	-0.4805	0.9472	44.3S	120.8W	61	331	221	05m00s
7090	355	0985 Jul 20	14:45:03	1644	-12547	100	T	-p	0.2923	1.0170	35.9N	26.6W	73	202	61	01m31s
7091	355	0986 Jan 13	05:08:47	1641	-12541	105	H	n-	0.2422	1.0011	7.2S	108.7E	76	163	4	00m07s
7092	355	0986 Jul 09	22:00:35	1638	-12535	110	A	p-	-0.4952	0.9651	7.0S	149.8W	60	14	144	04m22s
7093	355	0987 Jan 02	18:48:09	1636	-12529	115	T	p-	0.9056	1.0323	41.2N	108.9W	25	164	257	02m37s
7094	355	0987 Jun 28	23:17:31	1633	-12523	120	P	t-	-1.2690	0.5024	64.8S	172.8E	0	23		
7095	355	0987 Nov 23	23:08:42	1630	-12518	87	P	-t	-1.1533	0.7145	68.0S	28.3E	0	170		
7096	355	0988 May 18	13:13:08	1628	-12512	92	A	-p	0.7637	0.9929	70.2N	26.0W	40	164	39	00m31s
7097	355	0988 Nov 12	09:14:16	1625	-12506	97	A	-p	-0.4999	0.9636	49.2S	38.8E	60	11	152	03m23s
7098	355	0989 May 08	01:01:52	1622	-12500	102	T	nn	-0.0304	1.0523	16.8N	170.3E	88	350	173	05m01s
7099	355	0989 Nov 01	12:05:28	1619	-12494	107	A	nn	0.1989	0.9261	5.1S	3.9E	79	191	283	10m14s
7100	355	0990 Apr 27	17:28:21	1617	-12488	112	T	p-	-0.7599	1.0700	33.0S	65.2W	40	346	349	05m45s
7101	356	0990 Oct 21	11:18:41	1614	-12482	117	A	p-	0.8870	0.9211	47.9N	32.3E	27	201	644	08m09s
7102	356	0991 Mar 19	00:28:00	1612	-12477	84	P	-t	1.1249	0.7658	72.0N	87.6E	0	86		
7103	356	0991 Apr 17	10:00:06	1611	-12476	122	Pb	t-	-1.5013	0.0624	70.8S	87.9E	0	310		
7104	356	0991 Sep 11	01:47:57	1609	-12471	89	P	-t	-1.2447	0.5462	71.8S	77.3E	0	81		
7105	356	0991 Oct 10	14:31:21	1608	-12470	127	Pb	t-	1.5370	0.0321	71.5N	28.4E	0	242		
7106	356	0992 Mar 07	08:33:56	1606	-12465	94	A	-p	0.4248	0.9580	21.0N	52.9E	65	161	168	04m48s
7107	356	0992 Aug 30	15:48:52	1603	-12459	99	T	-p	-0.4826	1.0577	20.7S	59.7W	61	18	216	05m04s
7108	356	0993 Feb 24	09:45:28	1601	-12453	104	A	nn	-0.2971	0.9283	24.0S	49.0E	73	342	281	08m51s
7109	356	0993 Aug 20	08:33:40	1598	-12447	109	T	n-	0.2350	1.0654	24.0N	62.4E	76	196	220	05m37s
7110	356	0994 Feb 13	09:51:58	1595	-12441	114	As	t-	-0.9912	0.9303	74.3S	151.1E	6	249	-	04m33s
7111	356	0994 Aug 09	23:21:16	1592	-12435	119	T+	t-	0.9985	1.0017	70.5N	26.8W	0	318	-	-
7112	356	0995 Jan 04	03:13:44	1590	-12430	86	P	-t	1.1496	0.7206	67.7N	133.6E	0	174		
7113	356	0995 Jun 30	18:22:43	1587	-12424	91	A	-t	-0.9708	0.9465	53.9S	88.3W	13	360	854	05m11s
7114	356	0995 Dec 24	17:55:08	1585	-12418	96	T	-n	0.4518	1.0448	3.6N	80.1W	63	182	168	04m23s
7115	356	0996 Jun 18	19:01:57	1582	-12412	101	A	nn	-0.2142	0.9514	11.3N	97.9W	78	356	182	06m39s
7116	356	0996 Dec 13	09:21:13	1579	-12406	106	T	n-	-0.2176	1.0289	36.2S	44.3E	77	8	100	02m32s
7117	356	0997 Jun 07	22:20:33	1577	-12400	111	A	p-	0.5580	0.9853	56.8N	158.6W	56	164	63	01m15s
7118	356	0997 Dec 02	20:53:29	1574	-12394	116	A	p-	-0.9429	0.9663	78.7S	123.3E	19	112	374	02m02s
7119	356	0998 Apr 29	00:49:49	1572	-12389	83	P	-t	-1.2763	0.4887	62.2S	129.9W	0	306		
7120	356	0998 May 28	08:46:18	1571	-12388	121	P	t-	1.2820	0.4782	64.2N	93.7W	0	30		
7121	357	0998 Oct 23	06:43:47	1569	-12383	88	P	-t	1.3365	0.3902	61.6N	145.7E	0	241		
7122	357	0999 Apr 18	17:43:24	1566	-12377	93	T	-p	-0.5439	1.0700	17.3S	65.7W	57	334	268	05m47s
7123	357	0999 Oct 12	06:24:46	1564	-12371	98	A	-p	0.6349	0.9364	25.7N	105.8E	50	210	299	06m55s
7124	357	1000 Apr 07	09:21:38	1561	-12365	103	T	nn	0.1892	1.0348	18.7N	40.8E	79	151	119	03m01s
7125	357	1000 Sep 30	11:45:53	1558	-12359	108	A	nn	-0.1075	0.9919	10.6S	4.1E	84	29	29	00m47s
7126	357	1001 Mar 27	19:20:56	1556	-12353	113	A	p-	0.9775	0.9637	61.1N	178.0W	11	101	642	02m26s
7127	357	1001 Sep 20	00:19:08	1553	-12347	118	T	p-	-0.8111	1.0388	46.1S	144.9E	36	44	218	02m43s
7128	357	1002 Feb 15	04:30:46	1551	-12342	85	P	-t	-1.2256	0.5752	61.3S	127.0W	0	247		
7129	357	1002 Aug 11	08:28:52	1548	-12336	90	P	-t	1.0778	0.8675	61.8N	177.3E	0	301		
7130	357	1002 Sep 09	16:38:28	1548	-12335	128	P	t-	-1.4840	0.0911	61.1S	148.4W	0	83		
7131	357	1003 Feb 04	05:40:06	1545	-12330	95	A	-p	-0.4990	0.9517	41.5S	121.4E	60	328	202	04m31s
7132	357	1003 Jul 31	21:59:08	1543	-12324	100	H	-p	0.3621	1.0110	36.6N	132.5W	69	206	41	00m58s
7133	357	1004 Jan 24	13:37:07	1540	-12318	105	H	n-	0.2304	1.0060	5.5S	18.9W	77	160	21	00m36s
7134	357	1004 Jul 20	04:41:00	1537	-12312	110	A	p-	-0.4161	0.9618	3.7S	109.6E	65	18	151	04m42s
7135	357	1005 Jan 13	03:34:49	1535	-12306	115	T	p-	0.8978	1.0352	40.7N	115.1E	26	158	267	02m48s
7136	357	1005 Jul 09	05:42:18	1532	-12300	120	P	t-	-1.1837	0.6481	63.9S	66.5E	0	32		
7137	357	1005 Dec 04	07:52:26	1530	-12295	87	P	-t	-1.1589	0.7037	67.0S	114.2W	0	181		
7138	357	1006 May 29	20:07:43	1527	-12289	92	A	-p	0.8397	0.9955	79.8N	124.9W	33	171	29	00m18s
7139	357	1006 Nov 23	17:34:38	1525	-12283	97	A	-p	-0.5074	0.9596	52.3S	83.0W	59	6	171	03m43s
7140	357	1007 May 19	08:24:59	1522	-12277	102	Tm	nn	0.0409	1.0566	23.3N	58.6E	88	175	187	05m17s

Cat Num	Canon Plate	Calendar Date	TD of Greatest Eclipse	ΔT s	Luna Num	Saros Num	Ecl. Type	QLE	Gamma	Ecl. Mag.	Lat. °	Long. °	Sun Alt °	Sun Azm °	Path Width km	Central Line Dur.
7141	358	1007 Nov 12	20:00:23	1519	-12271	107	A	nn	0.1855	0.9233	8.8S	115.6W	79	188	294	10m49s
7142	358	1008 May 08	01:04:50	1517	-12265	112	T	p-	-0.6940	1.0734	25.1S	175.9E	46	350	330	06m22s
7143	358	1008 Oct 31	19:08:48	1514	-12259	117	A	p-	0.8676	0.9207	43.0N	90.9W	29	195	601	08m43s
7144	358	1009 Mar 29	08:11:00	1512	-12254	84	P	-t	1.1762	0.6719	71.8N	42.7W	0	72		
7145	358	1009 Apr 27	17:33:05	1511	-12253	122	P	t-	-1.4445	0.1700	70.1S	38.6W	0	322		
7146	358	1009 Sep 21	09:42:23	1509	-12248	89	P	-t	-1.2777	0.4860	71.9S	55.9W	0	95		
7147	358	1009 Oct 20	22:36:10	1509	-12247	127	P	t-	1.5123	0.0753	70.9N	106.7W	0	228		
7148	358	1010 Mar 18	16:00:00	1507	-12242	94	A	-p	0.4716	0.9581	28.1N	61.4W	62	160	172	04m36s
7149	358	1010 Sep 10	23:51:11	1504	-12236	99	T	-p	-0.5251	1.0557	27.5S	176.7E	58	20	215	04m42s
7150	358	1011 Mar 07	17:04:52	1501	-12230	104	A	nn	-0.2559	0.9311	17.3S	62.7W	75	342	266	08m44s
7151	358	1011 Aug 31	16:27:51	1499	-12224	109	T	n-	0.1851	1.0612	17.1N	58.0W	79	197	204	05m25s
7152	358	1012 Feb 24	17:31:58	1496	-12218	114	A	t-	-0.9593	0.9376	71.7S	11.9W	16	295	845	04m32s
7153	358	1012 Aug 20	06:50:51	1494	-12212	119	T	p-	0.9437	1.0086	73.0N	143.6E	19	244	91	00m32s
7154	358	1013 Jan 14	11:52:21	1491	-12207	86	P	-t	1.1531	0.7148	68.7N	7.9W	0	162		
7155	358	1013 Jul 11	00:53:58	1489	-12201	91	P	-t	-1.0554	0.8698	68.1S	162.5E	0	10		
7156	358	1014 Jan 04	02:45:41	1486	-12195	96	T	-n	0.4565	1.0462	5.0N	146.0E	63	178	173	04m29s
7157	358	1014 Jun 30	01:27:23	1484	-12189	101	A	np	-0.3013	0.9515	5.5N	165.0E	72	360	186	06m45s
7158	358	1014 Dec 24	18:08:45	1481	-12183	106	T	n-	-0.2153	1.0272	35.9S	85.3W	77	2	95	02m25s
7159	358	1015 Jun 19	05:08:27	1478	-12177	111	A	p-	0.4748	0.9894	52.0N	105.3E	61	173	43	00m57s
7160	358	1015 Dec 14	05:22:20	1476	-12171	116	A	p-	-0.9420	0.9633	82.5S	19.5W	19	127	407	02m14s
7161	359	1016 May 09	08:20:53	1474	-12166	83	P	-t	-1.3396	0.3670	62.9S	107.8E	0	315		
7162	359	1016 Jun 07	16:00:58	1473	-12165	121	P	t-	1.2042	0.6246	65.1N	147.4E	0	20		
7163	359	1016 Nov 02	14:30:40	1471	-12160	88	P	-t	1.3559	0.3582	62.3N	19.7E	0	232		
7164	359	1017 Apr 29	01:22:47	1469	-12154	93	T	-p	-0.6087	1.0706	18.7S	178.9E	52	337	286	05m56s
7165	359	1017 Oct 22	14:13:22	1466	-12148	98	A	-p	0.6573	0.9356	24.1N	13.1W	49	207	311	07m14s
7166	359	1018 Apr 18	16:54:52	1464	-12142	103	T	nn	0.1290	1.0344	19.5N	71.7W	82	154	117	03m03s
7167	359	1018 Oct 11	19:50:54	1461	-12136	108	A	nn	-0.0806	0.9923	13.5S	117.3W	85	27	27	00m45s
7168	359	1019 Apr 08	02:36:00	1458	-12130	113	A	p-	0.9222	0.9663	60.6N	87.3E	22	113	311	02m23s
7169	359	1019 Oct 01	08:34:20	1456	-12124	118	T	p-	-0.7781	1.0386	47.8S	20.9E	39	45	202	02m40s
7170	359	1020 Feb 26	11:58:42	1454	-12119	85	P	-t	-1.2592	0.5190	61.0S	111.9E	0	256		
7171	359	1020 Aug 21	16:12:38	1451	-12113	90	P	-t	1.1354	0.7563	61.4N	52.1E	0	292		
7172	359	1020 Sep 20	00:47:49	1451	-12112	128	P	t-	-1.4474	0.1625	61.0S	80.2E	0	92		
7173	359	1021 Feb 14	13:33:29	1449	-12107	95	A	-p	-0.5218	0.9568	38.5S	3.9E	58	326	182	04m01s
7174	359	1021 Aug 11	05:15:54	1446	-12101	100	H	-p	0.4287	1.0046	36.6N	120.8E	64	210	17	00m24s
7175	359	1022 Feb 03	22:00:47	1444	-12095	105	H	n-	0.2142	1.0113	3.4S	145.2W	78	156	40	01m06s
7176	359	1022 Jul 31	11:25:00	1441	-12089	110	A	pn	-0.3406	0.9580	1.5S	8.4E	70	21	161	05m03s
7177	359	1023 Jan 24	12:18:00	1439	-12083	115	T	p-	0.8869	1.0385	40.5N	19.7W	27	153	276	03m00s
7178	359	1023 Jul 20	12:11:29	1436	-12077	120	P	t-	-1.1015	0.7883	63.0S	40.6W	0	42		
7179	359	1023 Dec 15	16:36:13	1434	-12072	87	P	-t	-1.1640	0.6940	65.9S	103.9E	0	192		
7180	359	1024 Jun 09	03:04:03	1431	-12066	92	A	-p	0.9150	0.9971	87.6N	130.3W	23	271	25	00m10s
7181	360	1024 Dec 04	01:56:15	1429	-12060	97	A	-p	-0.5130	0.9561	54.2S	156.2E	59	359	187	04m00s
7182	360	1025 May 29	15:46:10	1426	-12054	102	T	nn	0.1145	1.0602	29.2N	51.8W	83	178	199	05m25s
7183	360	1025 Nov 23	03:59:16	1424	-12048	107	A	nn	0.1758	0.9211	11.5S	124.3E	80	184	303	11m14s
7184	360	1026 May 19	08:37:49	1421	-12042	112	T	p-	-0.6251	1.0758	17.8S	58.9E	51	355	314	06m52s
7185	360	1026 Nov 12	03:05:07	1419	-12036	117	A	p-	0.8527	0.9208	39.0N	145.2E	31	190	573	09m08s
7186	360	1027 Apr 09	15:45:07	1417	-12031	84	P	-t	1.2337	0.5664	71.4N	170.4W	0	59		
7187	360	1027 May 09	01:00:48	1416	-12030	122	P	t-	-1.3840	0.2848	69.3S	163.3W	0	334		
7188	360	1027 Oct 02	17:46:43	1414	-12025	89	P	-t	-1.3036	0.4389	71.7S	168.6E	0	109		
7189	360	1027 Nov 01	06:48:58	1414	-12024	127	P	t-	1.4930	0.1088	70.1N	116.7E	0	215		
7190	360	1028 Mar 28	23:15:25	1412	-12019	94	A	-p	0.5262	0.9579	35.8N	173.3W	58	159	179	04m24s
7191	360	1028 Sep 21	08:01:19	1409	-12013	99	T	-p	-0.5611	1.0535	34.0S	51.3E	56	22	212	04m19s
7192	360	1029 Mar 18	00:14:06	1407	-12007	104	A	nn	-0.2064	0.9342	10.2S	172.3W	78	342	250	08m32s
7193	360	1029 Sep 11	00:29:28	1404	-12001	109	T	n-	0.1422	1.0567	10.4N	179.6E	82	198	189	05m07s
7194	360	1030 Mar 07	01:01:54	1402	-11995	114	A	t-	-0.9185	0.9445	64.1S	147.6W	23	317	519	04m25s
7195	360	1030 Aug 31	14:27:31	1399	-11989	119	H	p-	0.8957	1.0044	64.8N	6.5E	26	223	34	00m18s
7196	360	1031 Jan 25	20:26:16	1397	-11984	86	P	-t	1.1606	0.7018	69.7N	148.9W	0	150		
7197	360	1031 Jul 22	07:27:48	1395	-11978	91	P	-t	-1.1369	0.7287	69.1S	52.1E	0	22		
7198	360	1032 Jan 15	11:33:38	1392	-11972	96	T	-n	0.4632	1.0479	7.4N	12.5E	62	173	179	04m35s
7199	360	1032 Jul 10	07:55:08	1390	-11966	101	A	-p	-0.3866	0.9513	1.1S	66.4E	67	4	193	06m46s
7200	360	1033 Jan 04	02:54:50	1387	-11960	106	T	n-	-0.2122	1.0260	34.5S	145.2E	78	357	91	02m21s

Cat Num	Canon Plate	Calendar Date	TD of Greatest Eclipse	ΔT s	Luna Num	Saros Num	Ecl. Type	QLE	Gamma	Ecl. Mag.	Lat. °	Long. °	Sun Alt °	Sun Azm °	Path Width km	Central Line Dur.
7201	361	1033 Jun 29	11:57:10	1385	-11954	111	A	p-	0.3912	0.9928	46.2N	6.8E	67	180	27	00m40s
7202	361	1033 Dec 24	13:51:14	1383	-11948	116	A	p-	-0.9418	0.9609	85.6S	179.6W	19	159	434	02m24s
7203	361	1034 May 20	15:48:37	1380	-11943	83	P	-t	-1.4058	0.2391	63.7S	14.0W	0	324		
7204	361	1034 Jun 18	23:16:23	1380	-11942	121	P	t-	1.1262	0.7724	66.0N	28.1E	0	10		
7205	361	1034 Nov 13	22:23:18	1378	-11937	88	P	-t	1.3704	0.3345	63.0N	107.9W	0	222		
7206	361	1035 May 10	08:58:31	1376	-11931	93	T	-p	-0.6762	1.0703	21.4S	64.1E	47	340	307	06m00s
7207	361	1035 Nov 02	22:09:33	1373	-11925	98	A	-p	0.6737	0.9352	22.7N	134.1W	47	203	320	07m32s
7208	361	1036 Apr 29	00:21:45	1371	-11919	103	Tm	nn	0.0642	1.0335	19.6N	177.6E	86	157	113	03m04s
7209	361	1036 Oct 22	04:03:26	1368	-11913	108	A	nn	-0.0593	0.9928	16.4S	119.5E	87	25	25	00m42s
7210	361	1037 Apr 18	09:42:40	1366	-11907	113	A	p-	0.8607	0.9679	60.7N	9.9W	30	121	225	02m21s
7211	361	1037 Oct 11	16:57:23	1363	-11901	118	T	p-	-0.7512	1.0382	50.3S	104.8W	41	45	191	02m36s
7212	361	1038 Mar 08	19:16:36	1361	-11896	85	P	-t	-1.3010	0.4487	60.9S	6.6W	0	265		
7213	361	1038 Sep 02	00:04:02	1359	-11890	90	P	-t	1.1865	0.6587	61.1N	74.9W	0	283		
7214	361	1038 Oct 01	09:05:32	1359	-11889	128	P	t-	-1.4177	0.2202	61.2S	53.4W	0	101		
7215	361	1039 Feb 25	21:18:55	1357	-11884	95	A	-p	-0.5521	0.9620	35.8S	112.0W	56	325	163	03m31s
7216	361	1039 Aug 22	12:38:24	1354	-11878	100	A	-p	0.4896	0.9980	35.8N	12.1E	60	213	8	00m10s
7217	361	1040 Feb 15	06:17:48	1352	-11872	105	H2	n-	0.1916	1.0169	1.0S	90.0E	79	154	59	01m35s
7218	361	1040 Aug 10	18:14:08	1349	-11866	110	A	nn	-0.2696	0.9539	0.4S	93.7W	74	24	174	05m24s
7219	361	1041 Feb 03	20:54:13	1347	-11860	115	T	p-	0.8704	1.0424	40.4N	152.0W	29	149	283	03m13s
7220	361	1041 Jul 30	18:45:44	1345	-11854	120	P	t-	-1.0226	0.9223	62.3S	148.7W	0	51		
7221	362	1041 Dec 26	01:19:06	1343	-11849	87	P	-t	-1.1695	0.6835	64.8S	37.3W	0	202		
7222	362	1042 Jun 20	10:01:14	1340	-11843	92	A	-t	0.9903	0.9958	71.9N	167.8W	7	339	126	00m13s
7223	362	1042 Dec 15	10:17:55	1338	-11837	97	A	-p	-0.5180	0.9533	54.8S	36.1E	59	351	200	04m14s
7224	362	1043 Jun 09	23:06:47	1335	-11831	102	T	nn	0.1893	1.0630	34.5N	161.0W	79	183	211	05m25s
7225	362	1043 Dec 04	12:01:48	1333	-11825	107	A	nn	0.1691	0.9196	13.4S	3.7E	80	179	309	11m28s
7226	362	1044 May 29	16:06:22	1331	-11819	112	T	p-	-0.5525	1.0775	11.1S	56.2W	56	359	300	07m12s
7227	362	1044 Nov 22	11:07:50	1328	-11813	117	A	p-	0.8426	0.9213	36.0N	20.3E	32	185	555	09m24s
7228	362	1045 Apr 19	23:10:13	1326	-11808	84	P	-t	1.2979	0.4491	70.7N	64.7E	0	46		
7229	362	1045 May 19	08:21:52	1326	-11807	122	P	t-	-1.3184	0.4089	68.3S	74.3E	0	346		
7230	362	1045 Oct 13	01:59:57	1324	-11802	89	P	-t	-1.3229	0.4038	71.2S	31.1E	0	122		
7231	362	1045 Nov 11	15:09:04	1323	-11801	127	P	t-	1.4792	0.1328	69.1N	21.0W	0	203		
7232	362	1046 Apr 09	06:20:59	1321	-11796	94	A	-p	0.5880	0.9576	44.0N	77.2E	54	158	190	04m11s
7233	362	1046 Oct 02	16:19:55	1319	-11790	99	T	-p	-0.5904	1.0512	40.3S	75.8W	54	23	209	03m58s
7234	362	1047 Mar 29	07:16:02	1317	-11784	104	A	nn	-0.1510	0.9373	2.9S	79.8E	81	343	236	08m15s
7235	362	1047 Sep 22	08:36:53	1314	-11778	109	T	nn	0.1046	1.0519	4.0N	55.8E	84	198	173	04m47s
7236	362	1048 Mar 17	08:26:35	1312	-11772	114	A	p-	-0.8726	0.9514	55.6S	88.7E	29	327	363	04m13s
7237	362	1048 Sep 10	22:09:40	1310	-11766	119	A	p-	0.8530	0.9995	56.8N	119.9W	31	214	4	00m02s
7238	362	1049 Feb 05	04:54:51	1308	-11761	86	P	-t	1.1733	0.6792	70.5N	70.9E	0	137		
7239	362	1049 Mar 06	16:00:57	1307	-11760	124	Pb	t-	-1.5374	0.0138	71.9S	47.7E	0	260		
7240	362	1049 Aug 01	14:05:43	1305	-11755	91	P	-t	-1.2142	0.5957	70.0S	59.8W	0	33		
7241	363	1050 Jan 25	20:16:08	1303	-11749	96	T	-p	0.4746	1.0499	10.6N	119.8W	62	170	188	04m41s
7242	363	1050 Jul 21	14:29:16	1301	-11743	101	A	-p	-0.4670	0.9506	8.2S	34.4W	62	7	204	06m42s
7243	363	1051 Jan 15	11:35:11	1298	-11737	106	T	n-	-0.2049	1.0252	31.9S	16.7E	78	352	88	02m18s
7244	363	1051 Jul 10	18:52:23	1296	-11731	111	A	p-	0.3120	0.9957	39.9N	95.1W	72	185	16	00m26s
7245	363	1052 Jan 04	22:16:17	1294	-11725	116	A	p-	-0.9395	0.9591	86.3S	14.2W	20	227	446	02m33s
7246	363	1052 May 30	23:13:57	1292	-11720	83	Pe	-t	-1.4743	0.1062	64.5S	135.3W	0	334		
7247	363	1052 Jun 29	06:33:32	1291	-11719	121	P	t-	1.0488	0.9196	67.0N	92.1W	0	0		
7248	363	1052 Nov 24	06:20:27	1289	-11714	88	P	-t	1.3809	0.3173	63.9N	123.1E	0	212		
7249	363	1053 May 20	16:28:31	1287	-11708	93	T	-p	-0.7481	1.0690	25.7S	49.6W	41	343	336	05m55s
7250	363	1053 Nov 13	06:13:19	1285	-11702	98	A	-p	0.6846	0.9355	21.6N	102.7E	47	199	324	07m44s
7251	363	1054 May 10	07:40:25	1283	-11696	103	T	nn	-0.0066	1.0319	18.8N	69.0E	90	332	108	03m02s
7252	363	1054 Nov 02	12:25:18	1280	-11690	108	A	nn	-0.0451	0.9937	19.2S	5.8W	87	22	22	00m38s
7253	363	1055 Apr 29	16:40:32	1278	-11684	113	A	p-	0.7927	0.9687	60.8N	105.2W	37	130	183	02m22s
7254	363	1055 Oct 23	01:27:47	1276	-11678	118	T	p-	-0.7301	1.0377	53.4S	128.1E	43	44	183	02m32s
7255	363	1056 Mar 19	02:27:00	1274	-11673	85	P	-t	-1.3488	0.3678	60.9S	123.2W	0	274		
7256	363	1056 Sep 12	08:01:37	1271	-11667	90	P	-t	1.2319	0.5725	60.9N	156.7E	0	274		
7257	363	1056 Oct 11	17:30:26	1271	-11666	128	P	t-	-1.3942	0.2656	61.4S	171.3E	0	110		
7258	363	1057 Mar 08	04:57:18	1269	-11661	95	A	-p	-0.5888	0.9675	33.6S	133.5E	54	325	142	03m01s
7259	363	1057 Sep 01	20:06:25	1267	-11655	100	A	-p	0.5448	0.9912	34.6N	98.6W	57	215	37	00m46s
7260	363	1058 Feb 25	14:28:43	1265	-11649	105	T	n-	0.1634	1.0229	1.5N	33.1W	81	152	79	02m05s

Cat Num	Canon Plate	Calendar Date	TD of Greatest Eclipse	ΔT s	Luna Num	Saros Num	Ecl. Type	QLE	Gamma	Ecl. Mag.	Lat. °	Long. °	Sun Alt °	Sun Azm °	Path Width km	Central Line Dur.
7261	364	1058 Aug 22	01:09:14	1262	-11643	110	A	nn	-0.2043	0.9496	0.4S	162.7E	78	27	188	05m47s
7262	364	1059 Feb 15	05:24:50	1260	-11637	115	T	p-	0.8492	1.0465	40.6N	77.7E	32	145	287	03m26s
7263	364	1059 Aug 11	01:27:30	1258	-11631	120	A	t-	-0.9493	0.9337	49.9S	125.8E	18	39	775	06m10s
7264	364	1060 Jan 06	09:59:33	1256	-11626	87	P	-t	-1.1766	0.6704	63.9S	177.5W	0	213		
7265	364	1060 Feb 04	21:21:56	1255	-11625	125	Pb	t-	1.5334	0.0080	62.1N	167.8E	0	122		
7266	364	1060 Jun 30	17:01:02	1254	-11620	92	P	-t	1.0642	0.8763	64.6N	82.7E	0	334		
7267	364	1060 Dec 25	18:38:02	1251	-11614	97	A	-p	-0.5234	0.9511	54.2S	83.7W	58	344	211	04m26s
7268	364	1061 Jun 20	06:28:17	1249	-11608	102	T	-n	0.2641	1.0651	38.9N	90.7E	74	189	221	05m20s
7269	364	1061 Dec 14	20:03:51	1247	-11602	107	A	nn	0.1623	0.9187	14.3S	116.7W	81	175	312	11m29s
7270	364	1062 Jun 09	23:34:05	1244	-11596	112	T	p-	-0.4793	1.0781	5.2S	170.2W	61	3	287	07m20s
7271	364	1062 Dec 03	19:13:41	1242	-11590	117	A	p-	0.8342	0.9223	33.8N	105.1W	33	180	534	09m26s
7272	364	1063 May 01	06:27:56	1240	-11585	84	P	-t	1.3668	0.3234	69.9N	57.8W	0	34		
7273	364	1063 May 30	15:39:34	1240	-11584	122	P	t-	-1.2508	0.5365	67.4S	46.7W	0	356		
7274	364	1063 Oct 24	10:20:36	1238	-11579	89	P	-t	-1.3370	0.3782	70.6S	107.7W	0	136		
7275	364	1063 Nov 22	23:33:39	1238	-11578	127	P	t-	1.4681	0.1517	68.1N	159.2W	0	191		
7276	364	1064 Apr 19	13:17:27	1236	-11573	94	A	-p	0.6564	0.9568	52.8N	30.1W	49	157	208	03m58s
7277	364	1064 Oct 13	00:46:25	1234	-11567	99	T	-p	-0.6133	1.0490	46.2S	156.0E	52	23	205	03m40s
7278	364	1065 Apr 08	14:06:48	1231	-11561	104	A	nn	-0.0864	0.9403	4.7N	25.3W	85	344	222	07m54s
7279	364	1065 Oct 02	16:52:55	1229	-11555	109	T	nn	0.0747	1.0471	2.0S	69.9W	86	197	157	04m24s
7280	364	1066 Mar 28	15:41:57	1227	-11549	114	A	p-	-0.8181	0.9582	46.7S	28.8W	35	335	262	03m57s
7281	365	1066 Sep 22	05:59:31	1225	-11543	119	A	p-	0.8173	0.9941	49.7N	115.7E	35	208	35	00m29s
7282	365	1067 Feb 16	13:16:39	1223	-11538	86	P	-t	1.1920	0.6453	71.2N	68.2W	0	124		
7283	365	1067 Mar 17	23:50:58	1222	-11537	124	P	t-	-1.4938	0.0904	71.9S	84.3W	0	274		
7284	365	1067 Aug 12	20:49:40	1221	-11532	91	P	-t	-1.2857	0.4734	70.8S	173.9W	0	46		
7285	365	1068 Feb 06	04:53:52	1218	-11526	96	T	-p	0.4899	1.0521	14.7N	108.8E	61	166	197	04m46s
7286	365	1068 Jul 31	21:07:35	1216	-11520	101	A	-p	-0.5441	0.9495	15.8S	137.0W	57	11	220	06m33s
7287	365	1069 Jan 25	20:11:55	1214	-11514	106	T	n-	-0.1952	1.0249	28.4S	111.7W	79	349	86	02m19s
7288	365	1069 Jul 21	01:51:43	1212	-11508	111	A	n-	0.2355	0.9979	33.2N	160.5E	76	189	8	00m13s
7289	365	1070 Jan 15	06:37:03	1209	-11502	116	A	p-	-0.9346	0.9580	83.3S	174.7E	20	272	440	02m42s
7290	365	1070 Jul 10	13:54:30	1207	-11496	121	T	t-	0.9739	1.0404	80.0N	133.7E	12	338	636	02m05s
7291	365	1070 Dec 05	14:20:57	1205	-11491	88	P	-t	1.3881	0.3054	64.8N	7.0W	0	202		
7292	365	1071 May 31	23:56:38	1203	-11485	93	T	-p	-0.8213	1.0667	31.6S	163.5W	35	347	381	05m38s
7293	365	1071 Nov 24	14:21:36	1201	-11479	98	A	-p	0.6922	0.9362	20.8N	21.8W	46	195	325	07m51s
7294	365	1072 May 20	14:54:52	1199	-11473	103	T	nn	-0.0801	1.0299	16.9N	38.6W	86	344	101	02m58s
7295	365	1072 Nov 12	20:52:43	1197	-11467	108	A	nn	-0.0348	0.9948	21.6S	132.2W	88	18	18	00m31s
7296	365	1073 May 09	23:30:35	1195	-11461	113	A	p-	0.7189	0.9690	60.3N	161.7E	44	140	160	02m27s
7297	365	1073 Nov 02	10:05:13	1192	-11455	118	T	p-	-0.7148	1.0373	56.8S	0.1E	44	42	178	02m29s
7298	365	1074 Mar 30	09:28:21	1191	-11450	85	P	-t	-1.4035	0.2745	61.0S	122.5E	0	282		
7299	365	1074 Apr 29	01:23:46	1190	-11449	123	Pb	t-	1.4965	0.1151	62.2N	39.8E	0	54		
7300	365	1074 Sep 23	16:06:44	1188	-11444	90	P	-t	1.2706	0.4998	60.9N	26.4E	0	265		
7301	366	1074 Oct 23	02:01:48	1188	-11443	128	P	t-	-1.3765	0.2999	61.9S	34.2E	0	119		
7302	366	1075 Mar 19	12:27:45	1186	-11438	95	A	-p	-0.6326	0.9728	32.2S	20.9E	51	325	123	02m32s
7303	366	1075 Sep 13	03:41:46	1184	-11432	100	A	-p	0.5929	0.9844	33.1N	148.1E	53	215	67	01m23s
7304	366	1076 Mar 07	22:31:53	1182	-11426	105	T	n-	0.1281	1.0290	3.9N	154.1W	83	151	99	02m34s
7305	366	1076 Sep 01	08:10:51	1180	-11420	110	A	nn	-0.1448	0.9452	1.2S	57.6E	82	28	204	06m13s
7306	366	1077 Feb 25	13:47:25	1178	-11414	115	T	p-	0.8214	1.0510	41.0N	49.7W	35	142	290	03m40s
7307	366	1077 Aug 21	08:17:11	1175	-11408	120	A	t-	-0.8817	0.9344	43.9S	27.0E	28	38	502	06m21s
7308	366	1078 Jan 16	18:34:30	1174	-11403	87	P	-t	-1.1878	0.6499	63.0S	44.1E	0	222		
7309	366	1078 Feb 15	05:47:57	1173	-11402	125	P	t-	1.5158	0.0407	61.6N	32.0E	0	113		
7310	366	1078 Jul 12	00:05:18	1171	-11397	92	P	-t	1.1351	0.7474	63.7N	33.3W	0	325		
7311	366	1079 Jan 06	02:55:27	1169	-11391	97	A	-p	-0.5307	0.9494	52.5S	156.6E	58	337	219	04m35s
7312	366	1079 Jul 01	13:51:08	1167	-11385	102	T	-n	0.3381	1.0663	42.3N	16.9W	70	195	230	05m12s
7313	366	1079 Dec 26	04:06:10	1165	-11379	107	A	nn	0.1559	0.9185	14.4S	123.0E	81	170	313	11m18s
7314	366	1080 Jun 20	07:00:13	1163	-11373	112	T	p-	-0.4047	1.0779	0.2S	76.9E	66	7	275	07m18s
7315	366	1080 Dec 14	03:22:42	1161	-11367	117	A	p-	0.8281	0.9239	32.4N	129.0E	34	174	512	09m16s
7316	366	1081 May 11	13:37:25	1159	-11362	84	P	-t	1.4413	0.1880	69.0N	177.6W	0	22		
7317	366	1081 Jun 09	22:52:36	1159	-11361	122	P	t-	-1.1800	0.6695	66.4S	166.1W	0	7		
7318	366	1081 Nov 03	18:49:33	1157	-11356	89	P	-t	-1.3451	0.3634	69.7S	112.0E	0	148		
7319	366	1081 Dec 03	08:03:27	1156	-11355	127	P	t-	1.4607	0.1642	67.0N	62.0E	0	179		
7320	366	1082 Apr 30	20:06:15	1155	-11350	94	A	-p	0.7301	0.9556	62.1N	135.6W	43	157	238	03m46s

Cat Num	Canon Plate	Calendar Date	TD of Greatest Eclipse	ΔT s	Luna Num	Saros Num	Ecl. Type	QLE	Gamma	Ecl. Mag.	Lat. °	Long. °	Sun Alt °	Sun Azm °	Path Width km	Central Line Dur.
7321	367	1082 Oct 24	09:19:34	1153	-11344	99	T	-p	-0.6308	1.0470	51.7S	27.3E	51	21	201	03m24s
7322	367	1083 Apr 19	20:52:17	1150	-11338	104	A	nn	-0.0173	0.9432	12.3N	128.8W	89	345	210	07m28s
7323	367	1083 Oct 14	01:15:15	1148	-11332	109	T	nn	0.0503	1.0424	7.4S	163.1E	87	196	142	04m00s
7324	367	1084 Apr 07	22:52:04	1146	-11326	114	A	p-	-0.7585	0.9649	38.1S	142.9W	40	340	192	03m35s
7325	367	1084 Oct 02	13:55:25	1144	-11320	119	A	p-	0.7874	0.9887	43.3N	8.6W	38	204	64	01m00s
7326	367	1085 Feb 26	21:32:13	1142	-11315	86	P	-t	1.2162	0.6007	71.7N	153.8E	0	110		
7327	367	1085 Mar 28	07:34:42	1142	-11314	124	P	t-	-1.4444	0.1790	71.8S	145.3E	0	287		
7328	367	1085 Aug 23	03:40:37	1140	-11309	91	P	-t	-1.3506	0.3635	71.4S	69.7E	0	59		
7329	367	1086 Feb 16	13:23:18	1138	-11303	96	T	-p	0.5120	1.0544	19.7N	20.9W	59	164	208	04m48s
7330	367	1086 Aug 12	03:55:33	1136	-11297	101	A	-p	-0.6134	0.9482	23.5S	117.3E	52	15	239	06m20s
7331	367	1087 Feb 06	04:40:57	1134	-11291	106	T	n-	-0.1801	1.0248	24.0S	121.1E	79	346	86	02m20s
7332	367	1087 Aug 01	08:58:39	1132	-11285	111	A	nn	0.1644	0.9996	26.2N	53.1E	80	192	1	00m02s
7333	367	1088 Jan 26	14:50:56	1130	-11279	116	A	p-	-0.9249	0.9575	78.9S	29.2E	22	293	415	02m50s
7334	367	1088 Jul 20	21:20:06	1128	-11273	121	T	t-	0.9023	1.0453	80.7N	83.3W	25	234	356	02m36s
7335	367	1088 Dec 15	22:22:23	1126	-11268	88	P	-t	1.3947	0.2944	65.9N	137.8W	0	192		
7336	367	1089 Jun 11	07:20:55	1124	-11262	93	T	-t	-0.8972	1.0629	40.1S	82.9E	26	351	469	05m05s
7337	367	1089 Dec 04	22:34:53	1122	-11256	98	A	-p	0.6963	0.9378	20.5N	147.6W	46	190	321	07m48s
7338	367	1090 May 31	22:03:44	1120	-11250	103	T	nn	-0.1573	1.0270	14.0N	145.2W	81	349	93	02m48s
7339	367	1090 Nov 24	05:25:50	1118	-11244	108	A	nn	-0.0286	0.9965	23.5S	100.3E	88	14	12	00m21s
7340	367	1091 May 21	06:14:32	1116	-11238	113	A	p-	0.6408	0.9687	58.7N	69.9E	50	150	146	02m37s
7341	368	1091 Nov 13	18:48:57	1114	-11232	118	T	p-	-0.7047	1.0371	60.3S	128.2W	45	38	175	02m26s
7342	368	1092 Apr 09	16:21:49	1112	-11227	85	P	-t	-1.4644	0.1696	61.4S	10.1E	0	291		
7343	368	1092 May 09	07:55:04	1111	-11226	123	P	t-	1.4182	0.2481	62.8N	67.4W	0	45		
7344	368	1092 Oct 04	00:18:28	1110	-11221	90	P	-t	1.3034	0.4387	61.1N	105.6W	0	256		
7345	368	1092 Nov 02	10:39:37	1109	-11220	128	P	t-	-1.3643	0.3234	62.4S	104.5W	0	129		
7346	368	1093 Mar 29	19:51:50	1108	-11215	95	A	-p	-0.6824	0.9782	31.6S	90.2W	47	326	103	02m02s
7347	368	1093 Sep 23	11:23:51	1106	-11209	100	A	-p	0.6346	0.9777	31.5N	32.4E	50	215	101	02m03s
7348	368	1094 Mar 19	06:28:14	1104	-11203	105	T	nn	0.0862	1.0350	6.1N	86.7E	85	151	118	03m04s
7349	368	1094 Sep 12	15:20:29	1102	-11197	110	A	nn	-0.0922	0.9408	2.8S	49.7W	85	29	220	06m41s
7350	368	1095 Mar 08	22:03:57	1100	-11191	115	T	p-	0.7883	1.0553	41.8N	174.8W	38	140	291	03m54s
7351	368	1095 Sep 01	15:14:39	1097	-11185	120	A	p-	-0.8197	0.9343	41.2S	76.0W	35	39	414	06m24s
7352	368	1096 Jan 28	03:04:32	1096	-11180	87	P	-t	-1.2026	0.6230	62.3S	92.9W	0	232		
7353	368	1096 Feb 26	14:07:35	1095	-11179	125	P	t-	1.4936	0.0820	61.3N	102.0W	0	104		
7354	368	1096 Jul 22	07:15:03	1094	-11174	92	P	-t	1.2025	0.6245	63.0N	150.4W	0	316		
7355	368	1096 Aug 20	18:35:35	1093	-11173	130	Pb	t-	-1.5110	0.0743	61.4S	164.7W	0	68		
7356	368	1097 Jan 16	11:07:06	1092	-11168	97	A	-p	-0.5420	0.9483	50.1S	37.2E	57	332	225	04m41s
7357	368	1097 Jul 11	21:17:18	1090	-11162	102	T	-n	0.4099	1.0667	44.5N	124.6W	66	201	239	05m01s
7358	368	1098 Jan 05	12:04:49	1088	-11156	107	A	nn	0.1464	0.9189	13.8S	3.7E	82	166	311	10m56s
7359	368	1098 Jul 01	14:28:20	1086	-11150	112	T	n-	-0.3320	1.0768	3.8N	35.8W	71	11	263	07m05s
7360	368	1098 Dec 25	11:30:26	1084	-11144	117	A	p-	0.8201	0.9263	31.4N	3.5E	35	169	483	08m53s
7361	369	1099 May 22	20:42:06	1082	-11139	84	Pe	-t	1.5185	0.0483	68.1N	64.4E	0	11		
7362	369	1099 Jun 21	06:05:15	1082	-11138	122	P	t-	-1.1092	0.8015	65.4S	75.1E	0	17		
7363	369	1099 Nov 15	03:24:19	1080	-11133	89	P	-t	-1.3496	0.3551	68.8S	29.2W	0	161		
7364	369	1099 Dec 14	16:35:04	1080	-11132	127	P	t-	1.4537	0.1757	65.9N	76.8W	0	169		
7365	369	1100 May 11	02:46:35	1078	-11127	94	A	-p	0.8100	0.9537	72.4N	119.9E	36	155	291	03m36s
7366	369	1100 Nov 03	17:59:41	1076	-11121	99	T	-p	-0.6430	1.0453	56.5S	101.5W	50	18	197	03m11s
7367	369	1101 Apr 30	03:29:14	1074	-11115	104	Am	nn	0.0594	0.9459	20.0N	130.3E	86	169	200	06m57s
7368	369	1101 Oct 24	09:45:16	1072	-11109	109	T	nn	0.0328	1.0378	12.3S	34.7E	88	194	127	03m37s
7369	369	1102 Apr 19	05:54:33	1070	-11103	114	A	p-	-0.6913	0.9714	29.5S	106.2E	46	344	141	03m07s
7370	369	1102 Oct 13	21:59:25	1068	-11097	119	A	p-	0.7648	0.9833	37.8N	133.9W	40	200	91	01m35s
7371	369	1103 Mar 10	05:40:52	1066	-11092	86	P	-t	1.2466	0.5440	71.9N	17.3E	0	96		
7372	369	1103 Apr 08	15:12:10	1066	-11091	124	P	t-	-1.3890	0.2797	71.4S	16.9E	0	301		
7373	369	1103 Sep 03	10:38:20	1064	-11086	91	P	-t	-1.4094	0.2646	71.8S	48.8W	0	72		
7374	369	1103 Oct 03	03:17:50	1064	-11085	129	Pb	t-	1.5318	0.0556	71.7N	157.1W	0	250		
7375	369	1104 Feb 27	21:46:57	1062	-11080	96	T	-p	0.5390	1.0568	25.3N	149.3W	57	161	221	04m49s
7376	369	1104 Aug 22	10:50:41	1060	-11074	101	A	-p	-0.6772	0.9466	31.5S	9.2E	47	18	264	06m04s
7377	369	1105 Feb 16	13:02:44	1058	-11068	106	T	nn	-0.1592	1.0249	18.9S	5.0W	81	344	86	02m23s
7378	369	1105 Aug 11	16:12:35	1056	-11062	111	H	nn	0.0982	1.0008	19.0N	56.6W	84	194	3	00m05s
7379	369	1106 Feb 05	22:58:18	1054	-11056	116	A	p-	-0.9106	0.9575	73.6S	106.6W	24	306	378	02m59s
7380	369	1106 Aug 01	04:51:33	1052	-11050	121	T	p-	0.8348	1.0481	70.8N	142.2E	33	214	292	03m00s

Cat Num	Canon Plate	Calendar Date	TD of Greatest Eclipse	ΔT s	Luna Num	Saros Num	Ecl. Type	QLE	Gamma	Ecl. Mag.	Lat. °	Long. °	Sun Alt °	Sun Azm °	Path Width km	Central Line Dur.
7381	370	1106 Dec 27	06:22:57	1051	-11045	88	P	-t	1.4019	0.2823	67.0N	91.3E	0	181		
7382	370	1107 Jun 22	14:45:52	1049	-11039	93	T	-t	-0.9722	1.0570	53.6S	32.0W	13	355	834	04m10s
7383	370	1107 Dec 16	06:49:58	1047	-11033	98	A	-p	0.6995	0.9398	20.8N	86.1E	45	185	312	07m35s
7384	370	1108 Jun 11	05:09:17	1045	-11027	103	T	-n	-0.2364	1.0235	10.0N	108.5E	76	353	82	02m32s
7385	370	1108 Dec 04	14:02:16	1043	-11021	108	A	nn	-0.0245	0.9986	24.7S	27.7W	88	9	5	00m08s
7386	370	1109 May 31	12:53:47	1041	-11015	113	A	p-	0.5596	0.9678	55.9N	21.8W	56	160	140	02m51s
7387	370	1109 Nov 24	03:36:12	1039	-11009	118	T	p-	-0.6974	1.0372	63.5S	104.6E	46	32	175	02m26s
7388	370	1110 Apr 20	23:08:10	1038	-11004	85	Pe	-t	-1.5310	0.0544	61.8S	100.7W	0	300		
7389	370	1110 May 20	14:22:27	1037	-11003	123	P	t-	1.3362	0.3876	63.6N	173.8W	0	36		
7390	370	1110 Oct 15	08:37:44	1036	-10998	90	P	-t	1.3297	0.3905	61.4N	120.6E	0	247		
7391	370	1110 Nov 13	19:22:20	1035	-10997	128	P	t-	-1.3568	0.3380	63.2S	115.3E	0	138		
7392	370	1111 Apr 10	03:09:37	1034	-10992	95	A	-p	-0.7383	0.9832	32.3S	160.1E	42	328	86	01m34s
7393	370	1111 Oct 04	19:12:31	1032	-10986	100	A	-p	0.6701	0.9712	30.1N	85.5W	48	213	136	02m44s
7394	370	1112 Mar 29	14:17:31	1030	-10980	105	T	nn	0.0378	1.0410	8.0N	30.5W	88	152	137	03m34s
7395	370	1112 Sep 22	22:38:26	1028	-10974	110	Am	nn	-0.0469	0.9365	4.8S	159.1W	87	29	237	07m13s
7396	370	1113 Mar 19	06:10:38	1026	-10968	115	T	p-	0.7471	1.0598	42.7N	63.6E	41	139	290	04m08s
7397	370	1113 Sep 11	22:22:41	1024	-10962	120	A	p-	-0.7658	0.9336	40.5S	177.7E	40	40	373	06m24s
7398	370	1114 Feb 07	11:26:45	1022	-10957	87	P	-t	-1.2234	0.5854	61.7S	132.3E	0	241		
7399	370	1114 Mar 08	22:17:27	1022	-10956	125	P	t-	1.4643	0.1367	61.1N	126.5E	0	95		
7400	370	1114 Aug 02	14:31:46	1021	-10951	92	P	-t	1.2648	0.5104	62.3N	91.0E	0	307		
7401	371	1114 Sep 01	01:57:49	1020	-10950	130	P	t-	-1.4527	0.1773	61.1S	75.7E	0	77		
7402	371	1115 Jan 27	19:12:32	1019	-10945	97	A	-p	-0.5578	0.9477	47.3S	81.6W	56	327	230	04m46s
7403	371	1115 Jul 23	04:47:53	1017	-10939	102	T	-p	0.4783	1.0663	45.6N	126.8E	61	207	246	04m50s
7404	371	1116 Jan 16	20:00:57	1015	-10933	107	A	nn	0.1350	0.9200	12.4S	115.2W	82	161	306	10m27s
7405	371	1116 Jul 11	21:56:21	1013	-10927	112	T	n-	-0.2594	1.0748	6.8N	148.0W	75	15	251	06m46s
7406	371	1117 Jan 04	19:38:51	1011	-10921	117	A	p-	0.8121	0.9292	31.0N	121.9W	35	164	450	08m19s
7407	371	1117 Jul 01	13:16:27	1009	-10915	122	P	t-	-1.0377	0.9337	64.5S	43.0W	0	26		
7408	371	1117 Nov 25	12:04:12	1008	-10910	89	P	-t	-1.3506	0.3531	67.8S	170.9W	0	172		
7409	371	1117 Dec 25	01:07:37	1007	-10909	127	P	t-	1.4469	0.1867	64.9N	144.7E	0	158		
7410	371	1118 May 22	09:22:22	1006	-10904	94	A	-t	0.8927	0.9508	83.9N	5.6E	26	141	406	03m27s
7411	371	1118 Nov 15	02:45:13	1004	-10898	99	T	-p	-0.6511	1.0439	60.4S	130.5E	49	13	194	03m01s
7412	371	1119 May 11	10:03:00	1002	-10892	104	A	nn	0.1386	0.9484	27.4N	30.9E	82	172	192	06m24s
7413	371	1119 Nov 04	18:19:38	1000	-10886	109	T	nn	0.0194	1.0336	16.4S	94.2W	89	191	113	03m14s
7414	371	1120 Apr 29	12:53:54	998	-10880	114	A	p-	-0.6205	0.9777	21.3S	2.9W	51	348	101	02m34s
7415	371	1120 Oct 24	06:09:13	996	-10874	119	A	p-	0.7478	0.9781	33.2N	100.0E	41	196	117	02m12s
7416	371	1121 Mar 20	13:41:29	995	-10869	86	P	-t	1.2840	0.4736	71.8N	117.2W	0	82		
7417	371	1121 Apr 18	22:43:16	994	-10868	124	P	t-	-1.3276	0.3929	70.7S	109.5W	0	314		
7418	371	1121 Sep 13	17:45:19	993	-10863	91	P	-t	-1.4600	0.1802	72.0S	170.0W	0	86		
7419	371	1121 Oct 13	10:56:27	992	-10862	129	P	t-	1.5060	0.1012	71.2N	74.2E	0	236		
7420	371	1122 Mar 10	06:01:29	991	-10857	96	T	-p	0.5736	1.0588	31.6N	84.2E	55	159	235	04m47s
7421	372	1122 Sep 02	17:56:11	989	-10851	101	A	-p	-0.7328	0.9449	39.3S	102.0W	43	22	295	05m48s
7422	372	1123 Feb 27	21:15:11	987	-10845	106	Tm	nn	-0.1313	1.0251	13.1S	129.3W	82	343	86	02m27s
7423	372	1123 Aug 22	23:36:13	985	-10839	111	H	nn	0.0391	1.0016	11.9N	169.3W	88	195	6	00m10s
7424	372	1124 Feb 17	06:56:07	983	-10833	116	A	p-	-0.8893	0.9581	67.5S	123.0E	27	316	335	03m08s
7425	372	1124 Aug 11	12:28:49	982	-10827	121	T	p-	0.7716	1.0497	61.4N	20.4E	39	209	259	03m19s
7426	372	1125 Jan 06	14:21:29	980	-10822	88	P	-t	1.4106	0.2676	68.1N	39.7W	0	170		
7427	372	1125 Jul 02	22:10:05	978	-10816	93	P	-t	-1.0469	0.9294	67.4S	151.5W	0	4		
7428	372	1125 Aug 01	05:15:09	978	-10815	131	Pb	t-	1.4666	0.1198	69.9N	109.5W	0	326		
7429	372	1125 Dec 26	15:05:32	976	-10810	98	A	-p	0.7027	0.9426	21.8N	40.3W	45	181	298	07m10s
7430	372	1126 Jun 22	12:12:40	975	-10804	103	T	-p	-0.3161	1.0193	5.0N	2.1E	72	357	69	02m09s
7431	372	1126 Dec 15	22:41:48	973	-10798	108	H	nn	-0.0228	1.0013	25.0S	156.4W	89	5	5	00m08s
7432	372	1127 Jun 11	19:28:58	971	-10792	113	A	p-	0.4756	0.9664	51.8N	114.2W	61	169	138	03m10s
7433	372	1127 Dec 05	12:27:08	969	-10786	118	T	p-	-0.6930	1.0377	65.9S	21.2W	46	23	176	02m28s
7434	372	1128 May 30	20:45:57	967	-10780	123	P	t-	1.2507	0.5336	64.5N	80.5E	0	26		
7435	372	1128 Oct 25	17:03:22	966	-10775	90	P	-t	1.3504	0.3528	61.9N	15.0W	0	238		
7436	372	1128 Nov 24	04:09:27	965	-10774	128	P	t-	-1.3531	0.3453	64.0S	26.2W	0	148		
7437	372	1129 Apr 20	10:21:11	964	-10769	95	A	-p	-0.8000	0.9879	34.5S	52.1E	37	330	69	01m08s
7438	372	1129 Oct 15	03:08:33	962	-10763	100	A	-p	0.6986	0.9651	28.7N	154.2E	45	210	172	03m27s
7439	372	1130 Apr 09	22:01:18	960	-10757	105	T	nn	-0.0160	1.0466	9.3N	146.1W	89	331	155	04m04s
7440	372	1130 Oct 04	06:03:42	958	-10751	110	A	nn	-0.0080	0.9324	7.2S	89.7E	89	26	253	07m48s

Cat Num	Canon Plate	Calendar Date	TD of Greatest Eclipse	ΔT s	Luna Num	Saros Num	Ecl. Type	QLE	Gamma	Ecl. Mag.	Lat. °	Long. °	Sun Alt °	Sun Azm °	Path Width km	Central Line Dur.
7441	373	1131 Mar 30	14:11:49	957	-10745	115	T	p-	0.7012	1.0639	43.9N	55.8W	45	139	289	04m22s
7442	373	1131 Sep 23	05:39:54	955	-10739	120	A	p-	-0.7188	0.9328	41.2S	68.9E	44	40	351	06m24s
7443	373	1132 Feb 18	19:41:35	953	-10734	87	P	-t	-1.2495	0.5380	61.3S	0.4W	0	250		
7444	373	1132 Mar 19	06:19:27	953	-10733	125	P	t-	1.4291	0.2024	61.1N	3.1W	0	86		
7445	373	1132 Aug 12	21:55:23	952	-10728	92	P	-t	1.3223	0.4048	61.8N	29.2W	0	298		
7446	373	1132 Sep 11	09:29:13	951	-10727	130	P	t-	-1.4007	0.2695	60.9S	46.1W	0	86		
7447	373	1133 Feb 07	03:10:08	950	-10722	97	A	-p	-0.5794	0.9473	44.3S	160.7E	54	325	235	04m49s
7448	373	1133 Aug 02	12:24:26	948	-10716	102	T	-p	0.5423	1.0652	45.8N	16.5E	57	212	252	04m38s
7449	373	1134 Jan 27	03:49:21	946	-10710	107	A	nn	0.1170	0.9217	10.6S	127.9E	83	158	298	09m54s
7450	373	1134 Jul 23	05:29:19	944	-10704	112	T	n-	-0.1910	1.0720	8.6N	99.0E	79	19	238	06m21s
7451	373	1135 Jan 16	03:43:17	943	-10698	117	A	p-	0.7997	0.9329	30.9N	114.0E	37	159	410	07m39s
7452	373	1135 Jul 12	20:28:48	941	-10692	122	T	t-	-0.9676	1.0179	51.5S	147.8W	14	25	248	01m25s
7453	373	1135 Dec 06	20:46:56	939	-10687	89	P	-t	-1.3503	0.3536	66.7S	47.1E	0	184		
7454	373	1136 Jan 05	09:38:55	939	-10686	127	P	t-	1.4383	0.2010	63.9N	6.9E	0	148		
7455	373	1136 Jun 01	15:52:58	937	-10681	94	A	-t	0.9790	0.9459	78.0N	124.5E	11	360	1063	03m18s
7456	373	1136 Nov 25	11:35:06	936	-10675	99	T	-p	-0.6557	1.0430	63.3S	4.0E	49	6	191	02m54s
7457	373	1137 May 21	16:30:23	934	-10669	104	A	nn	0.2236	0.9504	34.5N	66.0W	77	175	187	05m51s
7458	373	1137 Nov 15	03:00:21	932	-10663	109	Tm	nn	0.0116	1.0297	19.6S	135.7E	89	188	101	02m53s
7459	373	1138 May 10	19:48:33	930	-10657	114	A	p-	-0.5447	0.9835	13.5S	110.0W	57	352	70	01m58s
7460	373	1138 Nov 04	14:24:41	929	-10651	119	A	p-	0.7362	0.9732	29.4N	27.2W	42	191	141	02m51s
7461	374	1139 Mar 31	21:35:42	927	-10646	86	P	-t	1.3272	0.3913	71.6N	110.0E	0	68		
7462	374	1139 Apr 30	06:10:02	927	-10645	124	P	t-	-1.2616	0.5159	70.0S	125.8E	0	326		
7463	374	1139 Sep 25	01:00:38	925	-10640	91	P	-t	-1.5037	0.1081	72.0S	66.7E	0	100		
7464	374	1139 Oct 24	18:42:07	925	-10639	129	P	t-	1.4859	0.1369	70.6N	55.8W	0	223		
7465	374	1140 Mar 20	14:08:51	924	-10634	96	T	-p	0.6143	1.0607	38.6N	40.8W	52	157	251	04m42s
7466	374	1140 Sep 13	01:10:24	922	-10628	101	A	-p	-0.7820	0.9431	47.1S	144.0E	38	27	334	05m31s
7467	374	1141 Mar 10	05:19:43	920	-10622	106	T	nn	-0.0971	1.0254	6.9S	107.9E	84	342	87	02m30s
7468	374	1141 Sep 02	07:08:35	918	-10616	111	H	nn	-0.0136	1.0021	4.9N	75.7E	89	18	7	00m13s
7469	374	1142 Feb 27	14:45:00	917	-10610	116	A	p-	-0.8615	0.9589	60.8S	2.9W	30	323	293	03m17s
7470	374	1142 Aug 22	20:14:13	915	-10604	121	T	p-	0.7147	1.0504	52.9N	100.7W	44	207	238	03m36s
7471	374	1143 Jan 17	22:16:50	914	-10599	88	P	-t	1.4214	0.2491	69.1N	170.4W	0	158		
7472	374	1143 Jul 14	05:36:19	912	-10593	93	P	-t	-1.1194	0.7889	68.4S	85.4E	0	14		
7473	374	1143 Aug 12	12:57:36	911	-10592	131	P	t-	1.4088	0.2324	70.7N	121.9E	0	314		
7474	374	1144 Jan 06	23:19:57	910	-10587	98	A	-p	0.7075	0.9459	23.5N	166.6W	45	176	282	06m36s
7475	374	1144 Jul 02	19:15:50	908	-10581	103	H	-p	-0.3949	1.0145	0.8S	105.0W	67	1	54	01m39s
7476	374	1144 Dec 26	07:20:32	907	-10575	108	H	nn	-0.0195	1.0046	24.3S	75.1E	89	1	16	00m28s
7477	374	1145 Jun 22	02:02:44	905	-10569	113	A	p-	0.3909	0.9645	46.6N	151.7E	67	176	140	03m35s
7478	374	1145 Dec 15	21:18:46	903	-10563	118	T	p-	-0.6892	1.0387	67.0S	145.0W	46	12	180	02m32s
7479	374	1146 Jun 11	03:09:15	901	-10557	123	P	t-	1.1642	0.6817	65.4N	25.5W	0	17		
7480	374	1146 Nov 06	01:33:49	900	-10552	90	P	-t	1.3669	0.3233	62.6N	151.9W	0	228		
7481	375	1146 Dec 05	12:57:34	900	-10551	128	P	t-	-1.3507	0.3500	65.0S	168.3W	0	159		
7482	375	1147 May 01	17:28:51	898	-10546	95	A	-p	-0.8658	0.9918	38.7S	54.9W	30	333	57	00m45s
7483	375	1147 Oct 26	11:11:07	897	-10540	100	A	-p	0.7210	0.9595	27.6N	31.7E	44	207	207	04m11s
7484	375	1148 Apr 20	05:37:53	895	-10534	105	Tm	nn	-0.0765	1.0520	9.8N	100.3E	86	335	172	04m35s
7485	375	1148 Oct 14	13:38:06	893	-10528	110	A	nn	0.0231	0.9286	9.7S	23.9W	89	208	268	08m26s
7486	375	1149 Apr 09	22:04:02	891	-10522	115	T	p-	0.6479	1.0676	44.9N	171.9W	49	141	286	04m38s
7487	375	1149 Oct 03	13:08:11	890	-10516	120	A	p-	-0.6802	0.9320	43.0S	42.7W	47	40	339	06m24s
7488	375	1150 Mar 01	03:46:23	888	-10511	87	P	-t	-1.2835	0.4765	61.0S	130.6W	0	259		
7489	375	1150 Mar 30	14:11:34	888	-10510	125	P	t-	1.3867	0.2819	61.2N	130.1W	0	77		
7490	375	1150 Aug 24	05:28:34	887	-10505	92	P	-t	1.3729	0.3118	61.4N	151.6W	0	289		
7491	375	1150 Sep 22	17:12:01	886	-10504	130	P	t-	-1.3568	0.3471	60.9S	170.7W	0	95		
7492	375	1151 Feb 18	11:00:02	885	-10499	97	A	-p	-0.6070	0.9473	41.6S	44.2E	52	323	240	04m52s
7493	375	1151 Aug 13	20:06:10	883	-10493	102	T	-p	0.6024	1.0635	45.2N	95.7W	53	216	258	04m26s
7494	375	1152 Feb 07	11:32:55	882	-10487	107	A	nn	0.0950	0.9238	8.4S	12.0E	85	155	288	09m19s
7495	375	1152 Aug 02	13:04:59	880	-10481	112	T	nn	-0.1250	1.0685	9.5N	14.4W	83	23	225	05m55s
7496	375	1153 Jan 26	11:44:14	878	-10475	117	A	p-	0.7839	0.9372	31.1N	9.0W	38	155	367	06m53s
7497	375	1153 Jul 23	03:42:38	877	-10469	122	T	p-	-0.8994	1.0161	41.6S	105.0E	26	25	125	01m22s
7498	375	1153 Dec 17	05:32:01	875	-10464	89	P	-t	-1.3488	0.3560	65.7S	94.9W	0	194		
7499	375	1154 Jan 15	18:08:17	875	-10463	127	P	t-	1.4274	0.2192	63.1N	130.1W	0	138		
7500	375	1154 Jun 12	22:21:53	874	-10458	94	P	-t	1.0656	0.8530	66.1N	16.7E	0	350		

Cat Num	Canon Plate	Calendar Date	TD of Greatest Eclipse	ΔT s	Luna Num	Saros Num	Ecl. Type	QLE	Gamma	Ecl. Mag.	Lat. °	Long. °	Sun Alt °	Sun Azm °	Path Width km	Central Line Dur.
7501	376	1154 Dec 06	20:26:36	872	-10452	99	T	-p	-0.6593	1.0425	64.8S	121.0W	48	356	190	02m50s
7502	376	1155 Jun 01	22:57:44	870	-10446	104	A	np	0.3088	0.9520	41.1N	161.5W	72	180	185	05m19s
7503	376	1155 Nov 26	11:43:38	869	-10440	109	T	nn	0.0063	1.0262	21.9S	5.5E	90	184	89	02m34s
7504	376	1156 May 21	02:40:45	867	-10434	114	A	p-	-0.4656	0.9889	6.3S	144.3E	62	356	44	01m21s
7505	376	1156 Nov 14	22:44:29	865	-10428	119	A	p-	0.7287	0.9687	26.5N	155.2W	43	187	164	03m28s
7506	376	1157 Apr 11	05:23:00	864	-10423	86	P	-t	1.3767	0.2964	71.1N	20.7W	0	55		
7507	376	1157 May 10	13:32:54	864	-10422	124	P	t-	-1.1912	0.6486	69.1S	2.6E	0	338		
7508	376	1157 Oct 05	08:25:14	862	-10417	91	P	-t	-1.5396	0.0494	71.7S	58.7W	0	114		
7509	376	1157 Nov 04	02:33:58	862	-10416	129	P	t-	1.4710	0.1634	69.7N	173.3E	0	210		
7510	376	1158 Mar 31	22:07:25	861	-10411	96	T	-p	0.6623	1.0621	46.2N	163.9W	48	155	271	04m33s
7511	376	1158 Sep 24	08:36:09	859	-10405	101	A	-p	-0.8223	0.9415	54.6S	26.5E	34	31	378	05m15s
7512	376	1159 Mar 21	13:14:39	857	-10399	106	T	nn	-0.0557	1.0254	0.3S	12.7W	87	342	87	02m32s
7513	376	1159 Sep 13	14:50:28	856	-10393	111	H	nn	-0.0591	1.0023	1.9S	41.8W	87	18	8	00m15s
7514	376	1160 Mar 09	22:23:38	854	-10387	116	A	p-	-0.8259	0.9599	53.5S	124.9W	34	329	256	03m28s
7515	376	1160 Sep 02	04:07:30	852	-10381	121	T	p-	0.6640	1.0504	44.9N	137.1E	48	205	222	03m49s
7516	376	1161 Jan 28	06:05:42	851	-10376	88	P	-t	1.4375	0.2214	70.1N	59.8E	0	145		
7517	376	1161 Jul 24	13:04:53	849	-10370	93	P	-t	-1.1895	0.6540	69.4S	38.8W	0	26		
7518	376	1161 Aug 22	20:46:37	849	-10369	131	P	t-	1.3564	0.3340	71.3N	8.8W	0	301		
7519	376	1162 Jan 17	07:31:48	848	-10364	98	A	-p	0.7150	0.9499	26.1N	67.7E	44	172	262	05m54s
7520	376	1162 Jul 14	02:19:38	846	-10358	103	H	-p	-0.4722	1.0091	7.3S	147.1E	62	5	35	01m02s
7521	377	1163 Jan 06	15:58:40	844	-10352	108	H	nn	-0.0151	1.0084	22.6S	53.4W	89	357	29	00m51s
7522	377	1163 Jul 03	08:36:04	843	-10346	113	A	pn	0.3064	0.9620	40.5N	55.9E	72	182	145	04m06s
7523	377	1163 Dec 27	06:11:27	841	-10340	118	T	p-	-0.6860	1.0400	66.6S	91.5E	46	1	185	02m38s
7524	377	1164 Jun 21	09:30:36	840	-10334	123	P	t-	1.0754	0.8336	66.4N	131.2W	0	7		
7525	377	1164 Nov 16	10:09:43	838	-10329	90	P	-t	1.3785	0.3027	63.4N	69.7E	0	219		
7526	377	1164 Dec 15	21:48:04	838	-10328	128	P	t-	-1.3507	0.3502	66.0S	48.7E	0	169		
7527	377	1165 May 12	00:33:05	837	-10323	95	A	-p	-0.9349	0.9946	46.0S	160.4W	20	334	53	00m28s
7528	377	1165 Nov 05	19:19:17	835	-10317	100	A	-p	0.7379	0.9544	26.7N	92.6W	42	203	242	04m55s
7529	377	1166 May 01	13:11:03	833	-10311	105	T	nn	-0.1406	1.0567	9.5N	12.4W	82	338	189	05m06s
7530	377	1166 Oct 25	21:19:40	832	-10305	110	A	nn	0.0477	0.9253	12.2S	139.2W	87	205	282	09m05s
7531	377	1167 Apr 21	05:51:40	830	-10299	115	T	p-	0.5906	1.0709	45.8N	73.9E	54	145	284	04m53s
7532	377	1167 Oct 14	20:44:41	829	-10293	120	A	p-	-0.6477	0.9313	45.4S	156.0W	49	39	332	06m24s
7533	377	1168 Mar 11	11:43:09	827	-10288	87	P	-t	-1.3232	0.4045	60.9S	101.3E	0	268		
7534	377	1168 Apr 09	21:56:35	827	-10287	125	P	t-	1.3390	0.3713	61.5N	104.6E	0	68		
7535	377	1168 Sep 03	13:10:27	826	-10282	92	P	-t	1.4174	0.2298	61.2N	83.9E	0	280		
7536	377	1168 Oct 03	01:04:24	826	-10281	130	P	t-	-1.3197	0.4129	61.1S	62.2E	0	104		
7537	377	1169 Feb 28	18:38:38	824	-10276	97	A	-p	-0.6431	0.9473	39.3S	69.7W	50	322	248	04m55s
7538	377	1169 Aug 24	03:56:00	823	-10270	102	T	-p	0.6561	1.0612	44.1N	149.2E	49	219	263	04m15s
7539	377	1170 Feb 17	19:07:07	821	-10264	107	A	nn	0.0651	0.9264	6.1S	101.5W	86	152	277	08m46s
7540	377	1170 Aug 13	20:46:50	819	-10258	112	T	nn	-0.0646	1.0645	9.3N	129.3W	86	26	211	05m28s
7541	378	1171 Feb 06	19:38:30	818	-10252	117	A	p-	0.7618	0.9421	31.6N	129.7W	40	151	321	06m05s
7542	378	1171 Aug 03	11:00:18	816	-10246	122	T	p-	-0.8350	1.0126	36.2S	4.6W	33	28	77	01m06s
7543	378	1171 Dec 28	14:16:40	815	-10241	89	P	-t	-1.3484	0.3564	64.7S	123.7E	0	205		
7544	378	1172 Jan 27	02:33:05	815	-10240	127	P	t-	1.4121	0.2453	62.4N	94.3E	0	129		
7545	378	1172 Jun 23	04:47:38	813	-10235	94	P	-t	1.1539	0.7004	65.1N	90.0W	0	340		
7546	378	1172 Dec 17	05:20:06	812	-10229	99	T	-p	-0.6615	1.0426	64.6S	114.3E	48	345	190	02m49s
7547	378	1173 Jun 12	05:22:44	810	-10223	104	A	-p	0.3966	0.9531	47.0N	105.1E	66	186	187	04m51s
7548	378	1173 Dec 06	20:29:30	809	-10217	109	T	nn	0.0034	1.0234	23.2S	125.0W	90	172	80	02m17s
7549	378	1174 Jun 01	09:31:50	807	-10211	114	A	p-	-0.3843	0.9938	0.3N	39.7E	67	360	24	00m45s
7550	378	1174 Nov 26	07:08:31	806	-10205	119	A	p-	0.7250	0.9648	24.4N	76.0E	43	182	185	04m02s
7551	378	1175 Apr 22	13:04:58	804	-10200	86	P	-t	1.4309	0.1916	70.4N	149.4W	0	43		
7552	378	1175 May 21	20:53:01	804	-10199	124	P	t-	-1.1176	0.7882	68.1S	119.3W	0	349		
7553	378	1175 Oct 16	15:57:46	803	-10194	91	Pe	-t	-1.5690	0.0019	71.1S	174.3E	0	127		
7554	378	1175 Nov 15	10:31:20	803	-10193	129	P	t-	1.4605	0.1821	68.7N	41.6E	0	198		
7555	378	1176 Apr 11	05:59:31	801	-10188	96	T	-p	0.7156	1.0629	54.2N	74.2E	44	154	295	04m20s
7556	378	1176 Oct 04	16:11:30	800	-10182	101	A	-p	-0.8551	0.9400	61.7S	94.1W	31	37	428	05m00s
7557	378	1177 Mar 31	21:00:18	798	-10176	106	T	nn	-0.0073	1.0253	6.6N	131.2W	90	341	86	02m33s
7558	378	1177 Sep 23	22:42:22	797	-10170	111	H	nn	-0.0971	1.0025	8.3S	161.5W	84	18	9	00m16s
7559	378	1178 Mar 21	05:53:15	795	-10164	116	A	p-	-0.7838	0.9610	46.0S	116.4E	38	334	225	03m39s
7560	378	1178 Sep 13	12:08:37	794	-10158	121	T	p-	0.6196	1.0500	37.6N	13.3E	51	203	210	03m59s

Cat Num	Canon Plate	Calendar Date	TD of Greatest Eclipse	ΔT s	Luna Num	Saros Num	Ecl. Type	QLE	Gamma	Ecl. Mag.	Lat. °	Long. °	Sun Alt °	Sun Azm °	Path Width km	Central Line Dur.
7561	379	1179 Feb 08	13:48:49	792	-10153	88	P	-t	1.4582	0.1860	70.9N	69.2W	0	132		
7562	379	1179 Mar 10	07:39:51	792	-10152	126	Pb	t-	-1.5356	0.0536	72.0S	165.4E	0	265		
7563	379	1179 Aug 04	20:37:56	791	-10147	93	P	-t	-1.2552	0.5283	70.2S	164.7W	0	38		
7564	379	1179 Sep 03	04:42:14	791	-10146	131	P	t-	1.3096	0.4241	71.7N	141.7W	0	287		
7565	379	1180 Jan 28	15:39:21	789	-10141	98	A	-p	0.7267	0.9542	29.5N	57.2W	43	167	242	05m08s
7566	379	1180 Jul 24	09:25:08	788	-10135	103	H	-p	-0.5471	1.0031	14.4S	38.0E	57	9	13	00m21s
7567	379	1181 Jan 17	00:33:25	786	-10129	108	H	nn	-0.0073	1.0127	19.8S	178.5E	89	354	44	01m17s
7568	379	1181 Jul 13	15:11:38	785	-10123	113	A	nn	0.2244	0.9590	33.9N	42.0W	77	186	153	04m42s
7569	379	1182 Jan 06	15:00:32	783	-10117	118	T	p-	-0.6802	1.0419	64.4S	32.7W	47	351	192	02m48s
7570	379	1182 Jul 02	15:55:48	782	-10111	123	An	t-	0.9892	0.9368	74.6N	120.1E	7	355	-	03m50s
7571	379	1182 Nov 27	18:48:43	780	-10106	90	P	-t	1.3869	0.2883	64.3N	69.8W	0	209		
7572	379	1182 Dec 27	06:37:09	780	-10105	128	P	t-	-1.3505	0.3506	67.1S	94.4W	0	180		
7573	379	1183 May 23	07:34:59	779	-10100	95	P	-t	-1.0067	0.9797	63.9S	104.1E	0	327		
7574	379	1183 Nov 17	03:32:02	777	-10094	100	A	-p	0.7503	0.9500	26.2N	141.7E	41	198	274	05m38s
7575	379	1184 May 11	20:39:04	776	-10088	105	T	-n	-0.2094	1.0609	8.3N	123.9W	78	342	204	05m35s
7576	379	1184 Nov 05	05:09:12	774	-10082	110	A	nn	0.0659	0.9224	14.4S	103.7E	86	202	294	09m45s
7577	379	1185 May 01	13:30:57	773	-10076	115	T	p-	0.5264	1.0736	46.0N	37.2W	58	149	280	05m10s
7578	379	1185 Oct 25	04:31:56	771	-10070	120	A	p-	-0.6233	0.9308	48.4S	88.5E	51	37	328	06m24s
7579	379	1186 Mar 22	19:29:34	770	-10065	87	P	-t	-1.3708	0.3183	61.0S	24.2W	0	277		
7580	379	1186 Apr 21	05:32:46	770	-10064	125	P	t-	1.2847	0.4732	62.0N	18.6W	0	59		
7581	380	1186 Sep 14	21:01:17	769	-10059	92	P	-t	1.4555	0.1598	61.1N	42.7W	0	271		
7582	380	1186 Oct 14	09:06:01	769	-10058	130	P	t-	-1.2891	0.4670	61.4S	67.1W	0	113		
7583	380	1187 Mar 12	02:08:41	767	-10053	97	A	-p	-0.6856	0.9474	37.8S	178.2E	47	322	259	04m59s
7584	380	1187 Sep 04	11:52:36	766	-10047	102	T	-p	0.7044	1.0585	42.8N	31.5E	45	220	267	04m05s
7585	380	1188 Feb 29	02:34:01	764	-10041	107	A	nn	0.0292	0.9294	3.8S	146.9E	88	151	265	08m14s
7586	380	1188 Aug 24	04:32:58	763	-10035	112	Tm	nn	-0.0082	1.0598	8.4N	114.6E	90	31	197	05m01s
7587	380	1189 Feb 17	03:27:37	761	-10029	117	A	p-	0.7346	0.9475	32.3N	111.4E	43	148	276	05m18s
7588	380	1189 Aug 13	18:22:08	760	-10023	122	H	p-	-0.7744	1.0082	33.2S	115.4W	39	30	43	00m43s
7589	380	1190 Jan 07	22:59:43	759	-10018	89	P	-t	-1.3501	0.3528	63.7S	16.9W	0	215		
7590	380	1190 Feb 06	10:53:04	758	-10017	127	P	t-	1.3921	0.2801	61.8N	39.8W	0	119		
7591	380	1190 Jul 04	11:15:38	757	-10012	94	P	-t	1.2397	0.5528	64.2N	163.0E	0	331		
7592	380	1190 Dec 28	14:12:58	756	-10006	99	T	-p	-0.6648	1.0430	63.1S	11.2W	48	336	193	02m50s
7593	380	1191 Jun 23	11:49:25	754	-10000	104	A	-p	0.4833	0.9539	52.0N	13.2E	61	193	193	04m28s
7594	380	1191 Dec 18	05:15:22	753	-9994	109	T	nn	0.0008	1.0209	23.6S	104.7E	90	3	71	02m02s
7595	380	1192 Jun 11	16:23:44	751	-9988	114	A	p-	-0.3023	0.9981	6.0N	64.3W	72	4	7	00m14s
7596	380	1192 Dec 06	15:33:28	750	-9982	119	A	p-	0.7228	0.9614	23.1N	52.9W	44	178	203	04m30s
7597	380	1193 May 02	20:41:06	749	-9977	86	Pe	-t	1.4902	0.0765	69.6N	83.8E	0	31		
7598	380	1193 Jun 01	04:11:37	748	-9976	124	P	t-	-1.0418	0.9331	67.1S	119.8E	0	360		
7599	380	1193 Nov 25	18:32:28	747	-9970	129	P	t-	1.4531	0.1954	67.7N	90.3W	0	186		
7600	380	1194 Apr 22	13:44:30	746	-9965	96	T	-p	0.7748	1.0629	62.9N	46.7W	39	151	327	04m03s
7601	381	1194 Oct 15	23:56:12	744	-9959	101	A	-p	-0.8811	0.9389	68.2S	142.0E	28	43	481	04m45s
7602	381	1195 Apr 12	04:37:20	743	-9953	106	T	nn	0.0477	1.0248	13.6N	112.7E	87	165	84	02m29s
7603	381	1195 Oct 05	06:44:13	742	-9947	111	H	nn	-0.1277	1.0026	14.3S	76.6E	83	17	9	00m16s
7604	381	1196 Mar 31	13:11:09	740	-9941	116	A	p-	-0.7326	0.9621	38.1S	1.4E	43	339	200	03m52s
7605	381	1196 Sep 23	20:18:46	739	-9935	121	T	p-	0.5821	1.0491	30.9N	112.3W	54	201	199	04m06s
7606	381	1197 Feb 18	21:23:51	737	-9930	88	P	-t	1.4855	0.1390	71.5N	163.4E	0	119		
7607	381	1197 Mar 20	14:48:56	737	-9929	126	P	t-	-1.4880	0.1327	72.0S	43.8E	0	279		
7608	381	1197 Aug 15	04:16:07	736	-9924	93	P	-t	-1.3161	0.4129	71.0S	67.6E	0	50		
7609	381	1197 Sep 13	12:45:54	736	-9923	131	P	t-	1.2695	0.5009	71.9N	83.1E	0	273		
7610	381	1198 Feb 07	23:41:19	735	-9918	98	A	-p	0.7437	0.9590	33.9N	178.9E	42	163	221	04m20s
7611	381	1198 Aug 04	16:34:46	733	-9912	103	A	-p	-0.6176	0.9967	22.0S	72.8W	52	13	15	00m21s
7612	381	1199 Jan 28	09:05:27	732	-9906	108	H2	nn	0.0033	1.0174	16.2S	50.5E	90	156	60	01m45s
7613	381	1199 Jul 24	21:48:31	730	-9900	113	A	nn	0.1439	0.9557	26.8N	141.4W	82	190	163	05m21s
7614	381	1200 Jan 17	23:48:29	729	-9894	110	T	p-	-0.6731	1.0443	61.0S	159.1W	47	344	200	03m01s
7615	381	1200 Jul 12	22:22:50	728	-9888	123	A	t-	0.9039	0.9409	83.6N	101.5W	25	232	521	04m12s
7616	381	1200 Dec 08	03:30:03	726	-9883	90	P	-t	1.3933	0.2771	65.3N	149.9E	0	198		
7617	381	1201 Jan 06	15:24:30	726	-9882	128	P	t-	-1.3491	0.3532	68.1S	122.4E	0	191		
7618	381	1201 Jun 02	14:35:50	725	-9877	95	P	-t	-1.0802	0.8474	64.8S	11.0W	0	337		
7619	381	1201 Nov 27	11:48:14	724	-9871	100	A	-p	0.7592	0.9461	26.0N	15.0E	40	193	302	06m16s
7620	381	1202 May 23	04:06:00	722	-9865	105	T	-n	-0.2801	1.0643	6.0N	124.6E	74	346	219	06m02s

Cat Num	Canon Plate	Calendar Date	TD of Greatest Eclipse	ΔT s	Luna Num	Saros Num	Ecl. Type	QLE	Gamma	Ecl. Mag.	Lat. °	Long. °	Sun Alt °	Sun Azm °	Path Width km	Central Line Dur.
7621	382	1202 Nov 16	13:02:26	721	-9859	110	A	nn	0.0809	0.9201	16.2S	14.2W	85	198	303	10m23s
7622	382	1203 May 12	21:07:30	719	-9853	115	T	p-	0.4596	1.0755	45.5N	147.2W	62	155	275	05m26s
7623	382	1203 Nov 05	12:25:54	718	-9847	120	A	p-	-0.6037	0.9307	51.4S	27.9W	53	34	323	06m23s
7624	382	1204 Apr 02	03:06:26	717	-9842	87	P	-t	-1.4253	0.2196	61.2S	147.4W	0	286		
7625	382	1204 May 01	13:01:38	717	-9841	125	P	t-	1.2250	0.5852	62.5N	140.1W	0	50		
7626	382	1204 Sep 25	05:01:33	715	-9836	92	P	-t	1.4869	0.1021	61.1N	171.8W	0	262		
7627	382	1204 Oct 24	17:16:40	715	-9835	130	P	t-	-1.2650	0.5097	61.9S	161.2E	0	122		
7628	382	1205 Mar 22	09:27:32	714	-9830	97	A	-p	-0.7365	0.9473	37.3S	69.0E	42	323	278	05m04s
7629	382	1205 Sep 14	19:58:08	713	-9824	102	T	-p	0.7458	1.0556	41.4N	89.5W	42	220	270	03m55s
7630	382	1206 Mar 11	09:50:27	711	-9818	107	Am	nn	-0.0156	0.9326	1.8S	37.9E	89	331	252	07m47s
7631	382	1206 Sep 04	12:27:26	710	-9812	112	T	nn	0.0409	1.0549	6.8N	3.7W	88	208	181	04m36s
7632	382	1207 Feb 28	11:09:12	708	-9806	117	A	p-	0.7002	0.9534	33.2N	5.0W	45	145	232	04m32s
7633	382	1207 Aug 25	01:48:51	707	-9800	122	H	p-	-0.7186	1.0031	31.9S	132.6E	44	32	15	00m16s
7634	382	1208 Jan 19	07:39:49	706	-9795	89	P	-t	-1.3551	0.3431	62.9S	156.5W	0	225		
7635	382	1208 Feb 17	19:06:58	706	-9794	127	P	t-	1.3664	0.3257	61.3N	172.3W	0	110		
7636	382	1208 Jul 14	17:44:19	705	-9789	94	P	-t	1.3243	0.4078	63.3N	56.2E	0	321		
7637	382	1208 Aug 13	08:26:52	704	-9788	132	Pb	t-	-1.5227	0.0639	61.7S	8.8W	0	63		
7638	382	1209 Jan 07	23:03:30	703	-9783	99	T	-p	-0.6701	1.0439	60.6S	137.9W	48	328	197	02m54s
7639	382	1209 Jul 03	18:17:42	702	-9777	104	A	-p	0.5692	0.9540	55.8N	77.1W	55	202	204	04m11s
7640	382	1209 Dec 28	14:00:54	701	-9771	109	T	nn	-0.0018	1.0190	23.0S	25.5W	90	355	65	01m50s
7641	383	1210 Jun 22	23:17:17	699	-9765	114	H	n-	-0.2207	1.0018	10.8N	168.0W	77	8	6	00m12s
7642	383	1210 Dec 17	23:58:47	698	-9759	119	A	p-	0.7215	0.9585	22.5N	178.1E	44	173	217	04m51s
7643	383	1211 Jun 12	11:30:10	696	-9753	124	T	t-	-0.9649	1.0434	51.7S	3.4E	15	6	569	03m20s
7644	383	1211 Dec 07	02:36:45	695	-9747	129	P	t-	1.4484	0.2039	66.6N	137.5E	0	175		
7645	383	1212 May 02	21:22:43	694	-9742	96	T	-p	0.8394	1.0620	72.4N	168.7W	33	146	377	03m43s
7646	383	1212 Oct 26	07:50:26	693	-9736	101	A	-p	-0.9000	0.9382	74.2S	14.8E	25	51	533	04m33s
7647	383	1213 Apr 22	12:06:14	691	-9730	106	T	nn	0.1090	1.0239	20.7N	1.1W	84	167	82	02m23s
7648	383	1213 Oct 15	14:54:53	690	-9724	111	H	nn	-0.1519	1.0029	19.8S	47.0W	81	16	10	00m18s
7649	383	1214 Apr 11	20:20:49	689	-9718	116	A	p-	-0.6751	0.9629	30.2S	110.8W	47	343	180	04m05s
7650	383	1214 Oct 05	04:37:19	687	-9712	121	T	p-	0.5513	1.0480	24.8N	120.3E	56	198	190	04m11s
7651	383	1215 Mar 02	04:52:15	686	-9707	88	P	-t	1.5180	0.0828	71.8N	37.1E	0	105		
7652	383	1215 Mar 31	21:50:41	686	-9706	126	P	t-	-1.4344	0.2222	71.7S	75.8W	0	292		
7653	383	1215 Aug 26	11:59:14	685	-9701	93	P	-t	-1.3723	0.3075	71.5S	61.8W	0	63		
7654	383	1215 Sep 24	20:55:57	685	-9700	131	P	t-	1.2351	0.5664	71.9N	53.7W	0	260		
7655	383	1216 Feb 19	07:37:26	683	-9695	98	A	-p	0.7663	0.9641	39.1N	56.0E	40	159	200	03m33s
7656	383	1216 Aug 14	23:48:35	682	-9689	103	A	-p	-0.6832	0.9899	29.9S	174.6E	47	17	48	01m02s
7657	383	1217 Feb 07	17:30:25	681	-9683	108	T	nn	0.0204	1.0226	11.7S	76.2W	89	165	77	02m15s
7658	383	1217 Aug 04	04:30:50	679	-9677	113	Am	nn	0.0686	0.9520	19.4N	117.0E	86	192	176	06m01s
7659	383	1218 Jan 28	08:30:17	678	-9671	118	T	p-	-0.6613	1.0470	56.5S	73.1E	48	340	209	03m17s
7660	383	1218 Jul 24	04:55:35	677	-9665	123	A	p-	0.8225	0.9425	72.3N	135.3E	34	209	376	04m34s
7661	384	1218 Dec 19	12:10:11	676	-9660	90	P	-t	1.4002	0.2653	66.3N	9.4E	0	188		
7662	384	1219 Jan 18	00:06:47	676	-9659	128	P	t-	-1.3445	0.3619	69.1S	20.1W	0	203		
7663	384	1219 Jun 13	21:37:18	674	-9654	95	P	-t	-1.1537	0.7138	65.7S	126.7W	0	346		
7664	384	1219 Jul 13	08:23:41	674	-9653	133	Pb	t-	1.5337	0.0308	68.4N	137.2W	0	346		
7665	384	1219 Dec 08	20:06:08	673	-9648	100	A	-p	0.7661	0.9430	26.3N	112.3W	40	189	327	06m48s
7666	384	1220 Jun 02	11:28:52	672	-9642	105	T	-n	-0.3546	1.0670	2.6N	13.7E	69	350	234	06m24s
7667	384	1220 Nov 26	21:01:31	670	-9636	110	A	nn	0.0907	0.9185	17.4S	133.4W	85	193	311	10m57s
7668	384	1221 May 23	04:38:19	669	-9630	115	T	n-	0.3885	1.0767	43.9N	104.3E	67	161	269	05m43s
7669	384	1221 Nov 15	20:27:28	668	-9624	120	A	p-	-0.5900	0.9310	54.2S	145.0W	54	29	319	06m20s
7670	384	1222 Apr 13	10:33:58	667	-9619	87	Pe	-t	-1.4863	0.1092	61.5S	91.7E	0	295		
7671	384	1222 May 12	20:24:03	667	-9618	125	P	t-	1.1607	0.7054	63.2N	99.9E	0	41		
7672	384	1222 Oct 06	13:11:29	666	-9613	92	P	-t	1.5111	0.0576	61.4N	56.8E	0	253		
7673	384	1222 Nov 05	01:35:45	665	-9612	130	P	t-	-1.2474	0.5408	62.5S	27.2E	0	131		
7674	384	1223 Apr 02	16:37:47	664	-9607	97	A	-p	-0.7938	0.9471	38.2S	38.0W	37	324	310	05m09s
7675	384	1223 Sep 26	04:10:29	663	-9601	102	T	-p	0.7816	1.0525	40.2N	147.0E	38	219	272	03m46s
7676	384	1224 Mar 21	16:59:55	662	-9595	107	A	nn	-0.0663	0.9359	0.0S	69.1W	86	331	239	07m23s
7677	384	1224 Sep 14	20:27:49	660	-9589	112	T	nn	0.0847	1.0496	4.7N	123.7W	85	209	165	04m11s
7678	384	1225 Mar 10	18:43:38	659	-9583	117	A	p-	0.6590	0.9596	34.3N	118.9W	49	144	190	03m49s
7679	384	1225 Sep 04	09:21:58	658	-9577	122	A	p-	-0.6686	0.9977	31.9S	19.0E	48	34	11	00m12s
7680	384	1226 Jan 29	16:15:54	657	-9572	89	P	-t	-1.3640	0.3259	62.3S	65.2E	0	234		

Cat Num	Canon Plate	Calendar Date	TD of Greatest Eclipse	ΔT s	Luna Num	Saros Num	Ecl. Type	QLE	Gamma	Ecl. Mag.	Lat. °	Long. °	Sun Alt °	Sun Azm °	Path Width km	Central Line Dur.
7681	385	1226 Feb 28	03:15:05	656	-9571	127	P	t-	1.3351	0.3818	61.0N	56.8E	0	101		
7682	385	1226 Jul 26	00:17:26	655	-9566	94	P	-t	1.4044	0.2713	62.6N	51.5W	0	312		
7683	385	1226 Aug 24	15:25:45	655	-9565	132	P	t-	-1.4633	0.1684	61.3S	122.5W	0	72		
7684	385	1227 Jan 19	07:50:13	654	-9560	99	T	-p	-0.6788	1.0450	57.4S	94.3E	47	323	204	02m59s
7685	385	1227 Jul 15	00:51:05	653	-9554	104	A	-p	0.6512	0.9537	58.3N	167.2W	49	212	222	03m59s
7686	385	1228 Jan 08	22:42:54	652	-9548	109	H3	nn	-0.0068	1.0176	21.6S	155.1W	89	348	60	01m40s
7687	385	1228 Jul 03	06:13:46	650	-9542	114	H	nn	-0.1404	1.0049	14.6N	88.3E	82	13	17	00m32s
7688	385	1228 Dec 28	08:22:01	649	-9536	119	A	p-	0.7190	0.9563	22.5N	49.7E	44	168	227	05m04s
7689	385	1229 Jun 22	18:50:32	648	-9530	124	T	t-	-0.8886	1.0496	39.0S	109.7W	27	11	360	04m10s
7690	385	1229 Dec 17	10:40:21	647	-9524	129	P	t-	1.4431	0.2132	65.5N	6.1E	0	164		
7691	385	1230 May 14	04:56:10	645	-9519	96	T	-t	0.9078	1.0597	82.5N	52.4E	24	122	476	03m17s
7692	385	1230 Nov 06	15:52:58	644	-9513	101	A	-p	-0.9131	0.9380	79.6S	116.1W	24	60	574	04m21s
7693	385	1231 May 03	19:27:05	643	-9507	106	T	-n	0.1763	1.0224	27.8N	112.3W	80	169	78	02m11s
7694	385	1231 Oct 26	23:14:27	642	-9501	111	Hm	nn	-0.1694	1.0033	24.6S	172.2W	80	13	12	00m20s
7695	385	1232 Apr 22	03:20:16	640	-9495	116	A	p-	-0.6097	0.9636	22.2S	140.3E	52	346	165	04m18s
7696	385	1232 Oct 15	13:04:38	639	-9489	121	T	p-	0.5277	1.0469	19.5N	8.9W	58	196	183	04m14s
7697	385	1233 Mar 12	12:10:19	638	-9484	88	Pe	-t	1.5587	0.0121	72.0N	86.7W	0	91		
7698	385	1233 Apr 11	04:41:24	638	-9483	126	P	t-	-1.3717	0.3278	71.2S	167.7E	0	306		
7699	385	1233 Sep 05	19:49:30	637	-9478	93	P	-t	-1.4219	0.2156	71.9S	166.5E	0	77		
7700	385	1233 Oct 05	05:14:34	637	-9477	131	P	t-	1.2080	0.6174	71.6N	167.4E	0	246		
7701	386	1234 Mar 01	15:27:01	636	-9472	98	A	-p	0.7947	0.9693	45.2N	66.0W	37	155	180	02m49s
7702	386	1234 Aug 26	07:07:19	634	-9466	103	A	-p	-0.7438	0.9829	38.0S	60.0E	42	21	90	01m39s
7703	386	1235 Feb 19	01:50:49	633	-9460	108	Tm	nn	0.0419	1.0280	6.6S	157.8E	88	164	95	02m45s
7704	386	1235 Aug 15	11:17:23	632	-9454	113	A	nn	-0.0027	0.9481	11.9N	13.7E	90	28	191	06m40s
7705	386	1236 Feb 08	17:07:28	631	-9448	118	T	p-	-0.6454	1.0501	51.3S	55.5W	50	337	217	03m36s
7706	386	1236 Aug 03	11:33:03	629	-9442	123	A	p-	0.7441	0.9432	61.9N	29.8E	42	205	314	04m57s
7707	386	1236 Dec 29	20:49:58	628	-9437	90	P	-t	1.4069	0.2537	67.4N	131.4W	0	177		
7708	386	1237 Jan 28	08:44:50	628	-9436	128	P	t-	-1.3369	0.3759	70.0S	162.2W	0	216		
7709	386	1237 Jun 24	04:41:09	627	-9431	95	P	-t	-1.2260	0.5814	66.7S	116.7E	0	356		
7710	386	1237 Jul 23	15:20:43	627	-9430	133	P	t-	1.4562	0.1681	69.4N	106.7E	0	335		
7711	386	1237 Dec 19	04:23:11	626	-9425	100	A	-p	0.7728	0.9404	27.3N	120.5E	39	184	348	07m11s
7712	386	1238 Jun 13	18:53:33	625	-9419	105	T	-p	-0.4280	1.0689	1.7S	98.2W	65	354	248	06m38s
7713	386	1238 Dec 08	05:02:16	624	-9413	110	A	nn	0.0988	0.9175	17.9S	107.1E	84	188	315	11m23s
7714	386	1239 Jun 03	12:07:17	622	-9407	115	T	n-	0.3157	1.0771	41.3N	4.2W	71	167	263	05m58s
7715	386	1239 Nov 27	04:33:35	621	-9401	120	A	p-	-0.5795	0.9318	56.5S	98.1E	54	23	313	06m16s
7716	386	1240 May 23	03:41:19	620	-9395	125	P	t-	1.0929	0.8316	64.0N	19.0W	0	32		
7717	386	1240 Oct 16	21:30:05	619	-9390	92	Pe	-t	1.5295	0.0237	61.7N	76.9W	0	244		
7718	386	1240 Nov 15	10:01:04	619	-9389	130	P	t-	-1.2339	0.5649	63.3S	108.5W	0	141		
7719	386	1241 Apr 12	23:36:48	618	-9384	97	A	-t	-0.8597	0.9462	41.1S	141.6W	30	325	376	05m14s
7720	386	1241 Oct 06	12:32:01	616	-9378	102	T	-p	0.8103	1.0494	39.2N	20.3E	36	216	274	03m38s
7721	387	1242 Apr 01	23:59:52	615	-9372	107	A	nn	-0.1253	0.9393	1.1N	173.6W	83	332	227	07m04s
7722	387	1242 Sep 26	04:35:39	614	-9366	112	T	-n	0.1219	1.0443	2.2N	114.3E	83	209	149	03m48s
7723	387	1243 Mar 22	02:10:26	613	-9360	117	A	p-	0.6104	0.9659	35.5N	129.9E	52	144	152	03m08s
7724	387	1243 Sep 15	17:01:45	612	-9354	122	A	p-	-0.6249	0.9920	33.0S	96.0W	51	35	35	00m42s
7725	387	1244 Feb 10	00:45:37	611	-9349	89	P	-t	-1.3786	0.2979	61.7S	71.4W	0	244		
7726	387	1244 Mar 10	11:14:43	610	-9348	127	P	t-	1.2963	0.4525	60.9N	72.0W	0	92		
7727	387	1244 Aug 05	06:54:20	609	-9343	94	P	-t	1.4807	0.1421	62.0N	160.0W	0	303		
7728	387	1244 Sep 03	22:31:22	609	-9342	132	P	t-	-1.4095	0.2623	61.0S	122.1E	0	80		
7729	387	1245 Jan 29	16:32:07	608	-9337	99	T	-p	-0.6916	1.0465	54.0S	33.8W	46	320	213	03m05s
7730	387	1245 Jul 25	07:29:42	607	-9331	104	A	-p	0.7295	0.9528	59.6N	102.3E	43	222	251	03m52s
7731	387	1246 Jan 19	07:20:33	606	-9325	109	H	nn	-0.0150	1.0166	19.6S	76.3E	89	342	57	01m34s
7732	387	1246 Jul 14	13:15:19	605	-9319	114	H	nn	-0.0631	1.0074	17.2N	16.2W	86	18	26	00m46s
7733	387	1247 Jan 08	16:43:15	604	-9313	119	A	p-	0.7154	0.9547	23.1N	78.1W	44	163	234	05m09s
7734	387	1247 Jul 04	02:11:47	602	-9307	124	T	p-	-0.8122	1.0539	30.9S	137.9E	35	14	304	04m42s
7735	387	1247 Dec 28	18:44:27	601	-9301	129	P	t-	1.4383	0.2213	64.5N	125.1W	0	154		
7736	387	1248 May 24	12:24:47	600	-9296	96	T	-t	0.9801	1.0549	78.2N	170.9W	11	13	997	02m42s
7737	387	1248 Jun 22	19:09:36	600	-9295	134	Pb	t-	-1.5159	0.0223	65.2S	125.9W	0	20		
7738	387	1248 Nov 17	00:02:16	599	-9290	101	A	-p	-0.9210	0.9385	84.4S	107.9E	22	73	599	04m10s
7739	387	1249 May 14	02:41:14	598	-9284	106	T	-n	0.2482	1.0204	34.7N	139.0E	75	173	72	01m56s
7740	387	1249 Nov 06	07:41:11	597	-9278	111	H	nn	-0.1817	1.0041	28.7S	61.6E	79	10	14	00m24s

Cat Num	Canon Plate	Calendar Date	TD of Greatest Eclipse	ΔT s	Luna Num	Saros Num	Ecl. Type	QLE	Gamma	Ecl. Mag.	Lat. °	Long. °	Sun Alt °	Sun Azm °	Path Width km	Central Line Dur.
7741	388	1250 May 03	10:13:11	595	-9272	116	A	p-	-0.5397	0.9639	14.6S	33.6E	57	350	155	04m32s
7742	388	1250 Oct 26	21:37:26	594	-9266	121	T	p-	0.5085	1.0458	14.9N	139.2W	59	193	177	04m16s
7743	388	1251 Apr 22	11:26:32	593	-9260	126	P	t-	-1.3041	0.4423	70.5S	53.1E	0	319		
7744	388	1251 Sep 17	03:45:52	592	-9255	93	P	-t	-1.4660	0.1351	72.0S	33.1E	0	91		
7745	388	1251 Oct 16	13:39:20	592	-9254	131	P	t-	1.1863	0.6578	71.0N	27.4E	0	232		
7746	388	1252 Mar 11	23:08:36	591	-9249	98	A	-p	0.8306	0.9745	52.2N	172.8E	34	150	163	02m09s
7747	388	1252 Sep 05	14:32:18	590	-9243	103	A	-p	-0.7982	0.9757	46.2S	57.0W	37	26	143	02m11s
7748	388	1253 Mar 01	10:02:56	589	-9237	108	T	nn	0.0710	1.0336	0.8S	33.4E	86	163	113	03m15s
7749	388	1253 Aug 25	18:11:53	588	-9231	113	A	nn	-0.0671	0.9440	4.4N	91.8W	86	16	207	07m16s
7750	388	1254 Feb 19	01:36:18	586	-9225	118	T	p-	-0.6227	1.0534	45.4S	176.3E	51	337	225	03m59s
7751	388	1254 Aug 14	18:19:43	585	-9219	123	A	p-	0.6726	0.9433	52.6N	75.8W	47	204	282	05m23s
7752	388	1255 Jan 10	05:25:56	584	-9214	90	P	-t	1.4160	0.2376	68.5N	88.1E	0	165		
7753	388	1255 Feb 08	17:15:43	584	-9213	128	P	t-	-1.3244	0.3991	70.8S	57.0E	0	229		
7754	388	1255 Jul 05	11:47:02	583	-9208	95	P	-t	-1.2971	0.4503	67.7S	0.8W	0	7		
7755	388	1255 Aug 03	22:23:39	583	-9207	133	P	t-	1.3823	0.2996	70.2N	11.5W	0	323		
7756	388	1255 Dec 30	12:39:09	582	-9202	100	A	-p	0.7798	0.9385	28.9N	6.5W	39	179	365	07m23s
7757	388	1256 Jun 24	02:17:22	581	-9196	105	T	-p	-0.5023	1.0698	7.0S	149.4E	60	359	263	06m42s
7758	388	1256 Dec 18	13:04:38	580	-9190	110	A	nn	0.1055	0.9172	17.5S	12.7W	84	184	317	11m39s
7759	388	1257 Jun 13	19:33:21	579	-9184	115	T	n-	0.2409	1.0765	37.6N	112.8W	76	173	255	06m11s
7760	388	1257 Dec 07	12:44:36	577	-9178	120	A	p-	-0.5725	0.9332	57.9S	18.9W	55	15	305	06m09s
7761	389	1258 Jun 03	10:54:19	576	-9172	125	P	t-	1.0220	0.9628	64.9N	137.1W	0	22		
7762	389	1258 Nov 26	18:31:26	575	-9166	130	P	t-	-1.2239	0.5826	64.2S	114.3E	0	151		
7763	389	1259 Apr 24	06:28:48	574	-9161	97	A	-t	-0.9304	0.9445	46.9S	117.9E	21	325	548	05m15s
7764	389	1259 Oct 17	21:00:30	573	-9155	102	T	-p	0.8334	1.0464	38.4N	108.8W	33	213	274	03m30s
7765	389	1260 Apr 12	06:51:59	572	-9149	107	A	nn	-0.1907	0.9426	1.5N	83.9E	79	334	216	06m48s
7766	389	1260 Oct 06	12:50:25	571	-9143	112	T	-n	0.1527	1.0390	0.4S	9.6W	81	208	132	03m25s
7767	389	1261 Apr 01	09:30:56	570	-9137	117	A	p-	0.5560	0.9724	36.6N	21.0E	56	145	117	02m31s
7768	389	1261 Sep 26	00:48:31	569	-9131	122	A	p-	-0.5878	0.9863	35.0S	147.3E	54	36	59	01m12s
7769	389	1262 Feb 20	09:09:20	568	-9126	89	P	-t	-1.3986	0.2595	61.4S	153.7E	0	253		
7770	389	1262 Mar 21	19:08:33	567	-9125	127	P	t-	1.2522	0.5339	61.0N	160.7E	0	83		
7771	389	1262 Aug 16	13:39:00	567	-9120	94	Pe	-t	1.5501	0.0254	61.5N	89.8E	0	295		
7772	389	1262 Sep 15	05:45:07	566	-9119	132	P	t-	-1.3624	0.3436	60.9S	4.9E	0	89		
7773	389	1263 Feb 10	01:08:09	565	-9114	99	T	-p	-0.7093	1.0480	50.7S	161.5W	45	318	224	03m13s
7774	389	1263 Aug 05	14:14:43	564	-9108	104	A	-p	0.8030	0.9515	60.0N	10.2E	36	231	295	03m49s
7775	389	1264 Jan 30	15:52:26	563	-9102	109	H	nn	-0.0276	1.0159	17.1S	51.2W	88	338	55	01m29s
7776	389	1264 Jul 24	20:22:44	562	-9096	114	H	nn	0.0104	1.0093	18.8N	121.9W	89	198	32	00m56s
7777	389	1265 Jan 19	00:57:35	561	-9090	119	A	p-	0.7068	0.9538	23.9N	156.0E	45	159	234	05m08s
7778	389	1265 Jul 14	09:37:31	560	-9084	124	T	p-	-0.7388	1.0568	25.3S	25.1E	42	18	275	04m59s
7779	389	1266 Jan 08	02:44:29	559	-9078	129	P	t-	1.4301	0.2350	63.6N	105.2E	0	144		
7780	389	1266 Jun 04	19:50:32	558	-9073	96	P	-t	1.0541	0.9156	66.8N	60.4E	0	357		
7781	390	1266 Jul 04	02:38:30	558	-9072	134	P	t-	-1.4464	0.1578	64.3S	111.8E	0	29		
7782	390	1266 Nov 28	08:16:01	557	-9067	101	A	-p	-0.9262	0.9396	88.5S	67.7W	22	125	608	04m00s
7783	390	1267 May 25	09:49:28	556	-9061	106	T	-p	0.3243	1.0177	41.2N	32.9E	71	177	64	01m37s
7784	390	1267 Nov 17	16:14:34	555	-9055	111	H	-n	-0.1891	1.0052	31.8S	65.6W	79	6	18	00m30s
7785	390	1268 May 13	16:56:23	553	-9049	116	A	p-	-0.4622	0.9638	7.2S	69.9W	62	354	148	04m44s
7786	390	1268 Nov 06	06:18:16	552	-9043	121	T	p-	0.4959	1.0448	11.2N	88.8E	60	189	172	04m16s
7787	390	1269 May 02	18:03:15	551	-9037	126	P	t-	-1.2291	0.5699	69.7S	58.8W	0	331		
7788	390	1269 Sep 27	11:49:53	550	-9032	93	P	-t	-1.5034	0.0680	71.8S	102.2W	0	105		
7789	390	1269 Oct 26	22:11:15	550	-9031	131	P	t-	1.1708	0.6862	70.3N	113.8W	0	219		
7790	390	1270 Mar 23	06:43:33	549	-9026	98	A	-p	0.8726	0.9795	60.0N	50.7E	29	143	149	01m35s
7791	390	1270 Sep 16	22:04:26	548	-9020	103	A	-p	-0.8459	0.9686	54.3S	177.2W	32	31	211	02m38s
7792	390	1271 Mar 12	18:09:34	547	-9014	108	T	nn	0.1052	1.0392	5.4N	89.9W	84	162	132	03m44s
7793	390	1271 Sep 06	01:12:00	546	-9008	113	A	nn	-0.1263	0.9398	3.0S	161.0E	83	17	225	07m48s
7794	390	1272 Mar 01	09:59:32	545	-9002	118	T	p-	-0.5954	1.0569	39.2S	48.6E	53	337	232	04m24s
7795	390	1272 Aug 25	01:13:45	544	-8996	123	A	p-	0.6067	0.9430	43.9N	177.3E	52	203	264	05m50s
7796	390	1273 Jan 20	13:57:10	543	-8991	90	P	-t	1.4282	0.2159	69.5N	51.8W	0	153		
7797	390	1273 Feb 19	01:39:04	543	-8990	128	P	t-	-1.3066	0.4321	71.4S	82.5W	0	242		
7798	390	1273 Jul 15	18:58:15	542	-8985	95	P	-t	-1.3644	0.3257	68.7S	120.2W	0	18		
7799	390	1273 Aug 14	05:35:26	542	-8984	133	P	t-	1.3146	0.4205	71.0N	132.5W	0	310		
7800	390	1274 Jan 09	20:51:23	541	-8979	100	A	-p	0.7886	0.9372	31.3N	132.8W	38	174	380	07m26s

Cat Num	Canon Plate	Calendar Date	TD of Greatest Eclipse	ΔT s	Luna Num	Saros Num	Ecl. Type	QLE	Gamma	Ecl. Mag.	Lat. °	Long. °	Sun Alt °	Sun Azm °	Path Width km	Central Line Dur.
7801	391	1274 Jul 05	09:44:26	540	-8973	105	T	-p	-0.5742	1.0700	13.0S	35.5E	55	3	278	06m35s
7802	391	1274 Dec 29	21:04:54	539	-8967	110	A	nn	0.1138	0.9175	16.2S	132.2W	84	179	316	11m44s
7803	391	1275 Jun 25	02:59:56	538	-8961	115	T	nn	0.1668	1.0752	33.0N	137.5E	80	178	247	06m21s
7804	391	1275 Dec 18	20:56:34	537	-8955	120	A	p-	-0.5657	0.9352	58.0S	135.5W	55	7	294	06m00s
7805	391	1276 Jun 13	18:03:40	536	-8949	125	T	t-	0.9490	1.0202	82.2N	135.2E	18	42	226	01m07s
7806	391	1276 Dec 07	03:05:28	535	-8943	130	P	t-	-1.2165	0.5960	65.2S	24.2W	0	161		
7807	391	1277 May 04	13:11:57	534	-8938	97	A-	-t	-1.0071	0.9528	62.6S	34.0E	0	312	-	-
7808	391	1277 Oct 28	05:36:04	533	-8932	102	T	-p	0.8506	1.0438	37.9N	119.6E	31	209	273	03m23s
7809	391	1278 Apr 23	13:36:43	532	-8926	107	A	nn	-0.2623	0.9457	1.0N	16.7W	75	337	207	06m37s
7810	391	1278 Oct 17	21:13:01	530	-8920	112	T	-n	0.1762	1.0338	3.0S	135.4W	80	205	116	03m03s
7811	391	1279 Apr 12	16:44:05	529	-8914	117	A	p-	0.4945	0.9788	37.4N	85.3W	60	147	86	01m55s
7812	391	1279 Oct 07	08:42:21	528	-8908	122	A	p-	-0.5573	0.9805	37.6S	29.1E	56	35	82	01m42s
7813	391	1280 Mar 02	17:25:39	528	-8903	89	P	-t	-1.4251	0.2084	61.1S	20.7E	0	262		
7814	391	1280 Apr 01	02:54:21	527	-8902	127	P	t-	1.2008	0.6298	61.2N	35.4E	0	74		
7815	391	1280 Sep 25	13:06:53	526	-8896	132	P	t-	-1.3218	0.4131	61.0S	114.4W	0	98		
7816	391	1281 Feb 20	09:36:20	525	-8891	99	T	-p	-0.7337	1.0496	47.8S	72.1E	43	317	239	03m22s
7817	391	1281 Aug 15	21:07:56	524	-8885	104	A	-p	0.8702	0.9497	59.9N	84.2W	29	239	370	03m50s
7818	391	1282 Feb 10	00:17:59	523	-8879	109	H	nn	-0.0451	1.0156	14.3S	177.3W	87	334	54	01m26s
7819	391	1282 Aug 05	03:35:56	522	-8873	114	H	nn	0.0799	1.0107	19.4N	131.1E	85	204	37	01m01s
7820	391	1283 Jan 30	09:06:55	521	-8867	119	A	p-	0.6948	0.9533	25.1N	31.5E	46	155	232	05m02s
7821	392	1283 Jul 25	17:06:40	520	-8861	124	T	p-	-0.6677	1.0587	21.4S	88.1W	48	21	256	05m07s
7822	392	1284 Jan 19	10:41:02	519	-8855	129	P	t-	1.4194	0.2525	62.8N	23.3W	0	134		
7823	392	1284 Jun 15	03:13:07	518	-8850	96	P	-t	1.1301	0.7690	65.8N	60.8W	0	347		
7824	392	1284 Jul 14	10:08:18	518	-8849	134	P	t-	-1.3779	0.2915	63.4S	10.4W	0	39		
7825	392	1284 Dec 08	16:34:01	517	-8844	101	A	-p	-0.9284	0.9414	86.2S	56.2E	21	235	598	03m49s
7826	392	1285 Jun 04	16:53:58	516	-8838	106	H	-p	0.4024	1.0143	47.2N	70.8W	66	182	54	01m15s
7827	392	1285 Nov 28	00:51:28	515	-8832	111	H	-n	-0.1943	1.0068	33.9S	167.0E	79	1	24	00m39s
7828	392	1286 May 24	23:36:15	514	-8826	116	A	p-	-0.3825	0.9632	0.4S	171.9W	68	358	144	04m55s
7829	392	1286 Nov 17	15:03:22	513	-8820	121	T	p-	0.4865	1.0441	8.2N	43.9W	61	185	168	04m17s
7830	392	1287 May 14	00:35:25	512	-8814	126	P	t-	-1.1500	0.7052	68.8S	169.0W	0	342		
7831	392	1287 Oct 08	20:00:06	511	-8809	93	Pe	-t	-1.5351	0.0120	71.4S	121.2E	0	119		
7832	392	1287 Nov 07	06:48:15	511	-8808	131	P	t-	1.1600	0.7059	69.4N	104.2E	0	206		
7833	392	1288 Apr 02	14:11:28	510	-8803	98	A	-p	0.9211	0.9840	68.5N	76.4W	22	130	147	01m07s
7834	392	1288 Sep 27	05:43:53	509	-8797	103	A	-t	-0.8863	0.9616	62.1S	58.9E	27	39	301	03m00s
7835	392	1289 Mar 23	02:07:05	508	-8791	108	T	nn	0.1475	1.0448	12.1N	148.9E	81	162	151	04m10s
7836	392	1289 Sep 16	08:22:02	507	-8785	113	A	nn	-0.1768	0.9357	10.1S	51.5E	80	18	243	08m15s
7837	392	1290 Mar 12	18:13:59	506	-8779	118	T	p-	-0.5611	1.0604	32.7S	77.6W	56	339	238	04m52s
7838	392	1290 Sep 05	08:17:10	505	-8773	123	A	p-	0.5480	0.9424	35.8N	68.4E	57	201	253	06m17s
7839	392	1291 Jan 31	22:21:43	504	-8768	90	P	-t	1.4452	0.1856	70.4N	169.4E	0	140		
7840	392	1291 Mar 02	09:53:49	504	-8767	128	P	t-	-1.2826	0.4766	71.8S	139.8E	0	256		
7841	393	1291 Jul 27	02:14:27	503	-8762	95	P	-t	-1.4280	0.2075	69.6S	118.6E	0	29		
7842	393	1291 Aug 25	12:55:31	503	-8761	133	P	t-	1.2525	0.5314	71.6N	103.8E	0	297		
7843	393	1292 Jan 21	04:58:17	502	-8756	100	A	-p	0.8014	0.9363	34.6N	102.0E	36	169	395	07m17s
7844	393	1292 Jul 15	17:13:08	501	-8750	105	T	-p	-0.6450	1.0692	19.9S	79.6W	50	7	295	06m17s
7845	393	1293 Jan 09	05:03:32	500	-8744	110	A	nn	0.1233	0.9185	13.9S	108.4E	83	175	312	11m36s
7846	393	1293 Jul 05	10:26:45	499	-8738	115	T	nn	0.0933	1.0730	27.5N	26.7E	84	183	238	06m24s
7847	393	1293 Dec 29	05:09:12	499	-8732	120	A	p-	-0.5588	0.9379	56.8S	107.6E	56	358	279	05m48s
7848	393	1294 Jun 25	01:11:53	498	-8726	125	T	p-	0.8757	1.0195	84.5N	153.7E	28	168	140	01m11s
7849	393	1294 Dec 18	11:42:14	497	-8720	130	P	t-	-1.2108	0.6064	66.3S	163.8W	0	172		
7850	393	1295 May 15	19:48:34	496	-8715	97	P	-t	-1.0884	0.8131	63.3S	74.5W	0	321		
7851	393	1295 Nov 08	14:17:45	495	-8709	102	T	-p	0.8630	1.0414	37.5N	14.0W	30	204	271	03m17s
7852	393	1296 May 03	20:15:52	494	-8703	107	A	np	-0.3385	0.9485	0.5S	116.1W	70	340	200	06m27s
7853	393	1296 Oct 28	05:41:29	493	-8697	112	T	-n	0.1946	1.0289	5.5S	97.2E	79	203	100	02m41s
7854	393	1297 Apr 22	23:51:57	492	-8691	117	A	p-	0.4275	0.9850	37.7N	170.4E	64	150	58	01m22s
7855	393	1297 Oct 17	16:42:50	491	-8685	122	A	p-	-0.5330	0.9751	40.6S	90.4W	58	34	104	02m11s
7856	393	1298 Mar 14	01:35:41	490	-8680	89	P	-t	-1.4573	0.1459	61.1S	110.7W	0	271		
7857	393	1298 Apr 12	10:35:28	490	-8679	127	P	t-	1.1445	0.7357	61.5N	88.9W	0	66		
7858	393	1298 Oct 06	20:35:16	489	-8673	132	P	t-	-1.2867	0.4728	61.2S	124.7E	0	107		
7859	393	1299 Mar 03	17:57:31	488	-8668	99	T	-p	-0.7639	1.0510	45.5S	53.0W	40	317	257	03m30s
7860	393	1299 Aug 27	04:09:37	487	-8662	104	A	-p	0.9311	0.9474	60.0N	180.0E	21	248	526	03m53s

Cat Num	Canon Plate	Calendar Date	TD of Greatest Eclipse	ΔT s	Luna Num	Saros Num	Ecl. Type	QLE	Gamma	Ecl. Mag.	Lat. °	Long. °	Sun Alt °	Sun Azm °	Path Width km	Central Line Dur.
7861	394	1300 Feb 21	08:34:00	486	-8656	109	H	nn	-0.0698	1.0154	11.5S	58.8E	86	332	53	01m24s
7862	394	1300 Aug 15	10:57:25	485	-8650	114	Hm	nn	0.1434	1.0115	19.0N	22.0E	82	207	40	01m05s
7863	394	1301 Feb 09	17:06:50	484	-8644	119	A	p-	0.6757	0.9533	26.4N	90.3W	47	151	226	04m53s
7864	394	1301 Aug 05	00:42:42	483	-8638	124	T	p-	-0.6019	1.0597	19.1S	157.3E	53	24	242	05m07s
7865	394	1302 Jan 29	18:29:48	482	-8632	129	P	t-	1.4025	0.2805	62.1N	149.7W	0	125		
7866	394	1302 Jun 26	10:35:49	481	-8627	96	P	-t	1.2055	0.6240	64.8N	178.4E	0	337		
7867	394	1302 Jul 25	17:42:30	481	-8626	134	P	t-	-1.3129	0.4178	62.7S	133.4W	0	48		
7868	394	1302 Dec 20	00:53:51	480	-8621	101	A	-p	-0.9302	0.9438	82.1S	85.3W	21	251	578	03m37s
7869	394	1303 Jun 15	23:53:41	479	-8615	106	H	-p	0.4836	1.0103	52.5N	171.4W	61	189	41	00m52s
7870	394	1303 Dec 09	09:32:53	479	-8609	111	H	-n	-0.1964	1.0089	34.9S	39.0E	78	356	31	00m50s
7871	394	1304 Jun 04	06:09:36	478	-8603	116	A	nn	-0.2977	0.9622	5.9N	88.6E	73	2	144	05m04s
7872	394	1304 Nov 27	23:53:25	477	-8597	121	T	p-	0.4812	1.0438	6.2N	177.7W	61	181	167	04m17s
7873	394	1305 May 24	07:02:17	476	-8591	126	P	t-	-1.0658	0.8496	67.8S	82.8E	0	353		
7874	394	1305 Nov 17	15:30:56	475	-8585	131	P	t-	1.1540	0.7163	68.4N	38.4W	0	194		
7875	394	1306 Apr 13	21:34:09	474	-8580	98	A	-t	0.9745	0.9872	75.5N	130.6E	12	89	214	00m47s
7876	394	1306 Oct 08	13:29:27	473	-8574	103	A	-t	-0.9208	0.9547	69.4S	70.7W	22	51	428	03m18s
7877	394	1307 Apr 03	09:59:42	472	-8568	108	T	-n	0.1946	1.0501	18.9N	29.0E	79	163	169	04m33s
7878	394	1307 Sep 27	15:39:04	471	-8562	113	A	nn	-0.2211	0.9317	16.9S	59.6W	77	18	261	08m39s
7879	394	1308 Mar 23	02:21:00	470	-8556	118	T	p-	-0.5205	1.0638	25.9S	157.8E	58	341	243	05m21s
7880	394	1308 Sep 15	15:30:02	469	-8550	123	A	p-	0.4964	0.9417	28.2N	42.6W	60	200	247	06m43s
7881	395	1309 Feb 11	06:39:24	468	-8545	90	P	-t	1.4670	0.1465	71.1N	31.7E	0	127		
7882	395	1309 Mar 12	18:00:25	468	-8544	128	P	t-	-1.2527	0.5323	71.9S	3.9E	0	270		
7883	395	1309 Aug 06	09:38:14	467	-8539	95	Pe	-t	-1.4863	0.0990	70.4S	5.0W	0	42		
7884	395	1309 Sep 04	20:25:26	467	-8538	133	P	t-	1.1974	0.6300	72.0N	22.7W	0	283		
7885	395	1310 Jan 31	12:57:57	467	-8533	100	A	-p	0.8194	0.9358	38.9N	21.9W	35	164	415	07m01s
7886	395	1310 Jul 27	00:47:46	466	-8527	105	T	-p	-0.7111	1.0676	27.2S	163.0E	44	11	313	05m51s
7887	395	1311 Jan 20	12:57:37	465	-8521	110	A	nn	0.1365	0.9200	10.6S	10.1W	82	171	306	11m18s
7888	395	1311 Jul 16	17:55:04	464	-8515	115	Tm	nn	0.0216	1.0700	21.4N	85.4W	89	186	228	06m20s
7889	395	1312 Jan 09	13:20:03	463	-8509	120	A	p-	-0.5500	0.9413	54.3S	10.2W	56	352	261	05m33s
7890	395	1312 Jul 05	08:19:23	462	-8503	125	T	p-	0.8028	1.0171	75.4N	68.1E	36	191	99	01m08s
7891	395	1312 Dec 28	20:17:58	461	-8497	130	P	t-	-1.2038	0.6192	67.4S	56.4E	0	183		
7892	395	1313 May 26	02:19:28	460	-8492	97	P	-t	-1.1731	0.6672	64.2S	178.1E	0	331		
7893	395	1313 Nov 18	23:04:31	459	-8486	102	T	-p	0.8712	1.0395	37.4N	149.3W	29	199	268	03m13s
7894	395	1314 May 15	02:50:08	458	-8480	107	A	-p	-0.4192	0.9510	3.3S	145.5E	65	344	196	06m19s
7895	395	1314 Nov 08	14:15:05	457	-8474	112	T	-n	0.2080	1.0244	7.6S	31.3W	78	199	85	02m20s
7896	395	1315 May 04	06:54:52	457	-8468	117	A	p-	0.3556	0.9909	37.3N	67.7E	69	155	34	00m51s
7897	395	1315 Oct 29	00:49:59	456	-8462	122	A	p-	-0.5150	0.9698	43.8S	149.1E	59	32	126	02m40s
7898	395	1316 Mar 24	09:36:47	455	-8457	89	Pe	-t	-1.4970	0.0686	61.2S	120.1E	0	280		
7899	395	1316 Apr 22	18:08:42	455	-8456	127	P	t-	1.0812	0.8560	62.0N	148.8E	0	57		
7900	395	1316 Oct 17	04:12:33	454	-8450	132	P	t-	-1.2591	0.5193	61.6S	1.5E	0	116		
7901	396	1317 Mar 14	02:10:14	453	-8445	99	T	-p	-0.8008	1.0522	44.2S	175.9W	37	317	283	03m37s
7902	396	1317 Sep 06	11:21:05	452	-8439	104	An	-t	0.9843	0.9439	60.9N	87.6E	9	260	-	03m55s
7903	396	1318 Mar 03	16:42:11	451	-8433	109	H	-n	-0.1003	1.0153	8.8S	63.2W	84	331	53	01m24s
7904	396	1318 Aug 26	18:27:19	450	-8427	114	H	nn	0.2005	1.0120	17.9N	89.6W	78	209	42	01m06s
7905	396	1319 Feb 21	00:59:44	449	-8421	119	A	p-	0.6516	0.9537	28.0N	150.1E	49	148	218	04m42s
7906	396	1319 Aug 16	08:23:22	448	-8415	124	T	p-	-0.5396	1.0600	18.0S	41.9E	57	27	231	05m01s
7907	396	1320 Feb 10	02:13:16	448	-8409	129	P	t-	1.3813	0.3154	61.5N	85.4E	0	116		
7908	396	1320 Jul 06	17:58:24	447	-8404	96	P	-t	1.2804	0.4807	63.9N	57.9E	0	327		
7909	396	1320 Aug 05	01:20:21	447	-8403	134	P	t-	-1.2510	0.5375	62.1S	102.9E	0	57		
7910	396	1320 Dec 30	09:13:30	446	-8398	101	A	-p	-0.9327	0.9468	77.9S	141.3E	21	259	553	03m25s
7911	396	1321 Jun 26	06:52:55	445	-8392	106	H	-p	0.5641	1.0056	56.7N	90.3E	55	198	23	00m27s
7912	396	1321 Dec 19	18:15:02	444	-8386	111	H	-n	-0.1987	1.0115	34.9S	89.2W	78	351	40	01m04s
7913	396	1322 Jun 15	12:41:47	443	-8380	116	A	nn	-0.2127	0.9607	11.4N	9.8W	78	6	146	05m11s
7914	396	1322 Dec 09	08:44:26	442	-8374	121	T	n-	0.4767	1.0439	5.0N	48.4E	61	177	167	04m17s
7915	396	1323 Jun 04	13:27:55	441	-8368	126	As	t-	-0.9799	0.9383	56.1S	23.4W	11	3	-	05m59s
7916	396	1323 Nov 29	00:16:10	441	-8362	131	P	t-	1.1509	0.7215	67.3N	178.8E	0	183		
7917	396	1324 Apr 24	04:49:54	440	-8357	98	P	-t	1.0343	0.9272	70.3N	31.7W	0	38		
7918	396	1324 Oct 18	21:22:40	439	-8351	103	A	-t	-0.9481	0.9483	75.4S	149.9E	18	71	613	03m33s
7919	396	1325 Apr 13	17:44:29	438	-8345	108	T	-n	0.2487	1.0551	26.0N	88.8W	75	164	188	04m50s
7920	396	1325 Oct 07	23:05:32	437	-8339	113	A	nn	-0.2574	0.9281	23.2S	172.5W	75	17	279	08m57s

Cat Num	Canon Plate	Calendar Date	TD of Greatest Eclipse	ΔT s	Luna Num	Saros Num	Ecl. Type	QLE	Gamma	Ecl. Mag.	Lat. °	Long. °	Sun Alt °	Sun Azm °	Path Width km	Central Line Dur.
7921	397	1326 Apr 03	10:19:38	436	-8333	118	T	p-	-0.4731	1.0668	19.0S	35.3E	62	343	246	05m49s
7922	397	1326 Sep 26	22:53:53	435	-8327	123	A	p-	0.4531	0.9409	21.3N	156.1W	63	199	244	07m07s
7923	397	1327 Feb 22	14:47:53	435	-8322	90	P	-t	1.4956	0.0953	71.6N	104.3W	0	114		
7924	397	1327 Mar 24	01:56:52	435	-8321	128	P	t-	-1.2152	0.6020	71.8S	129.5W	0	284		
7925	397	1327 Sep 16	04:04:29	434	-8315	133	P	t-	1.1489	0.7168	72.1N	151.8W	0	270		
7926	397	1328 Feb 11	20:50:10	433	-8310	100	A	-p	0.8426	0.9356	44.1N	144.7W	32	159	442	06m38s
7927	397	1328 Aug 06	08:26:35	432	-8304	105	T	-p	-0.7736	1.0652	35.1S	43.6E	39	16	335	05m19s
7928	397	1329 Jan 30	20:45:47	431	-8298	110	A	nn	0.1543	0.9222	6.5S	127.7W	81	168	297	10m51s
7929	397	1329 Jul 27	01:26:16	430	-8292	115	T	nn	-0.0471	1.0662	14.9N	160.9E	87	11	217	06m08s
7930	397	1330 Jan 19	21:28:48	429	-8286	120	A	p-	-0.5391	0.9452	50.6S	129.0W	57	346	240	05m16s
7931	397	1330 Jul 16	15:26:57	429	-8280	125	T	p-	0.7307	1.0139	66.5N	35.7W	43	197	70	01m00s
7932	397	1331 Jan 09	04:53:22	428	-8274	130	P	t-	-1.1961	0.6333	68.4S	83.8W	0	194		
7933	397	1331 Jun 06	08:46:50	427	-8269	97	P	-t	-1.2597	0.5183	65.1S	71.4E	0	340		
7934	397	1331 Jul 05	22:46:38	427	-8268	135	Pb	t-	1.5532	0.0063	67.8N	12.6E	0	352		
7935	397	1331 Nov 30	07:54:51	426	-8263	102	T	-p	0.8766	1.0380	37.6N	74.2E	28	194	265	03m09s
7936	397	1332 May 25	09:20:33	425	-8257	107	A	-p	-0.5032	0.9531	7.3S	47.6E	60	347	197	06m10s
7937	397	1332 Nov 18	22:53:10	424	-8251	112	T	-n	0.2172	1.0202	9.2S	160.8W	78	195	71	02m01s
7938	397	1333 May 14	13:55:23	424	-8245	117	A	p-	0.2806	0.9964	36.1N	34.2W	74	160	13	00m20s
7939	397	1333 Nov 08	09:01:11	423	-8239	122	A	p-	-0.5012	0.9651	46.8S	28.4E	60	28	145	03m06s
7940	397	1334 May 04	01:39:14	422	-8233	127	P	t-	1.0149	0.9830	62.6N	27.0E	0	48		
7941	398	1334 Oct 28	11:56:18	421	-8227	132	P	t-	-1.2370	0.5560	62.1S	123.4W	0	126		
7942	398	1335 Mar 25	10:14:54	420	-8222	99	T	-p	-0.8444	1.0528	44.2S	63.3E	32	318	319	03m42s
7943	398	1335 Sep 17	18:41:54	419	-8216	104	P	-t	1.0305	0.9134	60.9N	12.4W	0	268		
7944	398	1336 Mar 14	00:40:15	419	-8210	109	H	-n	-0.1386	1.0152	6.4S	177.4E	82	331	52	01m23s
7945	398	1336 Sep 06	02:06:58	418	-8204	114	H	-n	0.2506	1.0122	16.2N	156.1E	75	210	43	01m07s
7946	398	1337 Mar 03	08:40:41	417	-8198	119	A	p-	0.6182	0.9543	29.5N	34.1E	52	146	207	04m32s
7947	398	1337 Aug 26	16:12:58	416	-8192	124	T	p-	-0.4842	1.0596	18.1S	75.7W	61	29	221	04m53s
7948	398	1338 Feb 20	09:47:30	415	-8186	129	P	t-	1.3524	0.3632	61.2N	37.0W	0	107		
7949	398	1338 Jul 18	01:22:26	414	-8181	96	P	-t	1.3535	0.3419	63.2N	62.6W	0	318		
7950	398	1338 Aug 16	09:03:11	414	-8180	134	P	t-	-1.1933	0.6482	61.6S	21.9W	0	66		
7951	398	1339 Jan 10	17:31:41	414	-8175	101	A	-p	-0.9371	0.9504	73.9S	10.9E	20	265	531	03m11s
7952	398	1339 Jul 07	13:50:33	413	-8169	106	H	-p	0.6451	1.0002	59.8N	5.5W	50	208	1	00m01s
7953	398	1339 Dec 31	02:57:36	412	-8163	111	H	-n	-0.2011	1.0147	33.9S	142.5E	78	346	52	01m20s
7954	398	1340 Jun 25	19:10:38	411	-8157	116	A	nn	-0.1253	0.9586	16.0N	106.7W	83	10	151	05m16s
7955	398	1340 Dec 19	17:37:50	410	-8151	121	T	n-	0.4741	1.0444	4.7N	86.0W	62	172	168	04m17s
7956	398	1341 Jun 14	19:52:10	409	-8145	126	A	t-	-0.8922	0.9433	39.6S	123.3W	27	7	465	06m25s
7957	398	1341 Dec 09	09:03:29	409	-8139	131	P	t-	1.1500	0.7229	66.3N	36.2E	0	172		
7958	398	1342 May 05	12:02:39	408	-8134	98	P	-t	1.0972	0.8150	69.4N	152.3W	0	26		
7959	398	1342 Oct 30	05:22:14	407	-8128	103	A	-t	-0.9696	0.9422	78.6S	5.8W	13	106	920	03m44s
7960	398	1343 Apr 25	01:24:16	406	-8122	108	T	-n	0.3077	1.0597	33.1N	155.2E	72	166	206	05m02s
7961	399	1343 Oct 19	06:39:25	405	-8116	113	A	-n	-0.2873	0.9247	29.0S	73.3E	73	16	296	09m12s
7962	399	1344 Apr 13	18:11:36	405	-8110	118	T	n-	-0.4200	1.0695	12.0S	85.4W	65	345	249	06m15s
7963	399	1344 Oct 07	06:26:57	404	-8104	123	A	p-	0.4170	0.9402	15.1N	88.4E	65	197	242	07m29s
7964	399	1345 Mar 04	22:47:14	403	-8099	90	Pe	-t	1.5306	0.0325	71.9N	121.7E	0	100		
7965	399	1345 Apr 03	09:44:58	403	-8098	128	P	t-	-1.1717	0.6830	71.5S	99.5E	0	297		
7966	399	1345 Sep 26	11:53:53	402	-8092	133	P	t-	1.1079	0.7902	72.0N	76.5E	0	255		
7967	399	1346 Feb 22	04:33:47	401	-8087	100	A	-p	0.8720	0.9354	50.4N	93.4E	29	154	488	06m11s
7968	399	1346 Aug 17	16:11:26	401	-8081	105	T	-p	-0.8312	1.0622	43.4S	78.6W	33	21	365	04m43s
7969	399	1347 Feb 11	04:27:03	400	-8075	110	A	nn	0.1778	0.9248	1.5S	116.0E	80	165	287	10m17s
7970	399	1347 Aug 07	09:01:38	399	-8069	115	T	nn	-0.1116	1.0618	8.1N	45.7E	84	13	204	05m48s
7971	399	1348 Jan 31	05:31:34	398	-8063	120	A	p-	-0.5226	0.9499	46.0S	111.9E	58	343	216	04m55s
7972	399	1348 Jul 26	22:37:09	397	-8057	125	H	p-	0.6616	1.0098	58.0N	143.3W	48	199	45	00m46s
7973	399	1349 Jan 19	13:24:42	396	-8051	130	P	t-	-1.1847	0.6546	60.5S	136.3E	0	206		
7974	399	1349 Jun 16	15:12:24	396	-8046	97	P	-t	-1.3468	0.3687	66.1S	35.3W	0	350		
7975	399	1349 Jul 16	05:25:45	396	-8045	135	P	t-	1.4782	0.1384	68.8N	98.7W	0	341		
7976	399	1349 Dec 10	16:46:27	395	-8040	102	T	-p	0.8811	1.0371	38.2N	62.7W	28	188	264	03m06s
7977	399	1350 Jun 05	15:49:57	394	-8034	107	A	-p	-0.5883	0.9547	12.4S	50.7W	54	351	204	05m59s
7978	399	1350 Nov 30	07:34:51	393	-8028	112	H3	-n	0.2227	1.0166	10.3S	68.8E	77	191	58	01m42s
7979	399	1351 May 25	20:52:03	393	-8022	117	H	nn	0.2015	1.0016	33.7N	135.5W	78	165	6	00m09s
7980	399	1351 Nov 19	17:18:04	392	-8016	122	A	p-	-0.4929	0.9608	49.4S	92.7W	60	23	163	03m32s

Cat Num	Canon Plate	Calendar Date	TD of Greatest Eclipse	ΔT s	Luna Num	Saros Num	Ecl. Type	QLE	Gamma	Ecl. Mag.	Lat. °	Long. °	Sun Alt °	Sun Azm °	Path Width km	Central Line Dur.
7981	400	1352 May 14	09:04:24	391	-8010	127	T	t-	0.9437	1.0427	73.6N	48.5W	19	81	441	02m18s
7982	400	1352 Nov 07	19:47:15	390	-8004	132	P	t-	-1.2209	0.5826	62.8S	109.7E	0	135		
7983	400	1353 Apr 04	18:11:06	389	-7999	99	T	-t	-0.8949	1.0527	46.0S	54.7W	26	319	383	03m41s
7984	400	1353 Sep 28	02:13:52	389	-7993	104	P	-t	1.0684	0.8481	61.0N	134.1W	0	259		
7985	400	1354 Mar 25	08:30:21	388	-7987	109	H	-n	-0.1829	1.0149	4.4S	60.0E	79	331	52	01m23s
7986	400	1354 Sep 17	09:54:40	387	-7981	114	H	-n	0.2947	1.0122	14.2N	39.3E	73	210	44	01m07s
7987	400	1355 Mar 14	16:13:55	386	-7975	119	A	p-	0.5792	0.9552	31.2N	79.3W	54	145	196	04m22s
7988	400	1355 Sep 07	00:09:07	385	-7969	124	T	p-	-0.4340	1.0586	19.2S	165.3E	64	31	212	04m43s
7989	400	1356 Mar 02	17:13:18	385	-7963	129	P	t-	1.3168	0.4225	60.9N	157.2W	0	98		
7990	400	1356 Jul 28	08:49:12	384	-7958	96	P	-t	1.4235	0.2101	62.5N	176.3E	0	309		
7991	400	1356 Aug 26	16:52:10	384	-7957	134	P	t-	-1.1410	0.7477	61.2S	148.1W	0	75		
7992	400	1357 Jan 21	01:46:45	383	-7952	101	A	-p	-0.9448	0.9543	70.1S	116.8W	19	269	517	02m56s
7993	400	1357 Jul 17	20:50:34	382	-7946	106	A	-p	0.7228	0.9942	61.5N	100.5W	43	219	29	00m26s
7994	400	1358 Jan 10	11:37:17	382	-7940	111	T	-n	-0.2065	1.0183	32.0S	14.4E	78	341	64	01m38s
7995	400	1358 Jul 07	01:41:45	381	-7934	116	Am	nn	-0.0404	0.9562	19.6N	156.6E	88	16	160	05m22s
7996	400	1358 Dec 31	02:29:35	380	-7928	121	T	n-	0.4701	1.0454	5.1N	140.0E	62	168	171	04m18s
7997	400	1359 Jun 26	02:16:31	379	-7922	126	A	p-	-0.8038	0.9463	29.9S	138.3E	36	11	330	06m30s
7998	400	1359 Dec 20	17:50:58	378	-7916	131	P	t-	1.1496	0.7231	65.2N	106.0W	0	161		
7999	400	1360 May 15	19:11:02	378	-7911	98	P	-t	1.1647	0.6932	68.5N	88.7E	0	15		
8000	400	1360 Jun 14	05:56:04	378	-7910	136	Pb	t-	-1.5227	0.0495	65.8S	78.2E	0	13		
8001	401	1360 Nov 09	13:27:09	377	-7905	103	As	-t	-0.9858	0.9366	76.8S	166.6W	9	145	-	03m53s
8002	401	1361 May 05	08:58:03	376	-7899	108	T	-n	0.3722	1.0635	40.2N	41.4E	68	169	224	05m07s
8003	401	1361 Oct 29	14:21:52	375	-7893	113	A	-n	-0.3101	0.9219	34.1S	42.1W	72	13	310	09m22s
8004	401	1362 Apr 25	01:56:16	375	-7887	118	T	n-	-0.3611	1.0717	5.2S	155.9E	69	348	249	06m37s
8005	401	1362 Oct 18	14:09:27	374	-7881	123	A	p-	0.3879	0.9397	9.6N	29.1W	67	194	241	07m48s
8006	401	1363 Apr 14	17:23:47	373	-7875	128	P	t-	-1.1212	0.7768	70.9S	28.7W	0	310		
8007	401	1363 Oct 07	19:52:55	372	-7869	133	P	t-	1.0741	0.8507	71.6N	57.4W	0	242		
8008	401	1364 Mar 04	12:06:40	372	-7864	100	A	-t	0.9095	0.9352	57.9N	28.8W	24	145	580	05m41s
8009	401	1364 Aug 28	00:03:01	371	-7858	105	T	-p	-0.8832	1.0584	51.9S	155.6E	28	28	409	04m06s
8010	401	1365 Feb 21	12:00:58	370	-7852	110	A	nn	0.2074	0.9279	4.1N	1.2E	78	164	276	09m38s
8011	401	1365 Aug 17	16:41:46	369	-7846	115	T	-n	-0.1716	1.0569	1.1N	71.2W	80	15	190	05m22s
8012	401	1366 Feb 10	13:29:55	369	-7840	120	A	p-	-0.5016	0.9549	40.6S	7.4W	60	341	190	04m32s
8013	401	1366 Aug 07	05:50:23	368	-7834	125	H	p-	0.5958	1.0051	49.7N	107.0E	53	200	22	00m26s
8014	401	1367 Jan 30	21:53:13	367	-7828	130	P	t-	-1.1704	0.6812	70.4S	3.5W	0	219		
8015	401	1367 Jun 27	21:36:10	366	-7823	97	P	-t	-1.4343	0.2188	67.1S	141.9W	0	0		
8016	401	1367 Jul 27	12:05:47	366	-7822	135	P	t-	1.4043	0.2679	69.7N	149.3E	0	330		
8017	401	1367 Dec 22	01:39:34	366	-7817	102	T	-p	0.8842	1.0366	39.2N	159.8E	28	182	265	03m03s
8018	401	1368 Jun 15	22:19:08	365	-7811	107	A	-p	-0.6736	0.9557	18.9S	149.8W	48	356	218	05m43s
8019	401	1368 Dec 10	16:17:17	364	-7805	112	H	-n	0.2270	1.0135	10.5S	61.6W	77	186	48	01m25s
8020	401	1369 Jun 05	03:49:31	363	-7799	117	H	nn	0.1222	1.0061	30.4N	122.5E	83	170	21	00m37s
8021	402	1369 Nov 30	01:36:58	363	-7793	122	A	p-	-0.4873	0.9570	51.4S	146.6E	61	17	179	03m55s
8022	402	1370 May 25	16:28:30	362	-7787	127	T	t-	0.8708	1.0497	76.2N	124.0W	29	117	338	02m51s
8023	402	1370 Nov 19	03:42:08	361	-7781	132	P	t-	-1.2082	0.6034	63.6S	18.4W	0	145		
8024	402	1371 Apr 16	02:00:13	360	-7776	99	T	-t	-0.9508	1.0512	50.7S	168.9W	18	319	545	03m30s
8025	402	1371 Oct 09	09:55:25	360	-7770	104	P	-t	1.0990	0.7952	61.3N	101.7E	0	250		
8026	402	1372 Apr 04	16:09:02	359	-7764	109	H	-n	-0.2359	1.0143	3.1S	54.4W	76	333	50	01m22s
8027	402	1372 Sep 27	17:53:15	358	-7758	114	H	-n	0.3305	1.0121	11.9N	80.4W	71	209	44	01m07s
8028	402	1373 Mar 24	23:35:23	357	-7752	119	A	p-	0.5311	0.9561	32.7N	170.9E	58	146	186	04m15s
8029	402	1373 Sep 17	08:14:16	357	-7746	124	T	n-	-0.3912	1.0573	21.0S	44.0E	67	31	204	04m33s
8030	402	1374 Mar 14	00:29:08	356	-7740	129	P	t-	1.2731	0.4957	60.9N	85.1E	0	89		
8031	402	1374 Aug 08	16:20:08	355	-7735	96	Pe	-t	1.4893	0.0875	61.9N	54.5E	0	300		
8032	402	1374 Sep 07	00:48:07	355	-7734	134	P	t-	-1.0949	0.8345	61.0S	84.0E	0	83		
8033	402	1375 Feb 01	09:57:38	355	-7729	101	A	-p	-0.9565	0.9586	66.8S	118.4E	16	271	525	02m39s
8034	402	1375 Jul 29	03:50:53	354	-7723	106	A	-p	0.7991	0.9876	62.3N	165.3E	37	231	72	00m54s
8035	402	1376 Jan 21	20:15:01	353	-7717	111	T	-n	-0.2141	1.0225	29.5S	113.5W	77	337	78	01m58s
8036	402	1376 Jul 17	08:13:14	352	-7711	116	A	nn	0.0439	0.9533	22.2N	60.3E	87	198	171	05m30s
8037	402	1377 Jan 10	11:19:31	352	-7705	121	T	n-	0.4646	1.0469	6.1N	6.5E	62	164	175	04m19s
8038	402	1377 Jul 06	08:43:28	351	-7699	126	A	p-	-0.7168	0.9484	22.8S	40.2E	44	15	269	06m24s
8039	402	1377 Dec 31	02:38:10	350	-7693	131	P	t-	1.1494	0.7234	64.3N	112.2E	0	151		
8040	402	1378 May 27	02:18:09	350	-7688	98	P	-t	1.2336	0.5674	67.5N	29.3W	0	4		

Cat Num	Canon Plate	Calendar Date	TD of Greatest Eclipse	ΔT s	Luna Num	Saros Num	Ecl. Type	QLE	Gamma	Ecl. Mag.	Lat. °	Long. °	Sun Alt °	Sun Azm °	Path Width km	Central Line Dur.
8041	403	1378 Jun 25	12:45:16	349	-7687	136	P	t-	-1.4392	0.1976	64.8S	34.2W	0	23		
8042	403	1378 Nov 20	21:36:04	349	-7682	103	A-	-t	-0.9981	0.9635	68.1S	45.5E	0	170	-	-
8043	403	1379 May 16	16:28:59	348	-7676	108	T	-p	0.4396	1.0668	47.0N	70.5W	64	172	243	05m07s
8044	403	1379 Nov 09	22:10:29	347	-7670	113	A	-n	-0.3275	0.9195	38.3S	158.0W	71	10	323	09m29s
8045	403	1380 May 05	09:34:58	347	-7664	118	T	n-	-0.2973	1.0732	1.5N	39.2E	73	351	249	06m52s
8046	403	1380 Oct 28	22:00:47	346	-7658	123	A	p-	0.3656	0.9395	4.9N	148.4W	69	191	240	08m01s
8047	403	1381 Apr 25	00:55:46	345	-7652	128	P	t-	-1.0659	0.8794	70.2S	154.8W	0	323		
8048	403	1381 Oct 18	04:00:20	344	-7646	133	P	t-	1.0464	0.9004	71.0N	167.1E	0	228		
8049	403	1382 Mar 15	19:30:25	344	-7641	100	A	-t	0.9536	0.9344	66.6N	156.5W	17	130	827	05m10s
8050	403	1382 Sep 08	08:02:24	343	-7635	105	T	-p	-0.9290	1.0541	60.5S	24.3E	21	38	487	03m29s
8051	403	1383 Mar 04	19:25:59	342	-7629	110	A	nn	0.2444	0.9312	10.4N	111.8W	76	162	265	08m56s
8052	403	1383 Aug 29	00:27:38	342	-7623	115	T	-n	-0.2262	1.0516	5.9S	170.2E	77	17	175	04m50s
8053	403	1384 Feb 21	21:20:45	341	-7617	120	A	p-	-0.4738	0.9605	34.6S	126.0W	62	340	162	04m05s
8054	403	1384 Aug 17	13:09:06	340	-7611	125	A	p-	0.5354	0.9999	41.7N	4.5W	57	200	1	00m01s
8055	403	1385 Feb 10	06:14:26	339	-7605	130	P	t-	-1.1498	0.7198	71.1S	142.1W	0	232		
8056	403	1385 Jul 08	04:02:10	339	-7600	97	Pe	-t	-1.5189	0.0745	68.1S	110.4E	0	11		
8057	403	1385 Aug 06	18:51:40	339	-7599	135	P	t-	1.3352	0.3878	70.6N	35.3E	0	318		
8058	403	1386 Jan 01	10:31:27	338	-7594	102	T	-p	0.8881	1.0366	40.8N	22.6E	27	177	269	03m01s
8059	403	1386 Jun 27	04:49:17	337	-7588	107	A	-p	-0.7583	0.9561	26.6S	110.0E	40	360	246	05m23s
8060	403	1386 Dec 22	01:00:27	337	-7582	112	H	-n	0.2300	1.0109	10.0S	167.7E	77	182	39	01m10s
8061	404	1387 Jun 16	10:46:23	336	-7576	117	H	nn	0.0416	1.0100	26.0N	19.8E	87	176	35	01m03s
8062	404	1387 Dec 11	09:58:30	335	-7570	122	A	p-	-0.4843	0.9539	52.4S	26.1E	61	10	193	04m16s
8063	404	1388 Jun 04	23:49:27	335	-7564	127	T	p-	0.7944	1.0552	74.2N	156.6E	37	148	302	03m20s
8064	404	1388 Nov 29	11:42:20	334	-7558	132	P	t-	-1.1999	0.6170	64.6S	148.1W	0	155		
8065	404	1389 Apr 26	09:42:22	333	-7553	99	P	-t	-1.0124	0.9944	62.2S	91.5E	0	307		
8066	404	1389 May 25	16:48:11	333	-7552	137	Pb	t-	1.4993	0.0549	64.4N	139.8E	0	29		
8067	404	1389 Oct 19	17:46:26	333	-7547	104	P	-t	1.1226	0.7545	61.7N	25.0W	0	241		
8068	404	1390 Apr 15	23:40:36	332	-7541	109	H	-n	-0.2940	1.0133	2.7S	167.0W	73	335	48	01m19s
8069	404	1390 Oct 09	02:00:26	331	-7535	114	H	-n	0.3598	1.0120	9.6N	157.3E	69	207	44	01m07s
8070	404	1391 Apr 05	06:47:41	330	-7529	119	A	p-	0.4761	0.9570	33.9N	64.2E	61	147	176	04m11s
8071	404	1391 Sep 28	16:26:31	330	-7523	124	T	n-	-0.3541	1.0557	23.4S	78.9W	69	31	195	04m23s
8072	404	1392 Mar 24	07:36:47	329	-7517	129	P	t-	1.2226	0.5809	60.9N	30.5W	0	80		
8073	404	1392 Sep 17	08:51:03	328	-7511	134	P	t-	-1.0548	0.9092	61.0S	45.5W	0	92		
8074	404	1393 Feb 11	18:02:26	328	-7506	101	A	-p	-0.9742	0.9628	64.1S	1.4W	12	270	618	02m22s
8075	404	1393 Aug 08	10:56:18	327	-7500	106	A	-t	0.8703	0.9804	62.3N	70.1E	29	241	140	01m22s
8076	404	1394 Feb 01	04:47:49	326	-7494	111	T	-n	-0.2268	1.0270	26.6S	119.2E	77	334	94	02m19s
8077	404	1394 Jul 28	14:48:17	326	-7488	116	A	nn	0.1249	0.9501	23.7N	36.5W	83	203	184	05m40s
8078	404	1395 Jan 21	20:05:24	325	-7482	121	T	n-	0.4555	1.0487	7.7N	126.0W	63	160	180	04m21s
8079	404	1395 Jul 17	15:14:16	324	-7476	126	A	p-	-0.6318	0.9497	17.7S	58.2W	51	18	234	06m12s
8080	404	1396 Jan 11	11:21:14	323	-7470	131	P	t-	1.1464	0.7287	63.4N	28.1W	0	141		
8081	405	1396 Jun 06	09:23:32	323	-7465	98	P	-t	1.3046	0.4365	66.5N	146.5W	0	354		
8082	405	1396 Jul 05	19:37:40	323	-7464	136	P	t-	-1.3568	0.3449	63.9S	147.2W	0	32		
8083	405	1396 Dec 01	05:48:03	322	-7459	103	A-	-t	-1.0074	0.9463	67.0S	88.6W	0	181	-	-
8084	405	1397 May 26	23:56:51	321	-7453	108	T	-p	0.5101	1.0692	53.4N	179.9W	59	178	263	05m01s
8085	405	1397 Nov 20	06:04:18	321	-7447	113	A	-n	-0.3407	0.9178	41.6S	85.7E	70	5	333	09m32s
8086	405	1398 May 16	17:08:41	320	-7441	118	T	n-	-0.2294	1.0741	7.7N	75.7W	77	355	247	06m59s
8087	405	1398 Nov 09	06:00:34	319	-7435	123	A	p-	0.3493	0.9397	1.1N	90.5E	70	188	238	08m07s
8088	405	1399 May 06	08:18:28	319	-7429	128	P	t-	-1.0035	0.9949	69.3S	82.1E	0	335		
8089	405	1399 Oct 29	12:17:08	318	-7423	133	P	t-	1.0256	0.9380	70.2N	29.8E	0	215		
8090	405	1400 Mar 26	02:43:41	317	-7418	100	A+	-t	1.0058	0.9506	72.0N	34.0E	0	72	-	-
8091	405	1400 Sep 18	16:09:43	317	-7412	105	T	-t	-0.9684	1.0490	68.7S	118.1W	14	57	679	02m53s
8092	405	1401 Mar 15	02:42:43	316	-7406	110	A	nn	0.2885	0.9347	17.2N	137.1E	73	162	253	08m12s
8093	405	1401 Sep 08	08:20:21	315	-7400	115	T	-n	-0.2746	1.0459	12.8S	49.0E	74	18	159	04m15s
8094	405	1402 Mar 04	05:06:39	315	-7394	120	A	p-	-0.4410	0.9665	28.2S	115.9E	64	340	134	03m34s
8095	405	1402 Aug 28	20:31:39	314	-7388	125	A	p-	0.4790	0.9943	34.0N	117.4W	61	200	23	00m33s
8096	405	1403 Feb 21	14:31:42	313	-7382	130	P	t-	-1.1253	0.7660	71.7S	79.8E	0	246		
8097	405	1403 Aug 18	01:41:42	313	-7376	135	P	t-	1.2697	0.5006	71.3N	80.4W	0	305		
8098	405	1404 Jan 12	19:20:46	312	-7371	102	T	-p	0.8945	1.0369	43.3N	114.3W	26	171	279	02m58s
8099	405	1404 Jul 07	11:22:59	311	-7365	107	A	-p	-0.8407	0.9558	35.9S	7.5E	33	5	299	05m00s
8100	405	1405 Jan 01	09:41:37	311	-7359	112	H	-n	0.2343	1.0089	8.6S	37.5E	77	177	32	00m57s

Cat Num	Canon Plate	Calendar Date	TD of Greatest Eclipse	ΔT s	Luna Num	Saros Num	Ecl. Type	QLE	Gamma	Ecl. Mag.	Lat. °	Long. °	Sun Alt °	Sun Azm °	Path Width km	Central Line Dur.
8101	406	1405 Jun 26	17:47:04	310	-7353	117	H	nn	-0.0370	1.0134	20.9N	84.7W	88	360	46	01m26s
8102	406	1405 Dec 21	18:17:58	309	-7347	122	A	p-	-0.4803	0.9514	52.2S	93.7W	61	3	204	04m35s
8103	406	1406 Jun 16	07:12:01	309	-7341	127	T	p-	0.7188	1.0596	69.4N	64.4E	44	168	283	03m48s
8104	406	1406 Dec 10	19:43:54	308	-7335	132	P	t-	-1.1928	0.6285	65.6S	81.5E	0	165		
8105	406	1407 May 07	17:17:22	307	-7330	99	P	-t	-1.0794	0.8660	62.9S	31.5W	0	316		
8106	406	1407 Jun 06	00:16:35	307	-7329	137	P	t-	1.4296	0.1902	65.2N	17.8E	0	19		
8107	406	1407 Oct 31	01:46:22	307	-7324	104	P	-t	1.1398	0.7250	62.3N	154.0W	0	232		
8108	406	1408 Apr 26	07:02:10	306	-7318	109	H	-p	-0.3595	1.0119	3.3S	82.8E	69	338	44	01m13s
8109	406	1408 Oct 19	10:16:59	305	-7312	114	H	-n	0.3820	1.0121	7.3N	32.6E	67	205	45	01m10s
8110	406	1409 Apr 15	13:49:19	305	-7306	119	A	p-	0.4130	0.9577	34.6N	39.1W	65	149	168	04m11s
8111	406	1409 Oct 09	00:48:09	304	-7300	124	T	n-	-0.3249	1.0539	26.2S	156.1E	71	30	188	04m15s
8112	406	1410 Apr 04	14:35:19	303	-7294	129	P	t-	1.1642	0.6799	61.2N	143.9W	0	71		
8113	406	1410 Sep 28	17:00:48	303	-7288	134	P	t-	-1.0206	0.9718	61.1S	176.7W	0	101		
8114	406	1411 Feb 23	02:01:36	302	-7283	101	As	-p	-0.9972	0.9654	61.6S	105.7W	1	258	-	02m05s
8115	406	1411 Aug 19	18:05:20	302	-7277	106	A	-t	0.9376	0.9724	62.2N	23.9W	20	253	284	01m52s
8116	406	1412 Feb 12	13:15:02	301	-7271	111	T	-n	-0.2446	1.0319	23.5S	7.0W	76	332	111	02m42s
8117	406	1412 Aug 07	21:27:46	300	-7265	116	A	nn	0.2018	0.9465	24.4N	134.5W	78	206	201	05m55s
8118	406	1413 Feb 01	04:47:05	300	-7259	121	T	n-	0.4429	1.0509	9.6N	102.6E	64	156	187	04m25s
8119	406	1413 Jul 27	21:50:24	299	-7253	126	A	p-	-0.5506	0.9506	14.1S	157.5W	57	22	214	05m58s
8120	406	1414 Jan 21	20:00:46	298	-7247	131	P	t-	1.1411	0.7384	62.6N	167.3W	0	131		
8121	407	1414 Jun 17	16:30:43	298	-7242	98	P	-t	1.3749	0.3061	65.5N	96.4E	0	344		
8122	407	1414 Jul 17	02:35:03	298	-7241	136	P	t-	-1.2770	0.4881	63.1S	99.0E	0	42		
8123	407	1414 Dec 12	14:01:50	297	-7236	103	A-	-t	-1.0145	0.9330	65.9S	137.3E	0	192	-	-
8124	407	1415 Jun 07	07:22:41	296	-7230	108	T	-p	0.5827	1.0708	59.2N	73.7E	54	185	284	04m51s
8125	407	1415 Dec 01	14:02:32	296	-7224	113	A	-n	-0.3503	0.9166	43.7S	30.7W	69	359	339	09m31s
8126	407	1416 May 27	00:38:48	295	-7218	118	T	nn	-0.1584	1.0742	13.5N	171.0E	81	359	244	06m56s
8127	407	1416 Nov 19	14:05:55	294	-7212	123	A	p-	0.3370	0.9404	1.8S	31.6W	70	184	234	08m05s
8128	407	1417 May 16	15:36:31	294	-7206	128	T	p-	-0.9378	1.0179	48.9S	46.3W	20	352	180	01m30s
8129	407	1417 Nov 08	20:41:02	293	-7200	133	P	t-	1.0097	0.9670	69.2N	108.6W	0	203		
8130	407	1418 Apr 06	09:48:10	293	-7195	100	P	-t	1.0643	0.8513	71.5N	86.1W	0	58		
8131	407	1418 Sep 30	00:24:11	292	-7189	105	T-	-t	-1.0021	1.0112	71.8S	63.7E	0	109	-	-
8132	407	1419 Mar 26	09:50:57	291	-7183	110	A	-p	0.3399	0.9383	24.5N	28.0E	70	161	243	07m25s
8133	407	1419 Sep 19	16:20:21	291	-7177	115	T	-n	-0.3162	1.0401	19.5S	72.1W	71	19	141	03m40s
8134	407	1420 Mar 14	12:42:57	290	-7171	120	A	p-	-0.3994	0.9727	21.4S	0.3W	66	341	106	02m59s
8135	407	1420 Sep 08	04:02:09	289	-7165	125	A	p-	0.4301	0.9885	26.7N	127.7E	64	200	45	01m10s
8136	407	1421 Mar 03	22:40:34	289	-7159	130	P	t-	-1.0933	0.8265	72.0S	56.6W	0	260		
8137	407	1421 Aug 28	08:38:54	288	-7153	135	P	t-	1.2101	0.6025	71.8N	161.6E	0	292		
8138	407	1422 Jan 23	04:05:41	287	-7148	102	T	-p	0.9044	1.0374	46.6N	109.5E	25	165	296	02m54s
8139	407	1422 Jul 18	18:00:59	287	-7142	107	A	-p	-0.9197	0.9545	47.2S	98.5W	23	11	427	04m35s
8140	407	1423 Jan 12	18:20:19	286	-7136	112	H	-n	0.2400	1.0073	6.2S	92.4W	76	173	26	00m48s
8141	408	1423 Jul 08	00:48:40	286	-7130	117	H2	nn	-0.1158	1.0161	15.0N	169.7E	83	4	55	01m45s
8142	408	1424 Jan 02	02:37:16	285	-7124	122	A	p-	-0.4768	0.9495	50.7S	146.2E	61	356	211	04m52s
8143	408	1424 Jun 26	14:34:25	284	-7118	127	T	p-	0.6425	1.0629	63.1N	36.6W	50	180	270	04m14s
8144	408	1424 Dec 21	03:46:29	284	-7112	132	P	t-	-1.1867	0.6384	66.7S	49.6W	0	176		
8145	408	1425 May 18	00:47:36	283	-7107	99	P	-t	-1.1498	0.7309	63.7S	153.5W	0	325		
8146	408	1425 Jun 16	07:44:06	283	-7106	137	P	t-	1.3592	0.3271	66.2N	104.4W	0	9		
8147	408	1425 Nov 10	09:54:28	282	-7101	104	P	-t	1.1506	0.7064	63.1N	74.8E	0	222		
8148	408	1426 May 07	14:17:32	282	-7095	109	H	-p	-0.4294	1.0100	5.0S	26.0W	65	341	38	01m03s
8149	408	1426 Oct 30	18:40:38	281	-7089	114	H	-n	0.3991	1.0123	5.2N	94.1W	66	202	46	01m13s
8150	408	1427 Apr 26	20:43:40	281	-7083	119	A	pn	0.3444	0.9583	34.7N	140.2W	70	153	161	04m15s
8151	408	1427 Oct 20	09:16:36	280	-7077	124	T	n-	-0.3009	1.0521	29.1S	29.6E	72	28	180	04m07s
8152	408	1428 Apr 14	21:25:47	279	-7071	129	P	t-	1.0987	0.7916	61.6N	104.6E	0	62		
8153	408	1428 Oct 09	01:18:10	279	-7065	134	Ts	t-	-0.9930	1.0281	63.0S	61.0E	5	101	-	01m30s
8154	408	1429 Mar 05	09:54:08	278	-7060	101	P	-t	-1.0266	0.9336	61.0S	129.6E	0	265		
8155	408	1429 Aug 30	01:20:30	277	-7054	106	A+	-t	0.9988	0.9782	61.0N	98.8W	0	282	-	-
8156	408	1430 Feb 22	21:35:50	277	-7048	111	T	-n	-0.2685	1.0369	20.5S	131.9W	74	330	128	03m05s
8157	408	1430 Aug 19	04:13:52	276	-7042	116	A	np	0.2729	0.9428	24.2N	125.7E	74	209	219	06m13s
8158	408	1431 Feb 12	13:21:50	276	-7036	121	T	n-	0.4245	1.0534	11.9N	26.9W	65	153	193	04m30s
8159	408	1431 Aug 08	04:33:00	275	-7030	126	A	p-	-0.4737	0.9509	12.0S	102.0E	62	25	201	05m45s
8160	408	1432 Feb 02	04:33:42	274	-7024	131	P	t-	1.1309	0.7571	62.0N	55.4E	0	122		

Cat Num	Canon Plate	Calendar Date	TD of Greatest Eclipse	ΔT s	Luna Num	Saros Num	Ecl. Type	QLE	Gamma	Ecl. Mag.	Lat. °	Long. °	Sun Alt °	Sun Azm °	Path Width km	Central Line Dur.
8161	409	1432 Jun 27	23:39:43	274	-7019	98	P	-t	1.4445	0.1760	64.6N	20.9W	0	334		
8162	409	1432 Jul 27	09:39:02	274	-7018	136	P	t-	-1.2011	0.6250	62.4S	16.3W	0	51		
8163	409	1432 Dec 22	22:14:13	273	-7013	103	P	-t	-1.0224	0.9188	64.9S	4.1E	0	203		
8164	409	1433 Jun 17	14:48:42	273	-7007	108	T	-p	0.6558	1.0714	64.0N	29.5W	49	196	309	04m38s
8165	409	1433 Dec 11	22:03:44	272	-7001	113	A	-n	-0.3579	0.9162	44.6S	147.3W	69	353	342	09m25s
8166	409	1434 Jun 07	08:05:20	271	-6995	118	T	nn	-0.0847	1.0735	18.7N	59.3E	85	3	239	06m45s
8167	409	1434 Nov 30	22:17:34	271	-6989	123	A	n-	0.3290	0.9416	3.7S	155.0W	71	180	229	07m54s
8168	409	1435 May 27	22:47:54	270	-6983	128	T	p-	-0.8670	1.0184	37.8S	160.2W	30	359	127	01m43s
8169	409	1435 Nov 20	05:12:02	269	-6977	133	A+	t-	0.9991	0.9868	68.2N	111.8E	0	191	-	-
8170	409	1436 Apr 16	16:42:11	269	-6972	100	P	-t	1.1306	0.7385	70.8N	157.0E	0	45		
8171	409	1436 Oct 10	08:47:28	268	-6966	105	P	-t	-1.0286	0.9594	71.3S	76.0W	0	123		
8172	409	1437 Apr 05	16:52:06	268	-6960	110	A	-p	0.3974	0.9419	32.1N	79.3W	66	162	233	06m39s
8173	409	1437 Sep 30	00:26:45	267	-6954	115	T	-n	-0.3519	1.0343	25.9S	164.7E	69	19	123	03m05s
8174	409	1438 Mar 25	20:14:23	266	-6948	120	A	p-	-0.3529	0.9790	14.5S	115.5W	69	342	80	02m21s
8175	409	1438 Sep 19	11:38:08	266	-6942	125	A	n-	0.3864	0.9826	19.8N	11.4E	67	199	66	01m51s
8176	409	1439 Mar 15	06:43:34	265	-6936	130	P	t-	-1.0559	0.8980	72.1S	168.3E	0	274		
8177	409	1439 Sep 08	15:42:20	265	-6930	135	P	t-	1.1555	0.6947	72.1N	41.6E	0	278		
8178	409	1440 Feb 03	12:45:48	264	-6925	102	T	-p	0.9183	1.0380	50.9N	26.5W	23	159	324	02m49s
8179	409	1440 Jul 29	00:45:41	264	-6919	107	As	-t	-0.9938	0.9505	66.1S	142.4E	5	27	-	04m02s
8180	409	1441 Jan 23	02:52:50	263	-6913	112	H	-n	0.2503	1.0062	2.9S	138.9E	76	170	22	00m40s
8181	410	1441 Jul 18	07:57:16	262	-6907	117	T	nn	-0.1896	1.0181	8.6N	61.6E	79	8	63	01m59s
8182	410	1442 Jan 12	10:51:53	262	-6901	122	A	p-	-0.4704	0.9481	48.0S	26.1E	62	350	216	05m06s
8183	410	1442 Jul 07	21:59:40	261	-6895	127	T	p-	0.5679	1.0654	56.2N	143.3W	55	187	261	04m39s
8184	410	1443 Jan 01	11:47:21	261	-6889	132	P	t-	-1.1793	0.6506	67.8S	179.2E	0	187		
8185	410	1443 May 29	08:13:09	260	-6884	99	P	-t	-1.2234	0.5897	64.6S	85.3E	0	334		
8186	410	1443 Jun 27	15:11:10	260	-6883	137	P	t-	1.2887	0.4640	67.2N	133.2E	0	359		
8187	410	1443 Nov 21	18:08:47	259	-6878	104	P	-t	1.1575	0.6946	63.9N	58.2W	0	213		
8188	410	1444 May 17	21:24:41	259	-6872	109	H	-p	-0.5052	1.0074	8.1S	133.0W	60	345	29	00m48s
8189	410	1444 Nov 10	03:12:20	258	-6866	114	H	-n	0.4102	1.0130	3.5N	137.2E	66	198	49	01m18s
8190	410	1445 May 07	03:29:38	258	-6860	119	A	nn	0.2692	0.9585	33.8N	121.2E	74	158	157	04m24s
8191	410	1445 Oct 30	17:52:12	257	-6854	124	T	n-	-0.2828	1.0505	32.0S	98.2W	73	25	174	04m01s
8192	410	1446 Apr 26	04:09:04	256	-6848	129	P	t-	1.0268	0.9147	62.1N	5.2W	0	54		
8193	410	1446 Oct 20	09:42:45	256	-6842	134	T	p-	-0.9718	1.0258	65.8S	56.8W	13	94	386	01m25s
8194	410	1447 Mar 16	17:39:01	255	-6837	101	P	-t	-1.0629	0.8715	61.0S	4.7E	0	274		
8195	410	1447 Sep 10	08:41:40	255	-6831	106	P	-t	1.0542	0.8785	60.8N	142.2E	0	274		
8196	410	1448 Mar 05	05:49:57	254	-6825	111	T	-n	-0.2984	1.0421	17.7S	104.7E	73	330	147	03m30s
8197	410	1448 Aug 29	11:07:05	254	-6819	116	A	-p	0.3380	0.9389	23.4N	23.8E	70	211	239	06m37s
8198	410	1449 Feb 22	21:50:09	253	-6813	121	T	n-	0.4008	1.0561	14.3N	154.6W	66	151	200	04m36s
8199	410	1449 Aug 18	11:24:05	252	-6807	126	A	p-	0.4030	0.9509	11.1S	0.5W	66	27	194	05m35s
8200	410	1450 Feb 12	13:01:23	252	-6801	131	P	t-	1.1169	0.7829	61.5N	80.5W	0	113		
8201	411	1450 Jul 09	06:51:32	251	-6796	98	Pe	-t	1.5125	0.0484	63.8N	138.5W	0	325		
8202	411	1450 Aug 07	16:48:49	251	-6795	136	P	t-	-1.1286	0.7560	61.8S	132.8W	0	60		
8203	411	1451 Jan 03	06:25:40	251	-6790	103	P	-t	-1.0306	0.9045	63.9S	128.6W	0	213		
8204	411	1451 Jun 28	22:15:28	250	-6784	108	T	-p	0.7287	1.0711	67.5N	128.8W	43	210	339	04m23s
8205	411	1451 Dec 23	06:05:20	249	-6778	113	A	-n	-0.3651	0.9164	44.3S	96.0E	68	347	342	09m16s
8206	411	1452 Jun 17	15:30:42	249	-6772	118	Tm	nn	-0.0102	1.0719	23.0N	51.3W	90	10	234	06m26s
8207	411	1452 Dec 11	06:31:53	248	-6766	123	A	n-	0.3224	0.9434	4.8S	81.1E	71	175	221	07m32s
8208	411	1453 Jun 07	05:56:44	248	-6760	128	T	p-	-0.7948	1.0175	29.4S	88.7E	37	3	99	01m45s
8209	411	1453 Nov 30	13:46:17	247	-6754	133	A	t-	0.9903	0.9842	60.4N	27.7W	7	179	469	01m14s
8210	411	1454 Apr 27	23:29:09	247	-6749	100	P	-t	1.2018	0.6169	70.0N	42.5E	0	33		
8211	411	1454 Oct 21	17:17:31	246	-6743	105	P	-t	-1.0499	0.9177	70.7S	143.0E	0	136		
8212	411	1455 Apr 16	23:44:01	245	-6737	110	A	-p	0.4628	0.9454	40.0N	176.1E	62	162	227	05m53s
8213	411	1455 Oct 11	08:40:44	245	-6731	115	T	-n	-0.3809	1.0286	31.8S	40.2E	67	18	104	02m31s
8214	411	1456 Apr 05	03:37:14	244	-6725	120	A	p-	-0.2980	0.9853	7.3S	131.5E	73	344	54	01m40s
8215	411	1456 Sep 29	19:22:26	244	-6719	125	A	n-	0.3503	0.9768	13.5N	106.7W	69	197	88	02m36s
8216	411	1457 Mar 25	14:38:16	243	-6713	130	P	t-	-1.0107	0.9845	71.9S	35.4E	0	288		
8217	411	1457 Sep 18	22:54:59	243	-6707	135	P	t-	1.1083	0.7737	72.1N	80.8W	0	264		
8218	411	1458 Feb 13	21:19:39	242	-6702	102	T	-t	0.9374	1.0385	56.3N	162.6W	20	151	375	02m41s
8219	411	1458 Aug 09	07:36:10	242	-6696	107	P	-t	-1.0636	0.8590	70.8S	20.6E	0	46		
8220	411	1459 Feb 03	11:20:41	241	-6690	112	H	-n	0.2638	1.0054	1.1N	11.1E	75	167	19	00m34s

Cat Num	Canon Plate	Calendar Date	TD of Greatest Eclipse	ΔT s	Luna Num	Saros Num	Ecl. Type	QLE	Gamma	Ecl. Mag.	Lat. °	Long. °	Sun Alt °	Sun Azm °	Path Width km	Central Line Dur.
8221	412	1459 Jul 29	15:10:11	240	-6684	117	T	nn	-0.2605	1.0196	1.8N	48.2W	75	11	69	02m07s
8222	412	1460 Jan 23	19:01:52	240	-6678	122	A	p-	-0.4607	0.9474	44.2S	94.0W	62	346	218	05m19s
8223	412	1460 Jul 18	05:27:53	239	-6672	127	T	p-	0.4954	1.0669	48.9N	106.4E	60	191	252	05m00s
8224	412	1461 Jan 11	19:46:21	239	-6666	132	P	t-	-1.1705	0.6651	68.9S	47.9E	0	198		
8225	412	1461 Jun 08	15:36:19	238	-6661	99	P	-t	-1.2986	0.4459	65.5S	35.5W	0	344		
8226	412	1461 Jul 07	22:39:29	238	-6660	137	P	t-	1.2191	0.5989	68.2N	10.0E	0	349		
8227	412	1461 Dec 02	02:27:45	238	-6655	104	P	-t	1.1614	0.6882	64.9N	167.3E	0	203		
8228	412	1462 May 29	04:28:02	237	-6649	109	H	-p	-0.5833	1.0042	12.4S	120.4E	54	349	18	00m28s
8229	412	1462 Nov 21	11:49:24	236	-6643	114	H	-n	0.4176	1.0139	2.2N	7.2E	65	194	52	01m26s
8230	412	1463 May 18	10:08:52	236	-6637	119	A	nn	0.1890	0.9584	31.9N	24.2E	79	163	154	04m38s
8231	412	1463 Nov 11	02:33:46	235	-6631	124	T	n-	-0.2696	1.0490	34.5S	133.0E	74	21	169	03m56s
8232	412	1464 May 06	10:46:58	235	-6625	129	A	t-	0.9502	0.9367	71.2N	72.7W	18	83	771	04m17s
8233	412	1464 Oct 30	18:13:13	234	-6619	134	T	p-	-0.9560	1.0225	68.9S	176.0E	17	94	267	01m14s
8234	412	1465 Mar 27	01:17:27	234	-6614	101	P	-t	-1.1052	0.7976	61.1S	118.7W	0	282		
8235	412	1465 Sep 20	16:10:43	233	-6608	106	P	-t	1.1020	0.7931	60.9N	21.2E	0	265		
8236	412	1466 Mar 16	13:57:13	233	-6602	111	T	-n	-0.3348	1.0471	15.3S	17.1W	70	330	165	03m56s
8237	412	1466 Sep 09	18:07:54	232	-6596	116	A	-p	0.3966	0.9351	22.2N	80.5W	67	212	260	07m05s
8238	412	1467 Mar 06	06:10:42	231	-6590	121	T	n-	0.3706	1.0588	16.7N	79.9E	68	150	207	04m44s
8239	412	1467 Aug 29	18:24:29	231	-6584	126	A	p-	-0.3391	0.9505	11.3S	105.2W	70	29	191	05m29s
8240	412	1468 Feb 23	21:18:55	230	-6578	131	P	t-	1.0953	0.8228	61.2N	146.3E	0	103		
8241	413	1468 Aug 18	00:08:08	230	-6572	136	P	t-	-1.0627	0.8753	61.3S	108.4E	0	68		
8242	413	1469 Jan 13	14:32:44	229	-6567	103	P	-t	-1.0420	0.8851	63.0S	100.3E	0	222		
8243	413	1469 Jul 09	05:44:22	229	-6561	108	T	-p	0.8000	1.0697	69.3N	134.7E	37	226	380	04m06s
8244	413	1470 Jan 02	14:05:56	228	-6555	113	A	-p	-0.3733	0.9173	43.1S	20.7W	68	341	339	09m02s
8245	413	1470 Jun 28	22:54:56	228	-6549	118	T	nn	0.0650	1.0695	26.4N	161.0W	86	192	227	06m02s
8246	413	1470 Dec 22	14:49:05	227	-6543	123	A	n-	0.3175	0.9458	4.9S	43.4W	72	171	210	07m02s
8247	413	1471 Jun 18	13:00:12	227	-6537	128	T	p-	-0.7189	1.0157	22.3S	19.7W	44	8	77	01m38s
8248	413	1471 Dec 11	22:25:20	226	-6531	133	A	t-	0.9849	0.9871	57.1N	165.0W	9	171	287	01m02s
8249	413	1472 May 08	06:07:58	226	-6526	100	P	-t	1.2791	0.4848	69.1N	69.4W	0	21		
8250	413	1472 Jun 06	20:20:31	225	-6525	138	Pb	t-	-1.5448	0.0209	66.4S	132.2W	0	7		
8251	413	1472 Nov 01	01:54:27	225	-6520	105	P	-t	-1.0657	0.8868	69.8S	0.9E	0	149		
8252	413	1473 Apr 27	06:30:57	224	-6514	110	A	-p	0.5328	0.9486	48.1N	73.3E	58	164	223	05m10s
8253	413	1473 Oct 21	17:01:28	224	-6508	115	T	-p	-0.4040	1.0230	37.3S	85.1W	66	16	86	02m00s
8254	413	1474 Apr 16	10:55:48	223	-6502	120	A	n-	-0.2387	0.9916	0.2S	19.7E	76	346	30	00m58s
8255	413	1474 Oct 11	03:12:17	223	-6496	125	A	n-	0.3195	0.9711	7.8N	134.1E	71	195	109	03m22s
8256	413	1475 Apr 05	22:27:42	222	-6490	130	T	t-	-0.9607	1.0310	60.5S	123.6W	15	327	386	02m08s
8257	413	1475 Sep 30	06:15:14	222	-6484	135	P	t-	1.0676	0.8411	71.9N	154.9E	0	250		
8258	413	1476 Feb 25	05:45:39	221	-6479	102	T	-t	0.9627	1.0386	63.1N	58.7E	15	140	491	02m29s
8259	413	1476 Aug 19	14:36:14	221	-6473	107	P	-t	-1.1260	0.7506	71.4S	97.8W	0	58		
8260	413	1477 Feb 13	19:40:23	220	-6467	112	H	-n	0.2833	1.0048	6.0N	115.0W	74	164	17	00m30s
8261	414	1477 Aug 08	22:30:57	220	-6461	117	T	-n	-0.3257	1.0206	5.2S	160.5W	71	14	74	02m10s
8262	414	1478 Feb 03	03:04:10	219	-6455	122	A	p-	-0.4455	0.9472	39.5S	146.4E	63	343	217	05m31s
8263	414	1478 Jul 29	13:01:17	219	-6449	127	T	p-	0.4269	1.0676	41.4N	6.8W	65	194	244	05m18s
8264	414	1479 Jan 23	03:39:45	218	-6443	132	P	t-	-1.1571	0.6875	69.9S	82.6W	0	211		
8265	414	1479 Jun 19	22:56:07	218	-6438	99	P	-t	-1.3756	0.2991	66.5S	155.9W	0	354		
8266	414	1479 Jul 19	06:09:16	217	-6437	137	P	t-	1.1509	0.7302	69.1N	114.1W	0	337		
8267	414	1479 Dec 13	10:50:14	217	-6432	104	P	-t	1.1630	0.6858	65.9N	31.5E	0	192		
8268	414	1480 Jun 08	11:26:08	217	-6426	109	H	-p	-0.6644	1.0002	18.0S	14.4E	48	353	1	00m02s
8269	414	1480 Dec 01	20:30:38	216	-6420	114	H2	-n	0.4218	1.0155	1.5N	123.9W	65	189	58	01m37s
8270	414	1481 May 28	16:42:59	215	-6414	119	Am	nn	0.1053	0.9577	28.8N	71.9W	84	168	155	04m57s
8271	414	1481 Nov 21	11:21:13	215	-6408	124	T	n-	-0.2617	1.0479	36.6S	3.3E	75	17	165	03m53s
8272	414	1482 May 17	17:19:00	214	-6402	129	A	t-	0.8681	0.9420	73.4N	137.0W	29	116	434	04m14s
8273	414	1482 Nov 11	02:49:49	214	-6396	134	T	p-	-0.9457	1.0189	72.5S	44.0E	18	97	203	01m03s
8274	414	1483 Apr 07	08:49:08	213	-6391	101	P	-t	-1.1536	0.7117	61.5S	119.5E	0	291		
8275	414	1483 Oct 01	23:47:15	213	-6385	106	P	-t	1.1431	0.7202	61.1N	101.7W	0	256		
8276	414	1484 Mar 26	21:56:47	212	-6379	111	T	-n	-0.3782	1.0521	13.5S	137.0W	68	331	185	04m22s
8277	414	1484 Sep 20	01:17:55	212	-6373	116	A	-p	0.4474	0.9313	20.8N	172.4E	63	211	283	07m39s
8278	414	1485 Mar 16	14:24:22	211	-6367	121	T	n-	0.3345	1.0615	19.1N	43.6W	70	149	213	04m53s
8279	414	1485 Sep 09	01:33:06	211	-6361	126	A	p-	-0.2811	0.9500	12.4S	148.0E	74	30	190	05m26s
8280	414	1486 Mar 06	05:30:00	210	-6355	131	P	t-	1.0689	0.8714	61.0N	14.7E	0	94		

Cat Num	Canon Plate	Calendar Date	TD of Greatest Eclipse	ΔT s	Luna Num	Saros Num	Ecl. Type	QLE	Gamma	Ecl. Mag.	Lat. °	Long. °	Sun Alt °	Sun Azm °	Path Width km	Central Line Dur.
8281	415	1486 Aug 29	07:34:56	210	-6349	136	P	t-	-1.0018	0.9856	61.0S	12.1W	0	77		
8282	415	1487 Jan 24	22:35:03	209	-6344	103	P	-t	-1.0566	0.8604	62.3S	29.5W	0	232		
8283	415	1487 Jul 20	13:15:36	209	-6338	108	T	-p	0.8696	1.0673	69.3N	39.6E	29	244	446	03m47s
8284	415	1488 Jan 13	22:03:45	208	-6332	113	A	-p	-0.3840	0.9188	41.0S	137.4W	67	336	333	08m45s
8285	415	1488 Jul 09	06:20:51	208	-6326	118	T	-n	0.1384	1.0663	28.9N	89.5E	82	197	219	05m36s
8286	415	1489 Jan 01	23:04:27	207	-6320	123	A	n-	0.3102	0.9489	4.3S	167.5W	72	166	197	06m24s
8287	415	1489 Jun 28	20:04:24	207	-6314	128	H3	p-	-0.6440	1.0130	16.8S	127.5W	50	12	58	01m23s
8288	415	1489 Dec 22	07:04:57	206	-6308	133	A	t-	0.9791	0.9904	54.6N	58.8E	11	164	175	00m47s
8289	415	1490 May 19	12:41:05	206	-6303	100	P	-t	1.3600	0.3462	68.1N	179.2W	0	10		
8290	415	1490 Jun 18	02:55:30	206	-6302	138	P	t-	-1.4661	0.1592	65.4S	118.7E	0	17		
8291	415	1490 Nov 12	10:36:45	205	-6297	105	P	-t	-1.0774	0.8639	68.8S	141.9W	0	161		
8292	415	1491 May 08	13:11:33	205	-6291	110	A	-p	0.6085	0.9514	56.5N	26.9W	52	166	225	04m30s
8293	415	1491 Nov 02	01:28:47	204	-6285	115	T	-p	-0.4209	1.0179	42.0S	148.9E	65	13	68	01m32s
8294	415	1492 Apr 26	18:07:10	204	-6279	120	A	nn	-0.1723	0.9976	6.8N	90.0W	80	349	8	00m16s
8295	415	1492 Oct 21	11:10:36	203	-6273	125	A	n-	0.2964	0.9657	2.8N	13.1E	73	193	129	04m08s
8296	415	1493 Apr 16	06:10:20	203	-6267	130	T	t-	-0.9042	1.0391	49.5S	107.3E	25	339	308	03m00s
8297	415	1493 Oct 10	13:43:35	202	-6261	135	P	t-	1.0334	0.8969	71.4N	28.8E	0	236		
8298	415	1494 Mar 07	14:04:20	202	-6256	102	Tn	-t	0.9940	1.0368	71.3N	99.2W	4	108	-	02m06s
8299	415	1494 Aug 30	21:44:35	201	-6250	107	P	-t	-1.1821	0.6529	71.8S	141.3E	0	72		
8300	415	1495 Feb 25	03:52:03	201	-6244	112	H	-n	0.3090	1.0044	11.5N	120.5E	72	163	16	00m27s
8301	416	1495 Aug 20	05:58:28	200	-6238	117	T	-n	-0.3862	1.0210	12.4S	85.1E	67	16	77	02m08s
8302	416	1496 Feb 14	10:59:31	200	-6232	122	A	p-	-0.4249	0.9474	34.2S	27.5E	65	341	213	05m41s
8303	416	1496 Aug 08	20:40:14	199	-6226	127	T	n-	0.3626	1.0675	33.9N	122.4W	69	196	236	05m30s
8304	416	1497 Feb 02	11:27:50	199	-6220	132	P	t-	-1.1393	0.7176	70.7S	147.5E	0	223		
8305	416	1497 Jun 30	06:16:37	198	-6215	99	P	-t	-1.4514	0.1556	67.5S	83.1E	0	4		
8306	416	1497 Jul 29	13:43:13	198	-6214	137	P	t-	1.0863	0.8539	70.0N	120.2E	0	326		
8307	416	1497 Dec 23	19:14:58	198	-6209	104	P	-t	1.1634	0.6856	67.0N	105.2W	0	181		
8308	416	1498 Jun 19	18:21:38	197	-6203	109	A	-p	-0.7466	0.9956	25.1S	91.8W	42	357	23	00m29s
8309	416	1498 Dec 13	05:15:08	197	-6197	114	T	-n	0.4242	1.0174	1.5N	104.3E	65	185	66	01m50s
8310	416	1499 Jun 08	23:13:39	196	-6191	119	A	nn	0.0195	0.9567	24.7N	167.8W	89	173	158	05m22s
8311	416	1499 Dec 02	20:11:32	196	-6185	124	T	n-	-0.2557	1.0471	37.9S	126.7W	75	12	162	03m51s
8312	416	1500 May 27	23:48:31	195	-6179	129	A	p-	0.7832	0.9461	71.5N	152.4E	38	143	320	04m13s
8313	416	1500 Nov 21	11:30:31	195	-6173	134	T	p-	-0.9393	1.0156	76.4S	91.4W	20	102	159	00m52s
8314	416	1501 Apr 17	16:15:52	194	-6168	101	P	-t	-1.2071	0.6155	61.9S	1.1W	0	300		
8315	416	1501 May 17	03:27:44	194	-6167	139	Pb	t-	1.5002	0.0905	63.7N	13.6W	0	35		
8316	416	1501 Oct 12	07:30:04	194	-6162	106	P	-t	1.1784	0.6585	61.4N	133.9E	0	247		
8317	416	1502 Apr 07	05:49:59	193	-6156	111	T	-n	-0.4276	1.0567	12.6S	104.7E	65	333	205	04m49s
8318	416	1502 Oct 01	08:36:17	193	-6150	116	A	-p	0.4913	0.9277	19.3N	62.8E	60	210	306	08m16s
8319	416	1503 Mar 27	22:28:20	192	-6144	121	T	n-	0.2904	1.0640	21.1N	164.1W	73	150	218	05m04s
8320	416	1503 Sep 20	08:52:38	192	-6138	126	A	n-	-0.2314	0.9494	14.1S	38.5E	77	30	190	05m27s
8321	417	1504 Mar 16	13:30:09	191	-6132	131	P	t-	1.0345	0.9348	61.0N	114.0W	0	86		
8322	417	1504 Sep 08	15:12:15	191	-6126	136	A	t-	-0.9486	0.9924	55.3S	102.6W	18	58	83	00m32s
8323	417	1505 Feb 04	06:29:08	190	-6121	103	P	-t	-1.0775	0.8255	61.7S	156.9W	0	241		
8324	417	1505 Jul 30	20:51:55	190	-6115	108	T	-t	0.9352	1.0635	67.9N	55.6W	20	263	593	03m25s
8325	417	1506 Jan 24	05:58:07	189	-6109	113	A	-p	-0.3979	0.9209	38.3S	106.1E	66	332	325	08m26s
8326	417	1506 Jul 20	13:46:58	189	-6103	118	T	-n	0.2112	1.0623	30.4N	19.8W	78	202	209	05m08s
8327	417	1507 Jan 13	07:20:10	189	-6097	123	A	n-	0.3024	0.9526	3.0S	68.3E	72	162	181	05m42s
8328	417	1507 Jul 10	03:06:33	188	-6091	128	H	p-	-0.5680	1.0095	12.4S	126.0E	55	16	40	01m01s
8329	417	1508 Jan 02	15:45:09	188	-6085	133	A	t-	0.9732	0.9941	52.8N	77.0W	13	157	92	00m28s
8330	417	1508 May 29	19:09:02	187	-6080	100	P	-t	1.4443	0.2019	67.1N	72.8E	0	360		
8331	417	1508 Jun 28	09:20:44	187	-6079	138	P	t-	-1.3860	0.2993	64.5S	10.4E	0	27		
8332	417	1508 Nov 22	19:24:23	187	-6074	105	P	-t	-1.0850	0.8489	67.8S	74.6E	0	173		
8333	417	1509 May 18	19:49:36	186	-6068	110	A	-p	0.6865	0.9539	64.9N	124.3W	46	171	233	03m56s
8334	417	1509 Nov 12	10:00:15	186	-6062	115	H	-p	-0.4338	1.0131	45.8S	23.2E	64	9	50	01m06s
8335	417	1510 May 08	01:16:15	185	-6056	120	H	nn	-0.1030	1.0033	13.5N	161.4E	84	352	12	00m22s
8336	417	1510 Nov 01	19:13:50	185	-6050	125	A	n-	0.2781	0.9607	1.5S	108.7W	74	190	148	04m54s
8337	417	1511 Apr 27	13:47:24	184	-6044	130	T	p-	-0.8425	1.0463	40.0S	14.7W	32	346	286	03m50s
8338	417	1511 Oct 21	21:19:49	184	-6038	135	A+	t-	1.0058	0.9416	70.7N	98.6W	0	223	-	-
8339	417	1512 Mar 17	22:14:35	183	-6033	102	P	-t	1.0322	0.9516	72.0N	110.2E	0	81		
8340	417	1512 Apr 16	06:22:25	183	-6032	140	Pb	t-	-1.5289	0.0003	70.6S	131.9E	0	314		

Cat Num	Canon Plate	Calendar Date	TD of Greatest Eclipse	ΔT s	Luna Num	Saros Num	Ecl. Type	QLE	Gamma	Ecl. Mag.	Lat. °	Long. °	Sun Alt °	Sun Azm °	Path Width km	Central Line Dur.
8341	418	1512 Sep 10	05:03:25	183	-6027	107	P	-t	-1.2305	0.5688	72.0S	17.4E	0	86		
8342	418	1513 Mar 07	11:54:03	182	-6021	112	H	-p	0.3421	1.0040	17.6N	1.8W	70	161	15	00m24s
8343	418	1513 Aug 30	13:35:52	182	-6015	117	T	-n	-0.4392	1.0211	19.5S	31.9W	64	18	80	02m03s
8344	418	1514 Feb 24	18:45:30	182	-6009	122	A	n-	-0.3974	0.9479	28.2S	90.0W	66	341	208	05m51s
8345	418	1514 Aug 20	04:25:15	181	-6003	127	T	n-	0.3032	1.0667	26.5N	119.9E	72	197	228	05m38s
8346	418	1515 Feb 13	19:08:19	181	-5997	132	P	t-	-1.1153	0.7580	71.5S	18.9E	0	237		
8347	418	1515 Jul 11	13:36:52	180	-5992	99	Pe	-t	-1.5262	0.0153	68.5S	38.2W	0	15		
8348	418	1515 Aug 09	21:21:25	180	-5991	137	P	t-	1.0258	0.9686	70.8N	7.1W	0	313		
8349	418	1516 Jan 04	03:38:41	180	-5986	104	P	-t	1.1652	0.6830	68.1N	117.7E	0	170		
8350	418	1516 Jun 30	01:15:15	179	-5980	109	A	-t	-0.8291	0.9899	33.8S	161.3E	34	2	64	01m03s
8351	418	1516 Dec 23	14:00:51	179	-5974	114	T	-n	0.4256	1.0199	2.2N	27.9W	65	181	75	02m05s
8352	418	1517 Jun 19	05:41:31	178	-5968	119	A	nn	-0.0683	0.9552	19.5N	96.1E	86	357	164	05m50s
8353	418	1517 Dec 13	05:04:13	178	-5962	124	T	n-	-0.2520	1.0468	38.2S	103.0E	75	6	161	03m52s
8354	418	1518 Jun 08	06:15:24	177	-5956	129	A	p-	0.6955	0.9496	67.0N	73.4E	46	162	259	04m13s
8355	418	1518 Dec 02	20:14:58	177	-5950	134	T	p-	-0.9365	1.0124	80.4S	128.7E	20	111	125	00m41s
8356	418	1519 Apr 28	23:35:43	177	-5945	101	P	-t	-1.2666	0.5070	62.5S	120.2W	0	309		
8357	418	1519 May 28	10:20:09	176	-5944	139	P	t-	1.4188	0.2342	64.6N	126.3W	0	26		
8358	418	1519 Oct 23	15:20:34	176	-5939	106	P	-t	1.2064	0.6096	61.9N	7.4E	0	238		
8359	418	1520 Apr 17	13:36:46	176	-5933	111	T	-p	-0.4825	1.0609	12.6S	12.2W	61	335	226	05m15s
8360	418	1520 Oct 11	16:03:20	175	-5927	116	A	-p	0.5277	0.9244	17.8N	49.4W	58	208	329	08m57s
8361	419	1521 Apr 07	06:26:06	175	-5921	121	T	n-	0.2414	1.0662	22.8N	77.3E	76	151	222	05m15s
8362	419	1521 Sep 30	16:21:42	174	-5915	126	A	nn	-0.1892	0.9489	16.2S	73.3W	79	29	191	05m30s
8363	419	1522 Mar 27	21:22:59	174	-5909	131	T	t-	0.9946	1.0076	62.0N	127.7E	4	84	347	00m26s
8364	419	1522 Sep 19	22:57:33	173	-5903	136	A	t-	-0.9011	0.9946	53.9S	146.1E	25	55	42	00m23s
8365	419	1523 Feb 15	14:16:44	173	-5898	103	P	-t	-1.1030	0.7827	61.2S	77.4E	0	250		
8366	419	1523 Aug 11	04:33:16	173	-5892	108	Tn	-t	0.9969	1.0558	62.7N	135.9W	2	294	-	02m44s
8367	419	1524 Feb 04	13:45:35	172	-5886	113	A	-p	-0.4176	0.9235	35.4S	9.3W	65	330	315	08m05s
8368	419	1524 Jul 30	21:17:39	172	-5880	118	T	-n	0.2797	1.0577	30.8N	130.2W	74	206	198	04m40s
8369	419	1525 Jan 23	15:31:21	171	-5874	123	A	n-	0.2897	0.9569	1.2S	54.8W	73	159	163	04m58s
8370	419	1525 Jul 20	10:11:04	171	-5868	128	H	p-	-0.4947	1.0054	9.3S	19.4E	60	19	21	00m35s
8371	419	1526 Jan 13	00:22:31	170	-5862	133	A	t-	0.9644	0.9985	51.0N	148.8E	15	151	19	00m07s
8372	419	1526 Jun 10	01:34:33	170	-5857	100	Pe	-t	1.5298	0.0557	66.1N	34.1W	0	350		
8373	419	1526 Jul 09	16:02:42	170	-5856	138	P	t-	-1.3063	0.4379	63.6S	97.8W	0	36		
8374	419	1526 Dec 04	04:14:39	170	-5851	105	P	-t	-1.0905	0.8382	66.7S	68.9W	0	184		
8375	419	1527 May 30	02:23:01	169	-5845	110	A	-p	0.7688	0.9556	73.4N	144.6E	39	180	255	03m28s
8376	419	1527 Nov 23	18:36:38	169	-5839	115	H	-p	-0.4422	1.0089	48.6S	102.5W	64	3	34	00m45s
8377	419	1528 May 18	08:21:05	168	-5833	120	H	nn	-0.0290	1.0085	19.9N	54.6E	88	356	29	00m56s
8378	419	1528 Nov 12	03:22:58	168	-5827	125	A	n-	0.2653	0.9562	4.9S	128.4E	75	186	166	05m36s
8379	419	1529 May 07	21:19:50	167	-5821	130	T	p-	-0.7760	1.0526	31.3S	133.1W	39	351	276	04m38s
8380	419	1529 Nov 01	05:04:11	167	-5815	135	An	t-	0.9846	0.9119	61.7N	122.8E	9	201	-	08m09s
8381	420	1530 Mar 29	06:16:37	167	-5810	102	P	-t	1.0769	0.8671	71.7N	24.4W	0	67		
8382	420	1530 Apr 27	14:07:20	166	-5809	140	P	t-	-1.4726	0.1083	69.9S	2.9E	0	327		
8383	420	1530 Sep 21	12:31:37	166	-5804	107	P	-t	-1.2718	0.4970	72.0S	108.9W	0	100		
8384	420	1531 Mar 18	19:47:21	166	-5798	112	H	-p	0.3818	1.0036	24.3N	122.1W	67	161	13	00m21s
8385	420	1531 Sep 10	21:21:52	165	-5792	117	T	-p	-0.4857	1.0208	26.4S	151.0W	61	20	81	01m56s
8386	420	1532 Mar 07	02:21:39	165	-5786	122	A	n-	-0.3625	0.9488	21.8S	154.3E	69	341	201	05m59s
8387	420	1532 Aug 30	12:17:45	164	-5780	127	T	n-	0.2500	1.0654	19.3N	0.0E	75	198	221	05m40s
8388	420	1533 Feb 24	02:42:10	164	-5774	132	P	t-	-1.0860	0.8077	71.9S	108.5W	0	251		
8389	420	1533 Aug 20	05:04:01	164	-5768	137	T	t-	0.9693	1.0479	73.7N	178.3E	13	257	678	02m40s
8390	420	1534 Jan 14	12:01:19	163	-5763	104	P	-t	1.1685	0.6778	69.1N	19.7W	0	158		
8391	420	1534 Jul 11	08:08:46	163	-5757	109	A	-t	-0.9104	0.9833	44.9S	52.5E	24	8	144	01m35s
8392	420	1535 Jan 03	22:45:49	162	-5751	114	T	-n	0.4285	1.0228	3.8N	160.1W	65	176	86	02m22s
8393	420	1535 Jun 30	12:08:20	162	-5745	119	A	nn	-0.1565	0.9533	13.5N	0.6W	81	1	173	06m19s
8394	420	1535 Dec 24	13:56:57	161	-5739	124	T	n-	-0.2482	1.0469	37.5S	27.4W	75	1	161	03m55s
8395	420	1536 Jun 18	12:43:21	161	-5733	129	A	p-	0.6079	0.9523	61.0N	13.4W	52	174	220	04m17s
8396	420	1536 Dec 13	04:59:20	161	-5727	134	T	p-	-0.9343	1.0098	84.5S	17.2W	20	125	97	00m33s
8397	420	1537 May 09	06:52:57	160	-5722	101	P	-t	-1.3289	0.3922	63.2S	121.2E	0	318		
8398	420	1537 Jun 07	17:14:05	160	-5721	139	P	t-	1.3373	0.3796	65.5N	120.2E	0	17		
8399	420	1537 Nov 02	23:17:01	160	-5716	106	P	-t	1.2286	0.5712	62.6N	120.7W	0	228		
8400	420	1538 Apr 28	21:17:31	159	-5710	111	T	-p	-0.5432	1.0645	13.7S	127.7W	57	338	249	05m38s

Cat Num	Canon Plate	Calendar Date	TD of Greatest Eclipse	ΔT s	Luna Num	Saros Num	Ecl. Type	QLE	Gamma	Ecl. Mag.	Lat. °	Long. °	Sun Alt °	Sun Azm °	Path Width km	Central Line Dur.
8401	421	1538 Oct 22	23:38:41	159	-5704	116	A	-p	0.5572	0.9214	16.6N	164.1W	56	205	351	09m41s
8402	421	1539 Apr 18	14:15:07	159	-5698	121	T	n-	0.1853	1.0680	23.7N	38.7W	79	154	225	05m28s
8403	421	1539 Oct 12	00:01:45	158	-5692	126	A	nn	-0.1551	0.9484	18.7S	172.2E	81	28	192	05m35s
8404	421	1540 Apr 07	05:04:30	158	-5686	131	T	p-	0.9462	1.0115	63.1N	34.7E	18	104	123	00m42s
8405	421	1540 Sep 30	06:54:11	157	-5680	136	A	p-	-0.8620	0.9960	54.6S	29.2E	30	54	27	00m17s
8406	421	1541 Feb 25	21:54:42	157	-5675	103	P	-t	-1.1360	0.7272	61.0S	45.8W	0	259		
8407	421	1541 Aug 21	12:20:07	157	-5669	108	P	-t	1.0541	0.9172	61.3N	102.0E	0	289		
8408	421	1541 Sep 19	20:34:01	157	-5668	146	Pb	t-	-1.5140	0.0378	61.1S	135.3E	0	95		
8409	421	1542 Feb 14	21:27:23	156	-5663	113	A	-p	-0.4424	0.9265	32.5S	123.8W	64	328	305	07m44s
8410	421	1542 Aug 11	04:51:06	156	-5657	118	T	-n	0.3454	1.0525	30.6N	118.6E	70	209	184	04m12s
8411	421	1543 Feb 03	23:38:52	155	-5651	123	A	n-	0.2735	0.9617	1.0N	177.0W	74	156	143	04m14s
8412	421	1543 Jul 31	17:16:23	155	-5645	128	H	p-	-0.4229	1.0007	7.3S	87.0W	65	22	3	00m05s
8413	421	1544 Jan 24	08:57:45	155	-5639	133	H	t-	0.9533	1.0035	49.7N	16.0E	17	146	40	00m16s
8414	421	1544 Jul 19	22:38:22	154	-5633	138	P	t-	-1.2281	0.5730	62.8S	153.9E	0	45		
8415	421	1544 Dec 14	13:06:28	154	-5628	105	P	-t	-1.0948	0.8297	65.6S	147.6E	0	195		
8416	421	1545 Jun 09	08:57:28	154	-5622	110	A	-p	0.8506	0.9567	81.2N	72.0E	31	208	303	03m06s
8417	421	1545 Dec 04	03:15:42	153	-5616	115	H	-p	-0.4480	1.0051	50.1S	132.1E	63	357	20	00m25s
8418	421	1546 May 29	15:24:40	153	-5610	120	H	nn	0.0470	1.0133	25.7N	51.0W	87	180	46	01m24s
8419	421	1546 Nov 23	11:35:42	152	-5604	125	A	n-	0.2561	0.9521	7.3S	4.9E	75	182	181	06m13s
8420	421	1547 May 19	04:48:58	152	-5598	130	T	p-	-0.7060	1.0581	23.5S	110.7E	45	356	270	05m22s
8421	422	1547 Nov 12	12:54:24	152	-5592	135	A	p-	0.9683	0.9106	55.5N	4.7W	14	191	1419	08m59s
8422	422	1548 Apr 08	14:10:08	151	-5587	102	P	-t	1.1282	0.7698	71.2N	156.5W	0	54		
8423	422	1548 May 07	21:46:52	151	-5586	140	P	t-	-1.4121	0.2250	69.0S	124.2W	0	338		
8424	422	1548 Oct 01	20:10:50	151	-5581	107	P	-t	-1.3049	0.4394	71.7S	122.3E	0	113		
8425	422	1549 Mar 29	03:30:55	150	-5575	112	H	-p	0.4285	1.0029	31.4N	120.1E	64	161	11	00m16s
8426	422	1549 Sep 21	05:16:24	150	-5569	117	T	-p	-0.5257	1.0205	33.2S	88.0E	58	21	82	01m49s
8427	422	1550 Mar 18	09:47:48	150	-5563	122	A	nn	-0.3200	0.9497	15.1S	40.8E	71	342	194	06m05s
8428	422	1550 Sep 10	20:17:38	149	-5557	127	T	n-	0.2029	1.0636	12.4N	121.8W	78	198	212	05m37s
8429	422	1551 Mar 07	10:05:18	149	-5551	132	P	t-	-1.0477	0.8730	72.2S	126.5E	0	265		
8430	422	1551 Aug 31	12:53:01	149	-5545	137	T	p-	0.9185	1.0460	65.7N	28.4E	23	226	391	02m52s
8431	422	1552 Jan 25	20:19:44	148	-5540	104	P	-t	1.1760	0.6655	70.0N	156.6W	0	146		
8432	422	1552 Jul 21	15:03:48	148	-5534	109	As	-t	-0.9893	0.9742	62.9S	64.6W	7	20	-	02m05s
8433	422	1553 Jan 14	07:28:09	147	-5528	114	T	-n	0.4340	1.0263	6.3N	68.3E	64	172	99	02m41s
8434	422	1553 Jul 10	18:36:34	147	-5522	119	A	np	-0.2430	0.9509	6.8N	98.5W	76	5	185	06m46s
8435	422	1554 Jan 03	22:49:38	147	-5516	124	T	n-	-0.2447	1.0474	35.8S	158.1W	76	355	163	04m00s
8436	422	1554 Jun 29	19:10:40	146	-5510	129	A	p-	0.5192	0.9546	54.0N	104.8W	58	182	195	04m22s
8437	422	1554 Dec 24	13:45:21	146	-5504	134	T	p-	-0.9341	1.0075	87.5S	159.2E	20	176	75	00m25s
8438	422	1555 May 20	14:06:06	146	-5499	101	P	-t	-1.3947	0.2696	64.0S	3.3E	0	328		
8439	422	1555 Jun 19	00:07:16	146	-5498	139	P	t-	1.2542	0.5290	66.5N	6.6E	0	7		
8440	422	1555 Nov 14	07:19:27	145	-5493	106	P	-t	1.2455	0.5423	63.3N	109.4E	0	219		
8441	423	1556 May 09	04:53:36	145	-5487	111	T	-p	-0.6079	1.0673	16.0S	117.8E	52	342	274	05m58s
8442	423	1556 Nov 02	07:22:13	145	-5481	116	A	-p	0.5798	0.9190	15.5N	78.9E	54	201	370	10m24s
8443	423	1557 Apr 28	21:59:05	144	-5475	121	T	nn	0.1251	1.0692	24.0N	153.1W	83	157	227	05m42s
8444	423	1557 Oct 22	07:49:28	144	-5469	126	A	nn	-0.1266	0.9482	21.1S	56.0E	83	25	192	05m40s
8445	423	1558 Apr 18	12:39:27	144	-5463	131	T	p-	0.8930	1.0132	64.1N	67.8W	26	114	100	00m50s
8446	423	1558 Oct 11	14:58:55	143	-5457	136	A	p-	-0.8289	0.9971	56.5S	90.3W	34	53	18	00m12s
8447	423	1559 Mar 09	05:23:01	143	-5452	103	P	-t	-1.1761	0.6598	60.8S	166.6W	0	268		
8448	423	1559 Sep 01	20:13:59	143	-5446	108	P	-t	1.1056	0.8172	61.1N	25.3W	0	280		
8449	423	1559 Oct 01	04:46:46	142	-5445	146	P	t-	-1.4772	0.1083	61.3S	3.4E	0	104		
8450	423	1560 Feb 26	05:00:44	142	-5440	113	A	-p	-0.4741	0.9299	29.9S	123.5E	62	327	294	07m22s
8451	423	1560 Aug 21	12:30:55	142	-5434	118	T	-p	0.4050	1.0469	29.7N	5.3E	66	212	170	03m44s
8452	423	1561 Feb 14	07:39:21	141	-5428	123	A	n-	0.2507	0.9670	3.4N	62.6E	75	153	122	03m30s
8453	423	1561 Aug 11	00:27:07	141	-5422	128	A	n-	-0.3564	0.9956	6.5S	165.5E	69	25	16	00m27s
8454	423	1562 Feb 03	17:27:33	141	-5416	133	T	t-	0.9373	1.0091	48.6N	114.5W	20	142	89	00m41s
8455	423	1562 Jul 31	05:16:46	140	-5410	138	P	t-	-1.1522	0.7034	62.2S	45.1E	0	54		
8456	423	1562 Dec 25	21:58:40	140	-5405	105	P	-t	-1.0990	0.8217	64.6S	4.6E	0	205		
8457	423	1563 Jun 20	15:30:55	140	-5399	110	A	-t	0.9338	0.9564	81.3N	55.3E	20	290	454	02m49s
8458	423	1563 Dec 15	11:55:49	139	-5393	115	H	-p	-0.4524	1.0020	50.3S	6.8E	63	350	8	00m10s
8459	423	1564 Jun 08	22:26:49	139	-5387	120	H2	nn	0.1253	1.0174	30.8N	155.4W	83	185	60	01m44s
8460	423	1564 Dec 03	19:52:06	139	-5381	125	A	nn	0.2504	0.9487	8.8S	119.2W	76	178	195	06m42s

Cat Num	Canon Plate	Calendar Date	TD of Greatest Eclipse	ΔT s	Luna Num	Saros Num	Ecl. Type	QLE	Gamma	Ecl. Mag.	Lat. °	Long. °	Sun Alt °	Sun Azm °	Path Width km	Central Line Dur.
8461	424	1565 May 29	12:15:00	138	-5375	130	T	p-	-0.6329	1.0629	16.5S	3.7W	51	0	266	05m57s
8462	424	1565 Nov 22	20:49:55	138	-5369	135	A	p-	0.9564	0.9092	51.4N	130.5W	16	184	1220	09m37s
8463	424	1566 Apr 19	21:56:01	138	-5364	102	P	-t	1.1855	0.6610	70.5N	73.9E	0	41		
8464	424	1566 May 19	05:21:00	138	-5363	140	P	t-	-1.3472	0.3507	68.1S	110.7E	0	350		
8465	424	1566 Oct 13	03:59:23	137	-5358	107	P	-t	-1.3312	0.3939	71.1S	8.6W	0	127		
8466	424	1567 Apr 09	11:04:08	137	-5352	112	H	-p	0.4830	1.0020	38.9N	4.9E	61	161	8	00m11s
8467	424	1567 Oct 02	13:20:27	137	-5346	117	T	-p	-0.5584	1.0200	39.7S	34.9W	56	22	82	01m42s
8468	424	1568 Mar 28	17:04:21	136	-5340	122	A	nn	-0.2701	0.9507	8.1S	70.5W	74	343	187	06m10s
8469	424	1568 Sep 21	04:25:02	136	-5334	127	T	n-	0.1619	1.0615	5.8N	114.6E	81	198	204	05m32s
8470	424	1569 Mar 17	17:21:18	136	-5328	132	A-	t-	-1.0033	0.9489	72.1S	3.1E	0	279	-	-
8471	424	1569 Sep 10	20:48:16	135	-5322	137	T	p-	0.8732	1.0428	57.4N	103.4W	29	215	293	02m55s
8472	424	1570 Feb 05	04:34:49	135	-5317	104	P	-t	1.1866	0.6475	70.9N	66.6E	0	133		
8473	424	1570 Aug 01	22:00:22	135	-5311	109	P	-t	-1.0655	0.8623	70.4S	171.9E	0	38		
8474	424	1571 Jan 25	16:07:36	135	-5305	114	T	-n	0.4422	1.0302	9.5N	62.8W	64	169	113	02m59s
8475	424	1571 Jul 22	01:07:18	134	-5299	119	A	-p	-0.3266	0.9481	0.5S	162.1E	71	9	201	07m08s
8476	424	1572 Jan 15	07:38:12	134	-5293	124	T	n-	-0.2380	1.0485	33.0S	71.6E	76	351	166	04m07s
8477	424	1572 Jul 10	01:42:42	134	-5287	129	A	p-	0.4338	0.9562	46.6N	159.7E	64	187	177	04m30s
8478	424	1573 Jan 03	22:28:35	133	-5281	134	H	p-	-0.9328	1.0058	85.9S	54.1W	21	258	57	00m20s
8479	424	1573 May 30	21:18:24	133	-5276	101	P	-t	-1.4619	0.1436	64.9S	114.6W	0	337		
8480	424	1573 Jun 29	07:03:36	133	-5275	139	P	t-	1.1724	0.6770	67.5N	108.2W	0	356		
8481	425	1573 Nov 24	15:24:46	133	-5270	106	P	-t	1.2591	0.5191	64.3N	21.4W	0	209		
8482	425	1574 May 20	12:25:42	132	-5264	111	T	-p	-0.6763	1.0694	19.7S	3.7E	47	346	305	06m09s
8483	425	1574 Nov 13	15:12:17	132	-5258	116	A	-p	0.5970	0.9171	14.8N	40.0W	53	197	387	11m03s
8484	425	1575 May 10	05:34:45	132	-5252	121	Tm	nn	0.0583	1.0697	23.1N	94.6E	87	162	227	05m56s
8485	425	1575 Nov 02	15:47:27	131	-5246	126	A	nn	-0.1061	0.9483	23.5S	62.6W	84	22	191	05m44s
8486	425	1576 Apr 28	20:04:44	131	-5240	131	T	p-	0.8328	1.0140	64.8N	168.0W	33	124	86	00m55s
8487	425	1576 Oct 21	23:13:06	131	-5234	136	A	p-	-0.8031	0.9981	59.2S	147.9E	36	51	11	00m08s
8488	425	1577 Mar 19	12:41:15	131	-5229	103	P	-t	-1.2235	0.5798	60.9S	75.2E	0	277		
8489	425	1577 Sep 12	04:15:22	130	-5223	108	P	-t	1.1507	0.7297	61.0N	154.4W	0	271		
8490	425	1577 Oct 11	13:08:02	130	-5222	146	P	t-	-1.4473	0.1654	61.6S	130.8W	0	113		
8491	425	1578 Mar 08	12:26:52	130	-5217	113	A	-p	-0.5120	0.9336	27.7S	12.3E	59	327	284	07m01s
8492	425	1578 Sep 01	20:15:08	130	-5211	118	T	-p	0.4602	1.0408	28.4N	109.6W	62	213	152	03m17s
8493	425	1579 Feb 25	15:34:47	129	-5205	123	A	n-	0.2229	0.9728	6.0N	56.4W	77	151	100	02m48s
8494	425	1579 Aug 22	07:41:32	129	-5199	128	A	n-	-0.2937	0.9901	6.6S	57.2E	73	27	36	01m00s
8495	425	1580 Feb 15	01:52:13	129	-5193	133	T	t-	0.9164	1.0151	47.9N	117.3E	23	138	127	01m07s
8496	425	1580 Aug 10	12:00:05	129	-5187	138	P	t-	-1.0802	0.8258	61.6S	64.7W	0	63		
8497	425	1581 Jan 05	06:49:58	128	-5182	105	P	-t	-1.1041	0.8121	63.7S	137.9W	0	216		
8498	425	1581 Jun 30	22:06:53	128	-5176	110	P	-t	1.0152	0.9454	64.2N	2.2W	0	331		
8499	425	1581 Dec 25	20:35:20	128	-5170	115	A	-p	-0.4567	0.9993	49.4S	118.5W	63	343	3	00m04s
8500	425	1582 Jun 20	05:30:27	128	-5164	120	T	nn	0.2032	1.0210	35.0N	100.8E	78	190	73	01m59s
8501	426	1582 Dec 25	04:08:39	127	-5158	125	A	nn	0.2457	0.9459	9.4S	116.8E	76	173	206	07m02s
8502	426	1583 Jun 19	19:39:32	127	-5152	130	T	p-	-0.5581	1.0667	10.4S	116.9W	56	4	262	06m23s
8503	426	1583 Dec 14	04:48:39	127	-5146	135	A	p-	0.9471	0.9083	48.5N	104.1E	18	177	1116	10m03s
8504	426	1584 May 10	05:35:06	126	-5141	102	P	-t	1.2478	0.5424	69.7N	53.5W	0	29		
8505	426	1584 Jun 08	12:52:25	126	-5140	140	P	t-	-1.2802	0.4805	67.1S	13.3W	0	0		
8506	426	1584 Nov 02	11:56:44	126	-5135	107	P	-t	-1.3510	0.3595	70.4S	141.1W	0	140		
8507	426	1585 Apr 29	18:28:58	126	-5129	112	H	-p	0.5436	1.0005	46.6N	107.7W	57	162	2	00m03s
8508	426	1585 Oct 22	21:33:25	126	-5123	117	T	-p	-0.5846	1.0196	45.7S	159.2W	54	21	82	01m35s
8509	426	1586 Apr 19	00:10:09	125	-5117	122	A	nn	-0.2120	0.9517	0.9S	179.1W	78	345	181	06m12s
8510	426	1586 Oct 12	12:40:32	125	-5111	127	T	n-	0.1278	1.0591	0.3S	10.8W	83	197	196	05m23s
8511	426	1587 Apr 08	00:27:05	125	-5105	132	A	t-	-0.9502	0.9271	60.5S	151.9W	18	325	889	06m26s
8512	426	1587 Oct 02	04:51:25	125	-5099	137	T	p-	0.8352	1.0387	50.0N	128.3E	33	208	235	02m51s
8513	426	1588 Feb 26	12:42:31	124	-5094	104	P	-t	1.2038	0.6178	71.5N	68.8W	0	119		
8514	426	1588 Aug 22	05:01:47	124	-5088	109	P	-t	-1.1364	0.7355	71.1S	53.5E	0	51		
8515	426	1589 Feb 15	00:42:20	124	-5082	114	T	-n	0.4545	1.0344	13.6N	167.1E	63	166	129	03m17s
8516	426	1589 Aug 11	07:41:04	124	-5076	119	A	-p	-0.4072	0.9450	8.2S	61.4E	66	12	221	07m24s
8517	426	1590 Feb 04	16:24:05	123	-5070	124	T	n-	-0.2293	1.0498	29.3S	58.8W	77	347	170	04m17s
8518	426	1590 Jul 31	08:17:39	123	-5064	129	A	p-	0.3503	0.9574	38.8N	61.6E	69	191	166	04m38s
8519	426	1591 Jan 25	07:09:22	123	-5058	134	H	p-	-0.9298	1.0047	81.9S	150.6E	21	283	45	00m16s
8520	426	1591 Jun 21	04:28:43	123	-5053	101	Pe	-t	-1.5311	0.0129	65.8S	127.7E	0	347		

Cat Num	Canon Plate	Calendar Date	TD of Greatest Eclipse	ΔT s	Luna Num	Saros Num	Ecl. Type	QLE	Gamma	Ecl. Mag.	Lat. °	Long. °	Sun Alt °	Sun Azm °	Path Width km	Central Line Dur.
8521	427	1591 Jul 20	14:02:08	123	-5052	139	P	t-	1.0911	0.8249	68.5N	136.0E	0	346		
8522	427	1591 Dec 15	23:33:56	122	-5047	106	P	-t	1.2690	0.5024	65.3N	153.5W	0	199		
8523	427	1592 Jun 09	19:55:49	122	-5041	111	T	-p	-0.7465	1.0705	24.7S	110.3W	42	350	344	06m11s
8524	427	1592 Dec 03	23:07:16	122	-5035	116	A	-p	0.6102	0.9159	14.5N	160.2W	52	193	401	11m36s
8525	427	1593 May 30	13:07:31	122	-5029	121	T	nn	-0.0106	1.0696	21.4N	17.0W	90	342	227	06m08s
8526	427	1593 Nov 22	23:52:06	121	-5023	126	A	nn	-0.0906	0.9488	25.4S	177.5E	85	18	189	05m46s
8527	427	1594 May 20	03:23:17	121	-5017	131	T	p-	0.7678	1.0141	64.9N	94.1E	40	136	76	00m58s
8528	427	1594 Nov 12	07:34:49	121	-5011	136	A	p-	-0.7829	0.9991	62.4S	25.1E	38	48	5	00m04s
8529	427	1595 Apr 09	19:50:05	121	-5006	103	P	-t	-1.2777	0.4879	61.1S	40.7W	0	286		
8530	427	1595 Oct 03	12:24:36	121	-5000	108	P	-t	1.1896	0.6546	61.1N	74.6E	0	262		
8531	427	1595 Nov 01	21:36:53	120	-4999	146	P	t-	-1.4233	0.2111	62.1S	93.1E	0	122		
8532	427	1596 Mar 28	19:43:19	120	-4994	113	A	-p	-0.5583	0.9373	26.3S	96.5W	56	328	275	06m41s
8533	427	1596 Sep 22	04:07:03	120	-4988	118	T	-p	0.5085	1.0346	26.8N	133.0E	59	213	134	02m50s
8534	427	1597 Mar 17	23:22:39	120	-4982	123	A	n-	0.1878	0.9788	8.4N	173.3W	79	151	77	02m08s
8535	427	1597 Sep 11	15:01:22	120	-4976	128	A	nn	-0.2363	0.9843	7.6S	52.4W	76	29	57	01m35s
8536	427	1598 Mar 07	10:10:01	119	-4970	133	T	p-	0.8893	1.0214	47.7N	8.2W	27	135	156	01m33s
8537	427	1598 Aug 31	18:48:48	119	-4964	138	A-	t-	-1.0126	0.9398	61.2S	175.6W	0	72	-	-
8538	427	1599 Jan 26	15:37:11	119	-4959	105	P	-t	-1.1125	0.7965	62.9S	80.9E	0	225		
8539	427	1599 Jul 22	04:45:15	119	-4953	110	P	-t	1.0949	0.8068	63.4N	111.2W	0	321		
8540	427	1600 Jan 16	05:12:46	118	-4947	115	A	-p	-0.4623	0.9972	47.4S	115.9E	62	337	11	00m14s
8541	428	1600 Jul 10	12:35:58	118	-4941	120	T	-n	0.2804	1.0238	38.2N	2.7W	74	196	84	02m08s
8542	428	1601 Jan 04	12:24:38	117	-4935	125	A	nn	0.2410	0.9437	9.1S	7.0W	76	169	214	07m13s
8543	428	1601 Jun 30	03:03:59	117	-4929	130	T	p-	-0.4826	1.0697	5.3S	130.7E	61	8	259	06m37s
8544	428	1601 Dec 24	12:50:31	116	-4923	135	A	p-	0.9402	0.9078	46.6N	21.5W	19	171	1051	10m14s
8545	428	1602 May 21	13:06:44	116	-4918	102	P	-t	1.3157	0.4132	68.8N	178.3W	0	18		
8546	428	1602 Jun 19	20:19:21	116	-4917	140	P	t-	-1.2097	0.6174	66.1S	135.7W	0	10		
8547	428	1602 Nov 13	20:03:05	115	-4912	107	P	-t	-1.3643	0.3363	69.5S	84.8E	0	153		
8548	428	1603 May 11	01:44:59	115	-4906	112	A	-p	0.6107	0.9987	54.7N	142.6E	52	163	6	00m07s
8549	428	1603 Nov 03	05:54:55	114	-4900	117	T	-p	-0.6041	1.0193	51.1S	75.6E	53	20	83	01m31s
8550	428	1604 Apr 29	07:07:21	114	-4894	122	A	nn	-0.1473	0.9525	6.3N	74.8E	82	347	176	06m12s
8551	428	1604 Oct 22	21:03:48	113	-4888	127	T	n-	0.1000	1.0567	5.9S	137.8W	84	195	188	05m12s
8552	428	1605 Apr 18	07:26:44	113	-4882	132	A	p-	-0.8918	0.9327	49.8S	89.9E	27	337	553	06m43s
8553	428	1605 Oct 12	12:59:58	112	-4876	137	T	p-	0.8022	1.0344	43.4N	0.6E	36	203	193	02m43s
8554	428	1606 Mar 08	20:45:39	112	-4871	104	P	-t	1.2253	0.5800	71.9N	156.4E	0	105		
8555	428	1606 Sep 02	12:07:23	111	-4865	109	P	-t	-1.2026	0.6182	71.7S	66.5W	0	64		
8556	428	1607 Feb 26	09:10:38	111	-4859	114	T	-n	0.4727	1.0388	18.4N	38.2E	62	163	147	03m34s
8557	428	1607 Aug 22	14:20:48	110	-4853	119	A	-p	-0.4824	0.9416	16.1S	41.4W	61	15	245	07m34s
8558	428	1608 Feb 16	01:03:28	110	-4847	124	T	n-	-0.2154	1.0515	24.8S	171.7E	77	345	175	04m29s
8559	428	1608 Aug 10	15:00:06	109	-4841	129	A	pn	0.2722	0.9581	31.0N	39.6W	74	194	158	04m46s
8560	428	1609 Feb 04	15:43:43	108	-4835	134	H	p-	-0.9224	1.0041	77.3S	7.2E	22	297	37	00m15s
8561	429	1609 Jul 30	21:07:08	108	-4829	139	P	t-	1.0140	0.9657	69.5N	17.9E	0	334		
8562	429	1609 Dec 26	07:43:34	107	-4824	106	P	-t	1.2776	0.4877	66.3N	73.9E	0	188		
8563	429	1610 Jun 21	03:23:00	107	-4818	111	T	-p	-0.8193	1.0705	31.5S	135.5E	35	354	400	05m59s
8564	429	1610 Dec 15	07:06:48	106	-4812	116	A	-p	0.6195	0.9153	14.7N	78.2E	52	188	409	11m56s
8565	429	1611 Jun 10	20:34:26	105	-4806	121	T	nn	-0.0836	1.0686	18.4N	127.6W	85	350	224	06m16s
8566	429	1611 Dec 04	08:03:43	105	-4800	126	A	nn	-0.0803	0.9498	26.9S	56.0E	85	13	185	05m44s
8567	429	1612 May 30	10:34:29	104	-4794	131	T	p-	0.6976	1.0135	63.6N	1.9W	45	149	65	00m58s
8568	429	1612 Nov 22	16:04:35	104	-4788	136	H	p-	-0.7691	1.0002	65.7S	98.4W	39	43	1	00m01s
8569	429	1613 Apr 20	02:49:29	103	-4783	103	P	-t	-1.3389	0.3839	61.5S	154.4W	0	295		
8570	429	1613 May 19	17:43:36	103	-4782	141	Pb	t-	1.5171	0.0712	63.3N	137.6E	0	41		
8571	429	1613 Oct 13	20:40:24	102	-4777	108	P	-t	1.2232	0.5902	61.3N	58.2W	0	253		
8572	429	1613 Nov 12	06:12:15	102	-4776	146	P	t-	-1.4048	0.2464	62.7S	44.9W	0	132		
8573	429	1614 Apr 09	02:52:58	102	-4771	113	A	-p	-0.6103	0.9411	25.7S	156.2E	52	329	268	06m22s
8574	429	1614 Oct 03	12:04:51	101	-4765	118	T	-p	0.5511	1.0282	25.2N	13.5E	56	212	113	02m22s
8575	429	1615 Mar 29	07:03:24	100	-4759	123	A	nn	0.1461	0.9851	10.7N	71.7E	82	151	53	01m28s
8576	429	1615 Sep 22	22:27:21	100	-4753	128	A	nn	-0.1849	0.9784	9.1S	163.5W	79	29	78	02m11s
8577	429	1616 Mar 17	18:21:45	99	-4747	133	T	p-	0.8568	1.0279	48.0N	131.4W	31	134	180	01m58s
8578	429	1616 Sep 11	01:44:06	99	-4741	138	A	t-	-0.9505	0.9319	54.1S	102.3E	18	55	807	05m42s
8579	429	1617 Feb 06	00:20:23	98	-4736	105	P	-t	-1.1241	0.7750	62.2S	59.0W	0	235		
8580	429	1617 Mar 07	10:05:36	98	-4735	143	Pb	t-	1.5110	0.0419	61.2N	48.6W	0	101		

Cat Num	Canon Plate	Calendar Date	TD of Greatest Eclipse	ΔT s	Luna Num	Saros Num	Ecl. Type	QLE	Gamma	Ecl. Mag.	Lat. °	Long. °	Sun Alt °	Sun Azm °	Path Width km	Central Line Dur.
8581	430	1617 Aug 01	11:29:44	97	-4730	110	P	-t	1.1702	0.6756	62.7N	138.3E	0	312		
8582	430	1618 Jan 26	13:46:44	97	-4724	115	A	-p	-0.4700	0.9955	44.7S	9.7W	62	333	18	00m23s
8583	430	1618 Jul 21	19:44:30	96	-4718	120	T	-n	0.3558	1.0260	40.4N	106.3W	69	201	94	02m13s
8584	430	1619 Jan 15	20:38:07	95	-4712	125	A	nn	0.2349	0.9422	8.1S	130.4W	76	165	220	07m16s
8585	430	1619 Jul 11	10:29:59	95	-4706	130	T	p-	-0.4077	1.0718	1.3S	18.6E	66	12	255	06m41s
8586	430	1620 Jan 04	20:51:05	94	-4700	135	A	p-	0.9321	0.9081	45.0N	146.5W	21	165	976	10m13s
8587	430	1620 May 31	20:33:45	93	-4695	102	P	-t	1.3868	0.2783	67.8N	58.5E	0	7		
8588	430	1620 Jun 30	03:46:25	93	-4694	140	P	t-	-1.1393	0.7535	65.1S	102.3E	0	20		
8589	430	1620 Nov 24	04:16:35	93	-4689	107	P	-t	-1.3729	0.3212	68.5S	50.6W	0	165		
8590	430	1621 May 21	08:53:44	92	-4683	112	A	-p	0.6828	0.9962	63.1N	36.1E	47	167	18	00m18s
8591	430	1621 Nov 13	14:23:13	91	-4677	117	T	-p	-0.6187	1.0194	55.8S	49.7W	52	16	84	01m28s
8592	430	1622 May 10	13:55:35	91	-4671	122	Am	nn	-0.0757	0.9531	13.5N	28.8W	86	350	172	06m07s
8593	430	1622 Nov 03	05:34:48	90	-4665	127	T	n-	0.0789	1.0544	10.7S	93.7E	86	193	180	05m01s
8594	430	1623 Apr 29	14:16:00	89	-4659	132	A	p-	-0.8244	0.9378	39.8S	20.4W	34	344	405	06m54s
8595	430	1623 Oct 23	21:17:10	88	-4653	137	T	p-	0.7770	1.0298	37.8N	128.0W	39	199	159	02m31s
8596	430	1624 Mar 19	04:40:36	88	-4648	104	P	-t	1.2540	0.5288	72.0N	23.5E	0	91		
8597	430	1624 Apr 17	17:16:18	88	-4647	142	Pb	t-	-1.5208	0.0582	71.2S	23.1W	0	306		
8598	430	1624 Sep 12	19:19:26	87	-4642	109	P	-t	-1.2625	0.5133	72.0S	171.5E	0	77		
8599	430	1624 Oct 12	08:53:55	87	-4641	147	Pb	t-	1.5466	0.0089	71.5N	109.9E	0	245		
8600	430	1625 Mar 08	17:32:39	86	-4636	114	T	-p	0.4965	1.0434	23.9N	89.4W	60	161	166	03m50s
8601	431	1625 Sep 01	21:06:57	86	-4630	119	A	-p	-0.5520	0.9380	24.2S	146.4W	56	18	274	07m37s
8602	431	1626 Feb 26	09:37:27	85	-4624	124	T	n-	-0.1971	1.0535	19.7S	42.7E	79	343	180	04m42s
8603	431	1626 Aug 21	21:47:42	84	-4618	129	A	nn	0.1975	0.9584	23.1N	142.9W	78	195	154	04m54s
8604	431	1627 Feb 16	00:13:31	84	-4612	134	H	p-	-0.9119	1.0040	72.3S	130.9W	24	307	34	00m15s
8605	431	1627 Aug 11	04:17:14	83	-4606	139	H	t-	0.9401	1.0001	77.7N	173.3W	19	253	1	00m00s
8606	431	1628 Jan 06	15:52:52	82	-4601	106	P	-t	1.2858	0.4739	67.4N	59.2W	0	177		
8607	431	1628 Jul 01	10:50:39	81	-4595	111	T	-t	-0.8917	1.0692	40.3S	20.0E	27	358	501	05m32s
8608	431	1628 Dec 25	15:08:47	81	-4589	116	A	-p	0.6265	0.9153	15.4N	44.0W	51	184	413	12m02s
8609	431	1629 Jun 21	03:59:24	80	-4583	121	T	-n	-0.1580	1.0670	14.5N	121.6E	81	354	221	06m20s
8610	431	1629 Dec 14	16:19:07	79	-4577	126	A	nn	-0.0725	0.9513	27.6S	66.2W	86	9	179	05m38s
8611	431	1630 Jun 10	17:41:07	79	-4571	131	H	p-	0.6244	1.0122	60.9N	98.3W	51	161	54	00m55s
8612	431	1630 Dec 04	00:38:59	78	-4565	136	H	p-	-0.7585	1.0017	68.7S	139.6E	40	36	9	00m07s
8613	431	1631 May 01	09:39:23	77	-4560	103	P	-t	-1.4070	0.2677	62.0S	94.2E	0	304		
8614	431	1631 May 31	00:25:37	77	-4559	141	P	t-	1.4433	0.1996	64.1N	27.6E	0	32		
8615	431	1631 Oct 25	05:04:15	76	-4554	108	P	-t	1.2502	0.5384	61.7N	167.0E	0	244		
8616	431	1631 Nov 23	14:53:44	76	-4553	146	P	t-	-1.3912	0.2723	63.5S	175.4E	0	141		
8617	431	1632 Apr 19	09:54:30	76	-4548	113	A	-p	-0.6694	0.9447	26.4S	50.8E	48	331	267	06m03s
8618	431	1632 Oct 13	20:09:39	75	-4542	118	T	-p	0.5873	1.0220	23.7N	108.2W	54	210	91	01m55s
8619	431	1633 Apr 08	14:37:06	74	-4536	123	A	nn	0.0976	0.9913	12.4N	41.2W	84	152	31	00m51s
8620	431	1633 Oct 03	06:00:37	73	-4530	128	Am	nn	-0.1405	0.9726	11.2S	83.4E	82	29	99	02m48s
8621	432	1634 Mar 29	02:25:11	73	-4524	133	T	p-	0.8169	1.0346	48.7N	108.6E	35	133	198	02m24s
8622	432	1634 Sep 22	08:47:04	72	-4518	138	A	p-	-0.8947	0.9300	51.5S	2.3E	26	51	572	06m03s
8623	432	1635 Feb 17	08:57:24	71	-4513	105	P	-t	-1.1407	0.7440	61.6S	162.7E	0	244		
8624	432	1635 Mar 18	18:24:53	71	-4512	143	P	t-	1.4813	0.0973	61.1N	177.7E	0	92		
8625	432	1635 Aug 12	18:20:10	71	-4507	110	P	-t	1.2412	0.5514	62.1N	26.6E	0	303		
8626	432	1636 Feb 06	22:14:33	70	-4501	115	A	-p	-0.4825	0.9943	41.6S	134.7W	61	329	23	00m29s
8627	432	1636 Aug 01	02:58:15	69	-4495	120	T	-p	0.4279	1.0275	41.5N	148.9E	64	207	103	02m15s
8628	432	1637 Jan 26	04:48:32	68	-4489	125	A	nn	0.2265	0.9412	6.4S	107.0E	77	161	223	07m12s
8629	432	1637 Jul 21	17:57:08	68	-4483	130	T	n-	-0.3335	1.0731	1.8N	93.4W	71	16	251	06m37s
8630	432	1638 Jan 15	04:51:53	67	-4477	135	A	p-	0.9242	0.9090	44.0N	88.9E	22	159	907	10m00s
8631	432	1638 Jun 12	03:55:44	66	-4472	102	Pe	-t	1.4614	0.1370	66.8N	62.9W	0	356		
8632	432	1638 Jul 11	11:11:52	66	-4471	140	P	t-	-1.0676	0.8917	64.2S	19.0W	0	30		
8633	432	1638 Dec 05	12:36:35	66	-4466	107	P	-t	-1.3768	0.3143	67.5S	173.0E	0	176		
8634	432	1639 Jan 04	04:56:19	65	-4465	145	Pb	t-	1.5650	0.0009	64.6N	80.0E	0	155		
8635	432	1639 Jun 01	15:55:16	65	-4460	112	A	-p	0.7597	0.9930	71.7N	65.3W	40	173	38	00m31s
8636	432	1639 Nov 24	22:58:55	64	-4454	117	T	-p	-0.6278	1.0197	59.6S	174.7W	51	10	87	01m27s
8637	432	1640 May 20	20:37:52	63	-4448	122	A	nn	0.0002	0.9533	20.4N	130.2W	90	179	171	06m00s
8638	432	1640 Nov 13	14:11:19	63	-4442	127	T	n-	0.0623	1.0522	14.8S	35.8W	87	189	173	04m50s
8639	432	1641 May 09	21:01:19	62	-4436	132	A	p-	-0.7532	0.9425	30.8S	127.3W	41	349	321	06m56s
8640	432	1641 Nov 03	05:40:09	61	-4430	137	T	p-	0.7570	1.0252	33.0N	102.5E	41	194	130	02m15s

Cat Num	Canon Plate	Calendar Date	TD of Greatest Eclipse	ΔT s	Luna Num	Saros Num	Ecl. Type	QLE	Gamma	Ecl. Mag.	Lat. °	Long. °	Sun Alt °	Sun Azm °	Path Width km	Central Line Dur.
8641	433	1642 Mar 30	12:29:29	61	-4425	104	P	-t	1.2884	0.4668	71.9N	108.0W	0	77		
8642	433	1642 Apr 29	00:29:43	60	-4424	142	P	t-	-1.4585	0.1660	70.6S	144.7W	0	318		
8643	433	1642 Sep 24	02:37:37	60	-4419	109	P	-t	-1.3163	0.4199	72.1S	47.6E	0	91		
8644	433	1642 Oct 23	16:48:36	60	-4418	147	P	t-	1.5221	0.0551	71.0N	22.5W	0	232		
8645	433	1643 Mar 20	01:47:19	59	-4413	114	T	-p	0.5271	1.0479	30.0N	144.6E	58	159	186	04m02s
8646	433	1643 Sep 13	04:01:21	58	-4407	119	A	-p	-0.6145	0.9343	32.3S	106.3E	52	21	307	07m35s
8647	433	1644 Mar 08	18:02:43	58	-4401	124	T	n-	-0.1717	1.0555	14.0S	84.7W	80	342	186	04m57s
8648	433	1644 Sep 01	04:45:28	57	-4395	129	A	nn	0.1307	0.9584	15.4N	110.9E	82	197	152	05m00s
8649	433	1645 Feb 26	08:35:06	56	-4389	134	H	p-	-0.8956	1.0043	66.7S	94.3E	26	316	34	00m17s
8650	433	1645 Aug 21	11:34:18	55	-4383	139	H	t-	0.8710	1.0040	68.2N	43.7E	29	222	28	00m16s
8651	433	1646 Jan 16	23:59:17	55	-4378	106	P	-t	1.2957	0.4574	68.5N	167.8E	0	166		
8652	433	1646 Jul 12	18:18:19	54	-4372	111	T	-t	-0.9641	1.0658	53.2S	98.0W	15	5	834	04m44s
8653	433	1647 Jan 05	23:10:59	53	-4366	116	A	-p	0.6336	0.9161	16.9N	166.5W	51	179	413	11m50s
8654	433	1647 Jul 02	11:21:21	53	-4360	121	T	-n	-0.2344	1.0643	9.6N	11.0E	77	359	217	06m15s
8655	433	1647 Dec 26	00:38:35	52	-4354	126	A	nn	-0.0675	0.9535	27.4S	170.6E	86	4	170	05m25s
8656	433	1648 Jun 21	00:43:22	51	-4348	131	H	p-	0.5483	1.0102	56.7N	164.0E	56	171	42	00m49s
8657	433	1648 Dec 14	09:17:55	51	-4342	136	H	p-	-0.7510	1.0035	70.9S	19.6E	41	25	18	00m14s
8658	433	1649 May 11	16:22:04	50	-4337	103	P	-t	-1.4801	0.1427	62.7S	15.7W	0	313		
8659	433	1649 Jun 10	07:02:37	50	-4336	141	P	t-	1.3657	0.3345	65.0N	81.5W	0	22		
8660	433	1649 Nov 04	13:35:08	49	-4331	108	P	-t	1.2716	0.4977	62.2N	30.2E	0	235		
8661	434	1649 Dec 03	23:40:37	49	-4330	146	P	t-	-1.3820	0.2896	64.4S	34.1E	0	151		
8662	434	1650 Apr 30	16:48:49	49	-4325	113	A	-p	-0.7347	0.9481	28.5S	52.9W	43	334	274	05m43s
8663	434	1650 Oct 25	04:21:25	48	-4319	118	T	-p	0.6170	1.0159	22.3N	127.9E	52	207	68	01m26s
8664	434	1651 Apr 19	22:04:37	47	-4313	123	A	nn	0.0433	0.9976	13.7N	152.4W	87	154	8	00m14s
8665	434	1651 Oct 14	13:40:56	46	-4307	128	A	nn	-0.1025	0.9668	13.5S	31.3W	84	28	120	03m27s
8666	434	1652 Apr 08	10:22:28	46	-4301	133	T	p-	0.7713	1.0412	49.6N	8.9W	39	135	213	02m49s
8667	434	1652 Oct 02	15:58:30	45	-4295	138	A	p-	-0.8458	0.9275	51.2S	102.7W	32	50	497	06m19s
8668	434	1653 Feb 27	17:28:50	44	-4290	105	P	-t	-1.1619	0.7043	61.3S	26.0E	0	253		
8669	434	1653 Mar 29	02:38:06	44	-4289	143	P	t-	1.4469	0.1622	61.2N	45.6E	0	83		
8670	434	1653 Aug 23	01:17:26	44	-4284	110	P	-t	1.3072	0.4356	61.6N	86.7W	0	295		
8671	434	1653 Sep 21	15:55:44	44	-4283	148	Pb	t-	-1.5450	0.0324	61.0S	149.7W	0	89		
8672	434	1654 Feb 17	06:36:38	43	-4278	115	A	-p	-0.4991	0.9933	38.3S	100.9E	60	327	27	00m34s
8673	434	1654 Aug 12	10:17:43	42	-4272	120	T	-p	0.4962	1.0285	41.7N	42.5E	60	211	110	02m16s
8674	434	1655 Feb 06	12:51:54	42	-4266	125	A	nn	0.2129	0.9408	4.3S	14.0W	78	157	224	07m03s
8675	434	1655 Aug 02	01:28:36	41	-4260	130	T	n-	-0.2625	1.0735	3.7N	154.0E	75	20	247	06m28s
8676	434	1656 Jan 26	12:48:10	40	-4254	135	A	p-	0.9122	0.9106	43.2N	34.1W	24	154	820	09m38s
8677	434	1656 Jul 21	18:39:48	40	-4248	140	T-	t-	-0.9983	1.0244	63.4S	140.7W	0	39	-	-
8678	434	1656 Dec 15	20:59:52	39	-4243	107	P	-t	-1.3790	0.3102	66.4S	36.3E	0	187		
8679	434	1657 Jan 14	13:08:11	39	-4242	145	P	t-	1.5547	0.0171	63.7N	52.7W	0	145		
8680	434	1657 Jun 11	22:52:09	39	-4237	112	A	-t	0.8395	0.9888	60.5N	153.7W	33	190	73	00m45s
8681	435	1657 Dec 05	07:39:36	38	-4231	117	T	-p	-0.6335	1.0205	62.1S	61.3E	50	3	91	01m29s
8682	435	1658 Jun 01	03:11:38	37	-4225	122	A	nn	0.0828	0.9532	27.0N	131.3E	85	178	172	05m49s
8683	435	1658 Nov 24	22:54:42	37	-4219	127	T	nn	0.0513	1.0502	18.0S	166.4W	87	186	167	04m40s
8684	435	1659 May 21	03:38:53	36	-4213	132	A	p-	-0.6747	0.9469	22.2S	129.2E	47	353	264	06m51s
8685	435	1659 Nov 14	14:10:08	35	-4207	137	T	p-	0.7432	1.0208	29.2N	28.2W	42	190	106	01m56s
8686	435	1660 Apr 09	20:10:11	35	-4202	104	P	-t	1.3301	0.3906	71.5N	122.9E	0	64		
8687	435	1660 May 09	07:36:45	35	-4201	142	P	t-	-1.3897	0.2868	69.7S	95.9E	0	331		
8688	435	1660 Oct 04	10:03:43	34	-4196	109	P	-t	-1.3629	0.3401	72.0S	78.2W	0	105		
8689	435	1660 Nov 03	00:50:39	34	-4195	147	P	t-	1.5038	0.0898	70.3N	156.2W	0	219		
8690	435	1661 Mar 30	09:55:24	34	-4190	114	T	-p	0.5634	1.0524	36.7N	20.2E	55	158	209	04m12s
8691	435	1661 Sep 23	11:02:34	33	-4184	119	A	-p	-0.6711	0.9306	40.3S	3.0W	48	23	347	07m29s
8692	435	1662 Mar 20	02:21:49	32	-4178	124	T	n-	-0.1414	1.0576	7.9S	149.1E	82	342	191	05m11s
8693	435	1662 Sep 12	11:50:45	32	-4172	129	A	nn	0.0694	0.9581	7.9N	2.6E	86	197	153	05m05s
8694	435	1663 Mar 09	16:48:41	31	-4166	134	H	p-	-0.8735	1.0049	60.5S	37.1W	29	323	35	00m21s
8695	435	1663 Sep 01	18:59:08	31	-4160	139	H	p-	0.8073	1.0065	58.6N	78.9W	36	212	38	00m29s
8696	435	1664 Jan 28	08:02:31	30	-4155	106	P	-t	1.3074	0.4376	69.6N	35.0E	0	154		
8697	435	1664 Jul 23	01:48:46	29	-4149	111	P	-t	-1.0343	0.9581	68.8S	134.7E	0	18		
8698	435	1664 Aug 21	08:58:23	29	-4148	149	Pb	t-	1.4870	0.0844	71.0N	173.8E	0	309		
8699	435	1665 Jan 16	07:11:51	29	-4143	116	A	-p	0.6420	0.9174	19.1N	71.2E	50	175	409	11m24s
8700	435	1665 Jul 12	18:44:06	28	-4137	121	T	-n	-0.3095	1.0611	3.9N	100.6W	72	3	211	06m02s

Cat Num	Canon Plate	Calendar Date	TD of Greatest Eclipse	ΔT s	Luna Num	Saros Num	Ecl. Type	QLE	Gamma	Ecl. Mag.	Lat. °	Long. °	Sun Alt °	Sun Azm °	Path Width km	Central Line Dur.
8701	436	1666 Jan 05	08:58:51	28	-4131	126	A	nn	-0.0624	0.9562	26.3S	47.1E	86	359	160	05m07s
8702	436	1666 Jul 02	07:42:30	27	-4125	131	H	p-	0.4704	1.0075	51.4N	64.4E	62	178	29	00m39s
8703	436	1666 Dec 25	17:59:16	27	-4119	136	H	p-	-0.7452	1.0058	71.6S	98.3W	42	11	30	00m24s
8704	436	1667 May 22	22:58:00	26	-4114	103	Pe	-t	-1.5574	0.0102	63.5S	124.0W	0	322		
8705	436	1667 Jun 21	13:36:07	26	-4113	141	P	t-	1.2858	0.4732	65.9N	170.1E	0	13		
8706	436	1667 Nov 15	22:12:06	26	-4108	108	P	-t	1.2880	0.4667	62.9N	108.2W	0	225		
8707	436	1667 Dec 15	08:29:59	26	-4107	146	P	t-	-1.3752	0.3024	65.3S	108.2W	0	162		
8708	436	1668 May 10	23:37:24	25	-4102	113	A	-p	-0.8049	0.9510	32.3S	155.4W	36	336	296	05m21s
8709	436	1668 Nov 04	12:40:05	25	-4096	118	H	-p	0.6401	1.0102	21.1N	1.8E	50	204	45	00m57s
8710	436	1669 Apr 30	05:26:07	24	-4090	123	H	nn	-0.0171	1.0036	14.1N	98.2E	89	334	13	00m22s
8711	436	1669 Oct 24	21:28:05	24	-4084	128	A	nn	-0.0710	0.9613	15.9S	147.7W	86	26	141	04m07s
8712	436	1670 Apr 19	18:12:20	23	-4078	133	T	p-	0.7191	1.0476	50.6N	123.3W	44	137	225	03m15s
8713	436	1670 Oct 13	23:19:00	23	-4072	138	A	p-	-0.8043	0.9247	52.4S	149.1E	36	49	467	06m34s
8714	436	1671 Mar 11	01:50:58	22	-4067	105	P	-t	-1.1906	0.6504	61.0S	108.3W	0	262		
8715	436	1671 Apr 09	10:41:25	22	-4066	143	P	t-	1.4047	0.2423	61.4N	84.1W	0	74		
8716	436	1671 Sep 03	08:23:57	22	-4061	110	P	-t	1.3664	0.3318	61.3N	157.8E	0	286		
8717	436	1671 Oct 02	23:13:22	22	-4060	148	P	t-	-1.4952	0.1177	61.0S	92.1E	0	98		
8718	436	1672 Feb 28	14:50:43	21	-4055	115	A	-p	-0.5218	0.9926	35.2S	22.0W	58	326	30	00m38s
8719	436	1672 Aug 22	17:44:06	21	-4049	120	T	-p	0.5594	1.0288	41.2N	66.2W	56	215	117	02m15s
8720	436	1673 Feb 16	20:49:18	20	-4043	125	A	nn	0.1950	0.9409	1.8S	133.5W	79	154	223	06m52s
8721	437	1673 Aug 12	09:04:05	20	-4037	130	T	n-	-0.1946	1.0731	4.6N	40.6E	79	23	242	06m15s
8722	437	1674 Feb 05	20:41:35	19	-4031	135	A	p-	0.8979	0.9129	42.8N	155.7W	26	149	736	09m09s
8723	437	1674 Aug 02	02:07:57	19	-4025	140	T	t-	-0.9295	1.0560	45.9S	120.8E	21	29	498	04m08s
8724	437	1674 Dec 27	05:27:32	18	-4020	107	P	-t	-1.3784	0.3108	65.4S	100.9W	0	198		
8725	437	1675 Jan 25	21:19:48	18	-4019	145	P	t-	1.5434	0.0346	62.9N	175.1E	0	135		
8726	437	1675 Jun 23	05:44:39	18	-4014	112	A	-t	0.9219	0.9835	84.1N	166.1W	22	282	154	01m01s
8727	437	1675 Dec 16	16:24:03	18	-4008	117	T	-p	-0.6367	1.0218	63.1S	62.0W	50	353	97	01m33s
8728	437	1676 Jun 11	09:42:37	17	-4002	122	A	nn	0.1673	0.9527	33.0N	34.6E	80	182	176	05m38s
8729	437	1676 Dec 05	07:42:08	17	-3996	127	T	nn	0.0435	1.0486	20.2S	62.5E	88	181	162	04m30s
8730	437	1677 May 31	10:13:53	16	-3990	132	A	p-	-0.5935	0.9510	14.4S	27.5E	53	358	223	06m36s
8731	437	1677 Nov 24	22:44:03	16	-3984	137	T	p-	0.7332	1.0166	26.3N	159.6W	43	186	84	01m36s
8732	437	1678 Apr 21	03:45:50	16	-3979	104	P	-t	1.3765	0.3049	71.0N	4.5W	0	51		
8733	437	1678 May 20	14:40:42	16	-3978	142	P	t-	-1.3172	0.4158	68.8S	22.1W	0	342		
8734	437	1678 Oct 15	17:36:58	15	-3973	109	P	-t	-1.4027	0.2730	71.6S	154.5E	0	119		
8735	437	1678 Nov 14	08:58:14	15	-3972	147	P	t-	1.4908	0.1148	69.4N	69.4E	0	206		
8736	437	1679 Apr 10	17:55:13	15	-3967	114	T	-p	0.6070	1.0565	43.8N	102.2W	52	157	233	04m17s
8737	437	1679 Oct 04	18:13:56	15	-3961	119	A	-p	-0.7191	0.9270	48.0S	114.9W	44	26	391	07m21s
8738	437	1680 Mar 30	10:32:01	14	-3955	124	T	nn	-0.1039	1.0595	1.5S	24.9E	84	343	197	05m25s
8739	437	1680 Sep 22	19:06:23	14	-3949	129	A	nn	0.0160	0.9578	0.7N	108.2W	89	198	153	05m08s
8740	437	1681 Mar 20	00:52:59	13	-3943	134	H	p-	-0.8445	1.0057	53.8S	165.3W	32	329	37	00m26s
8741	438	1681 Sep 12	02:33:12	13	-3937	139	H	p-	0.7504	1.0083	49.8N	161.1E	41	207	43	00m40s
8742	438	1682 Feb 07	15:59:21	13	-3932	106	P	-t	1.3238	0.4101	70.5N	96.8W	0	141		
8743	438	1682 Aug 03	09:21:11	13	-3926	111	P	-t	-1.1028	0.8246	69.7S	9.5E	0	30		
8744	438	1682 Sep 01	16:42:24	13	-3925	149	P	t-	1.4279	0.1978	71.5N	44.3E	0	296		
8745	438	1683 Jan 27	15:10:09	12	-3920	116	A	-p	0.6526	0.9195	22.1N	50.6W	49	171	401	10m44s
8746	438	1683 Jul 24	02:07:00	12	-3914	121	T	-p	-0.3838	1.0569	2.5S	147.1E	67	7	203	05m38s
8747	438	1684 Jan 16	17:18:53	12	-3908	126	A	nn	-0.0565	0.9596	24.2S	76.7W	87	355	147	04m43s
8748	438	1684 Jul 12	14:40:35	11	-3902	131	H	p-	0.3926	1.0041	45.2N	37.1W	67	184	16	00m23s
8749	438	1685 Jan 05	02:42:50	11	-3896	136	H	p-	-0.7409	1.0086	70.7S	143.1E	42	357	44	00m35s
8750	438	1685 Jul 01	20:06:07	11	-3890	141	P	t-	1.2030	0.6163	66.9N	62.2E	0	3		
8751	438	1685 Nov 26	06:54:43	11	-3885	108	P	-t	1.3000	0.4442	63.7N	111.8E	0	215		
8752	438	1685 Dec 25	17:22:35	11	-3884	146	P	t-	-1.3710	0.3102	66.4S	108.3E	0	172		
8753	438	1686 May 22	06:21:20	10	-3879	113	A	-p	-0.8791	0.9533	38.6S	103.3E	28	339	353	04m56s
8754	438	1686 Nov 15	21:05:00	10	-3873	118	H	-p	0.6578	1.0048	20.2N	126.0W	49	200	22	00m28s
8755	438	1687 May 11	12:42:28	10	-3867	123	H	nn	-0.0828	1.0094	13.6N	9.9W	85	339	33	00m57s
8756	438	1687 Nov 05	05:22:24	10	-3861	128	A	nn	-0.0460	0.9561	18.3S	94.3E	87	23	160	04m49s
8757	438	1688 Apr 30	01:57:34	10	-3855	133	T	p-	0.6621	1.0535	51.4N	124.4E	48	141	234	03m40s
8758	438	1688 Oct 24	06:46:41	9	-3849	138	A	p-	-0.7686	0.9221	54.4S	39.2E	39	47	453	06m49s
8759	438	1689 Mar 21	10:06:42	9	-3844	105	P	-t	-1.2245	0.5867	61.0S	119.0E	0	271		
8760	438	1689 Apr 19	18:39:23	9	-3843	143	P	t-	1.3581	0.3312	61.7N	147.5E	0	65		

Cat Num	Canon Plate	Calendar Date	TD of Greatest Eclipse	ΔT s	Luna Num	Saros Num	Ecl. Type	QLE	Gamma	Ecl. Mag.	Lat. °	Long. °	Sun Alt °	Sun Azm °	Path Width km	Central Line Dur.
8761	439	1689 Sep 13	15:39:22	9	-3838	110	P	-t	1.4191	0.2394	61.1N	40.2E	0	277		
8762	439	1689 Oct 13	06:40:02	9	-3837	148	P	t-	-1.4517	0.1920	61.2S	28.3W	0	107		
8763	439	1690 Mar 10	22:56:00	9	-3832	115	A	-p	-0.5512	0.9920	32.5S	143.0W	56	325	33	00m42s
8764	439	1690 Sep 03	01:17:47	9	-3826	120	T	-p	0.6173	1.0287	40.3N	177.4W	52	217	122	02m13s
8765	439	1691 Feb 28	04:37:41	9	-3820	125	A	nn	0.1701	0.9414	0.8N	109.3E	80	152	220	06m40s
8766	439	1691 Aug 23	16:45:57	9	-3814	130	T	n-	-0.1317	1.0720	4.5N	74.3W	82	26	236	06m01s
8767	439	1692 Feb 17	04:26:56	8	-3808	135	A	p-	0.8765	0.9159	42.4N	85.6E	28	145	644	08m36s
8768	439	1692 Aug 12	09:41:06	8	-3802	140	T	p-	-0.8649	1.0546	39.8S	8.6E	30	31	353	04m10s
8769	439	1693 Jan 06	13:55:33	8	-3797	107	P	-t	-1.3788	0.3097	64.4S	122.2E	0	208		
8770	439	1693 Feb 05	05:27:09	8	-3796	145	P	t-	1.5276	0.0597	62.2N	44.2E	0	125		
8771	439	1693 Jul 03	12:33:52	8	-3791	112	P	-t	1.0058	0.9718	64.8N	146.3E	0	336		
8772	439	1693 Dec 27	01:10:50	8	-3785	117	T	-p	-0.6387	1.0236	62.6S	174.3E	50	343	105	01m39s
8773	439	1694 Jun 22	16:08:45	8	-3779	122	A	np	0.2556	0.9517	38.4N	59.7W	75	187	183	05m27s
8774	439	1694 Dec 16	16:33:11	8	-3773	127	T	nn	0.0388	1.0475	21.3S	69.2W	88	176	158	04m22s
8775	439	1695 Jun 11	16:44:24	8	-3767	132	A	p-	-0.5077	0.9545	7.4S	72.2W	59	2	193	06m13s
8776	439	1695 Dec 06	07:23:18	8	-3761	137	T	p-	0.7280	1.0128	24.3N	67.9E	43	181	64	01m16s
8777	439	1696 May 01	11:15:19	8	-3756	104	P	-t	1.4286	0.2078	70.2N	129.8W	0	38		
8778	439	1696 May 30	21:41:23	8	-3755	142	P	t-	-1.2406	0.5534	67.8S	138.7W	0	353		
8779	439	1696 Oct 26	01:17:07	8	-3750	109	P	-t	-1.4361	0.2172	70.9S	25.9E	0	133		
8780	439	1696 Nov 24	17:10:41	8	-3749	147	P	t-	1.4822	0.1318	68.4N	65.6W	0	194		
8781	440	1697 Apr 21	01:49:22	8	-3744	114	T	-p	0.6559	1.0602	51.4N	136.9E	49	157	262	04m18s
8782	440	1697 Oct 15	01:33:41	8	-3738	119	A	-p	-0.7603	0.9236	55.5S	131.2E	40	29	441	07m12s
8783	440	1698 Apr 10	18:34:26	8	-3732	124	Tm	nn	-0.0599	1.0613	5.1N	97.3W	87	344	201	05m36s
8784	440	1698 Oct 04	02:31:25	8	-3726	129	A	nn	-0.0305	0.9573	6.2S	138.8E	88	17	155	05m10s
8785	440	1699 Mar 31	08:48:45	8	-3720	134	H	p-	-0.8089	1.0065	46.8S	69.7E	36	334	38	00m32s
8786	440	1699 Sep 23	10:16:12	8	-3714	139	H	p-	0.6999	1.0095	41.8N	40.7E	45	204	46	00m49s
8787	440	1700 Feb 18	23:49:35	8	-3709	106	P	-t	1.3451	0.3744	71.2N	132.4E	0	128		
8788	440	1700 Aug 14	16:59:06	8	-3703	111	P	-t	-1.1668	0.7000	70.6S	117.5W	0	42		
8789	440	1700 Sep 13	00:34:18	8	-3702	149	P	t-	1.3749	0.2996	71.9N	87.6W	0	283		
8790	440	1701 Feb 07	23:04:53	8	-3697	116	A	-p	0.6663	0.9219	25.9N	171.7W	48	167	393	09m55s
8791	440	1701 Aug 04	09:31:44	8	-3691	121	T	-p	-0.4559	1.0521	9.4S	33.7E	63	10	193	05m06s
8792	440	1702 Jan 28	01:37:10	8	-3685	126	A	nn	-0.0484	0.9636	21.2S	159.6E	87	351	132	04m14s
8793	440	1702 Jul 24	21:38:51	8	-3679	131	H	n-	0.3160	1.0001	38.4N	140.4W	71	188	1	00m01s
8794	440	1703 Jan 17	11:24:25	8	-3673	136	H2	p-	-0.7345	1.0120	67.9S	22.2E	42	347	61	00m50s
8795	440	1703 Jul 14	02:36:34	9	-3667	141	P	t-	1.1206	0.7580	67.9N	46.3W	0	352		
8796	440	1703 Dec 08	15:41:30	9	-3662	108	P	-t	1.3086	0.4281	64.6N	29.5W	0	205		
8797	440	1704 Jan 07	02:14:51	9	-3661	146	P	t-	-1.3669	0.3177	67.4S	35.5W	0	183		
8798	440	1704 Jun 02	13:02:36	9	-3656	113	A	-t	-0.9561	0.9542	49.1S	3.4E	16	341	578	04m26s
8799	440	1704 Nov 27	05:33:53	9	-3650	118	A	-p	0.6716	0.9999	19.7N	104.9E	48	196	1	00m01s
8800	440	1705 May 22	19:55:06	9	-3644	123	Hm	nn	-0.1525	1.0147	12.2N	117.0W	81	343	51	01m32s
8801	441	1705 Nov 16	13:23:06	9	-3638	128	A	nn	-0.0271	0.9514	20.4S	25.0W	88	19	178	05m31s
8802	441	1706 May 12	09:35:09	9	-3632	133	T	p-	0.5984	1.0591	51.5N	15.2E	53	147	242	04m06s
8803	441	1706 Nov 05	14:23:57	9	-3626	138	A	p-	-0.7407	0.9195	57.0S	72.6W	42	44	449	07m02s
8804	441	1707 Apr 02	18:12:25	9	-3621	105	P	-t	-1.2661	0.5082	61.1S	11.1W	0	280		
8805	441	1707 May 02	02:28:17	9	-3620	143	P	t-	1.3047	0.4339	62.2N	21.4E	0	56		
8806	441	1707 Sep 25	23:05:05	9	-3615	110	P	-t	1.4641	0.1603	61.1N	80.0W	0	268		
8807	441	1707 Oct 25	14:17:22	9	-3614	148	P	t-	-1.4161	0.2528	61.6S	151.3W	0	116		
8808	441	1708 Mar 22	06:51:37	9	-3609	115	A	-p	-0.5879	0.9913	30.4S	98.3E	54	326	37	00m46s
8809	441	1708 Sep 14	09:00:22	9	-3603	120	T	-p	0.6685	1.0281	39.2N	68.3E	48	218	126	02m10s
8810	441	1709 Mar 11	12:10:35	9	-3597	125	Am	nn	0.1394	0.9422	3.4N	5.9W	82	151	216	06m29s
8811	441	1709 Sep 04	00:32:26	9	-3591	130	T	nn	-0.0725	1.0703	3.7N	169.7E	86	28	229	05m47s
8812	441	1710 Feb 28	12:07:29	9	-3585	135	A	p-	0.8509	0.9194	42.5N	31.2W	31	141	562	08m00s
8813	441	1710 Aug 24	17:17:16	9	-3579	140	T	p-	-0.8031	1.0519	36.5S	105.1W	36	33	282	04m00s
8814	441	1711 Jan 18	22:23:38	9	-3574	107	P	-t	-1.3796	0.3075	63.5S	14.4W	0	218		
8815	441	1711 Feb 17	13:30:15	9	-3573	145	P	t-	1.5077	0.0919	61.6N	85.4W	0	116		
8816	441	1711 Jul 15	19:22:11	9	-3568	112	P	-t	1.0894	0.8216	63.9N	34.6E	0	327		
8817	441	1712 Jan 08	09:58:39	9	-3562	117	T	-p	-0.6406	1.0258	60.6S	49.2E	50	335	114	01m48s
8818	441	1712 Jul 03	22:34:57	9	-3556	122	A	-p	0.3434	0.9503	42.8N	152.7W	70	194	194	05m18s
8819	441	1712 Dec 28	01:24:54	9	-3550	127	T	nn	0.0346	1.0466	21.5S	159.0E	88	171	155	04m15s
8820	441	1713 Jun 22	23:15:39	9	-3544	132	A	p-	-0.4216	0.9576	1.3S	171.2W	65	6	170	05m45s

Cat Num	Canon Plate	Calendar Date	TD of Greatest Eclipse	ΔT s	Luna Num	Saros Num	Ecl. Type	QLE	Gamma	Ecl. Mag.	Lat. °	Long. °	Sun Alt °	Sun Azm °	Path Width km	Central Line Dur.
8821	442	1713 Dec 17	16:04:20	9	-3538	137	H	p-	0.7249	1.0094	23.1N	64.8W	43	176	47	00m56s
8822	442	1714 May 13	18:39:35	10	-3533	104	Pe	-t	1.4856	0.1007	69.4N	106.9E	0	26		
8823	442	1714 Jun 12	04:40:01	10	-3532	142	P	t-	-1.1610	0.6976	66.8S	105.8E	0	4		
8824	442	1714 Nov 07	09:04:34	10	-3527	109	P	-t	-1.4630	0.1730	70.1S	103.9W	0	145		
8825	442	1714 Dec 07	01:27:09	10	-3526	147	P	t-	1.4772	0.1420	67.4N	159.0E	0	183		
8826	442	1715 May 03	09:36:30	10	-3521	114	T	-p	0.7112	1.0632	59.4N	17.9E	44	157	295	04m14s
8827	442	1715 Oct 27	09:02:48	10	-3515	119	A	-p	-0.7939	0.9206	62.5S	15.5E	37	31	494	07m02s
8828	442	1716 Apr 22	02:28:33	10	-3509	124	T	nn	-0.0091	1.0625	11.8N	142.6E	90	343	205	05m43s
8829	442	1716 Oct 15	10:07:39	10	-3503	129	A	nn	-0.0687	0.9570	12.5S	23.5E	86	16	157	05m10s
8830	442	1717 Apr 11	16:34:40	10	-3497	134	H	p-	-0.7660	1.0072	39.5S	52.1W	40	339	39	00m39s
8831	442	1717 Oct 04	18:08:27	10	-3491	139	H	p-	0.6563	1.0104	34.6N	81.1W	49	201	47	00m56s
8832	442	1718 Mar 02	07:31:37	10	-3486	106	P	-t	1.3723	0.3285	71.8N	3.2E	0	114		
8833	442	1718 Aug 26	00:41:45	10	-3480	111	P	-t	-1.2267	0.5837	71.2S	113.7E	0	55		
8834	442	1718 Sep 24	08:34:20	10	-3479	149	P	t-	1.3282	0.3889	72.0N	138.3E	0	269		
8835	442	1719 Feb 19	06:52:57	10	-3474	116	A	-p	0.6856	0.9250	30.5N	68.6E	47	163	384	09m01s
8836	442	1719 Aug 15	16:59:51	10	-3468	121	T	-p	-0.5243	1.0466	16.8S	81.1W	58	13	181	04m27s
8837	442	1720 Feb 08	09:52:31	10	-3462	126	A	nn	-0.0375	0.9681	17.4S	36.1E	88	348	115	03m40s
8838	442	1720 Aug 04	04:38:15	10	-3456	131	A	nn	0.2409	0.9957	31.1N	114.8E	76	192	16	00m27s
8839	442	1721 Jan 27	20:05:11	10	-3450	136	T	p-	-0.7269	1.0158	64.0S	102.4W	43	340	79	01m07s
8840	442	1721 Jul 24	09:06:55	10	-3444	141	P	t-	1.0382	0.8990	68.9N	155.2W	0	341		
8841	443	1721 Dec 19	00:31:51	10	-3439	108	P	-t	1.3144	0.4172	65.7N	172.0W	0	195		
8842	443	1722 Jan 17	11:07:10	10	-3438	146	P	t-	-1.3629	0.3251	68.5S	179.9W	0	195		
8843	443	1722 Jun 13	19:40:19	10	-3433	113	P	-t	-1.0364	0.9083	65.2S	93.5W	0	340		
8844	443	1722 Dec 08	14:07:35	10	-3427	118	A	-p	0.6808	0.9955	19.5N	25.4W	47	191	21	00m28s
8845	443	1723 Jun 03	03:05:13	10	-3421	123	T	nn	-0.2251	1.0196	9.6N	136.1E	77	347	69	02m05s
8846	443	1723 Nov 27	21:28:16	10	-3415	128	A	nn	-0.0125	0.9471	22.0S	145.2W	89	14	195	06m12s
8847	443	1724 May 22	17:10:09	10	-3409	133	T	p-	0.5318	1.0640	50.8N	92.9W	58	154	247	04m33s
8848	443	1724 Nov 15	22:07:38	10	-3403	138	A	p-	-0.7183	0.9174	59.9S	175.0E	44	40	448	07m15s
8849	443	1725 Apr 13	02:11:23	10	-3398	105	P	-t	-1.3132	0.4193	61.4S	139.6W	0	289		
8850	443	1725 May 12	10:12:19	10	-3397	143	P	t-	1.2472	0.5447	62.8N	103.7W	0	47		
8851	443	1725 Oct 06	06:39:42	10	-3392	110	P	-t	1.5029	0.0923	61.2N	157.7E	0	259		
8852	443	1725 Nov 04	22:02:52	10	-3391	148	P	t-	-1.3861	0.3038	62.1S	83.5E	0	125		
8853	443	1726 Apr 02	14:38:16	10	-3386	115	A	-p	-0.6313	0.9906	29.2S	18.3W	51	327	42	00m52s
8854	443	1726 Sep 25	16:51:45	10	-3380	120	T	-p	0.7134	1.0273	38.0N	49.0W	44	218	129	02m07s
8855	443	1727 Mar 22	19:47:55	10	-3374	125	A	nn	0.0996	0.9432	5.7N	118.0W	84	151	211	06m20s
8856	443	1727 Sep 15	08:27:31	10	-3368	130	T	nn	-0.0202	1.0681	2.2N	51.4E	89	29	222	05m33s
8857	443	1728 Mar 10	19:38:56	10	-3362	135	A	p-	0.8172	0.9233	42.8N	144.6W	35	139	485	07m25s
8858	443	1728 Sep 04	00:59:22	10	-3356	140	T	p-	-0.7466	1.0484	35.0S	139.6E	41	34	236	03m44s
8859	443	1729 Jan 29	06:48:43	10	-3351	107	P	-t	-1.3838	0.2993	62.8S	149.9W	0	228		
8860	443	1729 Feb 27	21:27:02	10	-3350	145	P	t-	1.4817	0.1347	61.2N	146.6E	0	107		
8861	444	1729 Jul 26	02:10:40	11	-3345	112	P	-t	1.1718	0.6746	63.1N	76.9W	0	318		
8862	444	1729 Aug 24	13:48:31	11	-3344	150	Pb	t-	-1.5430	0.0067	61.7S	95.2W	0	66		
8863	444	1730 Jan 18	18:45:15	11	-3339	117	T	-p	-0.6440	1.0285	57.8S	77.4W	50	329	126	01m59s
8864	444	1730 Jul 15	04:59:09	11	-3333	122	A	-p	0.4325	0.9484	46.3N	115.9E	64	200	210	05m13s
8865	444	1731 Jan 08	10:17:44	11	-3327	127	Tm	nn	0.0313	1.0464	20.7S	27.0E	88	166	155	04m10s
8866	444	1731 Jul 04	05:46:25	11	-3321	132	A	p-	-0.3341	0.9602	3.8N	90.8E	71	10	153	05m15s
8867	444	1731 Dec 29	00:46:53	11	-3315	137	H	p-	0.7234	1.0065	22.7N	162.2E	44	171	32	00m39s
8868	444	1732 Jun 22	11:38:48	11	-3309	142	P	t-	-1.0800	0.8457	65.8S	9.3W	0	14		
8869	444	1732 Nov 17	16:58:51	11	-3304	109	P	-t	-1.4841	0.1389	69.2S	125.3E	0	158		
8870	444	1732 Dec 17	09:46:57	11	-3303	147	P	t-	1.4751	0.1470	66.3N	23.4E	0	172		
8871	444	1733 May 13	17:18:29	11	-3298	114	T	-p	0.7712	1.0656	67.9N	99.5W	39	157	339	04m06s
8872	444	1733 Nov 06	16:40:15	11	-3292	119	A	-p	-0.8208	0.9179	69.0S	101.2W	34	32	548	06m53s
8873	444	1734 May 03	10:15:56	11	-3286	124	T	nn	0.0472	1.0635	18.4N	24.6E	87	168	208	05m46s
8874	444	1734 Oct 26	17:53:28	11	-3280	129	A	nn	-0.0996	0.9567	18.2S	93.8W	84	14	159	05m08s
8875	444	1735 Apr 23	00:11:36	11	-3274	134	H	p-	-0.7164	1.0077	32.2S	171.0W	44	343	38	00m44s
8876	444	1735 Oct 16	02:10:34	11	-3268	139	H	p-	0.6202	1.0110	28.3N	155.2E	51	198	48	01m02s
8877	444	1736 Mar 12	15:05:55	11	-3263	106	P	-t	1.4049	0.2733	72.1N	124.5W	0	100		
8878	444	1736 Apr 11	07:18:07	11	-3262	144	Pb	t-	-1.5166	0.0748	71.5S	134.3E	0	298		
8879	444	1736 Sep 05	08:30:26	11	-3257	111	P	-t	-1.2817	0.4775	71.7S	17.1W	0	68		
8880	444	1736 Oct 04	16:41:34	11	-3256	149	P	t-	1.2874	0.4670	71.9N	2.4E	0	255		

Cat Num	Canon Plate	Calendar Date	TD of Greatest Eclipse	ΔT s	Luna Num	Saros Num	Ecl. Type	QLE	Gamma	Ecl. Mag.	Lat. °	Long. °	Sun Alt °	Sun Azm °	Path Width km	Central Line Dur.
8881	445	1737 Mar 01	14:35:17	11	-3251	116	A	-p	0.7099	0.9283	36.0N	50.1W	45	160	378	08m04s
8882	445	1737 Aug 26	00:32:08	11	-3245	121	T	-p	-0.5886	1.0407	24.4S	162.5E	54	17	167	03m44s
8883	445	1738 Feb 18	18:02:31	11	-3239	126	A	nn	-0.0211	0.9732	12.8S	86.7W	89	346	96	03m03s
8884	445	1738 Aug 15	11:40:12	11	-3233	131	A	nn	0.1688	0.9907	23.7N	8.4E	80	194	33	01m00s
8885	445	1739 Feb 08	04:41:13	11	-3227	136	T	p-	-0.7149	1.0203	59.2S	131.0E	44	336	99	01m27s
8886	445	1739 Aug 04	15:40:56	11	-3221	141	A	t-	0.9588	0.9408	79.9N	42.9E	16	280	801	03m59s
8887	445	1739 Dec 30	09:22:03	12	-3216	108	P	-t	1.3203	0.4062	66.7N	45.1E	0	184		
8888	445	1740 Jan 28	19:54:59	12	-3215	146	P	t-	-1.3555	0.3387	69.5S	36.2E	0	207		
8889	445	1740 Jun 24	02:18:54	12	-3210	113	P	-t	-1.1163	0.7697	66.2S	156.7E	0	350		
8890	445	1740 Dec 18	22:43:17	12	-3204	118	A	-p	0.6876	0.9917	19.9N	156.4W	46	187	40	00m53s
8891	445	1741 Jun 13	10:12:48	12	-3198	123	T	-n	-0.3007	1.0239	6.0N	29.4E	73	352	85	02m35s
8892	445	1741 Dec 08	05:38:00	12	-3192	128	A	nn	-0.0024	0.9434	23.0S	93.6E	90	6	209	06m51s
8893	445	1742 Jun 03	00:39:57	12	-3186	133	T	p-	0.4607	1.0683	49.0N	160.2E	62	161	251	05m00s
8894	445	1742 Nov 27	05:58:59	12	-3180	138	A	p-	-0.7019	0.9156	62.6S	62.2E	45	34	450	07m26s
8895	445	1743 Apr 24	10:00:10	12	-3175	105	P	-t	-1.3682	0.3152	61.8S	94.4E	0	298		
8896	445	1743 May 23	17:48:55	12	-3174	143	P	t-	1.1838	0.6672	63.5N	132.9E	0	38		
8897	445	1743 Oct 17	14:25:42	12	-3169	110	Pe	-t	1.5334	0.0387	61.5N	32.5E	0	250		
8898	445	1743 Nov 16	05:58:25	12	-3168	148	P	t-	-1.3634	0.3424	62.8S	44.4W	0	135		
8899	445	1744 Apr 12	22:15:24	12	-3163	115	A	-p	-0.6819	0.9895	29.1S	132.6W	47	329	49	00m59s
8900	445	1744 Oct 06	00:51:24	12	-3157	120	T	-p	0.7521	1.0263	37.0N	169.1W	41	216	132	02m04s
8901	446	1745 Apr 02	03:09:18	12	-3151	125	A	nn	0.0536	0.9444	7.7N	132.2E	87	152	205	06m13s
8902	446	1745 Sep 25	16:28:56	12	-3145	130	Tm	nn	0.0269	1.0655	0.3N	68.6W	88	209	214	05m21s
8903	446	1746 Mar 22	03:02:49	12	-3139	135	A	p-	0.7771	0.9277	43.5N	104.7E	39	138	419	06m51s
8904	446	1746 Sep 15	08:46:37	12	-3133	140	T	p-	-0.6948	1.0441	34.9S	23.0E	46	36	200	03m23s
8905	446	1747 Feb 09	15:11:18	12	-3128	107	P	-t	-1.3908	0.2860	62.1S	75.5E	0	237		
8906	446	1747 Mar 11	05:18:08	12	-3127	145	P	t-	1.4504	0.1872	61.0N	20.2E	0	98		
8907	446	1747 Aug 06	09:01:21	12	-3122	112	P	-t	1.2512	0.5339	62.4N	171.3E	0	309		
8908	446	1747 Sep 04	21:07:57	12	-3121	150	P	t-	-1.4880	0.1086	61.4S	146.1E	0	75		
8909	446	1748 Jan 30	03:29:13	13	-3116	117	T	-p	-0.6501	1.0316	54.4S	154.8E	49	324	140	02m12s
8910	446	1748 Jul 25	11:27:02	13	-3110	122	A	-p	0.5183	0.9461	48.7N	24.5E	59	207	231	05m12s
8911	446	1749 Jan 18	19:08:56	13	-3104	127	T	nn	0.0264	1.0465	19.1S	104.9W	89	161	155	04m07s
8912	446	1749 Jul 14	12:19:20	13	-3098	132	A	pn	-0.2476	0.9623	7.8N	7.2W	76	14	141	04m46s
8913	446	1750 Jan 08	09:28:43	13	-3092	137	H	p-	0.7217	1.0041	23.0N	29.3E	44	167	20	00m24s
8914	446	1750 Jul 03	18:38:52	13	-3086	142	P	t-	-0.9985	0.9956	64.8S	124.3W	0	23		
8915	446	1750 Nov 29	00:58:14	13	-3081	109	P	-t	-1.5004	0.1129	68.2S	6.2W	0	170		
8916	446	1750 Dec 28	18:06:51	13	-3080	147	P	t-	1.4737	0.1506	65.3N	111.8W	0	161		
8917	446	1751 May 25	00:55:16	13	-3075	114	T	-p	0.8359	1.0670	77.0N	144.7E	33	157	402	03m53s
8918	446	1751 Nov 18	00:26:00	13	-3069	119	A	-p	-0.8411	0.9159	74.9S	142.8E	32	31	597	06m45s
8919	446	1752 May 13	17:56:29	13	-3063	124	T	nn	0.1090	1.0637	24.9N	91.1W	84	171	210	05m42s
8920	446	1752 Nov 06	01:48:14	13	-3057	129	A	nn	-0.1239	0.9567	23.2S	147.4E	83	12	159	05m03s
8921	447	1753 May 03	07:39:40	13	-3051	134	H	p-	-0.6601	1.0079	24.9S	73.0E	49	347	36	00m48s
8922	447	1753 Oct 26	10:22:01	13	-3045	139	H	p-	0.5910	1.0115	22.7N	29.7E	54	195	49	01m08s
8923	447	1754 Mar 23	22:28:59	13	-3040	106	P	-t	1.4463	0.2032	72.1N	110.6E	0	86		
8924	447	1754 Apr 22	14:25:57	13	-3039	144	P	t-	-1.4631	0.1669	71.0S	14.0E	0	311		
8925	447	1754 Sep 16	16:25:41	14	-3034	111	P	-t	-1.3314	0.3821	71.9S	149.9W	0	82		
8926	447	1754 Oct 16	00:57:46	14	-3033	149	P	t-	1.2535	0.5314	71.5N	135.5W	0	241		
8927	447	1755 Mar 12	22:09:32	14	-3028	116	A	-p	0.7413	0.9319	42.2N	167.4W	42	156	375	07m07s
8928	447	1755 Sep 06	08:09:46	14	-3022	121	T	-p	-0.6478	1.0342	32.1S	44.3E	49	20	150	03m00s
8929	447	1756 Mar 01	02:07:09	14	-3016	126	A	nn	0.0006	0.9787	7.5S	151.4E	90	31	76	02m24s
8930	447	1756 Aug 25	18:46:17	14	-3010	131	Am	nn	0.1009	0.9853	16.1N	99.5W	84	196	52	01m38s
8931	447	1757 Feb 18	13:14:12	14	-3004	136	T	p-	-0.6999	1.0251	53.8S	2.9E	45	335	119	01m51s
8932	447	1757 Aug 14	22:16:45	14	-2998	141	A	p-	0.8807	0.9407	71.6N	113.5W	28	224	467	04m36s
8933	447	1758 Jan 09	18:13:42	14	-2993	108	P	-t	1.3251	0.3972	67.8N	98.7W	0	173		
8934	447	1758 Feb 08	04:40:52	14	-2992	146	P	t-	-1.3468	0.3549	70.4S	107.8W	0	220		
8935	447	1758 Jul 05	08:57:44	14	-2987	113	P	-t	-1.1961	0.6302	67.2S	46.4E	0	0		
8936	447	1758 Dec 30	07:20:12	14	-2981	118	A	-p	0.6929	0.9885	20.8N	72.2E	46	182	56	01m15s
8937	447	1759 Jun 24	17:20:59	14	-2975	123	T	-n	-0.3768	1.0275	1.4N	78.1W	68	356	101	02m59s
8938	447	1759 Dec 19	13:50:05	14	-2969	128	A	nn	0.0051	0.9404	23.3S	28.0W	90	193	221	07m25s
8939	447	1760 Jun 13	08:09:15	14	-2963	133	T	p-	0.3883	1.0719	46.0N	52.7E	67	168	254	05m27s
8940	447	1760 Dec 07	13:53:44	15	-2957	138	A	p-	-0.6881	0.9144	64.7S	49.4W	46	26	451	07m36s

Cat Num	Canon Plate	Calendar Date	TD of Greatest Eclipse	ΔT s	Luna Num	Saros Num	Ecl. Type	QLE	Gamma	Ecl. Mag.	Lat. °	Long. °	Sun Alt °	Sun Azm °	Path Width km	Central Line Dur.
8941	448	1761 May 04	17:43:11	15	-2952	105	P	-t	-1.4274	0.2031	62.4S	30.3W	0	307		
8942	448	1761 Jun 03	01:22:38	15	-2951	143	P	t-	1.1182	0.7939	64.4N	9.9E	0	29		
8943	448	1761 Nov 26	14:00:27	15	-2945	148	P	t-	-1.3451	0.3732	63.7S	174.2W	0	144		
8944	448	1762 Apr 24	05:42:10	15	-2940	115	A	-p	-0.7402	0.9881	30.3S	115.6E	42	331	61	01m08s
8945	448	1762 Oct 17	09:00:34	15	-2934	120	T	-p	0.7836	1.0253	36.2N	67.6E	38	214	135	02m02s
8946	448	1763 Apr 13	10:19:31	15	-2928	125	A	nn	-0.0010	0.9455	9.0N	25.3E	90	252	201	06m11s
8947	448	1763 Oct 07	00:39:04	15	-2922	130	T	nn	0.0666	1.0627	2.0S	169.1E	86	209	206	05m09s
8948	448	1764 Apr 01	10:17:15	15	-2916	135	A	p-	0.7288	0.9323	44.2N	2.5W	43	138	361	06m20s
8949	448	1764 Sep 25	16:41:43	15	-2910	140	T	p-	-0.6502	1.0394	36.0S	95.5W	49	37	171	03m01s
8950	448	1765 Feb 19	23:28:38	15	-2905	107	P	-t	-1.4028	0.2635	61.6S	57.7W	0	247		
8951	448	1765 Mar 21	13:01:45	15	-2904	145	P	t-	1.4120	0.2524	61.0N	104.3W	0	89		
8952	448	1765 Aug 16	15:54:02	15	-2899	112	P	-t	1.3279	0.3994	61.8N	59.2E	0	300		
8953	448	1765 Sep 15	04:32:34	15	-2898	150	P	t-	-1.4378	0.2009	61.1S	26.2E	0	84		
8954	448	1766 Feb 09	12:09:44	15	-2893	117	T	-p	-0.6598	1.0352	50.7S	26.6E	48	321	156	02m27s
8955	448	1766 Aug 05	17:56:58	15	-2887	122	A	-p	0.6023	0.9433	50.2N	67.0W	53	214	260	05m15s
8956	448	1767 Jan 30	03:56:55	15	-2881	127	T	nn	0.0190	1.0471	16.8S	123.9E	89	157	157	04m06s
8957	448	1767 Jul 25	18:55:48	16	-2875	132	A	nn	-0.1630	0.9638	10.8N	105.5W	81	18	132	04m21s
8958	448	1768 Jan 19	18:09:29	16	-2869	137	H	p-	0.7195	1.0022	23.9N	103.2W	44	162	11	00m13s
8959	448	1768 Jul 14	01:40:57	16	-2863	142	H	t-	-0.9176	1.0055	43.0S	137.4E	23	19	48	00m29s
8960	448	1768 Dec 09	09:01:39	16	-2858	109	P	-t	-1.5129	0.0932	67.1S	138.1W	0	181		
8961	449	1769 Jan 08	02:26:42	16	-2857	147	P	t-	1.4728	0.1530	64.3N	113.5E	0	151		
8962	449	1769 Jun 04	08:28:34	16	-2852	114	T	-t	0.9037	1.0671	87.3N	26.2E	25	153	521	03m36s
8963	449	1769 Nov 28	08:18:40	16	-2846	119	A	-p	-0.8559	0.9144	80.0S	32.0E	31	22	638	06m38s
8964	449	1770 May 25	01:30:12	16	-2840	124	T	-n	0.1760	1.0634	31.2N	155.6E	80	174	211	05m31s
8965	449	1770 Nov 17	09:51:53	16	-2834	129	A	nn	-0.1416	0.9571	27.3S	27.1E	82	9	158	04m56s
8966	449	1771 May 14	15:00:02	16	-2828	134	H	p-	-0.5980	1.0076	17.8S	40.4W	53	351	33	00m49s
8967	449	1771 Nov 06	18:41:02	16	-2822	139	H	p-	0.5676	1.0120	17.9N	97.3W	55	192	50	01m13s
8968	449	1772 Apr 03	05:43:53	16	-2817	106	P	-t	1.4935	0.1229	71.9N	12.3W	0	72		
8969	449	1772 May 02	21:26:41	16	-2816	144	P	t-	-1.4043	0.2683	70.2S	104.1W	0	323		
8970	449	1772 Sep 27	00:28:19	16	-2811	111	P	-t	-1.3751	0.2988	72.0S	75.4E	0	96		
8971	449	1772 Oct 26	09:21:18	16	-2810	149	P	t-	1.2255	0.5846	70.9N	85.1E	0	228		
8972	449	1773 Mar 23	05:36:58	16	-2805	116	A	-p	0.7785	0.9357	49.3N	76.2E	39	152	378	06m13s
8973	449	1773 Sep 16	15:52:23	16	-2799	121	T	-p	-0.7020	1.0275	39.9S	75.5W	45	23	130	02m18s
8974	449	1774 Mar 12	10:05:14	16	-2793	126	A	nn	0.0284	0.9845	1.7S	30.8E	88	162	55	01m43s
8975	449	1774 Sep 06	01:57:40	16	-2787	131	A	nn	0.0385	0.9797	8.7N	150.9E	88	197	72	02m20s
8976	449	1775 Mar 01	21:39:20	16	-2781	136	T	p-	-0.6783	1.0304	47.9S	124.8W	47	335	139	02m20s
8977	449	1775 Aug 26	04:59:40	16	-2775	141	A	p-	0.8088	0.9391	61.3N	132.0E	36	213	383	05m16s
8978	449	1776 Jan 21	03:02:27	16	-2770	108	P	-t	1.3318	0.3847	68.8N	117.6E	0	161		
8979	449	1776 Feb 19	13:20:11	16	-2769	146	P	t-	-1.3334	0.3800	71.1S	109.2E	0	233		
8980	449	1776 Jul 15	15:39:29	17	-2764	113	P	-t	-1.2739	0.4935	68.2S	65.1W	0	11		
8981	450	1776 Aug 14	05:22:56	17	-2763	151	Pb	t-	1.5357	0.0435	70.6N	123.5W	0	318		
8982	450	1777 Jan 09	15:55:35	17	-2758	118	A	-p	0.6988	0.9859	22.4N	58.9W	46	177	70	01m32s
8983	450	1777 Jul 05	00:29:29	17	-2752	123	T	-p	-0.4531	1.0305	4.2S	173.7E	63	0	115	03m17s
8984	450	1777 Dec 29	22:03:28	17	-2746	128	A	nn	0.0110	0.9380	22.7S	150.0W	90	183	231	07m53s
8985	450	1778 Jun 24	15:34:56	17	-2740	133	T	n-	0.3127	1.0746	41.8N	55.0W	72	175	255	05m52s
8986	450	1778 Dec 18	21:53:54	17	-2734	138	A	p-	-0.6788	0.9137	65.8S	160.6W	47	16	450	07m44s
8987	450	1779 May 16	01:17:39	17	-2729	105	Pe	-t	-1.4928	0.0796	63.0S	153.1W	0	316		
8988	450	1779 Jun 14	08:51:28	17	-2728	143	P	t-	1.0489	0.9276	65.3N	112.1W	0	19		
8989	450	1779 Dec 07	22:08:56	17	-2722	148	P	t-	-1.3315	0.3962	64.6S	54.2E	0	154		
8990	450	1780 May 04	13:00:42	17	-2717	115	A	-p	-0.8043	0.9861	33.3S	5.9E	36	334	81	01m21s
8991	450	1780 Oct 27	17:18:27	17	-2711	120	T	-p	0.8083	1.0244	35.6N	58.6W	36	210	138	02m00s
8992	450	1781 Apr 23	17:21:26	17	-2705	125	A	nn	-0.0620	0.9467	9.7N	79.2W	87	334	197	06m13s
8993	450	1781 Oct 17	08:55:59	17	-2699	130	T	-n	0.1007	1.0596	4.3S	45.1E	84	207	197	04m59s
8994	450	1782 Apr 12	17:24:47	17	-2693	135	A	p-	0.6745	0.9370	45.1N	107.1W	47	140	311	05m51s
8995	450	1782 Oct 07	00:43:19	17	-2687	140	T	p-	-0.6113	1.0344	37.9S	144.6E	52	37	144	02m37s
8996	450	1783 Mar 03	07:40:30	17	-2682	107	P	-t	-1.4200	0.2312	61.3S	170.5E	0	256		
8997	450	1783 Apr 01	20:38:39	17	-2681	145	P	t-	1.3671	0.3299	61.0N	132.8E	0	80		
8998	450	1783 Aug 27	22:52:06	17	-2676	112	P	-t	1.3991	0.2757	61.4N	54.1W	0	291		
8999	450	1783 Sep 26	12:04:17	17	-2675	150	P	t-	-1.3935	0.2814	61.1S	95.4W	0	93		
9000	450	1784 Feb 20	20:45:38	17	-2670	117	T	-p	-0.6739	1.0389	47.2S	101.5W	47	320	174	02m44s

Cat Num	Canon Plate	Calendar Date	TD of Greatest Eclipse	ΔT s	Luna Num	Saros Num	Ecl. Type	QLE	Gamma	Ecl. Mag.	Lat. °	Long. °	Sun Alt °	Sun Azm °	Path Width km	Central Line Dur.
9001	451	1784 Aug 16	00:31:53	17	-2664	122	A	-p	0.6819	0.9402	50.9N	159.8W	47	220	299	05m23s
9002	451	1785 Feb 09	12:40:41	17	-2658	127	T	nn	0.0080	1.0480	14.1S	6.6W	90	150	159	04m07s
9003	451	1785 Aug 05	01:37:22	17	-2652	132	A	nn	-0.0817	0.9650	12.7N	155.3E	85	22	127	04m01s
9004	451	1786 Jan 30	02:45:26	17	-2646	137	H	p-	0.7140	1.0009	25.1N	125.5E	44	158	5	00m05s
9005	451	1786 Jul 25	08:46:33	17	-2640	142	T	t-	-0.8384	1.0106	34.6S	30.8E	33	21	66	00m59s
9006	451	1786 Dec 20	17:07:24	17	-2635	109	P	-t	-1.5232	0.0772	66.0S	89.9E	0	192		
9007	451	1787 Jan 19	10:43:13	17	-2634	147	P	t-	1.4697	0.1591	63.4N	20.1W	0	141		
9008	451	1787 Jun 15	15:59:25	17	-2629	114	T	-t	0.9739	1.0648	78.7N	104.8E	12	346	998	03m09s
9009	451	1787 Dec 09	16:15:38	16	-2623	119	A	-p	-0.8675	0.9136	83.4S	62.7W	29	357	672	06m32s
9010	451	1788 Jun 04	08:59:31	16	-2617	124	T	-n	0.2465	1.0623	37.0N	44.4E	76	179	211	05m15s
9011	451	1788 Nov 27	18:02:54	16	-2611	129	A	nn	-0.1542	0.9579	30.4S	94.3W	81	4	155	04m46s
9012	451	1789 May 24	22:11:58	16	-2605	134	H	p-	-0.5297	1.0068	11.0S	151.0W	58	355	28	00m46s
9013	451	1789 Nov 17	03:08:35	16	-2599	139	H	p-	0.5504	1.0126	14.1N	133.9E	57	188	52	01m19s
9014	451	1790 Apr 14	12:48:15	16	-2594	106	Pe	-t	1.5487	0.0287	71.4N	132.1W	0	58		
9015	451	1790 May 14	04:17:21	16	-2593	144	P	t-	-1.3374	0.3840	69.4S	140.9E	0	335		
9016	451	1790 Oct 08	08:38:52	16	-2588	111	P	-t	-1.4122	0.2287	71.7S	61.3W	0	110		
9017	451	1790 Nov 06	17:53:11	16	-2587	149	P	t-	1.2044	0.6245	70.1N	55.8W	0	215		
9018	451	1791 Apr 03	12:55:13	16	-2582	116	A	-p	0.8236	0.9394	57.1N	39.5W	34	147	394	05m21s
9019	451	1791 Sep 27	23:42:30	16	-2576	121	T	-p	-0.7492	1.0206	47.6S	162.4E	41	27	106	01m38s
9020	451	1792 Mar 22	17:57:34	16	-2570	126	A	nn	0.0618	0.9905	4.5N	88.7W	86	162	33	01m02s
9021	452	1792 Sep 16	09:13:52	16	-2564	131	A	nn	-0.0191	0.9739	1.3N	39.9E	89	18	93	03m02s
9022	452	1793 Mar 12	06:00:07	16	-2558	136	T	p-	-0.6524	1.0359	41.7S	107.8E	49	336	158	02m51s
9023	452	1793 Sep 05	11:47:24	16	-2552	141	A	p-	0.7407	0.9370	51.7N	23.0E	42	207	347	06m02s
9024	452	1794 Jan 31	11:48:45	15	-2547	108	P	-t	1.3407	0.3680	69.8N	26.0W	0	149		
9025	452	1794 Mar 01	21:54:00	15	-2546	146	P	t-	-1.3155	0.4136	71.6S	32.9W	0	246		
9026	452	1794 Jul 26	22:24:27	15	-2541	113	P	-t	-1.3496	0.3599	69.1S	178.0W	0	22		
9027	452	1794 Aug 25	12:08:56	15	-2540	151	P	t-	1.4616	0.1709	71.3N	121.9E	0	305		
9028	452	1795 Jan 21	00:29:13	15	-2535	118	A	-p	0.7055	0.9837	24.8N	170.3E	45	173	81	01m44s
9029	452	1795 Jul 16	07:41:36	15	-2529	123	T	-p	-0.5274	1.0327	10.4S	63.8E	58	4	130	03m26s
9030	452	1796 Jan 10	06:14:52	15	-2523	128	A	nn	0.0179	0.9362	21.1S	88.3E	89	177	238	08m15s
9031	452	1796 Jul 04	23:02:54	15	-2517	133	T	n-	0.2385	1.0764	36.8N	164.6W	76	180	255	06m15s
9032	452	1796 Dec 29	05:54:58	15	-2511	138	A	p-	-0.6703	0.9136	65.5S	88.6E	48	5	446	07m51s
9033	452	1797 Jun 24	16:18:13	14	-2505	143	T	t-	0.9780	1.0570	77.2N	133.9E	11	17	975	02m47s
9034	452	1797 Dec 18	06:21:51	14	-2499	148	P	t-	-1.3208	0.4142	65.6S	79.0W	0	164		
9035	452	1798 May 15	20:10:32	14	-2494	115	A	-t	-0.8744	0.9832	38.6S	101.6W	29	336	121	01m36s
9036	452	1798 Nov 08	01:44:39	14	-2488	120	T	-p	0.8270	1.0237	35.1N	172.5E	34	206	141	01m59s
9037	452	1799 May 05	00:13:08	14	-2482	125	A	nn	-0.1310	0.9476	9.3N	178.9E	83	338	194	06m20s
9038	452	1799 Oct 28	17:21:46	14	-2476	130	T	-n	0.1274	1.0566	6.7S	81.3W	83	205	188	04m50s
9039	452	1800 Apr 24	00:24:00	13	-2470	135	A	p-	0.6125	0.9417	45.7N	151.3E	52	143	269	05m27s
9040	452	1800 Oct 18	08:51:53	13	-2464	140	T	p-	-0.5787	1.0293	40.3S	23.2E	54	36	120	02m14s
9041	453	1801 Mar 14	15:45:35	13	-2459	107	P	-t	-1.4434	0.1873	61.2S	40.6E	0	265		
9042	453	1801 Apr 13	04:08:06	13	-2458	145	P	t-	1.3152	0.4208	61.3N	11.7E	0	71		
9043	453	1801 Sep 08	05:54:40	13	-2453	112	P	-t	1.4657	0.1614	61.1N	168.5W	0	282		
9044	453	1801 Oct 07	19:42:34	13	-2452	150	P	t-	-1.3552	0.3505	61.2S	141.3E	0	102		
9045	453	1802 Mar 04	05:14:29	13	-2447	117	T	-p	-0.6943	1.0428	44.0S	131.5E	46	320	196	03m02s
9046	453	1802 Aug 28	07:12:00	13	-2441	122	A	-p	0.7569	0.9367	51.3N	105.7E	41	225	354	05m35s
9047	453	1803 Feb 21	21:18:46	13	-2435	127	T	nn	-0.0075	1.0492	11.1S	135.9W	90	337	163	04m09s
9048	453	1803 Aug 17	08:25:03	12	-2429	132	A	nn	-0.0048	0.9657	13.6N	54.7E	90	36	124	03m47s
9049	453	1804 Feb 11	11:16:33	12	-2423	137	H	p-	0.7053	1.0000	26.7N	4.5W	45	153	0	00m00s
9050	453	1804 Aug 05	15:57:13	12	-2417	142	T	p-	-0.7622	1.0144	29.3S	77.1W	40	24	75	01m20s
9051	453	1805 Jan 01	01:14:57	12	-2412	109	P	-t	-1.5315	0.0642	65.0S	42.1W	0	202		
9052	453	1805 Jan 30	18:57:01	12	-2411	147	P	t-	1.4651	0.1675	62.7N	152.8W	0	131		
9053	453	1805 Jun 26	23:27:40	12	-2406	114	P	-t	1.0462	0.9357	65.5N	9.9W	0	343		
9054	453	1805 Jul 26	06:14:19	12	-2405	152	Pb	t-	-1.4571	0.1405	63.2S	42.8E	0	42		
9055	453	1805 Dec 21	00:17:38	12	-2400	119	A	-p	-0.8751	0.9134	83.1S	143.8W	29	317	692	06m26s
9056	453	1806 Jun 16	16:24:27	12	-2394	124	T	-n	0.3204	1.0604	42.2N	64.6W	71	184	210	04m55s
9057	453	1806 Dec 10	02:19:40	12	-2388	129	A	nn	-0.1627	0.9591	32.4S	143.4E	80	360	151	04m32s
9058	453	1807 Jun 06	05:18:31	12	-2382	134	H	p-	-0.4577	1.0055	4.7S	100.4E	63	359	21	00m38s
9059	453	1807 Nov 29	11:42:09	12	-2376	139	H	p-	0.5377	1.0135	11.1N	3.9E	57	184	55	01m26s
9060	453	1808 May 25	11:02:35	12	-2370	144	P	t-	-1.2665	0.5064	68.4S	27.8E	0	347		

Cat Num	Canon Plate	Calendar Date	TD of Greatest Eclipse	ΔT s	Luna Num	Saros Num	Ecl. Type	QLE	Gamma	Ecl. Mag.	Lat. °	Long. °	Sun Alt °	Sun Azm °	Path Width km	Central Line Dur.
9061	454	1808 Oct 19	16:55:30	12	-2365	111	P	-t	-1.4443	0.1687	71.3S	160.8E	0	123		
9062	454	1808 Nov 18	02:30:03	12	-2364	149	P	t-	1.1874	0.6564	69.2N	162.6E	0	202		
9063	454	1809 Apr 14	20:07:11	12	-2359	116	A	-p	0.8742	0.9429	65.8N	157.3W	29	139	435	04m35s
9064	454	1809 Oct 09	07:38:42	12	-2353	121	T	-p	-0.7905	1.0137	55.1S	38.4E	37	30	77	01m02s
9065	454	1810 Apr 04	01:41:19	12	-2347	126	A	nn	0.1031	0.9967	11.1N	153.8E	84	163	12	00m21s
9066	454	1810 Sep 28	16:37:25	12	-2341	131	A	nn	-0.0696	0.9681	5.8S	72.8W	86	18	115	03m45s
9067	454	1811 Mar 24	14:12:13	12	-2335	136	T	p-	-0.6190	1.0416	35.2S	18.0W	52	338	176	03m27s
9068	454	1811 Sep 17	18:43:45	12	-2329	141	A	p-	0.6798	0.9345	43.0N	85.9W	47	204	330	06m51s
9069	454	1812 Feb 12	20:28:40	12	-2324	108	P	-t	1.3545	0.3422	70.7N	168.8W	0	136		
9070	454	1812 Mar 13	06:19:30	12	-2323	146	P	t-	-1.2913	0.4594	71.9S	173.3W	0	260		
9071	454	1812 Aug 07	05:15:50	12	-2318	113	P	-t	-1.4205	0.2343	70.0S	67.0E	0	34		
9072	454	1812 Sep 05	19:04:10	12	-2317	151	P	t-	1.3939	0.2874	71.8N	4.5E	0	292		
9073	454	1813 Feb 01	08:58:27	12	-2312	118	A	-p	0.7152	0.9820	27.9N	40.4E	44	169	91	01m53s
9074	454	1813 Jul 27	14:55:35	12	-2306	123	T	-p	-0.6006	1.0341	17.4S	47.4W	53	8	144	03m27s
9075	454	1814 Jan 21	14:24:47	12	-2300	128	A	nn	0.0253	0.9350	18.6S	33.4W	89	173	242	08m28s
9076	454	1814 Jul 17	06:30:29	12	-2294	133	T	n-	0.1641	1.0774	30.9N	84.7E	80	185	254	06m33s
9077	454	1815 Jan 10	13:57:06	12	-2288	138	A	p-	-0.6626	0.9143	63.7S	23.6W	48	355	438	07m55s
9078	454	1815 Jul 06	23:43:07	12	-2282	143	T	t-	0.9062	1.0593	88.1N	162.7W	25	192	470	03m13s
9079	454	1815 Dec 30	14:38:39	12	-2276	148	P	t-	-1.3129	0.4273	66.7S	146.4E	0	175		
9080	454	1816 May 27	03:13:24	12	-2271	115	A	-t	-0.9492	0.9791	48.0S	153.5E	18	338	238	01m54s
9081	455	1816 Nov 19	10:17:23	12	-2265	120	T	-p	0.8408	1.0233	35.0N	41.5E	33	202	145	02m00s
9082	455	1817 May 16	06:58:14	12	-2259	125	A	nn	-0.2049	0.9483	7.9N	78.5E	78	341	194	06m30s
9083	455	1817 Nov 09	01:53:53	12	-2253	130	T	-n	0.1487	1.0536	8.9S	150.9E	82	202	179	04m42s
9084	455	1818 May 05	07:15:49	12	-2247	135	A	p-	0.5440	0.9464	45.8N	52.5E	57	148	233	05m05s
9085	455	1818 Oct 29	17:07:10	12	-2241	140	T	p-	-0.5524	1.0241	43.1S	99.4W	56	34	98	01m51s
9086	455	1819 Mar 25	23:44:30	12	-2236	107	P	-t	-1.4722	0.1329	61.2S	87.9W	0	274		
9087	455	1819 Apr 24	11:31:59	12	-2235	145	P	t-	1.2579	0.5225	61.7N	108.0W	0	62		
9088	455	1819 Sep 19	13:03:47	12	-2230	112	Pe	-t	1.5258	0.0595	61.0N	75.6E	0	274		
9089	455	1819 Oct 19	03:27:17	12	-2229	150	P	t-	-1.3226	0.4085	61.5S	16.4E	0	111		
9090	455	1820 Mar 14	13:37:15	12	-2224	117	T	-p	-0.7199	1.0467	41.5S	5.7E	44	320	220	03m20s
9091	455	1820 Sep 07	13:59:58	11	-2218	122	A	-p	0.8251	0.9329	51.6N	8.7E	34	229	432	05m49s
9092	455	1821 Mar 04	05:50:13	11	-2212	127	T	nn	-0.0284	1.0506	8.0S	96.3E	88	333	168	04m14s
9093	455	1821 Aug 27	15:19:42	11	-2206	132	A	nn	0.0671	0.9661	13.6N	47.8W	86	207	123	03m38s
9094	455	1822 Feb 21	19:40:40	11	-2200	137	A	p-	0.6914	0.9996	28.6N	132.3W	46	150	2	00m02s
9095	455	1822 Aug 16	23:14:34	11	-2194	142	T	p-	-0.6904	1.0173	26.1S	173.5E	46	27	80	01m35s
9096	455	1823 Jan 12	09:20:12	11	-2189	109	P	-t	-1.5413	0.0484	64.0S	173.0W	0	212		
9097	455	1823 Feb 11	03:03:02	11	-2188	147	P	t-	1.4546	0.1856	62.0N	76.7E	0	122		
9098	455	1823 Jul 08	06:56:28	10	-2183	114	P	-t	1.1182	0.7958	64.6N	132.0W	0	333		
9099	455	1823 Aug 06	13:45:42	10	-2182	152	P	t-	-1.3871	0.2753	62.5S	79.3W	0	51		
9100	455	1824 Jan 01	08:21:09	10	-2177	119	A	-p	-0.8821	0.9139	79.9S	116.2E	28	295	705	06m21s
9101	456	1824 Jun 26	23:46:33	10	-2171	124	T	-p	0.3960	1.0578	46.6N	171.4W	66	190	207	04m31s
9102	456	1824 Dec 20	10:40:36	10	-2165	129	Am	nn	-0.1685	0.9610	33.3S	20.4E	80	354	144	04m15s
9103	456	1825 Jun 16	12:19:03	10	-2159	134	H	p-	-0.3812	1.0036	1.0N	6.0W	68	3	13	00m25s
9104	456	1825 Dec 09	20:21:45	9	-2153	139	H2	p-	0.5296	1.0148	9.2N	127.4W	58	180	60	01m34s
9105	456	1826 Jun 05	17:39:05	9	-2147	144	P	t-	-1.1887	0.6407	67.4S	82.5W	0	357		
9106	456	1826 Oct 31	01:20:38	9	-2142	111	P	-t	-1.4696	0.1222	70.6S	21.2E	0	136		
9107	456	1826 Nov 29	11:14:08	9	-2141	149	P	t-	1.1764	0.6770	68.2N	19.9E	0	191		
9108	456	1827 Apr 26	03:11:14	9	-2136	116	A	-p	0.9316	0.9458	74.8N	73.4E	21	118	559	03m53s
9109	456	1827 Oct 20	15:42:05	8	-2130	121	H	-p	-0.8251	1.0070	62.3S	87.6W	34	34	43	00m30s
9110	456	1828 Apr 14	09:19:38	8	-2124	126	Hm	nn	0.1498	1.0029	17.9N	37.7E	81	164	10	00m18s
9111	456	1828 Oct 09	00:07:47	8	-2118	131	A	nn	-0.1139	0.9623	12.5S	173.0E	83	17	137	04m26s
9112	456	1829 Apr 03	22:18:36	8	-2112	136	T	p-	-0.5803	1.0474	28.5S	142.6W	54	341	192	04m05s
9113	456	1829 Sep 28	01:46:53	8	-2106	141	A	p-	0.6243	0.9317	34.9N	164.3E	51	202	323	07m43s
9114	456	1830 Feb 23	05:04:13	7	-2101	108	P	-t	1.3716	0.3100	71.3N	49.0E	0	123		
9115	456	1830 Mar 24	14:38:43	7	-2100	146	P	t-	-1.2622	0.5148	72.0S	47.7E	0	274		
9116	456	1830 Aug 18	12:13:35	7	-2095	113	P	-t	-1.4866	0.1171	70.7S	50.2W	0	46		
9117	456	1830 Sep 17	02:08:12	7	-2094	151	P	t-	1.3325	0.3930	72.1N	115.6W	0	278		
9118	456	1831 Feb 12	17:21:45	7	-2089	118	A	-p	0.7288	0.9807	31.9N	88.3W	43	165	100	01m57s
9119	456	1831 Aug 07	22:15:59	7	-2083	123	T	-p	-0.6691	1.0349	24.9S	160.9W	48	12	158	03m20s
9120	456	1832 Feb 01	22:30:14	7	-2077	128	A	nn	0.0355	0.9344	15.3S	154.4W	88	169	245	08m35s

Cat Num	Canon Plate	Calendar Date	TD of Greatest Eclipse	ΔT s	Luna Num	Saros Num	Ecl. Type	QLE	Gamma	Ecl. Mag.	Lat. °	Long. °	Sun Alt °	Sun Azm °	Path Width km	Central Line Dur.
9121	457	1832 Jul 27	14:01:06	6	-2071	133	T	nn	0.0919	1.0776	24.5N	27.9W	85	188	252	06m46s
9122	457	1833 Jan 20	21:56:55	6	-2065	138	A	p-	-0.6530	0.9155	60.6S	137.4W	49	347	426	07m59s
9123	457	1833 Jul 17	07:08:02	6	-2059	143	T	p-	0.8348	1.0591	77.5N	92.5E	33	200	357	03m29s
9124	457	1834 Jan 09	22:55:31	6	-2053	148	P	t-	-1.3043	0.4418	67.8S	11.3E	0	186		
9125	457	1834 Jun 07	10:08:38	6	-2048	115	P	-t	-1.0291	0.9295	64.6S	55.4E	0	334		
9126	457	1834 Nov 30	18:56:35	6	-2042	120	T	-p	0.8498	1.0233	34.9N	91.6W	32	197	150	02m02s
9127	457	1835 May 27	13:35:42	6	-2036	125	A	np	-0.2846	0.9486	5.3N	20.2W	73	345	196	06m44s
9128	457	1835 Nov 20	10:31:58	6	-2030	130	T	-n	0.1649	1.0510	10.7S	21.6E	81	198	171	04m35s
9129	457	1836 May 15	14:01:39	5	-2024	135	A	p-	0.4700	0.9509	45.1N	44.4W	62	153	203	04m47s
9130	457	1836 Nov 09	01:29:26	5	-2018	140	T	p-	-0.5327	1.0191	46.1S	136.8E	58	31	77	01m28s
9131	457	1837 Apr 05	07:35:30	5	-2013	107	Pe	-t	-1.5081	0.0651	61.3S	145.6E	0	283		
9132	457	1837 May 04	18:48:28	5	-2012	145	P	t-	1.1934	0.6381	62.3N	133.9E	0	54		
9133	457	1837 Oct 29	11:19:24	5	-2006	150	P	t-	-1.2967	0.4542	61.9S	110.5W	0	120		
9134	457	1838 Mar 25	21:52:16	5	-2001	117	T	-p	-0.7525	1.0505	39.7S	118.3W	41	321	249	03m39s
9135	457	1838 Sep 18	20:55:56	5	-1995	122	A	-p	0.8868	0.9289	52.4N	90.6W	27	232	562	06m06s
9136	457	1839 Mar 15	14:13:42	5	-1989	127	T	nn	-0.0558	1.0520	5.1S	29.5W	87	331	172	04m20s
9137	457	1839 Sep 07	22:23:26	5	-1983	132	Am	nn	0.1325	0.9661	12.8N	152.7W	82	209	123	03m34s
9138	457	1840 Mar 04	03:58:22	5	-1977	137	A	p-	0.6728	0.9995	30.6N	101.7E	48	147	2	00m03s
9139	457	1840 Aug 27	06:37:32	5	-1971	142	T	p-	-0.6223	1.0195	24.3S	62.9E	51	29	83	01m45s
9140	457	1841 Jan 22	17:24:15	5	-1966	109	P	-t	-1.5516	0.0316	63.1S	56.6E	0	222		
9141	458	1841 Feb 21	11:03:56	5	-1965	147	P	t-	1.4406	0.2095	61.5N	52.4W	0	113		
9142	458	1841 Jul 18	14:25:14	5	-1960	114	P	-t	1.1903	0.6556	63.7N	106.2E	0	324		
9143	458	1841 Aug 16	21:20:24	5	-1959	152	P	t-	-1.3193	0.4059	61.9S	158.0E	0	60		
9144	458	1842 Jan 11	16:25:41	5	-1954	119	A	-p	-0.8882	0.9151	75.8S	1.4E	27	288	710	06m15s
9145	458	1842 Jul 08	07:06:27	6	-1948	124	T	-p	0.4727	1.0543	50.1N	83.6E	62	198	204	04m05s
9146	458	1842 Dec 31	19:04:24	6	-1942	129	A	nn	-0.1727	0.9634	33.1S	103.2W	80	349	135	03m54s
9147	458	1843 Jun 27	19:17:03	6	-1936	134	H	nn	-0.3037	1.0011	5.9N	111.0W	72	7	4	00m07s
9148	458	1843 Dec 21	05:03:26	6	-1930	139	T	p-	0.5227	1.0165	8.0N	101.0E	58	175	66	01m43s
9149	458	1844 Jun 16	00:13:22	6	-1924	144	P	t-	-1.1092	0.7778	66.4S	168.3E	0	8		
9150	458	1844 Nov 10	09:51:45	6	-1919	111	P	-t	-1.4902	0.0847	69.8S	119.3W	0	149		
9151	458	1844 Dec 09	20:01:39	6	-1918	149	P	t-	1.1682	0.6924	67.1N	123.0W	0	179		
9152	458	1845 May 06	10:09:00	6	-1913	116	An	-t	0.9945	0.9462	73.4N	110.6W	4	41	-	03m15s
9153	458	1845 Oct 30	23:51:58	6	-1907	121	H	-p	-0.8538	1.0005	69.1S	144.5E	31	39	3	00m02s
9154	458	1846 Apr 25	16:50:30	6	-1901	126	H	nn	0.2038	1.0088	24.8N	76.2W	78	165	31	00m53s
9155	458	1846 Oct 20	07:46:12	6	-1895	131	A	nn	-0.1506	0.9567	18.7S	57.3E	81	16	159	05m05s
9156	458	1847 Apr 15	06:16:13	7	-1889	136	T	p-	-0.5339	1.0530	21.6S	95.0E	58	343	206	04m44s
9157	458	1847 Oct 09	09:00:23	7	-1883	141	A	p-	0.5774	0.9290	27.7N	52.8E	55	199	323	08m35s
9158	458	1848 Mar 05	13:31:35	7	-1878	108	P	-t	1.3950	0.2662	71.8N	91.7W	0	109		
9159	458	1848 Apr 03	22:49:07	7	-1877	146	P	t-	-1.2264	0.5834	71.8S	89.0W	0	288		
9160	458	1848 Aug 28	19:18:22	7	-1872	113	Pe	-t	-1.5475	0.0090	71.3S	169.6W	0	59		
9161	459	1848 Sep 27	09:21:19	7	-1871	151	P	t-	1.2774	0.4875	72.2N	121.9E	0	264		
9162	459	1849 Feb 23	01:38:09	7	-1866	118	A	-p	0.7475	0.9796	36.7N	144.3E	41	161	108	01m58s
9163	459	1849 Aug 18	05:40:49	7	-1860	123	T	-p	-0.7343	1.0349	32.9S	83.5E	43	16	172	03m07s
9164	459	1850 Feb 12	06:29:37	7	-1854	128	A	nn	0.0503	0.9345	11.0S	85.6E	87	166	245	08m35s
9165	459	1850 Aug 07	21:33:54	7	-1848	133	T	nn	0.0215	1.0769	17.7N	141.8W	89	191	249	06m50s
9166	459	1851 Feb 01	05:54:27	7	-1842	138	A	p-	-0.6413	0.9175	56.4S	106.9E	50	342	409	08m01s
9167	459	1851 Jul 28	14:33:42	7	-1836	143	T	p-	0.7644	1.0577	68.0N	19.6W	40	201	296	03m41s
9168	459	1852 Jan 21	07:12:16	7	-1830	148	P	t-	-1.2948	0.4577	68.9S	124.3W	0	198		
9169	459	1852 Jun 17	16:59:50	7	-1825	115	P	-t	-1.1111	0.7828	65.6S	57.3W	0	344		
9170	459	1852 Dec 11	03:40:44	7	-1819	120	T	-p	0.8551	1.0237	35.2N	133.9E	31	191	156	02m05s
9171	459	1853 Jun 06	20:07:21	7	-1813	125	A	-p	-0.3686	0.9486	1.5N	117.9W	68	349	203	06m59s
9172	459	1853 Nov 30	19:15:39	7	-1807	130	T	-n	0.1763	1.0485	12.0S	109.0W	80	194	164	04m28s
9173	459	1854 May 26	20:42:53	7	-1801	135	A	p-	0.3918	0.9551	43.3N	140.1W	67	159	178	04m32s
9174	459	1854 Nov 20	09:56:58	7	-1795	140	H3	p-	-0.5179	1.0144	48.9S	12.7E	59	27	57	01m07s
9175	459	1855 May 16	02:01:12	7	-1789	145	P	t-	1.1249	0.7624	62.9N	16.6E	0	45		
9176	459	1855 Nov 09	19:17:51	7	-1783	150	P	t-	-1.2767	0.4892	62.5S	121.0E	0	129		
9177	459	1856 Apr 05	06:01:01	7	-1778	117	T	-p	-0.7906	1.0539	39.1S	119.2E	38	323	285	03m56s
9178	459	1856 Sep 29	03:59:44	7	-1772	122	A	-p	0.9420	0.9246	54.3N	169.1E	19	236	831	06m21s
9179	459	1857 Mar 25	22:29:38	7	-1766	127	T	-n	-0.0892	1.0534	2.4S	153.4W	85	331	177	04m28s
9180	459	1857 Sep 18	05:36:05	7	-1760	132	A	nn	0.1912	0.9659	11.6N	100.0E	79	210	125	03m34s

Cat Num	Canon Plate	Calendar Date	TD of Greatest Eclipse	ΔT s	Luna Num	Saros Num	Ecl. Type	QLE	Gamma	Ecl. Mag.	Lat. °	Long. °	Sun Alt °	Sun Azm °	Path Width km	Central Line Dur.
9181	460	1858 Mar 15	12:05:28	7	-1754	137	A	p-	0.6461	0.9996	32.7N	20.9W	50	145	2	00m02s
9182	460	1858 Sep 07	14:09:29	7	-1748	142	T	p-	-0.5609	1.0210	23.9S	49.8W	56	31	85	01m50s
9183	460	1859 Feb 03	01:22:42	7	-1743	109	Pe	-t	-1.5659	0.0077	62.4S	72.1W	0	232		
9184	460	1859 Mar 04	18:54:49	7	-1742	147	P	t-	1.4192	0.2461	61.2N	178.8W	0	103		
9185	460	1859 Jul 29	21:56:57	7	-1737	114	P	-t	1.2598	0.5205	63.0N	16.0W	0	315		
9186	460	1859 Aug 28	05:02:00	7	-1736	152	P	t-	-1.2569	0.5261	61.5S	33.7E	0	69		
9187	460	1860 Jan 23	00:27:31	8	-1731	119	A	-p	-0.8969	0.9168	71.8S	117.2W	26	286	719	06m07s
9188	460	1860 Jul 18	14:26:24	8	-1725	124	T	-p	0.5487	1.0500	52.5N	20.3W	56	205	198	03m39s
9189	460	1861 Jan 11	03:29:23	8	-1719	129	A	nn	-0.1766	0.9664	31.8S	132.7E	80	344	123	03m30s
9190	460	1861 Jul 08	02:10:26	8	-1713	134	A	nn	-0.2231	0.9979	10.0N	145.8E	77	12	7	00m14s
9191	460	1861 Dec 31	13:49:06	8	-1707	139	T	p-	0.5187	1.0186	7.8N	31.6W	59	171	74	01m55s
9192	460	1862 Jun 27	06:42:21	8	-1701	144	P	t-	-1.0252	0.9222	65.4S	60.8E	0	18		
9193	460	1862 Nov 21	18:29:48	7	-1696	111	P	-t	-1.5052	0.0580	68.8S	99.1E	0	161		
9194	460	1862 Dec 21	04:53:03	7	-1695	149	P	t-	1.1633	0.7016	66.0N	93.6E	0	168		
9195	460	1863 May 17	17:00:45	7	-1690	116	P	-t	1.0627	0.8606	69.2N	126.8E	0	22		
9196	460	1863 Nov 11	08:09:03	7	-1684	121	A	-p	-0.8760	0.9943	75.4S	15.1E	28	43	42	00m22s
9197	460	1864 May 06	00:16:48	6	-1678	126	H	-n	0.2622	1.0146	31.6N	171.5E	75	168	52	01m25s
9198	460	1864 Oct 30	15:30:31	6	-1672	131	A	nn	-0.1816	0.9514	24.3S	59.3W	79	14	181	05m41s
9199	460	1865 Apr 25	14:08:34	6	-1666	136	T	p-	-0.4826	1.0584	14.8S	25.8W	61	346	219	05m23s
9200	460	1865 Oct 19	16:21:14	5	-1660	141	A	p-	0.5366	0.9263	21.3N	60.2W	57	196	326	09m27s
9201	461	1866 Mar 16	21:51:25	5	-1655	108	P	-t	1.4241	0.2114	72.0N	129.2E	0	95		
9202	461	1866 Apr 15	06:51:40	5	-1654	146	P	t-	-1.1846	0.6637	71.4S	136.6E	0	302		
9203	461	1866 Oct 08	16:44:22	4	-1648	151	P	t-	1.2296	0.5693	71.9N	3.0W	0	250		
9204	461	1867 Mar 06	09:46:48	4	-1643	118	A	-p	0.7716	0.9787	42.3N	18.4E	39	157	118	01m57s
9205	461	1867 Aug 29	13:13:07	3	-1637	123	T	-p	-0.7940	1.0344	41.1S	34.9W	37	21	189	02m51s
9206	461	1868 Feb 23	14:21:31	3	-1631	128	A	nn	0.0706	0.9348	6.1S	33.0W	86	164	244	08m30s
9207	461	1868 Aug 18	05:12:10	2	-1625	133	Tm	nn	-0.0443	1.0756	10.6N	102.2E	88	14	245	06m47s
9208	461	1869 Feb 11	13:46:39	2	-1619	138	A	p-	-0.6251	0.9201	51.3S	9.7W	51	339	387	08m02s
9209	461	1869 Aug 07	22:01:05	1	-1613	143	T	p-	0.6960	1.0551	59.1N	133.2W	46	202	254	03m48s
9210	461	1870 Jan 31	15:26:25	1	-1607	148	P	t-	-1.2829	0.4781	69.9S	100.0E	0	210		
9211	461	1870 Jun 28	23:46:43	0	-1602	115	P	-t	-1.1949	0.6335	66.6S	169.4W	0	354		
9212	461	1870 Jul 28	11:02:31	0	-1601	153	Pb	t-	1.5044	0.0742	69.2N	170.9E	0	336		
9213	461	1870 Dec 22	12:27:33	-0	-1596	120	T	-p	0.8585	1.0248	35.7N	1.5W	31	186	165	02m11s
9214	461	1871 Jun 18	02:35:02	-1	-1590	125	A	-p	-0.4550	0.9481	3.5S	144.7E	63	353	214	07m14s
9215	461	1871 Dec 12	04:03:38	-1	-1584	130	T	-n	0.1836	1.0465	12.7S	119.4E	80	190	157	04m23s
9216	461	1872 Jun 06	03:20:03	-1	-1578	135	A	p-	0.3095	0.9590	40.5N	124.8E	72	166	157	04m20s
9217	461	1872 Nov 30	18:29:33	-2	-1572	140	H	p-	-0.5081	1.0099	51.2S	111.8W	59	22	40	00m47s
9218	461	1873 May 26	09:08:56	-2	-1566	145	P	t-	1.0513	0.8971	63.7N	99.6W	0	35		
9219	461	1873 Nov 20	03:22:52	-2	-1560	150	P	t-	-1.2625	0.5138	63.2S	9.5W	0	138		
9220	461	1874 Apr 16	14:00:53	-3	-1555	117	T	-p	-0.8364	1.0569	39.9S	0.9W	33	325	335	04m11s
9221	462	1874 Oct 10	11:13:33	-3	-1549	122	An	-t	0.9889	0.9193	58.6N	72.0E	7	244	-	06m28s
9222	462	1875 Apr 06	06:37:26	-3	-1543	127	T	-n	-0.1292	1.0547	0.2S	84.8E	83	332	182	04m37s
9223	462	1875 Sep 29	12:58:09	-4	-1537	132	A	nn	0.2427	0.9656	10.0N	10.1W	76	209	127	03m36s
9224	462	1876 Mar 25	20:05:06	-4	-1531	137	A	p-	0.6142	0.9999	34.8N	141.1W	52	144	1	00m01s
9225	462	1876 Sep 17	21:49:15	-4	-1525	142	T	p-	-0.5054	1.0220	24.6S	164.5W	60	32	86	01m53s
9226	462	1877 Mar 15	02:38:09	-4	-1519	147	P	t-	1.3924	0.2917	61.0N	56.7E	0	94		
9227	462	1877 Aug 09	05:30:24	-4	-1514	114	P	-t	1.3277	0.3889	62.3N	138.6W	0	306		
9228	462	1877 Sep 07	12:48:42	-4	-1513	152	P	t-	-1.1985	0.6382	61.2S	91.8W	0	78		
9229	462	1878 Feb 02	08:27:52	-5	-1508	119	A	-p	-0.9071	0.9191	67.9S	122.4E	24	286	729	05m59s
9230	462	1878 Jul 29	21:47:18	-5	-1502	124	T	-p	0.6232	1.0450	53.8N	124.0W	51	213	191	03m11s
9231	462	1879 Jan 22	11:53:08	-5	-1496	129	A	nn	-0.1824	0.9700	29.8S	8.5E	79	340	110	03m03s
9232	462	1879 Jul 19	09:04:32	-5	-1490	134	Am	nn	-0.1439	0.9942	13.0N	42.9E	82	16	20	00m39s
9233	462	1880 Jan 11	22:34:25	-5	-1484	139	T	p-	0.5136	1.0212	8.3N	164.1W	59	166	84	02m07s
9234	462	1880 Jul 07	13:10:28	-5	-1478	144	A	t-	-0.9406	0.9441	46.4S	33.4W	19	17	611	05m47s
9235	462	1880 Dec 02	03:11:33	-5	-1473	111	P	-t	-1.5172	0.0369	67.8S	42.9W	0	173		
9236	462	1880 Dec 31	13:45:01	-5	-1472	149	P	t-	1.1591	0.7096	65.0N	49.5W	0	158		
9237	462	1881 May 27	23:48:41	-5	-1467	116	P	-t	1.1345	0.7370	68.2N	13.3E	0	10		
9238	462	1881 Nov 21	16:31:10	-5	-1461	121	A	-p	-0.8931	0.9887	81.2S	114.5W	26	46	90	00m43s
9239	462	1882 May 17	07:36:27	-5	-1455	126	T	-n	0.3269	1.0200	38.4N	61.6E	71	171	72	01m50s
9240	462	1882 Nov 10	23:22:21	-6	-1449	131	A	-n	-0.2056	0.9465	29.2S	177.0W	78	11	201	06m14s

Cat Num	Canon Plate	Calendar Date	TD of Greatest Eclipse	ΔT s	Luna Num	Saros Num	Ecl. Type	QLE	Gamma	Ecl. Mag.	Lat. °	Long. °	Sun Alt °	Sun Azm °	Path Width km	Central Line Dur.
9241	463	1883 May 06	21:53:49	-6	-1443	136	T	p-	-0.4250	1.0634	8.1S	144.6W	65	349	229	05m58s
9242	463	1883 Oct 30	23:50:54	-6	-1437	141	A	p-	0.5030	0.9238	15.6N	174.9W	60	193	331	10m17s
9243	463	1884 Mar 27	06:02:11	-6	-1432	108	P	-t	1.4602	0.1436	72.0N	7.7W	0	81		
9244	463	1884 Apr 25	14:46:17	-6	-1431	146	P	t-	-1.1365	0.7563	70.7S	4.6E	0	315		
9245	463	1884 Oct 19	00:17:42	-6	-1425	151	P	t-	1.1892	0.6385	71.5N	130.2W	0	237		
9246	463	1885 Mar 16	17:45:43	-6	-1420	118	A	-p	0.8030	0.9778	48.9N	106.1W	36	153	132	01m55s
9247	463	1885 Sep 08	20:51:52	-6	-1414	123	T	-p	-0.8489	1.0332	49.6S	156.5W	32	27	211	02m31s
9248	463	1886 Mar 05	22:05:26	-6	-1408	128	A	nn	0.0970	0.9357	0.5S	150.1W	84	163	241	08m20s
9249	463	1886 Aug 29	12:55:23	-6	-1402	133	T	nn	-0.1059	1.0735	3.5N	15.3W	84	16	240	06m36s
9250	463	1887 Feb 22	21:33:04	-6	-1396	138	A	p-	-0.6040	0.9232	45.7S	126.5W	53	338	362	08m01s
9251	463	1887 Aug 19	05:32:05	-6	-1390	143	T	p-	0.6312	1.0518	50.6N	111.9E	51	202	221	03m50s
9252	463	1888 Feb 11	23:38:15	-6	-1384	148	P	t-	-1.2684	0.5029	70.7S	35.7W	0	223		
9253	463	1888 Jul 09	06:30:52	-6	-1379	115	P	-t	-1.2797	0.4832	67.6S	78.8E	0	4		
9254	463	1888 Aug 07	18:05:46	-6	-1378	153	P	t-	1.4369	0.1983	70.1N	53.0E	0	325		
9255	463	1889 Jan 01	21:16:50	-6	-1373	120	T	-p	0.8603	1.0262	36.7N	137.6W	30	181	175	02m17s
9256	463	1889 Jun 28	09:00:00	-6	-1367	125	A	-p	-0.5431	0.9471	9.6S	47.3E	57	357	232	07m22s
9257	463	1889 Dec 22	12:54:15	-6	-1361	130	T	-n	0.1888	1.0449	12.7S	12.8W	79	185	152	04m18s
9258	463	1890 Jun 17	09:55:05	-6	-1355	135	A	nn	0.2246	0.9625	36.5N	29.3E	77	172	140	04m09s
9259	463	1890 Dec 12	03:05:28	-6	-1349	140	H	p-	-0.5016	1.0059	52.8S	123.9E	60	15	24	00m28s
9260	463	1891 Jun 06	16:15:36	-6	-1343	145	A	t-	0.9754	0.9981	74.5N	163.8E	12	45	33	00m06s
9261	464	1891 Dec 01	11:31:08	-6	-1337	150	P	t-	-1.2515	0.5326	64.1S	140.9W	0	148		
9262	464	1892 Apr 26	21:55:20	-6	-1332	117	T	-p	-0.8870	1.0591	42.5S	119.4W	27	327	414	04m19s
9263	464	1892 Oct 20	18:36:06	-6	-1326	122	P	-t	1.0286	0.9054	61.4N	33.3W	0	247		
9264	464	1893 Apr 16	14:36:11	-6	-1320	127	T	-n	-0.1764	1.0556	1.3N	34.6W	80	334	186	04m47s
9265	464	1893 Oct 09	20:30:22	-6	-1314	132	A	nn	0.2866	0.9652	8.1N	123.0W	73	208	130	03m41s
9266	464	1894 Apr 06	03:53:41	-6	-1308	137	H	p-	0.5740	1.0001	36.7N	102.4E	55	144	1	00m01s
9267	464	1894 Sep 29	05:39:02	-6	-1302	142	T	p-	-0.4573	1.0226	26.1S	78.5E	63	32	85	01m55s
9268	464	1895 Mar 26	10:09:33	-6	-1296	147	P	t-	1.3565	0.3531	61.0N	64.8W	0	85		
9269	464	1895 Aug 20	13:09:16	-6	-1291	114	P	-t	1.3911	0.2665	61.8N	97.7E	0	297		
9270	464	1895 Sep 18	20:44:01	-6	-1290	152	P	t-	-1.1469	0.7369	61.0S	140.7E	0	86		
9271	464	1896 Feb 13	16:23:13	-6	-1285	119	A	-p	-0.9220	0.9218	64.6S	3.5E	22	287	761	05m48s
9272	464	1896 Aug 09	05:09:00	-6	-1279	124	T	-p	0.6964	1.0392	54.4N	132.2E	46	220	182	02m43s
9273	464	1897 Feb 01	20:15:15	-6	-1273	129	A	nn	-0.1903	0.9742	27.1S	115.7W	79	336	94	02m34s
9274	464	1897 Jul 29	15:56:58	-5	-1267	134	A	nn	-0.0640	0.9899	15.3N	59.0W	86	20	35	01m05s
9275	464	1898 Jan 22	07:19:12	-5	-1261	139	T	p-	0.5079	1.0244	9.5N	63.6E	59	162	96	02m21s
9276	464	1898 Jul 18	19:36:54	-4	-1255	144	A	p-	-0.8546	0.9450	35.7S	130.1W	31	19	385	06m11s
9277	464	1898 Dec 13	11:58:13	-4	-1250	111	P	-t	-1.5253	0.0231	66.8S	174.5E	0	184		
9278	464	1899 Jan 11	22:38:02	-4	-1249	149	P	t-	1.1558	0.7158	64.0N	167.5E	0	148		
9279	464	1899 Jun 08	06:33:43	-4	-1244	116	P	-t	1.2089	0.6076	67.2N	98.9W	0	360		
9280	464	1899 Dec 03	00:57:28	-3	-1238	121	A	-p	-0.9061	0.9836	86.6S	121.5E	25	43	140	01m01s
9281	465	1900 May 28	14:53:56	-2	-1232	126	T	-n	0.3943	1.0249	44.8N	46.5W	67	175	92	02m10s
9282	465	1900 Nov 22	07:19:43	-2	-1226	131	A	-n	-0.2245	0.9421	33.1S	64.8E	77	7	220	06m42s
9283	465	1901 May 18	05:33:48	-1	-1220	136	T	n-	-0.3626	1.0680	1.7S	98.4E	69	353	238	06m29s
9284	465	1901 Nov 11	07:28:21	-0	-1214	141	A	p-	0.4758	0.9216	10.8N	68.9E	62	190	336	11m01s
9285	465	1902 Apr 08	14:05:06	0	-1209	108	Pe	-t	1.5024	0.0643	71.7N	142.4W	0	67		
9286	465	1902 May 07	22:34:16	0	-1208	146	P	t-	-1.0831	0.8593	70.0S	125.1W	0	327		
9287	465	1902 Oct 31	08:00:18	1	-1202	151	P	t-	1.1556	0.6960	70.8N	100.8E	0	223		
9288	465	1903 Mar 29	01:35:23	2	-1197	118	A	-p	0.8413	0.9767	56.2N	130.3E	32	147	153	01m53s
9289	465	1903 Sep 21	04:39:52	2	-1191	123	T	-p	-0.8967	1.0316	58.0S	77.2E	26	35	241	02m12s
9290	465	1904 Mar 17	05:40:44	3	-1185	128	A	nn	0.1299	0.9367	5.6N	94.7E	82	162	237	08m07s
9291	465	1904 Sep 09	20:44:21	3	-1179	133	T	-n	-0.1625	1.0709	3.7S	134.5W	81	17	234	06m20s
9292	465	1905 Mar 06	05:12:26	4	-1173	138	A	p-	-0.5768	0.9269	39.5S	117.4E	55	338	334	07m58s
9293	465	1905 Aug 30	13:07:26	5	-1167	143	T	p-	0.5708	1.0477	42.5N	4.3W	55	202	192	03m46s
9294	465	1906 Feb 23	07:43:20	5	-1161	148	P	t-	-1.2479	0.5386	71.4S	170.3W	0	237		
9295	465	1906 Jul 21	13:14:19	6	-1156	115	P	-t	-1.3637	0.3355	68.6S	33.3W	0	15		
9296	465	1906 Aug 20	01:12:50	6	-1155	153	P	t-	1.3731	0.3147	70.8N	66.4W	0	313		
9297	465	1907 Jan 14	06:05:43	6	-1150	120	T	-p	0.8628	1.0281	38.3N	86.4E	30	175	189	02m25s
9298	465	1907 Jul 10	15:24:32	7	-1144	125	A	-p	-0.6313	0.9456	16.9S	50.9W	51	2	258	07m23s
9299	465	1908 Jan 03	21:45:22	8	-1138	130	T	-n	0.1934	1.0437	11.8S	145.1W	79	180	149	04m14s
9300	465	1908 Jun 28	16:29:51	8	-1132	135	A	nn	0.1389	0.9655	31.4N	67.2W	82	177	126	04m00s

Cat Num	Canon Plate	Calendar Date	TD of Greatest Eclipse	ΔT s	Luna Num	Saros Num	Ecl. Type	QLE	Gamma	Ecl. Mag.	Lat. °	Long. °	Sun Alt °	Sun Azm °	Path Width km	Central Line Dur.
9301	466	1908 Dec 23	11:44:28	9	-1126	140	H	n-	-0.4985	1.0024	53.4S	0.5W	60	8	10	00m12s
9302	466	1909 Jun 17	23:18:38	10	-1120	145	H	t-	0.8957	1.0065	82.9N	123.6E	26	110	51	00m24s
9303	466	1909 Dec 12	19:44:48	10	-1114	150	P	t-	-1.2456	0.5424	65.0S	86.0E	0	158		
9304	466	1910 May 09	05:42:13	11	-1109	117	T	-t	-0.9437	1.0600	48.2S	125.2E	19	328	594	04m15s
9305	466	1910 Nov 02	02:08:32	12	-1103	122	P	-t	1.0603	0.8515	61.9N	155.1W	0	238		
9306	466	1911 Apr 28	22:27:22	12	-1097	127	T	-n	-0.2294	1.0562	1.9N	151.9W	77	336	190	04m57s
9307	466	1911 Oct 22	04:13:02	13	-1091	132	A	-n	0.3224	0.9650	6.3N	121.4E	71	206	133	03m47s
9308	466	1912 Apr 17	11:34:22	14	-1085	137	H	p-	0.5280	1.0003	38.4N	11.3W	58	146	1	00m02s
9309	466	1912 Oct 10	13:36:14	14	-1079	142	T	p-	-0.4149	1.0229	28.1S	40.1W	65	32	85	01m55s
9310	466	1913 Apr 06	17:33:07	15	-1073	147	P	t-	1.3147	0.4244	61.2N	175.7E	0	77		
9311	466	1913 Aug 31	20:52:12	15	-1068	114	P	-t	1.4512	0.1513	61.5N	26.8W	0	288		
9312	466	1913 Sep 30	04:45:49	15	-1067	152	P	t-	-1.1005	0.8252	61.0S	11.6E	0	95		
9313	466	1914 Feb 25	00:13:01	16	-1062	119	A	-p	-0.9416	0.9248	62.1S	113.3W	19	287	839	05m35s
9314	466	1914 Aug 21	12:34:27	17	-1056	124	T	-p	0.7655	1.0328	54.5N	27.1E	40	227	170	02m14s
9315	466	1915 Feb 14	04:33:20	17	-1050	129	A	nn	-0.2024	0.9789	24.0S	120.7E	78	333	77	02m04s
9316	466	1915 Aug 10	22:52:25	18	-1044	134	A	nn	0.0124	0.9853	16.4N	161.4W	89	200	52	01m33s
9317	466	1916 Feb 03	16:00:21	18	-1038	139	T	p-	0.4987	1.0280	11.1N	67.7W	60	158	108	02m36s
9318	466	1916 Jul 30	02:06:10	19	-1032	144	A	p-	-0.7709	0.9447	29.0S	132.4E	39	22	313	06m24s
9319	466	1916 Dec 24	20:46:22	19	-1027	111	P	-t	-1.5321	0.0114	65.7S	32.1E	0	195		
9320	466	1917 Jan 23	07:28:31	19	-1026	149	P	t-	1.1508	0.7254	63.2N	25.6E	0	138		
9321	467	1917 Jun 19	13:16:21	20	-1021	116	P	-t	1.2857	0.4729	66.2N	150.1E	0	350		
9322	467	1917 Jul 19	02:42:42	20	-1020	154	Pb	t-	-1.5101	0.0863	63.7S	101.8E	0	36		
9323	467	1917 Dec 14	09:27:20	20	-1015	121	A	-t	-0.9157	0.9791	88.0S	124.7E	23	271	189	01m17s
9324	467	1918 Jun 08	22:07:43	20	-1009	126	T	-p	0.4658	1.0292	50.9N	152.0W	62	180	112	02m23s
9325	467	1918 Dec 03	15:22:02	21	-1003	131	A	-n	-0.2387	0.9383	36.1S	53.7W	76	3	236	07m06s
9326	467	1919 May 29	13:08:55	21	-997	136	T	n-	-0.2955	1.0719	4.4N	16.7W	73	356	244	06m51s
9327	467	1919 Nov 22	15:14:12	21	-991	141	A	p-	0.4549	0.9198	6.9N	48.9W	63	186	341	11m37s
9328	467	1920 May 18	06:14:55	21	-985	146	P	t-	-1.0239	0.9734	69.1S	107.7E	0	339		
9329	467	1920 Nov 10	15:52:15	22	-979	151	P	t-	1.1287	0.7420	69.9N	29.8W	0	211		
9330	467	1921 Apr 08	09:15:01	22	-974	118	A	-t	0.8869	0.9753	64.5N	5.6E	27	139	192	01m50s
9331	467	1921 Oct 01	12:35:58	22	-968	123	T	-p	-0.9383	1.0293	66.1S	56.1W	20	48	291	01m52s
9332	467	1922 Mar 28	13:05:26	23	-962	128	A	nn	0.1711	0.9381	12.3N	18.0W	80	162	233	07m50s
9333	467	1922 Sep 21	04:40:31	23	-956	133	T	-n	-0.2130	1.0678	10.7S	104.5E	78	18	226	05m59s
9334	467	1923 Mar 17	12:44:58	23	-950	138	A	p-	-0.5438	0.9310	33.0S	2.4E	57	339	305	07m51s
9335	467	1923 Sep 10	20:47:29	23	-944	143	T	p-	0.5149	1.0430	34.7N	121.8W	59	201	167	03m37s
9336	467	1924 Mar 05	15:44:20	24	-938	148	P	t-	-1.2232	0.5819	71.9S	55.6E	0	250		
9337	467	1924 Jul 31	19:58:20	24	-933	115	P	-t	-1.4459	0.1920	69.6S	146.0W	0	27		
9338	467	1924 Aug 30	08:23:00	24	-932	153	P	t-	1.3123	0.4245	71.5N	172.9E	0	300		
9339	467	1925 Jan 24	14:54:03	24	-927	120	T	-p	0.8661	1.0304	40.5N	49.6W	30	170	206	02m32s
9340	467	1925 Jul 20	21:48:42	24	-921	125	A	-p	-0.7193	0.9436	25.3S	150.0W	44	6	300	07m15s
9341	468	1926 Jan 14	06:36:58	24	-915	130	T	-n	0.1973	1.0430	10.1S	82.3E	79	176	147	04m11s
9342	468	1926 Jul 09	23:06:02	24	-909	135	A	nn	0.0538	0.9680	25.6N	165.1W	87	181	115	03m51s
9343	468	1927 Jan 03	20:22:53	24	-903	140	A	n-	-0.4956	0.9995	52.8S	124.8W	60	0	2	00m03s
9344	468	1927 Jun 29	06:23:27	24	-897	145	T	t-	0.8163	1.0128	78.1N	73.8E	35	167	77	00m50s
9345	468	1927 Dec 24	03:59:41	24	-891	150	P	t-	-1.2416	0.5490	66.1S	47.7W	0	169		
9346	468	1928 May 19	13:24:20	24	-886	117	T-	-t	-1.0048	1.0140	63.3S	22.5E	0	319	-	-
9347	468	1928 Jun 17	20:27:28	24	-885	155	Pb	t-	1.5107	0.0375	65.6N	70.6E	0	16		
9348	468	1928 Nov 12	09:48:24	24	-880	122	P	-t	1.0861	0.8078	62.6N	81.1E	0	229		
9349	468	1929 May 09	06:10:34	24	-874	127	T	-n	-0.2887	1.0562	1.6N	92.7E	73	339	193	05m07s
9350	468	1929 Nov 01	12:05:10	24	-868	132	A	-n	0.3514	0.9649	4.5N	3.1E	69	204	134	03m54s
9351	468	1930 Apr 28	19:03:34	24	-862	137	H	p-	0.4730	1.0003	39.4N	121.2W	62	149	1	00m01s
9352	468	1930 Oct 21	21:43:53	24	-856	142	T	p-	-0.3804	1.0230	30.5S	161.1W	67	31	84	01m55s
9353	468	1931 Apr 18	00:45:35	24	-850	147	P	t-	1.2643	0.5107	61.5N	58.9E	0	68		
9354	468	1931 Sep 12	04:41:25	24	-845	114	Pe	-t	1.5060	0.0471	61.2N	152.8W	0	280		
9355	468	1931 Oct 11	12:55:40	24	-844	152	P	t-	-1.0607	0.9005	61.2S	119.5W	0	104		
9356	468	1932 Mar 07	07:55:50	24	-839	119	A	-p	-0.9673	0.9277	60.7S	134.4E	14	285	1083	05m19s
9357	468	1932 Aug 31	20:03:41	24	-833	124	T	-p	0.8307	1.0257	54.5N	79.5W	34	232	155	01m45s
9358	468	1933 Feb 24	12:46:39	24	-827	129	A	nn	-0.2191	0.9841	20.8S	2.1W	77	331	58	01m32s
9359	468	1933 Aug 21	05:49:11	24	-821	134	A	nn	0.0869	0.9801	16.9N	95.9E	85	206	71	02m04s
9360	468	1934 Feb 14	00:38:41	24	-815	139	T	p-	0.4868	1.0321	13.2N	161.7E	61	155	123	02m53s

Cat Num	Canon Plate	Calendar Date	TD of Greatest Eclipse	ΔT s	Luna Num	Saros Num	Ecl. Type	QLE	Gamma	Ecl. Mag.	Lat. °	Long. °	Sun Alt °	Sun Azm °	Path Width km	Central Line Dur.
9361	469	1934 Aug 10	08:37:48	24	-809	144	A	p-	-0.6890	0.9436	24.5S	34.6E	46	25	280	06m33s
9362	469	1935 Jan 05	05:35:46	24	-804	111	Pe	-t	-1.5381	0.0013	64.7S	110.2W	0	205		
9363	469	1935 Feb 03	16:16:20	24	-803	149	P	t-	1.1438	0.7390	62.5N	115.4W	0	128		
9364	469	1935 Jun 30	19:59:46	24	-798	116	P	-t	1.3623	0.3375	65.2N	39.1E	0	340		
9365	469	1935 Jul 30	09:16:28	24	-797	154	P	t-	-1.4259	0.2315	62.9S	5.9W	0	45		
9366	469	1935 Dec 25	17:59:52	24	-792	121	A	-t	-0.9228	0.9752	83.5S	9.4E	22	258	234	01m30s
9367	469	1936 Jun 19	05:20:31	24	-786	126	T	-p	0.5389	1.0329	56.1N	104.7E	57	188	132	02m31s
9368	469	1936 Dec 13	23:28:12	24	-780	131	A	-n	-0.2493	0.9349	37.8S	172.6W	75	357	251	07m25s
9369	469	1937 Jun 08	20:41:02	24	-774	136	T	n-	-0.2253	1.0751	9.9N	130.5W	77	0	250	07m04s
9370	469	1937 Dec 02	23:05:45	24	-768	141	A	p-	0.4389	0.9184	4.0N	167.8W	64	182	344	12m00s
9371	469	1938 May 29	13:50:19	24	-762	146	T	t-	-0.9607	1.0552	52.7S	22.0W	16	354	675	04m05s
9372	469	1938 Nov 21	23:52:25	24	-756	151	P	t-	1.1077	0.7781	68.9N	162.0W	0	198		
9373	469	1939 Apr 19	16:45:53	24	-751	118	A	-t	0.9388	0.9731	73.1N	129.1W	20	118	285	01m49s
9374	469	1939 Oct 12	20:40:23	24	-745	123	T	-p	-0.9737	1.0266	72.8S	155.1E	12	74	418	01m32s
9375	469	1940 Apr 07	20:21:21	24	-739	128	A	nn	0.2190	0.9394	19.2N	128.5W	77	163	230	07m30s
9376	469	1940 Oct 01	12:44:06	25	-733	133	T	-n	-0.2573	1.0645	17.5S	18.2W	75	18	218	05m35s
9377	469	1941 Mar 27	20:08:08	25	-727	138	A	p-	-0.5025	0.9355	26.2S	110.9W	60	341	276	07m41s
9378	469	1941 Sep 21	04:34:03	25	-721	143	T	p-	0.4649	1.0379	27.3N	119.1E	62	200	143	03m22s
9379	469	1942 Mar 16	23:37:07	25	-715	148	P	t-	-1.1908	0.6393	72.2S	76.8W	0	264		
9380	469	1942 Aug 12	02:45:12	26	-710	115	Pe	-t	-1.5244	0.0561	70.4S	99.9E	0	39		
9381	470	1942 Sep 10	15:39:32	26	-709	153	P	t-	1.2571	0.5230	71.9N	50.0E	0	286		
9382	470	1943 Feb 04	23:38:10	26	-704	120	T	-p	0.8734	1.0331	43.6N	175.1E	29	165	229	02m39s
9383	470	1943 Aug 01	04:16:13	26	-698	125	A	-p	-0.8041	0.9409	34.8S	108.6E	36	11	367	06m59s
9384	470	1944 Jan 25	15:26:42	26	-692	130	T	-n	0.2025	1.0428	7.6S	50.2W	78	172	146	04m09s
9385	470	1944 Jul 20	05:43:13	27	-686	135	A	nn	-0.0314	0.9700	19.0N	95.7E	88	6	108	03m42s
9386	470	1945 Jan 14	05:01:43	27	-680	140	A	n-	-0.4937	0.9970	51.1S	110.3E	60	354	12	00m15s
9387	470	1945 Jul 09	13:27:45	27	-674	145	T	p-	0.7356	1.0180	70.0N	17.2W	42	184	92	01m15s
9388	470	1946 Jan 03	12:16:11	27	-668	150	P	t-	-1.2392	0.5529	67.1S	177.6E	0	180		
9389	470	1946 May 30	21:00:24	28	-663	117	P	-t	-1.0711	0.8865	64.1S	101.0W	0	328		
9390	470	1946 Jun 29	03:51:58	28	-662	155	P	t-	1.4361	0.1802	66.6N	50.8W	0	6		
9391	470	1946 Nov 23	17:37:12	28	-657	122	P	-t	1.1050	0.7758	63.4N	45.3W	0	219		
9392	470	1947 May 20	13:47:47	28	-651	127	T	-p	-0.3528	1.0557	0.2N	21.4W	69	343	196	05m13s
9393	470	1947 Nov 12	20:05:37	28	-645	132	A	-n	0.3743	0.9650	3.0N	117.4W	68	200	135	03m59s
9394	470	1948 May 09	02:26:04	28	-639	137	A	p-	0.4133	0.9999	39.8N	131.2E	65	153	0	00m00s
9395	470	1948 Nov 01	05:59:18	29	-633	142	T	n-	-0.3517	1.0231	33.1S	76.2E	69	28	84	01m56s
9396	470	1949 Apr 28	07:48:53	29	-627	147	P	t-	1.2068	0.6092	61.9N	55.7W	0	59		
9397	470	1949 Oct 21	21:13:01	29	-621	152	P	t-	-1.0270	0.9638	61.5S	107.5E	0	113		
9398	470	1950 Mar 18	15:32:01	29	-616	119	A-	-t	-0.9988	0.9620	60.9S	40.9E	0	268	-	-
9399	470	1950 Sep 12	03:38:47	29	-610	124	T	-t	0.8903	1.0182	54.8N	172.3E	27	236	134	01m14s
9400	470	1951 Mar 07	20:53:40	30	-604	129	A	-n	-0.2420	0.9896	17.7S	123.5W	76	330	38	00m59s
9401	471	1951 Sep 01	12:51:51	30	-598	134	A	nn	0.1557	0.9747	16.5N	8.5W	81	208	91	02m36s
9402	471	1952 Feb 25	09:11:35	30	-592	139	T	p-	0.4697	1.0366	15.6N	32.7E	62	152	138	03m09s
9403	471	1952 Aug 20	15:13:35	30	-586	144	A	p-	-0.6102	0.9420	21.7S	64.1W	52	27	264	06m40s
9404	471	1953 Feb 14	00:59:30	30	-580	149	P	t-	1.1331	0.7596	61.9N	104.9E	0	119		
9405	471	1953 Jul 11	02:44:14	30	-575	116	P	-t	1.4388	0.2015	64.3N	71.7W	0	331		
9406	471	1953 Aug 09	15:55:03	30	-574	154	P	t-	-1.3440	0.3729	62.2S	114.7W	0	54		
9407	471	1954 Jan 05	02:32:01	31	-569	121	A	-t	-0.9296	0.9720	79.1S	120.8W	21	260	278	01m42s
9408	471	1954 Jun 30	12:32:38	31	-563	126	T	-p	0.6135	1.0357	60.5N	4.2E	52	197	153	02m35s
9409	471	1954 Dec 25	07:36:42	31	-557	131	A	-n	-0.2576	0.9323	38.4S	68.2E	75	352	262	07m39s
9410	471	1955 Jun 20	04:10:42	31	-551	136	T	n-	-0.1528	1.0776	14.8N	117.0E	81	5	254	07m08s
9411	471	1955 Dec 14	07:02:25	31	-545	141	A	p-	0.4266	0.9176	2.1N	72.2E	65	178	346	12m09s
9412	471	1956 Jun 08	21:20:39	32	-539	146	T	p-	-0.8934	1.0581	40.8S	140.7W	26	0	429	04m45s
9413	471	1956 Dec 02	08:00:35	32	-533	151	P	t-	1.0923	0.8047	67.9N	64.6E	0	187		
9414	471	1957 Apr 30	00:05:28	32	-528	118	A+	-t	0.9992	0.9799	70.6N	40.3E	0	41	-	-
9415	471	1957 Oct 23	04:54:02	32	-522	123	T-	-t	-1.0022	1.0013	71.2S	23.1W	0	127	-	-
9416	471	1958 Apr 19	03:27:17	32	-516	128	A	np	0.2750	0.9408	26.5N	123.6E	74	164	228	07m07s
9417	471	1958 Oct 12	20:55:28	33	-510	133	T	-n	-0.2951	1.0608	24.0S	142.4W	73	18	209	05m11s
9418	471	1959 Apr 08	03:24:08	33	-504	138	A	p-	-0.4546	0.9401	19.1S	137.6E	63	343	247	07m26s
9419	471	1959 Oct 02	12:27:00	33	-498	143	T	n-	0.4207	1.0325	20.4N	1.4W	65	199	120	03m02s
9420	471	1960 Mar 27	07:25:07	33	-492	148	P	t-	-1.1537	0.7058	72.1S	151.9E	0	279		

Cat Num	Canon Plate	Calendar Date	TD of Greatest Eclipse	ΔT s	Luna Num	Saros Num	Ecl. Type	QLE	Gamma	Ecl. Mag.	Lat. °	Long. °	Sun Alt °	Sun Azm °	Path Width km	Central Line Dur.
9421	472	1960 Sep 20	22:59:56	33	-486	153	P	t-	1.2057	0.6139	72.1N	74.1W	0	273		
9422	472	1961 Feb 15	08:19:48	34	-481	120	T	-p	0.8830	1.0360	47.4N	40.0E	28	159	258	02m45s
9423	472	1961 Aug 11	10:46:47	34	-475	125	A	-p	-0.8859	0.9375	45.8S	4.0E	27	17	499	06m35s
9424	472	1962 Feb 05	00:12:38	34	-469	130	T	-n	0.2107	1.0430	4.2S	178.1E	78	169	147	04m08s
9425	472	1962 Jul 31	12:25:33	34	-463	135	Am	nn	-0.1130	0.9716	12.0N	5.7W	84	9	103	03m33s
9426	472	1963 Jan 25	13:37:12	35	-457	140	A	n-	-0.4898	0.9951	48.2S	15.0W	60	348	20	00m25s
9427	472	1963 Jul 20	20:36:13	35	-451	145	T	p-	0.6571	1.0224	61.7N	119.6W	49	191	101	01m40s
9428	472	1964 Jan 14	20:30:08	35	-445	150	P	t-	-1.2354	0.5591	68.2S	43.1E	0	191		
9429	472	1964 Jun 10	04:34:07	35	-440	117	P	-t	-1.1393	0.7545	65.0S	135.9E	0	338		
9430	472	1964 Jul 09	11:17:53	35	-439	155	P	t-	1.3623	0.3221	67.6N	172.9W	0	355		
9431	472	1964 Dec 04	01:31:54	36	-434	122	P	-t	1.1193	0.7518	64.3N	173.3W	0	209		
9432	472	1965 May 30	21:17:31	36	-428	127	T	-p	-0.4225	1.0544	2.5S	133.8W	65	347	198	05m15s
9433	472	1965 Nov 23	04:14:51	36	-422	132	A	-n	0.3906	0.9656	1.7N	119.8E	67	197	134	04m02s
9434	472	1966 May 20	09:39:02	37	-416	137	A	n-	0.3467	0.9991	39.2N	26.4E	70	158	3	00m05s
9435	472	1966 Nov 12	14:23:28	37	-410	142	T	n-	-0.3300	1.0234	35.6S	48.2W	71	25	84	01m57s
9436	472	1967 May 09	14:42:48	38	-404	147	P	t-	1.1422	0.7201	62.5N	168.1W	0	50		
9437	472	1967 Nov 02	05:38:56	38	-398	152	T-	t-	-1.0007	1.0126	62.0S	27.8W	0	122	-	-
9438	472	1968 Mar 28	23:00:30	38	-393	119	P	-t	-1.0370	0.8990	61.0S	79.8W	0	277		
9439	472	1968 Sep 22	11:18:46	39	-387	124	T	-t	0.9451	1.0099	56.2N	64.0E	19	240	104	00m40s
9440	472	1969 Mar 18	04:54:57	39	-381	129	A	-n	-0.2704	0.9954	14.8S	116.3E	74	330	16	00m26s
9441	473	1969 Sep 11	19:58:59	40	-375	134	A	nn	0.2201	0.9690	15.6N	114.1W	77	210	114	03m11s
9442	473	1970 Mar 07	17:38:30	40	-369	139	T	p-	0.4473	1.0414	18.2N	94.7W	63	150	153	03m28s
9443	473	1970 Aug 31	21:55:30	41	-363	144	A	p-	-0.5364	0.9400	20.3S	164.0W	57	29	258	06m47s
9444	473	1971 Feb 25	09:38:07	41	-357	149	P	t-	1.1188	0.7872	61.4N	33.5W	0	110		
9445	473	1971 Jul 22	09:31:55	42	-352	116	Pe	-t	1.5130	0.0689	63.5N	177.0E	0	321		
9446	473	1971 Aug 20	22:39:31	42	-351	154	P	t-	-1.2659	0.5080	61.7S	135.4E	0	63		
9447	473	1972 Jan 16	11:03:22	42	-346	121	A	-t	-0.9365	0.9692	74.9S	107.7E	20	263	321	01m53s
9448	473	1972 Jul 10	19:46:38	43	-340	126	T	-p	0.6872	1.0379	63.5N	94.2W	46	209	175	02m36s
9449	473	1973 Jan 04	15:46:21	43	-334	131	A	-n	-0.2644	0.9303	37.9S	51.2W	74	346	271	07m49s
9450	473	1973 Jun 30	11:38:41	44	-328	136	T	nn	-0.0785	1.0792	18.8N	5.6E	86	9	256	07m04s
9451	473	1973 Dec 24	15:02:44	44	-322	141	A	p-	0.4171	0.9174	1.1N	48.5W	65	174	345	12m02s
9452	473	1974 Jun 20	04:48:04	45	-316	146	T	p-	-0.8239	1.0592	32.1S	103.7E	34	5	344	05m09s
9453	473	1974 Dec 13	16:13:13	45	-310	151	P	t-	1.0797	0.8266	66.8N	69.4W	0	176		
9454	473	1975 May 11	07:17:33	46	-305	118	P	-t	1.0647	0.8636	69.7N	80.2W	0	28		
9455	473	1975 Nov 03	13:15:54	46	-299	123	P	-t	-1.0248	0.9588	70.4S	161.7W	0	141		
9456	473	1976 Apr 29	10:24:18	47	-293	128	A	-p	0.3378	0.9421	34.0N	18.3E	70	165	227	06m41s
9457	473	1976 Oct 23	05:13:45	47	-287	133	T	-n	-0.3270	1.0572	30.0S	92.3E	71	17	199	04m46s
9458	473	1977 Apr 18	10:31:30	48	-281	138	A	p-	-0.3990	0.9449	11.9S	28.3E	66	345	220	07m04s
9459	473	1977 Oct 12	20:27:27	48	-275	143	T	n-	0.3836	1.0269	14.1N	123.6W	67	197	99	02m37s
9460	473	1978 Apr 07	15:03:47	49	-269	148	P	t-	-1.1081	0.7883	71.9S	23.3E	0	293		
9461	474	1978 Oct 02	06:28:43	49	-263	153	P	t-	1.1616	0.6905	72.0N	159.6E	0	259		
9462	474	1979 Feb 26	16:55:06	50	-258	120	T	-p	0.8981	1.0391	52.1N	94.5W	26	153	298	02m49s
9463	474	1979 Aug 22	17:22:38	50	-252	125	A	-t	-0.9632	0.9329	59.6S	108.5W	15	29	953	06m03s
9464	474	1980 Feb 16	08:54:01	51	-246	130	T	-n	0.2224	1.0434	0.1S	47.1E	77	166	149	04m08s
9465	474	1980 Aug 10	19:12:21	51	-240	135	A	nn	-0.1915	0.9727	4.6N	108.9W	79	12	100	03m23s
9466	474	1981 Feb 04	22:09:24	51	-234	140	A	n-	-0.4838	0.9937	44.4S	140.8W	61	344	25	00m33s
9467	474	1981 Jul 31	03:46:37	52	-228	145	T	p-	0.5792	1.0258	53.3N	134.1E	54	195	108	02m02s
9468	474	1982 Jan 25	04:42:53	52	-222	150	P	t-	-1.2311	0.5663	69.3S	91.7W	0	203		
9469	474	1982 Jun 21	12:04:33	53	-217	117	P	-t	-1.2102	0.6168	65.9S	13.2E	0	347		
9470	474	1982 Jul 20	18:44:44	53	-216	155	P	t-	1.2886	0.4643	68.6N	64.2E	0	345		
9471	474	1982 Dec 15	09:32:09	53	-211	122	P	-t	1.1293	0.7350	65.3N	56.9E	0	199		
9472	474	1983 Jun 11	04:43:33	53	-205	127	T	-p	-0.4947	1.0524	6.2S	114.2E	60	351	199	05m11s
9473	474	1983 Dec 04	12:31:15	54	-199	132	A	-n	0.4015	0.9666	0.9N	4.7W	66	192	131	04m01s
9474	474	1984 May 30	16:45:41	54	-193	137	A	nn	0.2755	0.9980	37.5N	76.7W	74	163	7	00m11s
9475	474	1984 Nov 22	22:54:17	54	-187	142	T	n-	-0.3132	1.0237	37.8S	173.6W	72	21	85	02m00s
9476	474	1985 May 19	21:29:38	55	-181	147	P	t-	1.0720	0.8406	63.2N	81.1E	0	41		
9477	474	1985 Nov 12	14:11:27	55	-175	152	T	t-	-0.9795	1.0388	68.6S	142.6W	11	111	690	01m59s
9478	474	1986 Apr 09	06:21:22	55	-170	119	P	-t	-1.0822	0.8236	61.2S	161.4E	0	286		
9479	474	1986 Oct 03	19:06:15	55	-164	124	H	-t	0.9931	1.0000	59.9N	37.1W	5	252	1	00m00s
9480	474	1987 Mar 29	12:49:47	55	-158	129	H	-n	-0.3053	1.0013	12.3S	2.3W	72	331	5	00m08s

Cat Num	Canon Plate	Calendar Date	TD of Greatest Eclipse	ΔT s	Luna Num	Saros Num	Ecl. Type	QLE	Gamma	Ecl. Mag.	Lat. °	Long. °	Sun Alt °	Sun Azm °	Path Width km	Central Line Dur.
9481	475	1987 Sep 23	03:12:22	56	-152	134	A	-n	0.2787	0.9634	14.3N	138.4E	74	210	137	03m49s
9482	475	1988 Mar 18	01:58:56	56	-146	139	T	n-	0.4188	1.0464	20.7N	140.0E	65	149	169	03m46s
9483	475	1988 Sep 11	04:44:29	56	-140	144	A	p-	-0.4681	0.9377	20.0S	94.4E	62	31	258	06m57s
9484	475	1989 Mar 07	18:08:41	56	-134	149	P	t-	1.0981	0.8268	61.2N	169.8W	0	101		
9485	475	1989 Aug 31	05:31:47	57	-128	154	P	t-	-1.1928	0.6344	61.3S	23.6E	0	72		
9486	475	1990 Jan 26	19:31:24	57	-123	121	A	-t	-0.9457	0.9670	71.0S	22.2W	18	266	373	02m03s
9487	475	1990 Jul 22	03:03:07	57	-117	126	T	-p	0.7597	1.0391	65.2N	168.9E	40	222	201	02m33s
9488	475	1991 Jan 15	23:53:51	58	-111	131	A	-n	-0.2727	0.9290	36.4S	170.4W	74	341	277	07m53s
9489	475	1991 Jul 11	19:07:01	58	-105	136	Tm	nn	-0.0041	1.0800	22.0N	105.2W	90	30	258	06m53s
9490	475	1992 Jan 04	23:05:37	58	-99	141	A	p-	0.4091	0.9179	1.0N	169.7W	66	169	340	11m41s
9491	475	1992 Jun 30	12:11:22	59	-93	146	T	p-	-0.7512	1.0592	25.2S	9.5W	41	10	294	05m21s
9492	475	1992 Dec 24	00:31:41	59	-87	151	P	t-	1.0711	0.8422	65.7N	155.7E	0	165		
9493	475	1993 May 21	14:20:15	59	-82	118	P	-t	1.1372	0.7352	68.8N	162.3E	0	17		
9494	475	1993 Nov 13	21:45:51	60	-76	123	P	-t	-1.0411	0.9280	69.6S	58.3E	0	153		
9495	475	1994 May 10	17:12:26	60	-70	128	A	-p	0.4077	0.9431	41.5N	84.1W	66	168	230	06m13s
9496	475	1994 Nov 03	13:40:06	61	-64	133	T	-n	-0.3522	1.0535	35.4S	34.2W	69	15	189	04m23s
9497	475	1995 Apr 29	17:33:21	61	-58	138	A	p-	-0.3382	0.9497	4.8S	79.4W	70	348	196	06m37s
9498	475	1995 Oct 24	04:33:30	61	-52	143	T	n-	0.3518	1.0213	8.4N	113.2E	69	195	78	02m10s
9499	475	1996 Apr 17	22:38:12	62	-46	148	P	t-	-1.0580	0.8799	71.3S	104.0W	0	306		
9500	475	1996 Oct 12	14:03:04	62	-40	153	P	t-	1.1227	0.7575	71.7N	32.1E	0	245		
9501	476	1997 Mar 09	01:24:51	62	-35	120	T	-p	0.9183	1.0420	57.8N	130.7E	23	146	356	02m50s
9502	476	1997 Sep 02	00:04:48	63	-29	125	P	-t	-1.0352	0.8988	71.8S	114.3E	0	64		
9503	476	1998 Feb 26	17:29:27	63	-23	130	T	-n	0.2391	1.0441	4.7N	82.7W	76	164	151	04m09s
9504	476	1998 Aug 22	02:07:11	63	-17	135	A	nn	-0.2644	0.9734	3.0S	145.4E	75	14	99	03m14s
9505	476	1999 Feb 16	06:34:38	63	-11	140	A	n-	-0.4726	0.9928	39.8S	93.9E	62	342	29	00m40s
9506	476	1999 Aug 11	11:04:09	64	-5	145	T	p-	0.5062	1.0286	45.1N	24.3E	59	197	112	02m23s
9507	476	2000 Feb 05	12:50:27	64	1	150	P	t-	-1.2233	0.5795	70.2S	134.1E	0	215		
9508	476	2000 Jul 01	19:33:34	64	6	117	P	-t	-1.2821	0.4768	66.9S	109.5W	0	358		
9509	476	2000 Jul 31	02:14:08	64	7	155	P	t-	1.2166	0.6034	69.5N	59.9W	0	333		
9510	476	2000 Dec 25	17:35:57	64	12	122	P	-t	1.1367	0.7228	66.3N	74.1W	0	189		
9511	476	2001 Jun 21	12:04:46	64	18	127	T	-p	-0.5701	1.0495	11.3S	2.7E	55	355	200	04m57s
9512	476	2001 Dec 14	20:53:01	64	24	132	A	-n	0.4089	0.9681	0.6N	130.7W	66	188	126	03m53s
9513	476	2002 Jun 10	23:45:22	64	30	137	A	nn	0.1993	0.9962	34.5N	178.6W	78	169	13	00m23s
9514	476	2002 Dec 04	07:32:16	64	36	142	T	n-	-0.3020	1.0244	39.5S	59.6E	72	16	87	02m04s
9515	476	2003 May 31	04:09:22	64	42	147	An	t-	0.9960	0.9384	66.6N	24.5W	3	35	-	03m37s
9516	476	2003 Nov 23	22:50:22	64	48	152	T	t-	-0.9638	1.0379	72.7S	88.4E	15	111	495	01m57s
9517	476	2004 Apr 19	13:35:05	65	53	119	P	-t	-1.1335	0.7367	61.6S	44.3E	0	295		
9518	476	2004 Oct 14	03:00:23	65	59	124	P	-t	1.0348	0.9282	61.2N	153.7W	0	253		
9519	476	2005 Apr 08	20:36:51	65	65	129	H	-n	-0.3473	1.0074	10.6S	119.0W	70	332	27	00m42s
9520	476	2005 Oct 03	10:32:47	65	71	134	A	-p	0.3306	0.9576	12.9N	28.7E	71	209	162	04m32s
9521	477	2006 Mar 29	10:12:23	65	77	139	T	n-	0.3843	1.0515	23.2N	16.7E	67	149	184	04m07s
9522	477	2006 Sep 22	11:41:16	65	83	144	A	p-	-0.4062	0.9352	20.6S	9.1W	66	31	261	07m09s
9523	477	2007 Mar 19	02:32:57	65	89	149	P	t-	1.0728	0.8756	61.0N	55.5E	0	92		
9524	477	2007 Sep 11	12:32:24	66	95	154	P	t-	-1.1255	0.7507	61.0S	90.2W	0	80		
9525	477	2008 Feb 07	03:56:10	66	100	121	A	-t	-0.9570	0.9650	67.6S	150.5W	16	269	444	02m12s
9526	477	2008 Aug 01	10:22:12	66	106	126	T	-p	0.8307	1.0394	65.7N	72.3E	34	235	237	02m27s
9527	477	2009 Jan 26	07:59:45	66	112	131	A	-n	-0.2820	0.9282	34.1S	70.2E	73	337	280	07m54s
9528	477	2009 Jul 22	02:36:25	66	118	136	T	nn	0.0698	1.0799	24.2N	144.1E	86	198	258	06m39s
9529	477	2010 Jan 15	07:07:39	67	124	141	A	p-	0.4002	0.9190	1.6N	69.3E	66	165	333	11m08s
9530	477	2010 Jul 11	19:34:38	67	130	146	T	p-	-0.6788	1.0580	19.7S	121.9W	47	14	259	05m20s
9531	477	2011 Jan 04	08:51:42	67	136	151	P	t-	1.0627	0.8576	64.7N	20.8E	0	155		
9532	477	2011 Jun 01	21:17:18	67	141	118	P	-t	1.2130	0.6010	67.8N	46.8E	0	6		
9533	477	2011 Jul 01	08:39:30	67	142	156	Pb	t-	-1.4917	0.0971	65.2S	28.6E	0	21		
9534	477	2011 Nov 25	06:21:24	68	147	123	P	-t	-1.0536	0.9047	68.6S	82.4W	0	165		
9535	477	2012 May 20	23:53:54	68	153	128	A	-p	0.4828	0.9439	49.1N	176.3E	61	171	237	05m46s
9536	477	2012 Nov 13	22:12:55	68	159	133	T	-n	-0.3719	1.0500	40.0S	161.3W	68	11	179	04m02s
9537	477	2013 May 10	00:26:20	68	165	138	A	pn	-0.2694	0.9544	2.2N	175.5E	74	350	173	06m03s
9538	477	2013 Nov 03	12:47:36	68	171	143	H3	n-	0.3272	1.0159	3.5N	11.7W	71	192	58	01m40s
9539	477	2014 Apr 29	06:04:33	69	177	148	A-	t-	-1.0000	0.9868	70.6S	131.3E	0	319	-	-
9540	477	2014 Oct 23	21:45:39	69	183	153	P	t-	1.0908	0.8114	71.2N	97.2W	0	231		

Cat Num	Canon Plate	Calendar Date	TD of Greatest Eclipse	ΔT s	Luna Num	Saros Num	Ecl. Type	QLE	Gamma	Ecl. Mag.	Lat. °	Long. °	Sun Alt °	Sun Azm °	Path Width km	Central Line Dur.
9541	478	2015 Mar 20	09:46:47	69	188	120	T	-t	0.9454	1.0445	64.4N	6.6W	18	135	463	02m47s
9542	478	2015 Sep 13	06:55:19	69	194	125	P	-t	-1.1004	0.7875	72.1S	2.3W	0	77		
9543	478	2016 Mar 09	01:58:19	70	200	130	T	-n	0.2609	1.0450	10.1N	148.8E	75	162	155	04m09s
9544	478	2016 Sep 01	09:08:02	70	206	135	A	-n	-0.3330	0.9736	10.7S	37.8E	70	16	100	03m06s
9545	478	2017 Feb 26	14:54:33	70	212	140	A	n-	-0.4578	0.9922	34.7S	31.2W	63	340	31	00m44s
9546	478	2017 Aug 21	18:26:40	70	218	145	T	p-	0.4367	1.0306	37.0N	87.7W	64	198	115	02m40s
9547	478	2018 Feb 15	20:52:33	71	224	150	P	t-	-1.2116	0.5991	71.0S	0.6E	0	228		
9548	478	2018 Jul 13	03:02:16	71	229	117	P	-t	-1.3542	0.3365	67.9S	127.4E	0	8		
9549	478	2018 Aug 11	09:47:28	71	230	155	P	t-	1.1476	0.7368	70.4N	174.5E	0	321		
9550	478	2019 Jan 06	01:42:38	71	235	122	P	-t	1.1417	0.7145	67.4N	153.6E	0	178		
9551	478	2019 Jul 02	19:24:07	71	241	127	T	-p	-0.6466	1.0459	17.4S	109.0W	50	359	201	04m33s
9552	478	2019 Dec 26	05:18:53	72	247	132	A	-n	0.4135	0.9701	1.0N	102.3E	66	184	118	03m40s
9553	478	2020 Jun 21	06:41:15	72	253	137	Am	nn	0.1209	0.9940	30.5N	79.7E	83	174	21	00m38s
9554	478	2020 Dec 14	16:14:39	72	259	142	T	n-	-0.2939	1.0254	40.3S	67.9W	73	10	90	02m10s
9555	478	2021 Jun 10	10:43:07	72	265	147	A	t-	0.9152	0.9435	80.8N	66.8W	23	90	527	03m51s
9556	478	2021 Dec 04	07:34:38	73	271	152	T	p-	-0.9526	1.0367	76.8S	46.2W	17	115	419	01m54s
9557	478	2022 Apr 30	20:42:36	73	276	119	P	-t	-1.1901	0.6396	62.1S	71.5W	0	304		
9558	478	2022 Oct 25	11:01:20	73	282	124	P	-t	1.0701	0.8619	61.6N	77.4E	0	244		
9559	478	2023 Apr 20	04:17:56	73	288	129	H	-n	-0.3952	1.0132	9.6S	125.8E	67	334	49	01m16s
9560	478	2023 Oct 14	18:00:41	74	294	134	A	-p	0.3753	0.9520	11.4N	83.1W	68	208	187	05m17s
9561	479	2024 Apr 08	18:18:29	74	300	139	T	n-	0.3431	1.0566	25.3N	104.1W	70	149	198	04m28s
9562	479	2024 Oct 02	18:46:13	74	306	144	A	p-	-0.3509	0.9326	22.0S	114.5W	69	31	266	07m25s
9563	479	2025 Mar 29	10:48:36	75	312	149	P	t-	1.0405	0.9376	61.1N	77.1W	0	83		
9564	479	2025 Sep 21	19:43:04	75	318	154	P	t-	-1.0651	0.8550	60.9S	153.5E	0	89		
9565	479	2026 Feb 17	12:13:06	75	323	121	A	-t	-0.9743	0.9630	64.7S	86.8E	12	268	616	02m20s
9566	479	2026 Aug 12	17:47:06	75	329	126	T	-p	0.8977	1.0386	65.2N	25.2W	26	248	294	02m18s
9567	479	2027 Feb 06	16:00:48	76	335	131	A	-n	-0.2952	0.9281	31.3S	48.5W	73	334	282	07m51s
9568	479	2027 Aug 02	10:07:50	76	341	136	T	nn	0.1421	1.0790	25.5N	33.2E	82	202	258	06m23s
9569	479	2028 Jan 26	15:08:59	76	347	141	A	p-	0.3901	0.9208	3.0N	51.5W	67	161	323	10m27s
9570	479	2028 Jul 22	02:56:40	77	353	146	T	p-	-0.6056	1.0560	15.6S	126.7E	53	17	230	05m10s
9571	479	2029 Jan 14	17:13:48	77	359	151	P	t-	1.0553	0.8714	63.7N	114.2W	0	145		
9572	479	2029 Jun 12	04:06:13	77	364	118	P	-t	1.2943	0.4576	66.8N	66.2W	0	355		
9573	479	2029 Jul 11	15:37:19	77	365	156	P	t-	-1.4191	0.2303	64.3S	85.6W	0	30		
9574	479	2029 Dec 05	15:03:58	77	370	123	P	-t	-1.0609	0.8911	67.5S	135.7E	0	177		
9575	479	2030 Jun 01	06:29:13	78	376	128	A	-p	0.5626	0.9443	56.5N	80.1E	55	176	250	05m21s
9576	479	2030 Nov 25	06:51:37	78	382	133	T	-n	-0.3867	1.0468	43.6S	71.2E	67	7	169	03m44s
9577	479	2031 May 21	07:16:04	78	388	138	A	nn	-0.1970	0.9589	8.9N	71.7E	79	354	152	05m26s
9578	479	2031 Nov 14	21:07:31	79	394	143	H	n-	0.3078	1.0106	0.6S	137.6W	72	189	38	01m08s
9579	479	2032 May 09	13:26:42	79	400	148	A	t-	-0.9375	0.9957	51.3S	7.1W	20	345	44	00m22s
9580	479	2032 Nov 03	05:34:13	79	406	153	P	t-	1.0643	0.8554	70.4N	132.6E	0	218		
9581	480	2033 Mar 30	18:02:36	80	411	120	T	-t	0.9778	1.0462	71.3N	155.8W	11	111	781	02m37s
9582	480	2033 Sep 23	13:54:31	80	417	125	P	-t	-1.1583	0.6890	72.2S	121.2W	0	91		
9583	480	2034 Mar 20	10:18:45	80	423	130	T	-n	0.2894	1.0458	16.1N	22.2E	73	162	159	04m09s
9584	480	2034 Sep 12	16:19:28	81	429	135	A	-p	-0.3936	0.9736	18.2S	72.6W	67	18	102	02m58s
9585	480	2035 Mar 09	23:05:54	81	435	140	A	n-	-0.4368	0.9919	29.0S	154.9W	64	340	31	00m48s
9586	480	2035 Sep 02	01:56:46	81	441	145	T	p-	0.3727	1.0320	29.1N	158.0E	68	199	116	02m54s
9587	480	2036 Feb 27	04:46:49	82	447	150	P	t-	-1.1942	0.6286	71.6S	131.4W	0	242		
9588	480	2036 Jul 23	10:32:06	82	452	117	P	-t	-1.4250	0.1991	68.9S	3.6E	0	19		
9589	480	2036 Aug 21	17:25:45	82	453	155	P	t-	1.0825	0.8622	71.1N	47.0E	0	309		
9590	480	2037 Jan 16	09:48:55	82	458	122	P	-t	1.1477	0.7049	68.5N	20.8E	0	166		
9591	480	2037 Jul 13	02:40:36	83	464	127	T	-p	-0.7246	1.0413	24.8S	139.1E	43	3	201	03m58s
9592	480	2038 Jan 05	13:47:11	83	470	132	A	-n	0.4169	0.9728	2.1N	25.4W	65	179	107	03m18s
9593	480	2038 Jul 02	13:32:55	84	476	137	A	nn	0.0398	0.9911	25.4N	21.9W	88	179	31	01m00s
9594	480	2038 Dec 26	01:00:10	84	482	142	T	n-	-0.2881	1.0268	40.3S	164.0E	73	5	95	02m18s
9595	480	2039 Jun 21	17:12:54	84	488	147	A	p-	0.8312	0.9454	78.9N	102.1W	33	153	365	04m05s
9596	480	2039 Dec 15	16:23:46	85	494	152	T	p-	-0.9458	1.0356	80.9S	172.8E	18	123	380	01m51s
9597	480	2040 May 11	03:43:02	85	499	119	P	-t	-1.2529	0.5306	62.8S	174.4E	0	313		
9598	480	2040 Nov 04	19:09:02	85	505	124	P	-t	1.0993	0.8074	62.2N	53.4W	0	234		
9599	480	2041 Apr 30	11:52:21	86	511	129	T	-p	-0.4492	1.0189	9.6S	12.2E	63	337	72	01m51s
9600	480	2041 Oct 25	01:36:22	86	517	134	A	-p	0.4133	0.9467	9.9N	162.9E	66	206	213	06m07s

Cat Num	Canon Plate	Calendar Date	TD of Greatest Eclipse	ΔT s	Luna Num	Saros Num	Ecl. Type	QLE	Gamma	Ecl. Mag.	Lat. °	Long. °	Sun Alt °	Sun Azm °	Path Width km	Central Line Dur.
9601	481	2042 Apr 20	02:17:30	86	523	139	T	n-	0.2956	1.0614	27.0N	137.3E	73	151	210	04m51s
9602	481	2042 Oct 14	02:00:42	87	529	144	A	n-	-0.3030	0.9300	23.7S	137.8E	72	30	273	07m44s
9603	481	2043 Apr 09	18:57:49	87	535	149	T+	t-	1.0031	1.0095	61.3N	152.0E	0	74	-	-
9604	481	2043 Oct 03	03:01:49	88	541	154	A-	t-	-1.0102	0.9497	61.0S	35.3E	0	98	-	-
9605	481	2044 Feb 28	20:24:39	88	546	121	As	-t	-0.9954	0.9600	62.2S	25.6W	4	260	-	02m27s
9606	481	2044 Aug 23	01:17:02	88	552	126	T	-t	0.9613	1.0364	64.3N	120.4W	15	264	453	02m04s
9607	481	2045 Feb 16	23:56:07	89	558	131	A	-n	-0.3125	0.9285	28.3S	166.2W	72	331	281	07m47s
9608	481	2045 Aug 12	17:42:39	89	564	136	T	-n	0.2116	1.0774	25.9N	78.5W	78	206	256	06m06s
9609	481	2046 Feb 05	23:06:26	90	570	141	A	p-	0.3765	0.9232	4.8N	171.4W	68	157	310	09m42s
9610	481	2046 Aug 02	10:21:13	90	576	146	T	p-	-0.5350	1.0531	12.7S	15.2E	58	21	206	04m51s
9611	481	2047 Jan 26	01:33:18	90	582	151	P	t-	1.0450	0.8907	62.9N	111.7E	0	135		
9612	481	2047 Jun 23	10:52:31	91	587	118	P	-t	1.3766	0.3129	65.8N	178.0W	0	346		
9613	481	2047 Jul 22	22:36:17	91	588	156	P	t-	-1.3477	0.3604	63.4S	160.2E	0	40		
9614	481	2047 Dec 16	23:50:12	91	593	123	P	-t	-1.0661	0.8816	66.4S	6.6W	0	188		
9615	481	2048 Jun 11	12:58:53	92	599	128	A	-p	0.6468	0.9441	63.7N	11.5W	49	184	272	04m58s
9616	481	2048 Dec 05	15:35:27	92	605	133	T	-n	-0.3973	1.0440	46.1S	56.4W	66	1	160	03m28s
9617	481	2049 May 31	13:59:59	92	611	138	A	nn	-0.1187	0.9631	15.3N	29.9W	83	358	134	04m45s
9618	481	2049 Nov 25	05:33:48	93	617	143	H	n-	0.2943	1.0057	3.8S	95.2E	73	185	21	00m38s
9619	481	2050 May 20	20:42:50	94	623	148	H	t-	-0.8688	1.0038	40.1S	123.7W	29	352	27	00m21s
9620	481	2050 Nov 14	13:30:53	95	629	153	P	t-	1.0447	0.8874	69.5N	1.0E	0	206		
9621	482	2051 Apr 11	02:10:39	95	634	120	P	-t	1.0169	0.9849	71.6N	32.2E	0	63		
9622	482	2051 Oct 04	21:02:14	96	640	125	P	-t	-1.2094	0.6024	72.0S	117.7E	0	105		
9623	482	2052 Mar 30	18:31:53	97	646	130	T	-n	0.3238	1.0466	22.4N	102.5W	71	161	164	04m08s
9624	482	2052 Sep 22	23:39:10	98	652	135	A	-p	-0.4480	0.9734	25.7S	175.0E	63	20	106	02m51s
9625	482	2053 Mar 20	07:08:19	99	658	140	A	n-	-0.4089	0.9919	23.0S	83.0E	66	341	31	00m50s
9626	482	2053 Sep 12	09:34:09	100	664	145	T	n-	0.3140	1.0328	21.5N	41.7E	72	199	116	03m04s
9627	482	2054 Mar 09	12:33:40	101	670	150	P	t-	-1.1711	0.6678	72.0S	97.9E	0	256		
9628	482	2054 Aug 03	18:04:02	102	675	117	Pe	-t	-1.4941	0.0655	69.8S	121.3W	0	31		
9629	482	2054 Sep 02	01:09:34	102	676	155	P	t-	1.0215	0.9793	71.7N	82.3W	0	296		
9630	482	2055 Jan 27	17:54:05	103	681	122	P	-t	1.1550	0.6932	69.5N	112.2W	0	154		
9631	482	2055 Jul 24	09:57:50	104	687	127	T	-p	-0.8012	1.0359	33.3S	25.8E	37	8	202	03m17s
9632	482	2056 Jan 16	22:16:45	105	693	132	A	-n	0.4199	0.9759	3.9N	153.5W	65	175	95	02m52s
9633	482	2056 Jul 12	20:21:59	106	699	137	A	nn	-0.0426	0.9878	19.4N	123.7W	88	3	43	01m26s
9634	482	2057 Jan 05	09:47:52	107	705	142	T	n-	-0.2837	1.0287	39.2S	35.2E	73	359	102	02m29s
9635	482	2057 Jul 01	23:40:15	108	711	147	A	p-	0.7455	0.9464	71.5N	176.2W	41	177	298	04m22s
9636	482	2057 Dec 26	01:14:35	109	717	152	T	p-	-0.9405	1.0348	84.9S	21.8E	19	141	355	01m50s
9637	482	2058 May 22	10:39:25	110	722	119	P	-t	-1.3194	0.4141	63.5S	61.1E	0	322		
9638	482	2058 Jun 21	00:19:35	110	723	157	Pb	t-	1.4869	0.1260	65.9N	9.9E	0	13		
9639	482	2058 Nov 16	03:23:07	111	728	124	P	-t	1.1224	0.7644	62.9N	174.2E	0	225		
9640	482	2059 May 11	19:22:16	112	734	129	T	-p	-0.5080	1.0242	10.7S	100.4W	59	340	95	02m23s
9641	483	2059 Nov 05	09:18:15	113	740	134	A	-p	0.4454	0.9417	8.7N	47.1E	63	203	238	07m00s
9642	483	2060 Apr 30	10:10:00	114	746	139	T	n-	0.2422	1.0660	28.0N	20.9E	76	154	222	05m15s
9643	483	2060 Oct 24	09:24:10	115	752	144	A	nn	-0.2625	0.9277	25.8S	28.1E	75	28	281	08m06s
9644	483	2061 Apr 20	02:56:49	116	758	149	T	t-	0.9578	1.0475	64.5N	59.2E	16	97	559	02m37s
9645	483	2061 Oct 13	10:32:10	117	764	154	A	t-	-0.9639	0.9469	62.1S	54.4W	15	79	743	03m41s
9646	483	2062 Mar 11	04:26:16	118	769	121	P	-t	-1.0238	0.9331	61.0S	147.1W	0	263		
9647	483	2062 Sep 03	08:54:27	119	775	126	P	-t	1.0191	0.9749	61.3N	150.3E	0	286		
9648	483	2063 Feb 28	07:43:30	120	781	131	A	-p	-0.3360	0.9293	25.2S	77.7E	70	329	280	07m41s
9649	483	2063 Aug 24	01:22:11	121	787	136	T	-n	0.2771	1.0750	25.6N	168.4E	74	209	252	05m49s
9650	483	2064 Feb 17	07:00:23	122	793	141	A	p-	0.3597	0.9262	7.0N	69.7E	69	154	295	08m56s
9651	483	2064 Aug 12	17:46:06	123	799	146	T	p-	-0.4652	1.0495	10.9S	96.0W	62	24	184	04m28s
9652	483	2065 Feb 05	09:52:26	124	805	151	P	t-	1.0336	0.9123	62.2N	21.9W	0	125		
9653	483	2065 Jul 03	17:33:52	125	810	118	P	-t	1.4619	0.1638	64.8N	71.9E	0	336		
9654	483	2065 Aug 02	05:34:17	125	811	156	P	t-	-1.2759	0.4903	62.7S	46.5E	0	49		
9655	483	2065 Dec 27	08:39:56	126	816	123	P	-t	-1.0688	0.8769	65.4S	149.2W	0	198		
9656	483	2066 Jun 22	19:25:48	127	822	128	A	-p	0.7330	0.9435	70.1N	96.4W	43	198	309	04m40s
9657	483	2066 Dec 17	00:23:40	128	828	133	T	-n	-0.4043	1.0416	47.4S	175.8E	66	355	152	03m14s
9658	483	2067 Jun 11	20:42:26	129	834	138	A	nn	-0.0387	0.9670	21.0N	130.2W	88	2	119	04m05s
9659	483	2067 Dec 06	14:03:43	130	840	143	H	n-	0.2845	1.0011	6.0S	32.4W	74	181	4	00m08s
9660	483	2068 May 31	03:56:39	131	846	148	T	p-	-0.7970	1.0110	31.0S	123.2E	37	357	63	01m06s

Cat Num	Canon Plate	Calendar Date	TD of Greatest Eclipse	ΔT s	Luna Num	Saros Num	Ecl. Type	QLE	Gamma	Ecl. Mag.	Lat. °	Long. °	Sun Alt °	Sun Azm °	Path Width km	Central Line Dur.
9661	484	2068 Nov 24	21:32:30	132	852	153	P	t-	1.0299	0.9109	68.5N	131.1W	0	194		
9662	484	2069 Apr 21	10:11:09	133	857	120	P	-t	1.0624	0.8992	71.0N	101.3W	0	50		
9663	484	2069 May 20	17:53:18	133	858	158	Pb	t-	-1.4852	0.0879	68.8S	69.9W	0	342		
9664	484	2069 Oct 15	04:19:56	134	863	125	P	-t	-1.2524	0.5298	71.6S	5.5W	0	119		
9665	484	2070 Apr 11	02:36:09	135	869	130	T	-n	0.3652	1.0472	29.1N	135.1E	68	162	168	04m04s
9666	484	2070 Oct 04	07:08:57	136	875	135	A	-p	-0.4950	0.9731	32.8S	60.4E	60	21	110	02m44s
9667	484	2071 Mar 31	15:01:06	138	881	140	A	n-	-0.3739	0.9919	16.7S	37.0W	68	342	31	00m52s
9668	484	2071 Sep 23	17:20:28	139	887	145	T	n-	0.2620	1.0333	14.2N	76.7W	75	198	116	03m11s
9669	484	2072 Mar 19	20:10:31	140	893	150	P	t-	-1.1405	0.7199	72.2S	30.4W	0	270		
9670	484	2072 Sep 12	08:59:20	141	899	155	T	t-	0.9655	1.0558	69.8N	102.0E	14	240	732	03m13s
9671	484	2073 Feb 07	01:55:59	142	904	122	P	-t	1.1651	0.6768	70.5N	114.9E	0	141		
9672	484	2073 Aug 03	17:15:23	143	910	127	T	-t	-0.8763	1.0294	43.2S	89.4W	28	14	206	02m29s
9673	484	2074 Jan 27	06:44:15	144	916	132	A	-n	0.4251	0.9798	6.6N	78.8E	65	171	79	02m21s
9674	484	2074 Jul 24	03:10:32	145	922	137	A	nn	-0.1242	0.9838	12.8N	133.7E	83	7	58	01m57s
9675	484	2075 Jan 16	18:36:04	146	928	142	T	n-	-0.2799	1.0311	37.2S	94.1W	74	354	110	02m42s
9676	484	2075 Jul 13	06:05:44	147	934	147	A	p-	0.6583	0.9467	63.1N	95.2E	49	186	262	04m45s
9677	484	2076 Jan 06	10:07:27	148	940	152	T	p-	-0.9373	1.0342	87.2S	173.7W	20	203	340	01m49s
9678	484	2076 Jun 01	17:31:22	149	945	119	P	-t	-1.3897	0.2897	64.4S	51.2W	0	331		
9679	484	2076 Jul 01	06:50:43	149	946	157	P	t-	1.4005	0.2746	67.0N	98.1W	0	3		
9680	484	2076 Nov 26	11:43:01	150	951	124	P	-t	1.1401	0.7315	63.7N	40.1E	0	215		
9681	485	2077 May 22	02:46:05	151	957	129	T	-p	-0.5725	1.0290	13.1S	148.3E	55	343	119	02m54s
9682	485	2077 Nov 15	17:07:56	152	963	134	A	-p	0.4705	0.9371	7.8N	70.8W	62	199	262	07m54s
9683	485	2078 May 11	17:56:55	153	969	139	T	n-	0.1838	1.0701	28.1N	93.7W	79	158	232	05m40s
9684	485	2078 Nov 04	16:55:44	154	975	144	A	nn	-0.2285	0.9255	27.8S	83.3W	77	25	287	08m29s
9685	485	2079 May 01	10:50:13	155	981	149	T	p-	0.9081	1.0512	66.2N	46.3W	24	108	406	02m55s
9686	485	2079 Oct 24	18:11:21	156	987	154	A	t-	-0.9243	0.9484	63.4S	160.6W	22	72	495	03m39s
9687	485	2080 Mar 21	12:20:15	157	992	121	P	-t	-1.0578	0.8734	60.9S	85.9E	0	271		
9688	485	2080 Sep 13	16:38:09	158	998	126	P	-t	1.0723	0.8743	61.1N	25.8E	0	277		
9689	485	2081 Mar 10	15:23:31	159	1004	131	A	-p	-0.3653	0.9304	22.4S	36.7W	68	329	277	07m36s
9690	485	2081 Sep 03	09:07:31	160	1010	136	T	-n	0.3378	1.0720	24.6N	53.6E	70	211	247	05m33s
9691	485	2082 Feb 27	14:47:00	162	1016	141	A	p-	0.3361	0.9298	9.4N	47.1W	70	152	277	08m12s
9692	485	2082 Aug 24	01:16:21	163	1022	146	T	n-	-0.4004	1.0452	10.3S	151.8E	66	26	163	04m01s
9693	485	2083 Feb 16	18:06:36	164	1028	151	P	t-	1.0170	0.9433	61.6N	154.1W	0	116		
9694	485	2083 Jul 15	00:14:23	165	1033	118	Pe	-t	1.5465	0.0168	64.0N	37.7W	0	327		
9695	485	2083 Aug 13	12:34:41	165	1034	156	P	t-	-1.2064	0.6146	62.1S	67.5W	0	58		
9696	485	2084 Jan 07	17:30:24	166	1039	123	P	-t	-1.0715	0.8723	64.4S	68.5E	0	209		
9697	485	2084 Jul 03	01:50:26	167	1045	128	A	-p	0.8208	0.9421	75.0N	169.1W	35	222	377	04m25s
9698	485	2084 Dec 27	09:13:48	168	1051	133	T	-n	-0.4094	1.0396	47.3S	47.7E	66	349	146	03m04s
9699	485	2085 Jun 22	03:21:16	169	1057	138	A	nn	0.0452	0.9704	26.2N	131.3E	87	186	106	03m29s
9700	485	2085 Dec 16	22:37:48	170	1063	143	A	n-	0.2786	0.9971	7.3S	160.8W	74	176	10	00m19s
9701	486	2086 Jun 11	11:07:14	171	1069	148	T	p-	-0.7215	1.0174	23.2S	12.5E	44	2	86	01m48s
9702	486	2086 Dec 06	05:38:55	172	1075	153	P	p-	1.0194	0.9271	67.4N	96.2E	0	182		
9703	486	2087 May 02	18:04:42	173	1080	120	P	-t	1.1139	0.8011	70.3N	127.6E	0	37		
9704	486	2087 Jun 01	01:27:14	173	1081	158	P	t-	-1.4186	0.2146	67.8S	165.4E	0	354		
9705	486	2087 Oct 26	11:46:57	174	1086	125	P	-t	-1.2882	0.4696	71.0S	130.5W	0	132		
9706	486	2088 Apr 21	10:31:49	175	1092	130	T	-p	0.4135	1.0474	36.0N	15.1E	65	163	173	03m58s
9707	486	2088 Oct 14	14:48:05	177	1098	135	A	-p	-0.5349	0.9727	39.7S	56.0W	57	21	115	02m38s
9708	486	2089 Apr 10	22:44:42	178	1104	140	A	n-	-0.3319	0.9919	10.2S	154.8W	71	344	30	00m53s
9709	486	2089 Oct 04	01:15:23	179	1110	145	T	n-	0.2167	1.0333	7.4N	162.8E	77	197	115	03m14s
9710	486	2090 Mar 31	03:38:08	180	1116	150	P	t-	-1.1028	0.7843	72.1S	156.3W	0	284		
9711	486	2090 Sep 23	16:56:36	181	1122	155	T	t-	0.9157	1.0562	60.7N	40.5W	23	218	463	03m36s
9712	486	2091 Feb 18	09:54:40	182	1127	122	P	-t	1.1779	0.6558	71.2N	17.8W	0	128		
9713	486	2091 Aug 15	00:34:43	183	1133	127	T	-t	-0.9490	1.0216	55.6S	150.5E	18	23	236	01m38s
9714	486	2092 Feb 07	15:10:20	184	1139	132	A	-n	0.4322	0.9840	9.9N	48.7W	64	168	62	01m48s
9715	486	2092 Aug 03	09:59:33	185	1145	137	A	nn	-0.2044	0.9794	5.6N	30.3E	78	10	75	02m31s
9716	486	2093 Jan 27	03:22:16	186	1151	142	T	n-	-0.2737	1.0340	34.1S	136.4E	74	350	119	02m58s
9717	486	2093 Jul 23	12:32:04	187	1157	147	A	p-	0.5717	0.9463	54.6N	1.3E	55	191	241	05m11s
9718	486	2094 Jan 16	18:59:03	189	1163	152	T	p-	-0.9333	1.0342	84.8S	10.6W	21	267	329	01m51s
9719	486	2094 Jun 13	00:22:11	190	1168	119	P	-t	-1.4613	0.1618	65.3S	163.6W	0	341		
9720	486	2094 Jul 12	13:24:35	190	1169	157	P	t-	1.3150	0.4224	68.0N	152.8E	0	352		

Cat Num	Canon Plate	Calendar Date	TD of Greatest Eclipse	ΔT s	Luna Num	Saros Num	Ecl. Type	QLE	Gamma	Ecl. Mag.	Lat. °	Long. °	Sun Alt °	Sun Azm °	Path Width km	Central Line Dur.
9721	487	2094 Dec 07	20:05:56	191	1174	124	P	-t	1.1547	0.7046	64.7N	95.0W	0	205		
9722	487	2095 Jun 02	10:07:40	192	1180	129	T	-p	-0.6396	1.0332	16.7S	37.2E	50	347	145	03m18s
9723	487	2095 Nov 27	01:02:57	193	1186	134	A	-p	0.4903	0.9330	7.2N	169.8E	61	195	285	08m47s
9724	487	2096 May 22	01:37:14	194	1192	139	T	nn	0.1196	1.0737	27.3N	153.4E	83	162	241	06m06s
9725	487	2096 Nov 15	00:36:15	195	1198	144	A	nn	-0.2018	0.9237	29.7S	163.3E	78	22	294	08m53s
9726	487	2097 May 11	18:34:31	196	1204	149	T	p-	0.8516	1.0538	67.4N	149.5W	31	121	339	03m10s
9727	487	2097 Nov 04	02:01:25	197	1210	154	A	t-	-0.8926	0.9494	65.8S	86.8E	26	68	411	03m36s
9728	487	2098 Apr 01	20:02:31	198	1215	121	P	-t	-1.1005	0.7984	61.0S	38.1W	0	280		
9729	487	2098 Sep 25	00:31:16	199	1221	126	P	-t	1.1184	0.7871	61.1N	101.0W	0	268		
9730	487	2098 Oct 24	10:36:11	200	1222	164	Pb	t-	-1.5407	0.0056	61.8S	95.5W	0	116		
9731	487	2099 Mar 21	22:54:32	201	1227	131	A	-p	-0.4016	0.9318	20.0S	149.0W	66	329	275	07m32s
9732	487	2099 Sep 14	16:57:53	202	1233	136	T	-n	0.3942	1.0684	23.4N	62.8W	67	211	241	05m18s
9733	487	2100 Mar 10	22:28:11	203	1239	141	A	n-	0.3077	0.9338	12.0N	162.4W	72	151	257	07m29s
9734	487	2100 Sep 04	08:49:20	204	1245	146	T	n-	-0.3384	1.0402	10.5S	39.0E	70	28	142	03m32s
9735	487	2101 Feb 28	02:16:26	205	1251	151	An	t-	0.9964	0.9609	60.5N	80.0E	3	111	-	02m44s
9736	487	2101 Aug 24	19:37:03	206	1257	156	P	t-	-1.1392	0.7337	61.6S	178.2E	0	67		
9737	487	2102 Jan 19	02:21:30	207	1262	123	P	-t	-1.0741	0.8682	63.5S	73.6W	0	218		
9738	487	2102 Jul 15	08:15:14	208	1268	128	A	-t	0.9080	0.9398	75.9N	134.2E	24	261	539	04m14s
9739	487	2103 Jan 08	18:04:21	210	1274	133	T	-n	-0.4140	1.0381	46.1S	80.8W	65	342	140	02m57s
9740	487	2103 Jul 04	10:01:48	211	1280	138	Am	nn	0.1285	0.9734	30.3N	33.2E	82	191	96	02m57s
9741	488	2103 Dec 29	07:13:18	212	1286	143	A	n-	0.2747	0.9936	7.5S	70.5E	74	172	23	00m43s
9742	488	2104 Jun 22	18:16:21	213	1292	148	T	p-	-0.6438	1.0231	16.6S	96.8W	50	6	103	02m26s
9743	488	2104 Dec 17	13:48:27	214	1298	153	A+	p-	1.0120	0.9381	66.4N	36.6W	0	171	-	-
9744	488	2105 May 14	01:52:06	215	1303	120	P	-t	1.1708	0.6921	69.4N	1.4W	0	25		
9745	488	2105 Jun 12	08:58:11	215	1304	158	P	t-	-1.3489	0.3483	66.8S	41.9E	0	4		
9746	488	2105 Nov 06	19:23:02	216	1309	125	P	-t	-1.3168	0.4217	70.2S	102.7E	0	145		
9747	488	2106 May 03	18:19:20	217	1315	130	T	-p	0.4681	1.0472	43.1N	102.3W	62	164	177	03m47s
9748	488	2106 Oct 26	22:37:40	219	1321	135	A	-p	-0.5671	0.9725	45.9S	174.1W	55	20	119	02m32s
9749	488	2107 Apr 23	06:18:41	220	1327	140	A	nn	-0.2829	0.9918	3.6S	89.9E	74	346	30	00m56s
9750	488	2107 Oct 16	09:18:27	221	1333	145	T	n-	0.1778	1.0332	1.1N	40.6E	80	196	114	03m16s
9751	488	2108 Apr 11	10:55:37	222	1339	150	P	t-	-1.0573	0.8620	71.7S	80.5E	0	298		
9752	488	2108 Oct 05	01:01:20	223	1345	155	T	p-	0.8722	1.0551	52.5N	172.0W	29	209	371	03m50s
9753	488	2109 Mar 01	17:45:53	224	1350	122	P	-t	1.1972	0.6238	71.8N	149.1W	0	114		
9754	488	2109 Aug 26	07:57:26	225	1356	127	P	-t	-1.0178	0.9670	71.4S	5.1E	0	56		
9755	488	2110 Feb 18	23:31:35	227	1362	132	A	-n	0.4438	0.9888	14.1N	175.3W	64	165	44	01m12s
9756	488	2110 Aug 15	16:50:45	228	1368	137	A	-p	-0.2819	0.9746	2.0S	74.3W	74	13	94	03m07s
9757	488	2111 Feb 08	12:05:33	229	1374	142	T	n-	-0.2650	1.0374	30.2S	6.8E	74	346	130	03m17s
9758	488	2111 Aug 04	19:00:22	230	1380	147	A	p-	0.4867	0.9455	46.0N	95.3W	61	194	230	05m42s
9759	488	2112 Jan 29	03:49:52	231	1386	152	T	p-	-0.9292	1.0346	80.6S	163.8W	21	287	322	01m56s
9760	488	2112 Jun 24	07:09:53	232	1391	119	Pe	-t	-1.5356	0.0282	66.3S	84.4E	0	351		
9761	489	2112 Jul 23	19:58:32	233	1392	157	P	t-	1.2284	0.5725	69.0N	43.1E	0	341		
9762	489	2112 Dec 19	04:33:16	233	1397	124	P	-t	1.1648	0.6858	65.7N	128.4E	0	195		
9763	489	2113 Jun 13	17:26:00	235	1403	129	T	-p	-0.7097	1.0367	21.7S	73.8W	45	351	174	03m36s
9764	489	2113 Dec 08	09:03:27	236	1409	134	A	-p	0.5049	0.9296	7.1N	48.9E	60	191	304	09m35s
9765	489	2114 Jun 03	09:14:09	237	1415	139	T	nn	0.0525	1.0766	25.4N	41.3E	87	167	248	06m32s
9766	489	2114 Nov 27	08:24:15	238	1421	144	A	nn	-0.1815	0.9223	31.3S	48.4E	79	17	298	09m14s
9767	489	2115 May 24	02:13:56	239	1427	149	T	p-	0.7912	1.0557	67.8N	109.4E	37	134	301	03m24s
9768	489	2115 Nov 16	09:58:55	241	1433	154	A	p-	-0.8664	0.9503	68.7S	27.8W	30	63	365	03m32s
9769	489	2116 Apr 13	03:36:55	242	1438	121	P	-t	-1.1487	0.7138	61.3S	160.2W	0	289		
9770	489	2116 Oct 06	08:31:51	243	1444	126	P	-t	1.1589	0.7105	61.2N	130.4E	0	259		
9771	489	2116 Nov 04	18:50:09	243	1445	164	P	t-	-1.5103	0.0613	62.3S	132.3E	0	125		
9772	489	2117 Apr 02	06:15:20	244	1450	131	A	-p	-0.4459	0.9333	18.4S	101.1E	63	330	274	07m30s
9773	489	2117 Sep 26	00:55:42	245	1456	136	T	-p	0.4442	1.0645	21.9N	178.4E	64	211	233	05m03s
9774	489	2118 Mar 22	06:00:55	246	1462	141	A	n-	0.2719	0.9382	14.3N	84.7E	74	150	237	06m50s
9775	489	2118 Sep 15	16:28:26	248	1468	146	T	n-	-0.2823	1.0349	11.5S	75.2W	74	29	122	03m04s
9776	489	2119 Mar 11	10:19:19	249	1474	151	A	t-	0.9693	0.9694	56.7N	29.2W	14	120	451	02m13s
9777	489	2119 Sep 05	02:44:27	250	1480	156	P	t-	-1.0766	0.8431	61.2S	62.8E	0	75		
9778	489	2120 Jan 30	11:09:56	251	1485	123	P	-t	-1.0792	0.8594	62.7S	145.3E	0	228		
9779	489	2120 Jul 25	14:40:02	252	1491	128	An	-t	0.9948	0.9343	66.0N	90.4E	4	312	-	04m00s
9780	489	2121 Jan 19	02:54:15	253	1497	133	T	-n	-0.4190	1.0371	43.9S	150.1E	65	337	137	02m52s

Cat Num	Canon Plate	Calendar Date	TD of Greatest Eclipse	ΔT s	Luna Num	Saros Num	Ecl. Type	QLE	Gamma	Ecl. Mag.	Lat. °	Long. °	Sun Alt °	Sun Azm °	Path Width km	Central Line Dur.
9781	490	2121 Jul 14	16:42:39	255	1503	138	A	nn	0.2125	0.9758	33.6N	64.3W	78	197	88	02m32s
9782	490	2122 Jan 08	15:48:51	256	1509	143	A	n-	0.2713	0.9907	6.9S	58.2W	74	168	34	01m02s
9783	490	2122 Jul 04	01:25:31	257	1515	148	T	p-	-0.5649	1.0280	11.0S	154.7E	56	10	114	02m56s
9784	490	2122 Dec 28	22:00:56	258	1521	153	A+	p-	1.0072	0.9450	65.3N	169.8W	0	161	-	-
9785	490	2123 May 25	09:33:27	259	1526	120	P	-t	1.2325	0.5729	68.5N	128.2W	0	14		
9786	490	2123 Jun 23	16:26:12	260	1527	158	P	t-	-1.2763	0.4882	65.8S	80.3W	0	14		
9787	490	2123 Nov 18	03:07:26	261	1532	125	P	-t	-1.3389	0.3848	69.3S	25.5W	0	157		
9788	490	2124 May 14	01:59:10	262	1538	130	T	-p	0.5286	1.0464	50.3N	143.2E	58	167	182	03m34s
9789	490	2124 Nov 06	06:36:34	263	1544	135	A	-p	-0.5921	0.9724	51.6S	66.8E	53	18	123	02m26s
9790	490	2125 May 03	13:42:33	264	1550	140	A	nn	-0.2263	0.9915	3.0N	22.6W	77	349	31	00m59s
9791	490	2125 Oct 26	17:30:49	266	1556	145	T	n-	0.1461	1.0329	4.5S	83.6W	82	194	112	03m15s
9792	490	2126 Apr 22	18:04:22	267	1562	150	A-	t-	-1.0051	0.9514	71.1S	40.0W	0	311	-	-
9793	490	2126 Oct 16	09:12:51	268	1568	155	T	p-	0.8345	1.0534	45.3N	58.6E	33	203	319	04m00s
9794	490	2127 Mar 13	01:32:03	269	1573	122	P	-t	1.2208	0.5841	72.1N	80.4E	0	100		
9795	490	2127 Sep 06	15:24:17	270	1579	127	P	-t	-1.0822	0.8458	71.9S	120.1W	0	69		
9796	490	2128 Mar 01	07:48:32	271	1585	132	A	-n	0.4596	0.9940	18.9N	59.1E	63	163	24	00m37s
9797	490	2128 Aug 25	23:44:34	273	1591	137	A	-p	-0.3562	0.9694	9.8S	180.0E	69	15	117	03m41s
9798	490	2129 Feb 18	20:44:37	274	1597	142	T	n-	-0.2526	1.0411	25.6S	122.5W	75	344	142	03m38s
9799	490	2129 Aug 15	01:33:05	275	1603	147	A	p-	0.4055	0.9442	37.4N	165.8E	66	196	225	06m15s
9800	490	2130 Feb 08	12:35:23	276	1609	152	T	p-	-0.9212	1.0356	75.9S	51.8E	22	300	313	02m03s
9801	491	2130 Aug 04	02:38:44	278	1615	157	P	t-	1.1461	0.7158	69.9N	68.7W	0	330		
9802	491	2130 Dec 30	13:01:34	279	1620	124	P	-t	1.1730	0.6708	66.8N	8.8W	0	185		
9803	491	2131 Jun 25	00:43:16	280	1626	129	T	-p	-0.7813	1.0393	28.1S	174.7E	38	356	211	03m43s
9804	491	2131 Dec 19	17:06:51	281	1632	134	A	-p	0.5165	0.9267	7.6N	72.8W	59	186	321	10m14s
9805	491	2132 Jun 13	16:46:24	282	1638	139	Tm	nn	-0.0186	1.0788	22.3N	70.1W	89	350	255	06m55s
9806	491	2132 Dec 07	16:18:43	284	1644	144	A	nn	-0.1661	0.9215	32.2S	67.9W	80	13	301	09m33s
9807	491	2133 Jun 03	09:45:16	285	1650	149	T	p-	0.7247	1.0567	66.6N	10.7E	43	149	272	03m36s
9808	491	2133 Nov 26	18:05:55	286	1656	154	A	p-	-0.8473	0.9513	72.0S	143.5W	32	57	337	03m27s
9809	491	2134 Apr 24	10:59:59	287	1661	121	P	-t	-1.2052	0.6147	61.8S	80.5E	0	298		
9810	491	2134 May 23	23:01:18	287	1662	159	Pb	t-	1.5285	0.0308	63.7N	55.3E	0	37		
9811	491	2134 Oct 17	16:40:42	288	1667	126	P	-t	1.1931	0.6458	61.5N	0.4W	0	250		
9812	491	2134 Nov 16	03:12:08	288	1668	164	P	t-	-1.4857	0.1060	63.0S	2.1W	0	135		
9813	491	2135 Apr 13	13:27:05	290	1673	131	A	-p	-0.4973	0.9349	17.6S	6.5W	60	332	274	07m30s
9814	491	2135 Oct 07	09:00:03	291	1679	136	T	-p	0.4884	1.0603	20.3N	57.6E	61	210	224	04m50s
9815	491	2136 Apr 01	13:26:19	292	1685	141	A	nn	0.2295	0.9430	16.5N	26.0W	77	150	216	06m14s
9816	491	2136 Sep 26	00:12:14	293	1691	146	T	n-	-0.2309	1.0292	13.0S	169.4E	77	30	101	02m34s
9817	491	2137 Mar 21	18:16:38	294	1697	151	A	t-	0.9369	0.9769	55.6N	144.8W	20	121	233	01m40s
9818	491	2137 Sep 15	09:56:34	296	1703	156	P	t-	-1.0184	0.9436	61.0S	53.8W	0	84		
9819	491	2138 Feb 09	19:55:23	297	1708	123	P	-t	-1.0872	0.8453	62.1S	5.1E	0	238		
9820	491	2138 Aug 05	21:08:57	298	1714	128	P	-t	1.0781	0.8285	62.4N	9.2W	0	309		
9821	492	2139 Jan 30	11:42:25	299	1720	133	T	-n	-0.4255	1.0364	41.0S	20.7E	65	333	135	02m49s
9822	492	2139 Jul 25	23:26:33	301	1726	138	A	nn	0.2946	0.9778	35.8N	161.9W	73	202	83	02m13s
9823	492	2140 Jan 20	00:23:11	302	1732	143	A	n-	0.2676	0.9882	5.5S	173.4E	75	163	43	01m17s
9824	492	2140 Jul 14	08:36:11	303	1738	148	T	p-	-0.4861	1.0322	6.7S	46.5E	61	14	124	03m18s
9825	492	2141 Jan 08	06:12:38	304	1744	153	A+	p-	1.0024	0.9522	64.3N	57.7E	0	151	-	-
9826	492	2141 Jun 04	17:09:59	305	1749	120	P	-t	1.2981	0.4458	67.5N	106.7E	0	3		
9827	492	2141 Jul 03	23:53:38	306	1750	158	P	t-	-1.2029	0.6305	64.9S	158.0E	0	24		
9828	492	2141 Nov 28	10:59:33	307	1755	125	P	-t	-1.3552	0.3577	68.2S	155.0W	0	169		
9829	492	2142 May 25	09:32:37	308	1761	130	T	-p	0.5937	1.0449	57.4N	31.9E	53	171	187	03m17s
9830	492	2142 Nov 17	14:43:08	309	1767	135	A	-p	-0.6117	0.9727	56.4S	52.4W	52	14	124	02m19s
9831	492	2143 May 14	20:58:14	310	1773	140	Am	nn	-0.1638	0.9908	9.4N	132.7W	81	352	33	01m05s
9832	492	2143 Nov 07	01:51:16	312	1779	145	T	n-	0.1206	1.0326	9.4S	150.8E	83	191	111	03m14s
9833	492	2144 May 03	01:02:06	313	1785	150	A	t-	-0.9441	0.9363	53.6S	175.9W	19	341	727	06m09s
9834	492	2144 Oct 26	17:32:40	314	1791	155	T	p-	0.8037	1.0512	39.2N	71.2W	36	198	284	04m04s
9835	492	2145 Mar 23	09:09:38	315	1796	122	P	-t	1.2519	0.5311	72.1N	48.0W	0	86		
9836	492	2145 Sep 16	22:57:10	317	1802	127	P	-t	-1.1406	0.7368	72.1S	112.8E	0	83		
9837	492	2145 Oct 16	09:11:28	317	1803	165	Pb	t-	1.5190	0.0359	71.4N	101.7E	0	241		
9838	492	2146 Mar 12	15:58:15	318	1808	132	A	-p	0.4821	0.9995	24.4N	65.0W	61	161	2	00m03s
9839	492	2146 Sep 06	06:44:00	319	1814	137	A	-p	-0.4249	0.9639	17.8S	72.6E	65	18	143	04m13s
9840	492	2147 Mar 02	05:18:54	320	1820	142	T	n-	-0.2360	1.0452	20.5S	108.8E	76	343	155	04m02s

Cat Num	Canon Plate	Calendar Date	TD of Greatest Eclipse	ΔT s	Luna Num	Saros Num	Ecl. Type	QLE	Gamma	Ecl. Mag.	Lat. °	Long. °	Sun Alt °	Sun Azm °	Path Width km	Central Line Dur.
9841	493	2147 Aug 26	08:09:15	322	1826	147	A	pn	0.3271	0.9425	29.0N	65.2E	71	197	224	06m49s
9842	493	2148 Feb 19	21:18:00	323	1832	152	T	p-	-0.9111	1.0370	70.9S	88.3W	24	309	305	02m13s
9843	493	2148 Aug 14	09:22:21	324	1838	157	P	t-	1.0655	0.8562	70.7N	178.0E	0	318		
9844	493	2149 Jan 09	21:30:38	325	1843	124	P	-t	1.1802	0.6575	67.9N	146.7W	0	173		
9845	493	2149 Jul 05	07:59:34	327	1849	129	T	-p	-0.8544	1.0408	36.3S	62.4E	31	0	264	03m38s
9846	493	2149 Dec 30	01:13:04	328	1855	134	A	-p	0.5253	0.9245	8.6N	164.7E	58	182	334	10m42s
9847	493	2150 Jun 25	00:17:25	329	1861	139	T	nn	-0.0910	1.0802	18.3N	178.1E	85	356	260	07m14s
9848	493	2150 Dec 19	00:17:02	330	1867	144	A	nn	-0.1535	0.9211	32.3S	175.0E	81	8	302	09m46s
9849	493	2151 Jun 14	17:13:45	331	1873	149	T	p-	0.6561	1.0569	63.7N	89.4W	49	163	249	03m48s
9850	493	2151 Dec 08	02:18:31	332	1879	154	A	p-	-0.8320	0.9526	75.1S	103.1E	33	47	314	03m22s
9851	493	2152 May 04	18:14:02	333	1884	121	P	-t	-1.2679	0.5044	62.3S	36.8W	0	307		
9852	493	2152 Jun 03	06:11:19	333	1885	159	P	t-	1.4645	0.1478	64.5N	61.5W	0	28		
9853	493	2152 Oct 28	00:57:34	334	1890	126	P	-t	1.2213	0.5926	61.9N	133.3W	0	241		
9854	493	2152 Nov 26	11:41:08	334	1891	164	P	t-	-1.4665	0.1409	63.8S	138.4W	0	144		
9855	493	2153 Apr 23	20:29:24	335	1896	131	A	-p	-0.5557	0.9364	17.9S	111.8W	56	334	279	07m31s
9856	493	2153 Oct 17	17:12:18	336	1902	136	T	-p	0.5259	1.0560	18.8N	65.7W	58	208	214	04m36s
9857	493	2154 Apr 12	20:43:01	337	1908	141	A	nn	0.1794	0.9478	18.2N	134.2W	80	152	195	05m42s
9858	493	2154 Oct 07	08:03:50	338	1914	146	T	nn	-0.1867	1.0234	15.1S	52.1E	79	29	81	02m05s
9859	493	2155 Apr 02	02:06:34	339	1920	151	A	t-	0.8975	0.9844	55.6N	101.3E	26	123	123	01m07s
9860	493	2155 Sep 26	17:14:27	340	1926	156	A	t-	-0.9654	0.9593	58.6S	143.0W	15	68	570	02m55s
9861	494	2156 Feb 21	04:36:02	341	1931	123	P	-t	-1.0995	0.8230	61.6S	133.7W	0	247		
9862	494	2156 Aug 16	03:41:28	342	1937	128	P	-t	1.1584	0.6912	61.9N	116.1W	0	300		
9863	494	2157 Feb 09	20:25:36	343	1943	133	T	-p	-0.4358	1.0362	37.7S	108.4W	64	330	135	02m49s
9864	494	2157 Aug 05	06:14:19	344	1949	138	A	-p	0.3743	0.9792	37.1N	99.6E	68	207	80	01m59s
9865	494	2158 Jan 30	08:54:37	345	1955	143	A	n-	0.2620	0.9863	3.4S	45.5E	75	160	50	01m27s
9866	494	2158 Jul 25	15:49:17	346	1961	148	T	p-	-0.4087	1.0356	3.4S	61.8W	66	18	131	03m32s
9867	494	2159 Jan 19	14:23:26	347	1967	153	A+	p-	0.9974	0.9600	63.4N	74.2W	0	141	-	-
9868	494	2159 Jun 16	00:42:44	348	1972	120	P	-t	1.3668	0.3124	66.5N	17.0W	0	353		
9869	494	2159 Jul 15	07:20:50	348	1973	158	P	t-	-1.1288	0.7743	64.0S	36.7E	0	33		
9870	494	2159 Dec 09	18:58:33	349	1978	125	P	-t	-1.3663	0.3392	67.2S	74.4E	0	180		
9871	494	2160 Jun 04	16:58:36	350	1984	130	T	-p	0.6645	1.0428	64.5N	74.9W	48	178	192	02m58s
9872	494	2160 Nov 27	22:58:32	351	1990	135	A	-p	-0.6247	0.9734	60.1S	171.6W	51	8	123	02m12s
9873	494	2161 May 25	04:05:43	352	1996	140	A	nn	-0.0950	0.9898	15.7N	119.8E	85	355	36	01m12s
9874	494	2161 Nov 17	10:19:30	353	2002	145	T	n-	0.1012	1.0325	13.4S	23.6E	84	188	110	03m13s
9875	494	2162 May 14	07:52:46	355	2008	150	A	p-	-0.8775	0.9396	42.3S	72.8E	28	349	468	06m37s
9876	494	2162 Nov 07	01:59:40	356	2014	155	T	p-	0.7788	1.0489	34.1N	158.3E	39	193	258	04m05s
9877	494	2163 Apr 03	16:41:51	356	2019	122	P	-t	1.2876	0.4698	71.9N	175.0W	0	72		
9878	494	2163 Sep 28	06:34:34	358	2025	127	P	-t	-1.1943	0.6377	72.1S	15.6W	0	96		
9879	494	2163 Oct 27	17:20:52	358	2026	165	P	t-	1.4919	0.0888	70.8N	33.9W	0	227		
9880	494	2164 Mar 23	00:02:47	359	2031	132	H	-p	0.5095	1.0051	30.4N	172.1E	59	159	20	00m29s
9881	495	2164 Sep 16	13:48:20	360	2037	137	A	-p	-0.4885	0.9583	25.7S	36.3W	61	19	172	04m42s
9882	495	2165 Mar 12	13:45:50	361	2043	142	T	n-	-0.2130	1.0495	14.9S	18.8W	78	342	168	04m27s
9883	495	2165 Sep 05	14:52:45	362	2049	147	A	nn	0.2549	0.9406	20.7N	37.5W	75	198	227	07m22s
9884	495	2166 Mar 02	05:53:21	363	2055	152	T	p-	-0.8958	1.0388	65.4S	134.4E	26	317	294	02m26s
9885	495	2166 Aug 25	16:13:35	364	2061	157	An	t-	0.9901	0.9531	74.4N	41.5E	7	285	-	03m00s
9886	495	2167 Jan 21	05:56:25	365	2066	124	P	-t	1.1892	0.6413	68.9N	75.5E	0	162		
9887	495	2167 Jul 16	15:17:48	366	2072	129	T	-t	-0.9262	1.0410	46.8S	52.4W	22	6	368	03m19s
9888	495	2168 Jan 10	09:19:03	367	2078	134	A	-p	0.5337	0.9230	10.3N	42.1E	58	178	344	10m55s
9889	495	2168 Jul 05	07:45:23	368	2084	139	T	-n	-0.1660	1.0807	13.2N	66.4E	81	0	264	07m26s
9890	495	2168 Dec 29	08:19:33	369	2090	144	A	nn	-0.1444	0.9215	31.6S	56.7E	82	2	300	09m52s
9891	495	2169 Jun 25	00:37:09	370	2096	149	T	p-	0.5841	1.0562	59.2N	168.6E	54	173	229	03m58s
9892	495	2169 Dec 18	10:37:07	371	2102	154	A	p-	-0.8213	0.9544	77.3S	6.1W	34	31	295	03m15s
9893	495	2170 May 16	01:18:33	372	2107	121	P	-t	-1.3371	0.3831	63.0S	151.9W	0	316		
9894	495	2170 Jun 14	13:15:10	372	2108	159	P	t-	1.3963	0.2719	65.4N	177.1W	0	18		
9895	495	2170 Nov 08	09:23:07	373	2113	126	P	-t	1.2426	0.5524	62.5N	91.6E	0	232		
9896	495	2170 Dec 07	20:17:08	373	2114	164	P	t-	-1.4530	0.1653	64.7S	83.1E	0	154		
9897	495	2171 May 05	03:23:15	374	2119	131	A	-p	-0.6209	0.9378	19.4S	144.8E	51	337	289	07m32s
9898	495	2171 Oct 29	01:31:03	375	2125	136	T	-p	0.5577	1.0516	17.6N	169.1E	56	206	203	04m23s
9899	495	2172 Apr 23	03:53:15	377	2131	141	A	nn	0.1234	0.9528	19.2N	119.6E	83	154	174	05m12s
9900	495	2172 Oct 17	16:01:36	378	2137	146	H3	nn	-0.1484	1.0174	17.3S	66.6W	81	28	60	01m34s

Cat Num	Canon Plate	Calendar Date	TD of Greatest Eclipse	ΔT s	Luna Num	Saros Num	Ecl. Type	QLE	Gamma	Ecl. Mag.	Lat. °	Long. °	Sun Alt °	Sun Azm °	Path Width km	Central Line Dur.
9901	496	2173 Apr 12	09:49:40	379	2143	151	A	p-	0.8515	0.9919	56.2N	10.3W	31	126	53	00m35s
9902	496	2173 Oct 07	00:39:14	380	2149	156	A	p-	-0.9187	0.9558	57.8S	114.0E	23	62	402	03m17s
9903	496	2174 Mar 03	13:11:54	381	2154	123	P	-t	-1.1162	0.7924	61.3S	88.7E	0	256		
9904	496	2174 Apr 01	22:39:09	381	2155	161	Pb	t-	1.5107	0.0470	61.2N	103.8E	0	80		
9905	496	2174 Aug 27	10:19:55	382	2160	128	P	-t	1.2336	0.5629	61.4N	135.6E	0	291		
9906	496	2175 Feb 21	05:04:24	383	2166	133	T	-p	-0.4495	1.0362	34.2S	122.9E	63	328	135	02m50s
9907	496	2175 Aug 16	13:08:17	384	2172	138	A	-p	0.4497	0.9802	37.6N	0.5W	63	211	78	01m50s
9908	496	2176 Feb 10	17:21:21	385	2178	143	A	n-	0.2532	0.9849	0.9S	81.3W	75	156	55	01m34s
9909	496	2176 Aug 04	23:05:55	386	2184	148	T	p-	-0.3333	1.0383	1.3S	170.5W	71	21	136	03m40s
9910	496	2177 Jan 29	22:30:30	387	2190	153	An	p-	0.9897	0.9212	57.6N	165.1E	7	140	-	06m55s
9911	496	2177 Jun 26	08:13:28	388	2195	120	P	-t	1.4371	0.1758	65.5N	139.8W	0	343		
9912	496	2177 Jul 25	14:50:33	388	2196	158	P	t-	-1.0564	0.9149	63.2S	85.0W	0	43		
9913	496	2177 Dec 20	03:01:35	389	2201	125	P	-t	-1.3747	0.3251	66.1S	56.8W	0	191		
9914	496	2178 Jun 16	00:20:42	391	2207	130	T	-p	0.7378	1.0396	71.0N	175.3W	42	190	198	02m36s
9915	496	2178 Dec 09	07:20:02	392	2213	135	A	-p	-0.6338	0.9745	62.4S	69.9E	50	360	118	02m03s
9916	496	2179 Jun 05	11:05:36	393	2219	140	A	nn	-0.0209	0.9884	21.5N	15.0E	89	359	41	01m21s
9917	496	2179 Nov 28	18:54:18	394	2225	145	T	n-	0.0867	1.0325	16.5S	104.6W	85	184	110	03m12s
9918	496	2180 May 24	14:34:28	395	2231	150	A	p-	-0.8035	0.9422	32.6S	32.9W	36	354	359	06m59s
9919	496	2180 Nov 17	10:34:02	396	2237	155	T	p-	0.7605	1.0465	30.1N	26.5E	40	189	238	04m03s
9920	496	2181 Apr 14	00:04:05	397	2242	122	P	-t	1.3318	0.3931	71.5N	60.8E	0	59		
9921	497	2181 May 13	14:55:43	397	2243	160	Pb	t-	-1.5323	0.0510	69.4S	16.9W	0	335		
9922	497	2181 Oct 08	14:19:36	398	2248	127	P	-t	-1.2408	0.5529	71.9S	145.8W	0	110		
9923	497	2181 Nov 07	01:38:23	398	2249	165	P	t-	1.4718	0.1280	70.0N	170.9W	0	214		
9924	497	2182 Apr 03	07:59:43	399	2254	132	H	-p	0.5439	1.0108	36.9N	51.0E	57	159	44	00m58s
9925	497	2182 Sep 27	20:58:45	400	2260	137	A	-p	-0.5461	0.9527	33.5S	146.7W	57	21	205	05m05s
9926	497	2183 Mar 23	22:06:49	402	2266	142	T	n-	-0.1848	1.0540	8.9S	145.2W	79	342	181	04m54s
9927	497	2183 Sep 16	21:42:37	403	2272	147	A	nn	0.1877	0.9384	12.8N	141.9W	79	198	233	07m53s
9928	497	2184 Mar 12	14:22:32	404	2278	152	T	p-	-0.8755	1.0409	59.4S	0.2W	29	324	283	02m43s
9929	497	2184 Sep 04	23:11:00	405	2284	157	A	t-	0.9185	0.9576	67.1N	123.3W	23	227	393	03m12s
9930	497	2185 Jan 31	14:20:20	406	2289	124	P	-t	1.1991	0.6238	69.9N	62.4W	0	149		
9931	497	2185 Jul 26	22:38:16	407	2295	129	Ts	-t	-0.9967	1.0370	67.9S	178.5W	1	21	-	02m27s
9932	497	2186 Jan 20	17:23:44	408	2301	134	A	-p	0.5426	0.9221	12.8N	80.3W	57	174	350	10m53s
9933	497	2186 Jul 16	15:14:54	409	2307	139	T	-n	-0.2396	1.0805	7.4N	46.5W	76	4	267	07m29s
9934	497	2187 Jan 09	16:23:41	410	2313	144	A	nn	-0.1365	0.9224	30.0S	62.1W	82	358	296	09m51s
9935	497	2187 Jul 06	07:58:31	412	2319	149	T	p-	0.5109	1.0548	53.6N	63.8E	59	181	211	04m06s
9936	497	2187 Dec 29	18:59:03	413	2325	154	A	p-	-0.8126	0.9565	77.7S	111.2W	35	10	274	03m07s
9937	497	2188 May 26	08:15:53	414	2330	121	P	-t	-1.4109	0.2538	63.8S	94.6E	0	325		
9938	497	2188 Jun 24	20:14:39	414	2331	159	P	t-	1.3252	0.4008	66.4N	68.0E	0	8		
9939	497	2188 Nov 18	17:55:25	415	2336	126	P	-t	1.2591	0.5212	63.2N	45.5W	0	222		
9940	497	2188 Dec 18	04:56:59	415	2337	164	P	t-	-1.4420	0.1850	65.7S	56.6W	0	165		
9941	498	2189 May 15	10:08:34	416	2342	131	A	-p	-0.6928	0.9387	22.6S	43.3E	46	340	309	07m31s
9942	498	2189 Nov 08	09:57:28	417	2348	136	T	-p	0.5830	1.0474	16.5N	41.6E	54	202	192	04m10s
9943	498	2190 May 04	10:56:30	418	2354	141	A	nn	0.0608	0.9577	19.4N	15.4E	86	157	154	04m45s
9944	498	2190 Oct 29	00:05:50	419	2360	146	H	nn	-0.1161	1.0116	19.6S	173.2E	83	25	40	01m04s
9945	498	2191 Apr 23	17:26:06	420	2366	151	A	p-	0.7991	0.9993	57.0N	119.2W	37	130	4	00m03s
9946	498	2191 Oct 18	08:11:12	422	2372	156	A	p-	-0.8783	0.9516	58.7S	5.2E	28	59	365	03m39s
9947	498	2192 Mar 13	21:40:00	423	2377	123	P	-t	-1.1395	0.7491	61.1S	46.8W	0	265		
9948	498	2192 Apr 12	06:41:56	423	2378	161	P	t-	1.4678	0.1260	61.5N	25.4W	0	71		
9949	498	2192 Sep 06	17:05:08	424	2383	128	P	-t	1.3032	0.4444	61.2N	25.8E	0	282		
9950	498	2193 Mar 03	13:36:08	425	2389	133	T	-p	-0.4689	1.0365	30.9S	4.4W	62	327	137	02m53s
9951	498	2193 Aug 26	20:09:20	426	2395	138	A	-p	0.5200	0.9806	37.4N	102.9W	58	214	80	01m45s
9952	498	2194 Feb 21	01:41:31	427	2401	143	A	n-	0.2396	0.9840	1.9N	153.5E	76	154	58	01m38s
9953	498	2194 Aug 16	06:28:08	428	2407	148	T	n-	-0.2616	1.0403	0.2S	79.6E	75	24	139	03m44s
9954	498	2195 Feb 10	06:34:27	430	2413	153	An	p-	0.9797	0.9218	55.2N	41.6E	11	136	-	06m52s
9955	498	2195 Jul 07	15:41:21	430	2418	120	Pe	-t	1.5095	0.0353	64.6N	98.5E	0	333		
9956	498	2195 Aug 05	22:21:03	431	2419	158	Ts	t-	-0.9843	1.0618	56.1S	166.4E	9	40	-	04m03s
9957	498	2195 Dec 31	11:09:22	432	2424	125	P	-t	-1.3797	0.3166	65.1S	171.4E	0	202		
9958	498	2196 Jun 26	07:37:40	433	2430	130	T	-p	0.8149	1.0356	76.3N	97.0E	35	213	208	02m12s
9959	498	2196 Dec 19	15:47:09	434	2436	135	A	-p	-0.6387	0.9761	63.1S	48.6W	50	350	111	01m53s
9960	498	2197 Jun 15	17:59:33	435	2442	140	A	nn	0.0574	0.9864	26.8N	87.6W	87	184	48	01m32s

Cat Num	Canon Plate	Calendar Date	TD of Greatest Eclipse	ΔT s	Luna Num	Saros Num	Ecl. Type	QLE	Gamma	Ecl. Mag.	Lat. °	Long. °	Sun Alt °	Sun Azm °	Path Width km	Central Line Dur.
9961	499	2197 Dec 09	03:35:07	436	2448	145	T	n-	0.0769	1.0329	18.5S	126.0E	86	180	111	03m13s
9962	499	2198 Jun 04	21:11:35	437	2454	150	A	p-	-0.7260	0.9442	24.2S	135.7W	43	359	299	07m13s
9963	499	2198 Nov 28	19:12:46	439	2460	155	T	p-	0.7459	1.0442	26.9N	106.0W	42	184	221	03m58s
9964	499	2199 Apr 25	07:21:51	440	2465	122	P	-t	1.3799	0.3085	70.8N	61.7W	0	46		
9965	499	2199 May 24	21:42:07	440	2466	160	P	t-	-1.4596	0.1742	68.5S	130.1W	0	347		
9966	499	2199 Oct 19	22:10:26	441	2471	127	P	-t	-1.2817	0.4790	71.4S	82.9E	0	124		
9967	499	2199 Nov 18	10:01:01	441	2472	165	P	t-	1.4564	0.1583	69.1N	51.4E	0	202		
9968	499	2200 Apr 14	15:49:57	442	2477	132	T	-p	0.5847	1.0165	43.8N	68.3W	54	158	69	01m23s
9969	499	2200 Oct 09	04:16:21	443	2483	137	A	-p	-0.5972	0.9470	41.1S	101.3E	53	22	241	05m25s
9970	499	2201 Apr 04	06:19:57	444	2489	142	T	n-	-0.1495	1.0584	2.7S	90.2E	81	343	194	05m20s
9971	499	2201 Sep 28	04:41:51	446	2495	147	A	nn	0.1281	0.9361	5.2N	111.4E	83	198	240	08m21s
9972	499	2202 Mar 24	22:42:58	447	2501	152	T	p-	-0.8484	1.0431	52.9S	131.9W	32	330	271	03m03s
9973	499	2202 Sep 17	06:18:53	448	2507	157	A	t-	0.8546	0.9597	57.1N	114.2E	31	214	281	03m24s
9974	499	2203 Feb 12	22:38:35	449	2512	124	P	-t	1.2128	0.5998	70.8N	160.4E	0	136		
9975	499	2203 Aug 08	06:01:56	450	2518	129	P	-t	-1.0650	0.8898	70.1S	57.0E	0	34		
9976	499	2203 Sep 06	14:50:23	450	2519	167	Pb	t-	1.5374	0.0067	71.8N	69.4E	0	291		
9977	499	2204 Feb 02	01:25:26	451	2524	134	A	-p	0.5535	0.9218	16.0N	157.8E	56	170	353	10m38s
9978	499	2204 Jul 27	22:44:32	452	2530	139	T	-n	-0.3129	1.0793	1.0N	160.1W	72	8	269	07m22s
9979	499	2205 Jan 21	00:27:32	454	2536	144	A	nn	-0.1281	0.9241	27.5S	178.6E	82	353	289	09m42s
9980	499	2205 Jul 17	15:18:00	455	2542	149	T	p-	0.4367	1.0525	47.2N	43.0W	64	186	193	04m10s
9981	500	2206 Jan 10	03:24:08	456	2548	154	A	p-	-0.8060	0.9592	75.9S	140.5E	36	351	252	02m57s
9982	500	2206 Jun 07	15:05:59	457	2553	121	Pe	-t	-1.4894	0.1166	64.7S	17.3W	0	335		
9983	500	2206 Jul 07	03:10:26	457	2554	159	P	t-	1.2516	0.5335	67.4N	46.3W	0	358		
9984	500	2206 Dec 01	02:33:55	458	2559	126	P	-t	1.2711	0.4985	64.1N	175.7E	0	212		
9985	500	2206 Dec 30	13:40:30	458	2560	164	P	t-	-1.4337	0.1997	66.8S	162.3E	0	175		
9986	500	2207 May 27	16:47:47	459	2565	131	A	-p	-0.7692	0.9393	27.5S	57.0W	40	343	347	07m25s
9987	500	2207 Nov 20	18:30:26	461	2571	136	T	-p	0.6027	1.0434	15.8N	87.8W	53	198	180	03m56s
9988	500	2208 May 15	17:53:06	462	2577	141	A	nn	-0.0080	0.9625	18.7N	87.0W	90	334	136	04m19s
9989	500	2208 Nov 09	08:17:12	463	2583	146	H	nn	-0.0905	1.0059	21.8S	51.4E	85	22	20	00m34s
9990	500	2209 May 05	00:56:53	464	2589	151	H	p-	0.7413	1.0065	57.7N	134.4E	42	136	34	00m28s
9991	500	2209 Oct 29	15:50:20	465	2595	156	A	p-	-0.8445	0.9472	60.7S	106.3W	32	56	358	04m02s
9992	500	2210 Mar 26	06:01:57	466	2600	123	P	-t	-1.1680	0.6954	61.1S	179.2E	0	274		
9993	500	2210 Apr 24	14:39:19	467	2601	161	P	t-	1.4202	0.2148	61.9N	153.4W	0	62		
9994	500	2210 Sep 18	23:59:09	468	2606	128	P	-t	1.3657	0.3384	61.0N	86.2W	0	274		
9995	500	2211 Mar 15	22:01:40	469	2612	133	T	-p	-0.4931	1.0368	27.8S	130.6W	60	327	140	02m57s
9996	500	2211 Sep 08	03:17:18	470	2618	138	A	-p	0.5854	0.9808	36.9N	152.5E	54	216	83	01m43s
9997	500	2212 Mar 04	09:55:00	471	2624	143	A	nn	0.2211	0.9834	4.9N	30.1E	77	152	60	01m40s
9998	500	2212 Aug 27	13:56:17	473	2630	148	T	n-	-0.1940	1.0416	0.1S	31.7W	79	27	142	03m45s
9999	500	2213 Feb 21	14:30:14	474	2636	153	A	p-	0.9635	0.9230	53.4N	78.6W	15	133	1080	06m44s
10000	500	2213 Aug 17	05:56:32	475	2642	158	T	t-	-0.9161	1.0653	46.0S	60.3E	23	36	525	04m35s
10001	501	2214 Jan 11	19:17:52	476	2647	125	P	-t	-1.3848	0.3078	64.1S	39.7E	0	212		
10002	501	2214 Jul 08	14:52:45	477	2653	130	T	-t	0.8925	1.0303	78.1N	28.3E	26	253	230	01m46s
10003	501	2215 Jan 01	00:16:36	478	2659	135	A	-p	-0.6427	0.9783	62.3S	168.0W	50	340	101	01m41s
10004	501	2215 Jun 28	00:48:45	480	2665	140	A	nn	0.1388	0.9839	31.4N	172.0E	82	189	58	01m44s
10005	501	2215 Dec 21	12:20:08	481	2671	145	T	n-	0.0701	1.0336	19.5S	4.1W	86	175	114	03m14s
10006	501	2216 Jun 16	03:41:04	482	2677	150	A	p-	-0.6420	0.9458	16.7S	124.6E	50	3	260	07m20s
10007	501	2216 Dec 10	03:57:52	483	2683	155	T	p-	0.7367	1.0421	24.8N	120.2E	42	180	208	03m51s
10008	501	2217 May 06	14:31:15	484	2688	122	P	-t	1.4355	0.2100	70.0N	178.5E	0	33		
10009	501	2217 Jun 05	04:22:20	485	2689	160	P	t-	-1.3807	0.3094	67.5S	118.9E	0	357		
10010	501	2217 Oct 31	06:08:54	486	2694	127	P	-t	-1.3157	0.4185	70.7S	49.8W	0	137		
10011	501	2217 Nov 29	18:29:51	486	2695	165	P	t-	1.4464	0.1782	68.1N	87.3W	0	190		
10012	501	2218 Apr 25	23:33:14	487	2700	132	T	-p	0.6321	1.0219	51.1N	174.3E	51	158	96	01m43s
10013	501	2218 Oct 20	11:41:56	488	2706	137	A	-p	-0.6411	0.9416	48.4S	12.1W	50	23	280	05m41s
10014	501	2219 Apr 15	14:26:33	489	2712	142	T	n-	-0.1086	1.0628	3.7N	32.8W	84	344	207	05m45s
10015	501	2219 Oct 09	11:48:35	491	2718	147	A	nn	0.0744	0.9338	2.0S	3.0E	86	197	248	08m46s
10016	501	2220 Apr 04	06:56:42	492	2724	152	T	p-	-0.8162	1.0454	46.2S	99.0E	35	335	260	03m25s
10017	501	2220 Sep 27	13:35:07	493	2730	157	A	p-	0.7966	0.9609	48.0N	2.8W	37	207	232	03m36s
10018	501	2221 Feb 23	06:50:48	494	2735	124	P	-t	1.2305	0.5688	71.5N	24.2E	0	123		
10019	501	2221 Aug 18	13:30:39	495	2741	129	P	-t	-1.1295	0.7673	70.9S	67.8W	0	47		
10020	501	2221 Sep 16	22:25:14	495	2742	167	P	t-	1.4775	0.1170	72.1N	58.1W	0	278		

Cat Num	Canon Plate	Calendar Date	TD of Greatest Eclipse	ΔT s	Luna Num	Saros Num	Ecl. Type	QLE	Gamma	Ecl. Mag.	Lat. °	Long. °	Sun Alt °	Sun Azm °	Path Width km	Central Line Dur.
10021	502	2222 Feb 12	09:23:18	497	2747	134	A	-p	0.5669	0.9220	20.0N	36.7E	55	166	355	10m14s
10022	502	2222 Aug 08	06:17:05	498	2753	139	T	-n	-0.3837	1.0774	6.0S	84.9E	67	11	270	07m06s
10023	502	2223 Feb 01	08:29:43	499	2759	144	A	nn	-0.1180	0.9263	24.1S	59.2E	83	349	279	09m26s
10024	502	2223 Jul 28	22:38:03	500	2765	149	T	n-	0.3636	1.0495	40.2N	151.7W	68	190	176	04m09s
10025	502	2224 Jan 21	11:48:53	502	2771	154	A	p-	-0.7984	0.9626	72.4S	25.2E	37	339	227	02m46s
10026	502	2224 Jul 17	10:03:58	503	2777	159	P	t-	1.1767	0.6677	68.4N	160.6W	0	348		
10027	502	2224 Dec 11	11:17:51	504	2782	126	P	-t	1.2791	0.4834	65.0N	35.2E	0	202		
10028	502	2225 Jan 09	22:25:24	504	2783	164	P	t-	-1.4263	0.2125	67.8S	20.4E	0	187		
10029	502	2225 Jun 06	23:21:31	505	2788	131	A	-p	-0.8496	0.9392	34.6S	156.5W	32	347	425	07m10s
10030	502	2225 Dec 01	03:08:36	506	2794	136	T	-p	0.6178	1.0398	15.4N	141.4E	52	194	169	03m43s
10031	502	2226 May 27	00:45:11	508	2800	141	A	nn	-0.0810	0.9670	16.8N	171.5E	85	344	119	03m55s
10032	502	2226 Nov 20	16:34:56	509	2806	146	Hm	nn	-0.0711	1.0005	23.7S	71.7W	86	19	2	00m03s
10033	502	2227 May 16	08:21:31	510	2812	151	T	p-	0.6774	1.0135	57.7N	30.8E	47	144	63	00m59s
10034	502	2227 Nov 09	23:36:42	511	2818	156	A	p-	-0.8171	0.9429	63.3S	140.7E	35	53	364	04m24s
10035	502	2228 Apr 05	14:15:36	512	2823	123	P	-t	-1.2036	0.6279	61.3S	47.3E	0	283		
10036	502	2228 May 04	22:28:44	513	2824	161	P	t-	1.3659	0.3173	62.4N	80.4E	0	53		
10037	502	2228 Sep 29	07:02:08	514	2829	128	P	-t	1.4212	0.2445	61.1N	159.6E	0	265		
10038	502	2228 Oct 29	00:15:43	514	2830	166	Pb	t-	-1.5410	0.0477	61.9S	57.7E	0	119		
10039	502	2229 Mar 26	06:17:35	515	2835	133	T	-p	-0.5251	1.0371	25.5S	105.5E	58	328	144	03m02s
10040	502	2229 Sep 18	10:34:51	516	2841	138	A	-p	0.6439	0.9805	36.2N	44.8E	50	217	89	01m44s
10041	503	2230 Mar 15	18:00:26	517	2847	143	A	nn	0.1964	0.9831	7.9N	91.3W	79	151	61	01m40s
10042	503	2230 Sep 07	21:30:39	519	2853	148	T	n-	-0.1309	1.0424	0.7S	144.5W	82	28	143	03m44s
10043	503	2231 Mar 04	22:20:24	520	2859	153	A	p-	0.9430	0.9246	52.4N	163.0E	19	130	838	06m32s
10044	503	2231 Aug 28	13:35:31	521	2865	158	T	p-	-0.8506	1.0661	41.4S	52.2W	31	36	402	04m43s
10045	503	2232 Jan 23	03:27:39	522	2870	125	P	-t	-1.3891	0.3001	63.3S	91.9W	0	222		
10046	503	2232 Jul 18	22:04:56	524	2876	130	T	-t	0.9717	1.0229	72.4N	33.4W	13	299	348	01m14s
10047	503	2233 Jan 11	08:49:17	525	2882	135	A	-p	-0.6447	0.9811	60.0S	70.4E	50	333	88	01m28s
10048	503	2233 Jul 08	07:35:24	526	2888	140	A	nn	0.2215	0.9809	35.1N	73.1E	77	194	70	01m59s
10049	503	2233 Dec 31	21:07:37	527	2894	145	T	n-	0.0649	1.0348	19.5S	134.7W	86	170	117	03m18s
10050	503	2234 Jun 27	10:09:34	529	2900	150	A	p-	-0.5572	0.9468	10.3S	26.1E	56	8	235	07m18s
10051	503	2234 Dec 21	12:46:02	530	2906	155	T	p-	0.7299	1.0403	23.5N	14.1W	43	175	197	03m42s
10052	503	2235 May 17	21:36:41	531	2911	122	Pe	-t	1.4946	0.1044	69.1N	60.3E	0	22		
10053	503	2235 Jun 16	11:00:36	531	2912	160	P	t-	-1.2990	0.4502	66.5S	8.8E	0	8		
10054	503	2235 Nov 11	14:13:08	532	2917	127	P	-t	-1.3444	0.3682	69.9S	176.6E	0	150		
10055	503	2235 Dec 11	03:02:34	533	2918	165	P	t-	1.4400	0.1913	67.1N	133.6E	0	179		
10056	503	2236 May 06	07:11:03	534	2923	132	T	-p	0.6848	1.0269	58.7N	58.9E	46	159	126	01m59s
10057	503	2236 Oct 30	19:15:15	535	2929	137	A	-p	-0.6779	0.9365	55.2S	126.4W	47	23	321	05m54s
10058	503	2237 Apr 25	22:25:04	536	2935	142	T	nn	-0.0606	1.0668	10.1N	153.7W	87	346	219	06m05s
10059	503	2237 Oct 19	19:06:04	538	2941	147	A	nn	0.0295	0.9316	8.6S	107.6W	88	196	256	09m07s
10060	503	2238 Apr 15	15:01:45	539	2947	152	T	p-	-0.7772	1.0475	39.3S	27.3W	39	340	250	03m49s
10061	504	2238 Oct 08	21:01:18	540	2953	157	A	p-	0.7459	0.9618	40.1N	119.7W	41	202	206	03m47s
10062	504	2239 Mar 06	14:54:58	541	2958	124	P	-t	1.2541	0.5278	72.0N	110.6W	0	109		
10063	504	2239 Aug 29	21:05:15	543	2964	129	P	-t	-1.1897	0.6529	71.5S	165.5E	0	60		
10064	504	2239 Sep 28	06:09:02	543	2965	167	P	t-	1.4239	0.2160	72.1N	172.0E	0	264		
10065	504	2240 Feb 23	17:14:11	544	2970	134	A	-p	0.5859	0.9228	24.7N	83.0W	54	163	356	09m41s
10066	504	2240 Aug 18	13:52:25	545	2976	139	T	-p	-0.4522	1.0746	13.3S	31.3W	63	14	270	06m40s
10067	504	2241 Feb 11	16:28:39	546	2982	144	A	nn	-0.1046	0.9292	19.9S	60.0W	84	347	267	09m04s
10068	504	2241 Aug 08	05:59:21	548	2988	149	T	n-	0.2920	1.0457	32.9N	98.0E	73	193	159	04m02s
10069	504	2242 Jan 31	20:12:58	549	2994	154	A	p-	-0.7894	0.9665	67.9S	95.8W	38	333	197	02m31s
10070	504	2242 Jul 28	16:57:12	550	3000	159	P	t-	1.1020	0.8004	69.3N	84.8E	0	336		
10071	504	2242 Dec 22	20:06:40	551	3005	126	P	-t	1.2836	0.4750	66.0N	106.9W	0	192		
10072	504	2243 Jan 21	07:11:45	552	3006	164	P	t-	-1.4198	0.2238	68.9S	122.5W	0	198		
10073	504	2243 Jun 18	05:49:56	553	3011	131	A	-t	-0.9342	0.9380	45.6S	104.7E	20	351	652	06m41s
10074	504	2243 Dec 12	11:52:14	554	3017	136	T	-p	0.6284	1.0365	15.5N	9.0E	51	190	157	03m30s
10075	504	2244 Jun 06	07:33:12	555	3023	141	Am	nn	-0.1581	0.9712	13.8N	70.7E	81	349	105	03m31s
10076	504	2244 Dec 01	00:58:17	557	3029	146	A	nn	-0.0568	0.9955	25.1S	164.0E	87	14	16	00m27s
10077	504	2245 May 26	15:42:04	558	3035	151	T	p-	0.6089	1.0201	56.7N	71.4W	52	153	86	01m30s
10078	504	2245 Nov 20	07:29:36	559	3041	156	A	p-	-0.7955	0.9387	66.3S	27.1E	37	48	374	04m45s
10079	504	2246 Apr 16	22:23:24	560	3046	123	P	-t	-1.2445	0.5498	61.6S	83.2W	0	292		
10080	504	2246 May 16	06:14:10	561	3047	161	P	t-	1.3077	0.4284	63.1N	44.9W	0	44		

Cat Num	Canon Plate	Calendar Date	TD of Greatest Eclipse	ΔT s	Luna Num	Saros Num	Ecl. Type	QLE	Gamma	Ecl. Mag.	Lat. °	Long. °	Sun Alt °	Sun Azm °	Path Width km	Central Line Dur.
10081	505	2246 Oct 10	14:13:18	562	3052	128	P	-t	1.4705	0.1615	61.3N	43.4E	0	256		
10082	505	2246 Nov 09	07:47:03	562	3053	166	P	t-	-1.5082	0.1036	62.5S	63.8W	0	129		
10083	505	2247 Apr 06	14:26:51	563	3058	133	T	-p	-0.5624	1.0372	23.8S	16.9W	56	329	149	03m07s
10084	505	2247 Sep 29	18:01:05	564	3064	138	A	-p	0.6961	0.9801	35.6N	65.9W	46	216	96	01m47s
10085	505	2248 Mar 26	01:56:01	566	3070	143	A	nn	0.1643	0.9829	10.6N	150.1E	80	151	61	01m41s
10086	505	2248 Sep 18	05:13:07	567	3076	148	T	nn	-0.0738	1.0426	2.0S	100.6E	86	29	143	03m42s
10087	505	2249 Mar 15	06:00:45	568	3082	153	A	p-	0.9149	0.9266	52.0N	48.4E	23	128	666	06m18s
10088	505	2249 Sep 07	21:21:29	570	3088	158	T	p-	-0.7907	1.0656	39.4S	167.4W	38	37	343	04m42s
10089	505	2250 Feb 02	11:34:07	571	3093	125	P	-t	-1.3969	0.2864	62.5S	137.6E	0	231		
10090	505	2250 Jul 30	05:18:25	572	3099	130	P	-t	1.0490	0.9114	62.9N	124.7W	0	314		
10091	505	2250 Aug 28	13:51:18	572	3100	168	Pb	t-	-1.5278	0.0120	61.7S	96.6W	0	69		
10092	505	2251 Jan 22	17:21:41	573	3105	135	A	-p	-0.6480	0.9844	56.9S	53.2W	49	327	72	01m12s
10093	505	2251 Jul 19	14:18:46	575	3111	140	A	-p	0.3062	0.9773	38.0N	24.2W	72	200	85	02m16s
10094	505	2252 Jan 12	05:57:05	576	3117	145	T	n-	0.0608	1.0365	18.5S	94.0E	87	165	123	03m23s
10095	505	2252 Jul 07	16:34:12	577	3123	150	A	p-	-0.4686	0.9473	4.9S	70.6W	62	12	218	07m10s
10096	505	2252 Dec 31	21:37:06	579	3129	155	T	p-	0.7258	1.0389	23.1N	149.1W	43	170	189	03m33s
10097	505	2253 Jun 26	17:36:11	580	3135	160	P	t-	-1.2139	0.5981	65.5S	100.1W	0	18		
10098	505	2253 Nov 21	22:24:38	581	3140	127	P	-t	-1.3666	0.3297	68.9S	41.9E	0	162		
10099	505	2253 Dec 21	11:39:39	582	3141	165	P	t-	1.4374	0.1972	66.1N	6.0W	0	168		
10100	505	2254 May 17	14:43:39	583	3146	132	T	-p	0.7426	1.0315	66.7N	54.1W	42	161	160	02m09s
10101	506	2254 Nov 11	02:55:16	584	3152	137	A	-p	-0.7086	0.9317	61.4S	119.3E	45	21	363	06m05s
10102	506	2255 May 07	06:18:06	585	3158	142	T	nn	-0.0076	1.0706	16.4N	87.2E	90	346	230	06m22s
10103	506	2255 Oct 31	02:32:04	587	3164	147	A	nn	-0.0088	0.9295	14.5S	140.2E	89	11	264	09m24s
10104	506	2256 Apr 25	22:58:35	588	3170	152	T	p-	-0.7317	1.0495	32.3S	150.9W	43	344	240	04m14s
10105	506	2256 Oct 19	04:37:31	589	3176	157	A	p-	0.7025	0.9624	33.1N	122.3E	45	198	190	03m59s
10106	506	2257 Mar 16	22:51:29	590	3181	124	P	-t	1.2833	0.4770	72.2N	116.2E	0	95		
10107	506	2257 Apr 15	12:05:15	591	3182	162	Pb	t-	-1.5121	0.0633	71.3S	60.1E	0	302		
10108	506	2257 Sep 09	04:46:44	592	3187	129	P	-t	-1.2448	0.5480	71.9S	36.6E	0	73		
10109	506	2257 Oct 08	14:01:32	592	3188	167	P	t-	1.3765	0.3034	71.9N	40.0E	0	250		
10110	506	2258 Mar 06	00:58:23	593	3193	134	A	-p	0.6101	0.9239	30.2N	158.8E	52	160	359	09m04s
10111	506	2258 Aug 29	21:33:05	595	3199	139	T	-p	-0.5161	1.0712	20.9S	149.2W	59	17	269	06m09s
10112	506	2259 Feb 23	00:23:41	596	3205	144	A	nn	-0.0875	0.9326	15.0S	178.8W	85	345	253	08m36s
10113	506	2259 Aug 19	13:22:17	597	3211	149	T	nn	0.2226	1.0412	25.3N	13.6W	77	195	141	03m49s
10114	506	2260 Feb 12	04:34:24	599	3217	154	A	p-	-0.7776	0.9711	62.7S	140.2E	39	331	165	02m15s
10115	506	2260 Aug 07	23:51:13	600	3223	159	P	t-	1.0287	0.9293	70.2N	30.7W	0	325		
10116	506	2261 Jan 02	04:56:54	601	3228	126	P	-t	1.2873	0.4679	67.1N	110.2E	0	181		
10117	506	2261 Jan 31	15:55:00	601	3229	164	P	t-	-1.4107	0.2397	69.9S	94.8E	0	211		
10118	506	2261 Jun 28	12:16:28	603	3234	131	P	-t	-1.0198	0.9282	66.6S	6.0E	0	354		
10119	506	2261 Dec 22	20:38:50	604	3240	136	T	-p	0.6360	1.0337	16.1N	124.2W	50	185	147	03m17s
10120	506	2262 Jun 17	14:19:15	605	3246	141	A	nn	-0.2377	0.9750	9.8N	30.2W	76	353	92	03m08s
10121	507	2262 Dec 12	09:25:02	607	3252	146	A	nn	-0.0461	0.9910	25.8S	39.0E	87	10	32	00m56s
10122	507	2263 Jun 06	22:58:57	608	3258	151	T	p-	0.5366	1.0261	54.4N	173.1W	57	162	105	02m01s
10123	507	2263 Dec 01	15:28:45	609	3264	156	A	p-	-0.7794	0.9349	69.2S	85.8W	38	41	388	05m06s
10124	507	2264 Apr 27	06:21:41	611	3269	123	P	-t	-1.2931	0.4564	62.1S	148.5E	0	301		
10125	507	2264 May 26	13:52:07	611	3270	161	P	t-	1.2430	0.5526	63.9N	168.5W	0	35		
10126	507	2264 Oct 20	21:35:23	612	3275	128	P	-t	1.5111	0.0933	61.6N	75.7W	0	247		
10127	507	2264 Nov 19	15:28:13	612	3276	166	P	t-	-1.4830	0.1464	63.2S	172.0E	0	138		
10128	507	2265 Apr 16	22:26:19	613	3281	133	T	-p	-0.6073	1.0371	23.1S	136.8W	52	331	154	03m11s
10129	507	2265 Oct 10	01:37:34	615	3287	138	A	-p	0.7404	0.9796	35.1N	179.8W	42	215	105	01m51s
10130	507	2266 Apr 06	09:42:37	616	3293	143	Am	nn	0.1255	0.9829	12.9N	34.0E	83	151	61	01m42s
10131	507	2266 Sep 29	13:03:57	617	3299	148	T	nn	-0.0233	1.0425	3.7S	16.4W	89	28	142	03m40s
10132	507	2267 Mar 26	13:33:45	619	3305	153	A	p-	0.8810	0.9289	52.3N	63.7W	28	128	549	06m03s
10133	507	2267 Sep 19	05:12:14	620	3311	158	T	p-	-0.7348	1.0642	38.8S	75.9E	42	38	304	04m34s
10134	507	2268 Feb 13	19:39:32	621	3316	125	P	-t	-1.4059	0.2703	61.9S	7.5E	0	241		
10135	507	2268 Aug 09	12:32:05	623	3322	130	P	-t	1.1254	0.7684	62.2N	118.0E	0	305		
10136	507	2268 Sep 07	21:27:52	623	3323	168	P	t-	-1.4722	0.1194	61.4S	140.6E	0	78		
10137	507	2269 Feb 02	01:53:06	624	3328	135	A	-p	-0.6529	0.9883	53.2S	178.2W	49	323	54	00m54s
10138	507	2269 Jul 29	21:03:04	625	3334	140	A	-p	0.3893	0.9732	39.9N	121.3W	67	205	104	02m35s
10139	507	2270 Jan 22	14:46:29	627	3340	145	T	n-	0.0560	1.0385	16.7S	37.3W	87	161	130	03m29s
10140	507	2270 Jul 18	22:59:54	628	3346	150	A	p-	-0.3811	0.9474	0.7S	166.9W	68	16	208	06m57s

Cat Num	Canon Plate	Calendar Date	TD of Greatest Eclipse	ΔT s	Luna Num	Saros Num	Ecl. Type	QLE	Gamma	Ecl. Mag.	Lat. °	Long. °	Sun Alt °	Sun Azm °	Path Width km	Central Line Dur.
10141	508	2271 Jan 12	06:28:08	630	3352	155	T	p-	0.7217	1.0379	23.3N	76.0E	44	165	182	03m25s
10142	508	2271 Jul 08	00:13:02	631	3358	160	P	t-	-1.1284	0.7474	64.5S	151.1E	0	27		
10143	508	2271 Dec 03	06:40:47	632	3363	127	P	-t	-1.3843	0.2996	67.8S	93.4W	0	174		
10144	508	2272 Jan 01	20:17:51	632	3364	165	P	t-	1.4365	0.2000	65.1N	145.4W	0	158		
10145	508	2272 May 27	22:11:12	634	3369	132	T	-p	0.8053	1.0353	75.0N	163.2W	36	166	202	02m14s
10146	508	2272 Nov 21	10:42:52	635	3375	137	A	-p	-0.7327	0.9275	66.8S	5.9E	43	16	402	06m15s
10147	508	2273 May 17	14:04:31	636	3381	142	Tm	nn	0.0515	1.0738	22.5N	29.7W	87	173	240	06m31s
10148	508	2273 Nov 10	10:07:17	638	3387	147	A	nn	-0.0398	0.9278	19.6S	26.3E	88	10	272	09m34s
10149	508	2274 May 07	06:47:37	639	3393	152	T	p-	-0.6799	1.0510	25.5S	88.2E	47	348	230	04m37s
10150	508	2274 Oct 30	12:24:18	641	3399	157	A	p-	0.6667	0.9629	27.0N	2.4E	48	195	179	04m08s
10151	508	2275 Mar 28	06:37:50	642	3404	124	P	-t	1.3199	0.4133	72.2N	14.4W	0	81		
10152	508	2275 Apr 26	19:41:41	642	3405	162	P	t-	-1.4684	0.1423	70.7S	67.0W	0	315		
10153	508	2275 Sep 20	12:34:54	643	3410	129	P	-t	-1.2949	0.4527	72.0S	94.3W	0	87		
10154	508	2275 Oct 19	22:03:12	643	3411	167	P	t-	1.3358	0.3786	71.4N	93.9W	0	237		
10155	508	2276 Mar 16	08:34:03	645	3416	134	A	-p	0.6411	0.9253	36.4N	42.3E	50	158	362	08m23s
10156	508	2276 Sep 09	05:18:47	646	3422	139	T	-p	-0.5755	1.0671	28.5S	91.2E	55	20	266	05m33s
10157	508	2277 Mar 05	08:11:55	647	3428	144	A	nn	-0.0645	0.9366	9.5S	63.6E	86	343	236	08m04s
10158	508	2277 Aug 29	20:49:11	649	3434	149	T	nn	0.1573	1.0362	17.8N	126.7W	81	196	123	03m28s
10159	508	2278 Feb 22	12:52:48	650	3440	154	A	p-	-0.7628	0.9762	57.1S	14.8E	40	331	131	01m54s
10160	508	2278 Aug 19	06:46:23	652	3446	159	A	t-	0.9569	0.9712	75.8N	155.8E	16	257	367	01m53s
10161	509	2279 Jan 13	13:49:06	653	3451	126	P	-t	1.2899	0.4630	68.2N	33.7W	0	170		
10162	509	2279 Feb 12	00:37:06	653	3452	164	P	t-	-1.4003	0.2581	70.7S	48.3W	0	224		
10163	509	2279 Jul 09	18:41:13	654	3457	131	P	-t	-1.1065	0.7802	67.7S	100.7W	0	4		
10164	509	2280 Jan 03	05:28:11	656	3463	136	T	-p	0.6414	1.0314	17.2N	101.9E	50	180	138	03m04s
10165	509	2280 Jun 27	21:03:21	657	3469	141	A	nn	-0.3197	0.9784	4.6N	131.2W	71	357	81	02m45s
10166	509	2280 Dec 22	17:55:44	659	3475	146	A	nn	-0.0392	0.9870	25.8S	86.8W	88	5	46	01m23s
10167	509	2281 Jun 17	06:14:41	660	3481	151	T	p-	0.4621	1.0316	50.8N	84.2E	62	170	121	02m32s
10168	509	2281 Dec 11	23:31:24	661	3487	156	A	p-	-0.7667	0.9316	71.4S	163.7E	40	30	400	05m26s
10169	509	2282 May 08	14:15:16	663	3492	123	P	-t	-1.3458	0.3545	62.7S	21.3E	0	310		
10170	509	2282 Jun 06	21:28:19	663	3493	161	P	t-	1.1764	0.6815	64.8N	68.1E	0	25		
10171	509	2282 Nov 01	05:06:24	664	3498	128	Pe	-t	1.5448	0.0370	62.1N	163.0E	0	238		
10172	509	2282 Nov 30	23:15:23	664	3499	166	P	t-	-1.4625	0.1812	64.1S	46.1E	0	148		
10173	509	2283 Apr 28	06:18:21	666	3504	133	T	-p	-0.6581	1.0366	23.6S	105.0E	49	334	160	03m13s
10174	509	2283 Oct 21	09:23:11	667	3510	138	A	-p	0.7783	0.9790	34.9N	63.2E	39	212	116	01m56s
10175	509	2284 Apr 16	17:19:22	668	3516	143	A	nn	0.0792	0.9827	14.6N	79.2W	85	153	61	01m45s
10176	509	2284 Oct 09	21:03:48	670	3522	148	T	nn	0.0205	1.0420	5.7S	135.8W	89	209	140	03m39s
10177	509	2285 Apr 05	20:55:23	671	3528	153	A	p-	0.8379	0.9315	52.9N	171.4W	33	129	459	05m50s
10178	509	2285 Sep 29	13:11:38	673	3534	158	T	p-	-0.6859	1.0621	39.6S	42.9W	46	38	275	04m24s
10179	509	2286 Feb 24	03:39:23	674	3539	125	P	-t	-1.4203	0.2448	61.5S	121.0W	0	250		
10180	509	2286 Mar 25	20:37:48	674	3540	163	Pb	t-	1.5392	0.0472	61.0N	141.1E	0	86		
10181	510	2286 Aug 20	19:48:22	675	3545	130	P	-t	1.1987	0.6322	61.7N	0.2E	0	296		
10182	510	2286 Sep 19	05:10:04	676	3546	168	P	t-	-1.4214	0.2166	61.2S	16.5E	0	87		
10183	510	2287 Feb 13	10:21:25	677	3551	135	A	-p	-0.6613	0.9926	49.4S	56.3E	48	321	34	00m35s
10184	510	2287 Aug 10	03:47:42	678	3557	140	A	-p	0.4714	0.9686	41.0N	141.8E	62	210	127	02m56s
10185	510	2288 Feb 02	23:33:47	680	3563	145	T	n-	0.0492	1.0412	14.2S	168.4W	87	157	138	03m38s
10186	510	2288 Jul 29	05:25:23	681	3569	150	A	pn	-0.2930	0.9469	2.5N	97.4E	73	19	203	06m46s
10187	510	2289 Jan 22	15:19:25	683	3575	155	T	p-	0.7181	1.0374	24.3N	58.9W	44	161	178	03m18s
10188	510	2289 Jul 18	06:50:58	684	3581	160	P	t-	-1.0426	0.8980	63.6S	42.3E	0	37		
10189	510	2289 Dec 13	15:01:18	685	3586	127	P	-t	-1.3979	0.2767	66.8S	130.8E	0	185		
10190	510	2290 Jan 12	04:56:33	686	3587	165	P	t-	1.4365	0.2009	64.1N	75.4E	0	148		
10191	510	2290 Jun 08	05:35:49	687	3592	132	T	-p	0.8713	1.0382	83.8N	100.9E	29	182	265	02m14s
10192	510	2290 Dec 02	18:36:41	688	3598	137	A	-p	-0.7515	0.9237	70.9S	104.7W	41	7	439	06m23s
10193	510	2291 May 28	21:45:28	690	3604	142	T	nn	0.1153	1.0764	28.3N	144.5W	83	176	249	06m34s
10194	510	2291 Nov 21	17:50:53	691	3610	147	A	nn	-0.0644	0.9263	23.7S	88.9W	86	7	278	09m41s
10195	510	2292 May 17	14:29:33	693	3616	152	T	p-	-0.6224	1.0521	18.8S	30.3W	51	353	220	04m56s
10196	510	2292 Nov 09	20:20:07	694	3622	157	A	p-	0.6376	0.9635	22.0N	119.1W	50	191	171	04m14s
10197	510	2293 Apr 07	14:14:55	695	3627	124	P	-t	1.3632	0.3380	71.8N	142.5W	0	67		
10198	510	2293 May 07	03:09:47	696	3628	162	P	t-	-1.4186	0.2323	69.9S	168.5E	0	328		
10199	510	2293 Sep 30	20:31:28	697	3633	129	P	-t	-1.3386	0.3697	72.0S	132.7E	0	100		
10200	510	2293 Oct 30	06:13:45	697	3634	167	P	t-	1.3017	0.4416	70.7N	130.3E	0	223		

Cat Num	Canon Plate	Calendar Date	TD of Greatest Eclipse	ΔT s	Luna Num	Saros Num	Ecl. Type	QLE	Gamma	Ecl. Mag.	Lat. °	Long. °	Sun Alt °	Sun Azm °	Path Width km	Central Line Dur.
10201	511	2294 Mar 27	16:02:23	698	3639	134	A	-p	0.6776	0.9269	43.2N	72.6W	47	156	370	07m42s
10202	511	2294 Sep 20	13:09:58	700	3645	139	T	-p	-0.6300	1.0627	36.2S	29.9W	51	22	263	04m56s
10203	511	2295 Mar 16	15:54:34	701	3651	144	A	nn	-0.0362	0.9409	3.6S	53.0W	88	343	219	07m29s
10204	511	2295 Sep 10	04:20:19	703	3657	149	Tm	nn	0.0963	1.0307	10.3N	118.9E	84	197	104	03m01s
10205	511	2296 Mar 04	21:04:46	704	3663	154	A	p-	-0.7418	0.9819	51.1S	110.3W	42	333	95	01m31s
10206	511	2296 Aug 29	13:45:40	706	3669	159	A	p-	0.8888	0.9689	66.6N	15.0E	27	223	245	02m20s
10207	511	2297 Jan 23	22:39:47	707	3674	126	P	-t	1.2940	0.4550	69.2N	177.8W	0	158		
10208	511	2297 Feb 22	09:13:31	707	3675	164	P	t-	-1.3851	0.2853	71.4S	169.4E	0	237		
10209	511	2297 Jul 20	01:07:47	708	3680	131	P	-t	-1.1915	0.6346	68.7S	151.6E	0	15		
10210	511	2298 Jan 13	14:16:27	710	3686	136	T	-p	0.6474	1.0296	19.0N	31.9W	50	176	131	02m52s
10211	511	2298 Jul 09	03:49:02	711	3692	141	A	-p	-0.4012	0.9811	1.4S	126.5E	66	2	73	02m23s
10212	511	2299 Jan 03	02:27:43	713	3698	146	A	nn	-0.0341	0.9836	24.9S	146.9E	88	0	58	01m47s
10213	511	2299 Jun 28	13:27:43	714	3704	151	T	p-	0.3846	1.0365	46.0N	19.5W	67	176	133	03m03s
10214	511	2299 Dec 23	07:38:42	716	3710	156	A	p-	-0.7584	0.9288	72.5S	54.8E	40	16	413	05m45s
10215	511	2300 May 19	22:00:39	717	3715	123	P	-t	-1.4049	0.2399	63.4S	104.1W	0	319		
10216	511	2300 Jun 18	04:59:29	717	3716	161	P	t-	1.1056	0.8189	65.7N	54.5W	0	16		
10217	511	2300 Dec 12	07:09:43	719	3722	166	P	t-	-1.4473	0.2067	65.0S	81.9W	0	158		
10218	511	2301 May 09	14:00:59	720	3727	133	T	-p	-0.7161	1.0354	25.5S	11.0W	44	337	168	03m10s
10219	511	2301 Nov 01	17:19:33	721	3733	138	A	-p	0.8080	0.9786	34.8N	57.2W	36	209	126	02m01s
10220	511	2302 Apr 29	00:47:19	723	3739	143	A	nn	0.0263	0.9825	15.6N	170.0E	88	157	62	01m49s
10221	512	2302 Oct 22	05:11:16	724	3745	148	T	nn	0.0584	1.0413	7.8S	102.9E	87	207	139	03m38s
10222	512	2303 Apr 18	04:09:26	726	3751	153	A	p-	0.7889	0.9341	53.8N	83.7E	38	132	393	05m38s
10223	512	2303 Oct 11	21:17:25	727	3757	158	T	p-	-0.6424	1.0596	41.1S	163.2W	50	38	252	04m12s
10224	512	2304 Mar 07	11:34:24	729	3762	125	P	-t	-1.4389	0.2118	61.2S	111.8E	0	259		
10225	512	2304 Apr 06	04:00:21	729	3763	163	P	t-	1.4957	0.1189	61.2N	22.1E	0	77		
10226	512	2304 Sep 01	03:07:40	730	3768	130	P	-t	1.2684	0.5038	61.4N	118.2W	0	288		
10227	512	2304 Sep 30	12:58:17	730	3769	168	P	t-	-1.3760	0.3030	61.2S	109.0W	0	96		
10228	512	2305 Feb 24	18:46:09	732	3774	135	A	-p	-0.6732	0.9973	45.7S	69.3W	47	320	13	00m13s
10229	512	2305 Aug 21	10:35:44	733	3780	140	A	-p	0.5497	0.9637	41.5N	43.7E	56	214	155	03m21s
10230	512	2306 Feb 14	08:17:49	735	3786	145	T	nn	0.0394	1.0441	11.3S	61.0E	88	154	147	03m49s
10231	512	2306 Aug 10	11:55:10	736	3792	150	A	nn	-0.2083	0.9461	4.6N	1.0E	78	23	202	06m37s
10232	512	2307 Feb 04	00:08:01	738	3798	155	T	p-	0.7125	1.0373	25.7N	166.9E	44	156	176	03m12s
10233	512	2307 Jul 30	13:31:16	739	3804	160	A	t-	-0.9574	0.9602	50.0S	48.7W	16	30	501	03m37s
10234	512	2307 Dec 25	23:24:23	740	3809	127	P	-t	-1.4089	0.2585	65.7S	5.1W	0	195		
10235	512	2308 Jan 24	13:33:40	741	3810	165	P	t-	1.4358	0.2029	63.3N	63.0W	0	138		
10236	512	2308 Jun 19	12:57:53	742	3815	132	T	-t	0.9402	1.0396	84.1N	120.6E	19	313	401	02m08s
10237	512	2308 Dec 14	02:34:52	743	3821	137	A	-p	-0.7662	0.9207	73.4S	148.6E	40	353	470	06m31s
10238	512	2309 Jun 09	05:21:55	745	3827	142	T	-n	0.1833	1.0783	33.6N	102.7E	79	181	257	06m30s
10239	512	2309 Dec 03	01:42:05	746	3833	147	A	nn	-0.0832	0.9254	26.9S	154.6E	85	3	282	09m40s
10240	512	2310 May 29	22:04:50	748	3839	152	T	p-	-0.5599	1.0526	12.5S	146.5W	56	357	210	05m10s
10241	513	2310 Nov 22	04:24:19	749	3845	157	A	p-	0.6145	0.9642	17.9N	117.8E	52	187	164	04m16s
10242	513	2311 Apr 19	21:41:49	751	3850	124	P	-t	1.4139	0.2499	71.3N	92.3E	0	53		
10243	513	2311 May 19	10:28:46	751	3851	162	P	t-	-1.3621	0.3345	69.0S	46.9E	0	340		
10244	513	2311 Oct 13	04:36:09	752	3856	129	P	-t	-1.3762	0.2985	71.6S	2.1W	0	114		
10245	513	2311 Nov 11	14:33:19	753	3857	167	P	t-	1.2745	0.4919	69.9N	7.1W	0	211		
10246	513	2312 Apr 07	23:19:32	754	3862	134	A	-p	0.7231	0.9286	50.8N	174.7E	43	153	385	07m00s
10247	513	2312 Oct 01	21:08:26	755	3868	139	T	-p	-0.6783	1.0578	43.8S	152.9W	47	24	258	04m20s
10248	513	2313 Mar 27	23:29:31	757	3874	144	A	nn	-0.0011	0.9456	2.6N	167.9W	90	336	200	06m49s
10249	513	2313 Sep 21	11:57:00	758	3880	149	T	nn	0.0405	1.0249	3.0N	3.0E	88	198	85	02m30s
10250	513	2314 Mar 17	05:11:54	760	3886	154	A	p-	-0.7160	0.9880	44.9S	125.1E	44	335	60	01m03s
10251	513	2314 Sep 10	20:49:11	761	3892	159	A	p-	0.8247	0.9654	56.8N	103.3W	34	212	220	02m54s
10252	513	2315 Feb 05	07:29:49	763	3897	126	P	-t	1.2991	0.4453	70.1N	37.6E	0	145		
10253	513	2315 Mar 06	17:46:20	763	3898	164	P	t-	-1.3668	0.3187	71.9S	27.6E	0	251		
10254	513	2315 Aug 01	07:34:32	764	3903	131	P	-t	-1.2761	0.4898	69.6S	43.3E	0	27		
10255	513	2316 Jan 25	23:05:17	766	3909	136	T	-p	0.6526	1.0282	21.4N	166.0W	49	172	126	02m42s
10256	513	2316 Jul 20	10:36:18	767	3915	141	A	-p	-0.4819	0.9834	8.1S	23.1E	61	5	67	02m03s
10257	513	2317 Jan 14	10:59:38	769	3921	146	A	nn	-0.0298	0.9807	23.2S	20.5E	88	356	69	02m08s
10258	513	2317 Jul 09	20:42:40	770	3927	151	T	p-	0.3078	1.0406	40.4N	125.3W	72	182	143	03m32s
10259	513	2318 Jan 03	15:47:14	772	3933	156	A	p-	-0.7519	0.9265	71.9S	53.7W	41	1	422	06m02s
10260	513	2318 May 31	05:42:33	773	3938	123	Pe	-t	-1.4670	0.1192	64.2S	131.2E	0	329		

Cat Num	Canon Plate	Calendar Date	TD of Greatest Eclipse	ΔT s	Luna Num	Saros Num	Ecl. Type	QLE	Gamma	Ecl. Mag.	Lat. °	Long. °	Sun Alt °	Sun Azm °	Path Width km	Central Line Dur.
10261	514	2318 Jun 29	12:30:22	773	3939	161	P	t-	1.0340	0.9583	66.7N	177.3W	0	6		
10262	514	2318 Dec 23	15:07:26	775	3945	166	P	t-	-1.4346	0.2279	66.1S	148.9E	0	168		
10263	514	2319 May 20	21:37:23	776	3950	133	T	-p	-0.7786	1.0336	29.0S	125.8W	39	340	178	03m02s
10264	514	2319 Nov 13	01:24:39	778	3956	138	A	-p	0.8314	0.9784	35.0N	179.6E	34	205	136	02m04s
10265	514	2320 May 09	08:04:33	779	3962	143	A	nn	-0.0347	0.9820	15.6N	62.1E	88	337	64	01m56s
10266	514	2320 Nov 01	13:28:19	781	3968	148	Tm	nn	0.0888	1.0406	9.8S	20.8W	85	204	136	03m38s
10267	514	2321 Apr 28	11:12:59	783	3974	153	A	p-	0.7315	0.9367	54.5N	17.0W	43	136	341	05m30s
10268	514	2321 Oct 22	05:31:18	784	3980	158	T	p-	-0.6059	1.0567	43.3S	74.8E	52	37	233	04m00s
10269	514	2322 Mar 18	19:21:51	785	3985	125	P	-t	-1.4640	0.1671	61.1S	13.5W	0	268		
10270	514	2322 Apr 17	11:14:23	786	3986	163	P	t-	1.4446	0.2041	61.5N	94.9W	0	68		
10271	514	2322 Sep 12	10:32:06	787	3991	130	P	-t	1.3328	0.3865	61.1N	122.2E	0	279		
10272	514	2322 Oct 11	20:53:38	787	3992	168	P	t-	-1.3371	0.3763	61.4S	123.6E	0	105		
10273	514	2323 Mar 08	03:05:10	788	3997	135	H	-p	-0.6906	1.0023	42.4S	166.1E	46	320	11	00m11s
10274	514	2323 Sep 01	17:26:09	790	4003	140	A	-p	0.6253	0.9584	41.7N	55.3W	51	218	191	03m48s
10275	514	2324 Feb 25	16:57:32	792	4009	145	Tm	nn	0.0257	1.0475	8.1S	68.6W	89	152	158	04m02s
10276	514	2324 Aug 20	18:28:22	793	4015	150	A	nn	-0.1261	0.9449	5.7N	96.0W	83	25	205	06m33s
10277	514	2325 Feb 14	08:52:36	795	4021	155	T	p-	0.7038	1.0378	27.5N	33.9E	45	152	175	03m08s
10278	514	2325 Aug 09	20:16:24	796	4027	160	A	t-	-0.8749	0.9648	40.3S	146.1W	29	30	256	03m24s
10279	514	2326 Jan 05	07:49:43	798	4032	127	P	-t	-1.4177	0.2440	64.7S	141.2W	0	206		
10280	514	2326 Feb 03	22:08:49	798	4033	165	P	t-	1.4340	0.2068	62.6N	159.3E	0	128		
10281	515	2326 Jun 30	20:18:36	799	4038	132	P	-t	1.0107	0.9931	65.2N	37.3E	0	339		
10282	515	2326 Dec 25	10:36:53	801	4044	137	A	-p	-0.7774	0.9182	73.6S	43.3E	39	337	496	06m39s
10283	515	2327 Jun 20	12:55:01	802	4050	142	T	-n	0.2542	1.0795	38.3N	8.3W	75	186	265	06m21s
10284	515	2327 Dec 14	09:39:47	804	4056	147	A	nn	-0.0969	0.9250	28.8S	37.0E	84	358	284	09m34s
10285	515	2328 Jun 09	05:33:53	805	4062	152	T	p-	-0.4928	1.0524	6.7S	99.5E	60	1	199	05m15s
10286	515	2328 Dec 02	12:36:37	807	4068	157	A	p-	0.5974	0.9652	14.8N	6.9W	53	183	157	04m13s
10287	515	2329 Apr 30	04:59:58	808	4073	124	P	-t	1.4705	0.1514	70.6N	30.1W	0	40		
10288	515	2329 May 29	17:41:09	809	4074	162	P	t-	-1.3009	0.4449	68.1S	72.5W	0	351		
10289	515	2329 Oct 23	12:48:23	810	4079	129	P	-t	-1.4082	0.2383	71.1S	138.5W	0	127		
10290	515	2329 Nov 21	22:59:20	810	4080	167	P	t-	1.2521	0.5333	68.9N	145.4W	0	198		
10291	515	2330 Apr 19	06:29:25	811	4085	134	A	-p	0.7742	0.9302	59.0N	62.9E	39	151	412	06m19s
10292	515	2330 Oct 13	05:13:41	813	4091	139	T	-p	-0.7208	1.0528	51.2S	82.5E	44	27	251	03m46s
10293	515	2331 Apr 08	06:57:09	815	4097	144	A	nn	0.0408	0.9506	9.2N	79.0E	88	164	181	06m07s
10294	515	2331 Oct 02	19:39:16	816	4103	149	T	nn	-0.0097	1.0188	4.0S	114.2W	89	17	64	01m55s
10295	515	2332 Mar 27	13:11:34	818	4109	154	A	p-	-0.6831	0.9944	38.3S	2.0E	47	338	26	00m30s
10296	515	2332 Sep 21	03:59:10	819	4115	159	A	p-	0.7666	0.9613	47.9N	142.3E	40	207	217	03m34s
10297	515	2333 Feb 15	16:14:20	821	4120	126	P	-t	1.3087	0.4270	70.9N	106.2W	0	132		
10298	515	2333 Mar 17	02:10:53	821	4121	164	P	t-	-1.3417	0.3651	72.1S	112.5W	0	265		
10299	515	2333 Aug 11	14:06:48	822	4126	131	P	-t	-1.3558	0.3534	70.5S	66.9W	0	38		
10300	515	2333 Sep 10	05:42:00	823	4127	169	Pb	t-	1.5299	0.0592	72.0N	157.1W	0	286		
10301	516	2334 Feb 05	07:50:29	824	4132	136	T	-p	0.6603	1.0272	24.6N	60.8E	49	168	122	02m33s
10302	516	2334 Jul 31	17:26:33	825	4138	141	A	-p	-0.5608	0.9851	15.6S	81.8W	56	9	64	01m45s
10303	516	2335 Jan 25	19:29:43	827	4144	146	A	nn	-0.0247	0.9784	20.6S	105.9W	88	352	77	02m25s
10304	516	2335 Jul 21	03:57:49	829	4150	151	T	n-	0.2306	1.0440	34.0N	127.4E	76	186	151	03m58s
10305	516	2336 Jan 14	23:56:42	830	4156	156	A	p-	-0.7463	0.9250	69.6S	164.9W	41	349	427	06m19s
10306	516	2336 Jul 09	19:58:22	832	4162	161	T	t-	0.9598	1.0657	83.2N	49.4E	16	345	800	03m17s
10307	516	2337 Jan 02	23:09:44	833	4168	166	P	t-	-1.4252	0.2434	67.2S	18.0E	0	179		
10308	516	2337 May 31	05:05:56	835	4173	133	T	-t	-0.8470	1.0309	34.6S	121.2E	32	344	195	02m46s
10309	516	2337 Nov 23	09:37:55	836	4179	138	A	-p	0.8488	0.9786	35.5N	53.8E	32	200	142	02m05s
10310	516	2338 May 20	15:14:20	838	4185	143	A	nn	-0.1011	0.9812	14.5N	44.0W	84	342	67	02m07s
10311	516	2338 Nov 12	21:52:54	840	4191	148	T	nn	0.1131	1.0399	11.7S	146.4W	84	201	134	03m38s
10312	516	2339 May 09	18:08:04	841	4197	153	A	p-	0.6672	0.9392	54.7N	114.5W	48	143	300	05m24s
10313	516	2339 Nov 02	13:51:50	843	4203	158	T	p-	-0.5751	1.0536	45.8S	48.3W	55	34	215	03m47s
10314	516	2340 Mar 29	03:03:37	844	4208	125	P	-t	-1.4941	0.1131	61.2S	137.3W	0	277		
10315	516	2340 Apr 27	18:21:32	844	4209	163	P	t-	1.3873	0.3005	62.0N	149.8E	0	59		
10316	516	2340 Sep 22	18:01:34	846	4214	130	P	-t	1.3925	0.2793	61.1N	1.4E	0	270		
10317	516	2340 Oct 22	04:55:28	846	4215	168	P	t-	-1.3037	0.4387	61.7S	5.3W	0	114		
10318	516	2341 Mar 18	11:18:20	847	4220	135	H	-p	-0.7137	1.0075	39.8S	42.6E	44	321	36	00m36s
10319	516	2341 Sep 12	00:22:47	849	4226	140	A	-p	0.6950	0.9529	41.7N	156.4W	46	220	234	04m19s
10320	516	2342 Mar 08	01:32:14	851	4232	145	T	nn	0.0072	1.0511	4.9S	162.9E	90	149	169	04m16s

Cat Num	Canon Plate	Calendar Date	TD of Greatest Eclipse	ΔT s	Luna Num	Saros Num	Ecl. Type	QLE	Gamma	Ecl. Mag.	Lat. °	Long. °	Sun Alt °	Sun Azm °	Path Width km	Central Line Dur.
10321	517	2342 Sep 01	01:06:55	852	4238	150	A	nn	-0.0480	0.9434	6.1N	165.7E	87	28	209	06m34s
10322	517	2343 Feb 25	17:32:18	854	4244	155	T	p-	0.6913	1.0385	29.6N	97.7W	46	149	175	03m06s
10323	517	2343 Aug 21	03:07:05	855	4250	160	A	p-	-0.7957	0.9679	35.1S	112.8E	37	31	186	03m09s
10324	517	2344 Jan 16	16:13:41	857	4255	127	P	-t	-1.4270	0.2288	63.8S	83.5E	0	216		
10325	517	2344 Feb 15	06:37:58	857	4256	165	P	t-	1.4280	0.2178	62.0N	23.3E	0	119		
10326	517	2344 Jul 11	03:39:15	858	4261	132	P	-t	1.0818	0.8591	64.3N	82.3W	0	330		
10327	517	2344 Aug 09	11:59:05	859	4262	170	Pb	t-	-1.4974	0.0788	62.3S	52.2W	0	54		
10328	517	2345 Jan 04	18:40:23	860	4267	137	A	-p	-0.7872	0.9165	71.9S	64.6W	38	323	517	06m45s
10329	517	2345 Jun 30	20:26:17	862	4273	142	T	-n	0.3267	1.0797	42.1N	117.7W	71	192	272	06m07s
10330	517	2345 Dec 24	17:41:04	863	4279	147	Am	nn	-0.1081	0.9252	29.7S	81.1W	84	353	284	09m21s
10331	517	2346 Jun 20	12:58:44	865	4285	152	T	p-	-0.4224	1.0515	1.5S	12.7W	65	5	188	05m12s
10332	517	2346 Dec 13	20:55:36	867	4291	157	A	p-	0.5848	0.9665	12.8N	133.1W	54	178	149	04m04s
10333	517	2347 May 11	12:07:08	868	4296	124	Pe	-t	1.5351	0.0391	69.7N	149.1W	0	28		
10334	517	2347 Jun 10	00:44:42	868	4297	162	P	t-	-1.2329	0.5670	67.1S	170.8E	0	1		
10335	517	2347 Nov 03	21:09:19	870	4302	129	P	-t	-1.4337	0.1903	70.4S	83.4E	0	140		
10336	517	2347 Dec 03	07:33:33	870	4303	167	P	t-	1.2358	0.5635	67.9N	74.8E	0	187		
10337	517	2348 Apr 29	13:29:00	871	4308	134	A	-p	0.8338	0.9315	68.1N	48.8W	33	145	466	05m40s
10338	517	2348 Oct 23	13:26:56	873	4314	139	T	-p	-0.7564	1.0476	58.2S	43.6W	41	28	242	03m14s
10339	517	2349 Apr 18	14:16:52	874	4320	144	A	nn	0.0899	0.9557	16.0N	32.1W	85	165	162	05m23s
10340	517	2349 Oct 13	03:28:54	876	4326	149	H	nn	-0.0532	1.0126	10.6S	127.2E	87	16	43	01m18s
10341	518	2350 Apr 07	21:06:03	878	4332	154	H	p-	-0.6452	1.0011	31.7S	119.7W	50	340	5	00m06s
10342	518	2350 Oct 02	11:14:07	879	4338	159	A	p-	0.7131	0.9568	39.8N	28.7E	44	203	222	04m22s
10343	518	2351 Feb 27	00:56:12	881	4343	126	P	-t	1.3209	0.4037	71.5N	110.1E	0	119		
10344	518	2351 Mar 28	10:30:57	881	4344	164	P	t-	-1.3126	0.4195	72.1S	108.5E	0	279		
10345	518	2351 Aug 22	20:42:47	882	4349	131	P	-t	-1.4322	0.2228	71.2S	178.6W	0	51		
10346	518	2351 Sep 21	12:33:27	883	4350	169	P	t-	1.4664	0.1680	72.2N	86.2E	0	273		
10347	518	2352 Feb 16	16:32:06	884	4355	136	T	-p	0.6709	1.0266	28.5N	71.8W	48	164	121	02m24s
10348	518	2352 Aug 11	00:21:35	886	4361	141	A	-p	-0.6366	0.9862	23.6S	171.2E	50	13	63	01m32s
10349	518	2353 Feb 05	03:56:55	887	4367	146	A	nn	-0.0179	0.9766	17.1S	128.0E	89	349	84	02m38s
10350	518	2353 Jul 31	11:17:06	889	4373	151	T	n-	0.1559	1.0467	27.2N	17.8E	81	190	158	04m20s
10351	518	2354 Jan 25	08:03:20	891	4379	156	A	p-	-0.7388	0.9240	66.0S	80.6E	42	341	427	06m35s
10352	518	2354 Jul 21	03:28:22	892	4385	161	T	t-	0.8870	1.0697	81.4N	171.7E	27	221	499	03m51s
10353	518	2355 Jan 14	07:12:20	894	4391	166	P	t-	-1.4158	0.2588	68.2S	113.4W	0	190		
10354	518	2355 Jun 11	12:28:18	895	4396	133	T	-t	-0.9196	1.0269	43.3S	9.2E	23	348	233	02m18s
10355	518	2355 Dec 04	17:58:37	897	4402	138	A	-p	0.8609	0.9792	36.0N	74.4W	30	195	145	02m02s
10356	518	2356 May 30	22:15:18	899	4408	143	A	nn	-0.1735	0.9800	12.2N	148.0W	80	346	72	02m21s
10357	518	2356 Nov 23	06:24:55	900	4414	148	T	-n	0.1317	1.0394	13.2S	86.3E	83	197	133	03m40s
10358	518	2357 May 20	00:54:23	902	4420	153	A	p-	0.5961	0.9415	53.9N	151.0E	53	150	269	05m24s
10359	518	2357 Nov 12	22:20:23	904	4426	158	T	p-	-0.5514	1.0505	48.4S	172.7W	56	31	200	03m35s
10360	518	2358 Apr 09	10:37:39	905	4431	125	Pe	-t	-1.5309	0.0468	61.4S	100.7E	0	286		
10361	519	2358 May 09	01:21:14	905	4432	163	P	t-	1.3231	0.4097	62.6N	36.2E	0	50		
10362	519	2358 Oct 04	01:36:39	907	4437	130	P	-t	1.4464	0.1835	61.1N	120.7W	0	261		
10363	519	2358 Nov 02	13:04:00	907	4438	168	P	t-	-1.2765	0.4889	62.2S	136.1W	0	123		
10364	519	2359 Mar 29	19:24:46	908	4443	135	T	-p	-0.7429	1.0128	37.9S	79.3W	42	323	64	01m02s
10365	519	2359 Sep 23	07:24:42	910	4449	140	A	-p	0.7595	0.9471	41.9N	100.6E	40	221	291	04m53s
10366	519	2360 Mar 18	09:59:22	912	4455	145	T	nn	-0.0177	1.0549	1.8S	36.4E	89	331	181	04m33s
10367	519	2360 Sep 11	07:52:25	913	4461	150	Am	nn	0.0244	0.9415	5.7N	65.6E	89	208	217	06m41s
10368	519	2361 Mar 08	02:05:56	915	4467	155	T	p-	0.6743	1.0396	31.9N	132.7E	47	146	176	03m06s
10369	519	2361 Aug 31	10:04:30	917	4473	160	A	p-	-0.7211	0.9701	32.2S	9.7E	44	33	151	02m54s
10370	519	2362 Jan 27	00:36:00	918	4478	127	P	-t	-1.4368	0.2125	62.9S	51.1W	0	225		
10371	519	2362 Feb 25	15:02:03	918	4479	165	P	t-	1.4190	0.2344	61.6N	111.3W	0	109		
10372	519	2362 Jul 22	11:01:14	920	4484	132	P	-t	1.1522	0.7256	63.5N	157.9E	0	321		
10373	519	2362 Aug 20	19:18:10	920	4485	170	P	t-	-1.4239	0.2148	61.8S	170.7W	0	63		
10374	519	2363 Jan 16	02:15:07	922	4490	137	A	-p	-0.7955	0.9154	68.8S	177.6W	37	314	532	06m52s
10375	519	2363 Jul 12	03:55:03	923	4496	142	T	-p	0.4012	1.0792	45.0N	134.5E	66	198	279	05m51s
10376	519	2364 Jan 05	01:46:48	925	4502	147	A	nn	-0.1161	0.9259	29.4S	159.7E	83	348	281	09m03s
10377	519	2364 Jun 30	20:19:47	927	4508	152	T	n-	-0.3494	1.0499	2.9N	123.3W	70	9	176	05m00s
10378	519	2364 Dec 24	05:18:59	928	4514	157	A	p-	0.5752	0.9683	11.6N	99.8E	55	174	139	03m48s
10379	519	2365 Jun 20	07:44:13	930	4520	162	P	t-	-1.1623	0.6935	66.1S	55.7E	0	12		
10380	519	2365 Nov 14	05:37:33	931	4525	129	P	-t	-1.4540	0.1526	69.5S	55.9W	0	153		

Cat Num	Canon Plate	Calendar Date	TD of Greatest Eclipse	ΔT s	Luna Num	Saros Num	Ecl. Type	QLE	Gamma	Ecl. Mag.	Lat. °	Long. °	Sun Alt °	Sun Azm °	Path Width km	Central Line Dur.
10381	520	2365 Dec 13	16:12:42	932	4526	167	P	t-	1.2230	0.5872	66.8N	65.6W	0	176		
10382	520	2366 May 10	20:22:08	933	4531	134	A	-t	0.8981	0.9323	77.9N	169.5W	26	129	583	05m03s
10383	520	2366 Nov 03	21:46:04	935	4537	139	T	-p	-0.7868	1.0426	64.8S	170.2W	38	29	231	02m46s
10384	520	2367 Apr 29	21:30:03	936	4543	144	Am	nn	0.1451	0.9607	22.8N	141.2W	82	167	144	04m38s
10385	520	2367 Oct 24	11:25:04	938	4549	149	H	nn	-0.0902	1.0065	16.7S	7.3E	85	15	22	00m40s
10386	520	2368 Apr 18	04:51:38	940	4555	154	H	p-	-0.5992	1.0079	24.8S	120.8E	53	344	34	00m47s
10387	520	2368 Oct 12	18:37:20	942	4561	159	A	p-	0.6672	0.9522	32.5N	85.8W	48	199	233	05m13s
10388	520	2369 Mar 09	09:30:24	943	4566	126	P	-t	1.3392	0.3686	71.9N	32.2W	0	105		
10389	520	2369 Apr 07	18:42:11	943	4567	164	P	t-	-1.2763	0.4880	71.8S	28.1W	0	293		
10390	520	2369 Sep 02	03:25:56	945	4572	131	Pe	-t	-1.5027	0.1025	71.7S	67.4E	0	64		
10391	520	2369 Oct 01	19:33:31	945	4573	169	P	t-	1.4094	0.2651	72.1N	32.7W	0	259		
10392	520	2370 Feb 27	01:07:02	946	4578	136	T	-p	0.6865	1.0262	33.2N	157.0E	46	161	121	02m17s
10393	520	2370 Aug 22	07:22:21	948	4584	141	A	-p	-0.7082	0.9867	32.0S	62.0E	45	17	66	01m22s
10394	520	2371 Feb 16	12:18:49	950	4590	146	A	nn	-0.0075	0.9753	12.9S	2.7E	89	348	88	02m48s
10395	520	2371 Aug 11	18:38:04	951	4596	151	T	nn	0.0821	1.0487	19.9N	93.0W	85	192	162	04m36s
10396	520	2372 Feb 05	16:07:48	953	4602	156	A	p-	-0.7301	0.9237	61.5S	36.9W	43	336	422	06m50s
10397	520	2372 Jul 31	10:58:30	955	4608	161	T	p-	0.8144	1.0717	71.0N	45.5E	35	209	404	04m18s
10398	520	2373 Jan 24	15:14:59	957	4614	166	P	t-	-1.4062	0.2742	69.3S	114.5E	0	202		
10399	520	2373 Jun 21	19:45:29	958	4619	133	Ts	-t	-0.9954	1.0191	62.7S	100.1W	3	349	-	01m24s
10400	520	2373 Dec 15	02:25:55	960	4625	138	A	-p	0.8678	0.9803	36.7N	155.4E	29	190	141	01m56s
10401	521	2374 Jun 11	05:09:56	961	4631	143	A	-p	-0.2504	0.9784	8.8N	109.1E	76	351	79	02m39s
10402	521	2374 Dec 04	15:02:56	963	4637	148	T	-n	0.1455	1.0390	14.1S	42.4W	82	193	132	03m42s
10403	521	2375 May 31	07:34:33	965	4643	153	A	p-	0.5200	0.9436	52.0N	57.9E	58	158	243	05m26s
10404	521	2375 Nov 24	06:54:54	967	4649	158	T	p-	-0.5328	1.0474	50.7S	62.3E	58	26	186	03m23s
10405	521	2376 May 19	08:14:44	968	4655	163	P	t-	1.2528	0.5304	63.3N	76.1W	0	41		
10406	521	2376 Oct 14	09:18:28	970	4660	130	P	-t	1.4941	0.1003	61.4N	115.4E	0	252		
10407	521	2376 Nov 12	21:19:06	970	4661	168	P	t-	-1.2551	0.5279	62.8S	91.4E	0	132		
10408	521	2377 Apr 09	03:25:10	971	4666	135	T	-p	-0.7779	1.0180	37.1S	160.2E	39	325	96	01m28s
10409	521	2377 Oct 03	14:33:17	973	4672	140	A	-p	0.8178	0.9413	42.6N	4.7W	35	220	366	05m29s
10410	521	2378 Mar 29	18:20:23	975	4678	145	T	nn	-0.0480	1.0587	1.1N	88.6W	87	331	193	04m51s
10411	521	2378 Sep 22	14:45:48	977	4684	150	A	nn	0.0904	0.9396	4.8N	36.6W	85	209	225	06m54s
10412	521	2379 Mar 19	10:31:47	978	4690	155	T	p-	0.6512	1.0409	34.3N	5.6E	49	144	177	03m07s
10413	521	2379 Sep 11	17:09:32	980	4696	160	A	p-	-0.6518	0.9717	30.9S	95.4W	49	34	130	02m42s
10414	521	2380 Feb 07	08:54:01	982	4701	127	P	-t	-1.4496	0.1909	62.2S	175.7E	0	235		
10415	521	2380 Mar 07	23:17:52	982	4702	165	P	t-	1.4039	0.2615	61.3N	116.3E	0	100		
10416	521	2380 Aug 01	18:26:17	983	4707	132	P	-t	1.2207	0.5949	62.8N	37.7E	0	312		
10417	521	2380 Aug 31	02:44:39	984	4708	170	P	t-	-1.3553	0.3422	61.4S	69.2E	0	72		
10418	521	2381 Jan 26	10:46:38	985	4713	137	A	-p	-0.8064	0.9149	65.3S	66.8E	36	309	546	06m57s
10419	521	2381 Jul 22	11:25:02	987	4719	142	T	-p	0.4748	1.0777	46.9N	26.9E	61	205	285	05m32s
10420	521	2382 Jan 15	09:53:22	988	4725	147	A	nn	-0.1241	0.9274	28.1S	40.1E	83	343	275	08m40s
10421	522	2382 Jul 12	03:37:51	990	4731	152	T	n-	-0.2744	1.0475	6.5N	127.5E	74	13	164	04m41s
10422	522	2383 Jan 04	13:46:26	992	4737	157	A	p-	0.5682	0.9706	11.4N	28.1W	55	169	128	03m26s
10423	522	2383 Jul 01	14:37:42	994	4743	162	P	t-	-1.0870	0.8276	65.1S	57.5W	0	21		
10424	522	2383 Nov 25	14:13:32	995	4748	129	P	-t	-1.4683	0.1260	68.6S	163.5E	0	165		
10425	522	2383 Dec 25	00:57:04	995	4749	167	P	t-	1.2144	0.6033	65.8N	153.2E	0	165		
10426	522	2384 May 21	03:05:26	997	4754	134	A	-t	0.9701	0.9317	80.8N	0.8W	13	40	1115	04m28s
10427	522	2384 Nov 14	06:13:20	999	4760	139	T	-p	-0.8102	1.0377	70.9S	63.5E	36	28	217	02m22s
10428	522	2385 May 10	04:36:49	1000	4766	144	A	nn	0.2063	0.9657	29.5N	111.9E	78	169	126	03m53s
10429	522	2385 Nov 03	19:27:30	1002	4772	149	H	-n	-0.1212	1.0004	22.1S	113.5W	83	13	2	00m03s
10430	522	2386 Apr 29	12:32:25	1004	4778	154	H2	p-	-0.5483	1.0146	18.1S	2.9E	57	347	60	01m30s
10431	522	2386 Oct 24	02:06:43	1006	4784	159	A	p-	0.6268	0.9475	26.1N	158.8E	51	196	246	06m09s
10432	522	2387 Mar 20	17:59:08	1007	4789	126	P	-t	1.3624	0.3241	72.1N	173.3W	0	90		
10433	522	2387 Apr 19	02:47:06	1007	4790	164	P	t-	-1.2345	0.5677	71.3S	162.7W	0	306		
10434	522	2387 Oct 13	02:41:04	1009	4796	169	P	t-	1.3579	0.3523	71.8N	153.3W	0	245		
10435	522	2388 Mar 09	09:36:21	1011	4801	136	T	-p	0.7064	1.0260	38.5N	27.0E	45	158	124	02m10s
10436	522	2388 Sep 01	14:30:25	1012	4807	141	A	-p	-0.7744	0.9867	40.7S	50.1W	39	22	73	01m15s
10437	522	2389 Feb 26	20:33:52	1014	4813	146	A	nn	0.0078	0.9744	8.1S	121.3W	90	162	92	02m55s
10438	522	2389 Aug 22	02:05:53	1016	4819	151	T	nn	0.0133	1.0500	12.5N	153.9E	89	193	166	04m45s
10439	522	2390 Feb 16	00:06:58	1018	4825	156	A	p-	-0.7177	0.9239	56.4S	155.6W	44	335	411	07m06s
10440	522	2390 Aug 11	18:31:27	1019	4831	161	T	p-	0.7441	1.0724	61.3N	72.5W	42	206	353	04m41s

Cat Num	Canon Plate	Calendar Date	TD of Greatest Eclipse	ΔT s	Luna Num	Saros Num	Ecl. Type	QLE	Gamma	Ecl. Mag.	Lat. °	Long. °	Sun Alt °	Sun Azm °	Path Width km	Central Line Dur.
10441	523	2391 Feb 04	23:15:06	1021	4837	166	P	t-	-1.3944	0.2933	70.3S	17.6W	0	215		
10442	523	2391 Jul 03	02:58:53	1023	4842	133	P	-t	-1.0732	0.8664	67.1S	143.0E	0	358		
10443	523	2391 Aug 01	11:14:32	1023	4843	171	Pb	t-	1.4925	0.0766	69.6N	167.9E	0	332		
10444	523	2391 Dec 26	10:57:15	1024	4848	138	A	-p	0.8723	0.9820	37.6N	24.0E	29	184	131	01m46s
10445	523	2392 Jun 21	11:57:58	1026	4854	143	A	-p	-0.3319	0.9762	4.1N	7.3E	71	355	90	03m02s
10446	523	2392 Dec 14	23:46:26	1028	4860	148	T	-n	0.1550	1.0391	14.5S	172.4W	81	188	133	03m46s
10447	523	2393 Jun 10	14:08:41	1030	4866	153	A	p-	0.4389	0.9453	48.8N	34.4W	64	166	224	05m34s
10448	523	2393 Dec 04	15:34:35	1032	4872	158	T	p-	-0.5188	1.0445	52.6S	63.0W	58	20	174	03m13s
10449	523	2394 May 30	15:03:03	1033	4878	163	P	t-	1.1775	0.6609	64.1N	172.7E	0	32		
10450	523	2394 Oct 25	17:07:13	1035	4883	130	Pe	-t	1.5351	0.0298	61.8N	10.3W	0	243		
10451	523	2394 Nov 24	05:40:36	1035	4884	168	P	t-	-1.2398	0.5555	63.6S	43.0W	0	142		
10452	523	2395 Apr 20	11:17:15	1037	4889	135	T	-p	-0.8203	1.0230	37.7S	41.9E	35	327	134	01m52s
10453	523	2395 Oct 14	21:49:16	1038	4895	140	A	-p	0.8691	0.9354	44.0N	112.4W	29	219	471	06m07s
10454	523	2396 Apr 09	02:33:17	1040	4901	145	T	nn	-0.0851	1.0625	3.4N	148.7E	85	332	206	05m12s
10455	523	2396 Oct 02	21:48:07	1042	4907	150	A	nn	0.1493	0.9375	3.5N	141.2W	81	209	234	07m12s
10456	523	2397 Mar 29	18:49:52	1044	4913	155	T	p-	0.6221	1.0423	36.7N	118.9W	51	144	178	03m11s
10457	523	2397 Sep 22	00:23:55	1046	4919	160	A	p-	-0.5892	0.9728	30.9S	157.2E	54	34	118	02m34s
10458	523	2398 Feb 17	17:08:14	1047	4924	127	P	-t	-1.4648	0.1650	61.7S	43.5E	0	244		
10459	523	2398 Mar 19	07:27:08	1047	4925	165	P	t-	1.3844	0.2965	61.2N	14.4W	0	91		
10460	523	2398 Aug 13	01:53:37	1049	4930	132	P	-t	1.2877	0.4669	62.2N	82.9W	0	303		
10461	524	2398 Sep 11	10:16:34	1049	4931	170	P	t-	-1.2902	0.4632	61.2S	52.3W	0	81		
10462	524	2399 Feb 06	18:46:44	1051	4936	137	A	-p	-0.8180	0.9150	61.6S	51.0W	35	307	557	07m01s
10463	524	2399 Aug 02	18:55:14	1052	4942	142	T	-p	0.5482	1.0754	48.0N	80.4W	57	211	291	05m14s
10464	524	2400 Jan 26	18:00:10	1054	4948	147	A	nn	-0.1322	0.9295	26.0S	79.9W	82	339	267	08m13s
10465	524	2400 Jul 22	10:54:48	1056	4954	152	T	nn	-0.1992	1.0444	9.1N	19.0E	79	17	151	04m17s
10466	524	2401 Jan 14	22:15:20	1058	4960	157	A	p-	0.5617	0.9735	11.9N	156.4W	56	165	114	03m00s
10467	524	2401 Jul 11	21:29:20	1060	4966	162	P	t-	-1.0111	0.9620	64.2S	169.9W	0	31		
10468	524	2401 Dec 05	22:53:37	1061	4971	129	P	-t	-1.4797	0.1049	67.5S	22.4E	0	176		
10469	524	2402 Jan 04	09:42:28	1061	4972	167	P	t-	1.2064	0.6184	64.7N	12.2E	0	155		
10470	524	2402 Jun 01	09:44:38	1063	4977	134	P	-t	1.0452	0.8834	67.8N	135.9W	0	6		
10471	524	2402 Nov 25	14:45:41	1065	4983	139	T	-p	-0.8291	1.0332	76.2S	59.6W	34	22	202	02m02s
10472	524	2403 May 21	11:36:55	1066	4989	144	A	nn	0.2737	0.9705	36.1N	7.4E	74	173	110	03m10s
10473	524	2403 Nov 15	03:36:25	1068	4995	149	A	-n	-0.1461	0.9947	26.8S	124.9E	81	9	19	00m33s
10474	524	2404 May 09	20:05:45	1070	5001	154	T	p-	-0.4902	1.0212	11.4S	112.8W	61	350	83	02m14s
10475	524	2404 Nov 03	09:44:07	1072	5007	159	A	p-	0.5935	0.9430	20.5N	42.0E	53	193	260	07m05s
10476	524	2405 Mar 31	02:18:52	1073	5012	126	P	-t	1.3928	0.2654	71.9N	47.9E	0	76		
10477	524	2405 Apr 29	10:43:55	1074	5013	164	P	t-	-1.1858	0.6613	70.6S	65.3E	0	319		
10478	524	2405 Oct 23	09:58:55	1076	5019	169	P	t-	1.3142	0.4261	71.3N	83.8E	0	232		
10479	524	2406 Mar 20	17:57:23	1077	5024	136	T	-p	0.7327	1.0258	44.5N	101.3W	43	155	128	02m03s
10480	524	2406 Sep 12	21:45:23	1079	5030	141	A	-p	-0.8356	0.9862	49.6S	165.5W	33	27	88	01m11s
10481	525	2407 Mar 10	04:41:40	1081	5036	146	A	nn	0.0283	0.9739	2.7S	116.1E	88	163	93	02m59s
10482	525	2407 Sep 02	09:38:25	1082	5042	151	T	nn	-0.0517	1.0506	5.1N	39.2E	87	17	168	04m48s
10483	525	2408 Feb 27	07:59:40	1084	5048	156	A	p-	-0.7004	0.9249	50.8S	85.5E	45	334	394	07m22s
10484	525	2408 Aug 22	02:07:39	1086	5054	161	T	p-	0.6766	1.0720	52.3N	170.0E	47	204	317	05m00s
10485	525	2409 Feb 15	07:12:30	1088	5060	166	P	t-	-1.3802	0.3163	71.1S	149.6W	0	228		
10486	525	2409 Jul 13	10:09:33	1089	5065	133	P	-t	-1.1523	0.7186	68.1S	24.6E	0	9		
10487	525	2409 Aug 11	18:38:31	1090	5066	171	P	t-	1.4271	0.2021	70.4N	44.8E	0	320		
10488	525	2410 Jan 05	19:31:39	1091	5071	138	A	-p	0.8749	0.9842	38.8N	108.2W	29	179	116	01m31s
10489	525	2410 Jul 02	18:42:30	1093	5077	143	A	-p	-0.4152	0.9735	1.6S	94.4W	65	359	104	03m25s
10490	525	2410 Dec 26	08:33:58	1095	5083	148	T	-n	0.1613	1.0395	14.1S	56.6E	81	184	134	03m50s
10491	525	2411 Jun 21	20:37:43	1097	5089	153	A	pn	0.3537	0.9467	44.2N	126.9W	69	173	210	05m46s
10492	525	2411 Dec 16	00:19:07	1099	5095	158	T	p-	-0.5093	1.0419	53.6S	171.2E	59	13	163	03m04s
10493	525	2412 Jun 09	21:48:04	1100	5101	163	P	t-	1.0988	0.7983	65.0N	62.1E	0	22		
10494	525	2412 Dec 04	14:06:31	1102	5107	168	P	t-	-1.2288	0.5751	64.5S	178.7W	0	152		
10495	525	2413 Apr 30	19:03:57	1104	5112	135	T	-p	-0.8677	1.0274	40.0S	75.0W	30	330	183	02m13s
10496	525	2413 Oct 25	05:13:20	1106	5118	140	A	-p	0.9129	0.9298	46.2N	137.3E	24	218	628	06m43s
10497	525	2414 Apr 20	10:39:39	1107	5124	145	T	-n	-0.1279	1.0661	5.0N	27.7E	83	334	218	05m33s
10498	525	2414 Oct 14	04:58:50	1109	5130	150	A	nn	0.2015	0.9355	2.2N	111.8E	78	207	245	07m34s
10499	525	2415 Apr 10	02:59:35	1111	5136	155	T	p-	0.5866	1.0436	38.9N	119.6E	54	144	178	03m15s
10500	525	2415 Oct 03	07:47:48	1113	5142	160	A	p-	-0.5335	0.9736	31.8S	47.4E	58	34	110	02m27s

Cat Num	Canon Plate	Calendar Date	TD of Greatest Eclipse	ΔT s	Luna Num	Saros Num	Ecl. Type	QLE	Gamma	Ecl. Mag.	Lat. °	Long. °	Sun Alt °	Sun Azm °	Path Width km	Central Line Dur.
10501	526	2416 Feb 29	01:13:31	1115	5147	127	P	-t	-1.4865	0.1279	61.3S	86.2W	0	253		
10502	526	2416 Mar 29	15:24:54	1115	5148	165	P	t-	1.3563	0.3466	61.2N	142.2W	0	82		
10503	526	2416 Aug 23	09:26:38	1116	5153	132	P	-t	1.3505	0.3468	61.8N	155.2E	0	294		
10504	526	2416 Sep 21	17:57:51	1117	5154	170	P	t-	-1.2321	0.5716	61.1S	176.0W	0	89		
10505	526	2417 Feb 17	02:40:42	1118	5159	137	A	-p	-0.8345	0.9155	58.3S	168.3W	33	306	574	07m04s
10506	526	2417 Aug 13	02:28:06	1120	5165	142	T	-p	0.6189	1.0723	48.3N	171.4E	52	216	297	04m55s
10507	526	2418 Feb 06	02:04:04	1122	5171	147	A	nn	-0.1431	0.9322	23.3S	160.5E	82	336	256	07m43s
10508	526	2418 Aug 02	18:11:10	1124	5177	152	T	nn	-0.1242	1.0406	10.9N	89.0W	83	21	137	03m50s
10509	526	2419 Jan 26	06:44:37	1126	5183	157	A	p-	0.5550	0.9770	13.2N	75.2E	56	161	98	02m30s
10510	526	2419 Jul 23	04:16:45	1128	5189	162	A	t-	-0.9322	0.9753	45.6S	98.2E	21	24	242	02m17s
10511	526	2419 Dec 17	07:40:07	1129	5194	129	P	-t	-1.4865	0.0925	66.5S	119.7W	0	187		
10512	526	2420 Jan 15	18:30:39	1129	5195	167	P	t-	1.2004	0.6298	63.8N	129.2W	0	145		
10513	526	2420 Jun 11	16:17:02	1131	5200	134	P	-t	1.1256	0.7470	66.8N	115.6E	0	355		
10514	526	2420 Dec 05	23:23:52	1133	5206	139	T	-p	-0.8431	1.0290	80.2S	174.0W	32	6	185	01m44s
10515	526	2421 May 31	18:32:59	1135	5212	144	A	-p	0.3451	0.9750	42.4N	95.0W	70	177	95	02m32s
10516	526	2421 Nov 25	11:51:41	1136	5218	149	A	-n	-0.1652	0.9893	30.4S	2.4E	80	6	38	01m06s
10517	526	2422 May 21	03:34:51	1138	5224	154	T	p-	-0.4278	1.0275	5.0S	133.1E	65	354	103	02m56s
10518	526	2422 Nov 14	17:27:40	1140	5230	159	A	p-	0.5657	0.9386	15.9N	75.8W	55	189	275	08m01s
10519	526	2423 Apr 11	10:32:41	1142	5235	126	P	-t	1.4282	0.1970	71.6N	89.1W	0	63		
10520	526	2423 May 10	18:35:17	1142	5236	164	P	t-	-1.1323	0.7647	69.7S	64.8W	0	332		
10521	527	2423 Nov 03	17:24:51	1144	5242	169	P	t-	1.2769	0.4887	70.5N	40.5W	0	219		
10522	527	2424 Mar 31	02:10:10	1146	5247	136	T	-p	0.7652	1.0254	51.3N	131.9E	40	152	133	01m55s
10523	527	2424 Sep 23	05:09:46	1147	5253	141	A	-p	-0.8896	0.9853	58.6S	74.1E	27	36	114	01m08s
10524	527	2425 Mar 20	12:41:12	1149	5259	146	A	nn	0.0546	0.9735	3.1N	4.7W	87	162	95	03m02s
10525	527	2425 Sep 12	17:18:07	1151	5265	151	Tm	nn	-0.1113	1.0507	2.3S	77.5W	84	17	169	04m47s
10526	527	2426 Mar 09	15:44:45	1153	5271	156	A	p-	-0.6774	0.9262	44.7S	32.5W	47	336	374	07m38s
10527	527	2426 Sep 02	09:48:47	1155	5277	161	T	p-	0.6133	1.0709	43.8N	51.6E	52	203	291	05m14s
10528	527	2427 Feb 26	15:03:45	1157	5283	166	P	t-	-1.3607	0.3484	71.7S	79.3E	0	241		
10529	527	2427 Jul 24	17:18:10	1158	5288	133	P	-t	-1.2318	0.5709	69.1S	93.7W	0	20		
10530	527	2427 Aug 23	02:04:51	1159	5289	171	P	t-	1.3642	0.3222	71.1N	79.4W	0	308		
10531	527	2428 Jan 17	04:07:20	1160	5294	138	A	-p	0.8770	0.9870	40.5N	119.1E	28	173	96	01m13s
10532	527	2428 Jul 13	01:23:55	1162	5300	143	A	-p	-0.4998	0.9702	8.3S	163.9E	60	3	123	03m50s
10533	527	2429 Jan 05	17:22:56	1164	5306	148	T	-n	0.1666	1.0404	13.0S	74.9W	80	179	137	03m56s
10534	527	2429 Jul 02	03:04:30	1166	5312	153	A	pn	0.2668	0.9476	38.6N	139.6E	74	179	200	06m01s
10535	527	2429 Dec 26	09:07:20	1168	5318	158	T	n-	-0.5035	1.0397	53.7S	44.9E	60	6	155	02m57s
10536	527	2430 Jun 21	04:29:26	1170	5324	163	P	t-	1.0160	0.9438	66.0N	48.0W	0	13		
10537	527	2430 Dec 15	22:37:29	1172	5330	168	P	t-	-1.2227	0.5857	65.5S	44.0E	0	162		
10538	527	2431 May 12	02:43:30	1173	5335	135	T	-p	-0.9214	1.0310	44.8S	170.4E	22	332	267	02m27s
10539	527	2431 Nov 05	12:45:40	1175	5341	140	A	-t	0.9496	0.9242	49.5N	24.5E	18	216	902	07m15s
10540	527	2432 Apr 30	18:37:31	1177	5347	145	T	-n	-0.1780	1.0694	5.8N	91.0W	80	337	229	05m56s
10541	528	2432 Oct 24	12:19:58	1179	5353	150	A	nn	0.2455	0.9335	0.8N	2.1E	76	205	255	08m01s
10542	528	2433 Apr 20	11:01:32	1181	5359	155	T	p-	0.5450	1.0449	40.8N	0.9E	57	146	177	03m21s
10543	528	2433 Oct 13	15:20:16	1183	5365	160	A	p-	-0.4840	0.9742	33.4S	64.2W	61	33	104	02m23s
10544	528	2434 Mar 11	09:12:47	1184	5370	127	P	-t	-1.5121	0.0837	61.1S	145.6E	0	263		
10545	528	2434 Apr 09	23:15:24	1184	5371	165	P	t-	1.3232	0.4058	61.4N	91.7E	0	73		
10546	528	2434 Sep 03	17:04:08	1186	5376	132	P	-t	1.4099	0.2331	61.5N	32.2E	0	285		
10547	528	2434 Oct 03	01:46:03	1186	5377	170	P	t-	-1.1789	0.6705	61.1S	58.6E	0	98		
10548	528	2435 Feb 28	10:29:45	1188	5382	137	A	-p	-0.8546	0.9165	55.4S	75.0E	31	306	599	07m05s
10549	528	2435 Aug 24	10:03:12	1190	5388	142	T	-p	0.6875	1.0684	48.2N	62.2E	46	221	304	04m35s
10550	528	2436 Feb 17	10:05:24	1192	5394	147	A	nn	-0.1567	0.9355	20.2S	41.1E	81	333	243	07m12s
10551	528	2436 Aug 13	01:29:32	1194	5400	152	Tm	nn	-0.0517	1.0361	11.8N	162.8E	87	25	122	03m21s
10552	528	2437 Feb 05	15:11:25	1196	5406	157	A	p-	0.5453	0.9810	14.9N	52.5W	57	157	79	01m58s
10553	528	2437 Aug 02	11:06:01	1198	5412	162	A	p-	-0.8553	0.9741	37.4S	2.8W	31	26	175	02m33s
10554	528	2437 Dec 27	16:29:07	1199	5417	129	P	-t	-1.4920	0.0824	65.5S	98.1E	0	198		
10555	528	2438 Jan 26	03:17:37	1199	5418	167	P	t-	1.1929	0.6441	63.0N	90.1E	0	135		
10556	528	2438 Jun 22	22:46:47	1201	5423	134	P	-t	1.2079	0.6068	65.8N	8.2E	0	345		
10557	528	2438 Dec 17	08:05:40	1203	5429	139	T	-p	-0.8539	1.0254	81.7S	84.3E	31	336	168	01m30s
10558	528	2439 Jun 12	01:25:22	1205	5435	144	A	-p	0.4206	0.9791	48.2N	165.1E	65	183	82	01m59s
10559	528	2439 Dec 06	20:11:47	1207	5441	149	A	-n	-0.1794	0.9844	33.0S	120.5W	79	1	56	01m36s
10560	528	2440 May 31	10:58:15	1209	5447	154	T	p-	-0.3598	1.0334	1.0N	21.0E	69	358	121	03m33s

Cat Num	Canon Plate	Calendar Date	TD of Greatest Eclipse	ΔT s	Luna Num	Saros Num	Ecl. Type	QLE	Gamma	Ecl. Mag.	Lat. °	Long. °	Sun Alt °	Sun Azm °	Path Width km	Central Line Dur.
10561	529	2440 Nov 25	01:18:39	1211	5453	159	A	p-	0.5445	0.9347	12.2N	164.9E	57	185	290	08m51s
10562	529	2441 Apr 21	18:37:49	1212	5458	126	P	-t	1.4706	0.1149	71.0N	136.4E	0	50		
10563	529	2441 May 21	02:20:11	1213	5459	164	P	t-	-1.0733	0.8795	68.8S	167.3E	0	343		
10564	529	2441 Nov 14	00:59:17	1214	5465	169	P	t-	1.2459	0.5404	69.6N	166.3W	0	206		
10565	529	2442 Apr 11	10:14:04	1216	5470	136	T	-p	0.8046	1.0248	58.7N	6.2E	36	148	142	01m45s
10566	529	2442 Oct 04	12:43:00	1218	5476	141	A	-p	-0.9371	0.9838	67.2S	54.7W	20	50	166	01m08s
10567	529	2443 Mar 31	20:30:25	1220	5482	146	A	nn	0.0889	0.9734	9.3N	123.1W	85	163	95	03m02s
10568	529	2443 Sep 24	01:04:47	1222	5488	151	T	nn	-0.1656	1.0502	9.6S	164.1E	80	18	169	04m39s
10569	529	2444 Mar 19	23:21:38	1224	5494	156	A	p-	-0.6476	0.9280	38.3S	149.1W	49	337	351	07m53s
10570	529	2444 Sep 12	17:35:35	1226	5500	161	T	p-	0.5548	1.0688	35.7N	67.9W	56	202	268	05m23s
10571	529	2445 Mar 08	22:49:34	1228	5506	166	P	t-	-1.3361	0.3891	72.1S	50.9W	0	255		
10572	529	2445 Aug 04	00:27:22	1229	5511	133	P	-t	-1.3097	0.4272	70.0S	147.3E	0	31		
10573	529	2445 Sep 02	09:35:13	1230	5512	171	P	t-	1.3049	0.4344	71.7N	154.9E	0	295		
10574	529	2446 Jan 27	12:43:51	1231	5517	138	A	-p	0.8789	0.9903	42.7N	13.9W	28	168	72	00m53s
10575	529	2446 Jul 24	08:03:11	1233	5523	143	A	-p	-0.5854	0.9665	16.0S	61.9E	54	7	149	04m13s
10576	529	2447 Jan 17	02:14:03	1235	5529	148	T	-n	0.1703	1.0417	11.1S	152.9E	80	175	141	04m03s
10577	529	2447 Jul 13	09:29:35	1237	5535	153	A	nn	0.1786	0.9481	32.2N	45.1E	80	183	194	06m18s
10578	529	2448 Jan 06	17:57:07	1239	5541	158	T	n-	-0.4991	1.0380	52.6S	82.0W	60	358	147	02m51s
10579	529	2448 Jul 01	11:10:16	1241	5547	163	A	t-	0.9316	0.9620	87.6N	135.3W	21	25	389	02m26s
10580	529	2448 Dec 26	07:10:41	1243	5553	168	P	t-	-1.2190	0.5918	66.5S	94.2W	0	172		
10581	530	2449 May 22	10:19:15	1244	5558	135	T	-t	-0.9790	1.0328	54.4S	59.1E	11	332	567	02m24s
10582	530	2449 Nov 15	20:23:56	1246	5564	140	An	-t	0.9810	0.9186	54.9N	89.1W	10	214	-	07m35s
10583	530	2450 May 12	02:29:44	1248	5570	145	T	-n	-0.2330	1.0722	5.6N	151.7E	77	340	241	06m19s
10584	530	2450 Nov 04	19:49:31	1250	5576	150	A	nn	0.2828	0.9318	0.5S	109.9W	74	203	264	08m30s
10585	530	2451 May 01	18:53:37	1252	5582	155	T	p-	0.4958	1.0459	42.1N	114.3W	60	149	175	03m28s
10586	530	2451 Oct 24	23:03:09	1254	5588	160	A	p-	-0.4424	0.9746	35.3S	178.3W	64	31	101	02m21s
10587	530	2452 Mar 21	17:01:31	1256	5593	127	Pe	-t	-1.5455	0.0262	61.1S	20.1E	0	272		
10588	530	2452 Apr 20	06:54:27	1256	5594	165	P	t-	1.2819	0.4797	61.8N	31.5W	0	64		
10589	530	2452 Sep 14	00:49:17	1258	5599	132	P	-t	1.4635	0.1307	61.3N	92.5W	0	276		
10590	530	2452 Oct 13	09:44:05	1258	5600	170	P	t-	-1.1334	0.7553	61.3S	69.3W	0	107		
10591	530	2453 Mar 10	18:09:42	1260	5605	137	A	-p	-0.8820	0.9177	53.6S	39.1W	28	306	647	07m04s
10592	530	2453 Sep 03	17:43:48	1262	5611	142	T	-p	0.7513	1.0638	48.0N	49.1W	41	224	312	04m15s
10593	530	2454 Feb 27	18:01:47	1264	5617	147	A	nn	-0.1750	0.9393	17.0S	77.3W	80	331	228	06m40s
10594	530	2454 Aug 24	08:48:47	1266	5623	152	T	nn	0.0194	1.0310	11.9N	54.3E	89	205	105	02m50s
10595	530	2455 Feb 16	23:36:27	1268	5629	157	A	p-	0.5335	0.9857	17.1N	179.6W	58	154	59	01m25s
10596	530	2455 Aug 13	17:54:37	1270	5635	162	A	p-	-0.7781	0.9716	32.3S	104.2W	39	28	158	02m52s
10597	530	2456 Jan 08	01:21:04	1271	5640	129	P	-t	-1.4954	0.0760	64.5S	44.5W	0	208		
10598	530	2456 Feb 06	12:03:15	1272	5641	167	P	t-	1.1843	0.6607	62.3N	50.1W	0	125		
10599	530	2456 Jul 03	05:13:16	1273	5646	134	P	-t	1.2925	0.4621	64.9N	98.1W	0	336		
10600	530	2456 Dec 27	16:51:25	1275	5652	139	T	-p	-0.8614	1.0222	79.8S	22.0W	30	311	151	01m19s
10601	531	2457 Jun 22	08:16:13	1277	5658	144	A	-p	0.4979	0.9827	53.2N	67.4E	60	190	71	01m32s
10602	531	2457 Dec 17	04:35:27	1279	5664	149	A	-n	-0.1900	0.9799	34.4S	116.2E	79	356	73	02m04s
10603	531	2458 Jun 11	18:19:40	1281	5670	154	T	n-	-0.2891	1.0388	6.3N	90.0W	73	2	136	04m04s
10604	531	2458 Dec 06	09:14:46	1283	5676	159	A	p-	0.5280	0.9311	9.5N	44.7E	58	181	303	09m34s
10605	531	2459 May 03	02:35:54	1285	5681	126	Pe	-t	1.5188	0.0214	70.3N	4.3E	0	37		
10606	531	2459 Jun 01	09:59:50	1285	5682	164	T-	t-	-1.0097	1.0038	67.8S	41.3E	0	354	-	-
10607	531	2459 Nov 25	08:41:41	1287	5688	169	P	t-	1.2211	0.5818	68.6N	66.5E	0	194		
10608	531	2460 Apr 21	18:09:49	1289	5693	136	T	-p	0.8503	1.0236	66.8N	119.8W	31	142	154	01m34s
10609	531	2460 Oct 14	20:25:57	1291	5699	141	A	-p	-0.9775	0.9817	73.9S	156.1E	11	82	328	01m09s
10610	531	2461 Apr 11	04:10:36	1293	5705	146	A	nn	0.1300	0.9732	15.8N	120.8E	82	164	97	03m02s
10611	531	2461 Oct 04	09:00:22	1295	5711	151	T	-n	-0.2131	1.0495	16.5S	43.7E	78	18	168	04m30s
10612	531	2462 Mar 31	06:49:44	1297	5717	156	A	p-	-0.6111	0.9302	31.7S	96.3E	52	340	327	08m07s
10613	531	2462 Sep 24	01:28:08	1299	5723	161	T	p-	0.5014	1.0662	28.1N	171.4E	60	200	249	05m28s
10614	531	2463 Mar 20	06:27:54	1301	5729	166	P	t-	-1.3051	0.4410	72.3S	179.5W	0	270		
10615	531	2463 Aug 15	07:37:35	1302	5734	133	P	-t	-1.3853	0.2892	70.8S	27.4E	0	43		
10616	531	2463 Sep 13	17:10:28	1303	5735	171	P	t-	1.2504	0.5367	72.0N	27.6E	0	281		
10617	531	2464 Feb 07	21:17:16	1304	5740	138	A	-p	0.8840	0.9941	45.7N	146.4W	28	163	44	00m31s
10618	531	2464 Aug 03	14:43:00	1306	5746	143	A	-p	-0.6692	0.9621	24.5S	41.3W	48	11	184	04m32s
10619	531	2465 Jan 27	11:03:49	1308	5752	148	T	-n	0.1751	1.0435	8.3S	20.8E	80	171	147	04m11s
10620	531	2465 Jul 23	15:54:48	1310	5758	153	A	nn	0.0904	0.9482	25.0N	50.6W	85	187	191	06m35s

Cat Num	Canon Plate	Calendar Date	TD of Greatest Eclipse	ΔT s	Luna Num	Saros Num	Ecl. Type	QLE	Gamma	Ecl. Mag.	Lat. °	Long. °	Sun Alt °	Sun Azm °	Path Width km	Central Line Dur.
10621	532	2466 Jan 17	02:47:01	1312	5764	158	T	n-	-0.4953	1.0366	50.4S	150.3E	60	352	142	02m48s
10622	532	2466 Jul 12	17:50:51	1314	5770	163	A	t-	0.8461	0.9676	79.5N	65.8W	32	196	221	02m18s
10623	532	2467 Jan 06	15:46:09	1316	5776	168	P	t-	-1.2180	0.5934	67.6S	126.4E	0	183		
10624	532	2467 Jun 02	17:48:24	1318	5781	135	P	-t	-1.0425	0.9315	64.5S	50.8W	0	331		
10625	532	2467 Nov 27	04:10:21	1320	5787	140	A+	-t	1.0051	0.9434	63.7N	158.3E	0	216	-	-
10626	532	2468 May 22	10:15:11	1322	5793	145	T	-n	-0.2936	1.0744	4.2N	36.0E	73	344	252	06m41s
10627	532	2468 Nov 15	03:28:23	1324	5799	150	A	-n	0.3126	0.9304	1.5S	135.7E	72	199	273	08m59s
10628	532	2469 May 12	02:39:07	1326	5805	155	T	p-	0.4417	1.0466	42.6N	132.9E	64	153	172	03m36s
10629	532	2469 Nov 04	06:55:37	1328	5811	160	A	p-	-0.4081	0.9750	37.5S	65.7E	66	29	97	02m19s
10630	532	2470 May 01	14:25:40	1330	5817	165	P	t-	1.2347	0.5638	62.3N	152.9W	0	55		
10631	532	2470 Sep 25	08:39:57	1332	5822	132	Pe	-t	1.5130	0.0365	61.3N	141.4E	0	268		
10632	532	2470 Oct 24	17:49:29	1332	5823	170	P	t-	-1.0933	0.8298	61.7S	160.9E	0	116		
10633	532	2471 Mar 22	01:43:37	1334	5828	137	A	-p	-0.9141	0.9190	52.9S	151.3W	24	307	738	07m00s
10634	532	2471 Sep 15	01:29:11	1336	5834	142	T	-p	0.8109	1.0585	48.0N	162.2W	36	226	323	03m54s
10635	532	2472 Mar 10	01:52:11	1338	5840	147	A	nn	-0.1989	0.9436	13.9S	165.6E	78	331	212	06m08s
10636	532	2472 Sep 03	16:12:54	1340	5846	152	T	nn	0.0857	1.0255	11.3N	55.4W	85	208	87	02m19s
10637	532	2473 Feb 27	07:56:51	1342	5852	157	A	p-	0.5168	0.9907	19.6N	54.6E	59	151	37	00m53s
10638	532	2473 Aug 24	00:46:32	1344	5858	162	A	p-	-0.7043	0.9684	29.3S	153.6E	45	30	156	03m12s
10639	532	2474 Jan 18	10:12:11	1345	5863	129	P	-t	-1.5000	0.0673	63.6S	173.5E	0	218		
10640	532	2474 Feb 16	20:44:52	1346	5864	167	P	t-	1.1720	0.6841	61.7N	171.0E	0	116		
10641	533	2474 Jul 14	11:40:30	1347	5869	134	P	-t	1.3764	0.3182	64.0N	155.9E	0	327		
10642	533	2474 Aug 13	02:43:56	1348	5870	172	Pb	t-	-1.4827	0.1379	62.1S	85.2E	0	58		
10643	533	2475 Jan 08	01:37:52	1349	5875	139	T	-p	-0.8679	1.0196	76.2S	141.8W	29	299	136	01m09s
10644	533	2475 Jul 03	15:05:22	1351	5881	144	A	-p	0.5775	0.9858	57.3N	27.7W	54	199	62	01m11s
10645	533	2475 Dec 28	13:01:54	1354	5887	149	A	-n	-0.1977	0.9760	34.7S	7.6W	78	350	87	02m27s
10646	533	2476 Jun 22	01:38:29	1356	5893	154	T	n-	-0.2153	1.0435	11.1N	160.4E	78	6	149	04m25s
10647	533	2476 Dec 16	17:15:18	1358	5899	159	A	p-	0.5154	0.9282	7.7N	76.3W	59	176	314	10m04s
10648	533	2477 Jun 11	17:35:29	1360	5905	164	T	t-	-0.9423	1.0647	47.8S	80.8W	19	3	642	04m53s
10649	533	2477 Dec 05	16:32:05	1362	5911	169	P	t-	1.2019	0.6136	67.5N	62.0W	0	183		
10650	533	2478 May 03	01:55:59	1363	5916	136	T	-t	0.9034	1.0218	75.7N	107.7E	25	128	176	01m20s
10651	533	2478 Oct 26	04:18:22	1365	5922	141	P	-t	-1.0109	0.9645	71.0S	13.3W	0	132		
10652	533	2479 Apr 22	11:40:30	1367	5928	146	A	nn	0.1790	0.9731	22.5N	7.5E	80	165	98	03m01s
10653	533	2479 Oct 15	17:04:11	1369	5934	151	T	-n	-0.2538	1.0484	23.0S	78.3W	75	17	166	04m18s
10654	533	2480 Apr 10	14:07:46	1372	5940	156	A	p-	-0.5664	0.9326	24.8S	16.0W	55	342	303	08m18s
10655	533	2480 Oct 04	09:27:58	1374	5946	161	T	p-	0.4543	1.0631	21.1N	49.2E	63	198	231	05m26s
10656	533	2481 Mar 30	14:00:07	1376	5952	166	P	t-	-1.2685	0.5027	72.1S	53.5E	0	284		
10657	533	2481 Aug 25	14:49:25	1377	5957	133	P	-t	-1.4585	0.1568	71.4S	93.5W	0	56		
10658	533	2481 Sep 24	00:49:51	1378	5958	171	P	t-	1.1997	0.6307	72.1N	101.0W	0	268		
10659	533	2482 Feb 18	05:48:52	1379	5963	138	A	-p	0.8912	0.9982	49.3N	81.2E	27	157	14	00m09s
10660	533	2482 Aug 14	21:23:36	1381	5969	143	A	-p	-0.7515	0.9573	33.7S	145.8W	41	16	234	04m45s
10661	534	2483 Feb 07	19:51:56	1383	5975	148	T	-n	0.1817	1.0457	4.8S	111.2W	80	168	155	04m20s
10662	534	2483 Aug 03	22:21:18	1386	5981	153	A	nn	0.0030	0.9479	17.4N	147.7W	90	186	192	06m50s
10663	534	2484 Jan 28	11:35:53	1388	5987	158	T	n-	-0.4910	1.0358	47.2S	21.5E	60	347	138	02m48s
10664	534	2484 Jul 23	00:34:35	1390	5993	163	A	p-	0.7619	0.9720	68.9N	166.0W	40	198	156	02m10s
10665	534	2485 Jan 17	00:19:53	1392	5999	168	P	t-	-1.2162	0.5962	68.6S	13.0W	0	195		
10666	534	2485 Jun 13	01:16:18	1393	6004	135	P	-t	-1.1075	0.8095	65.4S	172.3W	0	341		
10667	534	2485 Jul 12	09:35:02	1394	6005	173	Pb	t-	1.4713	0.1259	68.0N	145.3W	0	352		
10668	534	2485 Dec 07	12:02:00	1395	6010	140	A+	-t	1.0242	0.9100	64.7N	31.2E	0	206	-	-
10669	534	2486 Jun 02	17:55:28	1398	6016	145	T	-n	-0.3587	1.0760	1.8N	78.7W	69	348	263	06m59s
10670	534	2486 Nov 26	11:15:08	1400	6022	150	A	-n	0.3363	0.9294	2.1S	19.2E	70	195	280	09m26s
10671	534	2487 May 23	10:16:15	1402	6028	155	T	n-	0.3811	1.0467	42.1N	22.6E	67	159	168	03m43s
10672	534	2487 Nov 15	14:57:35	1404	6034	160	A	p-	-0.3807	0.9756	39.5S	52.3W	67	25	94	02m16s
10673	534	2488 May 11	21:45:56	1406	6040	165	P	t-	1.1793	0.6626	62.9N	88.3E	0	46		
10674	534	2488 Nov 04	02:04:56	1408	6046	170	P	t-	-1.0609	0.8898	62.3S	28.4E	0	125		
10675	534	2489 Apr 01	09:07:55	1410	6051	137	A	-t	-0.9541	0.9200	54.4S	101.3E	17	305	997	06m50s
10676	534	2489 Sep 25	09:20:22	1412	6057	142	T	-p	0.8654	1.0527	48.6N	82.9E	30	227	341	03m32s
10677	534	2490 Mar 21	09:36:11	1414	6063	147	A	nn	-0.2288	0.9482	11.1S	50.0E	77	330	195	05m36s
10678	534	2490 Sep 14	23:40:38	1416	6069	152	T	-n	0.1483	1.0195	10.3N	166.2W	81	209	67	01m47s
10679	534	2491 Mar 10	16:11:57	1418	6075	157	A	p-	0.4952	0.9964	22.2N	69.6W	60	149	14	00m20s
10680	534	2491 Sep 04	07:41:15	1420	6081	162	A	p-	-0.6332	0.9646	27.7S	50.9E	51	32	161	03m34s

Cat Num	Canon Plate	Calendar Date	TD of Greatest Eclipse	ΔT s	Luna Num	Saros Num	Ecl. Type	QLE	Gamma	Ecl. Mag.	Lat. °	Long. °	Sun Alt °	Sun Azm °	Path Width km	Central Line Dur.
10681	535	2492 Jan 29	19:03:43	1422	6086	129	P	-t	-1.5046	0.0582	62.9S	31.6E	0	228		
10682	535	2492 Feb 28	05:22:53	1422	6087	167	P	t-	1.1568	0.7135	61.4N	33.1E	0	107		
10683	535	2492 Jul 24	18:08:32	1424	6092	134	P	-t	1.4594	0.1755	63.2N	49.8E	0	317		
10684	535	2492 Aug 23	09:17:49	1424	6093	172	P	t-	-1.4043	0.2723	61.6S	21.8W	0	67		
10685	535	2493 Jan 18	10:24:30	1426	6098	139	T	-p	-0.8742	1.0174	72.2S	90.8E	29	294	123	01m02s
10686	535	2493 Jul 13	21:56:36	1428	6104	144	A	-p	0.6562	0.9882	60.3N	121.4W	49	209	55	00m56s
10687	535	2494 Jan 07	21:30:21	1430	6110	149	A	-n	-0.2034	0.9727	33.7S	132.0W	78	345	100	02m46s
10688	535	2494 Jul 03	08:56:16	1432	6116	154	T	nn	-0.1397	1.0477	15.0N	51.7E	82	11	160	04m40s
10689	535	2494 Dec 28	01:19:28	1434	6122	159	A	p-	0.5061	0.9257	6.9N	161.8E	60	172	323	10m22s
10690	535	2495 Jun 23	01:08:06	1436	6128	164	T	t-	-0.8718	1.0696	37.2S	162.6E	29	8	464	05m39s
10691	535	2495 Dec 17	00:27:39	1438	6134	169	P	t-	1.1867	0.6389	66.4N	168.8E	0	172		
10692	535	2496 May 13	09:34:25	1440	6139	136	T	-t	0.9622	1.0185	81.0N	70.4W	15	65	243	01m02s
10693	535	2496 Nov 05	12:20:23	1442	6145	141	P	-t	-1.0373	0.9173	70.2S	146.4W	0	145		
10694	535	2497 May 02	19:01:52	1444	6151	146	A	-n	0.2341	0.9727	29.2N	103.2W	76	167	100	02m59s
10695	535	2497 Oct 26	01:15:23	1446	6157	151	T	-n	-0.2889	1.0472	29.1S	158.5E	73	16	164	04m06s
10696	535	2498 Apr 21	21:17:12	1448	6163	156	A	p-	-0.5148	0.9351	17.9S	125.9W	59	345	280	08m26s
10697	535	2498 Oct 15	17:34:44	1451	6169	161	T	n-	0.4131	1.0597	14.6N	74.5W	66	196	215	05m21s
10698	535	2499 Apr 10	21:22:39	1453	6175	166	P	t-	-1.2233	0.5797	71.8S	70.8W	0	298		
10699	535	2499 Sep 05	22:05:19	1454	6180	133	Pe	-t	-1.5273	0.0340	71.9S	144.2E	0	69		
10700	535	2499 Oct 05	08:36:23	1455	6181	171	P	t-	1.1554	0.7119	71.9N	128.6E	0	254		
10701	536	2500 Mar 01	14:14:47	1457	6186	138	H	-p	0.9038	1.0026	53.9N	50.7W	25	151	21	00m12s
10702	536	2500 Aug 26	04:08:16	1459	6192	143	A	-p	-0.8296	0.9518	43.8S	106.9E	34	21	313	04m53s
10703	536	2501 Feb 19	04:35:21	1461	6198	148	T	-n	0.1925	1.0483	0.6S	117.7E	79	165	163	04m31s
10704	536	2501 Aug 15	04:52:08	1463	6204	153	Am	nn	-0.0810	0.9471	9.5N	113.4E	85	13	195	07m01s
10705	536	2502 Feb 08	20:22:29	1465	6210	158	T	n-	-0.4851	1.0354	43.3S	107.9W	61	343	136	02m49s
10706	536	2502 Aug 04	07:19:53	1467	6216	163	A	p-	0.6779	0.9756	58.9N	91.6E	47	199	119	02m03s
10707	536	2503 Jan 29	08:53:33	1469	6222	168	P	t-	-1.2151	0.5979	69.6S	153.0W	0	207		
10708	536	2503 Jun 25	08:40:22	1471	6227	135	P	-t	-1.1759	0.6800	66.4S	66.7E	0	351		
10709	536	2503 Jul 24	16:46:37	1471	6228	173	P	t-	1.3926	0.2716	69.0N	95.8E	0	341		
10710	536	2503 Dec 19	19:59:21	1473	6233	140	P	-t	1.0385	0.8851	65.7N	97.7W	0	196		
10711	536	2504 Jun 14	01:31:03	1475	6239	145	T	-p	-0.4278	1.0769	1.9S	167.3E	65	352	275	07m10s
10712	536	2504 Dec 07	19:10:09	1477	6245	150	A	-n	0.3535	0.9289	2.3S	99.5W	69	191	284	09m46s
10713	536	2505 Jun 03	17:48:02	1479	6251	155	T	n-	0.3165	1.0464	40.5N	86.3W	71	165	163	03m50s
10714	536	2505 Nov 26	23:07:04	1482	6257	160	A	p-	-0.3588	0.9763	41.2S	171.6W	69	20	91	02m13s
10715	536	2506 May 24	04:59:35	1484	6263	165	P	t-	1.1192	0.7695	63.7N	29.1W	0	37		
10716	536	2506 Nov 16	10:27:38	1486	6269	170	P	t-	-1.0340	0.9396	63.0S	106.0W	0	135		
10717	536	2507 Apr 13	16:23:43	1488	6274	137	A-	-t	-1.0006	0.9539	61.3S	13.5E	0	289	-	-
10718	536	2507 Oct 07	17:18:18	1490	6280	142	T	-t	0.9141	1.0464	50.0N	34.0W	24	227	374	03m07s
10719	536	2508 Apr 01	17:13:22	1492	6286	147	A	nn	-0.2648	0.9532	8.7S	63.9W	75	331	177	05m06s
10720	536	2508 Sep 26	07:14:51	1494	6292	152	H	-n	0.2046	1.0134	9.0N	81.2E	78	209	47	01m14s
10721	537	2509 Mar 22	00:20:47	1496	6298	157	H	p-	0.4676	1.0023	24.8N	168.2E	62	148	9	00m12s
10722	537	2509 Sep 15	14:42:15	1498	6304	162	A	p-	-0.5679	0.9604	27.3S	53.4W	55	33	171	03m58s
10723	537	2510 Feb 10	03:51:56	1500	6309	129	P	-t	-1.5123	0.0430	62.2S	109.1W	0	238		
10724	537	2510 Mar 11	13:54:45	1500	6310	167	P	t-	1.1362	0.7531	61.1N	103.2W	0	98		
10725	537	2510 Aug 06	00:38:56	1502	6315	134	Pe	-t	1.5405	0.0362	62.6N	56.5W	0	309		
10726	537	2510 Sep 04	15:56:49	1503	6316	172	P	t-	-1.3292	0.4003	61.2S	130.0W	0	75		
10727	537	2511 Jan 30	19:09:33	1504	6321	139	T	-p	-0.8816	1.0157	68.1S	39.5W	28	293	114	00m57s
10728	537	2511 Jul 26	04:49:26	1506	6327	144	A	-p	0.7346	0.9899	62.1N	146.2E	42	220	52	00m45s
10729	537	2512 Jan 20	05:57:20	1509	6333	149	A	-n	-0.2096	0.9700	31.9S	103.6E	78	341	110	03m02s
10730	537	2512 Jul 14	16:14:11	1511	6339	154	T	nn	-0.0634	1.0510	18.1N	56.5W	86	15	170	04m47s
10731	537	2513 Jan 08	09:25:23	1513	6345	159	A	p-	0.4982	0.9240	7.0N	39.7E	60	168	329	10m25s
10732	537	2513 Jul 04	08:38:16	1515	6351	164	T	p-	-0.7992	1.0729	29.6S	48.0E	37	12	392	06m09s
10733	537	2513 Dec 28	08:28:05	1517	6357	169	P	t-	1.1748	0.6587	65.4N	38.8E	0	161		
10734	537	2514 May 25	17:04:32	1519	6362	136	P	-t	1.0272	0.9507	68.5N	123.2E	0	13		
10735	537	2514 Nov 17	20:31:22	1521	6368	141	P	-t	-1.0572	0.8818	69.3S	78.9E	0	157		
10736	537	2515 May 15	02:12:11	1523	6374	146	A	-p	0.2976	0.9722	36.0N	149.4E	72	170	104	02m57s
10737	537	2515 Nov 07	09:35:34	1525	6380	151	T	-n	-0.3169	1.0459	34.4S	33.9E	71	13	161	03m53s
10738	537	2516 May 03	04:17:47	1528	6386	156	A	p-	-0.4559	0.9377	10.9S	126.7E	63	349	259	08m28s
10739	537	2516 Oct 27	01:48:46	1530	6392	161	T	n-	0.3782	1.0560	8.8N	160.5E	68	194	199	05m11s
10740	537	2517 Apr 22	04:39:28	1532	6398	166	P	t-	-1.1726	0.6669	71.2S	166.8E	0	311		

Cat Num	Canon Plate	Calendar Date	TD of Greatest Eclipse	ΔT s	Luna Num	Saros Num	Ecl. Type	QLE	Gamma	Ecl. Mag.	Lat. °	Long. °	Sun Alt °	Sun Azm °	Path Width km	Central Line Dur.
10741	538	2517 Oct 16	16:28:59	1534	6404	171	P	t-	1.1165	0.7823	71.6N	3.1W	0	240		
10742	538	2518 Mar 12	22:37:02	1536	6409	138	T	-p	0.9200	1.0071	59.1N	176.7E	23	144	63	00m31s
10743	538	2518 Sep 06	10:55:41	1538	6415	143	A	-t	-0.9046	0.9458	54.9S	4.8W	25	30	467	04m54s
10744	538	2519 Mar 02	13:15:24	1540	6421	148	T	-n	0.2062	1.0511	4.2N	12.9W	78	164	173	04m42s
10745	538	2519 Aug 26	11:27:49	1542	6427	153	A	nn	-0.1610	0.9460	1.4N	12.7E	81	15	201	07m08s
10746	538	2520 Feb 20	05:04:06	1544	6433	158	T	n-	-0.4758	1.0353	38.7S	122.7E	61	341	135	02m54s
10747	538	2520 Aug 14	14:11:41	1547	6439	163	A	p-	0.5984	0.9784	49.7N	12.9W	53	200	96	01m57s
10748	538	2521 Feb 08	17:22:54	1549	6445	168	P	t-	-1.2116	0.6039	70.5S	67.4E	0	220		
10749	538	2521 Jul 05	16:04:53	1551	6450	135	P	-t	-1.2445	0.5492	67.4S	54.7W	0	1		
10750	538	2521 Aug 04	00:02:18	1551	6451	173	P	t-	1.3160	0.4141	69.9N	24.8W	0	329		
10751	538	2521 Dec 30	03:58:50	1553	6456	140	P	-t	1.0507	0.8642	66.8N	132.5E	0	185		
10752	538	2522 Jun 25	09:03:45	1555	6462	145	T	-p	-0.4991	1.0769	6.6S	53.5E	60	356	287	07m12s
10753	538	2522 Dec 19	03:10:40	1557	6468	150	A	-n	0.3668	0.9289	2.0S	140.4E	68	187	286	09m58s
10754	538	2523 Jun 15	01:12:30	1559	6474	155	T	n-	0.2464	1.0453	37.5N	166.2E	76	171	156	03m56s
10755	538	2523 Dec 08	07:24:54	1561	6480	160	A	n-	-0.3431	0.9774	42.4S	67.5E	70	15	86	02m08s
10756	538	2524 Jun 03	12:04:40	1564	6486	165	P	t-	1.0528	0.8873	64.5N	144.5W	0	28		
10757	538	2524 Nov 26	18:58:09	1566	6492	170	P	t-	-1.0134	0.9778	63.8S	117.4E	0	144		
10758	538	2525 Apr 23	23:30:15	1568	6497	137	P	-t	-1.0544	0.8646	61.8S	101.6W	0	298		
10759	538	2525 Oct 18	01:23:55	1570	6503	142	T	-t	0.9558	1.0396	52.7N	152.5W	17	227	450	02m39s
10760	538	2526 Apr 13	00:43:02	1572	6509	147	A	-n	-0.3077	0.9583	7.1S	175.8W	72	333	158	04m35s
10761	539	2526 Oct 07	14:54:21	1574	6515	152	H	-n	0.2557	1.0070	7.5N	33.0W	75	208	25	00m40s
10762	539	2527 Apr 02	08:23:26	1576	6521	157	H	p-	0.4341	1.0086	27.3N	48.1E	64	148	33	00m45s
10763	539	2527 Sep 26	21:48:45	1579	6527	162	A	p-	-0.5074	0.9559	27.9S	158.9W	59	33	183	04m25s
10764	539	2528 Feb 21	12:36:45	1580	6532	129	Pe	-t	-1.5232	0.0218	61.8S	111.1E	0	247		
10765	539	2528 Mar 21	22:20:29	1581	6533	167	P	t-	1.1103	0.8030	61.1N	122.0E	0	89		
10766	539	2528 Sep 14	22:42:54	1583	6539	172	P	t-	-1.2592	0.5192	61.0S	120.2E	0	84		
10767	539	2529 Feb 10	03:52:31	1585	6544	139	T	-p	-0.8908	1.0143	64.3S	170.7W	27	294	108	00m53s
10768	539	2529 Aug 05	11:45:36	1587	6550	144	A	-p	0.8109	0.9910	62.9N	53.9E	36	232	54	00m39s
10769	539	2530 Jan 30	14:23:10	1589	6556	149	A	-n	-0.2163	0.9678	29.3S	21.0W	77	337	119	03m14s
10770	539	2530 Jul 25	23:33:49	1591	6562	154	T	nn	0.0124	1.0538	20.2N	164.6W	89	196	178	04m50s
10771	539	2531 Jan 19	17:31:19	1594	6568	159	A	p-	0.4908	0.9228	7.9N	82.4W	61	163	332	10m17s
10772	539	2531 Jul 15	16:07:33	1596	6574	164	T	p-	-0.7256	1.0750	23.8S	65.5W	43	15	351	06m25s
10773	539	2532 Jan 08	16:31:05	1598	6580	169	P	t-	1.1645	0.6760	64.4N	91.4W	0	151		
10774	539	2532 Jun 05	00:28:58	1600	6585	136	P	-t	1.0962	0.8224	67.5N	1.3E	0	2		
10775	539	2532 Jul 04	08:54:58	1600	6586	174	Pb	t-	-1.4782	0.1040	64.9S	27.8E	0	24		
10776	539	2532 Nov 28	04:49:26	1602	6591	141	P	-t	-1.0722	0.8553	68.3S	57.0W	0	169		
10777	539	2533 May 25	09:15:50	1604	6597	146	A	-p	0.3660	0.9712	42.6N	44.8E	68	174	111	02m56s
10778	539	2533 Nov 17	18:03:10	1606	6603	151	T	-n	-0.3394	1.0448	38.9S	91.5W	70	10	159	03m43s
10779	539	2534 May 14	11:09:29	1609	6609	156	A	p-	-0.3896	0.9402	4.1S	21.9E	67	352	240	08m23s
10780	539	2534 Nov 07	10:10:07	1611	6615	161	T	n-	0.3495	1.0522	3.9N	34.0E	70	191	184	04m58s
10781	540	2535 May 03	11:47:37	1613	6621	166	P	t-	-1.1138	0.7691	70.4S	47.1E	0	324		
10782	540	2535 Oct 28	00:29:34	1615	6627	171	P	t-	1.0847	0.8388	70.9N	136.3W	0	227		
10783	540	2536 Mar 23	06:51:06	1617	6632	138	T	-p	0.9435	1.0115	65.3N	42.0E	19	133	121	00m46s
10784	540	2536 Sep 16	17:50:18	1619	6638	143	A	-t	-0.9727	0.9385	67.2S	131.1W	13	51	1025	04m48s
10785	540	2537 Mar 12	21:49:07	1622	6644	148	T	-n	0.2254	1.0542	9.5N	142.2W	77	162	184	04m53s
10786	540	2537 Sep 05	18:08:59	1624	6650	153	A	nn	-0.2368	0.9446	6.7S	89.6W	76	17	210	07m11s
10787	540	2538 Mar 02	13:41:10	1626	6656	158	T	n-	-0.4629	1.0357	33.6S	6.5W	62	340	135	03m01s
10788	540	2538 Aug 25	21:08:14	1628	6662	163	A	p-	0.5217	0.9806	40.7N	119.1W	58	200	81	01m52s
10789	540	2539 Feb 20	01:48:05	1631	6668	168	P	t-	-1.2050	0.6153	71.3S	71.7W	0	233		
10790	540	2539 Jul 16	23:27:49	1632	6673	135	P	-t	-1.3148	0.4143	68.3S	176.3W	0	12		
10791	540	2539 Aug 15	07:21:01	1633	6674	173	P	t-	1.2408	0.5548	70.8N	146.7W	0	317		
10792	540	2540 Jan 10	12:01:35	1635	6679	140	P	-t	1.0600	0.8483	67.9N	1.3E	0	174		
10793	540	2540 Jul 05	16:34:26	1637	6685	145	T	-p	-0.5722	1.0760	12.4S	60.6W	55	1	300	07m04s
10794	540	2540 Dec 29	11:15:59	1639	6691	150	A	-n	0.3765	0.9295	1.0S	19.0E	68	182	285	09m57s
10795	540	2541 Jun 25	08:33:57	1641	6697	155	T	nn	0.1743	1.0437	33.5N	58.6E	80	176	148	03m58s
10796	540	2541 Dec 18	15:48:55	1643	6703	160	A	n-	-0.3319	0.9788	42.8S	54.7W	70	9	80	02m01s
10797	540	2542 Jun 14	19:03:09	1646	6709	165	A	t-	0.9815	0.9737	74.8N	113.4E	10	30	540	01m30s
10798	540	2542 Dec 08	03:35:01	1648	6715	170	T-	t-	-0.9975	1.0072	64.7S	21.0W	0	154	-	-
10799	540	2543 May 05	06:29:19	1650	6720	137	P	-t	-1.1140	0.7648	62.3S	145.1E	0	307		
10800	540	2543 Oct 29	09:36:30	1652	6726	142	Tn	-t	0.9919	1.0316	58.7N	91.9E	6	232	-	02m02s

Cat Num	Canon Plate	Calendar Date	TD of Greatest Eclipse	ΔT s	Luna Num	Saros Num	Ecl. Type	QLE	Gamma	Ecl. Mag.	Lat. °	Long. °	Sun Alt °	Sun Azm °	Path Width km	Central Line Dur.
10801	541	2544 Apr 23	08:05:34	1654	6732	147	A	-n	-0.3575	0.9635	6.3S	74.0E	69	335	140	04m05s
10802	541	2544 Oct 17	22:41:14	1657	6738	152	H	-n	0.3001	1.0006	5.9N	149.2W	73	207	2	00m04s
10803	541	2545 Apr 12	16:19:46	1659	6744	157	H	p-	0.3942	1.0149	29.4N	69.8W	67	149	55	01m17s
10804	541	2545 Oct 07	05:01:15	1661	6750	162	A	p-	-0.4523	0.9514	29.1S	94.4E	63	33	197	04m54s
10805	541	2546 Apr 02	06:39:22	1663	6756	167	P	t-	1.0785	0.8647	61.2N	11.0W	0	80		
10806	541	2546 Sep 26	05:36:28	1666	6762	172	P	t-	-1.1945	0.6285	61.0S	8.5E	0	93		
10807	541	2547 Feb 21	12:29:30	1667	6767	139	T	-p	-0.9046	1.0132	61.1S	59.6E	25	295	106	00m50s
10808	541	2547 Aug 16	18:46:36	1670	6773	144	A	-p	0.8841	0.9910	63.0N	38.6W	28	244	67	00m37s
10809	541	2548 Feb 10	22:44:25	1672	6779	149	A	-n	-0.2262	0.9662	26.3S	144.9W	77	334	125	03m23s
10810	541	2548 Aug 05	06:56:36	1674	6785	154	Tm	nn	0.0862	1.0556	21.4N	86.6E	85	202	184	04m49s
10811	541	2549 Jan 30	01:34:51	1676	6791	159	A	p-	0.4815	0.9223	9.4N	156.1E	61	159	331	10m00s
10812	541	2549 Jul 25	23:37:26	1679	6797	164	T	p-	-0.6522	1.0761	19.5S	178.5W	49	19	322	06m30s
10813	541	2550 Jan 19	00:36:25	1681	6803	169	P	t-	1.1554	0.6914	63.5N	138.2E	0	141		
10814	541	2550 Jun 16	07:45:35	1683	6808	136	P	-t	1.1708	0.6840	66.4N	118.1W	0	352		
10815	541	2550 Jul 15	16:14:58	1683	6809	174	P	t-	-1.4089	0.2366	64.0S	91.6W	0	34		
10816	541	2550 Dec 09	13:15:41	1685	6814	141	P	-t	-1.0815	0.8390	67.2S	165.7E	0	180		
10817	541	2551 Jun 05	16:10:40	1687	6820	146	A	-p	0.4411	0.9699	49.0N	56.1W	64	179	121	02m55s
10818	541	2551 Nov 29	02:38:19	1690	6826	151	T	-n	-0.3559	1.0438	42.3S	142.3E	69	5	157	03m34s
10819	541	2552 May 24	17:54:09	1692	6832	156	A	pn	-0.3174	0.9425	2.5N	80.5W	72	355	224	08m09s
10820	541	2552 Nov 17	18:38:45	1694	6838	161	T	n-	0.3269	1.0485	0.3S	93.9W	71	188	170	04m42s
10821	542	2553 May 13	18:51:21	1696	6844	166	P	t-	-1.0504	0.8800	69.5S	70.8W	0	336		
10822	542	2553 Nov 07	08:35:22	1699	6850	171	P	t-	1.0577	0.8858	70.2N	89.7E	0	214		
10823	542	2554 Apr 03	15:00:51	1701	6855	138	T	-p	0.9713	1.0153	71.5N	102.2W	13	112	232	00m56s
10824	542	2554 Sep 28	00:50:14	1703	6861	143	P	-t	-1.0357	0.8994	72.3S	76.0E	0	97		
10825	542	2555 Mar 24	06:16:23	1705	6867	148	T	-n	0.2502	1.0574	15.2N	90.0E	75	162	195	05m04s
10826	542	2555 Sep 17	00:58:18	1707	6873	153	A	nn	-0.3061	0.9429	14.8S	165.9E	72	18	221	07m10s
10827	542	2556 Mar 12	22:11:21	1710	6879	158	T	n-	-0.4447	1.0362	28.1S	134.7W	63	340	135	03m10s
10828	542	2556 Sep 05	04:13:26	1712	6885	163	A	p-	0.4511	0.9823	32.2N	132.5E	63	200	70	01m48s
10829	542	2557 Mar 02	10:05:49	1714	6891	168	P	t-	-1.1932	0.6361	71.8S	150.5E	0	247		
10830	542	2557 Jul 27	06:54:08	1716	6896	135	P	-t	-1.3827	0.2835	69.3S	60.9E	0	23		
10831	542	2557 Aug 25	14:46:39	1717	6897	173	P	t-	1.1703	0.6870	71.4N	89.1E	0	304		
10832	542	2558 Jan 20	20:03:53	1718	6902	140	P	-t	1.0693	0.8326	69.0N	130.4W	0	162		
10833	542	2558 Jul 17	00:03:14	1721	6908	145	T	-p	-0.6466	1.0742	19.2S	175.0W	50	5	315	06m43s
10834	542	2559 Jan 09	19:24:29	1723	6914	150	A	-n	0.3841	0.9308	0.6N	103.4W	67	178	280	09m43s
10835	542	2559 Jul 06	15:50:37	1725	6920	155	Tm	nn	0.0992	1.0412	28.4N	48.9W	84	181	139	03m55s
10836	542	2559 Dec 30	00:17:19	1728	6926	160	A	n-	-0.3237	0.9808	42.2S	177.9W	71	3	72	01m50s
10837	542	2560 Jun 25	01:55:50	1730	6932	165	A	p-	0.9063	0.9754	86.4N	91.3E	25	111	211	01m35s
10838	542	2560 Dec 18	12:17:54	1732	6938	170	T	t-	-0.9868	1.0184	73.6S	153.6W	8	157	444	00m55s
10839	542	2561 May 15	13:20:16	1734	6943	137	P	-t	-1.1801	0.6534	63.1S	33.6E	0	316		
10840	542	2561 Nov 08	17:55:40	1736	6949	142	P	-t	1.0221	0.9660	62.5N	31.3W	0	231		
10841	543	2562 May 04	15:21:29	1739	6955	147	A	-p	-0.4133	0.9686	6.5S	34.5W	66	338	123	03m35s
10842	543	2562 Oct 29	06:34:40	1741	6961	152	A	-n	0.3382	0.9943	4.6N	92.7E	70	204	21	00m35s
10843	543	2563 Apr 24	00:08:31	1743	6967	157	T	p-	0.3474	1.0213	31.1N	174.8E	70	151	77	01m49s
10844	543	2563 Oct 18	12:21:39	1746	6973	162	A	p-	-0.4042	0.9467	30.7S	14.2W	66	31	213	05m26s
10845	543	2564 Apr 12	14:51:42	1748	6979	167	P	t-	1.0412	0.9373	61.5N	142.5W	0	71		
10846	543	2564 Oct 06	12:38:14	1750	6985	172	P	t-	-1.1361	0.7266	61.1S	105.2W	0	101		
10847	543	2565 Mar 03	21:01:39	1752	6990	139	T	-t	-0.9220	1.0121	58.7S	68.8W	22	296	107	00m46s
10848	543	2565 Aug 27	01:53:56	1754	6996	144	A	-t	0.9527	0.9900	63.0N	129.3W	17	258	117	00m39s
10849	543	2566 Feb 21	07:01:44	1757	7002	149	A	-n	-0.2388	0.9650	22.9S	91.8E	76	332	130	03m30s
10850	543	2566 Aug 16	14:22:25	1759	7008	154	T	nn	0.1581	1.0569	21.8N	22.8W	81	206	190	04m46s
10851	543	2567 Feb 10	09:35:51	1761	7014	159	A	p-	0.4703	0.9223	11.4N	35.2E	62	156	328	09m37s
10852	543	2567 Aug 06	07:09:09	1764	7020	164	T	p-	-0.5802	1.0762	16.5S	68.5E	54	22	299	06m26s
10853	543	2568 Jan 30	08:39:47	1766	7026	169	P	t-	1.1442	0.7106	62.7N	8.6E	0	131		
10854	543	2568 Jun 26	14:58:55	1768	7031	136	P	-t	1.2472	0.5426	65.5N	123.7E	0	342		
10855	543	2568 Jul 25	23:36:01	1768	7032	174	P	t-	-1.3408	0.3666	63.3S	149.0E	0	43		
10856	543	2568 Dec 19	21:47:01	1770	7037	141	P	-t	-1.0877	0.8284	66.2S	27.6E	0	191		
10857	543	2569 Jun 15	23:00:08	1773	7043	146	A	-p	0.5197	0.9680	54.8N	153.7W	58	186	135	02m56s
10858	543	2569 Dec 09	11:18:32	1775	7049	151	T	-n	-0.3687	1.0431	44.7S	15.8E	68	360	155	03m27s
10859	543	2570 Jun 05	00:32:17	1777	7055	156	A	nn	-0.2395	0.9446	8.7N	179.4E	76	359	211	07m48s
10860	543	2570 Nov 29	03:13:44	1779	7061	161	T	n-	0.3100	1.0449	3.4S	137.1E	72	184	158	04m25s

Cat Num	Canon Plate	Calendar Date	TD of Greatest Eclipse	ΔT s	Luna Num	Saros Num	Ecl. Type	QLE	Gamma	Ecl. Mag.	Lat. °	Long. °	Sun Alt °	Sun Azm °	Path Width km	Central Line Dur.
10861	544	2571 May 25	01:47:00	1782	7067	166	A	t-	-0.9794	0.9520	57.9S	169.4E	11	351	926	04m21s
10862	544	2571 Nov 18	16:49:14	1784	7073	171	P	t-	1.0379	0.9196	69.2N	45.7W	0	202		
10863	544	2572 Apr 13	23:02:08	1786	7078	138	P	-t	1.0068	0.9902	71.5N	81.8E	0	58		
10864	544	2572 Oct 08	07:58:20	1788	7084	143	P	-t	-1.0915	0.8031	72.0S	44.8W	0	111		
10865	544	2573 Apr 03	14:36:16	1791	7090	148	T	-n	0.2815	1.0606	21.4N	35.9W	74	162	207	05m13s
10866	544	2573 Sep 27	07:55:50	1793	7096	153	A	-p	-0.3690	0.9411	22.7S	59.4E	68	19	233	07m06s
10867	544	2574 Mar 24	06:34:21	1795	7102	158	T	n-	-0.4208	1.0371	22.3S	98.3E	65	341	137	03m21s
10868	544	2574 Sep 16	11:25:01	1798	7108	163	A	p-	0.3848	0.9835	24.0N	22.4E	67	199	63	01m45s
10869	544	2575 Mar 13	18:17:20	1800	7114	168	P	t-	-1.1771	0.6646	72.1S	13.8E	0	261		
10870	544	2575 Aug 07	14:21:38	1802	7119	135	P	-t	-1.4499	0.1541	70.1S	62.9W	0	35		
10871	544	2575 Sep 05	22:17:41	1802	7120	173	P	t-	1.1036	0.8125	71.9N	37.0W	0	291		
10872	544	2576 Feb 01	04:04:59	1804	7125	140	P	-t	1.0793	0.8161	70.0N	97.6E	0	150		
10873	544	2576 Jul 27	07:32:31	1807	7131	145	T	-p	-0.7203	1.0714	26.9S	69.5E	44	9	334	06m12s
10874	544	2577 Jan 20	03:35:00	1809	7137	150	A	-n	0.3901	0.9326	2.8N	133.7E	67	174	273	09m18s
10875	544	2577 Jul 16	23:05:23	1811	7143	155	T	nn	0.0230	1.0382	22.5N	156.8W	89	184	128	03m47s
10876	544	2578 Jan 09	08:49:00	1814	7149	160	A	n-	-0.3176	0.9831	40.7S	57.7E	71	357	63	01m37s
10877	544	2578 Jul 06	08:44:44	1816	7155	165	A	p-	0.8285	0.9753	78.8N	61.1E	34	183	159	01m45s
10878	544	2578 Dec 29	21:04:04	1818	7161	170	T	t-	-0.9781	1.0201	77.9S	61.6E	11	171	357	01m02s
10879	544	2579 May 26	20:04:28	1820	7166	137	P	-t	-1.2516	0.5320	63.9S	76.4W	0	325		
10880	544	2579 Nov 20	02:21:42	1823	7172	142	P	-t	1.0466	0.9182	63.3N	166.5W	0	222		
10881	545	2580 May 14	22:32:12	1825	7178	147	A	-p	-0.4743	0.9735	7.9S	141.9W	62	341	107	03m04s
10882	545	2580 Nov 08	14:34:19	1827	7184	152	A	-p	0.3704	0.9883	3.4N	27.1W	68	201	44	01m15s
10883	545	2581 May 04	07:51:50	1830	7190	157	T	n-	0.2951	1.0276	32.0N	61.3E	73	154	98	02m22s
10884	545	2581 Oct 28	19:49:23	1832	7196	162	A	n-	-0.3627	0.9422	32.6S	124.3W	69	29	228	06m00s
10885	545	2582 Apr 23	22:55:56	1834	7202	167	Tn	t-	0.9969	1.0462	62.8N	91.1E	2	65	-	02m17s
10886	545	2582 Oct 17	19:49:11	1837	7208	172	P	t-	-1.0846	0.8128	61.4S	138.8E	0	110		
10887	545	2583 Mar 15	05:25:52	1839	7213	139	T	-t	-0.9456	1.0109	57.4S	166.2E	19	297	115	00m42s
10888	545	2583 Sep 07	09:09:01	1841	7219	144	P	-t	1.0160	0.9596	61.3N	150.8E	0	282		
10889	545	2584 Mar 03	15:10:31	1844	7225	149	A	-n	-0.2580	0.9643	19.6S	29.8W	75	330	133	03m36s
10890	545	2584 Aug 26	21:54:18	1846	7231	154	T	-n	0.2258	1.0573	21.4N	134.0W	77	208	193	04m43s
10891	545	2585 Feb 20	17:31:56	1848	7237	159	A	p-	0.4550	0.9230	13.8N	84.3W	63	153	321	09m11s
10892	545	2585 Aug 16	14:42:33	1851	7243	164	T	p-	-0.5094	1.0753	14.7S	44.5W	59	25	281	06m16s
10893	545	2586 Feb 09	16:42:36	1853	7249	169	P	t-	1.1318	0.7319	62.0N	120.6W	0	122		
10894	545	2586 Jul 07	22:07:07	1855	7254	136	P	-t	1.3270	0.3957	64.5N	7.2E	0	332		
10895	545	2586 Aug 06	06:55:48	1855	7255	174	P	t-	-1.2718	0.4975	62.6S	30.2E	0	52		
10896	545	2586 Dec 31	06:23:26	1857	7260	141	P	-t	-1.0903	0.8243	65.1S	111.3W	0	202		
10897	545	2587 Jun 27	05:42:58	1860	7266	146	A	-p	0.6029	0.9656	59.9N	113.1E	53	195	156	02m58s
10898	545	2587 Dec 20	20:04:49	1862	7272	151	T	-n	-0.3767	1.0428	45.6S	111.5W	68	353	154	03m22s
10899	545	2588 Jun 15	07:06:20	1864	7278	156	A	nn	-0.1582	0.9463	14.3N	81.1E	81	4	200	07m21s
10900	545	2588 Dec 09	11:53:28	1867	7284	161	T	n-	0.2973	1.0416	5.6S	7.2E	73	179	146	04m07s
10901	546	2589 Jun 04	08:40:23	1869	7290	166	A	t-	-0.9054	0.9600	42.8S	58.6E	25	359	345	04m10s
10902	546	2589 Nov 29	01:08:10	1872	7296	171	P	t-	1.0227	0.9447	68.2N	178.3E	0	190		
10903	546	2590 Apr 25	06:57:46	1874	7301	138	P	-t	1.0476	0.9167	70.9N	50.1W	0	45		
10904	546	2590 Oct 19	15:13:18	1876	7307	143	P	-t	-1.1411	0.7180	71.5S	167.0W	0	124		
10905	546	2591 Apr 14	22:49:07	1878	7313	148	T	-n	0.3189	1.0637	27.7N	160.0W	71	163	220	05m19s
10906	546	2591 Oct 08	15:02:49	1881	7319	153	A	-p	-0.4245	0.9393	30.2S	49.1W	65	19	247	07m00s
10907	546	2592 Apr 03	14:48:32	1883	7325	158	T	n-	-0.3902	1.0378	16.3S	26.7W	67	342	137	03m32s
10908	546	2592 Sep 26	18:47:01	1886	7331	163	A	p-	0.3261	0.9844	16.4N	90.2W	71	198	58	01m42s
10909	546	2593 Mar 24	02:19:51	1888	7337	168	P	t-	-1.1544	0.7045	72.1S	120.7W	0	275		
10910	546	2593 Aug 17	21:53:04	1890	7342	135	Pe	-t	-1.5141	0.0303	70.8S	171.9E	0	47		
10911	546	2593 Sep 16	05:55:38	1890	7343	173	P	t-	1.0418	0.9285	72.2N	165.1W	0	277		
10912	546	2594 Feb 11	12:02:17	1892	7348	140	P	-t	1.0921	0.7951	70.9N	34.1W	0	137		
10913	546	2594 Aug 07	15:02:42	1895	7354	145	T	-p	-0.7928	1.0676	35.6S	47.4W	37	14	361	05m32s
10914	546	2595 Jan 31	11:44:03	1897	7360	150	A	-n	0.3981	0.9352	5.9N	10.9E	67	170	263	08m42s
10915	546	2595 Jul 28	06:18:18	1900	7366	155	T	nn	-0.0539	1.0343	15.8N	94.8E	87	9	116	03m30s
10916	546	2596 Jan 20	17:22:01	1902	7372	160	A	n-	-0.3119	0.9862	38.3S	67.6W	72	352	51	01m20s
10917	546	2596 Jul 16	15:30:49	1904	7378	165	A	p-	0.7487	0.9741	69.4N	32.6W	41	193	141	02m00s
10918	546	2597 Jan 09	05:52:34	1907	7384	170	T	t-	-0.9713	1.0219	80.8S	94.3W	13	195	334	01m08s
10919	546	2597 Jun 06	02:43:07	1909	7389	137	P	-t	-1.3272	0.4029	64.8S	174.7E	0	335		
10920	546	2597 Jul 05	17:53:16	1909	7390	175	Pb	t-	1.5370	0.0437	67.4N	99.2E	0	358		

Cat Num	Canon Plate	Calendar Date	TD of Greatest Eclipse	ΔT s	Luna Num	Saros Num	Ecl. Type	QLE	Gamma	Ecl. Mag.	Lat. °	Long. °	Sun Alt °	Sun Azm °	Path Width km	Central Line Dur.
10921	547	2597 Nov 30	10:54:08	1911	7395	142	P	-t	1.0654	0.8814	64.1N	56.4E	0	212		
10922	547	2598 May 26	05:36:09	1914	7401	147	A	-p	-0.5415	0.9782	10.6S	112.1E	57	345	91	02m34s
10923	547	2598 Nov 19	22:41:03	1916	7407	152	A	-p	0.3959	0.9825	2.6N	148.8W	67	198	67	01m57s
10924	547	2599 May 15	15:28:44	1918	7413	157	T	n-	0.2370	1.0337	32.0N	50.2W	76	158	117	02m56s
10925	547	2599 Nov 09	03:25:06	1921	7419	162	A	n-	-0.3282	0.9379	34.5S	123.9E	71	26	244	06m35s
10926	547	2600 May 05	06:53:54	1923	7425	167	T	t-	0.9474	1.0552	68.5N	2.2E	18	92	579	02m57s
10927	547	2600 Oct 29	03:09:34	1926	7431	172	P	t-	-1.0403	0.8863	61.8S	20.3E	0	119		
10928	547	2601 Mar 26	13:43:55	1928	7436	139	T	-t	-0.9740	1.0091	58.0S	45.6E	12	295	142	00m35s
10929	547	2601 Sep 18	16:30:38	1930	7442	144	P	-t	1.0746	0.8541	61.1N	32.1E	0	274		
10930	547	2602 Mar 15	23:13:25	1932	7448	149	A	-n	-0.2814	0.9638	16.4S	150.0W	74	330	136	03m41s
10931	547	2602 Sep 08	05:31:32	1935	7454	154	T	-n	0.2895	1.0572	20.6N	113.3E	73	210	196	04m39s
10932	547	2603 Mar 05	01:21:16	1937	7460	159	A	p-	0.4345	0.9243	16.3N	158.1E	64	151	312	08m45s
10933	547	2603 Aug 28	22:20:26	1940	7466	164	T	p-	-0.4425	1.0737	13.9S	158.4W	64	27	264	06m02s
10934	547	2604 Feb 22	00:40:35	1942	7472	169	P	t-	1.1150	0.7609	61.5N	111.5E	0	113		
10935	547	2604 Jul 19	05:14:31	1944	7477	136	P	-t	1.4062	0.2509	63.7N	108.8W	0	323		
10936	547	2604 Aug 17	14:18:38	1945	7478	174	P	t-	-1.2058	0.6216	62.0S	89.1W	0	61		
10937	547	2605 Jan 11	15:01:06	1947	7483	141	P	-t	-1.0928	0.8207	64.2S	110.0E	0	212		
10938	547	2605 Jul 08	12:23:21	1949	7489	146	A	-p	0.6873	0.9626	64.0N	24.0E	46	207	186	03m03s
10939	547	2606 Jan 01	04:53:56	1951	7495	151	T	-n	-0.3828	1.0428	45.3S	120.5E	67	347	155	03m20s
10940	547	2606 Jun 27	13:34:39	1954	7501	156	A	nn	-0.0720	0.9477	19.3N	15.0W	86	8	193	06m52s
10941	548	2606 Dec 21	20:37:56	1956	7507	161	T	n-	0.2888	1.0387	6.7S	123.7W	73	175	135	03m50s
10942	548	2607 Jun 16	15:28:35	1959	7513	166	A	p-	-0.8258	0.9664	32.6S	47.0W	34	4	216	03m47s
10943	548	2607 Dec 11	09:32:40	1961	7519	171	P	p-	1.0127	0.9607	67.1N	41.5E	0	179		
10944	548	2608 May 06	14:45:32	1963	7524	138	P	-t	1.0954	0.8288	70.1N	179.4W	0	33		
10945	548	2608 Jun 04	23:55:35	1964	7525	176	Pb	t-	-1.5249	0.0294	67.5S	168.1W	0	358		
10946	548	2608 Oct 30	22:37:25	1966	7530	143	P	-t	-1.1828	0.6469	70.8S	69.0E	0	137		
10947	548	2609 Apr 26	06:54:26	1968	7536	148	T	-n	0.3627	1.0665	34.2N	78.2E	69	164	233	05m23s
10948	548	2609 Oct 19	22:18:05	1971	7542	153	A	-p	-0.4734	0.9375	37.4S	159.2W	62	19	263	06m53s
10949	548	2610 Apr 15	22:55:08	1973	7548	158	T	n-	-0.3537	1.0387	10.2S	150.0W	69	344	138	03m44s
10950	548	2610 Oct 09	02:17:27	1975	7554	163	A	n-	0.2737	0.9849	9.2N	155.4E	74	197	55	01m41s
10951	548	2611 Apr 05	10:13:05	1978	7560	168	P	t-	-1.1251	0.7564	71.9S	107.3E	0	289		
10952	548	2611 Sep 28	13:41:25	1980	7566	173	T	t-	0.9859	1.0280	69.6N	39.6E	9	240	630	01m42s
10953	548	2612 Feb 23	19:55:50	1982	7571	140	P	-t	1.1076	0.7697	71.6N	165.6W	0	123		
10954	548	2612 Aug 18	22:35:27	1985	7577	145	T	-t	-0.8629	1.0629	45.2S	166.8W	30	20	407	04m45s
10955	548	2613 Feb 11	19:51:44	1987	7583	150	A	-n	0.4076	0.9382	9.6N	111.8W	66	167	250	08m00s
10956	548	2613 Aug 08	13:32:05	1990	7589	155	T	nn	-0.1292	1.0300	8.7N	14.5W	83	11	102	03m07s
10957	548	2614 Feb 01	01:55:16	1992	7595	160	A	n-	-0.3058	0.9897	35.0S	166.3E	72	348	38	01m00s
10958	548	2614 Jul 28	22:14:49	1995	7601	165	A	p-	0.6680	0.9721	60.2N	132.0W	48	196	135	02m21s
10959	548	2615 Jan 21	14:42:02	1997	7607	170	T	t-	-0.9651	1.0241	81.6S	99.5E	14	229	328	01m17s
10960	548	2615 Jun 18	09:17:56	1999	7612	137	P	-t	-1.4053	0.2689	65.7S	66.4E	0	344		
10961	549	2615 Jul 18	00:19:58	2000	7613	175	P	t-	1.4533	0.1869	68.4N	8.1W	0	348		
10962	549	2615 Dec 12	19:30:54	2002	7618	142	P	-t	1.0802	0.8524	65.1N	82.1W	0	202		
10963	549	2616 Jun 06	12:37:18	2004	7624	147	A	-p	-0.6115	0.9824	14.5S	6.3E	52	349	78	02m04s
10964	549	2616 Dec 01	06:53:19	2007	7630	152	A	-p	0.4156	0.9772	2.1N	88.0E	65	193	89	02m39s
10965	549	2617 May 26	23:01:04	2009	7636	157	T	n-	0.1741	1.0394	31.0N	160.5W	80	163	134	03m30s
10966	549	2617 Nov 20	11:07:13	2012	7642	162	A	n-	-0.2995	0.9339	36.2S	11.0E	72	22	258	07m11s
10967	549	2618 May 16	14:44:47	2014	7648	167	T	t-	0.8919	1.0612	70.9N	97.2W	27	109	447	03m24s
10968	549	2618 Nov 09	10:39:32	2016	7654	172	A-	t-	-1.0033	0.9474	62.4S	100.7W	0	128	-	-
10969	549	2619 Apr 06	21:51:02	2019	7659	139	P	-t	-1.0108	0.9781	61.2S	60.7W	0	283		
10970	549	2619 Sep 30	00:02:28	2021	7665	144	P	-t	1.1256	0.7620	61.2N	89.2W	0	265		
10971	549	2620 Mar 26	07:06:11	2023	7671	149	A	-p	-0.3125	0.9636	13.7S	92.2E	72	330	138	03m46s
10972	549	2620 Sep 18	13:15:47	2026	7677	154	T	-n	0.3476	1.0565	19.4N	1.6W	70	211	198	04m35s
10973	549	2621 Mar 15	09:03:08	2028	7683	159	A	p-	0.4080	0.9260	18.9N	42.7E	66	149	301	08m20s
10974	549	2621 Sep 08	06:02:19	2031	7689	164	T	n-	-0.3793	1.0713	14.1S	86.8E	68	29	249	05m45s
10975	549	2622 Mar 04	08:34:43	2033	7695	169	P	t-	1.0946	0.7963	61.2N	15.3W	0	104		
10976	549	2622 Jul 30	12:18:09	2035	7700	136	Pe	-t	1.4872	0.1039	63.0N	136.4E	0	314		
10977	549	2622 Aug 28	21:41:59	2036	7701	174	P	t-	-1.1408	0.7425	61.6S	151.5E	0	70		
10978	549	2623 Jan 22	23:41:24	2038	7706	141	P	-t	-1.0937	0.8199	63.3S	29.1W	0	222		
10979	549	2623 Jul 19	19:00:06	2040	7712	146	A	-p	0.7738	0.9589	66.8N	60.2W	39	223	235	03m10s
10980	549	2624 Jan 12	13:45:09	2043	7718	151	T	-n	-0.3874	1.0433	43.8S	8.4W	67	341	157	03m21s

Cat Num	Canon Plate	Calendar Date	TD of Greatest Eclipse	ΔT s	Luna Num	Saros Num	Ecl. Type	QLE	Gamma	Ecl. Mag.	Lat. °	Long. °	Sun Alt °	Sun Azm °	Path Width km	Central Line Dur.
10981	550	2624 Jul 07	20:02:10	2045	7724	156	Am	nn	0.0150	0.9487	23.4N	110.2W	89	191	188	06m24s
10982	550	2625 Jan 01	05:25:27	2048	7730	161	T	n-	0.2829	1.0361	6.8S	104.8E	74	171	126	03m32s
10983	550	2625 Jun 26	22:15:44	2050	7736	166	A	p-	-0.7444	0.9720	24.6S	150.9W	42	8	150	03m17s
10984	550	2625 Dec 21	18:00:12	2053	7742	171	P	p-	1.0058	0.9710	66.1N	95.5W	0	168		
10985	550	2626 May 17	22:28:40	2055	7747	138	P	-t	1.1476	0.7318	69.1N	53.1E	0	21		
10986	550	2626 Jun 16	07:13:26	2055	7748	176	P	t-	-1.4523	0.1621	66.5S	72.1E	0	8		
10987	550	2626 Nov 11	06:08:45	2057	7753	143	P	-t	-1.2180	0.5874	70.0S	56.1W	0	150		
10988	550	2627 May 07	14:52:04	2060	7759	148	T	-p	0.4129	1.0688	40.8N	41.0W	65	166	246	05m22s
10989	550	2627 Oct 31	05:43:51	2062	7765	153	A	-p	-0.5140	0.9358	44.0S	89.1E	59	18	278	06m44s
10990	550	2628 Apr 26	06:52:57	2065	7771	158	T	n-	-0.3105	1.0392	4.0S	89.1E	72	347	138	03m53s
10991	550	2628 Oct 19	09:57:24	2067	7777	163	A	n-	0.2284	0.9854	2.7N	39.0E	77	195	53	01m39s
10992	550	2629 Apr 15	17:56:38	2070	7783	168	P	t-	-1.0887	0.8209	71.5S	22.1W	0	303		
10993	550	2629 Oct 08	21:35:29	2072	7789	173	T	t-	0.9363	1.0312	59.0N	106.3W	20	214	302	02m10s
10994	550	2630 Mar 06	03:42:09	2075	7794	140	P	-t	1.1288	0.7350	72.1N	64.3E	0	109		
10995	550	2630 Aug 30	06:10:52	2077	7800	145	T	-t	-0.9302	1.0568	56.1S	68.9E	21	29	514	03m53s
10996	550	2631 Feb 23	03:55:11	2080	7806	150	A	-p	0.4211	0.9419	14.0N	126.3E	65	164	236	07m13s
10997	550	2631 Aug 19	20:47:03	2082	7812	155	T	-n	-0.2025	1.0249	1.2N	124.7W	78	14	86	02m36s
10998	550	2632 Feb 12	10:25:37	2085	7818	160	A	n-	-0.2969	0.9938	30.9S	40.1E	73	345	23	00m36s
10999	550	2632 Aug 08	04:59:27	2087	7824	165	A	p-	0.5886	0.9695	51.2N	126.5E	54	198	136	02m49s
11000	550	2633 Jan 31	23:31:25	2090	7830	170	T	t-	-0.9592	1.0266	79.9S	64.6W	16	261	332	01m27s
11001	551	2633 Jun 28	15:48:41	2092	7835	137	Pe	-t	-1.4864	0.1291	66.7S	41.2W	0	354		
11002	551	2633 Jul 28	06:45:14	2092	7836	175	P	t-	1.3679	0.3330	69.4N	115.6W	0	337		
11003	551	2633 Dec 23	04:12:15	2094	7841	142	P	-t	1.0909	0.8313	66.1N	137.9E	0	192		
11004	551	2634 Jun 17	19:34:22	2097	7847	147	A	-p	-0.6854	0.9862	19.8S	99.1W	47	353	67	01m36s
11005	551	2634 Dec 12	15:11:13	2099	7853	152	A	-p	0.4303	0.9723	2.2N	36.7W	64	189	110	03m19s
11006	551	2635 Jun 07	06:28:22	2102	7859	157	T	nn	0.1063	1.0447	28.8N	90.3E	84	168	150	04m04s
11007	551	2635 Dec 01	18:56:55	2104	7865	162	A	nn	-0.2774	0.9302	37.4S	103.5W	74	17	272	07m47s
11008	551	2636 May 26	22:30:53	2107	7871	167	T	p-	0.8322	1.0661	71.8N	165.5E	33	128	392	03m48s
11009	551	2636 Nov 19	18:16:59	2109	7877	172	As	t-	-0.9719	0.9164	70.9S	163.3E	13	113	-	05m33s
11010	551	2637 Apr 17	05:51:33	2111	7882	139	P	-t	-1.0525	0.9013	61.6S	170.8E	0	292		
11011	551	2637 Oct 10	07:42:11	2114	7888	144	P	-t	1.1709	0.6802	61.4N	147.6E	0	256		
11012	551	2638 Apr 06	14:50:17	2117	7894	149	A	-p	-0.3500	0.9635	11.6S	23.4W	69	331	140	03m52s
11013	551	2638 Sep 29	21:06:37	2119	7900	154	T	-n	0.4007	1.0554	18.1N	118.4W	66	210	198	04m31s
11014	551	2639 Mar 26	16:36:39	2122	7906	159	A	p-	0.3749	0.9281	21.4N	70.2W	68	149	288	07m58s
11015	551	2639 Sep 19	13:50:24	2124	7912	164	T	n-	-0.3212	1.0683	15.0S	29.4W	71	30	234	05m28s
11016	551	2640 Mar 14	16:21:09	2127	7918	169	P	t-	1.0672	0.8439	61.0N	140.1W	0	95		
11017	551	2640 Sep 08	05:10:48	2129	7924	174	P	t-	-1.0807	0.8532	61.3S	30.9E	0	79		
11018	551	2641 Feb 02	08:20:04	2131	7929	141	P	-t	-1.0971	0.8150	62.6S	167.5W	0	231		
11019	551	2641 Jul 30	01:35:56	2134	7935	146	A	-t	0.8602	0.9545	68.1N	140.0W	30	242	326	03m20s
11020	551	2642 Jan 22	22:36:33	2136	7941	151	T	-n	-0.3923	1.0443	41.4S	138.1W	67	337	160	03m24s
11021	552	2642 Jul 19	02:27:54	2139	7947	156	A	nn	0.1040	0.9493	26.6N	155.8E	84	197	187	06m00s
11022	552	2643 Jan 12	14:14:06	2142	7953	161	T	n-	0.2784	1.0341	6.0S	27.0W	74	166	119	03m18s
11023	552	2643 Jul 08	05:00:57	2144	7959	166	A	p-	-0.6602	0.9768	18.2S	106.7E	49	12	109	02m44s
11024	552	2644 Jan 02	02:31:07	2147	7965	171	A+	p-	1.0021	0.9759	65.0N	127.1E	0	157	-	-
11025	552	2644 May 28	06:05:58	2149	7970	138	P	-t	1.2051	0.6236	68.2N	72.3W	0	10		
11026	552	2644 Jun 26	14:29:01	2149	7971	176	P	t-	-1.3767	0.3016	65.5S	46.7W	0	18		
11027	552	2644 Nov 21	13:47:29	2151	7976	143	P	-t	-1.2468	0.5390	69.0S	177.5E	0	162		
11028	552	2645 May 17	22:43:18	2154	7982	148	T	-p	0.4686	1.0707	47.4N	157.7W	62	170	261	05m16s
11029	552	2645 Nov 10	13:18:36	2156	7988	153	A	-p	-0.5477	0.9344	49.9S	23.6W	57	15	293	06m35s
11030	552	2646 May 07	14:41:46	2159	7994	158	T	n-	-0.2602	1.0396	2.1N	29.4W	75	350	137	04m00s
11031	552	2646 Oct 30	17:46:59	2161	8000	163	A	n-	0.1902	0.9857	3.2S	79.4W	79	193	51	01m38s
11032	552	2647 Apr 27	01:30:49	2164	8006	168	P	t-	-1.0450	0.8980	70.8S	148.5W	0	316		
11033	552	2647 Oct 20	05:38:03	2167	8012	173	T	t-	0.8932	1.0324	50.5N	122.8E	26	204	243	02m27s
11034	552	2648 Mar 16	11:21:54	2169	8017	140	P	-t	1.1552	0.6917	72.3N	64.5W	0	95		
11035	552	2648 Sep 09	13:51:23	2171	8023	145	Ts	-t	-0.9929	1.0479	70.1S	79.1W	5	60	-	02m48s
11036	552	2649 Mar 05	11:55:21	2174	8029	150	A	-p	0.4378	0.9460	19.0N	5.2E	64	162	220	06m25s
11037	552	2649 Aug 30	04:03:55	2176	8035	155	T	-n	-0.2732	1.0194	6.5S	124.2E	74	16	69	02m01s
11038	552	2650 Feb 22	18:53:59	2179	8041	160	A	n-	-0.2856	0.9984	26.2S	86.4W	73	343	6	00m09s
11039	552	2650 Aug 19	11:45:23	2182	8047	165	A	p-	0.5110	0.9663	42.4N	23.7E	59	199	141	03m21s
11040	552	2651 Feb 12	08:17:28	2184	8053	170	T	t-	-0.9502	1.0298	76.6S	139.9E	18	284	332	01m41s

Cat Num	Canon Plate	Calendar Date	TD of Greatest Eclipse	ΔT s	Luna Num	Saros Num	Ecl. Type	QLE	Gamma	Ecl. Mag.	Lat. °	Long. °	Sun Alt °	Sun Azm °	Path Width km	Central Line Dur.
11041	553	2651 Aug 08	13:13:27	2187	8059	175	P	t-	1.2844	0.4757	70.3N	135.6E	0	325		
11042	553	2652 Jan 03	12:55:42	2189	8064	142	P	-t	1.0995	0.8144	67.2N	3.1W	0	181		
11043	553	2652 Jun 28	02:31:12	2191	8070	147	A	-p	-0.7603	0.9892	26.5S	154.7E	40	357	58	01m11s
11044	553	2652 Dec 22	23:31:17	2194	8076	152	A	-p	0.4424	0.9680	2.8N	162.0W	64	185	128	03m56s
11045	553	2653 Jun 17	13:53:05	2197	8082	157	T	nn	0.0356	1.0493	25.5N	18.8W	88	173	164	04m37s
11046	553	2653 Dec 12	02:51:33	2199	8088	162	A	nn	-0.2599	0.9271	38.0S	141.1E	75	12	284	08m21s
11047	553	2654 Jun 07	06:09:50	2202	8094	167	T	p-	0.7665	1.0703	70.6N	70.1E	40	148	358	04m12s
11048	553	2654 Dec 01	02:03:29	2204	8100	172	A	t-	-0.9477	0.9165	75.8S	53.4E	18	107	1021	05m41s
11049	553	2655 Apr 28	13:40:56	2207	8105	139	P	-t	-1.1024	0.8094	62.0S	45.1E	0	301		
11050	553	2655 May 27	22:51:50	2207	8106	177	Pb	t-	1.5050	0.0543	64.1N	62.4E	0	34		
11051	553	2655 Oct 21	15:32:13	2209	8111	144	P	-t	1.2084	0.6123	61.7N	21.7E	0	247		
11052	553	2656 Apr 16	22:23:12	2212	8117	149	A	-p	-0.3957	0.9633	10.5S	136.2W	67	333	143	04m00s
11053	553	2656 Oct 10	05:06:01	2214	8123	154	T	-n	0.4468	1.0539	16.7N	122.2E	63	209	197	04m28s
11054	553	2657 Apr 06	00:01:46	2217	8129	159	A	n-	0.3350	0.9305	23.6N	179.4E	70	149	274	07m38s
11055	553	2657 Sep 29	21:43:07	2219	8135	164	T	n-	-0.2675	1.0647	16.4S	146.6W	74	30	219	05m11s
11056	553	2658 Mar 26	00:02:23	2222	8141	169	P	t-	1.0350	0.9002	61.0N	96.4E	0	86		
11057	553	2658 Sep 19	12:42:52	2225	8147	174	P	t-	-1.0239	0.9564	61.1S	90.4W	0	87		
11058	553	2659 Feb 13	16:57:15	2227	8152	141	P	-t	-1.1020	0.8073	62.0S	54.6E	0	241		
11059	553	2659 Aug 10	08:11:51	2229	8158	146	A	-t	0.9454	0.9487	67.6N	146.8E	19	266	584	03m30s
11060	553	2660 Feb 03	07:27:31	2232	8164	151	T	-n	-0.3977	1.0457	38.2S	91.5E	66	333	165	03m30s
11061	554	2660 Jul 29	08:55:21	2235	8170	156	A	nn	0.1914	0.9495	28.9N	61.8E	79	202	189	05m42s
11062	554	2661 Jan 22	23:02:23	2237	8176	161	T	n-	0.2740	1.0324	4.4S	158.8W	74	162	113	03m04s
11063	554	2661 Jul 18	11:48:08	2240	8182	166	A	p-	-0.5766	0.9811	13.2S	4.5E	55	16	81	02m12s
11064	554	2662 Jan 12	11:02:26	2242	8188	171	A+	p-	0.9996	0.9787	64.1N	10.0W	0	147	-	-
11065	554	2662 Jun 08	13:38:43	2245	8193	138	P	-t	1.2666	0.5068	67.2N	163.9E	0	359		
11066	554	2662 Jul 07	21:43:23	2245	8194	176	P	t-	-1.2992	0.4460	64.6S	164.8W	0	27		
11067	554	2662 Dec 02	21:32:54	2247	8199	143	P	-t	-1.2698	0.5007	67.9S	50.2E	0	174		
11068	554	2663 May 29	06:28:21	2250	8205	148	T	-p	0.5295	1.0719	53.7N	88.7E	58	174	276	05m07s
11069	554	2663 Nov 21	21:02:01	2252	8211	153	A	-p	-0.5747	0.9333	54.9S	136.6W	55	11	305	06m26s
11070	554	2664 May 17	22:22:49	2255	8217	158	T	nn	-0.2040	1.0395	7.9N	145.4W	78	353	135	04m02s
11071	554	2664 Nov 10	01:46:10	2258	8223	163	A	n-	0.1591	0.9861	8.2S	160.4E	81	190	50	01m36s
11072	554	2665 May 07	08:55:09	2260	8229	168	As	t-	-0.9943	0.9668	66.3S	82.4E	4	334	-	02m35s
11073	554	2665 Oct 30	13:48:30	2263	8235	173	T	p-	0.8563	1.0330	43.6N	6.1W	31	198	215	02m40s
11074	554	2666 Mar 27	18:53:07	2265	8240	140	P	-t	1.1881	0.6371	72.2N	168.8E	0	81		
11075	554	2666 Sep 20	21:37:08	2268	8246	145	P	-t	-1.0506	0.9186	72.2S	136.0E	0	88		
11076	554	2666 Oct 20	05:56:28	2268	8247	183	Pb	t-	1.5196	0.0226	71.3N	154.0E	0	236		
11077	554	2667 Mar 16	19:47:40	2270	8252	150	A	-p	0.4613	0.9506	24.6N	114.3W	62	161	203	05m36s
11078	554	2667 Sep 10	11:25:05	2273	8258	155	H	-p	-0.3393	1.0134	14.2S	11.7E	70	18	49	01m22s
11079	554	2668 Mar 05	03:17:08	2276	8264	160	H	n-	-0.2697	1.0035	21.0S	147.7E	74	342	13	00m21s
11080	554	2668 Aug 29	18:33:48	2278	8270	165	A	p-	0.4360	0.9627	33.7N	80.2W	64	199	149	03m59s
11081	555	2669 Feb 22	17:00:38	2281	8276	170	T	p-	-0.9390	1.0333	72.2S	7.0W	20	300	333	01m58s
11082	555	2669 Aug 18	19:43:41	2283	8282	175	P	t-	1.2020	0.6161	71.1N	25.7E	0	313		
11083	555	2670 Jan 13	21:41:08	2286	8287	142	P	-t	1.1061	0.8013	68.3N	145.2W	0	170		
11084	555	2670 Jul 09	09:25:19	2288	8293	147	A	-p	-0.8379	0.9915	35.0S	48.0E	33	2	55	00m52s
11085	555	2671 Jan 03	07:55:01	2291	8299	152	A	-p	0.4505	0.9643	4.0N	71.7E	63	180	144	04m27s
11086	555	2671 Jun 28	21:15:30	2293	8305	157	T	nn	-0.0374	1.0534	21.1N	128.0W	88	357	177	05m07s
11087	555	2671 Dec 23	10:50:32	2296	8311	162	A	nn	-0.2466	0.9246	37.8S	24.7E	76	6	294	08m52s
11088	555	2672 Jun 17	13:46:16	2299	8317	167	T	p-	0.6987	1.0735	67.3N	28.6W	45	164	335	04m36s
11089	555	2672 Dec 11	09:56:21	2301	8323	172	A	p-	-0.9286	0.9165	80.5S	59.4W	21	102	874	05m47s
11090	555	2673 May 08	21:23:23	2304	8328	139	P	-t	-1.1574	0.7080	62.7S	79.1W	0	310		
11091	555	2673 Jun 07	06:25:46	2304	8329	177	P	t-	1.4460	0.1665	64.9N	60.3W	0	24		
11092	555	2673 Oct 31	23:29:55	2306	8334	144	P	-t	1.2404	0.5544	62.2N	106.2W	0	238		
11093	555	2674 Apr 28	05:47:47	2309	8340	149	A	-p	-0.4477	0.9631	10.2S	113.0E	63	336	147	04m09s
11094	555	2674 Oct 21	13:13:05	2312	8346	154	T	-p	0.4869	1.0522	15.4N	0.5E	61	207	196	04m25s
11095	555	2675 Apr 17	07:16:48	2314	8352	159	A	nn	0.2868	0.9331	25.3N	72.2E	73	151	259	07m23s
11096	555	2675 Oct 11	05:43:47	2317	8358	164	T	n-	-0.2206	1.0608	18.3S	94.2E	77	29	204	04m54s
11097	555	2676 Apr 05	07:35:27	2319	8364	169	An	t-	0.9952	0.9336	61.9N	17.3W	4	84	-	04m25s
11098	555	2676 Sep 29	20:20:52	2322	8370	174	T	t-	-0.9725	1.0089	60.1S	173.0E	13	73	134	00m33s
11099	555	2677 Feb 24	01:29:55	2324	8375	141	P	-t	-1.1113	0.7915	61.6S	81.9W	0	250		
11100	555	2677 Aug 20	14:50:18	2327	8381	146	P	-t	1.0277	0.9182	61.7N	81.5E	0	296		

Cat Num	Canon Plate	Calendar Date	TD of Greatest Eclipse	ΔT s	Luna Num	Saros Num	Ecl. Type	QLE	Gamma	Ecl. Mag.	Lat. °	Long. °	Sun Alt °	Sun Azm °	Path Width km	Central Line Dur.
11101	556	2678 Feb 13	16:14:59	2330	8387	151	T	-n	-0.4062	1.0475	34.8S	38.7W	66	330	172	03m38s
11102	556	2678 Aug 09	15:23:56	2332	8393	156	A	nn	0.2782	0.9492	30.4N	32.2W	74	206	194	05m30s
11103	556	2679 Feb 03	07:49:00	2335	8399	161	T	n-	0.2685	1.0312	2.1S	69.7E	74	159	109	02m54s
11104	556	2679 Jul 29	18:37:06	2338	8405	166	A	p-	-0.4933	0.9846	9.5S	97.5W	60	19	62	01m44s
11105	556	2680 Jan 23	19:32:56	2340	8411	171	An	p-	0.9970	0.9636	62.1N	144.2W	2	140	-	02m46s
11106	556	2680 Jun 18	21:08:08	2342	8416	138	P	-t	1.3315	0.3827	66.2N	41.5E	0	349		
11107	556	2680 Jul 18	04:58:41	2343	8417	176	P	t-	-1.2212	0.5923	63.7S	77.3E	0	37		
11108	556	2680 Dec 13	05:25:03	2345	8422	143	P	-t	-1.2874	0.4715	66.8S	78.3W	0	185		
11109	556	2681 Jun 08	14:07:31	2348	8428	148	T	-p	0.5953	1.0724	59.7N	21.1W	53	181	294	04m54s
11110	556	2681 Dec 02	04:53:55	2350	8434	153	A	-p	-0.5952	0.9326	58.6S	110.4E	53	5	314	06m18s
11111	556	2682 May 29	05:56:13	2353	8440	158	T	nn	-0.1419	1.0390	13.5N	100.9E	82	357	132	03m59s
11112	556	2682 Nov 21	09:54:31	2356	8446	163	A	n-	0.1351	0.9866	12.2S	38.4E	82	187	48	01m32s
11113	556	2683 May 18	16:10:08	2358	8452	168	A	p-	-0.9366	0.9706	49.7S	42.7W	20	349	305	02m44s
11114	556	2683 Nov 10	22:07:55	2361	8458	173	T	p-	0.8266	1.0331	37.9N	135.4W	34	193	198	02m49s
11115	556	2684 Apr 07	02:17:17	2363	8463	140	P	-t	1.2265	0.5732	71.9N	44.0E	0	67		
11116	556	2684 Oct 01	05:28:03	2366	8469	145	P	-t	-1.1036	0.8161	72.1S	4.5E	0	101		
11117	556	2684 Oct 30	14:11:22	2367	8470	183	P	t-	1.4864	0.0875	70.6N	17.5E	0	223		
11118	556	2685 Mar 27	03:35:09	2369	8475	150	A	-p	0.4895	0.9554	30.7N	127.5E	61	160	185	04m48s
11119	556	2685 Sep 20	18:50:12	2371	8481	155	H	-p	-0.4011	1.0071	22.0S	101.7W	66	19	27	00m42s
11120	556	2686 Mar 16	11:34:58	2374	8487	160	H	n-	-0.2486	1.0090	15.4S	22.6E	76	342	32	00m54s
11121	557	2686 Sep 10	01:25:58	2377	8493	165	A	p-	0.3646	0.9587	25.3N	174.7E	68	199	160	04m41s
11122	557	2687 Mar 06	01:38:13	2379	8499	170	T	p-	-0.9230	1.0374	66.9S	148.6W	22	312	330	02m19s
11123	557	2687 Aug 30	02:19:46	2382	8505	175	P	t-	1.1238	0.7490	71.7N	86.2W	0	300		
11124	557	2688 Jan 25	06:24:18	2384	8510	142	P	-t	1.1141	0.7860	69.3N	72.8E	0	158		
11125	557	2688 Jul 19	16:22:31	2387	8516	147	A	-p	-0.9136	0.9926	45.6S	61.4W	24	8	64	00m41s
11126	557	2689 Jan 13	16:18:41	2390	8522	152	A	-p	0.4578	0.9612	5.9N	54.7W	63	176	158	04m52s
11127	557	2689 Jul 09	04:36:31	2392	8528	157	Tm	nn	-0.1123	1.0568	15.8N	122.4E	84	2	188	05m31s
11128	557	2690 Jan 02	18:52:25	2395	8534	162	A	nn	-0.2360	0.9226	36.6S	92.6W	76	1	301	09m17s
11129	557	2690 Jun 28	21:18:07	2398	8540	167	T	p-	0.6272	1.0759	62.2N	131.3W	51	176	317	05m00s
11130	557	2690 Dec 22	17:55:30	2401	8546	172	A	p-	-0.9140	0.9168	85.1S	166.7W	23	89	795	05m52s
11131	557	2691 May 20	04:55:09	2403	8551	139	P	-t	-1.2203	0.5922	63.4S	159.1E	0	320		
11132	557	2691 Jun 18	13:52:32	2403	8552	177	P	t-	1.3818	0.2885	65.9N	178.4E	0	15		
11133	557	2691 Nov 12	07:38:14	2405	8557	144	P	-t	1.2646	0.5105	62.8N	123.2E	0	229		
11134	557	2691 Dec 11	21:05:53	2406	8558	182	Pb	t-	-1.5501	0.0114	65.1S	76.1E	0	158		
11135	557	2692 May 08	13:02:03	2408	8563	149	A	-p	-0.5074	0.9627	11.2S	4.8E	59	339	155	04m21s
11136	557	2692 Oct 31	21:27:40	2411	8569	154	T	-p	0.5212	1.0503	14.4N	123.3W	58	204	193	04m23s
11137	557	2693 Apr 27	14:23:45	2414	8575	159	A	nn	0.2320	0.9359	26.4N	32.5W	76	154	245	07m12s
11138	557	2693 Oct 21	13:50:36	2416	8581	164	T	n-	-0.1793	1.0566	20.3S	26.3W	80	27	189	04m37s
11139	557	2694 Apr 16	15:01:10	2419	8587	169	A	t-	0.9489	0.9422	63.2N	105.8W	18	103	679	04m05s
11140	557	2694 Oct 11	04:04:20	2422	8593	174	H	p-	-0.9263	1.0054	59.5S	67.4E	22	65	49	00m21s
11141	558	2695 Mar 07	09:58:56	2424	8598	141	P	-t	-1.1242	0.7693	61.3S	142.5E	0	259		
11142	558	2695 Aug 31	21:32:02	2427	8604	146	P	-t	1.1064	0.7816	61.3N	27.3W	0	287		
11143	558	2696 Feb 25	00:58:48	2429	8610	151	T	-n	-0.4179	1.0496	31.1S	168.7W	65	329	180	03m48s
11144	558	2696 Aug 19	21:57:56	2432	8616	156	A	-p	0.3608	0.9485	31.1N	127.6W	69	210	201	05m24s
11145	558	2697 Feb 13	16:33:04	2435	8622	161	T	n-	0.2612	1.0305	0.7N	61.3W	75	156	106	02m46s
11146	558	2697 Aug 09	01:29:22	2437	8628	166	A	p-	-0.4116	0.9877	6.9S	160.1E	66	23	47	01m20s
11147	558	2698 Feb 03	04:01:19	2440	8634	171	An	p-	0.9933	0.9625	59.1N	85.7E	5	135	-	02m52s
11148	558	2698 Jun 30	04:35:43	2442	8639	138	P	-t	1.3983	0.2539	65.3N	80.1W	0	339		
11149	558	2698 Jul 29	12:15:55	2443	8640	176	P	t-	-1.1437	0.7385	63.0S	40.9W	0	46		
11150	558	2698 Dec 24	13:21:08	2445	8645	143	P	-t	-1.3014	0.4485	65.8S	152.7E	0	195		
11151	558	2699 Jun 19	21:42:32	2448	8651	148	T	-p	0.6645	1.0720	64.9N	126.6W	48	191	314	04m38s
11152	558	2699 Dec 13	12:52:50	2451	8657	153	A	-p	-0.6106	0.9325	60.9S	2.3W	52	356	320	06m08s
11153	558	2700 Jun 09	13:23:20	2453	8663	158	Tm	nn	-0.0753	1.0379	18.7N	10.5W	86	1	128	03m49s
11154	558	2700 Dec 02	18:09:37	2456	8669	163	A	n-	0.1159	0.9874	15.3S	84.8W	83	183	45	01m26s
11155	558	2701 May 29	23:16:46	2459	8675	168	A	p-	-0.8726	0.9720	39.4S	155.4W	29	356	206	02m54s
11156	558	2701 Nov 22	06:34:43	2462	8681	173	T	p-	0.8026	1.0331	33.3N	94.4E	36	188	187	02m56s
11157	558	2702 Apr 19	09:30:34	2464	8686	140	P	-t	1.2736	0.4942	71.4N	77.6W	0	53		
11158	558	2702 Oct 13	13:25:52	2467	8692	145	P	-t	-1.1504	0.7261	71.8S	128.5W	0	115		
11159	558	2702 Nov 11	22:34:30	2467	8693	183	P	t-	1.4601	0.1390	69.8N	120.6W	0	210		
11160	558	2703 Apr 08	11:13:59	2469	8698	150	A	-p	0.5256	0.9605	37.2N	11.4E	58	159	167	04m01s

Cat Num	Canon Plate	Calendar Date	TD of Greatest Eclipse	ΔT s	Luna Num	Saros Num	Ecl. Type	QLE	Gamma	Ecl. Mag.	Lat. °	Long. °	Sun Alt °	Sun Azm °	Path Width km	Central Line Dur.
11161	559	2703 Oct 03	02:21:25	2472	8704	155	H	-p	-0.4570	1.0006	29.6S	143.4E	63	20	2	00m03s
11162	559	2704 Mar 27	19:45:56	2475	8710	160	H	n-	-0.2211	1.0148	9.5S	101.1W	77	342	52	01m29s
11163	559	2704 Sep 21	08:23:52	2477	8716	165	A	nn	0.2985	0.9545	17.3N	68.2E	73	198	173	05m26s
11164	559	2705 Mar 17	10:10:57	2480	8722	170	T	p-	-0.9031	1.0419	61.0S	74.0E	25	321	327	02m44s
11165	559	2705 Sep 10	08:59:54	2483	8728	175	P	t-	1.0484	0.8767	72.2N	160.4E	0	286		
11166	559	2706 Feb 05	15:07:13	2485	8733	142	P	-t	1.1218	0.7713	70.3N	69.9W	0	145		
11167	559	2706 Jul 31	23:20:38	2488	8739	147	A	-t	-0.9889	0.9911	63.0S	178.8W	7	20	240	00m41s
11168	559	2707 Jan 26	00:42:05	2491	8745	152	A	-p	0.4646	0.9587	8.5N	178.7E	62	172	169	05m08s
11169	559	2707 Jul 21	11:58:10	2493	8751	157	T	nn	-0.1871	1.0593	9.8N	11.7E	79	6	199	05m48s
11170	559	2708 Jan 15	02:56:17	2496	8757	162	A	nn	-0.2277	0.9212	34.6S	149.2E	77	356	306	09m38s
11171	559	2708 Jul 10	04:49:16	2499	8763	167	T	p-	0.5551	1.0774	56.1N	121.9E	56	183	302	05m22s
11172	559	2709 Jan 03	01:57:50	2502	8769	172	A	p-	-0.9017	0.9175	87.5S	147.9E	25	14	737	05m55s
11173	559	2709 May 31	12:21:17	2504	8774	139	P	-t	-1.2869	0.4697	64.2S	38.6E	0	329		
11174	559	2709 Jun 29	21:16:23	2504	8775	177	P	t-	1.3157	0.4139	66.8N	57.5E	0	4		
11175	559	2709 Nov 23	15:53:41	2507	8780	144	P	-t	1.2837	0.4759	63.6N	9.5W	0	219		
11176	559	2709 Dec 23	05:25:45	2507	8781	182	P	t-	-1.5350	0.0371	66.1S	58.5W	0	168		
11177	559	2710 May 20	20:07:03	2510	8786	149	A	-p	-0.5738	0.9620	13.6S	101.3W	55	343	166	04m34s
11178	559	2710 Nov 13	05:50:18	2512	8792	154	T	-p	0.5489	1.0486	13.6N	110.5E	57	201	191	04m20s
11179	559	2711 May 09	21:21:41	2515	8798	159	A	nn	0.1701	0.9385	26.5N	134.7W	80	157	231	07m05s
11180	559	2711 Nov 02	22:05:02	2518	8804	164	T	n-	-0.1448	1.0523	22.3S	148.6W	82	25	175	04m21s
11181	560	2712 Apr 27	22:19:08	2521	8810	169	A	t-	0.8956	0.9492	64.2N	156.8E	26	113	417	03m41s
11182	560	2712 Oct 22	11:55:16	2523	8816	174	H	p-	-0.8873	1.0009	60.7S	45.3W	27	62	6	00m03s
11183	560	2713 Mar 18	18:21:32	2526	8821	141	P	-t	-1.1428	0.7362	61.2S	8.6E	0	268		
11184	560	2713 Sep 12	04:18:12	2528	8827	146	P	-t	1.1807	0.6536	61.1N	137.1W	0	279		
11185	560	2714 Mar 08	09:37:13	2531	8833	151	T	-n	-0.4342	1.0520	27.6S	62.3E	64	328	189	04m01s
11186	560	2714 Sep 01	04:36:50	2534	8839	156	A	-p	0.4397	0.9474	31.3N	135.5E	64	212	213	05m24s
11187	560	2715 Feb 26	01:11:19	2537	8845	161	T	n-	0.2495	1.0302	3.7N	169.2E	76	153	105	02m42s
11188	560	2715 Aug 21	08:26:53	2539	8851	166	A	p-	-0.3333	0.9900	5.4S	56.7E	70	25	37	01m03s
11189	560	2716 Feb 15	12:26:26	2542	8857	171	An	p-	0.9874	0.9617	57.4N	43.8W	8	130	-	02m55s
11190	560	2716 Jul 11	12:01:43	2545	8862	138	Pe	-t	1.4666	0.1219	64.4N	159.1E	0	330		
11191	560	2716 Aug 09	19:35:30	2545	8863	176	P	t-	-1.0674	0.8829	62.3S	159.5W	0	55		
11192	560	2717 Jan 04	21:20:53	2547	8868	143	P	-t	-1.3123	0.4307	64.7S	23.3E	0	206		
11193	560	2717 Jul 01	05:13:30	2550	8874	148	T	-p	0.7368	1.0707	69.2N	133.9E	42	206	342	04m20s
11194	560	2717 Dec 24	20:58:06	2553	8880	153	A	-p	-0.6210	0.9329	61.5S	115.6W	51	347	321	06m00s
11195	560	2718 Jun 20	20:43:15	2556	8886	158	T	nn	-0.0034	1.0362	23.3N	119.4W	90	18	122	03m34s
11196	560	2718 Dec 14	02:32:46	2558	8892	163	A	n-	0.1027	0.9885	17.3S	150.5E	84	178	41	01m17s
11197	560	2719 Jun 10	06:16:10	2561	8898	168	A	p-	-0.8032	0.9726	30.8S	96.0E	36	1	165	03m04s
11198	560	2719 Dec 03	15:08:00	2564	8904	173	T	p-	0.7837	1.0331	29.9N	36.8W	38	183	180	03m01s
11199	560	2720 Apr 29	16:37:17	2566	8909	140	P	-t	1.3257	0.4061	70.7N	163.1E	0	41		
11200	560	2720 Oct 23	21:30:13	2569	8915	145	P	-t	-1.1917	0.6475	71.2S	97.2E	0	129		
11201	561	2720 Nov 22	07:03:47	2570	8916	183	P	t-	1.4388	0.1806	68.9N	100.5E	0	198		
11202	561	2721 Apr 18	18:47:26	2572	8921	150	A	-p	0.5665	0.9657	44.1N	103.1W	55	159	150	03m17s
11203	561	2721 Oct 13	09:57:38	2575	8927	155	A	-p	-0.5077	0.9940	37.0S	27.6E	59	20	24	00m34s
11204	561	2722 Apr 08	03:51:03	2577	8933	160	T	n-	-0.1881	1.0208	3.5S	136.6E	79	343	72	02m06s
11205	561	2722 Oct 02	15:28:06	2580	8939	165	A	nn	0.2384	0.9501	9.7N	39.7W	76	197	188	06m12s
11206	561	2723 Mar 28	18:35:42	2583	8945	170	T	p-	-0.8766	1.0465	54.6S	59.8W	28	329	321	03m13s
11207	561	2723 Sep 21	15:48:55	2586	8951	175	An	t-	0.9799	0.9291	70.0N	12.0E	11	242	-	05m24s
11208	561	2724 Feb 16	23:45:25	2588	8956	142	P	-t	1.1327	0.7511	71.1N	148.0E	0	132		
11209	561	2724 Aug 11	06:23:26	2591	8962	147	P	-t	-1.0610	0.8797	70.5S	56.2E	0	39		
11210	561	2725 Feb 05	09:02:01	2594	8968	152	A	-p	0.4734	0.9567	11.8N	52.9E	62	169	178	05m17s
11211	561	2725 Jul 31	19:21:13	2597	8974	157	T	-n	-0.2611	1.0612	3.1N	100.0W	75	9	208	05m57s
11212	561	2726 Jan 25	10:59:24	2599	8980	162	A	nn	-0.2189	0.9206	31.7S	30.7E	77	351	308	09m52s
11213	561	2726 Jul 21	12:17:48	2602	8986	167	T	p-	0.4807	1.0780	49.1N	12.9E	61	189	288	05m43s
11214	561	2727 Jan 14	10:03:51	2605	8992	172	A	p-	-0.8920	0.9189	83.8S	75.5E	26	325	691	05m57s
11215	561	2727 Jun 11	19:39:01	2607	8997	139	P	-t	-1.3590	0.3372	65.2S	80.2W	0	339		
11216	561	2727 Jul 11	04:36:05	2608	8998	177	P	t-	1.2468	0.5441	67.8N	62.7W	0	354		
11217	561	2727 Dec 05	00:17:21	2610	9003	144	P	-t	1.2968	0.4521	64.4N	144.5W	0	209		
11218	561	2728 Jan 03	13:50:32	2611	9004	182	P	t-	-1.5233	0.0567	67.2S	165.2E	0	179		
11219	561	2728 May 31	03:03:54	2613	9009	149	A	-p	-0.6458	0.9608	17.4S	154.2E	50	346	185	04m48s
11220	561	2728 Nov 23	14:20:46	2616	9015	154	T	-p	0.5701	1.0468	13.1N	17.8W	55	197	188	04m17s

Cat Num	Canon Plate	Calendar Date	TD of Greatest Eclipse	ΔT s	Luna Num	Saros Num	Ecl. Type	QLE	Gamma	Ecl. Mag.	Lat. °	Long. °	Sun Alt °	Sun Azm °	Path Width km	Central Line Dur.
11221	562	2729 May 20	04:11:51	2619	9021	159	A	nn	0.1017	0.9412	25.6N	125.3E	84	161	219	07m01s
11222	562	2729 Nov 13	06:26:10	2621	9027	164	T	n-	-0.1162	1.0480	24.1S	87.7E	83	22	161	04m05s
11223	562	2730 May 09	05:30:52	2624	9033	169	A	p-	0.8363	0.9560	65.0N	60.6E	33	124	291	03m16s
11224	562	2730 Nov 02	19:52:57	2627	9039	174	A	p-	-0.8541	0.9959	62.8S	160.6W	31	59	28	00m17s
11225	562	2731 Mar 30	02:38:14	2629	9044	141	P	-t	-1.1669	0.6925	61.2S	123.8W	0	277		
11226	562	2731 Apr 28	13:11:25	2630	9045	179	Pb	t-	1.5160	0.0518	62.1N	125.4W	0	59		
11227	562	2731 Sep 23	11:10:46	2632	9050	146	P	-t	1.2491	0.5366	61.0N	111.6E	0	270		
11228	562	2732 Mar 18	18:10:35	2635	9056	151	T	-n	-0.4552	1.0544	24.4S	65.8W	63	328	200	04m15s
11229	562	2732 Sep 11	11:21:54	2638	9062	156	A	-p	0.5138	0.9461	31.1N	36.6E	59	214	227	05m29s
11230	562	2733 Mar 08	09:44:50	2641	9068	161	T	n-	0.2343	1.0301	7.0N	40.8E	76	151	104	02m39s
11231	562	2733 Aug 31	15:30:17	2643	9074	166	A	n-	-0.2587	0.9920	4.9S	48.1W	75	27	29	00m49s
11232	562	2734 Feb 25	20:45:26	2646	9080	171	A	p-	0.9774	0.9615	56.2N	170.5W	11	126	674	02m55s
11233	562	2734 Aug 21	02:59:13	2649	9086	176	Ts	t-	-0.9937	1.0332	59.4S	89.8E	5	56	-	02m08s
11234	562	2735 Jan 16	05:21:36	2652	9091	143	P	-t	-1.3223	0.4143	63.8S	106.0W	0	216		
11235	562	2735 Jul 12	12:43:11	2654	9097	148	T	-p	0.8101	1.0682	71.7N	40.5E	36	225	381	03m59s
11236	562	2736 Jan 05	05:06:12	2657	9103	153	A	-p	-0.6300	0.9340	60.6S	130.0E	51	338	318	05m50s
11237	562	2736 Jul 01	03:59:45	2660	9109	158	T	nn	0.0707	1.0339	27.1N	133.3E	86	190	114	03m15s
11238	562	2736 Dec 24	11:00:57	2663	9115	163	A	n-	0.0927	0.9900	18.2S	24.7E	85	173	35	01m06s
11239	562	2737 Jun 20	13:08:00	2666	9121	168	A	p-	-0.7282	0.9726	23.4S	9.4W	43	5	143	03m14s
11240	562	2737 Dec 13	23:47:43	2668	9127	173	T	p-	0.7697	1.0332	27.4N	169.3W	39	178	176	03m03s
11241	563	2738 May 10	23:34:31	2671	9132	140	P	-t	1.3856	0.3042	69.8N	46.7E	0	28		
11242	563	2738 Jun 09	15:23:02	2671	9133	178	Pb	t-	-1.5105	0.0905	67.1S	41.3W	0	2		
11243	563	2738 Nov 04	05:42:27	2674	9138	145	P	-t	-1.2259	0.5827	70.5S	38.5W	0	142		
11244	563	2738 Dec 03	15:40:13	2674	9139	183	P	t-	1.4237	0.2102	67.9N	39.6W	0	187		
11245	563	2739 Apr 30	02:11:56	2677	9144	150	A	-p	0.6157	0.9708	51.4N	145.0E	52	160	133	02m37s
11246	563	2739 Oct 24	17:41:47	2679	9150	155	A	-p	-0.5510	0.9874	43.9S	89.4W	56	20	53	01m08s
11247	563	2740 Apr 18	11:49:23	2682	9156	160	T	n-	-0.1487	1.0268	2.7N	15.9E	81	345	92	02m43s
11248	563	2740 Oct 12	22:38:24	2685	9162	165	A	nn	0.1837	0.9456	2.5N	148.8W	79	196	204	06m59s
11249	563	2741 Apr 08	02:54:50	2688	9168	170	T	p-	-0.8453	1.0513	47.9S	169.2E	32	335	317	03m46s
11250	563	2741 Oct 01	22:44:42	2691	9174	175	A	t-	0.9163	0.9303	58.5N	120.9W	23	215	652	06m14s
11251	563	2742 Feb 27	08:19:28	2693	9179	142	P	-t	1.1468	0.7250	71.7N	6.3E	0	118		
11252	563	2742 Aug 22	13:29:55	2696	9185	147	P	-t	-1.1308	0.7534	71.2S	63.0W	0	51		
11253	563	2743 Feb 16	17:18:53	2699	9191	152	A	-p	0.4842	0.9553	15.8N	72.4W	61	166	185	05m20s
11254	563	2743 Aug 12	02:47:40	2702	9197	157	T	-n	-0.3329	1.0623	4.0S	146.9E	71	12	216	05m56s
11255	563	2744 Feb 05	19:00:30	2705	9203	162	A	nn	-0.2086	0.9205	28.0S	88.1W	78	348	308	10m01s
11256	563	2744 Jul 31	19:48:25	2707	9209	167	T	p-	0.4082	1.0778	41.8N	98.6W	66	192	276	05m59s
11257	563	2745 Jan 24	18:10:38	2710	9215	172	A	p-	-0.8825	0.9207	78.8S	40.0W	28	318	646	05m58s
11258	563	2745 Jun 22	02:51:30	2713	9220	139	P	-t	-1.4345	0.1992	66.1S	162.0E	0	348		
11259	563	2745 Jul 21	11:53:46	2713	9221	177	P	t-	1.1767	0.6759	68.8N	177.0E	0	343		
11260	563	2745 Dec 15	08:47:08	2716	9226	144	P	-t	1.3057	0.4358	65.4N	78.7E	0	199		
11261	564	2746 Jan 13	22:18:11	2716	9227	182	P	t-	-1.5133	0.0731	68.3S	27.6E	0	190		
11262	564	2746 Jun 11	09:53:44	2718	9232	149	A	-p	-0.7226	0.9591	22.8S	50.8E	44	350	214	04m59s
11263	564	2746 Dec 04	22:57:39	2721	9238	154	T	-p	0.5864	1.0454	13.0N	147.9W	54	192	186	04m15s
11264	564	2747 May 31	10:54:36	2724	9244	159	A	nn	0.0271	0.9436	23.5N	27.0E	88	167	208	07m01s
11265	564	2747 Nov 24	14:54:29	2727	9250	164	T	nn	-0.0940	1.0438	25.7S	37.6W	84	18	147	03m49s
11266	564	2748 May 19	12:36:24	2730	9256	169	A	p-	0.7709	0.9624	65.1N	33.2W	39	136	213	02m53s
11267	564	2748 Nov 13	03:56:37	2733	9262	174	A	p-	-0.8266	0.9908	65.5S	83.1E	34	54	57	00m37s
11268	564	2749 Apr 09	10:47:47	2735	9267	141	P	-t	-1.1976	0.6362	61.5S	105.6E	0	286		
11269	564	2749 May 08	20:49:43	2736	9268	179	P	t-	1.4607	0.1506	62.7N	111.5E	0	50		
11270	564	2749 Oct 03	18:09:52	2738	9273	146	P	-t	1.3115	0.4305	61.1N	1.4W	0	261		
11271	564	2750 Mar 30	02:35:25	2741	9279	151	T	-p	-0.4832	1.0570	21.8S	168.1E	61	329	212	04m31s
11272	564	2750 Sep 22	18:14:45	2744	9285	156	A	-p	0.5817	0.9445	30.9N	64.7W	54	214	246	05m40s
11273	564	2751 Mar 19	18:10:35	2747	9291	161	T	n-	0.2131	1.0303	10.2N	85.4W	78	151	105	02m38s
11274	564	2751 Sep 11	22:41:05	2750	9297	166	A	n-	-0.1895	0.9934	5.2S	154.6W	79	29	24	00m40s
11275	564	2752 Mar 08	04:57:54	2752	9303	171	A	p-	0.9627	0.9618	55.6N	65.5E	15	123	511	02m52s
11276	564	2752 Aug 31	10:28:18	2755	9309	176	T	t-	-0.9236	1.0394	49.4S	6.1W	22	44	339	02m46s
11277	564	2753 Jan 26	13:23:39	2758	9314	143	P	-t	-1.3311	0.3997	63.0S	124.7E	0	225		
11278	564	2753 Jul 22	20:10:02	2761	9320	148	T	-t	0.8853	1.0646	72.2N	46.7W	27	249	458	03m35s
11279	564	2754 Jan 15	13:18:44	2764	9326	153	A	-p	-0.6358	0.9356	58.3S	13.1E	50	331	310	05m39s
11280	564	2754 Jul 12	11:11:56	2766	9332	158	T	nn	0.1479	1.0308	30.2N	27.8E	81	195	105	02m52s

Cat Num	Canon Plate	Calendar Date	TD of Greatest Eclipse	ΔT s	Luna Num	Saros Num	Ecl. Type	QLE	Gamma	Ecl. Mag.	Lat. °	Long. °	Sun Alt °	Sun Azm °	Path Width km	Central Line Dur.
11281	565	2755 Jan 04	19:33:47	2769	9338	163	A	n-	0.0860	0.9920	17.9S	102.2W	85	169	28	00m52s
11282	565	2755 Jul 01	19:55:36	2772	9344	168	A	p-	-0.6502	0.9719	17.2S	112.6W	49	10	132	03m24s
11283	565	2755 Dec 25	08:32:19	2775	9350	173	T	p-	0.7595	1.0335	26.0N	57.3E	40	173	174	03m05s
11284	565	2756 May 21	06:26:50	2778	9355	140	P	-t	1.4490	0.1955	68.8N	67.7W	0	17		
11285	565	2756 Jun 19	21:57:07	2778	9356	178	P	t-	-1.4326	0.2230	66.1S	149.9W	0	12		
11286	565	2756 Nov 14	13:59:51	2780	9361	145	P	-t	-1.2556	0.5271	69.6S	174.9W	0	154		
11287	565	2756 Dec 14	00:20:29	2781	9362	183	P	t-	1.4120	0.2331	66.8N	179.8E	0	176		
11288	565	2757 May 10	09:32:11	2783	9367	150	A	-p	0.6690	0.9758	58.9N	35.1E	48	162	116	02m01s
11289	565	2757 Nov 04	01:31:50	2786	9373	155	A	-p	-0.5886	0.9811	50.4S	153.2E	54	18	83	01m39s
11290	565	2758 Apr 29	19:40:31	2789	9379	160	T	nn	-0.1026	1.0328	8.9N	102.7W	84	347	111	03m18s
11291	565	2758 Oct 24	05:56:58	2792	9385	165	A	nn	0.1364	0.9412	4.0S	100.4E	82	194	220	07m44s
11292	565	2759 Apr 19	11:05:50	2795	9391	170	T	p-	-0.8071	1.0560	41.0S	41.2E	36	340	311	04m22s
11293	565	2759 Oct 13	05:50:16	2798	9397	175	A	p-	0.8601	0.9301	49.0N	122.5E	30	205	509	07m00s
11294	565	2760 Mar 09	16:45:54	2800	9402	142	P	-t	1.1667	0.6887	72.1N	133.9W	0	104		
11295	565	2760 Apr 08	03:22:07	2801	9403	180	Pb	t-	-1.5099	0.0484	71.7S	151.0W	0	294		
11296	565	2760 Sep 01	20:44:19	2803	9408	147	P	-t	-1.1947	0.6372	71.7S	175.4E	0	64		
11297	565	2760 Oct 01	08:36:13	2804	9409	185	Pb	t-	1.5442	0.0176	72.1N	139.6E	0	259		
11298	565	2761 Feb 27	01:29:39	2806	9414	152	A	-p	0.4993	0.9543	20.4N	163.7E	60	163	191	05m17s
11299	565	2761 Aug 22	10:17:10	2809	9420	157	T	-n	-0.4025	1.0626	11.5S	32.4E	66	15	223	05m47s
11300	565	2762 Feb 16	02:58:17	2812	9426	162	A	nn	-0.1959	0.9211	23.6S	153.3E	79	345	304	10m04s
11301	566	2762 Aug 12	03:19:41	2815	9432	167	T	n-	0.3366	1.0766	34.2N	148.5E	70	195	263	06m11s
11302	566	2763 Feb 05	02:16:47	2818	9438	172	A	p-	-0.8721	0.9233	73.6S	162.2W	29	318	596	05m58s
11303	566	2763 Jul 03	09:58:23	2820	9443	139	Pe	-t	-1.5132	0.0562	67.1S	45.2E	0	359		
11304	566	2763 Aug 01	19:10:33	2821	9444	177	P	t-	1.1066	0.8069	69.7N	56.5E	0	332		
11305	566	2763 Dec 26	17:23:22	2823	9449	144	P	-t	1.3098	0.4283	66.4N	60.1W	0	189		
11306	566	2764 Jan 25	06:48:13	2824	9450	182	P	t-	-1.5049	0.0865	69.3S	111.1W	0	202		
11307	566	2764 Jun 21	16:37:03	2826	9455	149	A	-p	-0.8039	0.9568	30.2S	51.7W	36	354	265	05m06s
11308	566	2764 Dec 15	07:40:02	2829	9461	154	T	-p	0.5984	1.0443	13.3N	80.4E	53	188	184	04m12s
11309	566	2765 Jun 10	17:31:59	2832	9467	159	Am	nn	-0.0520	0.9459	20.2N	70.4W	87	350	200	07m02s
11310	566	2765 Dec 04	23:28:46	2835	9473	164	T	nn	-0.0771	1.0398	26.7S	164.2W	85	13	134	03m33s
11311	566	2766 May 30	19:36:02	2838	9479	169	A	p-	0.6996	0.9686	63.8N	125.4W	45	149	158	02m29s
11312	566	2766 Nov 24	12:07:07	2841	9485	174	A	p-	-0.8054	0.9858	68.4S	33.6W	36	48	85	00m59s
11313	566	2767 Apr 20	18:51:20	2843	9490	141	P	-t	-1.2335	0.5694	61.8S	23.6W	0	295		
11314	566	2767 May 20	04:23:19	2844	9491	179	P	t-	1.4007	0.2593	63.4N	10.7W	0	41		
11315	566	2767 Oct 15	01:16:22	2846	9496	146	P	-t	1.3673	0.3366	61.4N	116.2W	0	253		
11316	566	2768 Apr 09	10:54:27	2849	9502	151	T	-p	-0.5162	1.0595	19.9S	43.4E	59	331	225	04m48s
11317	566	2768 Oct 03	01:16:28	2852	9508	156	A	-p	0.6427	0.9428	30.7N	168.8W	50	214	269	05m54s
11318	566	2769 Mar 30	02:29:32	2855	9514	161	T	n-	0.1866	1.0307	13.2N	150.3E	79	151	105	02m40s
11319	566	2769 Sep 22	05:59:23	2858	9520	166	A	nn	-0.1254	0.9944	6.1S	96.9E	83	29	20	00m33s
11320	566	2770 Mar 19	13:02:32	2861	9526	171	A	p-	0.9422	0.9624	55.6N	55.5W	19	122	401	02m48s
11321	567	2770 Sep 11	18:04:10	2864	9532	176	T	t-	-0.8583	1.0422	45.8S	115.7W	31	43	269	03m00s
11322	567	2771 Feb 06	21:22:12	2866	9537	143	P	-t	-1.3429	0.3802	62.3S	3.5W	0	235		
11323	567	2771 Aug 03	03:38:34	2869	9543	148	T	-t	0.9590	1.0590	69.6N	129.5W	16	277	704	03m05s
11324	567	2772 Jan 26	21:31:18	2872	9549	153	A	-p	-0.6426	0.9378	55.2S	105.7W	50	326	300	05m26s
11325	567	2772 Jul 22	18:22:15	2875	9555	158	T	-n	0.2259	1.0272	32.4N	76.7W	77	200	95	02m27s
11326	567	2773 Jan 15	04:08:33	2878	9561	163	A	n-	0.0801	0.9945	16.8S	130.4E	86	164	19	00m35s
11327	567	2773 Jul 12	02:38:32	2881	9567	168	A	p-	-0.5687	0.9707	12.0S	146.0E	55	13	127	03m35s
11328	567	2774 Jan 04	17:20:31	2884	9573	173	T	p-	0.7521	1.0342	25.4N	76.9W	41	168	174	03m07s
11329	567	2774 Jun 01	13:10:10	2886	9578	140	Pe	-t	1.5196	0.0738	67.8N	179.3W	0	6		
11330	567	2774 Jul 01	04:24:54	2887	9579	178	P	t-	-1.3490	0.3657	65.1S	103.6E	0	22		
11331	567	2774 Nov 25	22:25:15	2889	9584	145	P	-t	1.2785	0.4846	68.6S	47.4E	0	166		
11332	567	2774 Dec 25	09:06:23	2890	9585	183	P	t-	1.4050	0.2469	65.8N	38.4E	0	165		
11333	567	2775 May 21	16:45:20	2892	9590	150	A	-p	0.7292	0.9804	66.7N	71.6W	43	165	102	01m31s
11334	567	2775 Nov 15	09:29:02	2895	9596	155	A	-p	-0.6195	0.9750	56.1S	35.8E	51	15	114	02m07s
11335	567	2776 May 10	03:25:50	2898	9602	160	T	nn	-0.0507	1.0386	14.9N	140.5E	87	350	130	03m50s
11336	567	2776 Nov 03	13:23:19	2901	9608	165	A	nn	0.0956	0.9369	9.9S	11.8W	85	192	236	08m25s
11337	567	2777 Apr 29	19:10:09	2904	9614	170	T	p-	-0.7632	1.0607	34.1S	84.2W	40	345	307	05m00s
11338	567	2777 Oct 23	13:04:19	2907	9620	175	A	p-	0.8102	0.9294	40.9N	7.9E	36	199	448	07m45s
11339	567	2778 Mar 21	01:06:37	2909	9625	142	P	-t	1.1908	0.6446	72.3N	87.0E	0	90		
11340	567	2778 Apr 19	11:29:53	2910	9626	180	P	t-	-1.4770	0.1100	71.2S	74.0E	0	307		

Cat Num	Canon Plate	Calendar Date	TD of Greatest Eclipse	ΔT s	Luna Num	Saros Num	Ecl. Type	QLE	Gamma	Ecl. Mag.	Lat. °	Long. °	Sun Alt °	Sun Azm °	Path Width km	Central Line Dur.
11341	568	2778 Sep 13	04:05:03	2912	9631	147	P	-t	-1.2541	0.5291	72.0S	51.8E	0	78		
11342	568	2778 Oct 12	16:03:27	2913	9632	185	P	t-	1.4902	0.1127	71.8N	14.2E	0	245		
11343	568	2779 Mar 10	09:33:37	2915	9637	152	A	-p	0.5193	0.9537	25.7N	41.3E	59	161	196	05m10s
11344	568	2779 Sep 02	17:52:25	2918	9643	157	T	-p	-0.4676	1.0622	19.2S	83.8W	62	17	230	05m31s
11345	568	2780 Feb 27	10:51:54	2921	9649	162	A	nn	-0.1801	0.9221	18.6S	35.1E	80	343	299	10m03s
11346	568	2780 Aug 22	10:53:34	2924	9655	167	T	n-	0.2672	1.0747	26.5N	34.2E	74	196	251	06m16s
11347	568	2781 Feb 15	10:20:50	2927	9661	172	A	p-	-0.8595	0.9265	68.1S	74.0E	30	321	543	05m57s
11348	568	2781 Aug 12	02:27:38	2930	9667	177	P	t-	1.0373	0.9353	70.6N	64.8W	0	320		
11349	568	2782 Jan 06	02:02:17	2933	9672	144	P	-t	1.3126	0.4228	67.5N	159.9E	0	178		
11350	568	2782 Feb 04	15:16:43	2933	9673	182	P	t-	-1.4944	0.1034	70.2S	109.9E	0	215		
11351	568	2782 Jul 02	23:15:18	2936	9678	149	A	-t	-0.8886	0.9534	40.2S	154.2W	27	359	373	05m06s
11352	568	2782 Dec 26	16:26:47	2939	9684	154	T	-p	0.6070	1.0435	14.1N	52.4W	53	184	183	04m10s
11353	568	2783 Jun 22	00:05:19	2942	9690	159	A	nn	-0.1347	0.9477	15.7N	167.4W	82	355	194	07m04s
11354	568	2783 Dec 16	08:07:08	2945	9696	164	T	nn	-0.0640	1.0362	27.0S	68.3E	86	8	122	03m18s
11355	568	2784 Jun 10	02:32:16	2948	9702	169	A	p-	0.6245	0.9744	60.9N	141.7E	51	161	118	02m07s
11356	568	2784 Dec 04	20:23:18	2951	9708	174	A	p-	-0.7896	0.9810	71.2S	149.1W	38	39	110	01m19s
11357	568	2785 May 01	02:47:04	2953	9713	141	P	-t	-1.2764	0.4886	62.3S	151.0W	0	304		
11358	568	2785 May 30	11:50:35	2954	9714	179	P	t-	1.3346	0.3806	64.2N	131.5W	0	31		
11359	568	2785 Oct 25	08:31:14	2956	9719	146	P	-t	1.4158	0.2555	61.8N	126.8E	0	244		
11360	568	2785 Nov 24	01:44:26	2957	9720	184	Pb	t-	-1.5497	0.0271	63.6S	24.0E	0	142		
11361	569	2786 Apr 20	19:04:46	2959	9725	151	T	-p	-0.5565	1.0617	18.9S	79.1W	56	333	240	05m05s
11362	569	2786 Oct 14	08:28:06	2962	9731	156	A	-p	0.6961	0.9410	30.8N	84.0E	46	212	296	06m11s
11363	569	2787 Apr 10	10:38:57	2965	9737	161	T	nn	0.1525	1.0310	15.7N	28.7E	81	152	106	02m43s
11364	569	2787 Oct 03	13:27:29	2968	9743	166	A	nn	-0.0685	0.9950	7.5S	14.0W	86	29	18	00m29s
11365	569	2788 Mar 29	20:59:26	2971	9749	171	A	p-	0.9159	0.9633	56.3N	173.5W	23	121	326	02m43s
11366	569	2788 Sep 22	01:45:51	2974	9755	176	T	p-	-0.7971	1.0439	44.4S	131.7E	37	42	238	03m08s
11367	569	2789 Feb 17	05:19:18	2976	9760	143	P	-t	-1.3557	0.3587	61.7S	131.1W	0	244		
11368	569	2789 Aug 13	11:07:04	2979	9766	148	P	-t	1.0325	0.9580	62.1N	146.1E	0	302		
11369	569	2789 Sep 11	18:18:50	2980	9767	186	Pb	t-	-1.4846	0.0862	61.4S	165.2W	0	81		
11370	569	2790 Feb 06	05:44:15	2982	9772	153	A	-p	-0.6495	0.9407	51.5S	133.7E	49	323	286	05m12s
11371	569	2790 Aug 03	01:30:57	2986	9778	158	T	-p	0.3044	1.0228	33.9N	179.4E	72	205	81	02m00s
11372	569	2791 Jan 26	12:45:16	2988	9784	163	A	nn	0.0751	0.9975	14.8S	2.3E	86	160	9	00m15s
11373	569	2791 Jul 23	09:20:33	2992	9790	168	A	p-	-0.4869	0.9689	8.0S	45.6E	61	17	127	03m46s
11374	569	2792 Jan 16	02:09:57	2995	9796	173	T	p-	0.7454	1.0353	25.6N	148.8E	42	164	177	03m09s
11375	569	2792 Jul 11	10:53:02	2998	9802	178	P	t-	-1.2650	0.5092	64.2S	2.8W	0	31		
11376	569	2792 Dec 06	06:55:37	3000	9807	145	P	-t	-1.2973	0.4500	67.6S	91.0W	0	177		
11377	569	2793 Jan 04	17:54:33	3001	9808	183	P	t-	1.4001	0.2568	64.8N	103.1W	0	154		
11378	569	2793 May 31	23:54:30	3003	9813	150	A	-p	0.7933	0.9846	74.7N	173.3W	37	172	90	01m06s
11379	569	2793 Nov 25	17:32:25	3006	9819	155	A	-p	-0.6447	0.9693	60.9S	80.9W	50	10	145	02m32s
11380	569	2794 May 21	11:05:18	3009	9825	160	T	nn	0.0070	1.0441	20.7N	25.6E	89	175	147	04m16s
11381	570	2794 Nov 14	20:57:54	3012	9831	165	Am	nn	0.0620	0.9329	14.8S	125.4W	87	189	251	09m02s
11382	570	2795 May 11	03:06:46	3015	9837	170	T	p-	-0.7126	1.0649	27.2S	153.0E	44	350	302	05m37s
11383	570	2795 Nov 03	20:29:04	3018	9843	175	A	p-	0.7683	0.9285	34.2N	107.5W	40	194	416	08m26s
11384	570	2796 Mar 31	09:18:22	3021	9848	142	P	-t	1.2216	0.5883	72.1N	49.7W	0	76		
11385	570	2796 Apr 29	19:28:58	3021	9849	180	P	t-	-1.4377	0.1839	70.5S	58.4W	0	320		
11386	570	2796 Sep 23	11:33:43	3024	9854	147	P	-t	-1.3078	0.4310	72.1S	74.0W	0	91		
11387	570	2796 Oct 22	23:40:08	3024	9855	185	P	t-	1.4426	0.1965	71.3N	113.1W	0	232		
11388	570	2797 Mar 20	17:29:20	3027	9860	152	A	-p	0.5454	0.9533	31.6N	79.2W	57	159	202	05m00s
11389	570	2797 Sep 13	01:33:06	3030	9866	157	T	-p	-0.5286	1.0611	26.9S	158.4E	58	19	235	05m11s
11390	570	2798 Mar 09	18:37:54	3033	9872	162	A	nn	-0.1580	0.9238	13.1S	81.7W	81	343	291	09m57s
11391	570	2798 Sep 02	18:30:51	3036	9878	167	T	n-	0.2008	1.0719	18.8N	81.4W	78	197	238	06m14s
11392	570	2799 Feb 26	18:21:15	3039	9884	172	A	p-	-0.8432	0.9304	62.3S	49.8W	32	324	484	05m54s
11393	570	2799 Aug 23	09:46:32	3042	9890	177	T	t-	0.9698	1.0204	75.2N	127.2E	13	263	300	01m11s
11394	570	2800 Jan 17	10:43:57	3044	9895	144	P	-t	1.3141	0.4200	68.6N	18.7E	0	166		
11395	570	2800 Feb 15	23:44:06	3045	9896	182	P	t-	-1.4823	0.1232	71.1S	29.5W	0	228		
11396	570	2800 Jul 13	05:50:34	3047	9901	149	A	-t	-0.9747	0.9483	56.1S	101.0E	12	6	893	04m52s
11397	570	2801 Jan 06	01:17:30	3050	9907	154	T	-p	0.6127	1.0432	15.5N	173.7E	52	179	182	04m07s
11398	570	2801 Jul 02	06:34:26	3053	9913	159	A	nn	-0.2210	0.9492	10.2N	95.8E	77	359	191	07m03s
11399	570	2801 Dec 26	16:50:37	3056	9919	164	T	nn	-0.0555	1.0328	26.6S	60.4W	87	4	111	03m04s
11400	570	2802 Jun 21	09:25:30	3060	9925	169	A	p-	0.5461	0.9798	56.5N	46.9E	57	171	86	01m46s

Cat Num	Canon Plate	Calendar Date	TD of Greatest Eclipse	ΔT s	Luna Num	Saros Num	Ecl. Type	QLE	Gamma	Ecl. Mag.	Lat. °	Long. °	Sun Alt °	Sun Azm °	Path Width km	Central Line Dur.
11401	571	2802 Dec 16	04:44:25	3063	9931	174	A	p-	-0.7785	0.9765	73.1S	97.6E	39	27	134	01m39s
11402	571	2803 May 12	10:37:20	3065	9936	141	P	-t	-1.3244	0.3976	63.0S	82.7E	0	313		
11403	571	2803 Jun 10	19:15:11	3066	9937	179	P	t-	1.2654	0.5091	65.1N	108.1E	0	22		
11404	571	2803 Nov 05	15:54:34	3068	9942	146	P	-t	1.4572	0.1871	62.3N	7.6E	0	234		
11405	571	2803 Dec 05	09:36:34	3069	9943	184	P	t-	-1.5312	0.0602	64.5S	103.0W	0	152		
11406	571	2804 May 01	03:09:24	3071	9948	151	T	-p	-0.6018	1.0636	19.0S	159.6E	53	336	257	05m21s
11407	571	2804 Oct 24	15:48:02	3074	9954	156	A	-p	0.7433	0.9392	31.2N	26.0W	42	210	327	06m30s
11408	571	2805 Apr 20	18:41:09	3077	9960	161	T	nn	0.1129	1.0313	17.7N	90.7W	83	154	106	02m46s
11409	571	2805 Oct 13	21:04:29	3080	9966	166	A	nn	-0.0181	0.9954	9.1S	127.2W	89	26	16	00m27s
11410	571	2806 Apr 10	04:45:31	3083	9972	171	A	p-	0.8815	0.9643	57.3N	72.8E	28	122	269	02m39s
11411	571	2806 Oct 03	09:36:12	3086	9978	176	T	p-	-0.7426	1.0447	44.5S	16.3E	42	42	219	03m10s
11412	571	2807 Feb 28	13:10:08	3089	9983	143	P	-t	-1.3736	0.3288	61.4S	102.9E	0	253		
11413	571	2807 Aug 24	18:39:28	3092	9989	148	P	-t	1.1023	0.8227	61.7N	24.5E	0	293		
11414	571	2807 Sep 23	02:08:10	3093	9990	186	P	t-	-1.4296	0.1933	61.3S	69.2E	0	90		
11415	571	2808 Feb 17	13:53:37	3095	9995	153	A	-p	-0.6601	0.9441	47.8S	12.9E	48	321	271	04m56s
11416	571	2808 Aug 13	08:40:55	3098	10001	158	T	-p	0.3810	1.0178	34.5N	75.3E	67	209	66	01m32s
11417	571	2809 Feb 05	21:20:58	3101	10007	163	H	nn	0.0684	1.0011	12.1S	125.8W	86	157	4	00m06s
11418	571	2809 Aug 02	15:59:43	3104	10013	168	A	p-	-0.4032	0.9665	4.9S	53.6W	66	21	131	03m58s
11419	571	2810 Jan 26	11:00:54	3107	10019	173	T	p-	0.7398	1.0369	26.5N	14.1E	42	159	182	03m11s
11420	571	2810 Jul 22	17:18:37	3110	10025	178	P	t-	-1.1780	0.6579	63.4S	108.1W	0	40		
11421	572	2810 Dec 17	15:31:45	3113	10030	145	P	-t	-1.3111	0.4248	66.5S	129.7E	0	188		
11422	572	2811 Jan 16	02:44:41	3113	10031	183	P	t-	1.3971	0.2629	63.9N	115.2E	0	144		
11423	572	2811 Jun 12	06:58:46	3116	10036	150	A	-p	0.8623	0.9880	82.8N	102.4E	30	195	84	00m47s
11424	572	2811 Dec 07	01:42:19	3119	10042	155	A	-p	-0.6641	0.9640	64.3S	163.6E	48	2	175	02m55s
11425	572	2812 May 31	18:39:58	3122	10048	160	Tm	nn	0.0694	1.0493	26.2N	87.3W	86	178	164	04m36s
11426	572	2812 Nov 25	04:39:02	3125	10054	165	A	nn	0.0341	0.9292	18.9S	120.0E	88	185	266	09m33s
11427	572	2813 May 21	10:57:38	3128	10060	170	T	p-	-0.6571	1.0688	20.7S	32.4E	49	354	297	06m11s
11428	572	2813 Nov 14	04:02:23	3131	10066	175	A	p-	0.7331	0.9276	28.6N	136.0E	43	190	398	09m04s
11429	572	2814 Apr 11	17:21:36	3134	10071	142	P	-t	1.2589	0.5204	71.7N	176.0E	0	62		
11430	572	2814 May 11	03:19:42	3134	10072	180	P	t-	-1.3923	0.2696	69.7S	171.9E	0	332		
11431	572	2814 Oct 04	19:10:43	3137	10077	147	P	-t	-1.3554	0.3440	71.9S	158.2E	0	105		
11432	572	2814 Nov 03	07:26:40	3138	10078	185	P	t-	1.4021	0.2678	70.6N	117.6E	0	219		
11433	572	2815 Apr 01	01:17:07	3140	10083	152	A	-p	0.5774	0.9532	38.0N	162.2E	55	158	208	04m48s
11434	572	2815 Sep 24	09:20:27	3143	10089	157	T	-p	-0.5844	1.0596	34.6S	38.7E	54	22	240	04m48s
11435	572	2816 Mar 20	02:17:34	3146	10095	162	A	nn	-0.1307	0.9259	7.2S	162.7E	82	342	281	09m48s
11436	572	2816 Sep 13	02:13:14	3149	10101	167	T	nn	0.1390	1.0686	11.1N	161.5E	82	198	226	06m06s
11437	572	2817 Mar 09	02:16:54	3152	10107	172	A	p-	-0.8224	0.9348	56.2S	172.9W	34	328	425	05m49s
11438	572	2817 Sep 02	17:07:37	3155	10113	177	T	p-	0.9047	1.0186	66.5N	23.1W	25	225	150	01m14s
11439	572	2818 Jan 27	19:25:58	3158	10118	144	P	-t	1.3157	0.4167	69.6N	123.1W	0	154		
11440	572	2818 Feb 26	08:07:44	3159	10119	182	P	t-	-1.4667	0.1492	71.7S	168.5W	0	241		
11441	573	2818 Jul 24	12:24:20	3161	10124	149	P	-t	1.0608	0.8615	69.2S	11.7W	0	20		
11442	573	2819 Jan 17	10:08:37	3164	10130	154	T	-p	0.6180	1.0433	17.4N	39.6E	52	175	184	04m04s
11443	573	2819 Jul 13	13:03:16	3167	10136	159	A	np	-0.3075	0.9502	3.8N	1.7W	72	3	192	06m58s
11444	573	2820 Jan 07	01:36:11	3170	10142	164	Tm	nn	-0.0495	1.0300	25.4S	170.2E	87	359	102	02m51s
11445	573	2820 Jul 01	16:17:03	3174	10148	169	A	p-	0.4652	0.9847	50.8N	50.3W	62	179	61	01m24s
11446	573	2820 Dec 26	13:08:54	3177	10154	174	A	p-	-0.7709	0.9725	73.7S	13.7W	39	11	156	01m59s
11447	573	2821 May 22	18:21:01	3179	10159	141	P	-t	-1.3779	0.2949	63.8S	42.0W	0	323		
11448	573	2821 Jun 21	02:35:50	3180	10160	179	P	t-	1.1922	0.6463	66.1N	11.6W	0	12		
11449	573	2821 Nov 15	23:26:43	3182	10165	146	P	-t	1.4912	0.1314	63.0N	114.0W	0	225		
11450	573	2821 Dec 15	17:34:54	3183	10166	184	P	t-	-1.5177	0.0845	65.5S	128.0E	0	162		
11451	573	2822 May 12	11:04:34	3185	10171	151	T	-p	-0.6549	1.0650	20.4S	40.6E	49	339	278	05m34s
11452	573	2822 Nov 04	23:19:12	3188	10177	156	A	-p	0.7817	0.9376	31.9N	139.4W	38	207	362	06m49s
11453	573	2823 May 02	02:34:05	3192	10183	161	Tm	nn	0.0662	1.0314	18.9N	152.6E	86	157	106	02m51s
11454	573	2823 Oct 25	04:50:58	3195	10189	166	A	nn	0.0252	0.9957	10.8S	117.2E	89	207	15	00m26s
11455	573	2824 Apr 20	12:23:21	3198	10195	171	A	p-	0.8410	0.9654	58.7N	37.5W	32	125	228	02m35s
11456	573	2824 Oct 13	17:34:17	3201	10201	176	T	p-	-0.6942	1.0449	45.7S	101.0W	46	41	205	03m10s
11457	573	2825 Mar 10	20:56:44	3204	10206	143	P	-t	-1.3948	0.2930	61.1S	21.8W	0	262		
11458	573	2825 Sep 04	02:13:41	3207	10212	148	P	-t	1.1700	0.6920	61.3N	97.3W	0	284		
11459	573	2825 Oct 03	10:02:41	3207	10213	186	P	t-	-1.3789	0.2916	61.3S	57.7W	0	99		
11460	573	2826 Feb 27	22:01:00	3210	10218	153	A	-p	-0.6729	0.9480	44.2S	108.3W	47	320	254	04m38s

Cat Num	Canon Plate	Calendar Date	TD of Greatest Eclipse	ΔT s	Luna Num	Saros Num	Ecl. Type	QLE	Gamma	Ecl. Mag.	Lat. °	Long. °	Sun Alt °	Sun Azm °	Path Width km	Central Line Dur.
11461	574	2826 Aug 24	15:52:15	3213	10224	158	H	-p	0.4557	1.0123	34.6N	29.4W	63	212	47	01m03s
11462	574	2827 Feb 17	05:54:44	3216	10230	163	H	nn	0.0594	1.0052	9.0S	106.3E	87	154	18	00m30s
11463	574	2827 Aug 13	22:41:10	3219	10236	168	A	pn	-0.3219	0.9637	3.0S	152.9W	71	24	138	04m12s
11464	574	2828 Feb 06	19:50:44	3222	10242	173	T	p-	0.7327	1.0388	28.0N	120.3W	43	155	188	03m15s
11465	574	2828 Aug 01	23:45:36	3225	10248	178	P	t-	-1.0917	0.8055	62.6S	146.4E	0	49		
11466	574	2828 Dec 28	00:10:20	3228	10253	145	P	-t	-1.3228	0.4038	65.4S	9.7W	0	199		
11467	574	2829 Jan 26	11:34:12	3228	10254	183	P	t-	1.3939	0.2693	63.1N	26.0W	0	135		
11468	574	2829 Jun 22	14:01:25	3231	10259	150	A	-p	0.9335	0.9904	83.4N	97.9E	21	296	97	00m35s
11469	574	2829 Dec 17	09:56:19	3234	10265	155	A	-p	-0.6793	0.9594	66.1S	49.5E	47	351	202	03m15s
11470	574	2830 Jun 12	02:09:55	3237	10271	160	T	nn	0.1365	1.0538	31.1N	161.8E	82	182	180	04m50s
11471	574	2830 Dec 06	12:27:18	3240	10277	165	A	nn	0.0124	0.9261	21.8S	4.1E	89	181	278	09m57s
11472	574	2831 Jun 01	18:42:34	3244	10283	170	T	p-	-0.5964	1.0720	14.6S	86.0W	53	358	292	06m39s
11473	574	2831 Nov 25	11:44:06	3247	10289	175	A	p-	0.7040	0.9270	24.1N	18.1E	45	185	386	09m32s
11474	574	2832 Apr 22	01:15:31	3249	10294	142	P	-t	1.3031	0.4397	71.1N	44.5E	0	48		
11475	574	2832 May 21	11:02:59	3250	10295	180	P	t-	-1.3411	0.3662	68.7S	44.7E	0	344		
11476	574	2832 Oct 15	02:57:22	3252	10300	147	P	-t	-1.3962	0.2695	71.6S	28.3E	0	119		
11477	574	2832 Nov 13	15:23:04	3253	10301	185	P	t-	1.3684	0.3271	69.7N	13.5W	0	206		
11478	574	2833 Apr 11	08:54:12	3256	10306	152	A	-p	0.6177	0.9531	45.0N	46.3E	52	157	217	04m34s
11479	574	2833 Oct 04	17:14:56	3259	10312	157	T	-p	-0.6347	1.0576	42.2S	82.6W	50	23	244	04m23s
11480	574	2834 Mar 31	09:48:16	3262	10318	162	A	nn	-0.0959	0.9284	1.0S	49.1E	84	343	270	09m32s
11481	575	2834 Sep 24	10:01:22	3265	10324	167	T	nn	0.0823	1.0647	3.8N	42.9E	85	198	212	05m51s
11482	575	2835 Mar 20	10:06:06	3268	10330	172	A	p-	-0.7955	0.9398	49.7S	65.6E	37	332	365	05m41s
11483	575	2835 Sep 14	00:33:18	3271	10336	177	T	p-	0.8441	1.0151	56.9N	147.8W	32	213	96	01m07s
11484	575	2836 Feb 08	04:08:21	3274	10341	144	P	-t	1.3175	0.4130	70.5N	94.2E	0	141		
11485	575	2836 Mar 08	16:28:20	3274	10342	182	P	t-	-1.4480	0.1810	72.1S	52.8E	0	255		
11486	575	2836 Aug 03	18:56:34	3277	10347	149	P	-t	-1.1473	0.7116	70.1S	121.4W	0	31		
11487	575	2837 Jan 27	19:01:17	3280	10353	154	T	-p	0.6223	1.0438	20.0N	95.0W	51	171	187	04m02s
11488	575	2837 Jul 23	19:31:12	3283	10359	159	A	-p	-0.3949	0.9508	3.5S	99.8W	67	7	196	06m47s
11489	575	2838 Jan 17	10:23:36	3286	10365	164	T	nn	-0.0450	1.0277	23.3S	40.1E	87	355	94	02m39s
11490	575	2838 Jul 12	23:08:16	3290	10371	169	A	p-	0.3828	0.9891	44.3N	149.7W	67	184	42	01m03s
11491	575	2839 Jan 06	21:36:05	3293	10377	174	A	p-	-0.7660	0.9690	72.6S	126.1W	40	356	175	02m17s
11492	575	2839 Jun 03	02:01:17	3295	10382	141	P	-t	-1.4350	0.1847	64.6S	166.2W	0	332		
11493	575	2839 Jul 02	09:56:30	3296	10383	179	P	t-	1.1177	0.7870	67.1N	131.8W	0	2		
11494	575	2839 Nov 27	07:05:15	3299	10388	146	P	-t	1.5198	0.0847	63.8N	122.6E	0	216		
11495	575	2839 Dec 27	01:35:57	3299	10389	184	P	t-	-1.5066	0.1044	66.5S	1.9W	0	172		
11496	575	2840 May 22	18:55:22	3302	10394	151	T	-p	-0.7118	1.0657	23.1S	77.7W	44	343	303	05m41s
11497	575	2840 Nov 15	06:59:00	3305	10400	156	A	-p	0.8135	0.9363	32.9N	104.4E	35	203	399	07m05s
11498	575	2841 May 12	10:18:35	3308	10406	161	T	nn	0.0129	1.0312	19.1N	38.2E	89	162	105	02m55s
11499	575	2841 Nov 04	12:47:01	3311	10412	166	A	nn	0.0615	0.9959	12.4S	0.7W	87	203	14	00m25s
11500	575	2842 May 01	19:50:36	3314	10418	171	A	p-	0.7926	0.9664	59.9N	143.2W	37	131	197	02m33s
11501	576	2842 Oct 25	01:41:24	3317	10424	176	T	p-	-0.6527	1.0447	47.5S	139.8E	49	39	195	03m09s
11502	576	2843 Mar 22	04:34:12	3320	10429	143	P	-t	-1.4236	0.2443	61.1S	144.3W	0	271		
11503	576	2843 Apr 20	22:09:23	3321	10430	181	Pb	t-	1.5548	0.0206	61.8N	108.5E	0	64		
11504	576	2843 Sep 15	09:54:00	3323	10435	148	P	-t	1.2325	0.5724	61.2N	139.4E	0	276		
11505	576	2843 Oct 14	18:05:30	3324	10436	186	P	t-	-1.3350	0.3763	61.6S	173.3E	0	108		
11506	576	2844 Mar 10	06:02:50	3326	10441	153	A	-p	-0.6913	0.9523	41.0S	131.5E	46	321	236	04m18s
11507	576	2844 Sep 03	23:05:38	3330	10447	158	H	-p	0.5278	1.0063	34.5N	135.0W	58	214	25	00m32s
11508	576	2845 Feb 27	14:25:03	3333	10453	163	H	nn	0.0467	1.0098	5.7S	20.8W	87	152	34	00m55s
11509	576	2845 Aug 24	05:23:05	3336	10459	168	A	nn	-0.2411	0.9603	1.8S	107.9E	76	26	148	04m30s
11510	576	2846 Feb 17	04:37:38	3339	10465	173	T	p-	0.7233	1.0413	29.8N	106.3E	43	151	196	03m20s
11511	576	2846 Aug 13	06:14:04	3342	10471	178	A-	t-	-1.0057	0.9522	62.0S	40.8E	0	58	-	-
11512	576	2847 Jan 08	08:52:24	3345	10476	145	P	-t	-1.3314	0.3886	64.5S	149.6W	0	209		
11513	576	2847 Feb 06	20:22:50	3346	10477	183	P	t-	1.3904	0.2760	62.4N	166.7W	0	125		
11514	576	2847 Jul 03	21:02:28	3348	10482	150	P	-t	1.0066	0.9775	64.9N	33.4E	0	336		
11515	576	2847 Dec 28	18:13:42	3351	10488	155	A	-p	-0.6911	0.9552	66.1S	64.7W	46	340	227	03m34s
11516	576	2848 Jun 22	09:37:37	3355	10494	160	T	-n	0.2062	1.0578	35.3N	52.3E	78	187	195	04m57s
11517	576	2848 Dec 16	20:21:20	3358	10500	165	A	nn	-0.0045	0.9233	23.7S	112.7W	90	358	289	10m13s
11518	576	2849 Jun 12	02:21:56	3361	10506	170	T	p-	-0.5310	1.0747	9.0S	157.7E	58	2	286	07m00s
11519	576	2849 Dec 05	19:34:20	3364	10512	175	A	p-	0.6814	0.9266	20.7N	101.4W	47	181	377	09m51s
11520	576	2850 May 03	09:01:02	3367	10517	142	P	-t	1.3537	0.3475	70.3N	84.3W	0	36		

Cat Num	Canon Plate	Calendar Date	TD of Greatest Eclipse	ΔT s	Luna Num	Saros Num	Ecl. Type	QLE	Gamma	Ecl. Mag.	Lat. °	Long. °	Sun Alt °	Sun Azm °	Path Width km	Central Line Dur.
11521	577	2850 Jun 01	18:39:24	3367	10518	180	P	t-	-1.2846	0.4728	67.8S	80.3W	0	355		
11522	577	2850 Oct 26	10:52:49	3370	10523	147	P	-t	-1.4306	0.2068	71.0S	103.5W	0	132		
11523	577	2850 Nov 24	23:27:55	3370	10524	185	P	t-	1.3407	0.3758	68.7N	146.1W	0	194		
11524	577	2851 Apr 22	16:22:32	3373	10529	152	A	-p	0.6644	0.9529	52.4N	67.3W	48	157	230	04m20s
11525	577	2851 Oct 16	01:17:27	3376	10535	157	T	-p	-0.6786	1.0553	49.5S	154.4E	47	25	248	04m00s
11526	577	2852 Apr 10	17:12:01	3379	10541	162	A	nn	-0.0555	0.9310	5.4N	62.8W	87	344	258	09m13s
11527	577	2852 Oct 04	17:54:44	3383	10547	167	Tm	nn	0.0305	1.0604	3.3S	76.9W	88	197	198	05m31s
11528	577	2853 Mar 30	17:49:21	3386	10553	172	A	p-	-0.7631	0.9452	43.1S	54.2W	40	336	309	05m29s
11529	577	2853 Sep 24	08:03:33	3389	10559	177	H3	p-	0.7880	1.0107	48.1N	92.4E	38	207	59	00m52s
11530	577	2854 Feb 18	12:46:44	3392	10564	144	P	-t	1.3232	0.4021	71.2N	48.0W	0	128		
11531	577	2854 Mar 20	00:41:36	3392	10565	182	P	t-	-1.4228	0.2247	72.2S	84.3W	0	270		
11532	577	2854 Aug 15	01:31:02	3395	10570	149	P	-t	-1.2310	0.5673	70.9S	127.9E	0	43		
11533	577	2855 Feb 08	03:51:22	3398	10576	154	T	-p	0.6288	1.0448	23.2N	131.0E	51	167	191	04m00s
11534	577	2855 Aug 04	02:02:02	3401	10582	159	A	-p	-0.4802	0.9510	11.3S	160.5E	61	10	204	06m32s
11535	577	2856 Jan 28	19:09:09	3405	10588	164	T	nn	-0.0395	1.0258	20.4S	89.9W	88	351	88	02m30s
11536	577	2856 Jul 23	06:01:10	3408	10594	169	A	p-	0.3009	0.9929	37.2N	108.7E	72	188	26	00m43s
11537	577	2857 Jan 17	06:04:05	3411	10600	174	A	p-	-0.7626	0.9660	70.0S	117.8E	40	345	191	02m34s
11538	577	2857 Jun 13	09:35:05	3414	10605	141	Pe	-t	-1.4973	0.0637	65.5S	70.9E	0	342		
11539	577	2857 Jul 12	17:14:23	3414	10606	179	P	t-	1.0403	0.9341	68.1N	108.3E	0	351		
11540	577	2857 Dec 07	14:52:22	3417	10611	146	P	-t	1.5412	0.0502	64.8N	3.2W	0	206		
11541	578	2858 Jan 06	09:41:23	3417	10612	184	P	t-	-1.4993	0.1173	67.6S	133.5W	0	183		
11542	578	2858 Jun 03	02:37:58	3420	10617	151	T	-p	-0.7750	1.0656	27.6S	165.7E	39	347	338	05m38s
11543	578	2858 Nov 26	14:48:33	3423	10623	156	A	-p	0.8377	0.9354	34.1N	14.9W	33	198	435	07m17s
11544	578	2859 May 23	17:54:35	3426	10629	161	T	nn	-0.0467	1.0305	18.2N	74.0W	87	343	103	02m58s
11545	578	2859 Nov 15	20:52:40	3430	10635	166	A	nn	0.0901	0.9962	13.8S	121.1W	85	200	13	00m23s
11546	578	2860 May 12	03:09:22	3433	10641	171	A	p-	0.7377	0.9673	60.7N	114.9E	42	138	173	02m33s
11547	578	2860 Nov 04	09:55:28	3436	10647	176	T	p-	-0.6168	1.0441	49.6S	19.3E	52	35	186	03m07s
11548	578	2861 Apr 01	12:06:04	3439	10652	143	P	-t	-1.4566	0.1880	61.2S	94.6E	0	280		
11549	578	2861 May 01	05:17:17	3439	10653	181	P	t-	1.5031	0.1069	62.3N	6.9W	0	55		
11550	578	2861 Sep 25	17:38:14	3442	10658	148	P	-t	1.2912	0.4607	61.2N	15.1E	0	267		
11551	578	2861 Oct 25	02:14:31	3443	10659	186	P	t-	-1.2964	0.4504	61.9S	42.7E	0	117		
11552	578	2862 Mar 21	13:59:08	3445	10664	153	A	-p	-0.7148	0.9570	38.5S	12.4E	44	322	218	03m55s
11553	578	2862 Sep 15	06:23:08	3449	10670	158	A	-p	0.5956	0.9999	34.2N	117.9E	53	215	0	00m01s
11554	578	2863 Mar 10	22:51:08	3452	10676	163	H	nn	0.0299	1.0147	2.3S	146.9W	88	151	50	01m21s
11555	578	2863 Sep 04	12:09:14	3455	10682	168	A	nn	-0.1646	0.9567	1.6S	7.7E	81	28	159	04m50s
11556	578	2864 Feb 28	13:20:36	3458	10688	173	T	p-	0.7105	1.0442	32.0N	25.8W	45	147	205	03m26s
11557	578	2864 Aug 23	12:47:18	3461	10694	178	A	t-	-0.9231	0.9378	48.1S	34.4W	22	41	586	05m47s
11558	578	2865 Jan 18	17:34:21	3464	10699	145	P	-t	-1.3398	0.3737	63.6S	70.9E	0	219		
11559	578	2865 Feb 17	05:07:43	3465	10700	183	P	t-	1.3843	0.2874	61.9N	53.7E	0	116		
11560	578	2865 Jul 14	04:03:03	3467	10705	150	P	-t	1.0808	0.8446	64.0N	80.8W	0	326		
11561	579	2866 Jan 08	02:33:07	3471	10711	155	A	-p	-0.7007	0.9518	64.5S	179.3E	45	331	248	03m51s
11562	579	2866 Jul 03	17:03:16	3474	10717	160	T	-n	0.2785	1.0610	38.7N	55.8W	74	193	209	04m59s
11563	579	2866 Dec 28	04:18:59	3477	10723	165	A	nn	-0.0184	0.9213	24.4S	129.9E	89	352	298	10m19s
11564	579	2867 Jun 23	09:57:35	3480	10729	170	T	p-	-0.4622	1.0766	4.1S	43.0E	62	6	279	07m10s
11565	579	2867 Dec 17	03:31:29	3484	10735	175	A	p-	0.6635	0.9266	18.4N	137.7E	48	176	369	09m55s
11566	579	2868 May 13	16:37:07	3486	10740	142	P	-t	1.4111	0.2430	69.4N	149.9E	0	24		
11567	579	2868 Jun 12	02:08:55	3487	10741	180	P	t-	-1.2230	0.5889	66.8S	157.0E	0	5		
11568	579	2868 Nov 05	18:57:09	3489	10746	147	P	-t	-1.4586	0.1556	70.2S	123.1E	0	145		
11569	579	2868 Dec 05	07:40:55	3490	10747	185	P	t-	1.3188	0.4144	67.6N	79.9E	0	183		
11570	579	2869 May 02	23:40:23	3493	10752	152	A	-p	0.7192	0.9525	60.4N	178.2W	44	157	250	04m05s
11571	579	2869 Oct 26	09:28:07	3496	10758	157	T	-p	-0.7160	1.0528	56.5S	30.1E	44	25	250	03m38s
11572	579	2870 Apr 22	00:24:56	3499	10764	162	A	nn	-0.0061	0.9340	12.0N	171.8W	90	342	246	08m47s
11573	579	2870 Oct 16	01:55:41	3502	10770	167	T	nn	-0.0147	1.0557	9.9S	161.9E	89	15	184	05m07s
11574	579	2871 Apr 11	01:25:37	3506	10776	172	A	p-	-0.7242	0.9510	36.3S	172.1W	43	340	258	05m12s
11575	579	2871 Oct 05	15:39:10	3509	10782	177	H	p	0.7370	1.0057	40.2N	26.4W	42	203	29	00m30s
11576	579	2872 Feb 29	21:22:44	3512	10787	144	P	-t	1.3315	0.3864	71.7N	169.8E	0	114		
11577	579	2872 Mar 30	08:50:29	3512	10788	182	P	t-	-1.3933	0.2766	72.1S	139.8E	0	284		
11578	579	2872 Aug 25	08:07:26	3515	10793	149	P	-t	-1.3122	0.4279	71.5S	16.1E	0	56		
11579	579	2872 Sep 23	22:50:02	3515	10794	187	Pb	t-	1.5378	0.0363	72.1N	62.5W	0	267		
11580	579	2873 Feb 18	12:39:50	3518	10799	154	T	-p	0.6369	1.0461	27.1N	2.8W	50	164	198	03m59s

Cat Num	Canon Plate	Calendar Date	TD of Greatest Eclipse	ΔT s	Luna Num	Saros Num	Ecl. Type	QLE	Gamma	Ecl. Mag.	Lat. °	Long. °	Sun Alt °	Sun Azm °	Path Width km	Central Line Dur.
11581	580	2873 Aug 14	08:34:15	3521	10805	159	A	-p	-0.5646	0.9506	19.8S	59.8E	55	14	218	06m12s
11582	580	2874 Feb 08	03:54:12	3525	10811	164	T	nn	-0.0340	1.0244	16.9S	139.7E	88	348	83	02m23s
11583	580	2874 Aug 03	12:56:53	3528	10817	169	A	n-	0.2203	0.9961	29.7N	5.1E	77	191	14	00m24s
11584	580	2875 Jan 28	14:30:34	3531	10823	174	A	p-	-0.7588	0.9636	66.3S	2.2W	40	337	203	02m49s
11585	580	2875 Jul 24	00:34:50	3535	10829	179	T	t-	0.9640	1.0388	81.5N	47.8W	15	306	514	02m04s
11586	580	2875 Dec 18	22:44:21	3537	10834	146	P	-t	1.5581	0.0230	65.8N	130.6W	0	196		
11587	580	2876 Jan 17	17:47:01	3538	10835	184	P	t-	-1.4928	0.1288	68.7S	94.3E	0	195		
11588	580	2876 Jun 13	10:16:45	3541	10840	151	T	-p	-0.8414	1.0645	33.8S	49.4E	32	351	391	05m25s
11589	580	2876 Dec 06	22:44:54	3544	10846	156	A	-p	0.8570	0.9349	35.5N	136.5W	31	193	468	07m22s
11590	580	2877 Jun 03	01:23:34	3547	10852	161	T	nn	-0.1114	1.0294	16.2N	175.3E	84	348	100	02m58s
11591	580	2877 Nov 26	05:06:29	3550	10858	166	A	nn	0.1130	0.9967	14.8S	116.6E	84	196	12	00m21s
11592	580	2878 May 23	10:17:43	3554	10864	171	A	p-	0.6749	0.9679	60.3N	17.1E	47	148	156	02m36s
11593	580	2878 Nov 15	18:18:28	3557	10870	176	T	p-	-0.5879	1.0435	51.7S	102.6W	54	31	179	03m04s
11594	580	2879 Apr 12	19:28:29	3560	10875	143	P	-t	-1.4973	0.1183	61.5S	24.2W	0	289		
11595	580	2879 May 12	12:15:42	3560	10876	181	P	t-	1.4435	0.2069	62.9N	120.0W	0	46		
11596	580	2879 Oct 07	01:29:04	3563	10881	148	P	-t	1.3441	0.3612	61.3N	110.8W	0	258		
11597	580	2879 Nov 05	10:30:50	3563	10882	186	P	t-	-1.2640	0.5122	62.5S	89.8W	0	126		
11598	580	2880 Mar 31	21:48:29	3566	10887	153	A	-p	-0.7447	0.9619	36.8S	105.1W	42	324	201	03m32s
11599	580	2880 Sep 25	13:45:31	3569	10893	158	A	-p	0.6583	0.9932	34.0N	9.1E	49	216	31	00m36s
11600	580	2881 Mar 21	07:10:48	3573	10899	163	T	nn	0.0071	1.0201	0.9N	88.6E	90	151	68	01m49s
11601	581	2881 Sep 14	18:58:55	3576	10905	168	Am	nn	-0.0914	0.9527	1.9S	93.3W	85	29	174	05m15s
11602	581	2882 Mar 10	21:58:32	3579	10911	173	T	p-	0.6934	1.0475	34.4N	156.2W	46	145	215	03m34s
11603	581	2882 Sep 03	19:25:39	3583	10917	178	A	p-	-0.8438	0.9390	42.7S	129.0W	32	39	408	05m52s
11604	581	2883 Jan 30	02:15:42	3585	10922	145	P	-t	-1.3484	0.3583	62.8S	68.1W	0	229		
11605	581	2883 Feb 28	13:48:03	3586	10923	183	P	t-	1.3751	0.3046	61.5N	84.7W	0	106		
11606	581	2883 Jul 25	11:05:22	3589	10928	150	P	-t	1.1544	0.7116	63.3N	164.7E	0	317		
11607	581	2883 Aug 23	22:19:19	3589	10929	188	Pb	t-	-1.5524	0.0010	61.7S	152.0E	0	67		
11608	581	2884 Jan 19	10:53:53	3592	10934	155	A	-p	-0.7088	0.9489	61.7S	60.4E	45	324	265	04m07s
11609	581	2884 Jul 14	00:27:39	3595	10940	160	T	-n	0.3523	1.0635	41.3N	162.8W	69	199	222	04m58s
11610	581	2885 Jan 07	12:20:24	3598	10946	165	A	nn	-0.0289	0.9197	23.9S	11.5E	88	348	304	10m20s
11611	581	2885 Jul 03	17:29:55	3602	10952	170	T	n-	-0.3905	1.0777	0.1N	70.2W	67	10	272	07m11s
11612	581	2885 Dec 27	11:34:23	3605	10958	175	A	p-	0.6500	0.9270	17.1N	15.6E	49	172	360	09m46s
11613	581	2886 May 25	00:04:54	3608	10963	142	P	-t	1.4742	0.1283	68.5N	26.8E	0	12		
11614	581	2886 Jun 23	09:33:12	3608	10964	180	P	t-	-1.1577	0.7117	65.8S	36.1E	0	15		
11615	581	2886 Nov 17	03:10:24	3611	10969	147	P	-t	-1.4801	0.1162	69.3S	12.0W	0	157		
11616	581	2886 Dec 16	16:00:53	3612	10970	185	P	t-	1.3018	0.4443	66.5N	55.3W	0	172		
11617	581	2887 May 14	06:50:27	3614	10975	152	A	-p	0.7793	0.9518	69.0N	73.3E	38	157	283	03m52s
11618	581	2887 Nov 06	17:45:34	3618	10981	157	T	-p	-0.7479	1.0502	63.0S	94.6W	41	25	252	03m18s
11619	581	2888 May 02	07:31:10	3621	10987	162	Am	nn	0.0488	0.9369	18.6N	81.1E	87	169	235	08m17s
11620	581	2888 Oct 26	10:03:09	3624	10993	167	T	nn	-0.0541	1.0509	16.0S	39.5E	87	14	169	04m42s
11621	582	2889 Apr 21	08:53:56	3628	10999	172	A	p-	-0.6779	0.9570	29.3S	72.4E	47	344	211	04m49s
11622	582	2889 Oct 15	23:21:08	3631	11005	177	H	p-	0.6918	1.0004	33.1N	145.6W	46	199	2	00m02s
11623	582	2890 Mar 12	05:52:26	3634	11010	144	P	-t	1.3454	0.3600	72.0N	28.9E	0	100		
11624	582	2890 Apr 10	16:51:03	3634	11011	182	P	t-	-1.3561	0.3430	71.7S	6.2E	0	298		
11625	582	2890 Sep 05	14:49:00	3637	11016	149	P	-t	-1.3883	0.2981	72.0S	97.5W	0	69		
11626	582	2890 Oct 05	05:59:26	3638	11017	187	P	t-	1.4844	0.1307	72.0N	176.6E	0	253		
11627	582	2891 Mar 01	21:22:05	3640	11022	154	T	-p	0.6501	1.0477	31.7N	135.2W	49	161	207	03m58s
11628	582	2891 Aug 25	15:12:39	3644	11028	159	A	-p	-0.6441	0.9498	28.6S	43.3W	50	17	238	05m52s
11629	582	2892 Feb 19	12:34:51	3647	11034	164	T	nn	-0.0255	1.0234	12.6S	9.9E	88	346	80	02m18s
11630	582	2892 Aug 13	19:55:10	3650	11040	169	A	nn	0.1410	0.9988	21.9N	100.0W	82	194	4	00m08s
11631	582	2893 Feb 07	22:54:55	3654	11046	174	A	p-	-0.7539	0.9617	61.9S	125.1W	41	334	211	03m04s
11632	582	2893 Aug 03	07:55:25	3657	11052	179	T	t-	0.8874	1.0448	76.0N	120.7E	27	226	328	02m40s
11633	582	2893 Dec 29	06:42:03	3660	11057	146	Pe	-t	1.5706	0.0028	66.8N	100.1E	0	185		
11634	582	2894 Jan 28	01:52:44	3660	11058	184	P	t-	-1.4865	0.1397	69.7S	38.5W	0	207		
11635	582	2894 Jun 24	17:49:14	3663	11063	151	T	-t	-0.9127	1.0620	42.9S	66.3W	24	355	502	04m55s
11636	582	2894 Dec 18	06:49:29	3666	11069	156	A	-p	0.8700	0.9350	37.0N	99.4E	29	188	492	07m20s
11637	582	2895 Jun 14	08:45:51	3670	11075	161	T	-n	-0.1811	1.0278	12.9N	65.9E	80	352	96	02m55s
11638	582	2895 Dec 07	13:27:23	3673	11081	166	A	nn	0.1306	0.9974	15.3S	7.4W	83	191	9	00m17s
11639	582	2896 Jun 02	17:19:15	3676	11087	171	A	p-	0.6071	0.9683	58.5N	79.0W	52	158	144	02m42s
11640	582	2896 Nov 26	02:48:19	3680	11093	176	T	p-	-0.5647	1.0427	53.6S	134.7E	55	26	173	03m02s

Cat Num	Canon Plate	Calendar Date	TD of Greatest Eclipse	ΔT s	Luna Num	Saros Num	Ecl. Type	QLE	Gamma	Ecl. Mag.	Lat. °	Long. °	Sun Alt °	Sun Azm °	Path Width km	Central Line Dur.
11641	583	2897 Apr 23	02:43:17	3682	11098	143	Pe	-t	-1.5438	0.0380	61.9S	141.2W	0	298		
11642	583	2897 May 22	19:05:57	3683	11099	181	P	t-	1.3773	0.3186	63.7N	128.7E	0	37		
11643	583	2897 Oct 17	09:25:31	3686	11104	148	P	-t	1.3918	0.2723	61.6N	121.9E	0	249		
11644	583	2897 Nov 15	18:54:00	3686	11105	186	P	t-	-1.2375	0.5623	63.1S	135.9E	0	135		
11645	583	2898 Apr 12	05:31:42	3689	11110	153	A	-p	-0.7801	0.9669	36.2S	138.9E	38	326	185	03m06s
11646	583	2898 Oct 06	21:13:41	3692	11116	158	A	-p	0.7154	0.9864	34.1N	101.8W	44	215	67	01m13s
11647	583	2899 Apr 01	15:24:34	3696	11122	163	T	nn	-0.0212	1.0255	3.8N	34.3W	89	331	87	02m17s
11648	583	2899 Sep 26	01:55:26	3699	11128	168	A	nn	-0.0243	0.9486	2.7S	163.9E	89	29	189	05m44s
11649	583	2900 Mar 22	06:30:52	3702	11134	173	T	p-	0.6716	1.0510	36.9N	75.3E	48	143	224	03m44s
11650	583	2900 Sep 15	02:09:43	3706	11140	178	A	p-	-0.7686	0.9394	40.1S	133.1E	40	39	340	05m52s
11651	583	2901 Feb 10	10:54:17	3709	11145	145	P	-t	-1.3590	0.3395	62.1S	153.8E	0	238		
11652	583	2901 Mar 11	22:22:36	3709	11146	183	P	t-	1.3617	0.3296	61.3N	138.5E	0	97		
11653	583	2901 Aug 05	18:10:19	3712	11151	150	P	-t	1.2266	0.5801	62.6N	49.9E	0	308		
11654	583	2901 Sep 04	05:19:41	3712	11152	188	P	t-	-1.4766	0.1347	61.3S	38.7E	0	75		
11655	583	2902 Jan 30	19:12:16	3715	11157	155	A	-p	-0.7182	0.9466	58.3S	60.1W	44	319	280	04m21s
11656	583	2902 Jul 26	07:52:48	3719	11163	160	T	-p	0.4260	1.0651	42.9N	90.4E	65	205	235	04m54s
11657	583	2903 Jan 19	20:22:19	3722	11169	165	A	nn	-0.0391	0.9189	22.6S	107.2W	88	343	308	10m12s
11658	583	2903 Jul 16	01:00:45	3725	11175	170	T	n-	-0.3177	1.0780	3.4N	177.5E	71	14	265	07m04s
11659	583	2904 Jan 08	19:40:31	3729	11181	175	A	p-	0.6386	0.9281	16.7N	107.2W	50	167	348	09m24s
11660	583	2904 Jun 05	07:24:49	3732	11186	142	Pe	-t	1.5428	0.0040	67.5N	93.8W	0	2		
11661	584	2904 Jul 04	16:52:58	3732	11187	180	P	t-	-1.0890	0.8402	64.8S	83.3W	0	25		
11662	584	2904 Nov 28	11:32:06	3735	11192	147	P	-t	-1.4959	0.0874	68.3S	148.6W	0	169		
11663	584	2904 Dec 28	00:26:56	3735	11193	185	P	t-	1.2891	0.4669	65.5N	168.5E	0	161		
11664	584	2905 May 25	13:49:06	3738	11198	152	A	-t	0.8482	0.9505	78.6N	32.5W	32	157	346	03m39s
11665	584	2905 Nov 18	02:11:36	3742	11204	157	T	-p	-0.7731	1.0477	68.9S	141.3E	39	21	252	03m01s
11666	584	2906 May 14	14:27:44	3745	11210	162	A	nn	0.1121	0.9398	25.2N	23.0W	83	171	225	07m41s
11667	584	2906 Nov 07	18:18:03	3748	11216	167	T	-n	-0.0869	1.0461	21.3S	84.1W	85	11	154	04m15s
11668	584	2907 May 03	16:15:29	3752	11222	172	A	p-	-0.6251	0.9632	22.4S	40.9W	51	348	170	04m20s
11669	584	2907 Oct 28	07:09:47	3755	11228	177	A	p-	0.6527	0.9949	26.8N	94.3E	49	195	23	00m31s
11670	584	2908 Mar 23	14:18:02	3758	11233	144	P	-t	1.3632	0.3262	72.1N	111.2W	0	86		
11671	584	2908 Apr 22	00:46:33	3758	11234	182	P	t-	-1.3143	0.4189	71.1S	125.7W	0	311		
11672	584	2908 Sep 16	21:33:42	3761	11239	149	P	-t	-1.4611	0.1750	72.2S	147.8E	0	83		
11673	584	2908 Oct 16	13:13:45	3762	11240	187	P	t-	1.4352	0.2170	71.6N	54.7E	0	240		
11674	584	2909 Mar 13	06:00:58	3765	11245	154	T	-p	0.6663	1.0495	36.8N	93.1E	48	158	219	03m56s
11675	584	2909 Sep 05	21:55:40	3768	11251	159	A	-p	-0.7198	0.9486	37.7S	148.5W	44	21	269	05m31s
11676	584	2910 Mar 02	21:10:45	3771	11257	164	T	nn	-0.0136	1.0228	7.9S	119.0W	89	344	78	02m15s
11677	584	2910 Aug 26	02:59:33	3775	11263	169	H	nn	0.0660	1.0009	14.0N	152.8E	86	195	3	00m06s
11678	584	2911 Feb 20	07:15:04	3778	11269	174	A	p-	-0.7465	0.9604	57.0S	110.7E	41	332	215	03m19s
11679	584	2911 Aug 15	15:20:21	3781	11275	179	T	p-	0.8136	1.0488	65.5N	5.9W	35	213	280	03m10s
11680	584	2912 Feb 09	09:55:08	3785	11281	184	P	t-	-1.4779	0.1542	70.6S	171.1W	0	219		
11681	585	2912 Jul 06	01:20:07	3788	11286	151	Ts	-t	-0.9849	1.0568	58.5S	176.0E	9	1	-	03m59s
11682	585	2912 Aug 04	08:04:45	3788	11287	189	Pb	t-	1.5114	0.0308	70.0N	137.2W	0	328		
11683	585	2912 Dec 29	14:58:37	3791	11292	156	A	-p	0.8798	0.9356	38.8N	26.2W	28	183	507	07m10s
11684	585	2913 Jun 25	16:01:38	3794	11298	161	T	-n	-0.2551	1.0255	8.5N	42.4W	75	357	90	02m45s
11685	585	2913 Dec 18	21:55:06	3798	11304	166	Am	nn	0.1430	0.9985	15.2S	133.1W	82	187	5	00m10s
11686	585	2914 Jun 15	00:12:40	3801	11310	171	A	p-	0.5337	0.9681	55.2N	174.3W	57	167	136	02m52s
11687	585	2914 Dec 08	11:24:42	3805	11316	176	T	p-	-0.5469	1.0422	55.0S	11.2E	57	19	169	03m02s
11688	585	2915 Jun 04	01:48:52	3808	11322	181	P	t-	1.3051	0.4410	64.6N	19.1E	0	28		
11689	585	2915 Oct 29	17:29:53	3811	11327	148	P	-t	1.4323	0.1978	62.1N	7.5W	0	240		
11690	585	2915 Nov 28	03:24:22	3811	11328	186	P	t-	-1.2177	0.5994	63.9S	0.5W	0	145		
11691	585	2916 Apr 23	13:07:22	3814	11333	153	A	-p	-0.8223	0.9718	37.1S	24.8E	34	329	173	02m39s
11692	585	2916 Oct 18	04:47:37	3818	11339	158	A	-p	0.7665	0.9794	34.6N	145.3E	40	213	111	01m54s
11693	585	2917 Apr 12	23:31:15	3821	11345	163	T	nn	-0.0560	1.0312	6.2N	155.2W	87	333	106	02m48s
11694	585	2917 Oct 07	08:58:15	3824	11351	168	A	nn	0.0371	0.9443	3.8S	59.5E	88	209	206	06m19s
11695	585	2918 Apr 02	14:55:22	3828	11357	173	T	p-	0.6433	1.0547	39.3N	50.4W	50	142	233	03m55s
11696	585	2918 Sep 26	09:02:12	3831	11363	178	A	p-	-0.6997	0.9391	39.2S	32.5E	45	38	308	05m54s
11697	585	2919 Feb 21	19:29:46	3834	11368	145	P	-t	-1.3716	0.3168	61.7S	16.6E	0	247		
11698	585	2919 Mar 23	06:50:58	3835	11369	183	P	t-	1.3440	0.3627	61.3N	3.2E	0	88		
11699	585	2919 Aug 17	01:18:50	3837	11374	150	P	-t	1.2963	0.4526	62.1N	65.7W	0	300		
11700	585	2919 Sep 15	12:26:49	3838	11375	188	P	t-	-1.4052	0.2611	61.2S	76.3W	0	84		

Cat Num	Canon Plate	Calendar Date	TD of Greatest Eclipse	ΔT s	Luna Num	Saros Num	Ecl. Type	QLE	Gamma	Ecl. Mag.	Lat. °	Long. °	Sun Alt °	Sun Azm °	Path Width km	Central Line Dur.
11701	586	2920 Feb 11	03:28:36	3841	11380	155	A	-p	-0.7285	0.9449	54.6S	178.1E	43	317	293	04m34s
11702	586	2920 Aug 05	15:19:10	3844	11386	160	T	-p	0.4991	1.0660	43.8N	16.6W	60	210	248	04m48s
11703	586	2921 Jan 30	04:24:58	3848	11392	165	A	nn	-0.0482	0.9187	20.3S	133.7E	87	339	309	10m01s
11704	586	2921 Jul 26	08:29:29	3851	11398	170	T	n-	-0.2434	1.0775	5.8N	66.2E	76	18	258	06m50s
11705	586	2922 Jan 19	03:50:04	3854	11404	175	A	p-	0.6295	0.9296	17.2N	129.3E	51	163	335	08m53s
11706	586	2922 Jul 16	00:09:45	3858	11410	180	P	t-	-1.0185	0.9713	64.0S	158.3E	0	35		
11707	586	2922 Dec 09	19:59:54	3861	11415	147	P	-t	-1.5074	0.0662	67.3S	73.9E	0	180		
11708	586	2923 Jan 08	08:55:51	3861	11416	185	P	t-	1.2780	0.4864	64.5N	32.0E	0	151		
11709	586	2923 Jun 05	20:41:46	3864	11421	152	A	-t	0.9210	0.9483	89.5N	149.9E	22	84	498	03m28s
11710	586	2923 Nov 29	10:43:53	3867	11427	157	T	-p	-0.7936	1.0454	73.8S	20.2E	37	14	251	02m47s
11711	586	2924 May 24	21:17:54	3871	11433	162	A	nn	0.1805	0.9426	31.5N	124.7W	79	175	216	07m02s
11712	586	2924 Nov 18	02:38:44	3874	11439	167	T	-n	-0.1143	1.0413	25.8S	151.6E	83	8	139	03m47s
11713	586	2925 May 13	23:30:12	3878	11445	172	A	p-	-0.5657	0.9693	15.6S	152.0W	55	352	133	03m45s
11714	586	2925 Nov 07	15:05:14	3881	11451	177	A	p-	0.6202	0.9894	21.4N	26.9W	52	192	47	01m08s
11715	586	2926 Apr 03	22:34:55	3884	11456	144	P	-t	1.3882	0.2785	71.9N	111.0E	0	72		
11716	586	2926 May 03	08:33:18	3885	11457	182	P	t-	-1.2647	0.5099	70.4S	105.2E	0	324		
11717	586	2926 Sep 28	04:26:34	3887	11462	149	Pe	-t	-1.5263	0.0655	72.2S	31.0E	0	97		
11718	586	2926 Oct 27	20:36:57	3888	11463	187	P	t-	1.3936	0.2895	71.0N	69.0W	0	227		
11719	586	2927 Mar 24	14:32:03	3891	11468	154	T	-p	0.6886	1.0514	42.6N	36.8W	46	156	233	03m54s
11720	586	2927 Sep 17	04:46:05	3894	11474	159	A	-p	-0.7897	0.9470	47.1S	103.2E	38	26	314	05m10s
11721	587	2928 Mar 13	05:39:54	3898	11480	164	T	nn	0.0034	1.0225	2.6S	113.3E	90	161	77	02m13s
11722	587	2928 Sep 05	10:09:23	3901	11486	169	H	nn	-0.0052	1.0024	6.1N	43.8E	90	19	8	00m16s
11723	587	2929 Mar 02	15:29:16	3905	11492	174	A	p-	-0.7349	0.9597	51.7S	13.5W	42	333	214	03m32s
11724	587	2929 Aug 25	22:47:43	3908	11498	179	T	p-	0.7413	1.0515	55.6N	124.4W	42	207	254	03m37s
11725	587	2930 Feb 19	17:54:40	3911	11504	184	P	t-	-1.4671	0.1719	71.3S	56.3E	0	233		
11726	587	2930 Jul 17	08:47:32	3914	11509	151	P	-t	-1.0593	0.9063	68.5S	53.0E	0	13		
11727	587	2930 Aug 15	15:36:52	3915	11510	189	P	t-	1.4430	0.1642	70.8N	97.7E	0	316		
11728	587	2931 Jan 09	23:12:05	3918	11515	156	A	-p	0.8864	0.9369	40.7N	153.2W	27	177	510	06m52s
11729	587	2931 Jul 06	23:13:24	3921	11521	161	T	-p	-0.3316	1.0226	3.1N	150.5W	71	1	81	02m30s
11730	587	2931 Dec 30	06:28:24	3925	11527	166	A	nn	0.1511	1.0000	14.5S	99.7E	81	182	0	00m00s
11731	587	2932 Jun 25	07:00:01	3928	11533	171	A	p-	0.4557	0.9676	50.5N	90.0E	63	175	131	03m06s
11732	587	2932 Dec 18	20:06:44	3931	11539	176	T	p-	-0.5336	1.0418	55.4S	113.0W	57	11	166	03m01s
11733	587	2933 Jun 14	08:26:15	3935	11545	181	P	t-	1.2282	0.5718	65.5N	89.5W	0	18		
11734	587	2933 Nov 09	01:40:31	3938	11550	148	P	-t	1.4676	0.1339	62.6N	138.6W	0	231		
11735	587	2933 Dec 08	12:00:20	3938	11551	186	P	t-	-1.2025	0.6275	64.9S	138.6W	0	155		
11736	587	2934 May 04	20:36:37	3941	11556	153	A	-p	-0.8706	0.9764	39.7S	87.5W	29	331	168	02m12s
11737	587	2934 Oct 29	12:28:43	3945	11562	158	A	-p	0.8111	0.9727	35.6N	29.9E	36	210	163	02m35s
11738	587	2935 Apr 24	07:32:14	3948	11568	163	Tm	nn	-0.0964	1.0368	7.9N	85.5E	85	335	124	03m20s
11739	587	2935 Oct 18	16:07:51	3952	11574	168	A	nn	0.0925	0.9401	5.0S	46.7W	85	207	223	06m59s
11740	587	2936 Apr 12	23:13:37	3955	11580	173	T	p-	0.6097	1.0584	41.6N	173.6W	52	143	240	04m08s
11741	588	2936 Oct 06	16:02:39	3958	11586	178	A	p-	-0.6368	0.9385	39.4S	70.1W	50	38	290	05m57s
11742	588	2937 Mar 04	03:58:34	3961	11591	145	P	-t	-1.3893	0.2851	61.3S	118.8W	0	257		
11743	588	2937 Apr 02	15:10:19	3962	11592	183	P	t-	1.3195	0.4083	61.4N	129.7W	0	79		
11744	588	2937 Aug 27	08:32:10	3965	11597	150	P	-t	1.3627	0.3306	61.7N	177.6E	0	291		
11745	588	2937 Sep 25	19:42:19	3965	11598	188	P	t-	-1.3394	0.3777	61.1S	166.7E	0	93		
11746	588	2938 Feb 21	11:39:31	3968	11603	155	A	-p	-0.7427	0.9436	51.1S	56.6E	42	316	306	04m46s
11747	588	2938 Aug 16	22:49:03	3972	11609	160	T	-p	0.5697	1.0660	43.9N	124.7W	55	215	261	04m42s
11748	588	2939 Feb 10	12:23:38	3975	11615	165	A	nn	-0.0607	0.9191	17.6S	15.2E	86	336	307	09m45s
11749	588	2939 Aug 06	15:59:27	3979	11621	170	T	n-	-0.1702	1.0761	7.3N	45.1W	80	22	250	06m33s
11750	588	2940 Jan 30	11:59:57	3982	11627	175	A	p-	0.6198	0.9319	18.3N	5.8E	52	159	319	08m15s
11751	588	2940 Jul 26	07:23:06	3986	11633	180	T	t-	-0.9456	1.0244	48.0S	60.3E	18	28	256	01m56s
11752	588	2940 Dec 20	04:34:38	3988	11638	147	P	-t	-1.5142	0.0534	66.2S	64.9W	0	191		
11753	588	2941 Jan 18	17:28:26	3989	11639	185	P	t-	1.2692	0.5021	63.6N	105.1W	0	141		
11754	588	2941 Jun 16	03:25:38	3992	11644	152	A+	-t	1.0004	0.9657	66.4N	43.9W	0	351	-	-
11755	588	2941 Dec 09	19:23:14	3995	11650	157	T	-p	-0.8082	1.0434	77.0S	95.1W	36	358	248	02m36s
11756	588	2942 Jun 05	04:00:12	3999	11656	162	A	nn	0.2556	0.9452	37.5N	136.5E	75	179	209	06m22s
11757	588	2942 Nov 29	11:06:48	4002	11662	167	T	-n	-0.1353	1.0367	29.3S	26.2E	82	4	124	03m21s
11758	588	2943 May 25	06:39:59	4006	11668	172	A	p-	-0.5014	0.9753	9.2S	98.8E	60	355	102	03m04s
11759	588	2943 Nov 18	23:06:18	4009	11674	177	A	p-	0.5931	0.9840	17.0N	148.9W	54	188	70	01m48s
11760	588	2944 Apr 14	06:47:04	4012	11679	144	P	-t	1.4176	0.2219	71.4N	25.3W	0	58		

Cat Num	Canon Plate	Calendar Date	TD of Greatest Eclipse	ΔT s	Luna Num	Saros Num	Ecl. Type	QLE	Gamma	Ecl. Mag.	Lat. °	Long. °	Sun Alt °	Sun Azm °	Path Width km	Central Line Dur.
11761	589	2944 May 13	16:15:49	4013	11680	182	P	t-	-1.2108	0.6101	69.5S	22.3W	0	336		
11762	589	2944 Nov 07	04:06:22	4016	11686	187	P	t-	1.3571	0.3525	70.2N	166.4E	0	214		
11763	589	2945 Apr 03	22:56:50	4019	11691	154	T	-p	0.7164	1.0532	49.0N	165.4W	44	154	251	03m50s
11764	589	2945 Sep 27	11:43:51	4023	11697	159	A	-p	-0.8539	0.9451	56.6S	9.2W	31	33	387	04m50s
11765	589	2946 Mar 24	14:02:24	4026	11703	164	T	nn	0.0255	1.0224	3.0N	12.9W	89	163	76	02m13s
11766	589	2946 Sep 16	17:27:21	4030	11709	169	H	nn	-0.0706	1.0036	1.6S	67.3W	86	17	12	00m23s
11767	589	2947 Mar 13	23:36:01	4033	11715	174	A	p-	-0.7178	0.9594	46.1S	137.0W	44	334	210	03m46s
11768	589	2947 Sep 06	06:21:54	4037	11721	179	T	p-	0.6740	1.0534	46.5N	117.8E	47	205	238	04m01s
11769	589	2948 Mar 02	01:48:24	4040	11727	184	P	t-	-1.4521	0.1967	71.9S	75.3W	0	247		
11770	589	2948 Jul 27	16:14:08	4043	11732	151	P	-t	-1.1339	0.7620	69.5S	69.8W	0	24		
11771	589	2948 Aug 25	23:11:43	4044	11733	189	P	t-	1.3768	0.2930	71.4N	28.5W	0	303		
11772	589	2949 Jan 20	07:27:37	4046	11738	156	A	-p	0.8919	0.9388	43.0N	79.2E	27	172	504	06m26s
11773	589	2949 Jul 17	06:21:27	4050	11744	161	T	-p	-0.4099	1.0189	3.2S	101.6E	66	5	71	02m06s
11774	589	2950 Jan 09	15:04:40	4053	11750	166	H	nn	0.1574	1.0020	13.0S	28.4W	81	178	7	00m13s
11775	589	2950 Jul 06	13:42:14	4057	11756	171	A	p-	0.3743	0.9665	44.7N	6.5W	68	181	130	03m25s
11776	589	2950 Dec 30	04:53:21	4060	11762	176	T	p-	-0.5243	1.0417	54.9S	121.7E	58	3	164	03m03s
11777	589	2951 Jun 25	14:58:47	4064	11768	181	P	t-	1.1468	0.7107	66.5N	162.7E	0	8		
11778	589	2951 Nov 20	09:57:41	4067	11773	148	P	-t	1.4970	0.0814	63.4N	88.5E	0	222		
11779	589	2951 Dec 19	20:40:59	4067	11774	186	P	t-	-1.1917	0.6471	65.9S	81.8E	0	165		
11780	589	2952 May 15	03:59:11	4070	11779	153	A	-p	-0.9249	0.9803	44.9S	162.2E	22	334	182	01m46s
11781	590	2952 Nov 08	20:16:40	4074	11785	158	A	-p	0.8489	0.9660	37.2N	87.8W	32	207	227	03m18s
11782	590	2953 May 04	15:24:35	4077	11791	163	T	nn	-0.1443	1.0424	8.6N	31.5W	82	338	143	03m54s
11783	590	2953 Oct 28	23:25:40	4081	11797	168	A	nn	0.1405	0.9359	6.1S	155.1W	82	205	241	07m44s
11784	590	2954 Apr 24	07:23:39	4084	11803	173	T	p-	0.5696	1.0622	43.6N	66.2E	55	145	246	04m23s
11785	590	2954 Oct 17	23:12:15	4088	11809	178	A	p-	-0.5810	0.9377	40.4S	175.0W	54	36	280	06m03s
11786	590	2955 Mar 15	12:21:15	4091	11814	145	P	-t	-1.4112	0.2456	61.2S	107.5E	0	266		
11787	590	2955 Apr 13	23:22:16	4091	11815	183	P	t-	1.2899	0.4636	61.6N	99.1E	0	70		
11788	590	2955 Sep 07	15:51:40	4094	11820	150	P	-t	1.4246	0.2164	61.4N	59.5E	0	282		
11789	590	2955 Oct 07	03:06:31	4095	11821	188	P	t-	-1.2798	0.4834	61.2S	47.5E	0	101		
11790	590	2956 Mar 03	19:46:05	4098	11826	155	A	-p	-0.7598	0.9428	47.8S	64.6W	40	316	318	04m57s
11791	590	2956 Aug 27	06:20:57	4101	11832	160	T	-p	0.6387	1.0653	43.8N	126.2E	50	218	274	04m34s
11792	590	2957 Feb 20	20:20:31	4105	11838	165	A	nn	-0.0746	0.9201	14.5S	103.0W	86	333	303	09m28s
11793	590	2957 Aug 16	23:30:11	4108	11844	170	T	nn	-0.0978	1.0739	8.0N	156.4W	84	25	241	06m13s
11794	590	2958 Feb 09	20:08:50	4112	11850	175	A	p-	0.6088	0.9347	20.0N	117.4W	52	155	301	07m33s
11795	590	2958 Aug 06	14:36:40	4115	11856	180	T	p-	-0.8736	1.0235	40.0S	45.9W	29	28	161	01m58s
11796	590	2958 Dec 31	13:13:25	4118	11861	147	P	-t	-1.5186	0.0450	65.2S	155.8E	0	201		
11797	590	2959 Jan 30	02:01:14	4119	11862	185	P	t-	1.2599	0.5187	62.8N	118.1E	0	132		
11798	590	2959 Jun 27	10:05:42	4122	11867	152	P	-t	1.0817	0.8254	65.4N	153.5W	0	341		
11799	590	2959 Dec 21	04:06:06	4125	11873	157	T	-p	-0.8202	1.0417	77.9S	155.0E	35	337	246	02m28s
11800	590	2960 Jun 15	10:38:45	4129	11879	162	A	np	0.3340	0.9474	42.9N	40.0E	70	185	205	05m45s
11801	591	2960 Dec 09	19:39:39	4133	11885	167	T	-n	-0.1517	1.0323	31.7S	99.8W	81	359	111	02m57s
11802	591	2961 Jun 04	13:43:19	4136	11891	172	A	p-	-0.4308	0.9811	3.1S	8.2W	64	359	74	02m21s
11803	591	2961 Nov 29	07:14:06	4140	11897	177	A	p-	0.5722	0.9789	13.5N	87.7E	55	184	92	02m28s
11804	591	2962 Apr 25	14:50:32	4143	11902	144	P	-t	1.4546	0.1507	70.8N	158.9W	0	45		
11805	591	2962 May 24	23:51:19	4143	11903	182	P	t-	-1.1502	0.7238	68.5S	147.3W	0	347		
11806	591	2962 Nov 18	11:43:40	4147	11909	187	P	t-	1.3271	0.4039	69.3N	40.3E	0	202		
11807	591	2963 Apr 15	07:12:58	4150	11914	154	T	-p	0.7513	1.0547	55.9N	68.0E	41	152	273	03m44s
11808	591	2963 Oct 08	18:51:34	4153	11920	159	A	-p	-0.9105	0.9428	66.0S	129.0W	24	45	514	04m32s
11809	591	2964 Apr 03	22:16:19	4157	11926	164	T	nn	0.0540	1.0224	8.9N	137.0W	87	163	76	02m13s
11810	591	2964 Sep 27	00:52:08	4160	11932	169	Hm	nn	-0.1310	1.0043	9.2S	179.9E	82	18	15	00m27s
11811	591	2965 Mar 24	07:35:07	4164	11938	174	A	p-	-0.6953	0.9595	40.2S	100.9E	46	337	202	03m58s
11812	591	2965 Sep 16	14:01:24	4167	11944	179	T	p-	0.6110	1.0543	37.9N	0.5W	52	202	225	04m20s
11813	591	2966 Mar 13	09:35:29	4171	11950	184	P	t-	-1.4320	0.2297	72.2S	154.3E	0	261		
11814	591	2966 Aug 07	23:40:32	4174	11955	151	P	-t	-1.2079	0.6193	70.3S	166.9E	0	30		
11815	591	2966 Sep 06	06:50:40	4174	11956	189	P	t	1.3142	0.4145	71.9N	156.3W	0	290		
11816	591	2967 Jan 31	15:44:49	4177	11961	156	A	-p	0.8961	0.9413	45.6N	49.1W	26	166	490	05m55s
11817	591	2967 Jul 28	13:27:22	4181	11967	161	H3	-p	-0.4892	1.0147	10.4S	6.5W	61	8	58	01m37s
11818	591	2968 Jan 20	23:43:38	4184	11973	166	H	nn	0.1618	1.0044	10.8S	157.3W	81	174	16	00m29s
11819	591	2968 Jul 16	20:21:17	4188	11979	171	A	nn	0.2909	0.9651	37.9N	104.0W	73	186	132	03m48s
11820	591	2969 Jan 09	13:43:10	4192	11985	176	T	p-	-0.5176	1.0420	53.3S	4.9W	59	356	164	03m06s

Cat Num	Canon Plate	Calendar Date	TD of Greatest Eclipse	ΔT s	Luna Num	Saros Num	Ecl. Type	QLE	Gamma	Ecl. Mag.	Lat. °	Long. °	Sun Alt °	Sun Azm °	Path Width km	Central Line Dur.
11821	592	2969 Jul 05	21:27:17	4195	11991	181	P	t-	1.0617	0.8563	67.5N	55.6E	0	358		
11822	592	2969 Nov 30	18:20:54	4198	11996	148	P	-t	1.5210	0.0393	64.2N	46.2W	0	212		
11823	592	2969 Dec 30	05:25:34	4199	11997	186	P	t-	-1.1847	0.6598	66.9S	59.2W	0	176		
11824	592	2970 May 26	11:17:06	4202	12002	153	A	-t	-0.9834	0.9826	55.7S	55.7E	10	334	362	01m26s
11825	592	2970 Nov 20	04:10:31	4205	12008	158	A	-p	0.8810	0.9597	39.2N	152.3E	28	203	305	03m59s
11826	592	2971 May 15	23:12:11	4209	12014	163	T	-n	-0.1967	1.0476	8.4N	147.3W	79	341	161	04m27s
11827	592	2971 Nov 09	06:51:08	4212	12020	168	A	nn	0.1815	0.9320	7.1S	94.6E	80	201	258	08m32s
11828	592	2972 May 04	15:26:42	4216	12026	173	T	p-	0.5235	1.0657	45.0N	51.1W	58	149	251	04m40s
11829	592	2972 Oct 28	06:30:35	4219	12032	178	A	p-	-0.5318	0.9369	41.9S	78.3E	58	33	274	06m10s
11830	592	2973 Mar 25	20:35:44	4222	12037	145	P	-t	-1.4393	0.1949	61.2S	24.2W	0	275		
11831	592	2973 Apr 24	07:25:04	4223	12038	183	P	t-	1.2534	0.5318	62.1N	29.9W	0	61		
11832	592	2973 Sep 17	23:18:46	4226	12043	150	P	-t	1.4812	0.1119	61.3N	60.5W	0	273		
11833	592	2973 Oct 17	10:40:42	4227	12044	188	P	t-	-1.2272	0.5769	61.5S	74.1W	0	110		
11834	592	2974 Mar 15	03:43:12	4230	12049	155	A	-p	-0.7841	0.9422	45.4S	176.4E	38	317	335	05m07s
11835	592	2974 Sep 07	13:59:21	4233	12055	160	T	-p	0.7028	1.0638	43.5N	14.9E	45	220	289	04m25s
11836	592	2975 Mar 04	04:11:02	4237	12061	165	A	nn	-0.0939	0.9217	11.3S	140.1E	85	332	297	09m10s
11837	592	2975 Aug 28	07:03:32	4240	12067	170	Tm	nn	-0.0279	1.0709	8.0N	91.8E	88	28	231	05m53s
11838	592	2976 Feb 21	04:15:07	4244	12073	175	A	p-	0.5951	0.9382	22.1N	120.2E	53	152	279	06m49s
11839	592	2976 Aug 16	21:49:43	4247	12079	180	T	p-	-0.8016	1.0210	35.2S	152.8W	36	30	117	01m48s
11840	592	2977 Jan 10	21:56:20	4250	12084	147	P	-t	-1.5202	0.0413	64.3S	15.9E	0	211		
11841	593	2977 Feb 09	10:33:46	4251	12085	185	P	t-	1.2500	0.5366	62.1N	18.4W	0	122		
11842	593	2977 Jul 07	16:38:50	4254	12090	152	P	-t	1.1677	0.6768	64.5N	99.0E	0	332		
11843	593	2977 Dec 31	12:54:33	4258	12096	157	T	-p	-0.8278	1.0405	76.1S	40.7E	34	319	244	02m23s
11844	593	2978 Jun 26	17:12:36	4261	12102	162	A	-p	0.4165	0.9493	47.7N	53.9W	65	191	205	05m12s
11845	593	2978 Dec 21	04:17:20	4265	12108	167	T	-n	-0.1638	1.0284	32.9S	133.5E	80	354	98	02m34s
11846	593	2979 Jun 15	20:44:09	4268	12114	172	A	p-	-0.3570	0.9866	2.4N	113.9W	69	3	51	01m39s
11847	593	2979 Dec 10	15:27:25	4272	12120	177	A	p-	0.5563	0.9740	11.1N	36.6W	56	179	112	03m07s
11848	593	2980 May 05	22:48:34	4275	12125	144	Pe	-t	1.4963	0.0697	70.0N	69.5E	0	33		
11849	593	2980 Jun 04	07:22:42	4275	12126	182	P	t-	-1.0854	0.8465	67.5S	89.2E	0	358		
11850	593	2980 Nov 28	19:27:28	4279	12132	187	P	t-	1.3026	0.4456	68.3N	86.7W	0	190		
11851	593	2981 Apr 25	15:22:39	4282	12137	154	T	-p	0.7917	1.0560	63.3N	57.6W	37	150	303	03m36s
11852	593	2981 Oct 19	02:08:17	4286	12143	159	A	-p	-0.9600	0.9400	74.1S	93.7E	16	72	820	04m14s
11853	593	2982 Apr 15	06:21:40	4289	12149	164	T	nn	0.0890	1.0223	14.9N	101.1E	85	164	76	02m12s
11854	593	2982 Oct 08	08:26:58	4293	12155	169	H	nn	-0.1838	1.0047	16.4S	65.0E	79	17	17	00m29s
11855	593	2983 Apr 04	15:25:41	4296	12161	174	A	p-	-0.6666	0.9599	34.1S	19.3W	48	340	193	04m11s
11856	593	2983 Sep 27	21:47:41	4300	12167	179	T	p-	0.5531	1.0547	30.0N	120.0W	56	200	216	04m36s
11857	593	2984 Mar 23	17:14:44	4304	12173	184	P	t-	-1.4059	0.2730	72.3S	25.8E	0	275		
11858	593	2984 Aug 18	07:08:25	4307	12178	151	P	-t	-1.2800	0.4810	71.1S	42.6E	0	48		
11859	593	2984 Sep 16	14:34:20	4307	12179	189	P	t-	1.2556	0.5277	72.1N	74.3E	0	276		
11860	593	2985 Feb 11	00:00:02	4310	12184	156	A	-p	0.9028	0.9444	48.9N	177.3W	25	160	477	05m19s
11861	594	2985 Aug 07	20:31:50	4314	12190	161	H	-p	-0.5686	1.0097	18.3S	115.0W	55	12	41	01m02s
11862	594	2986 Jan 31	08:22:37	4317	12196	166	H	-n	0.1669	1.0075	7.8S	73.5E	80	170	26	00m48s
11863	594	2986 Jul 28	02:58:21	4321	12202	171	A	nn	0.2064	0.9630	30.6N	157.6E	78	189	137	04m16s
11864	594	2987 Jan 20	22:33:24	4325	12208	176	T	p-	-0.5111	1.0427	50.6S	132.7W	59	350	166	03m13s
11865	594	2987 Jul 17	03:54:36	4328	12214	181	A	t-	0.9751	0.9372	80.0N	66.4W	12	333	1130	04m01s
11866	594	2987 Dec 12	02:50:04	4331	12219	148	Pe	-t	1.5396	0.0074	65.2N	177.4E	0	202		
11867	594	2988 Jan 10	14:12:58	4332	12220	186	P	t-	-1.1806	0.6671	68.0S	158.6E	0	187		
11868	594	2988 Jun 05	18:28:53	4335	12225	153	P	-t	-1.0476	0.9018	64.8S	51.6W	0	335		
11869	594	2988 Nov 30	12:11:10	4339	12231	158	A	-t	0.9066	0.9538	41.7N	29.8E	25	198	398	04m38s
11870	594	2989 May 26	06:52:44	4342	12237	163	T	-n	-0.2555	1.0525	7.0N	98.6E	75	345	179	05m00s
11871	594	2989 Nov 19	14:25:04	4346	12243	168	A	nn	0.2155	0.9283	7.8S	18.0W	78	198	275	09m23s
11872	594	2990 May 15	23:22:03	4349	12249	173	T	p-	0.4710	1.0689	45.4N	165.7W	62	154	254	04m58s
11873	594	2990 Nov 08	13:59:19	4353	12255	178	A	p-	-0.4905	0.9360	43.5S	30.7W	60	30	272	06m19s
11874	594	2991 Apr 06	04:43:03	4356	12260	145	P	-t	-1.4726	0.1346	61.4S	154.2W	0	284		
11875	594	2991 May 05	15:20:42	4357	12261	183	P	t-	1.2116	0.6100	62.6N	157.2W	0	52		
11876	594	2991 Sep 29	06:52:19	4360	12266	150	Pe	-t	1.5333	0.0156	61.4N	178.0E	0	265		
11877	594	2991 Oct 28	18:23:00	4360	12267	188	P	t-	-1.1802	0.6604	61.9S	162.1E	0	119		
11878	594	2992 Mar 25	11:34:16	4363	12272	155	A	-p	-0.8128	0.9419	43.8S	58.8E	35	318	358	05m17s
11879	594	2992 Sep 17	21:42:08	4367	12278	160	T	-p	0.7636	1.0617	43.5N	98.1W	40	221	307	04m16s
11880	594	2993 Mar 14	11:55:59	4371	12284	165	A	nn	-0.1176	0.9238	8.1S	24.5E	83	331	289	08m53s

Cat Num	Canon Plate	Calendar Date	TD of Greatest Eclipse	ΔT s	Luna Num	Saros Num	Ecl. Type	QLE	Gamma	Ecl. Mag.	Lat. °	Long. °	Sun Alt °	Sun Azm °	Path Width km	Central Line Dur.
11881	595	2993 Sep 07	14:40:11	4374	12290	170	T	nn	0.0387	1.0673	7.4N	21.0W	88	208	220	05m33s
11882	595	2994 Mar 03	12:17:48	4378	12296	175	A	p-	0.5777	0.9422	24.5N	1.0W	55	149	256	06m06s
11883	595	2994 Aug 28	05:05:38	4381	12302	180	T	p-	-0.7327	1.0176	32.5S	99.3E	43	32	87	01m31s
11884	595	2995 Jan 22	06:39:24	4384	12307	147	P	-t	-1.5225	0.0363	63.5S	123.8W	0	221		
11885	595	2995 Feb 20	19:02:58	4385	12308	185	P	t-	1.2366	0.5608	61.6N	153.9W	0	113		
11886	595	2995 Jul 18	23:11:40	4388	12313	152	P	-t	1.2531	0.5297	63.7N	8.1W	0	322		
11887	595	2995 Aug 17	13:03:11	4389	12314	190	Pb	t-	-1.5542	0.0036	62.0S	60.2W	0	61		
11888	595	2996 Jan 11	21:44:38	4392	12319	157	T	-p	-0.8345	1.0397	72.9S	81.5W	33	308	243	02m20s
11889	595	2996 Jul 06	23:44:03	4395	12325	162	A	-p	0.5013	0.9508	51.6N	145.6W	60	199	208	04m44s
11890	595	2996 Dec 31	12:58:17	4399	12331	167	T	-n	-0.1729	1.0249	32.9S	6.2E	80	349	86	02m14s
11891	595	2997 Jun 26	03:41:44	4403	12337	172	A	p-	-0.2793	0.9916	7.2N	141.9E	74	8	31	01m00s
11892	595	2997 Dec 20	23:45:15	4406	12343	177	A	p-	0.5449	0.9696	9.6N	161.9W	57	175	130	03m40s
11893	595	2998 Jun 15	14:49:27	4410	12349	182	P	t-	-1.0158	0.9792	66.5S	32.5W	0	9		
11894	595	2998 Dec 10	03:18:31	4414	12355	187	P	t-	1.2838	0.4773	67.2N	145.0E	0	179		
11895	595	2999 May 06	23:23:57	4417	12360	154	T	-p	0.8388	1.0566	71.5N	177.3E	33	146	345	03m25s
11896	595	2999 Oct 30	09:34:33	4420	12366	159	A-	-t	-1.0023	0.9586	70.9S	84.7W	0	137	-	-
11897	595	3000 Apr 26	14:18:06	4424	12372	164	T	-n	0.1310	1.0222	21.1N	18.4W	82	166	76	02m11s
11898	595	3000 Oct 19	16:10:16	4428	12378	169	H	nn	-0.2303	1.0049	23.1S	51.6W	77	16	17	00m29s

Astropixels Publications

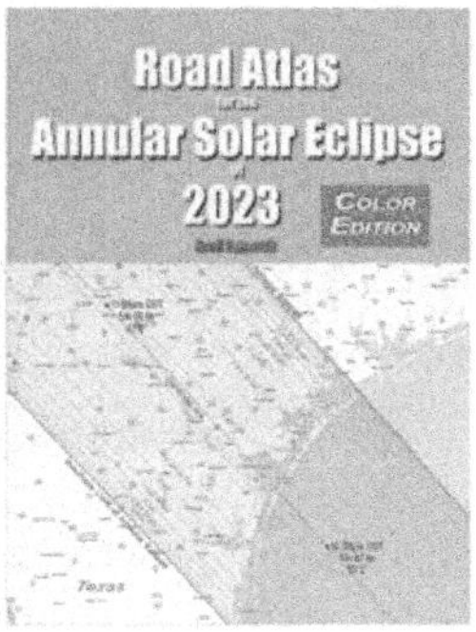

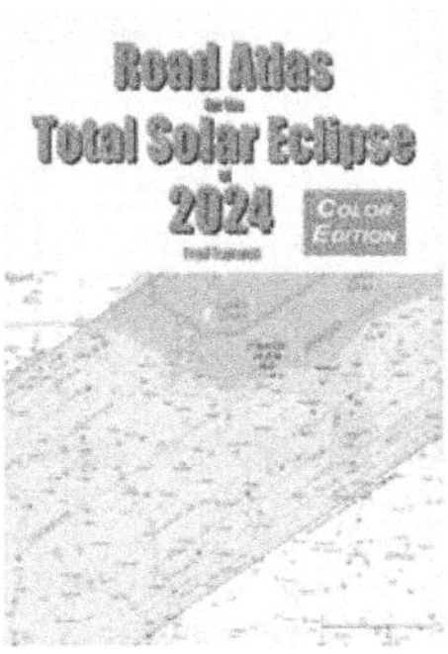

Below is a partial list of some of the books Fred Espenak has written through Astropixels Publishing:

21st Century Canon of Solar Eclipses

The complete guide to every solar eclipse occurring from 2001 tom 2100 (224 eclipses in all). It includes information and maps for all total, annular, hybrid, and partial eclipses. A special world atlas shows detailed full page maps of all central eclipse paths (total, annular and hybrid).

Road Atlas for the Annular Solar Eclipse of 2023

Detailed road maps of the entire eclipse path from the western USA, through Mexico, Central and South America. Information printed on the maps makes it easy to estimate the duration of annularity from any location in the eclipse path. This is the next annular solar eclipse visible from the USA.

Atlas of Central Solar Eclipses in the USA

When was the last total eclipse through the USA and when is the next? How often do they happen? What total eclipse tracks passed across the USA during the 17th, 18th, and 19th centuries, etc., and what states did they include? And how often is a total solar eclipse visible from each of the 50 states? The Atlas of Central Solar Eclipses in the USA answers all of these questions and more with hundreds of maps and tables.

Road Atlas for the Total Solar Eclipse of 2024

This book contains detailed road maps of the entire eclipse path from Mexico, through the USA and Canada. Information printed on the maps makes it easy to estimate the duration of totality from any location in the eclipse path. This is the next total solar eclipse visible from the USA.

21st Century Canon of Lunar Eclipses

The complete guide to every lunar eclipse occurring from 2001 tom 2100 (228 eclipses in all). It includes information and full page maps for all total, partial, and penumbral eclipses. The predictions use a new model for Earth's elliptical shadows.

For information on these books and more, visit *Astropixels Publishing*: *http://Astropixels.com/pubs/index.html*

All Astropixels Publications are available from Amazon.com

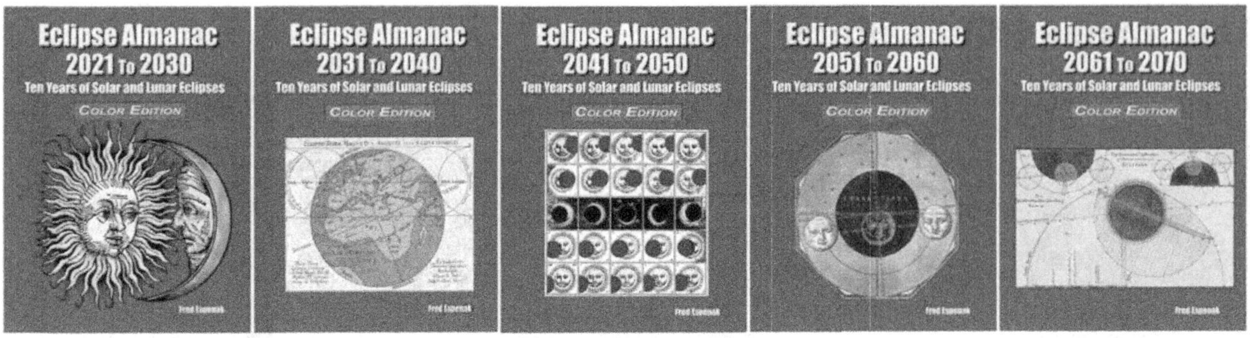

The ***Eclipse Almanac*** contains maps and diagrams of every solar and lunar eclipse over a ten-year period. This permits the reader to look ahead and easily determine when and where each of these events will be seen. Particular details about each eclipse are included as well as a 25-year table looking further into the future.

Section 1 covers solar eclipses, while Section 2 is devoted to lunar eclipses. Explanations are given for each type of eclipse, and the visual appearance of each one is included. Section 3 tabulates the date and time of the Moon's phases over a decade. New Moon and Full Moon phases that coincide with solar and lunar eclipses, are identified.

The ***Eclipse Almanac*** series consists of five volumes, each available in a color, black and white, and Kindle editions. Each volume covers a single decade:

1) Eclipse Almanac 2021 to 2030
2) Eclipse Almanac 2031 to 2040
3) Eclipse Almanac 2041 to 2050
4) Eclipse Almanac 2051 to 2060
5) Eclipse Almanac 2061 to 2070

For information on these books and more, visit *Astropixels Publishing*: *http://astropixels.com/pubs/index.html*

All Astropixels Publications are available from Amazon.com

The **Thousand Year Canon of Solar Eclipses 1501 to 2500** contains maps and data for each of the 2,389 solar eclipses occurring over the ten century period centered on the present era.

The primary content of the Thousand Year Canon is a comprehensive catalog listing the essential characteristics of each eclipse and an atlas of 2,389 maps depicting the geographic regions of visibility of each eclipse.

Available in either black & white or color editions.

The **Thousand Year Canon of Lunar Eclipses 1501 to 2500** contains diagrams, maps and data for each of the 2,424 lunar eclipses occurring over the ten century period centered on the present era.

Appendix A is a comprehensive catalog listing the essential characteristics of each eclipse. Appendix B is an atlas depicting the path of the Moon through Earth's shadows and maps identifying the geographic regions of visibility of each of the 2,424 eclipses.

Available in either black & white or color editions.

All Astropixels Publications are available from Amazon.com

Made in the USA
Monee, IL
05 November 2026

67811476R10151